*RNA and*
*PROTEIN SYNTHESIS*

**SELECTED METHODS IN ENZYMOLOGY SERIES**

**Series Editors**

Sidney P. Colowick
*Vanderbilt University*
*School of Medicine*
*Nashville, Tennessee*

Nathan O. Kaplan
*Department of Chemistry*
*University of California at San Diego*
*La Jolla, California*

**Kivie Moldave** (editor). RNA and Protein Synthesis, 1981

# RNA *and* PROTEIN SYNTHESIS

*Edited by*

KIVIE MOLDAVE

Department of Biological Chemistry
*California College of Medicine*
*University of California*
*Irvine, California*

ACADEMIC PRESS 1981

*A Subsidiary of Harcourt Brace Jovanovich, Publishers*

New York London Toronto Sydney San Francisco

ACADEMIC PRESS, INC.
111 Fifth Avenue, New York, New York 10003

*United Kingdom Edition published by*
ACADEMIC PRESS, INC. (LONDON) LTD.
24/28 Oval Road, London NW1 7DX

ISBN 0-12-504180-2

PRINTED IN THE UNITED STATES OF AMERICA

81 82 83 84 9 8 7 6 5 4 3 2 1

# Table of Contents

Section II. Aminoacyl-tRNA Synthetases



Section III. Polypeptide Chain Initiation

Section IV. Polypeptide Chain Elongation



Section V. Polypeptide Chain Termination



Section VI. Ribosomes

## List of Contributors

Article numbers are in parentheses following the names of contributors. Affiliations listed are current.

GEORGE ACS (6), *Department of Biochemistry and Pediatrics, Mount Sinai School of Medicine, New York, New York 10029*

IRENE L. ANDRULIS (9), *Department of Medical Genetics, University of Toronto, Toronto, Ontario, Canada*

STUART M. ARFIN (9), *Department of Biological Chemistry, College of Medicine, University of California, Irvine, California 92717*

A. L. BEAUDET (45), *Department of Internal Medicine, Baylor College of Medicine, Houston, Texas 77030*

ROB BENNE (28), *Department of Biochemistry, State University of Leiden, Leiden, The Netherlands*

CLAIRE H. BIRGE (34), *Department of Medicine, The Jewish Hospital of St. Louis, St. Louis, Missouri 63110*

LAWRENCE BITTE (62), *Department of Biochemistry, University of Oregon Medical School, Portland, Oregon 97201*

JAMES W. BODLEY (38), *Department of Biochemistry, University of Minnesota, Minneapolis, Minnesota 55455*

KALLOL K. BOSE (5), *Department of Chemistry, University of Nebraska, Lincoln, Nebraska 68105*

NATHAN BROT (38), *Department of Biochemistry, Roche Institute of Molecular Biology, Nutley, New Jersey 07110*

MARIANNE L. BROWN-LUEDI (28), *Department of Biological Chemistry, School of Medicine, University of California, Davis, California 95616*

C. T. CASKEY (44), (45), *Department of Medicine, Baylor College of Medicine, 1200 Moursund Avenue, Houston, Texas 77030*

F. CHAPEVILLE (13), *Institut de Biologie Moléculaire, Faculté des Sciences, Paris, France*

NANDO K. CHATTERJEE (5), *Department of Chemistry, University of Nebraska, Lincoln, Nebraska 68105*

S. CHOUSTERMAN (13), *Institut de Biologie Moléculaire, Faculté des Sciences, Paris, France*

W. W. CLELAND (25), *Department of Biochemistry, University of Wisconsin, Madison, Wisconsin 53706*

B. S. COOPERMAN (57), *Department of Chemistry, University of Pennsylvania, Philadelphia, Pennsylvania 19104*

GARY R. CRAVEN (65), *Laboratory of Molecular Biology, University of Wisconsin-Madison, Madison, Wisconsin 53706*

ALBERT E. DAHLBERG (48), *Division of Biology and Medicine, Brown University, Providence, Rhode Island 02912*

MURRAY P. DEUTSCHER (19), (23), *Department of Biochemistry, University of Connecticut Health Center, Farmington, Connecticut 06105*

RALPH F. DI CAMELLI (60), *Department of Biochemistry, The University of Chicago, Chicago, Illinois 60637*

JAN DIJK (52), *Max-Planck-Institut für Molekulare Genetik, Abteilung Wittmann, Ihnestrasse 63-73, D-1000 Berlin 33 (Dahlem), Federal Republic of Germany*

FERDINAND DOHME (50), *Althouse Laboratory, The Pennsylvania State University, University Park, Pennsylvania 16802*

BERNARD S. DUDOCK (20), *Department of Biochemistry, State University of New York at Stony Brook, Stony Brook, New York 11794*

B. Z. EGAN (2), *Chemical Technology Division, Oak Ridge National Laboratory, Oak Ridge, Tennessee 37830*

ELIZABETH ANN EIGNER (24), *142 Marlboro Street, Boston, Massachusetts 02178*

CHARLES EIL (63), *Department of Biochemistry, The University of Chicago, Chicago, Illinois 60637*

JOHN L. FAKUNDING (27), *Department of Biological Chemistry, School of Medicine, University of California, Davis, California 95616*

I. C. GILLAM (4), *Department of Biochemistry, University of British Columbia, Vancouver, British Columbia, Canada*

AMANDA M. GILLUM (14), *Department of Pathology, Stanford University School of Medicine, Palo Alto, California 94305*

EMANUEL GOLDMAN (18), *Department of Microbiology, New Jersey Medical School, 100 Bergen Street, Newark, New Jersey 07103*

R. A. GOLDMAN (57), *Department of Chemistry, University of Pennsylvania, Philadelphia, Pennsylvania 19104*

J. GOLDSTEIN (44), *Department of Internal Medicine, University of Washington School of Medicine, Seattle, Washington 98105*

FRED GOLINI (34), *Department of Microbiology, Health Sciences Center, State University of New York at Stony Brook, Stony Brook, New York 11794*

P. G. GRANT (57), *Department of Chemistry, University of Pennsylvania, Philadelphia, Pennsylvania 19104*

ROBERT GREENBERG (20), *Department of Biochemistry, State University of New York at Stony Brook, Stony Brook, New York 11794*

CLAUDIO GUALERZI (58), *Max-Planck-Institut für Molekulare Genetik, Abteilung Wittmann, Ihnestrasse 63-73, D-1000 Berlin 33 (Dahlem), Federal Republic of Germany*

NABA K. GUPTA (5), *Department of Chemistry, University of Nebraska, Lincoln, Nebraska 68105*

ARNOLD E. HAMPEL (3), *Department of Biological Sciences, Northern Illinois University, DeKalb, Illinois 60115*

S. J. S. HARDY (51), *Department of Genetics, University of Leicester, Leicester, England*

G. WESLEY HATFIELD (18), *Department of Medical Microbiology, College of Medicine, University of California, Irvine, California 92717*

GARY M. HATHAWAY (35), *Department of Biochemistry, University of California, Riverside, California 92502*

D. E. HEATHERLY (2), *Chemical Technology Division, Oak Ridge National Laboratory, Oak Ridge, Tennessee 37830*

EDGAR C. HENSHAW (30), *University of Rochester Cancer Center, Rochester, New York 14642*

C. F. HEREDIA (42), *Instituto de Enzimología del C.S.I.C., Facultad de Medicina de la Universidad Autónoma, Madrid, Spain*

JOHN W. B. HERSHEY (27), (28), *Department of Biological Chemistry, School of Medicine, University of California, Davis, California 95616*

G. A. HOWARD (54), *Department of Biological Chemistry, School of Medicine, University of California, Davis, California 95616*

KENTARO IWASAKI (40), *Institute of Medical Science, University of Tokyo, Takanawa, Minato-ku, Tokyo 108, Japan*

DAVID KABAT (62), *Department of Biochemistry, University of Oregon Medical School, Portland, Oregon 97201*

RAYMOND KAEMPFER (46), *Hebrew University-Hadassah Medical School, Jerusalem, Israel*

YOSHITO KAZIRO (40), *Institute of Medical Science, University of Tokyo, Takanawa, Minato-ku, Tokyo 108, Japan*

A. D. KELMERS (1), (2), *Chemical Technology Division, Oak Ridge National Laboratory, Oak Ridge, Tennessee 37830*

JAMES W. KENNY (56), *Department of Biological Chemistry, School of Medicine, University of California, Davis, California 95616*

MARILYN KOZAK (32), *Department of Biological Sciences, University of Pittsburgh, Pittsburgh, Pennsylvania 15260*

C. G. KURLAND (51), *The Wallemberg Laboratory, University of Uppsala, Uppsala, Sweden*

JOHN M. LAMBERT (56), *Department of Biological Chemistry, School of Medicine, University of California, Davis, California 95616*

ALAN M. LAMBOWITZ (66), *The Edward A. Doisy Department of Biochemistry, St. Louis University Medical School, St. Louis, Missouri 63104*

YEHUDA LAPIDOT (10), (11), *Department of Biological Chemistry, The Hebrew University of Jerusalem, Jerusalem, Israel*

PHILIP LEDER (17), (37), *National Institute of Child Health and Human Development, National Institutes of Health, Bethesda, Maryland 20014*

FRITZ LIPMANN (6), (41), *The Rockefeller University, New York, New York 10021*

JENNY LITTLECHILD (52), *Max-Planck-Institut für Molekulare Genetik, Abteilung Wittmann, Ihnestrasse 63-73, D-1000 Berlin 33 (Dahlem), Federal Republic of Germany*

ROBERT B. LOFTFIELD (24), *Department of Biochemistry, University of New Mexico School of Medicine, Albuquerque, New Mexico 87131*

C. V. LOWRY (49), *Laboratory of Genetics, University of Wisconsin, Madison, Wisconsin 53706*

M. A. LUDDY (57), *Department of Chemistry, University of Pennsylvania, Philadelphia, Pennsylvania 19104*

TINA S. LUNDAK (35), *Department of Biochemistry, University of California, Riverside, California 92502*

LJUBICA MAUSER (29), *Department of Biochemistry, New York University Medical Center, New York, New York 10016*

ALAN H. MEHLER (26), *Department of Biochemistry, Medical College of Wisconsin, Milwaukee, Wisconsin 53706*

MATTHEW MESELSON (46), *The Biological Laboratories, Harvard University, Cambridge, Massachusetts 02138*

CHRISTIAN F. MIDELFORT (26), *Department of Biochemistry, Medical College of Wisconsin, Milwaukee, Wisconsin 53706*

GREGORY MILMAN (44), *Department of Biochemistry, The Johns Hopkins University, School of Hygiene and Public Health, Baltimore, Maryland 21205*

A. MINNELLA (57), *Department of Chemistry, University of Pennsylvania, Philadelphia, Pennsylvania 19104*

S. MIZUSHIMA (49), *Faculty of Agriculture, Nagoya University, Chikusa, Nagoya, Japan*

G. MORA (51), *Department of Microbiology, The Catholic University of Chile, Santiago, Chile*

LYDA NEELEMAN (33), *Department of Biochemistry, State University of Leiden, 2300 RA Leiden, The Netherlands*

A. W. NICHOLSON (57), *Department of Chemistry, University of Pennsylvania, Philadelphia, Pennsylvania 19104*

KNUD H. NIERHAUS (36), (50), *Max-Planck-Institut für Molekulaire Genetik, Abteilung Wittmann, 1 Berlin-Dahlem 33, Federal Republic of Germany*

M. NOMURA (49), *Laboratory of Genetics, University of Wisconsin, Madison, Wisconsin 53706*

G. DAVID NOVELLI (1), *Biology Division, Oak Ridge National Laboratory, Oak Ridge, Tennessee 37830*

JAMES OFENGAND (21), *Department of Biochemistry, Roche Institute of Molecular Biology, Nutley, New Jersey 07110*

KIKUO OGATA (55), *Department of Biochemistry, Niigata University School of Medicine, Niigata, Japan*

R. L. PEARSON (1), *Chemical Technology Division, Oak Ridge National Laboratory, Oak Ridge, Tennessee*

SIDNEY PESTKA (12), (59), *Department of Biochemistry, Roche Institute of Molecular Biology, Nutley, New Jersey 07110*

CYNTHIA L. PON (58), *Max-Planck-Institut für Molekulare Genetik, Abteilung Wittmann, Ihnestrasse 63-73, D-1000 Berlin 33 (Dahlem), Federal Republic of Germany*

UTTAM L. RAJBHANDARY (14), *Department of Biology, Massachusetts Institute of Technology, Cambridge, Massachusetts 02139*

SARA RAPPOPORT (10), (11), *Department of Biological Chemistry, The Hebrew University of Jerusalem, Jerusalem, Israel*

JOANNE W. RAVEL (39), *Clayton Foundation Biochemical Institute, The University of Texas at Austin, Austin, Texas 78712*

SCOTT A. REINES (15), *Department of Psychiatry, Montefiore Hospital and Medical Center, Bronx, New York 10467*

DIETMAR RICHTER (41), *Physiologisch-Chemisches Institut der Universität Hamburg, Abteilung Zellbiochemie, Hamburg, West Germany*

WILLIAM G. ROBINSON (29), *Department of Biochemistry, New York University Medical Center, New York, New York 10016*

BRUCE RUEFER (3), *Department of Biological Sciences, Northern Illinois University, DeKalb, Illinois 60115*

A. SANDOVAL (42), *Instituto de Enzimología del C.S.I.C., Facultad de Medicina de la Universidad Autónoma, Madrid, Spain*

DANIEL V. SANTI (25), *Department of Biochemistry and Biophysics, University of California School of Medicine, San Francisco, California 94143*

ASOK K. SARKAR (16), *Department of Developmental Biology and Cancer, Albert Einstein College of Medicine, Bronx, New York 10461*

PAUL R. SCHIMMEL (22), *Department of Biology, Massachusetts Institute of Technology, Cambridge, Massachusetts 02139*

HUBERT J. P. SCHOEMAKER (22), *Corning Medical Center, Medfield, Massachusetts 02052*

LADONNE H. SCHULMAN (15), (16), *Department of Developmental Biology and Cancer, Albert Einstein College of Medicine, Bronx, New York 10461*

JEAN SCHWARZBAUER (65), *Laboratory of Molecular Biology, University of Wisconsin-Madison, Madison, Wisconsin 53706*

E. SCOLNICK (44), *National Cancer Institute, National Institutes of Health, Bethesda, Maryland 20014*

AARON J. SHATKIN (32), *Department of Cell Biology, Roche Institute of Molecular Biology, Nutley, New Jersey 07110*

CORINNE C. SHERTON (60), (61), *Department of Biochemistry, The University of Chicago, Chicago, Illinois 60637*

ROSE ANN L. SHOREY (39), *Clayton Foundation Biochemical Institute and the Department of Home Economics, The University of Texas at Austin, Austin, Texas 78712*

MELVIN SILBERKLANG (14), *Department of Biochemistry and Biophysics, University of California, San Francisco, California 94143*

LAWRENCE SKOGERSON (43), *Department of Biochemistry, The Medical College of Wisconsin, Milwaukee, Wisconsin 53233*

MATHIAS SPRINZL (8), *Max-Planck-Institut für Experimentelle Medizin, Abteilung Chemie, Hermann-Rein-Strasse 3, D-3400 Göttingen, Federal Republic of Germany*

JOAN ARGETSINGER STEITZ (31), *Department of Molecular Biophysics and Biochemistry, Yale University, New Haven, Connecticut 06510*

HANS STERNBACH (8), *Max-Planck-Institut für Experimentelle Medizin, Abteilung Chemie, Hermann-Rein-Strasse 3, D-3400 Göttingen, Federal Republic of Germany*

W. A. STRYCHARZ (57), *Institute for Enzyme Research, University of Wisconsin, Madison, Wisconsin 53705*

M. P. STULBERG (1), *Biology Division, Oak Ridge National Laboratory, Oak Ridge, Tennessee 37830*

PAUL S. SYPHERD (47), *Department of Medical Microbiology, California College of Medicine, University of California, Irvine, California 92717*

S. M. TAHARA (35), *Department of Biochemistry, University of California, Riverside, California 92502*

W. P. TATE (45), *Department of Internal Medicine, Baylor College of Medicine, Houston, Texas 77030*

G. M. TENER (4), *Department of Biochemistry, University of British Columbia, Vancouver, British Columbia, Canada*

KAZUO TERAO (55), *Department of Biochemistry, Niigata University School of Medicine, Niigata, Japan*

ROBERT E. THACH (34), *Department of Biology, Washington University, St. Louis, Missouri 63130*

R. TOMPKINS (44), *Department of Medicine, Dartmouth Medical School, Hanover, New Hampshire 03755*

A. TORAÑO (42), *Instituto de Enzimologia del C.S.I.C., Facultad de Medicina de la Universidad Autónoma, Madrid, Spain*

P. TRAUB (49), *Max Planck Institut für Zell Biologie, Bremerhaven, Germany*

ROBERT R. TRAUT (54), (56), *Department of Biological Chemistry, School of Medicine, University of California, Davis, California 95616*

JOLINDA A. TRAUGH (35), *Department of Biochemistry, University of California, Riverside, California 92502*

BENJAMIN V. TREADWELL (29), *Department of Biochemistry, New York University*

*Medical Center, New York, New York 10016*

LOUS VAN VLOTEN-DOTING (33), *Department of Biochemistry, State University of Leiden, 2300 RA Leiden, The Netherlands*

ROBERT W. WEBSTER, JR. (25), *Department of Biochemistry and Biophysics, University of California School of Medicine, San Francisco, California 94143*

H. O. WEEREN (1), *Chemical Technology Division, Oak Ridge National Laboratory, Oak Ridge, Tennessee*

J. F. WEISS (1), *Chemical Technology Division, Oak Ridge National Laboratory, Oak Ridge, Tennessee 37830*

HERBERT WEISSBACH (38), *Department of Biochemistry, Roche Institute of Molecular Biology, Nutley, New Jersey 07110*

H. G. WITTMANN (53), *Max-Planck-Institut für Molekulare Genetik, Abteilung Wittmann, 1 Berlin-Dahlem 33, Germany*

JOHN W. WIREMAN (47), *Department of Microbiology, University of Minnesota Medical School, Minneapolis, Minnesota*

RICHARD WOLFENDEN (7), *Frick Chemical Laboratory, Princeton University, Princeton, New Jersey 08540*

IRA G. WOOL (60), (61), (63), *Department of Biochemistry, University of Chicago, Chicago, Illinois 60637*

PETER WURMBACH (36), *Max-Planck-Institut für Molekulaire Genetik,. Abteilung Wittmann, 1 Berlin-Dahlem 33, Federal Republic of Germany*

JOAN YANOV (27), *Department of Biological Chemistry, School of Medicine, University of California, Davis, California 95616*

ROBERT A. ZIMMERMANN (64), *Department of Biochemistry, University of Massachusetts, Amherst, Massachusetts 01003*

## Foreword

The *Methods in Enzymology* series, which was originally published as a four-volume treatise over twenty-five years ago, has now grown to over 80 volumes. It has become more and more difficult for an individual investigator to locate particular methods of interest, especially in rapidly developing fields in which pertinent information now appears in many volumes of the series. Although individual and cumulative indexes are provided, the task of information retrieval is still formidable. We have, therefore, undertaken to provide such investigators and their students with a single volume work in a given area of interest, compiled by selection of the most essential and widely used procedures published in volumes of *Methods in Enzymology* in that particular area. The aim is to permit the individual investigator or student to have conveniently at hand all of the basic methodology in that field at relatively low cost. The articles, which are selected by the editors in that area, will be unabridged. A new Subject Index will be prepared for each volume of "Selected Methods in Enzymology."

It is our intention that one volume of "Selected Methods" will be derived from five to six related volumes of equivalent size in the parent series. This volume, which is the first of the "Selected Methods" series, deals with RNA and Protein Synthesis, and is comprised of articles selected by Dr. Moldave from volumes of the *Methods in Enzymology* series for which he served as an editor. We hope that this experiment in publication proves useful to the broad audience for this new series.

SIDNEY P. COLOWICK
NATHAN O. KAPLAN

# Preface

One of the considerations in selecting articles from the *Methods in Enzymology* series volumes for this volume was to provide a comprehensive compendium of methods for the assay, characterization, isolation purification, etc., of various organelles, enzymes, nucleic acids, translational factors, and other components and reactions that participate in protein synthesis. A large number of equally important articles, however, had to be omitted. In many cases, the decision was an arbitrary one made by the Editor, and does not reflect on the quality or the reproducibility of the procedures that were not included; indeed, users of this volume are also referred to those articles for modifications, alternative procedures, and, in some cases, topics not covered in this volume. Where a cross reference is given to a volume and paper in this series, it refers to the *Methods in Enzymology* series. Where only volumes and paper numbers are referred to, the volumes too are those in the *Methods in Enzymology* series.

I would like to take this opportunity to again thank those who have contributed so generously to these works on RNA and protein synthesis.

KIVIE MOLDAVE

# Section I

# Transfer RNA

# [1] Reversed-Phase Chromatography Systems for Transfer Ribonucleic Acids—Preparatory-Scale Methods[1]

*By* A. D. KELMERS, H. O. WEEREN, J. F. WEISS, R. L. PEARSON, M. P. STULBERG, and G. DAVID NOVELLI

Reversed-phase chromatography (RPC) is a system in which a water-immiscible organic extractant is present as a film on an inert support and an aqueous solution, passed through the column, develops the chromatogram. Column chromatographic techniques are inherently simple with regard to the apparatus involved (a length of glass pipe) and can readily be scaled up for production applications or scaled down for microanalytical use by changing the column size.

In designing these RPC systems for tRNA separations,[2-4] it was assumed that the tRNA's could be considered as long-chain polyphosphates, and thus an anion-exchange type of column would be needed. Problems of diffusion within conventional anion-exchange resin beads are avoided, since all the exchange sites are on the surface of the inert diatomaceous earth material of relatively high surface area employed as the solid support. The latter is acid washed and then treated with dimethyldichlorosilane to yield a hydrophobic surface of minimum surface activity. This support is then coated with a water-insoluble quaternary ammonium salt of high molecular weight which functions as the active extractant. A variety of quaternary ammonium compounds that meet these fundamental criteria are commercially available, and several different ones have proved useful for the separation of tRNA's.

A simple model of the mechanism of tRNA mobility on these RPC columns is anion exchange controlled by mass action. The tRNA's are applied to the column in the chloride form in a dilute sodium chloride solution and chloride ions bound to the quaternary ammonium extractant exchange for tRNA phosphate anionic sites; the tRNA's are thus retained on the column with essentially zero mobility. At higher sodium

[1]Research sponsored by the National Institute of General Medical Sciences, National Institutes of Health and the U. S. Atomic Energy Commission under contract with Union Carbide Corporation, Nuclear Division.

[2]A. D. Kelmers, G. David Novelli, and M. P. Stulberg, *J. Biol. Chem.* **240**, 3979 (1965).

[3]J. F. Weiss and A. D. Kelmers, *Biochemistry* **6**, 2507 (1967).

[4]J. F. Weiss, R. L. Pearson, and A. D. Kelmers, *Biochemistry* **7**, 3479 (1968).




ISBN 0-12-504180-2

chloride concentrations, mass action then favors chloride binding with the quaternary ammonium compound; the tRNA's are thus released from the support to the aqueous phase and eluted from the column. If step elution (loading at a low sodium chloride concentration followed by elution at a high sodium chloride concentration) is employed, the tRNA's are eluted as a group with little separation. However, if the sodium chloride concentration increases in a continuous manner, gradient elution, each tRNA transfers from the immobile quaternary ammonium compound to the mobile aqueous phase at a characteristic sodium chloride concentration determined by the specific chromatographic conditions.

Several factors regulate the separation of individual tRNA's during such gradient elution. Since most tRNA's have similar molecular weights, and thus nearly equivalent numbers of phosphate groups, it is unlikely that differences in size or total number of phosphates is a major factor in determining the elution sequence. The tRNA's possess a considerable degree of secondary and tertiary structure that would restrict the number of phosphates available for interaction. Solution conditions known to affect the structure of tRNA's, such as temperature and magnesium ion concentration, affect the elution position from RPC columns. Further, the order of elution of tRNA's from polyacrylamide gel columns, where the controlling factor is effective size (not molecular weight), is the reverse of the order from reversed-phase columns.[5] These results are consistent with the concept that tRNA elution from reversed-phase columns is controlled by the availability of phosphate groups for interaction with the quaternary ammonium exchange sites. Thus, the more tightly structured tRNA's would be eluted first (at low sodium chloride concentrations), and the more flexible, loosely structured tRNA's would be eluted later (at higher sodium chloride concentrations).

A number of commercially available quaternary ammonium compounds have proved useful in these RPC systems, either deposited as a solid on the surface of the diatomaceous earth or dissolved in a water-immiscible inert diluent. Each of these systems, while in general similar, exhibits certain differences in the elution order, sodium chloride concentration, sharpness of peaks, etc., that give each system certain advantages or disadvantages in specific applications. Three RPC systems are described in this report. They are:

RPC-2, tricaprylylmethylammonium chloride (Aliquat 336, General Mills, Kankakee, Illinois) dissolved in tetrafluorotetrachloropro-

[5] B. Z. Egan, R. W. Rhear, and A. D. Kelmers, *Biochim. Biophys. Acta* 174, 23 (1969).

pane[6] (Peninsular ChemResearch, Gainesville, Florida) on Chromosorb W, acid washed, dimethyldichlorosilane-treated, 100/120 mesh (Johns Manville Corp., New York, New York)

RPC-3, trioctylpropylammonium bromide (Eastman Organic Chemicals, Rochester, New York) on Chromosorb G, acid washed, dimethyldichlorosilane-treated, 100/120 mesh

RPC-4, dimethyldicocoammonium chloride[7] (Adogen 462, Ashland Chemical Co., Columbus, Ohio) on Chromosorb G, acid washed, dimethyldichlorosilane-treated, 100/120 mesh

The RPC-4 system is similar to the first system prepared,[2] RPC-1, in that it employs the same quaternary ammonium chloride compound, however, the use of isoamyl acetate as a solvent has been discontinued. The RPC-4 system gives the same chromatographic elution sequence, but with sharper peaks and more reproducible results than the RPC-1 system. Hence, that original system is no longer recommended.

The general effect of experimental variables, such as pH, temperature, load, magnesium ion concentration, are discussed in the next section; a detailed description follows of several specific examples of the application of RPC columns to both small- and large-scale separation and recovery of purified tRNA's. A review of the published applications is given in the final section.

## General Considerations

### Preparation of the Packing

It is important to obtain an even distribution of the quaternary ammonium salt extractant on the surface of the diatomaceous earth. Several different techniques have been devised to achieve this with the different RPC systems. For RPC-3 or RPC-4, a chloroform solution of the quaternary ammonium salt is blended with the diatomaceous earth and the chloroform is allowed to evaporate while the mixture is evenly mixed. For small batches of packing, this can be achieved with a spatula and glass dish; for larger batches, a rotary blending drum with provisions for airflow was constructed. In the case of RPC-2, the quaternary ammonium salt is dissolved in a high-boiling fluorocarbon diluent. Since it is essential in this case to retain the diluent, the solution is added from a dropping funnel and quickly blended with the diatomaceous earth by

[6] Tetrafluorotetrachloropropane is no longer supplied by DuPont Chemical Co. as Freon 214.

[7] The term "coco" refers to straight-chain saturated alkyl groups derived from coconut oil. They range from $C_8$ to $C_{18}$, but $C_{12}$ and $C_{14}$ are the major chains.

hand stirring. Particularly with the RPC-2 system, and also to a lesser extent with RPC-3 and RPC-4, achievement of maximum resolution of the tRNA's is dependent upon an even coating of the support.

On several occasions, both in our laboratories and in others, batches of various RPC packings have given poor or no resolution. In almost every case the problem has been traced to the diatomaceous earth support. If either the acid washing or dimethyldichlorosilane treatment is incomplete, or if excessive grinding or crushing of the diatomaceous earth has occurred so that untreated surfaces are broken open, the presence of active sites on the surface of the diatomaceous earth reduces the chromatographic resolution or even binds tRNA's irreversibly. We have found that poor batches of diatomaceous earth can be redeemed by carrying out the acid wash and dimethyldichlorosilane treatments as described by Horning *et al.*[8]

For small-scale or analytical RPC-2 columns, Chromosorb W, AW-DMCS, 100/120 mesh, gives the most satisfactory results, possibly since its larger surface area permits a more even distribution of the fluorocarbon diluent. For RPC-3 or RPC-4 either Chromosorb G, or Chromosorb W, both AW, DMCS, 100/120-mesh, give equivalent chromatographic results; for small-scale columns, either diatomaceous earth can be used. Chromosorb G is recommended, however, because its greater physical strength resists breakage during handling, with the resultant exposure of active sites. In larger (2.5–10 cm diameter) columns, better results are obtained with the G form of diatomaceous earth, since its greater density and harder particles permit a more even packing with concomitant reduced diffusion mixing in the column and thus sharper peaks.

All the RPC system columns can be reused after a high strength sodium chloride "bump" to remove residual material from the column followed by reequilibration with the loading solution. The number of times a column can be used depends on the specific RPC system, load of tRNA applied and the quantity of contaminants in the crude tRNA applied to the column. In some cases, RPC-2 and RPC-4 columns, where the quaternary ammonium salt has a very low aqueous solubility, have been used for 25–50 chromatographic runs when low loadings (50–100 $A_{260}$) were applied to a 1 × 240-cm column. At the other extreme, in production runs where loads of 200,000 $A_{260}$ units were applied to a 5 × 240 cm RPC-4 column, the chromatographic resolution and tRNA specific activity decreased with subsequent runs on the column, presumably due to buildup of nonelutable contaminants from the feed.

[8] E. C. Horning, W. S. A. VandenHeuvel, and B. G. Creech, *Methods Biochem. Anal.* 11, 82–83, (1963).

With the RPC-3 system, the quaternary ammonium compound used has a low but significant solubility in the dilute sodium chloride solutions, so that a maximum of four to six runs can be made before the column is depleted of extractant.

The diatomaceous earth from an exhausted or discarded RPC-3 or RPC-4 column can be used to make fresh packing following a simple washing procedure. This results in considerable economy, since the diatomaceous earth is the most expensive ingredient. The used packing is washed with water, then several volumes of ethanol to remove the quaternary ammonium salt, and finally with acetone. After it has been air-dried overnight to remove the last traces of acetone, it can be reused in the preparation of RPC-3 or RPC-4 packing. No loss in chromatographic resolution or tRNA specific activity has been observed with Chromosorb G reused many times in these systems. Chromosorb W is more subject to physical breakage, and we recommend it be reused only once or twice. The RPC-2 system has proved to be more sensitive to the nature of the Chromosorb W surface than any of the other RPC systems, and in order to reuse diatomaceous earth in this system it is necessary to repeat the acid wash and dimethyldichlorosilane treatment steps[8] after the water, alcohol, and acetone washes.

## Column Geometry and Preparation

A wide range of sizes and shapes of jacketed glass columns can be used with these RPC systems. The major factors controlling the tRNA resolution on these columns are the gradient volume and shape (linear or concave) and eluent solution conditions (pH, magnesium concentration, etc.) rather than column geometry. This is different from classical elution chromatography where column geometry is a dominant variable. Equally good tRNA separations have been achieved, for example, with RPC-3 systems of 1 × 240 cm, 2.5 × 100 cm, or 10 × 100 cm column dimensions, although the volume of the gradient must be changed corresponding to the total column volume. The column to be selected for a given experiment is determined by the load of tRNA to be chromatogrammed, and by the apparatus available. Excellent results have been obtained with commercially available jacketed glass columns 100 cm in length (Pharmacia), and no advantage is obtained by using the longer, 240 cm, columns specially fabricated for some of the earlier experimental work.[2,3]

The RPC-2 system is more sensitive to experimental conditions than the other systems, and for maximum resolution the inner wall of the glass column should be silane treated. By making the wall hydrophobic, wall effects in the long, thin columns ($1 \times 240$ cm) are reduced, and active sites that can bind small quantities of tRNA's on the glass surface are neutralized. The column is filled with 1 *M* NaOH and allowed to stand overnight at room temperature. It is then rinsed thoroughly with water and dried by blowing air through it. The circulating water in the jacket is heated to roughly 50° to help dry the column and speed the reaction in the next step. The column is then filled with a 2% solution of dimethyldichlorosilane in anhydrous benzene and allowed to stand for 2–3 hours at about 50°. The column is then drained and again dried with flowing air.

The RPC columns are prepared by filling the jacketed column at 37° with a buffer solution and slurrying the prepared packing with the same buffer. The packing is then poured rapidly as a thick slurry into the top of the column, while the excess buffer is allowed to run out at the bottom of the column. With packing prepared on the Chromosorb G support, the poured packing settles to its final volume simply by gravity compaction. The dense, hard particles of Chromosorb G cannot be further compacted significantly by mechanical or hydrodynamic means. When the commercial Pharmacia columns with movable end pieces are used, after the desired quantity of packing has been poured and has settled, the upper end-piece is pressed firmly down against the packing bed, tightened, and the column is ready to use. Pressurization of the packing as described in our earlier publications[2,3] is of, at best, marginal advantage in increasing tRNA resolution with columns packed with Chromosorb G support. The Chromosorb W support is less dense and does not compact as well by gravity when poured into columns in the same manner. In order to pack these columns to maximum density, and thereby reduce the liquid void volume and concomitant band spreading or back-diffusion, pressure compaction should be used. Either mechanical pressure using the movable end-pieces in the commercial columns or pressurization of the eluent solution by means of a positive displacement pump to 50–80 psi will cause a substantial compaction of the packing in the column. Additional packing is then added to the column, and pressurization is repeated until the column is filled to the desired height; by this technique the amount of Chromosorb W packing in the column can be substantially increased, and the column void volume correspondingly decreased.

## Temperature

The distribution coefficient for the tRNA's between the mobile aqueous phase and stationary packing is a function of the temperature.

At increased temperatures the tRNA's bind more tightly and elute at higher NaCl concentrations. This is consistent with the concept that higher temperatures partially "melt" the tRNA structure and make more phosphate groups available for interaction with the quaternary ammonium sites on the packing. Thus, it is extremely important that the temperature not change during the chromatographic run or from one run to the next run to prevent band spreading or lack of reproducibility of results. Jacketed glass columns with a thermostatically regulated circulating fluid are mandatory for maximum resolution. The temperature control should be to at least ±0.1° or better. Short-time temperature fluctuations larger than this can be observed as a sine-wave superimposed on the UV trace of the column eluent.

It has been established empirically that optimal resolution is obtained at 37°. Whether it is coincidental that this is also the physiological temperature for *E. coli* is problematical. Satisfactory resolution still can be achieved down to about 25°, but at low temperatures (4°) resolution is substantially decreased. At 50° sharp peaks are obtained, but a large fraction of some of the tRNA's are denatured under these conditions. Thus, 37° is the temperature of choice for all the RPC systems.

With the large (5 and 10 cm) diameter preparatory-scale columns, it is necessary to preheat the eluent solution to 37° before it enters the column. Heat transfer through these large-diameter packed beds is poor, and without the preheater, a temperature gradient cone is established at the inlet end of the column which significantly diminishes the chromatographic resolution.

## Column Load

The load of tRNA, amount of tRNA per unit volume, that can be bound to the RPC columns and then subsequently successfully chromatogrammed varies substantially depending upon the specific conditions. Much of the experimental work with RPC-2, RPC-3, and RPC-4 has been done at loadings of 5–10 $A_{260}/cm^3$ of column packing, equivalent to loads of 1000–2000 total $A_{260}$ units on a 1 × 240-cm column. This is well below the maximum capacity of the packings, however, and in large-scale preparatory work with RPC-3 and RPC-4 columns, loads of 32 $A_{260}/cm^3$ are routinely chromatogrammed. This represents a total load of 250,000 $A_{260}$ units of tRNA on a 10 × 100-cm column. Even this does not represent the maximum load attainable, and under certain conditions even higher loadings can be handled. Some RPC-3 runs have been made at loads as high as 50 $A_{260}/cm^3$.

Since the tRNA distribution coefficient is a function of concentration

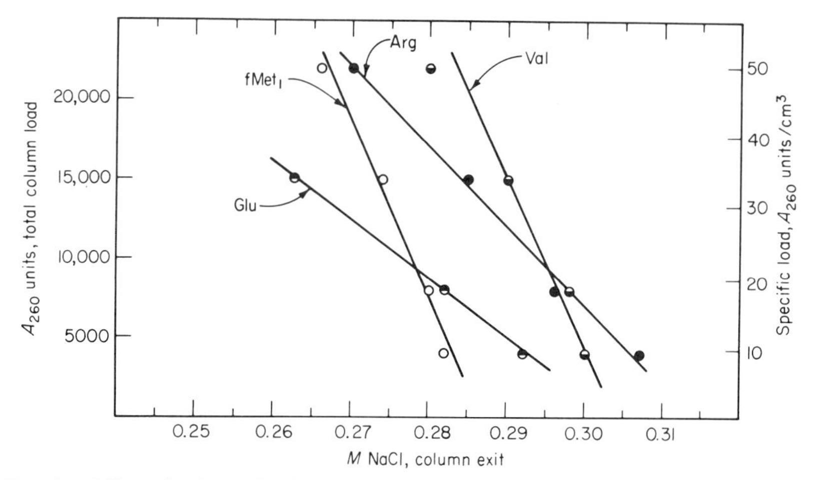


FIG. 1. Effect of column load on elution position for an RPC-3 column. Dimensions, 2.5 × 90 cm; feed, *E. coli* K12 MO7 tRNA as indicated in 150–500 ml of equilibration solution, 0.20 $M$ NaCl, 10 m$M$ $MgCl_2$, 2m$M$ $Na_2S_2O_3$, 10 m$M$ $NaC_2H_3O_2$ buffer at pH 5; flow rate, 8 ml/min; elution gradient, 0.26–0.36 $M$ NaCl, 10 m$M$ $MgCl_2$, 2 m$M$ $Na_2S_2O_3$, 10 m$M$ $NaC_2H_3O_2$ buffer pH 5, total volume 4–12 liters; temperature, 37°.

as well as temperature, the elution position of the tRNA's is a function of the column load. In general, the higher the load, the lower the NaCl concentration at which a given tRNA will be eluted. Particularly with the RPC-2 and RPC-3 systems, substantial changes in the NaCl concentration range of the elution gradient must be made to compensate for changes in load. Much smaller load effects are noted with the RPC-4 system. With this system, some tRNA's seem to elute at a characteristic NaCl concentration over wide changes in load or other experimental conditions.

The relative elution sequence of various tRNA's as well as their position with regard to the concentration of NaCl in the RPC-3 system is a function of load. Data for a series of runs on a 2.5 × 90 cm RPC-3 column are shown in Fig. 1. The relative elution positions change with load. For example, with a total load of 5000 $A_{260}$ units, the elution sequence of tRNA's is fMet$_1$, Glu, Val, Arg; while with a 20,000 $A_{260}$ total load, the elution sequence is Glu, fMet$_1$, Arg, then Val. This differential sensitivity to load affords a useful variable in obtaining selected purification of specific tRNA's. It also introduces an additional variable to be reevaluated when scaling-up a fractionation scheme.

### Relationship of Gradient Slope to Chromatographic Development

With these reversed-phase chromatographic systems, a complex relationship exists between the rate of increase of the sodium chloride concentration in the elution gradient and chromatographic development and degree of separation of adjacent tRNA's. In general, the chromatographic resolution is controlled by the way in which the elution gradient develops and is independent of column geometry. This, of course, is not true of classical elution chromatography. The distribution coefficient for each individual tRNA at a given point within the column and the equilibration rate between the fixed quaternary ammonium exchanger and the mobile sodium chloride solution actually establish the chromatographic development; however, they cannot readily be measured and cannot simply be related to a measure of separation. Through practical experience, however, some empirical guide lines have been established for some of the RPC systems. It is interesting to note, also, that each RPC system appears to be somewhat unique in the way that it responds to the interaction of column load and gradient development so that generalizations cannot be made.

A field plot showing areas of good, intermediate, and poor resolution of the two phenylalanine tRNA's on RPC-1 and RPC-4 columns is shown in Fig. 2. The criterion for degree of resolution in Fig. 2 was the ability to resolve the two phenylalanine tRNA's into two well-defined peaks. The ordinates of the graph were empirically established; no other method of plotting reduced the data to fields of comparable resolution. These fields represent data from both RPC-1 and RPC-4 columns with dimensions of 1 inch by 3 feet, 1 inch by 8 feet, 2 inches by 3 feet, and 2 inches by 8 feet. The total load of phenylalanine tRNA's applied to the column, regardless of column size, interacted with the linear gradient slope as shown in the figure. This figure may be used to establish experimental conditions for the resolution of the two phenylalanine tRNA's from *E. coli* B or *E. coli* K12. For example, to chromatograph a load of $10^6$ pmoles of phenylalanine tRNA on a 5 × 100 cm column (cross-sectional area of 19.6 cm$^2$) in order to achieve good separation, the elution gradient should be at least 1.9 liters/cm$^2$ for each 0.1 *M* NaCl increase in concentration, or 37.2 liters per 0.1 *M* NaCl increase. If the gradient is steeper (total volume decreased), poorer resolution will be achieved. For example, with 0.95 liter/cm$^2$ or 18.6 liters per 0.1 *M* NaCl increase, the two phenylalanine tRNA peaks will not be resolved. It also illustrates the factors to be manipulated in optimizing the separation of other pairs of adjacent tRNA's in RPC-4 chromatograms.

A field plot for the RPC-3 system is shown in Fig. 3. In this case specific column load is plotted against gradient slope. This plot represents data from RPC-3 columns 1 inch by 3 feet, 1 inch by 8 feet, 2 inches by 3 feet, 2 inches by 8 feet, and 4 inches by 3 feet. Surprisingly,

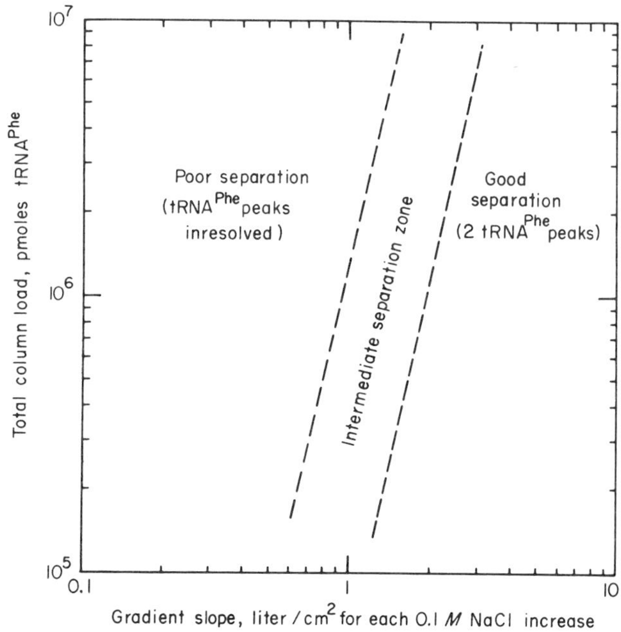


FIG. 2. Field plot relating degree of resolution to total column load and gradient slope for RPC-4 columns. Column geometries, 1 inch × 3 feet, 1 inch × 8 feet, 2 inches × 3 feet, 2 inches × 8 feet; elution gradient flow rate, 1.6 ml/min cm²; temperature, 37°.

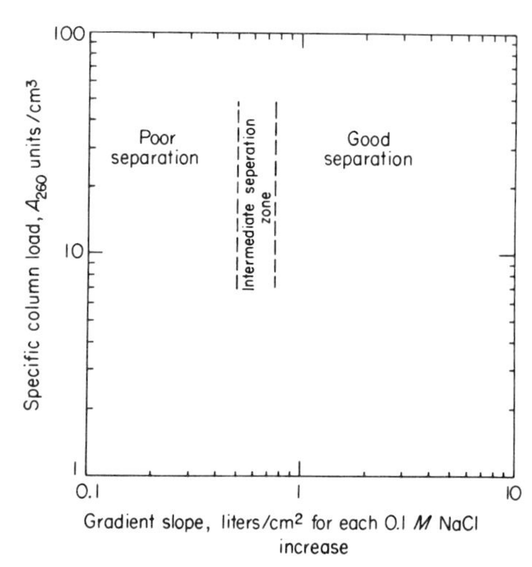


FIG. 3. Field plot relating degree of resolution to specific column load and gradient slope for RPC-3 columns. Column geometries, 1 inch × 3 feet, 1 inch × 8 feet, 2 inches × 3 feet, 2 inches × 8 feet, and 4 inches × 3 feet; flow rate 1.6 ml/min cm²; temperature, 37°.

in this system resolution appears to be unaffected by column load within the range tested and is entirely controlled by gradient slope.

## Eluent pH

The eluent pH affects the elution sequence of the tRNA's. This is probably due to changes in the number of protonated bases in the tRNA, which change with pH, and thus modulate the ionic interaction between the tRNA and the quaternary ammonium chloride. A typical plot showing the shift of elution position of tRNA's with pH on an RPC-3 column is shown in Fig. 4. Some tRNA's, such as formylmethionine-1 or arginine, show little change in elution position from pH 4.5 to pH 8. Others, such as valine or formylmethionine-2 tRNA's display a more pronounced dependence on elution pH, although in opposite directions. Glutamic acid tRNA showed the most remarkable effect, shifting backward in the chromatogram from pH 8 to pH5, and then abruptly shifting forward at pH 4.5. Thus, the tRNA elution sequence at pH 4.5 was glutamic acid, formylmethionine-1 and -2, valine, and arginine, while at pH 8 it was formylmethionine-1, glutamic acid, valine, formylmethionine-2, and arginine tRNA's. In all the RPC systems, advantage can be taken of this parameter to shift the chro-

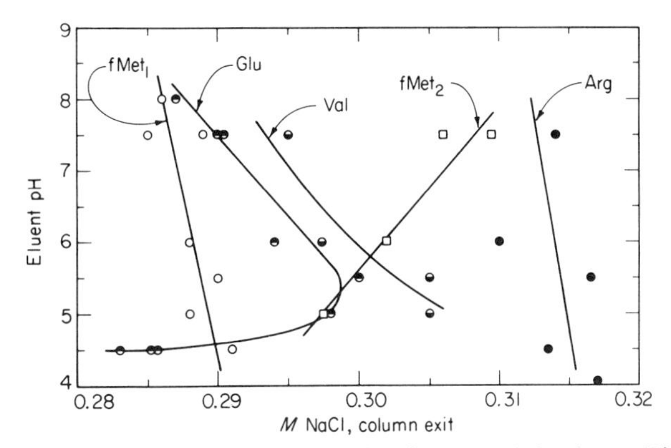


FIG. 4. Effect of eluent pH on elution position for an RPC-3 column. Dimension, 2.5 × 90 cm; feed, 4000 $A_{260}$ units of *Escherichia coli* K12 MO7 tRNA in 150 ml of equilibration solution, 0.20 *M* NaCl, 10 m*M* $MgCl_2$, 2 m*M* $Na_2S_2O_3$, 10 m*M* $NaC_2H_3O_2$ buffer at pH 4.5, 5, 5.5, 6, and 10 m*M* Tris·HCl buffer at pH 7.5 and 8; flow rate, 8 ml/min; elution gradient 0.26 to 0.36 *M* NaCl, 10 m*M* $MgCl_2$, 2 m*M* $Na_2S_2O_3$, and 10 m*M* buffer as above, total volume 4 liters; temperature, 37°.

matographic position of adjacent tRNA's so as to favor the separation of a particular tRNA.

Not all tRNA's are equally stable at all pH levels. Figure 5 shows the recovery of phenylalanine and tyrosine tRNA's after RPC-1 chromatography at various pH levels. Over a range of pH 4.5–9.5, recovery of phenylalanine tRNA averaged 71% and showed no trend. The tyrosine tRNA, however, displayed a different behavior pattern. From pH 6.5–8.5 the recovery averaged 76%, while at higher or lower pH levels it decreased sharply. At pH 4.5 in various runs the tyrosine tRNA recovery ranged from 0 to 20%. Solutions of the crude tRNA stored for several days at room temperature in the same pH 4.5 solution did not show a loss of tyrosine acceptance. The inactivation of tyrosine tRNA occurred on the RPC-1 column, possibly because of confor-

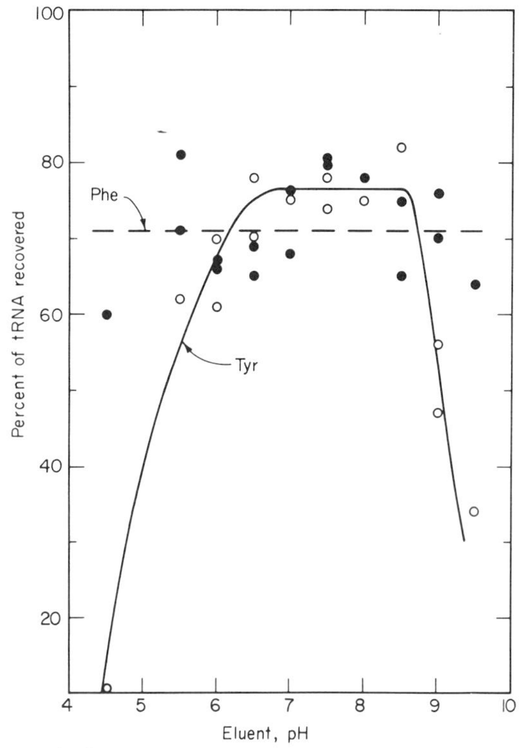


FIG. 5. Recovery of phenylalanine and tyrosine tRNA's on an RPC-1 column. Dimensions, 1 inch × 3 feet; feed, 3500 $A_{260}$ units of crude *Escherichia coli* W tRNA in 500 ml of 0.50 *M* NaCl, 10 m*M* $MgCl_2$, and 10 m*M* sodium acetate buffer at pH 4.5, 5.5, 6, and 20 m*M* Tris·HCl buffer at pH 6.5, 7.0, 7.5, 8.0, 8.5, 9.0, and 9.5; flow rate 8 ml/min; elution gradient, 0.55 *M* to 0.65 *M* NaCl, 10 m*M* $MgCl_2$ plus buffers indicated above, total volume 2.5 liters; temperature, 37°. ● = phenylanine tRNA; ○ = tyrosine tRNA.

mational changes or chemical sensitization while the tyrosine tRNA was dissolved in the immobile organic phase. Other specific tRNA's may be similarly labile under certain conditions, and each must be evaluated individually.

### Presence of Magnesium Ion

The presence or absence of 10 m$M$ magnesium ion in the eluent solutions affects both the elution sequence and recovery of certain tRNA's. In order to obtain a high recovery after RPC, certain tRNA's require the presence of magnesium ion, while others display an effect only on chromatographic position but no loss of recovery in the absence of magnesium. Furthermore, 10 m$M$ $Mg^{2+}$ shifts the entire elution range of the tRNA's to a lower NaCl concentration. These observations are consistent with the concept that tRNA elution from RPC columns is controlled by the availability of tRNA phosphate groups for interaction with the quaternary ammonium sites. Thus, the presence of magnesium ion, which is known to increase tRNA tertiary structure, makes the tRNA's easier to elute.

An example of the effect of magnesium ion on an RPC-3 chromatogram is shown in Fig. 6. With 10 m$M$ $MgCl_2$ present in the eluent, the major leucine tRNA eluted before the major valine tRNA near the front of the chromatogram and a sharp phenylalanine tRNA peak was formed near the rear of the chromatogram. The elution range was about 0.20–0.28 $M$ NaCl. In the absence of magnesium, the elution order of tRNA's in the front reversed, the valine tRNA now preceding the leucine tRNA while the phenylalanine tRNA gave a broad peak with poor recovery. Also, the elution range was shifted to about 0.30–0.38 $M$ NaCl.

As with pH changes, addition or removal of magnesium ion can be used to shift the chromatographic position of neighboring tRNA's, and advantage can be taken of this to favor the purification of desired tRNA's.

### Flow Rate

The rate at which the eluent gradient solution can be pumped through the column determines the length of time required to develop the chromatogram. With all the systems, a relationship exists between the flow rate and elution gradient volume required to obtain satisfactory chromatographic resolution. In general, the faster the flow rate, the larger volume of eluent (shallower gradient slope) is required. Much of the preparatory-scale work with the RPC-3 and RPC-4 systems has been carried out with flow rates of 1–2 ml/min/cm², although in other tests rates as high as 15 ml/min/cm² have routinely been used. In

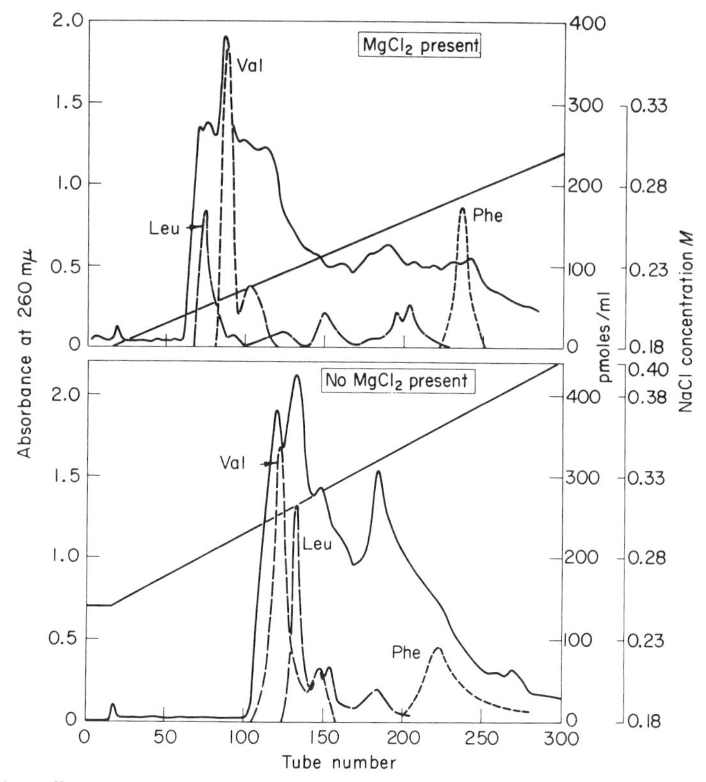


FIG. 6. Effect of 10 m$M$ magnesium chloride on tRNA elution position with an RPC-2 column. Dimensions, 1 × 240 cm; feed, 100 mg of crude *Escherichia coli* B tRNA in 2 ml of equilibration solution, 0.18 or 0.25 $M$ NaCl, 0 or 10 m$M$ $MgCl_2$ and 10 m$M$ Tris·HCl buffer at pH 7.0; flow rate, 2 ml/min; elution gradient, 0.18 to 0.30 or 0.25 to 0.40 $M$ NaCl plus other constituents as shown, total volume, 3 liters; temperature, 37°.

general, the time required for a chromatographic experiment can be shortened by increasing the flow rate, since the gradient volume increase is less than the flow rate increase. In addition, the interaction of flow rate and elution volume are not completely independent of column geometry effects; thus, to some degree, the best conditions for a given situation must be empirically established.

The RPC-2 system is much more sensitive to flow rate effects, possibly because in this case a discrete liquid ion-exchange phase exists and diffusion through the interface or in the organic phase may be rate limiting. Most of the experimental work with this system has been carried out with 1 × 240 cm columns at flow rates of 0.5–1.5 ml/min. Higher flow rates cause a significant loss of tRNA resolution.

## Protection with Reducing Agents

Crude tRNA from *E. coli* contains on the average about one 4-thiouridine per tRNA plus additional amounts of other thionucleotides. In order to protect these thionucleotides from oxidation during reversed-phase chromatography or storage, reducing agents are added to all chromatographic solutions. The addition of 2 m*M* sodium thiosulfate to eluent solutions for phenylalanine tRNA recovery on RPC-4 columns or for formylmethionine tRNA's on RPC-3 columns, as shown in Table I, increased overall recovery, increased the specific activity (amino acid acceptance) of the peak, and in tests with $^{35}S$-labeled crude tRNA, increased the sulfur content of the peak. Also, addition of thiosulfate sharpened the peak, improving chromatographic resolution. From these results it is concluded that the addition of sodium thiosulfate "protects" the tRNA's. Crude tRNA from many other organisms also contain thionucleotides, although not at as high a level as *E. coli*. Thus, thiosulfate protection should also be of advantage with tRNA's from sources other than *E. coli*. Only yeast tRNA's seem to be free of thionucleotides.

TABLE I
EFFECT OF THIOSULFATE "PROTECTION"

| tRNA | RPC system | $Na_2S_2O_3$[a] | Recovery (%) | Amino acid accepting activity (picomoles/$A_{260}$) | $^{35}S$ content (cpm/$A_{260}$) | Peak shape ($h/w_{1/2}$) |
|---|---|---|---|---|---|---|
| Phe | 4 | − | 70 | 543 | 1737 | 3.2 |
| | | + | 90 | 780 | 2025 | 5.6 |
| Met | 3 | − | 76 | 1278 | 1451 | 8.9 |
| | | + | 93 | 1608 | 1603 | 11.4 |

[a] Plus sign indicates 2 m*M* $Na_2S_2O_3$ added to chromatographic solutions.

In the RPC-2 system, it was found that the tricaprylylmethyammonium chloride selectively bound the thiosulfate anion in preference to tRNA's and the column was rapidly saturated with thiosulfate and rendered inoperable. Therefore, in place of thiosulfate a nonionic reducing agent, 10 m*M* $\beta$-mercaptoethanol, is added to eluent solutions for the RPC-2 system.

## Purity of Feed tRNA

In order to obtain the desired chromatographic resolution, the feed tRNA sample must be free of certain contaminants. Ribonuclease, of course, must be absent. It has been found that DNA at levels of 1–3% in the feed tRNA will rapidly poison a reversed-phase column; the resolution decreases in subsequent runs on the column. Ribosomal

RNA fragments or other higher molecular weight RNA's must also be low in the feed tRNA. These larger RNA's bind to the packing and cannot be efficiently eluted with 1 or 1.5 *M* NaCl solutions; again, chromatographic resolution and product purity are sharply reduced.

## Specific Examples

### RPC-2 System

The following experimental procedure details the separation of 100 mg of crude *E. coli* B tRNA on a 1 × 240 cm RPC-2 column.

*Preparation of Organic Phase.* The organic phase is prepared by dissolving 5 ml of Aliquat 336 in 95 ml of tetrachlorotetrafluoropropane. This solution, in a separatory funnel, is washed successively with about 2 volumes each of 1 *M* NaOH, 1 *M* HCl, and 0.5 *M* NaCl to remove any water-soluble contaminants. The wash solution and organic phase are shaken for approximately 2 minutes and then allowed to stand until phase disengagement is nearly complete. After the 0.5 *M* NaCl wash, the organic is passed through filter paper to clarify it by removing any dispersed aqueous droplets. The prepared organic phase is stored in a brown glass bottle until needed.

*Preparation of Column Packing.* Chromosorb W, AW-DMCS 100–120 mesh, 100 g, is placed in a plastic or glass beaker of sufficient size so that it can be easily and rapidly stirred with a flat-bladed ceramic spatula. Metal spatulas cannot be used because of the abrasive nature of the Chromosorb. Then, 56 ml of the prepared organic, placed in a dropping funnel suspended above the beaker or tray, is dropped onto the Chromosorb in a fine stream while the solid is constantly stirred with the spatula. About 10 minutes is required for addition of the organic; longer times should be avoided since excessive evaporation of the tetrachlorotetrafluoropropane may occur. The packing is then placed in a plastic bottle, capped tightly, and mechanically rotated for 2 hours to ensure even distribution of the organic on the Chromosorb. Rotation should be rapid enough only to ensure blending of the solids; violent shaking may fracture the Chromosorb particles and expose untreated surfaces. After preparation, the packing should weigh about 190–200 g. If the weight is below 180 g, as a result of evaporation of the diluent, additional tetrachlorotetrafluoropropane should be added and the blending step repeated. The prepared packing should be stored in a tightly capped glass bottle until needed.

*Column Preparation.* A jacketed glass column, 1 × 240 cm, is connected to a circulating water thermostat at 37° and filled with the initial equilibration solution, 0.18 *M* NaCl, 10 m*M* $MgCl_2$, 10 m*M* β-mercaptoethanol, and 10 m*M* Tris. HCl buffer at pH 7.0. The prepared packing is slurried with a portion of the initial equilibration solution and allowed to stand for a few minutes to permit entrapped air to escape. The packing slurry is then poured into the column and allowed to settle, with maximum possible aqueous flow through the column to facilitate dense packing of the bed. The bed is then further compacted by shutting off the stopcock on the bottom of the column, pumping initial equilibration solution at 5–10 ml/minute with a piston pump (Milton Roy Mini Pump) equipped with a pressure gauge isolated by a Teflon-coated all-PVC diaphram (Mansfield and Green, Type B) to a pressure of 60–80 psi, and then abruptly opening the stopcock. Chromatographic resolution is improved by this compaction procedure. Also, during normal column operation, the needle-valve Teflon stopcock on the bottom of the column is restricted to maintain a positive pressure of 10–15 psi in the column to prevent degassing in the column of air dissolved in the aqueous solutions.

*Chromatographic Operation.* The column is equilibrated before use by pumping through at least 500 ml of initial equilibration solution. Then 100 mg (approximately 2000 $A_{260}$) of *E. coli* B crude tRNA[9] is dissolved in 2–3 ml of initial equilibration solution and pipetted onto the column. A 3-liter concave gradient from 0.20 *M* to 0.36 *M* NaCl, containing 10 m*M* $MgCl_2$, 10 m*M* β-mercaptoethanol, and 10 m*M* Tris. HCl pH 7, is pumped at 0.5 ml/minute. A 9-chamber gradient generator (Phoenix Precision Instrument Co.) is employed to produce the gradient. The concave gradient[3] affords a more even distribution of the tRNA's. The column effluent is monitored at 260 mμ (Beckman Instruments, DB spectrophotometer and log recorder), and 15-ml fractions are collected in a refrigerated fraction collector (LKB Products). Selected fractions are subsequently assayed for amino acid acceptance.[2,10] The column is then washed to remove any residual material with 500 ml of a solution containing 0.5 *M* NaCl, 10 m*M* $MgCl_2$, 10 m*M* β-mercaptoethanol, 10 m*M* Tris. HCl pH 7 at 0.5 ml/min. After re-equilibration with the initial solution, the column may be reused. The columns may be used for many runs, depending on the purity of the crude tRNA applied. As a precaution, columns are stored at 4° between experiments to minimize growth of extraneous microorganisms, although this never appeared to be a problem.

The results of a run performed under these conditions is shown in

[9] C. W. Hancher, E. F. Phares, G. D. Novelli, and A. D. Kelmers, *Biotechnol. Bioeng.*, 11, 1055 (1969).

[10] I. B. Rubin, A. D. Kelmers, and G. Goldstein, *Anal. Biochem.* **20**, 533 (1967).

Fig. 7. A relatively even distribution of tRNA's from the front to the rear of the chromatogram is seen. Also, multiple peaks were developed for many tRNA's, for example, 5 for leucine, and 2 for methionine, arginine, valine, isoleucine, histidine, and alanine tRNA's.

## RPC-3 System

The experimental procedure for the purification of formylmethionine and arginine tRNA's from *E. coli* K12 MO on a production size RPC-3 column is given in this section. Approximately 200,000 $A_{260}$ units of tRNA that had been previously partially purified by DEAE-cellulose chromatography are fractionated in one run on a 10 × 100 cm RPC-3 column.

*Preparation of Column Packing.* In order to prepare large quantities (greater than 800-g batches) of RPC packings, a "packing making machine" was constructed from a 12-inch long section of 8-inch diameter glass pipe with removable Teflon end plates held in place by modified pipe flanges. Four 1.5-inch Teflon baffles are attached to the interior of the glass pipe. The assembled glass pipe and end plates are mounted horizontally on trunnions on a frame and chain-driven by an air motor so that it rotates at approximately 10 rpm. Connections are provided through the trunnions for the addition of liquids or air and for air discharge. Chromosorb G, AW-DMCS, 100–120 mesh, 2000 g, is placed in the machine, and it is assembled as described above. Sixty grams of trioctylpropylammonium bromide is dissolved in 2 liters of chloroform, and this solution is added by gravity flow to the Chromosorb while it is tumbling in the rotating packing maker. A flow rate of roughly 30 ml/minute is maintained for the addition of the chloroform solution. The inlet and outlet vents are then closed with stoppers, and blending is continued for about an hour to ensure thorough distribution of the organic throughout the Chromosorb. The chloroform is then evaporated by removing the stoppers and blowing filtered dry air at about 5 cfm through the rotating packing maker until the packing is completely dry. The test for completion is the absence of chloroform odor in the exit air. The prepared dry packing is stored in a tightly covered plastic container until needed.

*Column Preparation.* A 10 × 100 cm jacketed glass column (Pharmacia) is employed. The jacket is connected to a circulating thermostatted water bath which maintains the column jacket at 37°. The column is about half filled with the initial equilibration solution, 0.20 *M* NaCl, 10 m*M* $MgCl_2$, 2 m*M* $Na_2S_2O_3$, and 10 m*M* $NaC_2H_3O_2$ buffer at pH 4.5. Several hundred milliliters of the prepared dry column packing is

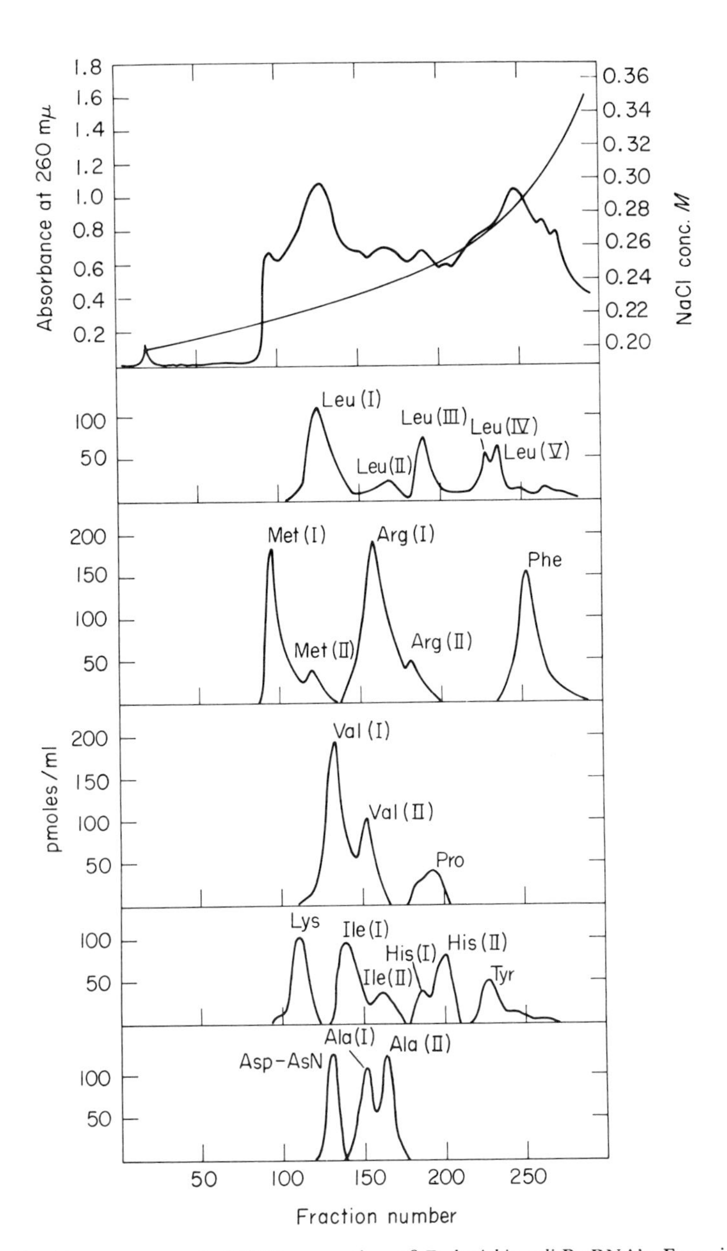


FIG. 7. Preparatory-scale RPC-2 separation of *Escherichia coli* B tRNA's. Experimental details in Specific Examples section.

slurried with additional equilibration solution in a beaker and then poured into the column and allowed to settle. The valve on the bottom of the column is then opened, and the excess equilibration solution is drained through the settled packing down to the level of the packing. Care is taken not to drain the solution below the packing level at any time because this would allow air to get into the packed bed. Excess solution is drained from the column, then more packing is slurried in a beaker, poured into the column, and allowed to settle; the excess aqueous again is withdrawn. The bed is then firmly compacted by tamping with a glass or plastic rod with a flat bottom of about ¾-inch diameter. Addition of slurried packing and tamping are continued until the column is filled, and the upper column end-piece is positioned against the top of the packed bed.

*Column Operation.* The column is equilibrated before use by pumping through at least one column volume of equilibration solution at 32 ml/minute with a positive displacement piston pump (Milton Roy Mini-Pump). The connecting lines are ⅛-inch polyethylene tubing with Teflon fittings. The solution feed to the column is preheated to 37° by connecting the pump discharge to a 50-foot coil of tubing in the thermostatted water bath and then through a jacketed length of tubing connected to the column inlet. The columns are generally operated down-flow. This is more convenient than up-flow since if a small amount of air enters the inlet line, it collects at the top of the column and can be removed between runs. Approximately 200,000 $A_{260}$ units of *E. coli* K12 MO crude tRNA (partially concentrated for methionine tRNA's by DEAE-cellulose chromatography[9]) are dissolved in the equilibration solution at a concentration of 60 $A_{260}$ units/ml or lower (3.3 liters or greater feed solution volume). This feed solution is pumped on the column at 32 ml/minute. A linear elution gradient with a volume of 45 liters from 235 m$M$ NaCl to 305 m$M$ NaCl is generated and pumped through the column, again at 32 ml/minute. The eluent also contains 10 m$M$ $MgCl_2$, 2 m$M$ $Na_2S_2O_3$, and 10 m$M$ $NaC_2H_3O_2$ at pH 4.5. The gradient generating system consists of a stirred 30-gal plastic tank initially containing 22.5 liters of the dilute sodium chloride concentration solution into which the concentrated solution is pumped at 16 ml/minute from a second unstirred 30-gal plastic tank initially containing 22.5 liters of the concentrated solution. The elution gradient is pumped at 32 ml/min from the stirred tank. The column effluent is monitored at 260 m$\mu$ with a double beam spectrophotometer (Beckman DB) equipped with a quartz flow cell. Fractions of 160 ml are collected in a large-sized 250-tube fraction collector,[11] and selected

[11] C. W. Hancher, *Biotechnol. Bioeng.* **10**, 681 (1968).

fractions subsequently assayed for amino acid acceptance. The column is then washed with 8 liters of 1.5 *M* NaCl, 10 m*M* $MgCl_2$, 2 m*M* $Na_2S_2O_3$, and 10 m*M* $NaC_2H_3O_2$ at pH 4.5 to remove residual material. After a subsequent wash with the equilibration solution, the column is ready for reuse. The effective life of a column is approximately 5 runs because of a low but appreciable aqueous solubility of the quaternary ammonium compound. After the fifth run, the packing is removed and the column is refilled with fresh packing.

A chromatogram from such an experiment is shown in Fig. 8. The bulk of the formylmethionine tRNA's, 73%, are eluted after a small glutamic acid peak and before the arginine tRNA peak, with a methionine accepting activity of 1360 pmoles/$A_{260}$ unit. About 47% of the arginine tRNA was also recovered with an arginine acceptance of 905 pmoles/$A_{260}$. This was subsequently further purified by rechromatography on RPC-4 columns at pH 7.0 to an activity of 1266 pmoles/$A_{260}$.

The procedure and results described in this section are for large-scale runs for the production of purified formylmethionine and arginine tRNA's. Smaller scale runs have been described previously.[4]

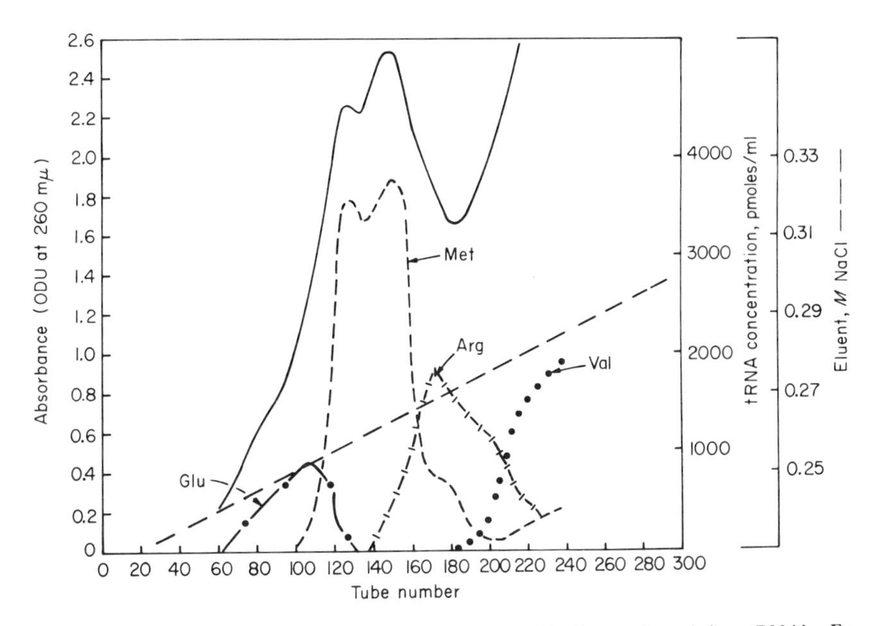


FIG. 8. Large-scale RPC-3 recovery of formylmethionine and arginine tRNA's. Experimental details in Specific Examples section.

**RPC-4 System**

The following experimental procedure details the separation of 200,000 $A_{260}$ unit lots of *E. coli* K12 crude tRNA's on a 5 × 240 cm RPC-4 column for the preparation of purified phenylalanine tRNA's.

*Recrystallization of Adogen 462.* The quaternary ammonium extractant, dimethyldicocoammonium chloride is purchased as an ~50% solution (Adogen 462) from Ashland Chemical Company. It must be recrystallized to a white powder before use. About 200 ml of Adogen 462 is dissolved in 2 liters of acetone and filtered to remove insoluble contaminants. The filtrate in a tightly stoppered flask is placed in an explosion-proof freezer (approximately −20°) overnight to crystallize. The crystals are recovered by vacuum filtration on a fritted-glass funnel and then are recrystallized at least once more, or until white crystals are obtained. These are air-dried overnight until free of acetone.

*Preparation of Column Packing.* Chromosorb G, 2000 g, acid washed, dimethyldichlorosilane-treated, 100–120 mesh size, is placed in the "packing-making machine" described under the RPC-3 system. Then, 22.5 g of recrystallized dimethyldicocoammonium chloride is dissolved in 2 liters of chloroform, blended with the Chromosorb, and dried as previously described. The prepared dry packing is stored in a tightly covered container until needed.

*Column Preparation.* A jacketed glass column, 5 × 240 cm, was constructed in our glass shop using 2-inch diameter glass pipe as the chromatographic column. This large-scale separation could also be carried out using a commercial 10 × 100 cm jacketed column (Pharmacia). The jacket is connected to a thermostatted circulating water bath at 37°. The column is partially filled with the initial equilibration solution, 0.52 *M* NaCl, 10 m*M* $MgCl_2$, 2 m*M* $Na_2S_2O_3$, and 20 m*M* Tris-HCl buffer at pH 8.0. Several hundred milliliters of prepared column packing is slurried with equilibration solution in a beaker and poured into the column and allowed to settle. The valve on the bottom of the column is opened, and the excess solution is drained through the packing down to the level of the packing. Then, more dry packing is slurried with equilibration solution and poured into the column, and the excess solution is withdrawn. Care is taken that the solution level does not fall below the level of the packing and allow air to get into the packed bed. A glass or plastic rod with a flat bottom of about ¾-inch diameter is used to firmly tamp the packing down to tightly compact the bed. Alternate filling and tamping is continued until the column is full.

*Column Operation.* The column is equilibrated before use by pumping through at least one column volume of equilibration solution at 32 ml/

minute with a positive displacement piston pump (Milton Roy Mini-Pump). The tubing is ⅛-inch polyethylene with Teflon fittings. The feed solution to the column is preheated as described for the RPC-3 column. Approximately 200,000 $A_{260}$ units of *E. coli* K12 MO crude tRNA, partially concentrated in phenylalanine tRNA[9] by DEAE-cellulose chromatography is dissolved in 500 ml of equilibration solution and pumped onto the column at 32 ml/minute. This is followed by 15 liters of equilibration solution. Most of the tRNA's do not bind to the column at this salt concentration and are eluted in this solution. A linear elution gradient of 56-liter volume is generated using the apparatus described in the RPC-3 section. The linear gradient goes from 0.52 to 0.60 *M* NaCl, and the solutions also contain 10 m*M* $MgCl_2$, 2 m*M* $Na_2S_2O_3$, and 20 m*M* Tris·HCl buffer at pH 8.0. The column effluent is monitored with a double-beam spectrophotometer (Beckman DB) and collected in fractions of 160 ml in a modified large-size fraction collector.[11] Selected fractions are assayed for amino acid acceptance. After completion of the gradient, 4 liters of a washing solution, 1.5 *M* NaCl, 10 m*M* $MgCl_2$, 2 m*M* $Na_2S_2O_3$, and 20 m*M* Tris. HCl buffer at pH 8.0 are pumped through the column to remove residual materials. After reequilibration, the column is ready to be used again. As long as the sodium chloride concentration is kept above 0.45 *M*, the Adogen 462 displays very low aqueous solubility and the column may be used again many times provided the feed tRNA is low in contaminants, such as DNA, that bind irreversibly and "poison" the column.

The results of a typical run are shown in Fig. 9. The two phenylalanine tRNA's are partially resolved and are well separated from other tRNA's. About 14% of the phenylalanine tRNA in the feed was pooled as $tRNA_1^{Phe}$ product with a phenylalanine acceptance activity (purity) of 880 pmoles/$A_{260}$ unit and 24% as $tRNA_2^{Phe}$ product with an acceptance activity of 1042 pmoles/$A_{260}$. The tyrosine tRNA also yielded a pool of 780 pmoles tyrosine acceptance/$A_{260}$ unit.

Small-scale RPC-4 experiments have been reported previously,[4] and, in general, the RPC-4 system is very similar to the RPC-1 system used for the purification of phenylalanine tRNA from *E. coli* B[12] and leucine tRNA.[13]

## Applications

In order to apply these reversed-phase chromatographic systems to a specific problem, the individual experimenter is faced with the task of establishing conditions for each of the experimental variables described

[12] A. D. Kelmers, *J. Biol. Chem.* **241**, 3540 (1966).

[13] A. D. Kelmers, *Biochem. Biophys. Res. Commun.* **25**, 562 (1966).

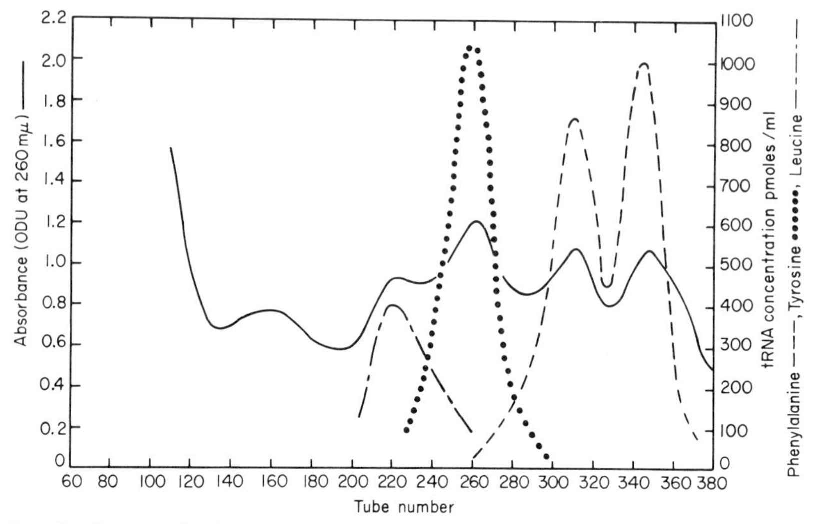


FIG. 9. Large-scale RPC-4 recovery of phenylalanine and tyrosine tRNA's. Experimental details in Specific Examples section.

under General Considerations. A number of specific examples for the separation and purification of *E. coli* tRNA's are given above, and for similar experiments it is possible to decide quickly on the conditions required for the purification of these *E. coli* tRNA's. For other tRNA's or for organisms other than *E. coli*, the investigator will have to spend more time optimizing the chromatographic conditions for his specific case.

Several useful generalizations about the RPC systems can be made. The RPC-2 system has the most even distribution of tRNA's throughout the chromatogram from the front to the rear. This makes this system particularly useful in studying the heterogeneity of tRNA's, as described in volume XX [3–5]. Another advantage is the relatively long life of a column due to the very low aqueous solubility of the fluorocarbon diluent and quaternary ammonium chloride. The RPC-2 system has proved less useful for large-scale preparatory work because of the flow rate limitations with this system. It also is more sensitive to variations in lots of Chromosorb W and to changes in the nature of the feed tRNA and, at preparatory-scale level, has proved difficult to operate so as to obtain consistent results. On some occasions, for extended periods of time, it has been impossible to obtain satisfactory resolution of 50- or 100-mg samples of tRNA on RPC-2 columns even though excellent results with tracer-level quantities, as described in this volume [3–5], were being simultaneously obtained. Despite extensive investigations, the reasons for these failures remain unknown. Thus, in general, the RPC-2 system should be considered for preparatory-scale work only when it is clear that the desired separations cannot be obtained on RPC-3 or RPC-4. Excellent preparatory-scale RPC-2 chromatograms, at loads of 4000–5000 $A_{260}$ units of crude tRNA, have recently been published by Anderson.[14]

The RPC-3 system has proved to be particularly useful for separation of tRNA's near the front of the chromatogram; tRNA's can be readily resolved with this system which do not separate in any other systems. Also, the RPC-3 system has proved to be highly dependable and reproducible. The only disadvantage is the low but appreciable solubility of the quaternary ammonium compound, which limits the life of the columns.

The RPC-4 system is most useful for the separation of tRNA's that run near the rear of the chromatogram, but tends to give less separation of tRNA's near the front of the chromatogram. Results with this column also tend to be highly reproducible, and the column can be used for many runs provided contaminants such as DNA are not present in the feed in appreciable quantities.

The RPC systems have been used for the separation and purification of a variety of tRNA's from different organisms. A listing of the published results is given in Table II, as a guide to the prospective investigator.

TABLE II

COMPILATION OF PUBLISHED REVERSED-PHASE CHROMATOGRAPHY APPLICATIONS

| Source | tRNA | Reference |
|---|---|---|
| *Escherichia coli* | Ala | *a* |
| | Ala, Val | *b* |
| | Arg | *c* |
| | Arg | *d* |
| | Leu | *e* |
| | Leu | *f* |
| | Leu | *g* |
| | fMet | *h* |
| | fMet | *i* |
| | Phe | *j* |
| | Phe | *k* |
| | Tyr | *l* |
| | Several | *m* |
| | Several | *n* |
| | Several | *o* |
| | Several | *p* |
| | Several | *q* |
| Yeast | Ala | *r* |
| | His | *s* |
| | Several | *t* |

[14] W. F. Anderson, *Biochemistry* **8**, 3687 (1969).

TABLE II (*Continued*)

| Source | tRNA | Reference |
|---|---|---|
| *Bacillus subtilis* | His | *u* |
| *Neurospora crassa* | Several | *v* |
| *Euglena* | Several | *w* |
| Wheat germ | Phe | *x* |
| Mouse liver | Leu | *y* |
| Mouse plasma cell tumor | Several | *z* |
| | Several | *aa* |
| | Several | *bb* |
| Chicken erythrocytes | Several | *cc* |
| Rabbit reticulocytes | Several | *dd* |
| Rabbit liver | Ala | *a* |
| Soybean | Leu | *ee* |
| Human spleen | Ala | *a* |

[a] W. F. Anderson, *Biochemistry* **8**, 3687 (1969).
[b] V. Z. Holten and K. B. Jacobson, *Arch. Biochem. Biophys.* **129**, 283 (1969).
[c] W. F. Anderson, *Proc. Nat. Acad. Sci. U. S.* **63**, 566 (1969).
[d] T. Leisinger and H. J. Vogel, *Biochim. Biophys. Acta* **182**, 572 (1969).
[e] L. C. Waters and G. D. Novelli, *Proc. Nat. Acad. Sci. U. S.* **57**, 979 (1967).
[f] A. D. Kelmers, G. D. Novelli, and M. P. Stulberg, *J. Biol. Chem.* **240**, 3979 (1965).
[g] L. C. Waters and G. D. Novelli, *Biochem. Biophys. Res. Commun.* **32**, 971 (1968).
[h] L. Shugart, B. Chastain, and G. D. Novelli, *Biochem. Biophys. Res. Commun.* **37**, 305 (1969).
[i] L. Shugart, B. Chastain, and G. D. Novelli, *Biochim. Biophys. Acta* **186**, 384 (1969).
[j] A. D. Kelmers, *J. Biol. Chem.* **241**, 3540 (1966).
[k] L. Shugart, G. D. Novelli, and M. P. Stulberg, *Biochim. Biophys. Acta* **157**, 83 (1968).
[l] M. L. Gefter and R. L. Russell, *J. Mol. Biol.* **39**, 145 (1969).
[m] S. Nishimura, F. Harada, V. Narushima, and T. Seno, *Biochim. Biophys. Acta* **142**, 133 (1967).
[n] J. F. Weiss, R. L. Pearson, and A. D. Kelmers, *Biochemistry* **7**, 3749 (1968).
[o] J. F. Weiss and A. D. Kelmers, *Biochemistry* **6**, 2507 (1967).
[p] A. D. Kelmers, G. D. Novelli, and M. P. Stulberg, *J. Biol. Chem.* **249**, 3979 (1965).
[q] F. J. Kull and K. B. Jacobson, *Proc. Nat. Acad. Sci. U. S.* **62**, 1137 (1969).
[r] C. R. Merril, *Biopolymers* **6**, 1727 (1968).
[s] K. Takeishi, S. Nishimura, and T. Ukita, *Biochim. Biophys. Acta* **145**, 605 (1967).
[t] G. Dirheimer and B. Kuntzel, *Bull. Soc. Pharm. Strasbourg* **9**, 97 (1966).
[u] M. P. Stulberg, K. R. Isham, and A. Stevens, *Biochim. Biophys. Acta* **186**, 297 (1969).
[v] J. L. Epler, *Biochemistry* **8**, 2285 (1969).
[w] W. E. Barnett, C. J. Pennington, and S. A. Fairfield, *Proc. Nat. Acad. Sci. U. S.*, **63**, 1261 (1969).
[x] B. S. Dudock, G. Katz, E. K. Taylor, and R. W. Holley, *Proc. Nat. Acad. Sci. U. S.* **62**, 941 (1969).
[y] J. F. Mushinski and M. Potter, *Biochemistry* **8**, 1684 (1969).
[z] W.-K. Yang and G. D. Novelli, *Proc. Nat. Acad. Sci. U. S.* **59**, 208 (1968).
[aa] W.-K. Yang and G. D. Novelli, *in* "Nucleic Acids in Immunology" (O. J. Plescia and W. Braun, eds.), p. 644. Springer, New York, 1968.
[bb] W.-K. Yang and G. D. Novelli, *Biochem. Biophys. Res. Commun.* **31**, 534 (1968).
[cc] J. C. Lee and V. M. Ingram, *Science* **158**, 1330 (1967).
[dd] W. F. Anderson and J. M. Gilbert, *Biochem. Biophys. Res. Commun.* **26**, 456 (1969).
[ee] M. B. Anderson and J. H. Cherry, *Proc. Nat. Acad. Sci. U. S.* **62**, 202 (1969).

# [2] Miniature Reversed-Phase Chromatography Systems for the Rapid Resolution of Transfer RNA's and Ribosomal RNA's[1]

*By* A. D. KELMERS, D. E. HEATHERLY, and B. Z. EGAN

Reversed-phase chromatography (RPC) techniques[2] utilizing Plaskon-supported[3] quaternary ammonium chloride extractants have been adapted for the rapid (30–60 minute) resolution of small quantities of tracer-labeled transfer RNA's[4] and for the analytical scale resolution of ribosomal RNA's.[5] These new methods using miniature chromatographic columns offer the advantages of much smaller sample sizes and more rapid operation than previous methods.

## Methods

### *Preparation of Chromatographic Support*

The RPC-5 system[3] utilizing Adogen 464 [methyltrialkyl($C_8$-$C_{10}$)-ammonium chloride obtained from Ashland Chemical Company, Columbus, Ohio] supported on Plaskon (polychlorotrifluoroethylene obtained as Plaskon CTFE 2300 powder from Allied Chemical Company, Morristown, New Jersey) is prepared by modifications of the published methods.[3] Four milliliters of Adogen 464 is dissolved in 200 ml of chloroform and vigorously mixed with 100 g of Plaskon powder. Mixing can be accomplished by use of a jar on a shaking table for approximately 1 hour or by vigorous mechanical stirring for shorter periods of time. Aggregates of Plaskon are dissociated during the mixing. The mixture is then placed in a glass tray and blended with a ceramic spatula while the chloroform is completely evaporated. Continuous blending during the latter stages of the chloroform evaporation assures even coating of each particle of Plaskon with Adogen 464. The dry RPC-5 packing may be stored indefinitely. RPC-5 packing is now also commercially available.

[1] Research sponsored by the National Institute of General Medical Sciences, National Institutes of Health, under contract with the U.S. Atomic Energy Commission and Union Carbide Corporation, Nuclear Division, at Oak Ridge National Laboratory, Oak Ridge, Tennessee.

[2] A. D. Kelmers, H. O. Weeren, J. F. Weiss, R. L. Pearson, M. P. Stulberg, and G. D. Novelli, this series, Vol. 20, p. 9.

[3] R. L. Pearson, J. F. Weiss, and A. D. Kelmers, *Biochim. Biophys. Acta* **228,** 770 (1971).

[4] A. D. Kelmers and D. E. Heatherly, *Anal. Biochem.* **44,** 486 (1971).

[5] B. Z. Egan and A. D. Kelmers, *Prep. Biochem.* **2,** 265 (1972).




ISBN 0-12-504180-2

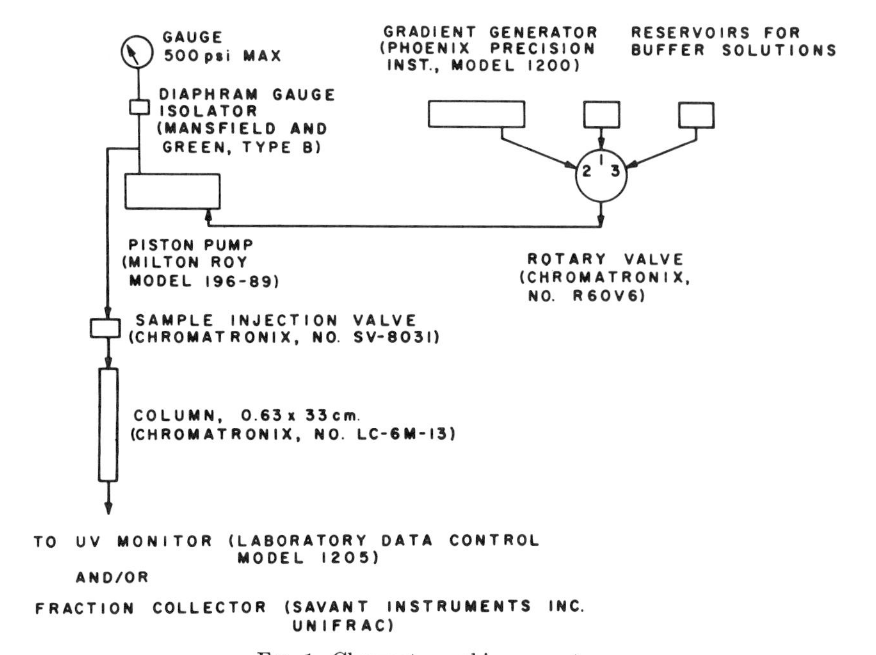


FIG. 1. Chromatographic apparatus.

*Chromatographic Apparatus*

The chromatographic apparatus used is shown in Fig. 1. The column and all solutions are at room temperature. For gradient volumes of 50 ml or larger, a nine-chamber Phoenix gradient generator can be used. For volumes of 50 ml or less, a generator with two conical chambers (Buchler Instruments, Inc.) can be used. The flow of eluent through the column is maintained at a constant flow rate by a positive displacement pump, and the pressure (usually 200–400 psi depending on the flow rate) is monitored with a gauge isolated from the system by a gauge protector (Mansfield and Green, Model B) constructed of polyvinyl chloride with a Teflon-coated diaphragm. Tubing on the inlet side of the pump is 0.031 inch ID Teflon (Chromatronix No. T063031), and on the discharge side is 0.012 inch ID Teflon (Chromatronix No. T063012). The column effluent is connected to a fraction collector that discharges directly into scintillation vials for tracer-labeled tRNA's or to a high-sensitivity ultraviolet monitor for analysis of ribosomal RNA's.

*Chromatographic Experiments*

To prepare a chromatographic column, the dry RPC-5 packing is slurried with an excess of the final gradient solution and degassed. This is accomplished by stirring the mixture with a magnetic stirrer in a flask connected to an aspirator. The slurry is then poured into the column and compacted by flow under pressure. After the column is equilibrated, the sample is applied with the sample injection valve. The pump is not stopped during the run; changing from equilibration to gradient solutions is accomplished by means of the rotary selection valve. After completion of a chromatographic run the column is reequilibrated and can be used repeatedly. Details of typical runs are given in the figure legends.

## Results

A typical chromatographic run showing the resolution of the five *E. coli* leucyl-tRNA's is presented in Fig. 2. The run required only 30 minutes, and sharp resolution of even the minor leucine-2 tRNA can be seen. The chromatographic conditions, given in the figure legend, are applicable for the resolution of both bacterial and mammalian tracer-labeled tRNA's.[4]

Experiments were made to show the minimum quantity of tracer-labeled tRNA that can be practically resolved. A total of 50 cpm of [$^{14}$C]-phenylalanyl-tRNA, equivalent to $3.5 \times 10^{-9}$ g, or $6 \times 10^{-5}$ $A_{260}$ unit,[6]

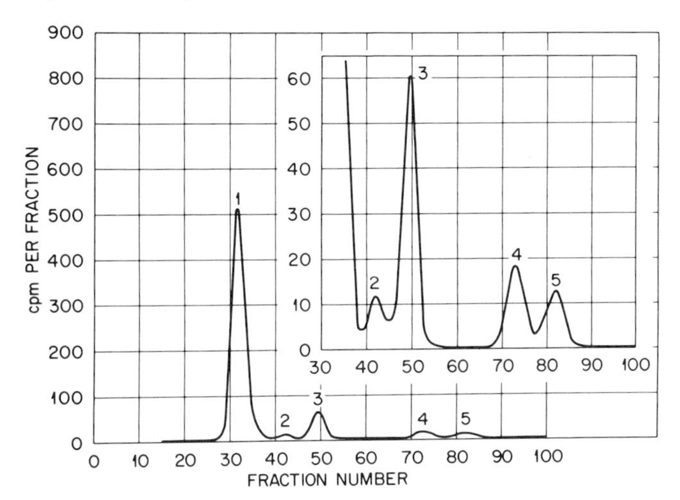


FIG. 2. Resolution of *Escherichia coli* leucyl-tRNA's. The minor peaks are shown in expanded detail in the inset. Approximately 0.10 $A_{260}$ unit of crude tRNA aminoacylated with L-[$^{14}$C]leucine (A. D. Kelmers *et al.*, this series, Vol. 20, p. 9), approximately 5000 cpm, was applied to a column equilibrated with a solution containing 0.5 *M* NaCl plus buffer constituents. The 100-ml volume elution gradient was from 0.5 to 0.8 *M* NaCl plus buffer constituents (10 m*M* $MgCl_2$, 10 m*M* $NaC_2H_3O_2$ at pH 4.5, and 2 m*M* $\beta$-mercaptoethanol). The column was operated at 3.3 ml per minute, and 1-minute fractions were collected.

[6] One $A_{260}$ unit is the quantity of RNA that, when dissolved in 1 ml of $H_2O$, gives an absorbance of 1 at 260 nm in a 1-cm cell.

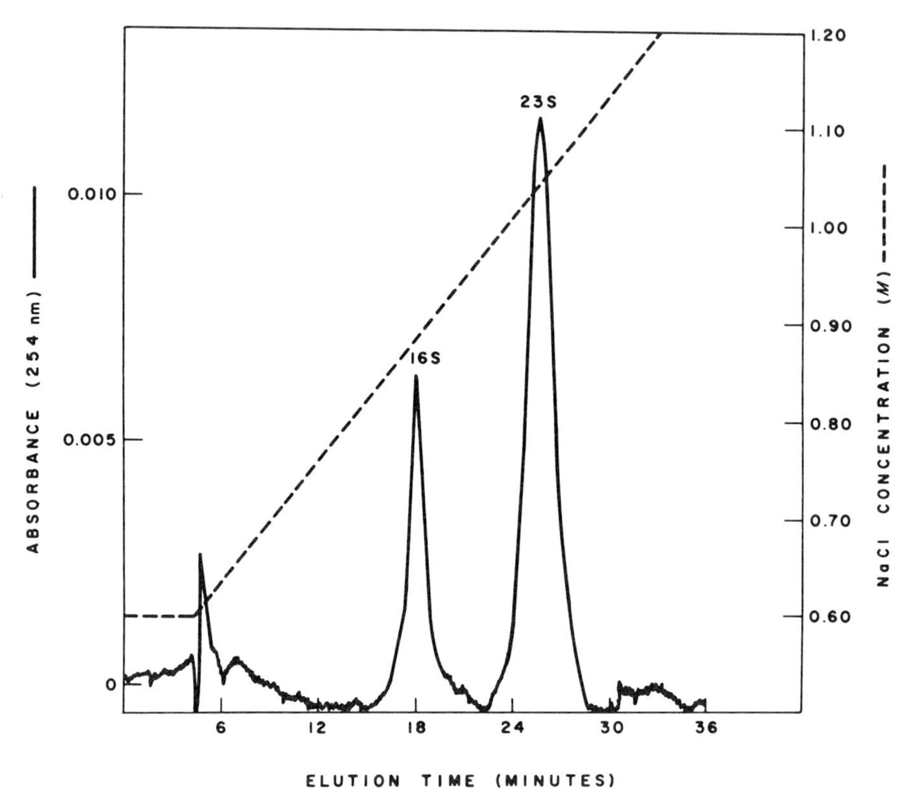


FIG. 3. Resolution of *Escherichia coli* ribosomal RNA's. Approximately 0.1 $A_{260}$ unit of rRNA was applied to a column equilibrated with 0.6 $M$ NaCl plus buffer constituents. A 30-ml volume gradient from 0.6 to 1.2 $M$ NaCl plus buffer constituents (10 m$M$ $MgCl_2$, and 50 m$M$ Tris·HCl at pH 7.3) was pumped through the column at 1 ml per minute, and the column effluent was continuously monitored and recorded.

was chromatographed and resolved in a single peak. In these tests 3 $A_{260}$ units of crude *E. coli* tRNA were added as a carrier.

Figure 3 shows a chromatogram obtained with a sample of about 0.1 $A_{260}$ unit of ribosomal RNA's, prepared as described.[7] This is a much smaller quantity of material than that usually required by other methods. The quantity of 16 S and 23 S rRNA can be calculated from the peak area, and, since the method is nondestructive, the fractions can be collected for additional tests.

[7] B. Z. Egan, J. E. Caton, and A. D. Kelmers, *Biochemistry* **10,** 1890 (1971).

# [3] Transfer RNA Isoacceptors in Cultured Chinese Hamster Ovary Cells[1]

*By* ARNOLD E. HAMPEL[2] and BRUCE RUEFER

Multiplicity of tRNA species for a single amino acid is a commonly observed phenomenon.[3] Multiple tRNA isoacceptors are required for many amino acids because of degeneracy of the genetic code;[4] however, the number of isoacceptors often seen is far greater than that required by the code. Because of this observation, it has often been suggested that tRNA is involved in roles other than classical protein synthesis.[5]

For several mammalian tRNAs there has been preservation of identical base sequences between species. The pattern of communality of tRNA species in mammals argues for the selection of a single versatile model mammalian system to completely catalog the isoacceptor profiles as the first step in systematically defining all their cellular functions. The Chinese hamster ovary (CHO) cell line is especially well suited for this purpose because it is easily grown in a partially defined media and amenable to the study of a variety of cellular and genetic problems.[6]

We have shown that multiple isoacceptors exist for all CHO tRNAs except $tRNA^{Trp}$, which has a single isoaccepting species. A total of 77 tRNA isoaccepting species in CHO cells are found by the RPC-5 chromatographic method, which represents a minimal number with more likely to be present. Variations between tRNA isoacceptors specific for a given amino acid would be expected to be due to both structural differences and the degree of base modification.

## Methods

*Cell Culture.* A hypodiploid (modal chromosome number of 21) line of Chinese hamster ovary cells[7] is grown in Ham's F-10 medium supple-

[1] This research was supported by NIH Grant GM 19506 and Research Career Development Award 1 KO4 GM 70424.

[2] Work performed in the Departments of Biological Sciences and Chemistry, Northern Illinois University, DeKalb, Illinois.

[3] R. M. Kathari and M. W. Taylor, *J. Chromatogr.* **86,** 289 (1973).

[4] D. Söll, J. Cherayil, D. Jones, R. Faulkner, A. Hampel, R. Bock, and H. G. Khorana, *Cold Spring Harbor Symp. Quant. Biol.* **32,** 51 (1966).

[5] U. Littauer and H. Inouye, *Annu. Rev. Biochem.* **42,** 439 (1973).

[6] C. R. Richmond, D. F. Peterson, and P. F. Mullaney, "Mammalian Cells: Probes and Problems." Technical Information Center, U.S. Energy Res. and Devel. Admin., 1975.

[7] J. Tjio and T. T. Puck, *J. Exp. Med.* **108,** 259 (1958).



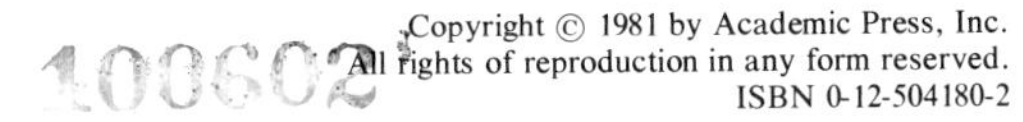



ISBN 0-12-504180-2

mented with 15% calf serum (prepared in our laboratory), 100 units of penicillin, and 100 $\mu$g of streptomycin per milliliter.

Cells are grown in 4-liter spinner flasks (Bellco) and harvested during logarithmic growth at a density of 3 to 5 $\times$ $10^5$ cells/ml by the addition of 500 ml of 0.25 *M* sucrose ($-20°$) to the suspension and centrifuged (4°) at 1000 *g* for 10 min in 250-ml polypropylene centrifuge bottles (Nalge) and resuspended in cold 0.25 *M* sucrose. Cells ($10^9$) are washed once and resuspended in 9 ml of buffer A: 0.1 *M* KCl, 10 m*M* Tris·HCl (pH 7.5 at 25°), 1 m*M* $MgCl_2$, and 0.1 m*M* dithiothreitol.

*Crude Aminoacyl-tRNA Synthetase Preparation.* A crude aminoacyl-tRNA synthetase preparation containing activities for all 20 amino acids is prepared from cells fresh or stored frozen at $-80°$. The cells ($10^9$) in 18 ml of buffer A are broken by the addition of 2 ml of 10% Nonidet P-40 (Particle Data Laboratories) with gentle hand mixing for 30 min on ice. Nuclei and cell debris are removed by centrifuging for 30 min at 1000 $g_{av}$ at 4°. The supernatant is centrifuged at 145,000 *g* at 4° for 2.5 hr in one A-211 centrifuge tube (International Equipment Co.) through a 1.5-ml 34% sucrose pad containing buffer A with KCl at 10 m*M* (buffer B). The crude supernatant is dialyzed overnight at 4° against 2 liters of buffer B with 40% glycerol. The final enzyme preparation containing 50 $A_{280}$ units/ml with 7.5 mg of protein per milliliter is stored at $-20°$. Protein is determined by a modification of the method of Lowry as described.[8]

*Preparation of Chinese Hamster Ovary tRNA.* Supernatant prepared by removing all ribosomes from broken cells through centrifugation of $10^9$ cells in 20 ml of buffer A, for 2.5 hr through a 1.5-ml 34% sucrose pad at 145,000 *g* at 4° is shaken with an equal volume of phenol for 15 min at room temperature. The emulsion is centrifuged at 4000 *g* for 20 min, the aqueous phase is carefully removed, and the preparation is phenol-treated twice more. The final aqueous phase is precipitated with 2 volumes of 100% ethanol and allowed to precipitate overnight at $-20°$. This is centrifuged; the pellet is redissolved in buffer A and precipitated twice more with ethanol. The final tRNA pellet is redissolved in buffer B without glycerol.

The long period of time involved in the tRNA preparation is adequate for removal of endogenous amino acids acylated to tRNA because of the lability of the aminoacyl-tRNA ester linkage at pH 7.5. This avoids using the high alkaline pH stripping conditions that destroy certain minor bases in tRNA and even produce a limited number of phosphodiester cleavages. The level of charging of $tRNA^{Val}$, which contains the most stable ami-

[8] A. Hampel and M. D. Enger, *J. Mol. Biol.* **79**, 285 (1973).

noacyl-tRNA ester bond, is used as a reference. An average preparation accepted 46 pmol of valine/$A_{260}$ unit of tRNA, which corresponds to 3.5% of the total heterogeneous tRNA preparation being available for valine aminoacylation.

The—CCA termini of these preparations have been shown to be intact by their inability to incorporate ATP when assayed with *Escherichia coli* ATP, CTP-tRNA nucleotidyltransferase according to Hampel *et al.*[9] Yeast tRNA is used as a control to verify enzymic activity.

*Aminoacylation of tRNA.* Two $A_{260}$ units of CHO tRNA are aminoacylated at 37° in 1 ml of incubation mix containing 20 m*M* Tris·HCl, 15 m*M* $MgCl_2$, 0.1 m*M* EDTA, 5 m*M* ATP, 0.75 m*M* CTP, and 200 μl of crude aminoacyl-tRNA synthetase (50 $A_{280}$/ml) all at pH 7.5. The incubation mix contains 10–20 μ*M* $^{14}C$- or $^{3}H$-labeled amino acid, which is about a 100-fold molar excess over tRNA. The other nonradioactive 19 amino acids are present at 10 μ*M* to overcome possible spurious charging by radioactive contaminants and to prevent misacylation of noncognate tRNAs. Enough aminoacyl-tRNA synthetase is present to allow completion of reaction in 10 min. Incubation is terminated after 20 min by the addition of 1/20 volume of 2 *M* Na acetate (pH 5.0). An equal volume of phenol is added, and the mixture is shaken at room temperature for 6 min; phases are separated by centrifugation at 2000 *g* for 20 min at 25°, and the aqueous phase is removed. The phenol step is repeated twice with volume restored by 0.1 *M* Na acetate, pH 5.0, when necessary; 16 $A_{260}$ units of carrier yeast tRNA are then added, and the tRNA is washed 3 times by ethanol precipitation. The last tRNA precipitate is lyophilized to dryness and redissolved in 200 μl of the respective RPC-5 starting buffer.

*RPC-5 Chromatography.* Reversed-phase chromatography is done using method B of the RPC-5 system described by Pearson *et al.*[10] The Plaskon 2300 CTFE powder (polychlorotrifloroethylene) used was a generous gift from Allied Chemical Corp., Morristown, New Jersey. The powder (300 g) is mixed at very low speed in a Waring blender with 12 ml of Tricaprylylmonomethyl-ammonium chloride (Ashland Chemical Company, Columbus, Ohio) dissolved in 450 ml of chloroform. After blending 1–2 hr, dry nitrogen gas is blown through the mix while blending until the chloroform is driven off. The average particle size of the resulting resin is 10 μm.

[9] A. Hampel, A. Saponara, R. Walters, and M. D. Enger, *Biochim. Biophys. Acta* **269,** 428 (1972).

[10] R. Pearson, J. Weiss, and A. D. Kelmers, *Biochim. Biophys. Acta* **228,** 770 (1971).

Chromatography is carried out using a modification of the high-pressure system of Kelmers and Heatherly.[11] Columns, 0.64 × 20 cm, are poured in a 25° water-jacketed LC-6M-13 high-pressure glass column (Laboratory Data Control) using a 1:1 slurry in starting buffer and maintaining a flow rate of near 2 ml/min with pressure up to 500 psi from a Milton Roy Model 396 minipump. Sample (200 $\mu$l) is added and gradient changes are made with an 8-part sample injection valve (SV-8031 Laboratory Data Control).

Elution is at 500 psi with 100-ml linear gradients of various NaCl concentrations (examples are given in figure legends) in 10 m*M* $MgCl_2$, 0.4 m*M* dithiothreitol, and 10 m*M* Na acetate, pH 4.5. Temperature is maintained at 25° using a Lauda constant-temperature bath. Fractions are collected at 30-sec intervals with pressures near 500 psi and flow rates near 1.85 ml/min. All flow rates are normalized to this rate. The column void volume is tube 7, where all free amino acids are eluted, and gradients are started at tube 10. Effluent is monitored at 260 nm using a 1-mm flow cell in a Hitachi 124 spectrophotometer. All samples (approximately 0.9 ml) are collected directly in scintillation vials with $^{14}C$ radioactivity determined by liquid scintillation counting using 10 ml of Bray's solution[12] for $^{14}C$ single-label samples or 10 ml of aquasol (New England Nuclear) for $^{14}C$, $^{3}H$ double-label samples.

## Identification of Multiple Isoacceptors

*Asp, Asn, His, and Tyr Specific tRNA Isoacceptors.* Figure 1 shows the isoacceptor profiles for tRNAs specific for Asp, Asn, His, and Tyr. $tRNA^{Asp}$ shows five isoacceptor peaks. Peaks 1, 2, 3, and 5 are small, and peak 4, the major peak, constitutes most of the $tRNA^{Asp}$. $tRNA^{Asn}$ contains a very minor peak 1, the largest 2, and another major peak 3. A significant shoulder is present on peak 3 representing a later-eluting peak 4. $tRNA^{His}$ contains three major peaks with an additional very small early-eluting peak.

The isoacceptor profiles of $tRNA^{Asp}$ from CHO cells correspond to pattern II of Gallagher *et al.*[13] This pattern is most commonly found in rapidly growing cells rather than in more quiescent cells and is characterized by a larger peak 4. By analogy with SV40-transformed 3T3 cells[14] the minor peaks 1 and 3 would be expected to contain the modified G base Q, 7-(4,5 *cis*-dihydroxy-l-cyclopenten-3-ylaminomethyl)-7-deaza-

[11] A. D. Kelmers and D. E. Heatherly, *Anal. Biochem.* **44,** 486 (1971).

[12] G. A. Bray, *Anal. Biochem.* **1,** 279 (1960).

[13] R. E. Gallagher, R. C. Ting, and R. Gallo, *Biochim. Biophys. Acta* **272,** 568 (1972).

[14] J. R. Katze, *Biochim. Biophys. Acta* **383,** 131 (1975).

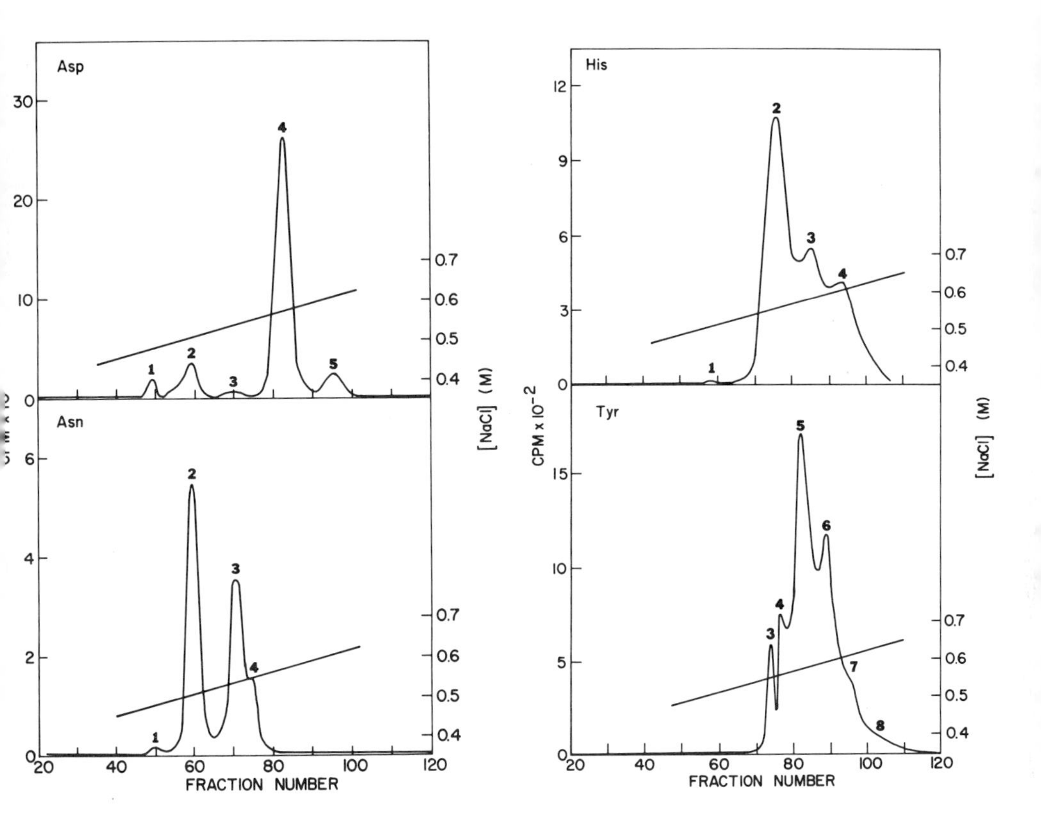


FIG. 1. Isoacceptor profiles of tRNA specific for Asp, Asn, His, and Tyr. NaCl gradients were 0.4 to 0.7 *M*. Radioactive amino acid specific activities: [$^{14}$C]Asp, 218 Ci/mol; [$^{14}$C]Asn 100 Ci/mol; [$^{14}$C]His, 388 Ci/mol; [$^{14}$C]Tyr, 522 Ci/mol.

guanosine,[15] while peaks 2 and 4 would contain the unmodified G. Q base is found as the 5′-terminal base of the anticodon of certain isoacceptors specific for Asp, Asn, His, and Tyr. These tRNAs correspond to the codons -$A_C^U$ with Q preferentially recognizing U.

*Glu, Gln, and Lys Specific tRNA Isoacceptors.* Figure 2 shows the isoacceptor profiles for Glu, Gln, and Lys specific tRNAs. $tRNA^{Glu}$ contains three peaks, the major species peak 1 eluting quite early at fraction 22 followed by a smaller peak 2 and another major peak 3. $tRNA^{Gln}$ contains two major peaks. $tRNA^{Lys}$ has four characteristic peaks. Peaks are numbered following the convention of Ortwerth and

[15] H. Kasai, Z. Ohashi, F. Harada, S. Nishimura, W. Oppenheimer, P. Crain, J. Liehr, D. von Minden, and J. McCloskey, *Biochemistry* **14,** 4198 (1975).

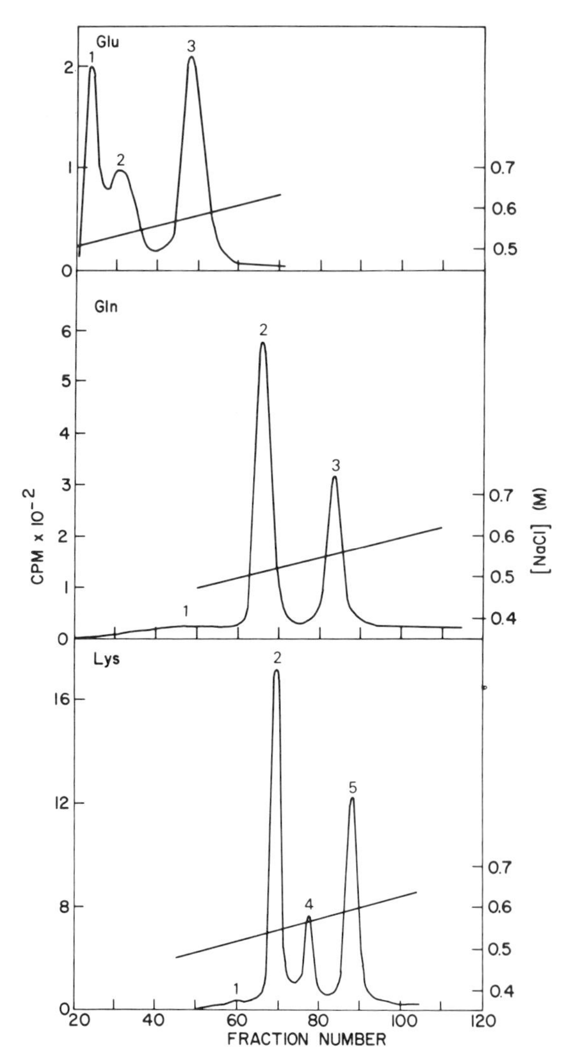


FIG. 2. Isoacceptor profiles of tRNA specific for Glu, Gln, and Lys. NaCl gradients were: 0.4 to 0.7 $M$ for Gln-tRNA$^{Gln}$ and Lys-tRNA$^{Lys}$; 0.5 to 0.8 $M$ for Glu-tRNA$^{Glu}$. Radioactive amino acid specific activities: [$^{14}$C]Glu, 250 Ci/mol; [$^{14}$C]Gln, 46 Ci/mol; [$^{14}$C]Lys, 50 Ci/mol.

Liu,[16] who reported a variable peak 3, which is not seen here. A minor peak 1 is followed by the major peak 2. Peak 5 is the second largest peak, with a smaller peak 4 between peaks 2 and 5. Peak 4 is significant and

[16] B. J. Ortwerth and L. Liu, *Biochemistry* **12,** 3978 (1973).

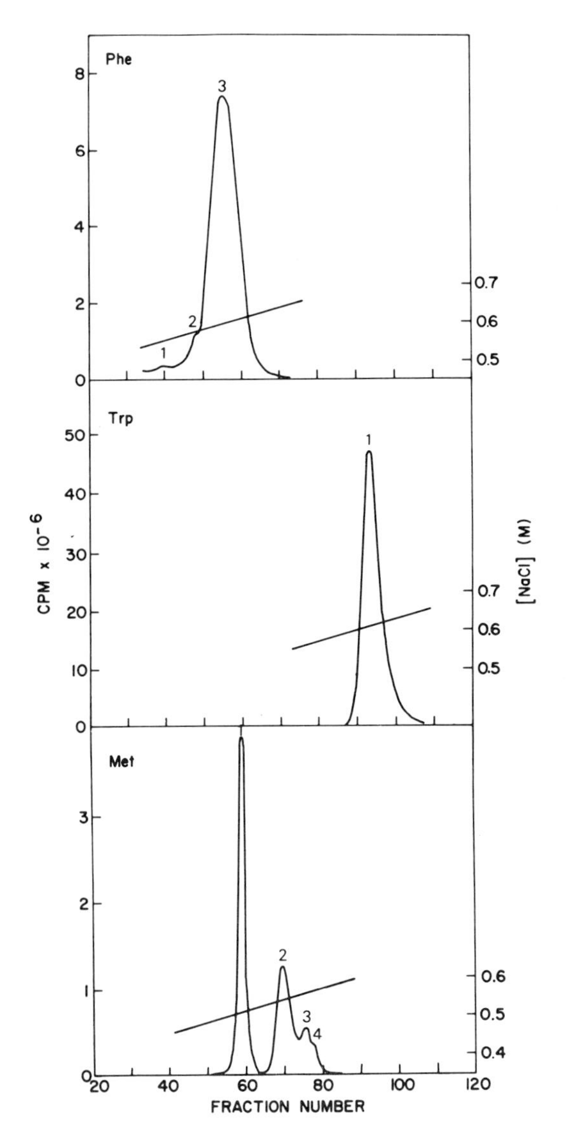


FIG. 3. Isoacceptor profiles of tRNA specific for Phe, Trp, and Met. NaCl gradients were 0.4 to 0.7 *M* for Trp-tRNA$^{Trp}$ and Met-tRNA$^{Met}$; 0.5 to 0.8 *M* for Phe-tRNA$^{Phe}$. Radioactive amino acid specific activities: [$^{14}$C]Phe, 50 Ci/mol; [$^{14}$C]Trp, 566 Ci/mol; [$^{14}$C]Met, 50 Ci/mol.

characteristic of proliferating cells, as quiescent cells normally show an insignificant peak 4.[17]

[17] H. Juarez, D. Juarez, C. Hedgcoth, and B. J. Ortwerth, *Nature (London)* **254,** 359 (1975).

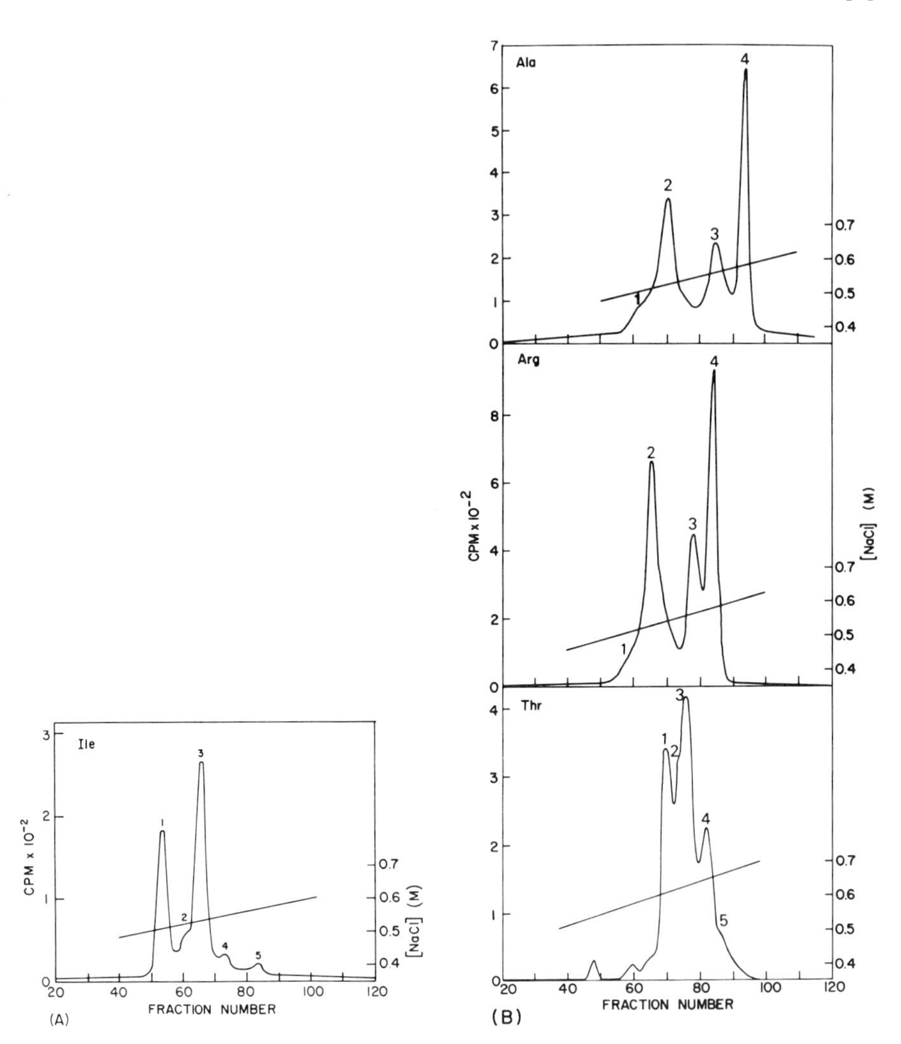


FIG. 4. Isoacceptor profiles of tRNA specific for Ile, Ala, Arg, Thr, Cys, Gly, Val, Pro, Ser, and Leu. NaCl gradients used were: 0.45 to 0.65 *M* for Ile-$tRNA^{Il}$ 0.4 to 0.7 *M* for Ala-$tRNA^{Ala}$, Arg-$tRNA^{Arg}$, Thr-$tRNA^{Thr}$, Gly-$tRNA^{Gly}$, Val-$tRNA^{Val}$, and Ser-$tRNA^{Ser}$; 0.5 to 0.8 *M* for Cys-$tRNA^{Cys}$, Pro-$tRNA^{Pro}$, and Leu-$tRNA^{Leu}$. Radioactive amino acid specific activities: [$^{14}$C]Ala, 50 Ci/mol; [$^{14}$]Arg, 240 Ci/mol; [$^{14}$C]Val, 50 Ci/mol; [$^{14}$C]Pro, 290 Ci/mol; [$^{14}$C]Ser, 50 Ci/mol; [$^{14}$C]Leu, 342 Ci/mol; [$^{14}$C]Thr, 50 Ci/mol; [$^{14}$C]Cys, 264 Ci/mol; [$^{14}$C]Gly, 112 Ci/mol, [$^{14}$C]Ile, 312 Ci/mol.

Certain isoacceptors for tRNAs specific for the amino acids Glu, Gln, and Lys contain a periodate-sensitive thiolated N base, 5-methylamino-

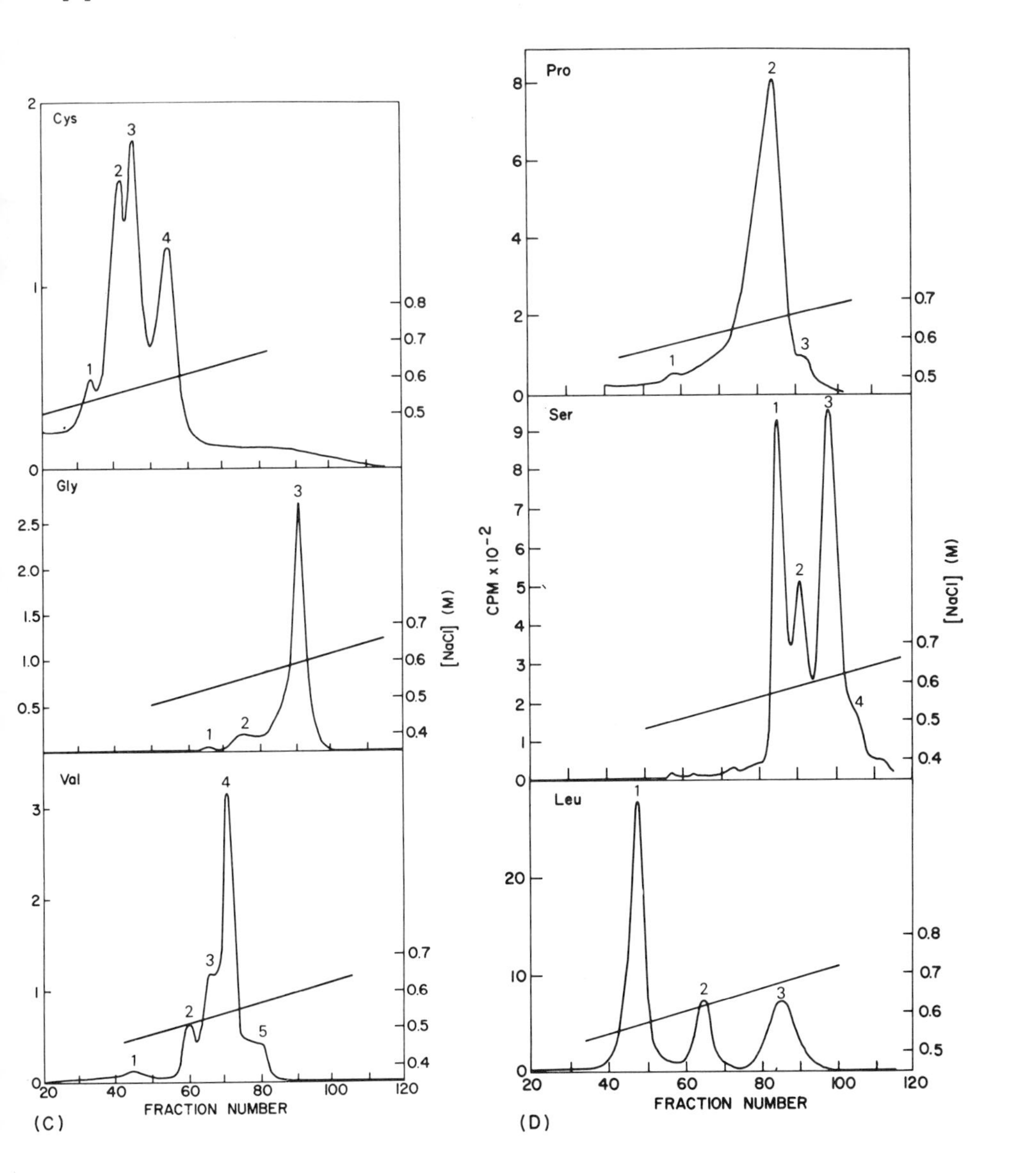


methyl-2-thiouracil, as the first base at the 5′ end of the anticodon.[18] These tRNAs recognize the codons $-A_{A}^{G}$ with N base preferentially recognizing A.

[18] Z. Ohashi, M. Saneyoshi, F. Harada, H. Hara, and S. Nishimura, *Biochem. Biophys. Res. Commun.* **40,** 866 (1970).

*Phe, Trp, and Met Specific tRNA Isoacceptors.* Figure 3 shows the isoacceptor profiles for Phe, Trp, and Met specific tRNAs. $tRNA^{Phe}$ has three peaks, early-eluting minor peaks 1 and 2 and a later-eluting major peak 3.

$tRNA^{Trp}$ shows a single peak which is similar to that for higher vertebrates, where only one isoacceptor exists, and its primary sequence is identical in rat, human, and avian cells with the possible exception of the degree of base modification. $tRNA^{Trp}$ in avian cells is spot 1, the primer for initiation of Rous sarcoma viral DNA synthesis.[19]

$tRNA^{Met}$ shows 4 distinct isoacceptors. Elution profiles correspond to peak 1 as the eukaryotic cytoplasmic initiator tRNA and peaks 2, 3, and 4 as noninitiator $tRNA_m^{Met}$. The CHO $tRNA^{Met}$ isoacceptors correspond to those of Krebs II mouse ascites tumor cells and chick embryo.[20] Relative amounts of the three $tRNA_m^{Met}$ species are considerably different, however, in CHO cells than in ascites and chick embryo. Peak 4 is the largest in ascites and chick while peak 2 is a minor peak. The opposite is true in CHO cells where peak 2 is the major $tRNA_m^{Met}$ and peak 4 is minor.

*Ile, Ala, Arg, Thr, Cys, Gly, Val, Pro, Ser, and Leu, Specific tRNA Isoacceptors.* Figure 4 shows the isoacceptor patterns of the tRNAs specific for the remaining 10 amino acids. $tRNA^{Ile}$ from proliferating CHO cells contains five reproducible peaks, and $tRNA^{Ala}$ shows 4 peaks. Four peaks are seen for $tRNA^{Arg}$, and $tRNA^{Thr}$ has 5 reproducible peaks preceded by two variable minor peaks. $tRNA^{Cys}$ shows 4 peaks, and $tRNA^{Gly}$ shows a single major peak preceded by 2 minor species. $tRNA^{Val}$ shows five distinct peaks. Three peaks are observed with $tRNA^{Pro}$, with peak 2 predominant and peaks 1 and 3 minor. $tRNA^{Ser}$ from CHO cells contains three major peaks with peak 4 a shoulder on peak 3.

$tRNA^{Leu}$ from CHO cells consistently shows three characteristic peaks. Because of the redundancy of the genetic code for leucine at least three species are required. Six are seen in most normal mammalian tissues,[21] hence the presence of only three in CHO cells is intriguing.

All tRNA isoacceptor profiles were obtained at 25°. In addition, several other parameters of separation were varied, i.e., pH, salt concentration, and temperature (5°, 25°, 37°, and 45°) with no conditions giving superior separation to those presented herein. It must be emphasized that the number of tRNA species identified in CHO cells (77) represents a minimal number, and no doubt more exist.

[19] F. Harada, R. Sawyer, and J. Dahlberg, *J. Biol. Chem.* **250,** 3487 (1975).
[20] K. Elder and A. Smith, *Proc. Natl. Acad. Sci. U.S.A.* **70,** 2823 (1973).
[21] W. K. Yang and G. D. Novelli, this series, Vol. 20, p. 44.

**Summary**

Methods are presented for separating and identifying the various tRNA isoacceptors from cultured Chinese hamster ovary cells with conditions for optimal resolution of the multiple tRNA isoacceptors determined for reversed-phase chromatography (RPC-5). Multiple isoacceptors for 19 amino acid-specific tRNAs were seen, only $tRNA^{Trp}$ showing a single species. The minimum number of tRNA isoacceptors shown by this method was 77. These chromatographic patterns serve as a useful reference for determining various tRNA functions in this model mammalian cell culture system. Because of the interspecies similarities between mammalian tRNAs, a model system such as CHO is very useful in studies directed toward understanding nonclassical roles for tRNA. The emergence of CHO as an outstanding somatic cell genetic system adds greatly to its usefulness along these lines. A catalog of tRNA isoacceptors in a defined mammalian system represents an essential first step in the elucidation of possible tRNA involvement in mammalian metabolism and cell regulation.

# [4] The Use of BD-Cellulose in Separating Transfer RNA's[1]

*By* I. C. GILLAM and G. M. TENER

Esterification of the hydroxyl groups of DEAE-cellulose by benzoyl residues leads to a product with altered physical properties. BD-cellulose[2] is an ion-exchanger with an increased affinity for lipophilic and particularly for aromatic groups, which is useful in the preparative and analytical fractionation of tRNA's by various schemes.[3] The first of these is ion-exchange chromatography in which the increased lipophilic attractions allow considerable purification of many species of tRNA. Certain species of tRNA, which contain in their structures a single lipophilic-substituted adenine having cytokinin activity, are

[1]Supported by research grants from the National Institutes of Health (GM 14007).
[2]Abbreviations used; BD-cellulose, benzoylated DEAE-cellulose; UV, ultraviolet; EDTA, ethylenediamine tetracetate; $A_{260}$ unit, the quantity that, dissolved in 1 ml of solvent, has an $A_{260}$ of 1.
[3]I. Gillam, S. Millward, D. Blew, M. von Tigerstrom, E. Wimmer, and G. M. Tener, *Biochemistry* 6, 3043 (1967).



ISBN 0-12-504180-2

retarded relative to most other tRNA's on columns of BD-cellulose.[4] In certain higher organisms the principal phenylalanine tRNA ($tRNA^{Phe}$) contains an even more lipophilic base of unknown structure, which causes this tRNA to be held to such columns much more strongly than any other. This is the basis of a simple means of purification of $tRNA^{Phe}$.[5] Strong salt solution is used to elute all other tRNA's from BD-cellulose, and salt solution containing ethanol or other organic solvent removes the $tRNA^{Phe}$.

Esterification of aromatic amino acids to their acceptors causes these to be bound more strongly to BD-cellulose and to require addition of ethanol to the eluent in order to release them.[6] In this manner, those phenylalanine tRNA's that are not strongly bound in the uncharged state may be purified. This approach, with the use of a gradient of ethanol in salt solution to elute the Phe-$tRNA^{Phe}$, is the method of choice even where the uncharged $tRNA^{Phe}$ is held strongly, for it eliminates a certain amount of nonaccepting material otherwise eluted with the desired product.[7] The essence of this method is the use of the great specificity of an aminoacyl-tRNA synthetase to attach an aromatic aminoacyl group only to the species of interest in the whole tRNA mixture and then to utilize the attraction of this aromatic group for BD-cellulose to separate acceptors for this one amino acid from all others.

By exploiting the potential reactivity of the amino group of an aminoacyl ester to substitute it with an aromatic residue, this method may be extended to other tRNA's.[8] The preferred reagent for the substitution is the 2-naphthoxyacetyl ester of *N*-hydroxysuccinimide, although appropriately active derivatives of other aromatic acids would undoubtedly be suitable.

## Preparation of BD-Cellulose

BD-cellulose is available commercially[9] or is easily prepared. The two commonest problems encountered in the preparation are the incomplete reaction caused by the presence of water in the reaction mixture

[4] D. J. Armstrong, F. Skoog, L. Kirkegaard, A. Hampel, R. M. Bock, I. Gillam, and G. M. Tener, *Proc. Nat. Acad. Sci. U. S.* **63**, 504 (1969).

[5] E. Wimmer, I. H. Maxwell, and G. M. Tener, *Biochemistry* 7, 2623 (1968).

[6] I. H. Maxwell, E. Wimmer, and G. M. Tener, *Biochemistry* 7, 2629 (1968).

[7] M. Litt, *Biochem. Biophys. Res. Commun.* **32**, 507 (1968).

[8] I. Gillam, D. Blew, R. C. Warrington, M. von Tigerstrom, and G. M. Tener, *Biochemistry* 7, 3459 (1968).

[9] Schwarz BioResearch, Orangeburg, New York; Regis Chemical Co. Inc., 1101 N. Franklin Street, Chicago, Illinois; Serva Entwicklungslabor, Heidelberg, Germany.

(usually due to incomplete drying of the DEAE-cellulose or by the use of wet pyridine) and the burning of the material by heating it too fast and mixing it too little.

*Materials*

*DEAE-Cellulose.* Commercially prepared material of the usual capacity of 1 meq/g dry weight is used without washing or removing fines. Samples prepared from cotton linters or from wood cellulose are equally suitable, but crosslinked ("microgranular") DEAE-cellulose gives a product that does not give good chromatographic separations. As DEAE-cellulose powder may contain up to 15% by weight of water, it is essential that it be dried before use. This is done by heating it in a vacuum over (60°–80°, approximately 100 torr, overnight) preferably with a desiccant ($P_2O_5$) or with a slight stream of air to remove the water driven out of the sample.

*Pyridine.* Reagent grade pyridine from a freshly opened bottle of stated negligible water content is suitable. Batches of unknown water content should be dried over calcium hydride and filtered.

*Benzoyl Chloride.* Fresh reagent grade material is used; other grades should be redistilled.

*Method*

Glassware is dried overnight at 110°. A 5-liter round-bottomed flask is fitted with a reflux condenser and drying tube. Pyridine, 2.5 liters, is put into the flask and 100 g of dried DEAE-cellulose is added slowly through a powder funnel. It is essential to add the DEAE-cellulose slowly enough and with sufficient shaking to avoid the formation of unwetted lumps that will not react and may burn.

Benzoyl chloride (284 ml, 4.0 eq/mole of anhydroglucose) is added through a long-stemmed funnel,[10] and the mixture is swirled to mix the contents thoroughly. The flask is placed in an electric heating mantle and moderate heating is started. As the contents warm up, the color changes through red shades to brown and eventually to a very dark brown. While heating, it is essential that the reaction mixture be stirred often enough to prevent the DEAE-cellulose from layering onto the bottom of the flask and burning. Mixing is most simply accomplished by lifting the flask (with one gloved hand supporting the bottom) and swirling the contents. As reaction begins to occur, it will be seen that the DEAE-cellulose fibers start to swell and eventually they dissolve. Heating

[10]Benzoyl chloride left on the ground-glass joint of the flask may cause the condenser to become cemented into place.

to boiling should take 45–60 minutes; after this time, provided no lumps had been allowed to form, there is little danger of burning. The mixture is left to boil gently under reflux for a further 45 minutes with occasional swirling. At the end of this period the contents of the flask should be a dark, viscous solution. When swirled up the sides of the flask and allowed to drain, very few fibrous particles should be detectable.

The flask is left to cool for about 30 minutes. (Note that when white crystals begin to appear in the solution they will grow fast enough to make it difficult to pour out the contents before they solidify. If such crystals are seen, the mixture should be slowly reheated before any attempt is made to pour out the solution).

A tub of cold water (20–30 liters; a plastic garbage can serves well) is set up with a mechanical stirrer in a fume hood or well-ventilated area. The brown, viscous solution of BD-cellulose in pyridine is poured in a steady stream into the water, where the rapid cooling causes it to set into a long rope, or to break into flakes.[11] As the product precipitates it becomes pale in color. After about an hour the water, now containing much pyridine and some of the colored materials, may be siphoned off and replaced. Stirring should be vigorous enough to break the now brittle rope into smaller pieces.

*Washing and Sizing the Product*

BD-cellulose prepared as above is likely to contain, besides pyridine and pyridinium chloride (which are water-soluble), benzoic acid, benzoic anhydride, and colored materials of unknown constitution. The last are relatively insoluble in water and are strongly held to the BD-cellulose. Hence it is desirable to wash the product first with water, then with ethanol, and finally with ethanolic salt solution.

Washing with water is continued until the major part of the pyridine has been removed, when the product is washed by stirring in 95% ethanol. The intensely colored ethanol is separated by aspiration through a tube with the end covered by a cloth filter. After several washings with 5 liter batches of ethanol (preferably left to stand overnight), the product is clean enough to be prepared for final washing. When wet with ethanol it is rather soft and dries too easily, losing its desirable porous character. It is therefore suspended in 5 liters of water before grinding. Grinding is perhaps best done in two stages, the first to a coarse grade merely to increase the rate of diffusion in washing the material; the second is to the final size for chromatography.

BD-cellulose prepared as described is a porous, light tan solid, brittle

[11]The residue in the flask may be washed out by dissolving it in methylene chloride or chloroform.

but not very hard. Thus it is readily broken into smaller crumbs on rubbing through a metal sieve. Alternatively a mechanical blender may be used, but only with great caution as it will rapidly degrade the material into particles too fine to allow reasonable flow rates in columns. The slower procedure of grinding through a stainless steel mesh is preferred. Initially a No. 10 mesh (2-mm opening) is useful. The particles washed through the sieve by water are allowed to settle and then washed in turn with 95% ethanol, 2 *M* NaCl in 50% ethanol, and finally when the $A_{260}$ of the washings falls to about 0.05, a wash with 2 *M* NaCl is given. The material is then ready for the final grinding, which is done as described above. The final mesh size depends upon the use to which the product is to be put. For batchwise separations using "steps" of eluent concentration, a fast flow rate is desirable and a No. 30 mesh (0.6-mm opening) is suitable. For chromatographic separations with a continuous gradient of eluent concentration, finer particles are desirable. Material passing a No. 50 mesh (0.3-mm opening) gives good separations in larger columns. In narrow columns, resolution can often be dramatically improved by using even finer material, for example, that passing a No. 100 mesh (0.15-mm opening). In this case it is essential to de-fine the product carefully to avoid problems with high back-pressure in columns. Whatever size of sieve is used, the product should be rid of fine particles by being suspended in water and allowed to settle briefly; the upper layer is decanted off. It should finally be left for some time in a buffered solution of EDTA to remove metal ions introduced in the sieving process. Where critical chromatographic separations are to be made, it is desirable to check the resolving power of a new batch of BD-cellulose by using it for chromatography of a standard sample of tRNA (usually a good preparation of yeast tRNA).

*Storage and Regeneration of BD-Cellulose*

BD-cellulose, particularly when contaminated with traces of nucleic acids, is a substrate for microbial growth. Sodium chloride alone does not inhibit the growth of all forms, and material left for more than a day or so should be kept in 2 *M* NaCl containing 20% (v/v) ethanol. Columns should be washed with this solution after use to remove traces of strongly adsorbed material.[12] The ethanol causes BD-cellulose to swell and soften slightly, and such columns washed free of ethanol will give poor chromatographic resolution unless they are repacked. Stored BD-cellulose should be washed free of ethanol with 2 *M* NaCl before packing into a column. Other organic solvents, particularly toluene and

[12]But note that high molecular weight RNA's, e.g., ribosomal RNA, are insoluble in this solution and hence are not eluted.

chloroform, should not be used as preservatives, as they will destroy the structure of BD-cellulose; nor should strongly acid or alkaline solutions be used to regenerate it.

**Chromatographic Methods**

Glass columns are usually rinsed with 1% dichlorodimethylsilane in benzene, then heated to give a nonwettable surface. All solutions are routinely degassed by allowing them to boil briefly in a Büchner flask under vacuum from an aspirator. (This precludes much trouble due to gas bubbles blocking the pump, the siphons, and the flow-cell of the UV monitor). The column is then filled from below with 2 *M* NaCl solution by suction at the top, carefully regulated to avoid bubbles. BD-cellulose, after de-fining as described above, is suspended as a 25–50% (v/v) slurry in degassed 2 *M* NaCl solution and poured into a packing funnel attached to the top of the column. The BD-cellulose is allowed to pack by gravity until 1–2 cm of packed bed have formed, and then the column is allowed to flow. As the bed rises and the flow rate decreases, air pressure (5 psi, 0.35 $kg/cm^2$) is applied to the top of the packing funnel. A pad of tissue paper moistened with alcohol and stroked firmly down the sides of the column causes vibrations, which rapidly settle the packing. The bed is allowed to rise almost to the top of the column and then a chromatographic pump is attached to wash with 2 *M* NaCl at a slightly greater rate than is to be used for the chromatography. This settles the packing further and removes UV-absorbing material. When the $A_{260}$ of the effluent has reached an acceptably low level (about 0.025), the column is washed with the starting solution for chromatography, which is continued until the effluent is identical with the influent, checked by pH and conductivity or refractive index.

*The Choice of Conditions for Chromatography*

Nonionic interactions occur between tRNA's and DEAE-cellulose and depend largely upon the secondary structure of the particular tRNA. In general the more open the configuration of the tRNA the stronger are the secondary interactions. The indirect effects of pH, concentration of divalent cation, temperature, and chaotropic agents, such as urea, formamide, and alcohols, upon these secondary interactions have been used to obtain partial separations between specific tRNA's by chromatography on DEAE-cellulose and DEAE-Sephadex.[13,14] The secondary interactions between BD-cellulose and tRNA's are stronger but are affected in much the same way by variation of these conditions. By varying them systematically, it should be possible to find

[13] R. M. Bock and J. D. Cherayil, Vol. XII [85].

[14] P. L. Bergquist, B. C. Baguley, and R. K. Ralph, Vol. XII [88].

sets of conditions that will allow any species of tRNA to be purified completely by repeated chromatography on BD-cellulose. In general the initial separation is performed around neutrality, at room temperature (about 20°), and in the presence of magnesium ion (about 0.01 *M*). Selected fractions are then rechromatographed at a different temperature[15] or at low pH (3.5–4.0) in the presence of EDTA. Initial separation on BD-cellulose has been combined successfully with rechromatography on DEAE-Sephadex[16,17] to purify several species of tRNA, and combination with countercurrent distribution or reversed-phase column chromatography is likely to be just as successful. Alternatively, the derivatization method (described below) followed by chromatography may be used.

The majority of tRNA's elute from columns of BD-cellulose between 0.5 and 1.1 *M* NaCl (containing 0.01 *M* $Mg^{2+}$). To allow tRNA applied to the column to reach equilibrium under the conditions of chromatography, it is desirable to start elution with a lower concentration of salt than will displace the earliest tRNA's, usually 0.4 *M*.

The desirability of buffered eluents is a debatable point for chromatography around neutrality. It is so difficult to prevent microbial growth (and production of nucleases) in buffers, and elution patterns apparently vary so little in the range pH 5–8, that we have usually preferred to omit buffer salts. For chromatography of aminoacyl-tRNA, whether derivatized or not, buffering around pH 4.5 is desirable to stabilize the aminoacyl linkage. Some separations of uncharged tRNA's are best performed at low pH, as mentioned above, and suitable buffers must be used.

*Loading and Running the Column*

BD-cellulose has a high capacity for tRNA. Figure 1 illustrates the separation of 5 g (70,000 $A_{260}$ units) of tRNA on a column 3.2 × 110 cm (885 $cm^3$ bed volume) giving good resolution with the $A_{260}$ of the effluent reaching above 20 at peaks. This approaches the capacity of such a column.

*Assay of Fractions for Biological Activity.*

Fractions eluted from chromatographic columns may be assayed for capacity to accept a given amino acid by several procedures.[18–20] In the

[15] K. L. Roy and D. Söll, personal communication, (1968).
[16] S. Cory, S. K. Dube, B. F. C. Clark, and K. A. Marcker, *Fed. Eur. Biochem. Soc. Lett.* 1, 259 (1968).
[17] S. Nishimura and I. B. Weinstein, *Biochemistry* **8**, 832 (1969).
[18] J. D. Cherayil, A. Hampel, and R. M. Bock, Vol. XII [106a].
[19] F. J. Bollum, Vol. XII [106b].
[20] J. F. Scott, Vol. XII [106c].

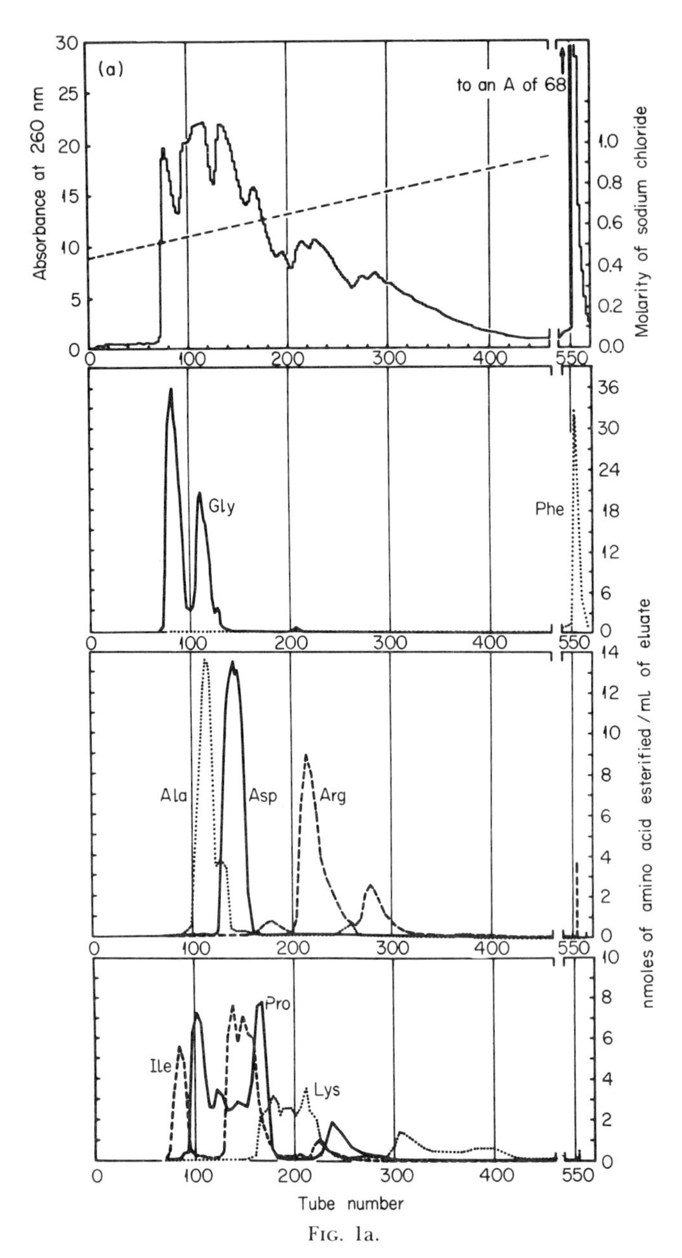


FIG. 1a.

case of each amino acid it is necessary to check that the concentration of salt and buffer present in the fractions is not strongly inhibitory to the enzymatic reaction. Where inhibition is found, it can be minimized by appropriate dilution of the fractions before assay or by use of a

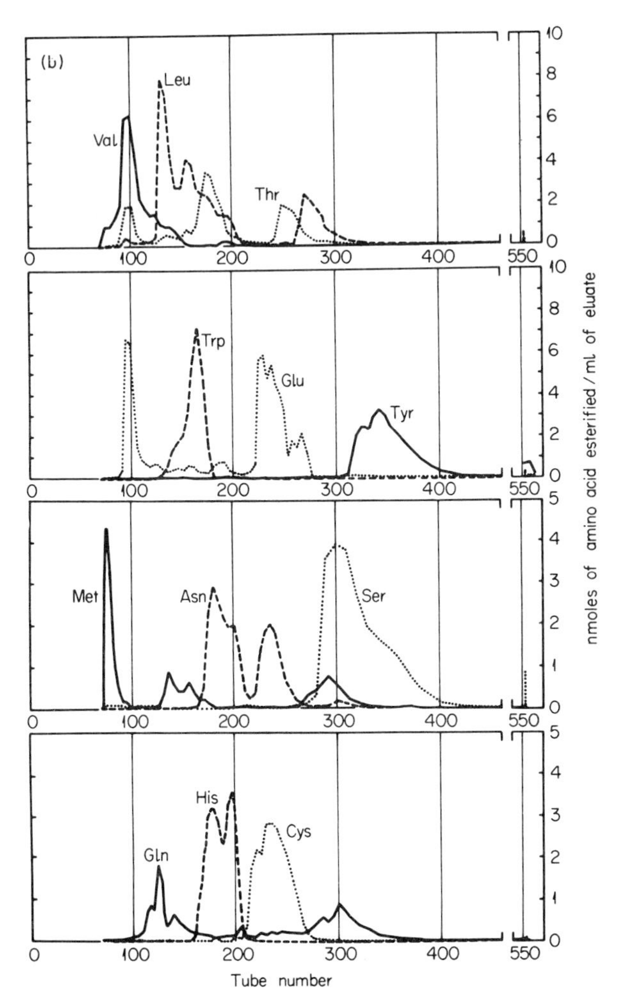


FIG. 1. Fractionation of 5 g (70,000 $A_{260}$ units) of tRNA from brewer's yeast (Boehringer) on a column (3.2 × 110 cm) of BD-cellulose. The sample was applied in 500 ml of 0.45 *M* NaCl plus 10 m*M* $MgSO_4$ and eluted with the indicated (dashed line) gradient of NaCl solution (a total of 10 liters) containing 10 m*M* $MgSO_4$. Fractions were 20 ml/15 min. At the end of the gradient (tube 497) elution was continued with 1.0 *M* NaCl plus 10 m*M* $MgSO_4$ containing 10% (v/v) methoxyethanol. Total recovery, about 66,000 $A_{260}$ units. Fractions were assayed for amino acid acceptor activity as described in the text. From I. Gillam, S. Millward, D. Blew, M. von Tigerstrom, E. Wimmer, and G. M. Tener [*Biochemistry* 6, 3043 (1967)], reprinted with permission of the American Chemical Society.

method[18] that allows many samples to be washed free of salt and assayed together. Amino acid acceptor activities found in this way are always minimal values. For the use of these methods to be valid in any study comparing relative quantities of acceptors in different fractions it must be shown that the maximal acceptor activity is being measured. The aminoacylation of each fraction plotted against time of reaction should show a rapid rise to a stable plateau, and this final value should be proportional to the amount of tRNA present. The fact that even iso-accepting tRNA's from the same organism show different behavior with the same aminoacyl-tRNA synthetase makes these precautions necessary. The results of assays for acceptor activity for various amino acids in the eluate from a column loaded with crude tRNA of brewer's yeast are shown in Fig. 1.

*Recovery and Rechromatography of tRNA*

Fractions having $A_{260}$ of 2–5 or greater give quantitative recovery of tRNA by precipitation with two volumes of 95% ethanol at 0°. For more dilute samples, it is usually advantageous to concentrate the solutions before precipitation. Fractions containing a particular acceptor may then be rechromatographed on BD-cellulose under different conditions. Figure 2 illustrates the purification by this method of $tRNA^{Asp}$ from the column of Fig. 1.

## Isolation of Specific tRNA's by Derivatization

Crude tRNA is first stripped of that fraction which remains bound to BD-cellulose in the presence of 1 *M* NaCl. This fraction may be eluted with salt containing 9.5% ethanol and is called the "ethanol fraction." In some organisms, notably in yeasts and some other higher organisms, this fraction contains all the acceptor activity for phenylalanine plus traces of other activities and some inactive material. If this fraction is not of interest, it may be left in the crude tRNA and recovered in the final chromatography.

The tRNA recovered (crude tRNA minus the ethanol fraction) is esterified with the single chosen amino acid, using a preparation of aminoacyl-tRNA synthetase, and the introduced amino acid is then derivatized with a reactive aromatic ester such as the *N*-hydroxysuccinimide ester of 2-naphthoxyacetic acid. The mixture of unesterified tRNA and *N*-2-naphthoxyacetylaminoacyl-tRNA is adsorbed to a column of BD-cellulose, buffered at pH 4.5 to stabilize the ester. tRNA is eluted with 1 *M* NaCl except for a slight "tail" of material not rapidly desorbed. This may then be released by a brief wash with 1 *M* NaCl in 4.7% ethanol. The derivatized aminoacyl-tRNA is finally eluted with 1 *M* NaCl in 19% ethanol. The substituted aminoacyl ester is hydrolyzed,[21]

[21] P. S. Sarin and P. C. Zamecnik, *Biochim. Biophys. Acta* 91, 653 (1964).

and the purified tRNA is recovered. Generally this product will consist of a family of isoaccepting tRNA's, not distinguishable, except perhaps under special conditions, by the aminoacyl-tRNA synthetase used. These isoacceptors may be purified finally by chromatography on BD-cellulose, as described above, or by other chromatographic methods.

*Limitations of the Method and a Cause of Error*

The method described is believed to be generally applicable, but quantitative recovery of a given family of isoaccepting tRNA's may not be possible because of incomplete enzymatic charging, incomplete derivatization, and loss of the amino acid because of the marked lability of the ester bond. Other factors can affect the purity of the final product, especially when one is isolating a tRNA that occurs in only trace amounts. During the derivatization of the amino group of the aminoacyl-tRNA, there is a small amount of reaction (usually less than 1%) of the ester with other functions in the tRNA. This appears to be relatively nonspecific and thus gives a product contaminated with trace amounts of various other tRNA's. These can be removed by further chromatography. Also, there may be a small amount of nonspecific adsorption of uncharged tRNA's to BD-cellulose. These are normally removed with solution C (below), but any traces remaining will contaminate the derivatized tRNA. They are usually eliminated by further chromatography. A potential source of erroneous conclusions is the tendency of tRNA's to form specific dimers[22] that have acceptor activity and are separable from the monomers on BD-cellulose. Formation of dimers is favored by precipitation of tRNA from solutions in which purified tRNA is present in high concentration. Such dimers can be converted to monomers by appropriate treatment before chromatography.

*Materials*

BD-cellulose is prepared as described above. Material passing a No. 30 mesh (0.6-mm opening) is used after washing to an effluent $A_{260}$ about 0.025 with 2 *M* NaCl in 19% (v/v) ethanol.

Solutions A-E contain 10 m*M* $MgCl_2$ plus 10 m*M* AcOH, pH 4.5 (NaOH) together with the following additions: Solution A, 0.3 *M* NaCl; Solution B, 1.0 *M* NaCl; Solution C, 1.0 *M* NaCl plus 4.7% (v/v) EtOH; Solution D, 1.0 *M* NaCl plus 9.5% (v/v) EtOH; Solution E, 1.0 *M* NaCl plus 19% (v/v) EtOH. Solutions C–E are 5, 10, and 20% (v/v) of 95% EtOH. *N*-Hydroxysuccinimide is commercially available.[23]

[22] J. S. Loehr and E. B. Keller, *Proc. Nat. Acad. Sci. U. S.* **61**, 1115 (1968).

[23] Aldrich Chemical Co. Inc., 2371 N. 30th St., Milwaukee, Wisconsin; Pierce Chemical Co., P.O. Box 117, Rockford, Illinois; Eastman Organic Chemicals, Rochester, New York.

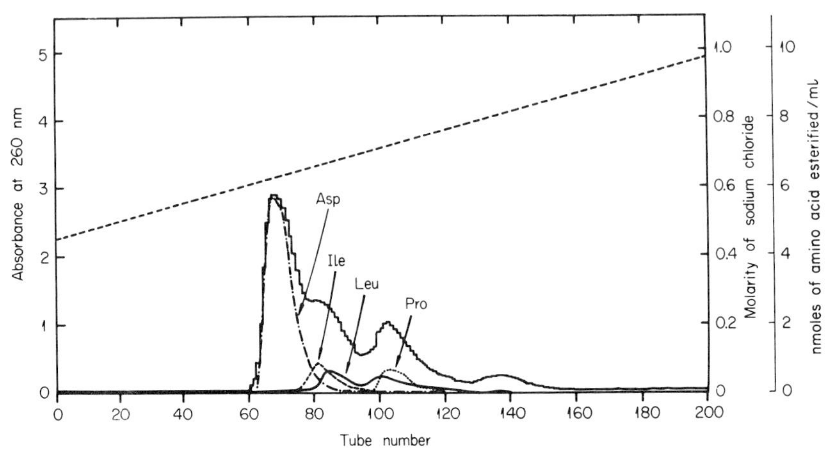


FIG. 2. Rechromatography of a sample (25 mg) of tRNA[Asp] (recovered from fractions 126–161 of Fig. 1) at pH 4.0 on a column (0.9 × 97 cm) of BD-cellulose. The sample was applied in 3 ml of 0.45 *M* NaCl plus 10 m*M* $MgCl_2$ plus 50 m*M* acetate buffer (pH 4.0, NaOH). Elution was with the indicated (dotted line) gradient of NaCl containing 10 m*M* $MgCl_2$ plus 50 m*M* acetate buffer (pH 4.0, NaOH) (total volume 1.2 liters). Fractions were 5 ml/7 min. Assays for amino acid acceptor activity were performed as described in the text after neutralization of the buffer. From I. Gillam, S. Millward, D. Blew, M. von Tigerstrom, E. Wimmer, and G. M. Tener [*Biochemistry* 6, 3043 (1967)], reprinted with permission of the American Chemical Society.

*Preparation of 2-Naphthoxyacetyl Ester of N-Hydroxysuccinimide*

2-Naphthoxyacetic acid (2.02 g, 10 mmoles) and *N*-hydroxysuccinimide (1.15 g, 10 mmoles) are dissolved in dry dioxane (40 ml) in a stoppered flask. To this is added dicyclohexylcarbodiimide[24] (2.10 g, 10 mmoles), and the flask is shaken to dissolve the reagent. Almost immediately, fine crystals of dicyclohexylurea begin to precipitate. The reaction mixture is left to stand at room temperature for about 2 hours, and the crystals are filtered off and discarded. The filtrate is evaporated under reduced pressure to a syrup which is crystallized from 1-propanol. The melting point of the product is 146.5–147°, and the yield after crystallization is 2.35 g (79%).

*Removal of the Ethanol Fraction from Crude tRNA*

A column (2.5 × 37 cm) of BD-cellulose is packed in 2 *M* NaCl solution. (Columns for stepwise elution require less care in packing than do

[24] *CAUTION:* Dicyclohexylcarbodiimide is a toxic irritant to which many people develop severe allergic reactions. It is slightly volatile and should be used with precautions which include gloves, eye protection and good ventilation. Spilled material and that on spatulas, etc., may be rapidly destroyed with dilute acetic acid.

columns for elution with continuous gradients of concentration of salt.) The column is washed with solution A until equilibrated and then loaded with crude tRNA (1 g) in solution A (50 ml). The column is washed with solution A until the effluent has $A_{260}$ less than 0.01. (This fraction is kept until the isolation has been completed successfully and then discarded. It usually contains chiefly ATP.) The bulk of the tRNA is eluted with solution B at a flow rate about 2 ml/minute. Fractions containing the main peak of absorbancy are pooled, and the tRNA is recovered by precipitation with ethanol. Elution with solution B is continued until $A_{260}$ of the effluent drops to 0.3. Elution with solution D releases the ethanol fraction, which is recovered by precipitation. The column is finally washed with solution E to remove unwanted material. It may be prepared for use again by washing with solution A. Any large cracks in the packing can be filled by stirring up the BD-cellulose. In this case repacking is not necessary.

*Aminoacylation of Crude tRNA*

The enzymatic aminoacylation of crude tRNA with the single amino acid of interest is performed essentially as described earlier in this series.[20,25,26] Crude or purified preparations of aminoacyl-tRNA synthetase may be used. An important requirement is that the enzyme contain no free amino acids. This can be assured by passing the preparation through Sephadex G-25 immediately before use.

The yield of purified tRNA depends upon the efficiency of aminoacylation obtained. Due to the relatively rapid hydrolysis of aminoacyl-tRNA under the conditions of incubation, the extent of aminoacylation is increased as the rate of synthesis is increased. Thus it is well worthwhile to spend some time examining the effects of some variables on the rate of reaction. These studies are done on a small scale using isotopically labeled amino acid as described.[18–21] After the concentration of enzyme, the factors usually of most importance are the concentrations of amino acid, ATP, and free $Mg^{2+}$ (the last will be affected by the concentration of ATP). The pH and the buffer used may require investigation. (Note that Tris buffer may be undesirable as it catalyzes hydrolysis of aminoacyl-tRNA.[21]) In some cases, addition of ammonium or potassium ions, chelating agent or disulfide reducing agent may be stimulatory. Completion of the reaction within 5–10 minutes at 30° may be taken to indicate almost quantitative aminoacylation. The reaction of the main part of the tRNA with [$^{12}$C]amino acid is then performed

[25] A. Kaji, Vol. XII [152].
[26] G. von Ehrenstein, Vol. XII [76].

under the conditions found. At the same time, a small sample is labeled with radioactive amino acid in a separate incubation. The solutions are adjusted to pH 4.5, and the RNA is precipitated and washed with ethanol.

*Derivatization of Aminoacyl-tRNA*

In order to undergo nucleophilic substitution the amino group of an aminoacyl ester must be uncharged. The reaction must therefore be performed around pH 8. Under the right conditions reaction is rapid and hydrolysis of the aminoacyl ester is slight. The pH should be lowered as soon as the derivatization is complete to minimize this. The preferred reagent, the 2-naphthoxyacetyl ester of *N*-hydroxysuccinimide, is less soluble in aqueous systems than the corresponding phenoxyacetyl ester, which has also been used. Most difficulties in obtaining complete derivatization seem to be due to failure to get enough of the ester into solution and can be overcome by increasing the content of organic solvent (tetrahydrofuran) in the mixture. Such failures lead to the appearance of radioactivity from labeled aminoacyl-tRNA in the 1 *M* salt wash of BD-cellulose—that is, together with the bulk of the tRNA.

*N-2-Naphthoxyacetylaminoacyl-tRNA*

Aminoacyl-tRNA derived from 1 g of crude tRNA and mixed with the sample of radioactively labeled aminoacyl-tRNA is dissolved in 50 ml of 0.1 *M* triethanolamine hydrochloride containing 0.01 *M* $MgCl_2$ (pH 4.3) at 0°. The cloudy solution is centrifuged to remove insoluble material (protein). To the stirred solution, kept at 0°, is added 2-naphthoxyacetyl ester of *N*-hydroxysuccinimide (60 mg) dissolved in dry tetrahydrofuran (5 ml). Much of the reagent precipitates. The pH of the solution is rapidly adjusted to 8.0 by the addition of 1 *N* NaOH, and stirring at 0° is continued. After 5 minutes an equal quantity of acylating agent in tetrahydrofuran is added together with any NaOH required to maintain pH 8.0. After a total of 10 minutes, the solution is adjusted to pH 4.5 with glacial acetic acid and the tRNA is recovered by precipitation with ethanol (two volumes). The precipitate is washed twice with cold 95% ethanol and finally drained as completely as possible.

*Separation of N-2-Naphthoxyacetylaminoacyl-tRNA from Uncharged tRNA's*

The derivatized tRNA prepared from 1 g of crude tRNA is dissolved in solution A (200 ml). Samples are taken for determination of the total content of $A_{260}$ units and of radioactivity, and the remainder is

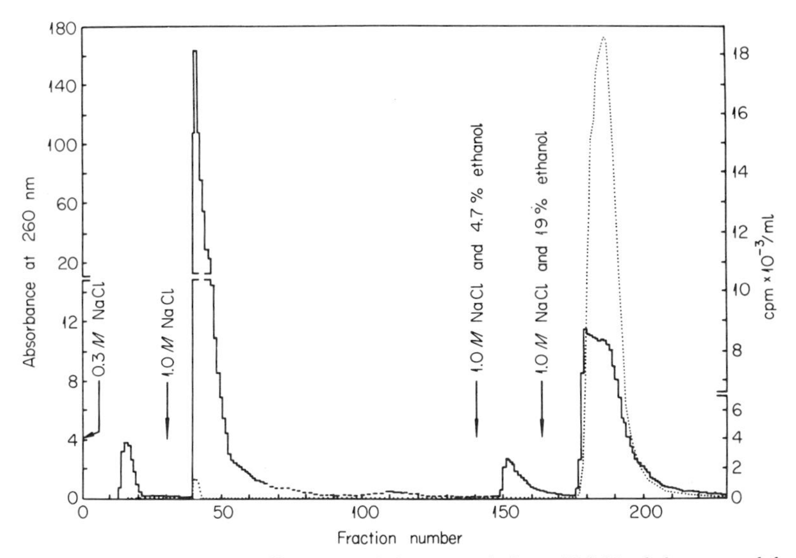


FIG. 3. Isolation of $tRNA^{Gly}$ by naphthoxyacetylation. *N*-2-Naphthoxyacetylglycyl-tRNA prepared from 1 g of mixed tRNA's was applied to a column (2.5 × 37 cm) of BD-cellulose in Solution A. The column was washed with the same solution. At the indicated points, elution with Solutions B, C, and E was started. Solid line, $A_{260}$; dotted line, radioactivity from *N*-2-naphthoxyacetyl-[$^{14}$C]glycyl-tRNA. From I. Gillam, D. Blew, R. C. Warrington, M. von Tigerstrom, and G. M. Tener [*Biochemistry* 7, 3459 (1968)], reprinted with the permission of the American Chemical Society.

applied to the column of BD-cellulose (2.5 × 37 cm) used above and equilibrated with solution A. The sample is rinsed in, and the column briefly washed with, the same solution, fractions being collected in the usual way. A small peak of material elutes, and when this has been collected the liquid above the column packing is replaced with solution B. Elution is continued until most of the tRNA has eluted (until $A_{260}$ has dropped to about 0.2). The eluent is then replaced with solution C. As soon as the small peak of UV-absorbing material has emerged and $A_{260}$ has dropped again to about 0.2, the eluent is replaced with solution E. Elution is continued with solution E, fractions being collected as before. Samples of fractions of interest are taken into vials containing Bray's solution[27] for determination of radioactivity in the scintillation counter.

Results of a typical experiment, the isolation of $tRNA^{Gly}$ from crude tRNA of brewer's yeast are shown in Fig. 3. The bulk of the tRNA elutes with solution B and contains only a trace of radioactivity due to unsubstituted glycyl-tRNA. A small peak of UV-absorbing material elutes with solution C, again containing negligible radioactivity. La-

[27] G. Bray, *Anal. Biochem.* 1, 279 (1960).

beled *N*-2-naphthoxyacetylglycyl-tRNA is eluted with solution E. It is usually found that an overloaded column works as well as the one shown. Overloading causes solution A to wash off a considerable amount of tRNA, but this usually contains no radioactivity, so that *N*-2-naphthoxyacetylaminoacyl-tRNA still elutes only with solution E.

*Recovery of Purified tRNA*

tRNA is precipitated from suitably pooled fractions, centrifuged, and washed well with 95% ethanol. The recovered *N*-2-naphthoxyacetylaminoacyl-tRNA is hydrolyzed[21] to yield the purified isoaccepting tRNA's by incubation in 1.8 *M* Tris·HCl buffer, pH 8.0, for 2 hours at 23° and then reprecipitated and washed. The completion of hydrolysis may be checked by determining that no radioactivity remains in the precipitated tRNA.

*Separation of Individual Species of Acceptor tRNA*

The purified isoaccepting tRNA's may be separated by chromatography on BD-cellulose as described above. If aggregation is suspected to have occurred the tRNA at a concentration less than 1 mg/ml should be heated (60°, 5 minutes) in 10 m*M* EDTA, pH 6, before application to the column.[22]

## [5] Fractionation of Rabbit Liver Methionyl-tRNA Species

*By* KALLOL K. BOSE, NANDO K. CHATTERJEE, and NABA K. GUPTA

As in bacterial protein synthesis, a specific methionine-tRNA species ($tRNA_f^{Met}$) in eukaryotic protein synthesis preferentially recognizes the terminal methionine codon (AUG or GUG) in a genetic message and initiates polypeptide chain synthesis with methionine at the N-terminal position of the peptide chains.[1-8] Another methionine-tRNA species ($tRNA_m^{Met}$), also present in all the eukaryotic organisms studied, specifically recognizes the internal methionine codon (AUG) and transfers methionine

[1] N. K. Gupta, N. K. Chatterjee, K. K. Bose, S. Bhaduri, and A. Chung, *J. Mol. Biol.* **54,** 145 (1970).

[2] S. Bhaduri, N. K. Chatterjee, K. K. Bose, and N. K. Gupta, *Biochem. Biophys. Res. Commun.* **40,** 402 (1970).

[3] A. E. Smith and K. A. Marcker, *Nature* (*London*) **226,** 607 (1970).

[4] J. C. Brown and A. E. Smith, *Nature* (*London*) **226,** 610 (1970).

[5] D. Housman, M. Jacobs-Lorena, U. L. RajBhandary, and H. F. Lodish, *Nature* (*London*) **247,** 913 (1970).





ISBN 0-12-504180-2

into the internal positions of the synthesized polypeptides. In most cases, the eukaryotic initiator $tRNA_f^{Met}$ species can be charged with *Escherichia coli* synthetase and formylated at the $\alpha$-amino position with *E. coli* transformylase and formyltetrahydrofolate (except wheat germ[8] and brine shrimp embryo[9] initiator Met-$tRNA_i^{Met}$ species, which are not formylated with *E. coli* transformylase). However, there is no convincing evidence of the presence of Met-$tRNA_f^{Met}$ transformylase activity in eukaryotic cell cytoplasm, and the available evidence suggests that these cells probably use the Met-$tRNA_f^{Met}$ species in the nonformylated form.

In this article, we describe the fractionation[1,2] of rabbit liver methionine tRNA species. At least three methionine acceptor activities ($tRNA_{m_1}^{Met}$, $tRNA_{m_2}^{Met}$, and $tRNA_f^{Met}$) were obtained by DEAE-Sephadex chromatography of crude liver tRNA. Only one of these $tRNA^{Met}$ species ($tRNA_f^{Met}$) could be charged with *E. coli* synthetase and formylated with *E. coli* transformylase. All three $tRNA^{Met}$ species were further purified by passage through reverse-phase chromatography (RPC 3)[10] and were obtained free from cross-contamination by the other methionyl tRNA species. The $tRNA_f^{Met}$ species was obtained in approximately 80% purity as judged by its methionine acceptor activity (1 $A_{260}$ unit pure tRNA = 1.6 nmole).

## General Procedures

### *Preparation of Crude Rabbit Liver tRNA*

Frozen rabbit liver, 480 g, was cut into small pieces and suspended in 900 ml of a cold buffer containing 0.1 *M* Tris·HCl, pH 7.5, 0.35 *M* sucrose, 3 m*M* $MgCl_2$, and 30 m*M* KCl. The suspension was then homogenized in a Waring Blendor for 10 minutes in the cold. The homogenate was centrifuged at 25,000 *g* for 15 minutes in a Sorvall RC2-B centrifuge. The supernatant (approximately 800 ml) was decanted and was further centrifuged at 100,000 *g* for 2 hours in a Spinco Model L2-65B ultracentrifuge. The 100,000 *g* supernatant was adjusted to pH 5.0 by slow addition of cold 1 *M* acetic acid. The solution was stirred for 15 minutes in the cold and was centrifuged at 20,000 *g* for 15 minutes. The precipitate was suspended in 250 ml of 20 m*M* Tris·HCl buffer, pH 7.5; and was mixed with an equal volume of phenol (88%, Analytical Reagent, Mallinckrodt Chemical Works). The mixture was stirred at room temperature for 1 hour and

[6] J. Ilan and J. Ilan, *Biochim. Biophys. Acta* **224,** 614 (1970).

[7] J. P. Leis and E. B. Keller, *Biochem. Biophys. Res. Commun.* **40,** 416 (1970).

[8] K. Ghosh, A. Grishko, and H. P. Ghosh, *Biochem. Biophys. Res. Commun.* **42,** 462 (1971).

[9] M. Zasloff and S. Ochoa, *Proc. Nat. Acad. Sci. U.S.* **68,** 3059 (1971).

[10] L. Shugart, B. Chastain, and G. D. Novelli, *Biochim. Biophys. Acta* **186,** 384 (1969).

was centrifuged at 20,000 *g* for 15 minutes at room temperature. The aqueous layer was carefully decanted. The phenol layer was reextracted twice with 100 ml of 20 m*M* Tris·HCl (pH 7.5) buffer. To the combined aqueous layer (400 ml) was added 40 ml of 20% potassium acetate followed by 880 ml of cold 95% ethanol. The suspension was kept overnight at −20° and was centrifuged at 25,000 *g* for 20 minutes. The precipitate was extracted twice with 20 ml of 1 *M* NaCl. The combined NaCl extract (approximately 40 ml) was mixed with an equal volume of 1 *M* Tris·HCl, pH 9, and the mixture was incubated at 37° for 45 minutes. The solution was then dialyzed against 6 liters of deionized water at 5° for 24 hours, with two changes of water after 8 hours. The dialyzed tRNA preparation (approximately 100 ml, 2200 $A_{260}$ unit) was then applied to a DEAE-cellulose (DE-11, Whatman) column (1.5 × 30 cm) equilibrated with 10 m*M* Tris·HCl buffer, pH 7.5. The column was then thoroughly washed with 500 ml of a buffer containing 10 m*M* Tris·HCl, pH 7.5, and 0.1 *M* NaCl. The tRNA was then eluted from the column with 1 *M* NaCl solution. Fractions containing 260 nm of absorbing material were pooled and thoroughly dialyzed against cold deionized water. The dialyzed tRNA preparation (approximately 2000 $A_{260}$ units) was dried by lyophilization and stored at −20°. The $tRNA_f^{Met}$ activity in the crude tRNA preparation was approximately 5 pmoles per $A_{260}$ unit.

*Preparation of Crude E. coli Aminoacyl-tRNA Synthetases*

Crude *E. coli* aminoacyl-tRNA synthetases were prepared from *E. coli* B cells (grown to 1/0.5 log phase in minimal media, purchased from Grain Processing Co., Muscatine, Iowa) according to the procedure of Muench and Berg.[11] The cells (24 g) were ground with 3 times their weight of alumina 305 (Sigma Chemicals), and the paste was suspended in 50 ml of a solution containing 10 m*M* Tris·HCl, pH 8.0, 10 m*M* $MgCl_2$, and 10% glycerol. The suspension was centrifuged at 25,000 *g* for 15 minutes to remove alumina and cell debris. The supernatant was then centrifuged at 100,000 *g* for 2 hours. The 100,000 *g* supernatant was applied to a DEAE-cellulose (DE-11, Whatman) column (2.6 × 22 cm) previously equilibrated with a buffer containing 20 m*M* potassium phosphate (pH 7.5) 20 m*M* 2-mercaptoethanol, 1 m*M* $MgCl_2$, and 10% glycerol, and the column was thoroughly washed with the same buffer. The mixture of aminoacyl-tRNA synthetases was eluted with a solution containing 0.25 *M* potassium phosphate, pH 6.5, 20 m*M* 2-mercaptoethanol, 1 m*M* $MgCl_2$, and 10% glycerol. The protein peak was pooled and dialyzed against a solution containing 20 m*M* potas-

[11] K. Muench and P. Berg, "Procedures in Nucleic Acid Research" (G. L. Cantoni and D. R. Davies, eds.), p. 375. Harper, New York, 1966.

sium phosphate, pH 7.0, 40 m$M$ 2-mercaptoethanol, 10% glycerol, for 12 hours and then it was further concentrated by dialyzing against the same buffer containing 50% glycerol for 12 hours. The concentrated enzyme solution was stored at $-20°$.

*Preparation of Rabbit Reticulocyte Aminoacyl-tRNA Synthetases*

Crude rabbit reticulocyte aminoacyl-tRNA synthetases were prepared from 100,000 $g$ reticulocyte supernatant by the procedure described for the preparation of *E. coli* synthetases. The concentrated enzyme solution was stored in 50% glycerol at 0°. The methionyl-tRNA synthetase activity in the preparation is very labile and cannot be stored for more than 1 week.

*Assay for Methionine Acceptor Activities*

The reaction mixture contained in a total volume of 0.125 ml: 80 m$M$ sodium cacodylate, pH 7.2, 8 m$M$ magnesium acetate, 8 m$M$ KCl, 1 m$M$ ATP, 4 m$M$ 2-mercaptoethanol, 0.5 nmole of [$^{14}$C]methionine (200 mCi/mmole, New England Nuclear), 0.5–2 $A_{260}$ units of crude rabbit liver tRNA, and either 0.2 mg of reticulocyte synthetase or 50 $\mu$g of *E. coli* synthetase. The reaction mixture was incubated at 37° for 20 minutes. After the incubation period, a 0.1-ml aliquot of the incubation mixture was spotted on a filter paper disk (Whatman No. 3, 2.3 cm diameter). The filter paper disks were washed 3 times in cold trichloroacetic acid (5%), once in ethanol:ether (1:1), and once in ether. The filter paper disks were dried and then counted for radioactivity in toluene-Omnifluor scintillation solution (New England Nuclear) using a Packard liquid scintillation counter (Tri-Carb III).

## Fractionation of Crude Rabbit Liver tRNA by DEAE-Sephadex Chromatography

DEAE-Sephadex A-50 (Pharmacia Fine Chemicals, Inc. 3.5 meq/g, particle size 40–120 $\mu$) beads were soaked in deionized water overnight and were then washed several times with a buffer containing 0.375 $M$ NaCl, 10 m$M$ $MgCl_2$ and 20 m$M$ Tris·HCl, pH 7.2. A column (2.8 cm × 95 cm) was slowly packed with this suspension of DEAE-Sephadex and was equilibrated by passing 2 liters of the above buffer. Crude tRNA (2 g, 33,000 $A_{260}$ units) dissolved in 35 ml of water was slowly adsorbed on the column at a flow rate of about 0.5 ml per minute. The column was then eluted with a linear gradient (0.375 $M$ to 0.425 $M$) of NaCl (total volume of the gradient was 8 liters), containing 20 m$M$ Tris·HCl (pH 7.2) and 10 m$M$ $MgCl_2$. Approximately 75-ml fractions were collected every 150 minutes and were then dialyzed against distilled water. The total amount of tRNA present in these fractions was determined by measuring $A_{260}$

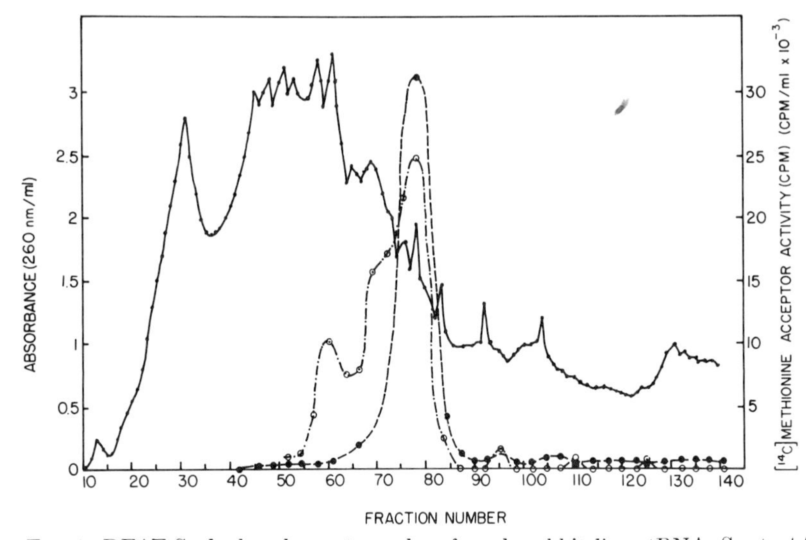


FIG. 1. DEAE-Sephadex chromotography of crude rabbit liver tRNA. See text for details. ●——●, Absorbance (260 nm); ●- - -●, methionine acceptor activity (*Escherichia coli* enzyme); ○- · - · ○, methionine acceptor activity (reticulocyte enzyme).

of the fractions, and the concentration of methionyl-tRNA in the individual fraction was determined by assaying for methionine acceptor activity using either reticulocyte synthetase or the *E. coli* enzyme. Three distinct methionine acceptor activity species designated as $tRNA^{Met}_{m_1}$, $tRNA^{Met}_{m_2}$, and $tRNA^{Met}_{f}$ were observed when the fractions were charged with reticulocyte enzyme. Only one of these species ($tRNA^{Met}_{f}$) could be charged with *E. coli* synthetase. The same Met-$tRNA^{Met}_{f}$ species could also be formylated with *E. coli* transformylase and formyltetrahydrofolate to form *N*-formylmethionyl-$tRNA^{Met}_{f}$ (f-Met-$tRNA^{Met}_{f}$). The extent of formylation of this species was above 99%.

The extent of methionine incorporation into the $tRNA^{Met}_{f}$ species with the reticulocyte enzyme (Fig. 1) was significantly less than that with the *E. coli* enzyme (Fig. 1). This is due to the fact that in these experiments using the same part of different tRNA fractions, the reticulocyte enzyme became limiting. When freshly prepared reticulocyte synthetase and limiting $tRNA^{Met}_{f}$ fractions were used, the maximum charging of the $tRNA^{Met}_{f}$ fractions was the same with both *E. coli* and reticulocyte enzymes.

## Further Purification of the Methionyl-tRNA Species ($tRNA^{Met}_{m_1}$, $tRNA^{Met}_{m_2}$ and $tRNA^{Met}_{f}$) by Reversed-Phase Column Chromatography

The three methionyl-tRNA species obtained by DEAE-Sephadex chromatography as above ($tRNA^{Met}_{m_1}$, pooled fractions 57–63; $tRNA^{Met}_{m_2}$, pooled fractions 66–74; and $tRNA^{Met}_{f}$, pooled fractions 75–81) were further purified by reversed phase chromatography (RPC 3).[10] Chromosorb G (Johns-Manville) was coated with trioctylpropylammonium bromide (Eastman Organic Chemicals) following the procedure of Shugart *et al.*[10] The adsorbant was then suspended in a buffer containing 10 m*M* sodium acetate (pH 4.75), 10 m*M* $MgCl_2$ and 0.275 *M* NaCl and was packed in a column (2 cm × 110 cm). The column was washed with 500 ml of a buffer containing 10 m*M* sodium acetate (pH 4.75) 10 m*M* $MgCl_2$ and 1 *M* NaCl and then equilibrated with the same buffer containing 0.275 *M* NaCl. The three methionyl-tRNA fractions after DEAE-Sephadex chromatography were pooled separately ($tRNA^{Met}_{m_1}$, total 890 $A_{260}$ units; $tRNA^{Met}_{m_2}$, total 900 $A_{260}$ units; $tRNA^{Met}_f$ total 430 $A_{260}$ units) and concentrated to approximately 15 ml. Each fraction was then added separately to different identical reversed-phase columns, and the columns were eluted with a linear gradient (total volume 2 liters) of 0.275 *M* to 0.575 *M* NaCl in a buffer containing 10 m*M* sodium acetate, pH 4.75, and 10 m*M* $MgCl_2$. The column was operated at room temperature (23°) and at a flow rate of 1.3 ml per minute. Approximately 30-ml fractions were collected. The fractions were thoroughly dialyzed against deionized water. The total optical densities of the fractions at 260 nm and the methionine acceptor activities using both *E. coli* enzyme and reticulocyte enzyme were determined as before. The results obtained in each case are shown in Fig. 2.

Figure 2A shows the optical densities and the charging patterns of the fractions obtained with $tRNA^{Met}_{m_1}$ preparation. One sharp major peak (fractions 10–20) and two minor peaks (fractions 38–45 and fractions 52–58) were observed. However, none of these methionyl tRNA species could be charged with *E. coli* synthetase.

The chromatographic profile with $tRNA^{Met}_{m_2}$ showed two major peaks (fractions 9–12 and 13–19). The first peak could be charged with both reticulocyte and *E. coli* enzyme and was, therefore, $tRNA^{Met}_f$ contaminant in the original $tRNA^{Met}_{m2}$ preparation. The second peak was charged by only reticulocyte synthetase and was apparently free from contamination with $tRNA^{Met}_f$. The purity of the $tRNA^{Met}_{m2}$ peak fraction as judged by its acceptor activity was approximately 30%.

The bottom curve (Fig. 2) shows the chromatographic profile of $tRNA^{Met}_f$ fractions. A sharp symmetrical peak was observed which eluted at the beginning of the optical density profile and was charged by both *E. coli* and reticulocyte enzymes. The purity of the $tRNA^{Met}_f$ peak fraction was approximately 80%.

### Some Properties of the Methionyl-tRNA Species

Only the $tRNA^{Met}_f$ species was charged with *E. coli* synthetase. Attempts to charge other methionine-tRNA species ($tRNA^{Met}_{m_1}$ and $tRNA^{Met}_{m_2}$) with the *E. coli* enzyme under various conditions of magnesium ion and

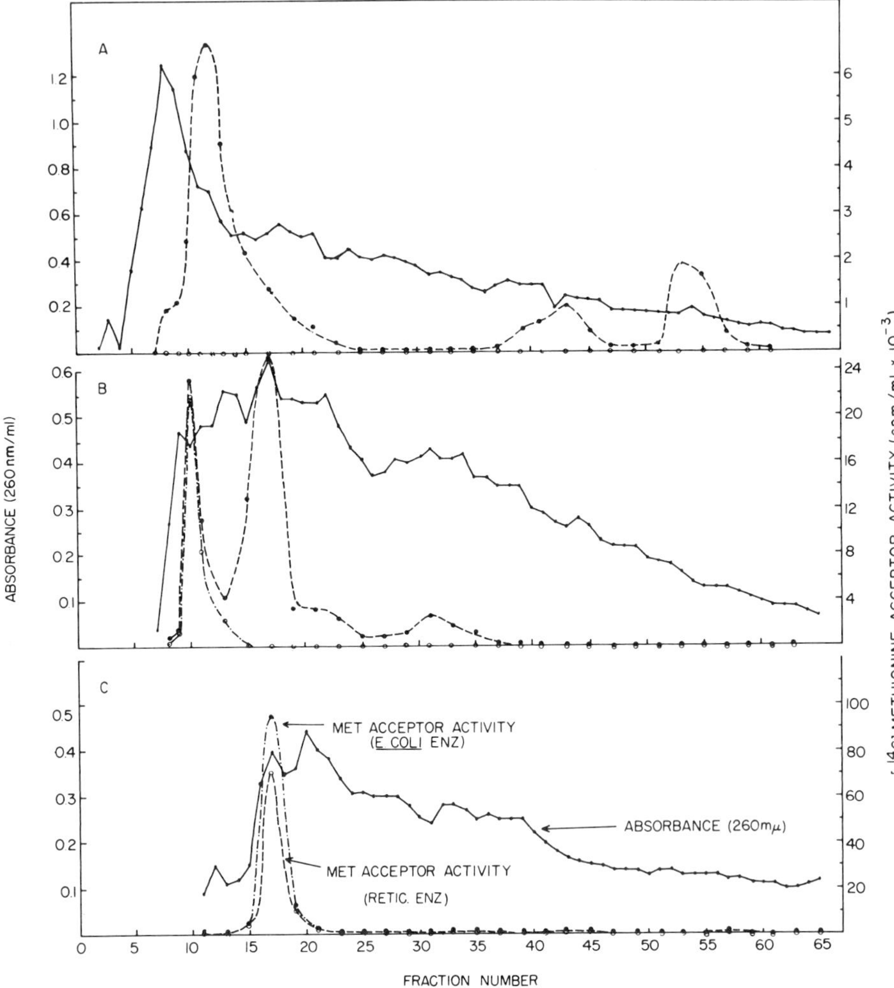


FIG. 2. Purification of $tRNA^{Met}_{m_1}$, $tRNA^{Met}_{m_2}$, and $tRNA^{Met}_{f}$ by reversed-phase chromatography. See text for details. (A) $tRNA^{Met}_{m_1}$, DEAE-Sephadex fraction 57–63; (B) $tRNA^{Met}_{m_2}$, fractions 66–74; (C) $tRNA^{Met}_{f}$, fraction 75–81. See text for details. ●——●, Absorbance (260 nm); ●- - -●, methionine acceptor activity (reticulocyte enzyme); ○-·-·○, methionine acceptor activity (*Escherichia coli* enzyme).

salt concentrations were unsuccessful. This fact has been used to charge preferentially the $tRNA^{Met}_{f}$ species present in crude tRNA preparations.

DEAE-Sephadex column chromatography separated two $tRNA^{Met}_{m}$ species ($tRNA^{Met}_{m_1}$ and $tRNA^{Met}_{m_2}$). These two $tRNA^{Met}_{m}$ species differed somewhat in their efficiency of translation of poly[r(A-U-G)] messenger.[1] However, they behaved similarly during charging by reticulocyte synthetase (i.e., their response to pH, $Mg^{2+}$ ion and salt concentrations were identical) and their codon recognition pattern was also similar.

The coding properties of the fractionated $tRNA^{Met}$ species were determined in a cell-free system from rabbit reticulocytes using poly[r(A-U-G)] and poly[r(U-G)] messengers.[12,13]

Only Met-$tRNA_f^{Met}$ recognized the terminal AUG and GUG codons and initiated polypeptide chain synthesis by transfer of methionine into the N-terminal positions of polypeptides synthesized in response to poly-[r(A-U-G)] and poly[r(U-G)] messengers. The preferential recognition of the terminal methionine codons (AUG and GUG) by Met-$tRNA_f^{Met}$ required the presence of peptide chain initiation factor(s) and low $Mg^{2+}$.

Both Met-$tRNA_f^{Met}$ and Met-$tRNA_m^{Met}$ recognized the internal AUG codons and transferred methionine into the internal positions of polymethionine products synthesized in response to poly[r(A-U-G)] messenger. However, in these transfer reactions, the apparent $K_m$ of poly[r(A-U-G)] was significantly higher with Met-$tRNA_f^{Met}$ than with Met-$tRNA_m^{Met}$. When both Met-$tRNA^{Met}$ species were present in excess, Met-$tRNA_m^{Met}$ preferentially transferred methionine into polypeptides in response to internal AUG codons. The recognition of the internal methionine codon by Met-$tRNA_f^{Met}$ has also been reported by Drews *et al.*[14,15] and has also been observed in *E. coli* protein synthesis.[15–17]

Met-$tRNA_m^{Met}$ and not Met-$tRNA_f^{Met}$ recognized the internal GUG codons. However, this recognition of internal GUG codons by Met-$tRNA_m^{Met}$ required a higher $Mg^{2+}$ concentration.

[12] N. K. Chatterjee, K. K. Bose, C. L. Woodley, and N. K. Gupta, *Biochem. Biophys. Res. Commun.* **43,** 771 (1971).

[13] N. K. Gupta, N. K. Chatterjee, C. L. Woodley, and K. K. Bose, *J. Biol. Chem.* **246,** 7460 (1971).

[14] J. Drews, C. Hogenauer, F. Unger, and R. Weil, *Biochem. Biophys. Res. Commun.* **43,** 905 (1971).

[15] J. Drews, H. Grasmuk, and R. Weil, *Eur. J. Biochem.* **29,** 119 (1972).

[16] K. K. Bose and N. K. Gupta (unpublished observation).

[17] H. P. Ghosh and K. Ghosh, *Biochem. Biophys. Res. Commun.* **49,** 550 (1972).

# [6] RNase Hydrolysis of Aminoacyl-sRNA to Aminoacyl Adenosine

*By* GEORGE ACS and FRITZ LIPMANN

The method described here was introduced as a means of exploring the manner of attachment of amino acids to soluble RNA (sRNA).[1] Digestion with pancreatic RNase of amino acid-charged sRNA rather cleanly releases the 3′-terminal adenosine with the amino acid linked to it. Figure 1 shows the specific cleavage adjacent to a pyrimidine nucleo-

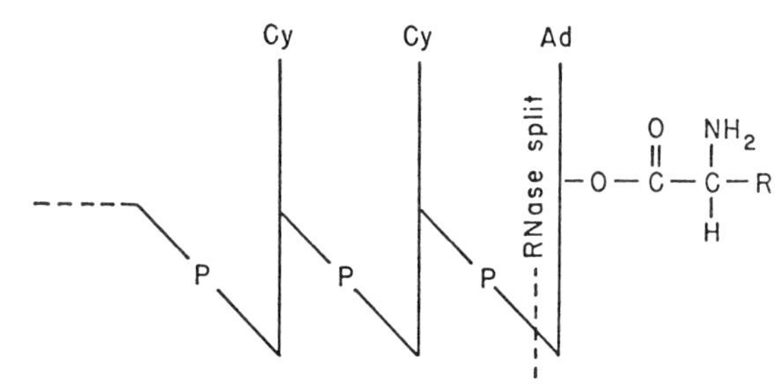


FIG. 1. Illustration of the action of RNase on aminoacyl-sRNA. Cleavage at 5′-position of phosphate bridge between cytidylic acid and terminal aminoacyl adenosine.

tide, and bears out the finding of Hecht *et al.*[2] that the amino acid attachment to sRNA requires the uniform 3′-terminal sequence of two cytidylic acids followed by an adenylic acid.

Brief exposure to pH 9 liberates the amino acid from adenosine (or sRNA). Periodate, however, does not cause a cleavage of the aminoacyl derivative but reacts after amino acid removal. This characterizes the linkage as an ester bond to one of the adjacent 2′ or 3′ OH of the ribose in adenosine.[1-3] To establish the point of attachment was shown to be rather complicated because of the ease of transacylation between the two positions.[4-6]

[1] H. G. Zachau, G. Acs, and F. Lipmann, *Proc. Natl. Acad. Sci. U.S.* **44**, 885 (1958). We use sRNA throughout; tRNA, which stands for transfer RNA, is frequently used synonymously.

[2] L. I. Hecht, M. L. Stephenson, and P. C. Zamecnik, *Proc. Natl. Acad. Sci. U.S.* **45**, 505 (1959).

[3] J. Preiss, P. Berg, E. J. Ofengand, F. H. Bergmann, and M. Dieckmann, *Proc. Natl. Acad. Sci. U.S.* **45**, 319 (1959).

[4] D. H. Rammler and H. G. Khorana, *J. Am. Chem. Soc.* **85**, 1997 (1963).

[5] R. Woldfenden, D. H. Rammler, and F. Lipmann, *Biochemistry* **3**, 329 (1964).

[6] C. S. McLaughlin and V. M. Ingram, *Biochemistry* **4**, 1448 (1965).



ISBN 0-12-504180-2

The isolation of aminoacyl adenosine from a fresh organ, rat liver, will be described as an example of the general procedure.

### Isolation of Aminoacyl Adenosine from a RNase Digest of Aminoacyl-sRNA of Fresh Rat Liver

Sixty grams of fresh rat livers was homogenized in 300 ml of 8.5% sucrose for 1 minute at 0° in a Waring blendor. The homogenate was centrifuged at 100,000 *g* for 2 hours, and the supernatant fraction was shaken with an equal volume of 90% phenol at 4° for 30 minutes. The mixture was centrifuged at 10,000 *g* for 20 minutes, the water layer was removed, and the phenol layer was reextracted with 100 ml of 0.001 *M* Tris buffer, pH 7.2. The combined water layers were dialyzed overnight against excess water and lyophilized. The dried powder was dissolved in 10 ml of water; 1 ml of 20% potassium acetate was added, and the RNA was precipitated with 2 volumes of alcohol. The precipitate containing the aminoacyl-sRNA was allowed to settle for 2 hours at —20°. The heavy, flocculent precipitate was removed by centrifugation at 3000 *g* for 10 minutes, dissolved in 5 ml of water, and spun again; a small residue was rejected, and the supernatant fluid was dialyzed overnight against distilled water. The yield was 3000 OD units at 260 m$\mu$.

### Digestion with Pancreatic RNase

To the dialyzed supernatant containing the aminoacyl sRNA's in 5 ml, 100 $\mu$g of five-times recrystallized RNase, kindly supplied to us by Dr. M. Kunitz, in 0.001 *M* acetate buffer, pH 5.6, was added and the mixture was incubated at room temperature for 30 minutes; the low pH was chosen to prevent hydrolysis of amino acid ester bond. After incubation, the pH was lowered to 3.5 and the solution was passed through a Dowex 1-formate column (2 cm $\times$ 30 cm). This column retains the acidic mono- and oligonucleotides while the adenosyl amino acids are eluted with about 3 column volumes of water. The eluate was concentrated by freeze drying and was dissolved in 3 ml of water. It contained 57 OD units at 260 m$\mu$ at pH 7, and showed the characteristic spectrum for adenosine at neutrality with an absorption maximum at 259 m$\mu$. In acid the maximum absorption was at 257 m$\mu$.

### Separation of Aminoacyl Adenosine

The eluate was subjected to electrophoresis using 0.05 *M* ammonium acetate buffer, pH 3.2, and applying 11 V/cm for 16 hours. The electropherogram showed two ultraviolet quenching regions; one was electrophoretically identical with adenosine, the other moved faster toward the negative pole. The eluate of this latter spot also showed the ultraviolet spectrum of adenosine, but when tested with periodate on the paper according to Gordon *et al.*,[7] it proved to be negative. Alkali hydrolysis of the eluate yielded a mixture of amino acids. Almost all amino acids were found to be present on determination with the Moore and Stein method.[8]

In many cases, sRNA charged with radioactive amino acids has been degraded by RNase to prove terminal esterification of sRNA. For separation of the aminoacyl adenosine from free adenosine, a DEAE-cellulose column was used; the latter eluted later than the former.[5] The isolation of polylysyl adenosine from polylysyl-sRNA by RNase digestion was described by Bretscher.[9] In his experiments, a $^{14}C$-marker on the adenosine of sRNA was introduced by first stripping the terminal adenylic acid from sRNA and then reconstituting it with $^{14}C$-ATP. Most recently, Marcker and Sanger[10] used RNase digestion to confirm the formylation of sRNA-bound methionine in *Escherichia coli* and yeast extracts.

### Attempts to Identify the Site of Esterification

Chromatographic separation of 2′- and 3′-*O*-valyladenosine was described by McLaughlin and Ingram,[6] and an equilibrium ratio of 3:1 in favor of the 3′-compound was found, in agreement with Wolfenden *et al.*[5] The half-time of 3′, 2′ migration is estimated to be 0.1 second at pH 7.5; this fast acyl migration makes a determination of the initial site of attachment seemingly impossible. A higher ratio of 95% 3′ to 5% 2′ isomer under somewhat different conditions has been reported by Sonnenbichler *et al.*[11] The reasons for obtaining such a high figure for the 3′ isomer are not quite clear. An apparently important but indirect argument for a 3′ attachment of the amino acid appears to be that it is so attached in puromycin, and that the 2′ isomer of this inhibitor has been found to be inactive.[12] The inhibition of protein synthesis by puromycin is attributed to premature polypeptide chain termination by competition with the aminoacyl adenosine at the terminal of aminoacyl-sRNA.[13]

[7] H. T. Gordon, W. Thornburg, and L. N. Werum, *Anal. Chem.* **28,** 849 (1956).
[8] S. Moore and W. H. Stein, *J. Biol. Chem.* **211,** 907 (1954).
[9] M. S. Bretscher, *J. Mol. Biol.* **12,** 913 (1965).
[10] K. Marcker and F. Sanger, *J. Mol. Biol.* **8,** 835 (1964).
[11] J. Sonnenbichler, H. Feldmann, and H. G. Zachau, *Z. Physiol. Chem.* **341,** 249 (1965).
[12] D. Nathans and A. Neidle, *Nature* **197,** 1076 (1963).
[13] A. J. Morris and R. S. Schweet, *Biochim. Biophys. Acta* **47,** 415 (1961).

**Characteristics of Alkali Hydrolysis and Hydroxyaminolysis of Aminoacyl- and Peptidyl-sRNA**

The acylated RNA is rather sensitive to alkali;[14-16] at pH 9 at 30° in 0.1 *M* Tris, the amino acids are removed from the sRNA in 15 minutes. Gilbert[17] showed that polyphenylalanyl-sRNA is more resistant to alkali than is phenylalanyl-sRNA. The half-life of phenylalanyl-sRNA at 30° and pH 8.8 is 15 minutes. The polyphenylalanyl-sRNA half-life under the same conditions is 100 minutes. The greater stability of peptidyl-sRNA was first observed by Simon *et al.*[18] with chemically prepared polypeptidyl sRNA. It is due to the greater distance between the amino terminal and the ester link, which lowers the dissociation constant of the terminal carboxyl. Recently, Bretscher[19] confirmed polylysyl sRNA to be more stable to alkali than lysyl-sRNA. The alkaline decomposition of different aminoacyl-sRNA's was tested by Coles *et al.*[20] and found to be quite different—most stable with valyl-sRNA and very unstable with aspartyl and glutamyl derivatives.

REACTIVITY OF AMINO ACID DERIVATIVES WITH 1 *M* HYDROXYLAMINE AT pH 5.5 AND 0°[a]

| Compound[b] | % Decomposition after 1 hour |
|---|---|
| Leucine ethyl ester | 0 |
| 2′/3′-leucine ester of AMP | 28 |
| Leucyl-$^{14}$C-RNA | 41 |
| Leucyl-AMP anhydride | 100 |

[a] From H. G. Zachau, G. Acs, and F. Lipmann, *Proc. Natl. Acad. Sci. U.S.* **44,** 885 (1958).

[b] The leucine ester of AMP was prepared according to T. H. Wieland and G. Pfleiderer [*Advan. Enzymol.* **19,** 235 (1957)] and the leucyl-AMP anhydride by a modification of the procedure of P. Berg [*Federation Proc.* **16,** 152 (1957)]. With the synthetic compounds, hydroxamate formation was measured as described by F. Lipmann and L. C. Tuttle [*J. Biol. Chem.* **159,** 21 (1945)]. With leucyl-$^{14}$C-RNA, the liberation of radioactivity was followed.

[14] M. B. Hoagland, M. L. Stephenson, J. F. Scott, L. I. Hecht, and P. C. Zamecnik, *J. Biol. Chem.* **231,** 241 (1958).

[15] D. Nathans and F. Lipmann, *Proc. Natl. Acad. Sci.* **47,** 497 (1961).

[16] P. S. Sarin and P. C. Zamecnik, *Biochim. Biophys. Acta* **91,** 653 (1964).

[17] W. Gilbert, *J. Mol. Biol.* **6,** 389 (1963).

[18] S. Simon, U. Z. Littauer, and E. Katchalski, *Biochim. Biophys. Acta* **80,** 169 (1964).

[19] M. S. Bretscher, *J. Mol. Biol.* **7,** 446 (1963).

[20] N. Coles, M. W. Bukenberger, and A. Meister, *Biochemistry* **1,** 317 (1962).

The aminoacyl sRNA shows a rather high reactivity to hydroxylamine.[1] At pH 5.5 and 0°, 41% of the RNA amino acid is decomposed in 1 hour. However, compared to an acid anhydride, the reaction is quite slow. Under identical conditions, leucyl-AMP anhydride is almost immediately decomposed. The relative resistance to hydroxylamine was an early clue to the nature of the link, it obviously not being an acid anhydride. The table lists the reactivity of a number of compounds.

# [7] Determination of Adenosine- and Aminoacyl Adenosine-Terminated sRNA Chains by Ion-Exclusion Chromatography

*By* RICHARD WOLFENDEN

Hydrolysis of soluble RNA (sRNA) by pancreatic ribonuclease produces mainly nucleotides and oligonucleotides, which bear negative charges in the neutral pH range. In addition, a free nucleoside or aminoacyl nucleoside is released from each chain, to which it was terminally attached by a 5′-phosphodiester bond.[1] The free nucleosides are not ionized at pH values between 5 and 8. Nucleosides esterified with neutral or basic amino acids bear net positive charge below pH 8 because of the presence of protonated amino groups. When the products of hydrolysis of soluble RNA, partially esterified with such amino acids, are subjected to column chromatography on an anion exchange material such as DEAE-cellulose, the positively charged aminoacyl nucleosides are eluted quantitatively with the solvent front by ion exclusion. This peak is closely followed by a peak or peaks containing the free nucleosides, which are weakly adsorbed. Under the present conditions nucleotides and oligonucleotides are completely retained on the column; they could be eluted, if desired, with more concentrated buffers.[2]

The present procedure was developed as a method for recovering chain ends rapidly and quantitatively for structure determination.[3] Accordingly elution is performed with dilute triethylammonium formate, a volatile buffer which can be removed by freeze-drying. Hydrolysis and separation require 15–30 minutes, and the pH selected for both is low enough to prevent appreciable saponification of the amino acid esters

[1] H. G. Zachau, G. Acs, and F. Lipmann, *Proc. Natl. Acad. Sci. U.S.* **44**, 885 (1958); J. Preiss, P. Berg, E. J. Ofengand, F. H. Bergmann, and M. Dieckmann, *ibid.* **45**, 319 (1959).
[2] B. G. Lane and G. C. Butler, *Can. J. Biochem. Physiol.* **37**, 1329 (1959).
[3] R. Wolfenden, D. H. Rammler, and F. Lipmann, *Biochemistry* **3**, 329 (1964).



ISBN 0-12-504180-2

during this time.[4] Thus, comparison of the ultraviolet absorbancy of the nucleoside ester and free nucleoside peaks provides a simple measure of the total degree of esterification of a sample of soluble RNA, without requiring radioactivity measurements.

*Reagents*

Triethylammonium formate buffer, 1 *M* and 0.02 *M*, pH 4.6

sRNA esterified enzymatically with amino acids. This material should be freed of amino acids and other low molecular weight contaminants by dialysis or, more rapidly, by gel filtration on Sephadex G-25 (coarse grade; Pharmacia, Inc.). Either procedure should be performed so as to render the sRNA in a concentrated form (e.g., 50 mg/ml) in 0.02 *M* triethylammonium formate buffer, pH 4.6 (footnote 5)

Pancreatic ribonuclease (3× recrystallized; Worthington Co.), dissolved in water to a concentration of 10 mg/ml

DEAE-cellulose ("Selectacel;" Schleicher and Schuell Co.). This material should be washed with alkali and acid,[6] and then washed with ten volumes of 1 *M* buffer and twenty volumes of 0.02 *M* buffer, pH 4.6. A column prepared from this material will release negligible ultraviolet-absorbing material; the optical density of the eluent at 260 mμ should not exceed 0.003.

## Method

A solution of soluble RNA (2 ml of 0.02 *M* triethylammonium formate buffer, containing 100 mg of RNA) is incubated for 5 minutes at room temperature with ribonuclease (0.5 mg in 0.05 ml of water).

The mixture is cooled in ice then transferred in the cold to a column of DEAE-cellulose (100-ml bed volume, preequilibrated with and containing 0.02 *M* triethylammonium formate buffer, and packed under a pressure of 5 psi). After application of this sample, any drops adhering

[4] R. Wolfenden, *Biochemistry* **2**, 1090 (1963).

[5] Enzymatic synthesis of aminoacyl adenosine esters and their isolation by phenol extraction and ethanol precipitation may be performed by a variety of methods, one of which is described in footnote 3. Unless otherwise noted, steps after enzymatic synthesis should be performed in the cold, and at a pH no greater than 5.5, to minimize ester hydrolysis. The molarity and pH of the buffer employed in the present procedure can be varied over a considerable range, and if the object is end-group analysis rather than isolation, sodium or potassium acetate buffers are satisfactory. Triethylammonium formate, which can be removed by lyophilization, lends itself to the isolation of salt-free end groups.

[6] E. A. Peterson and H. A. Sober, Vol. V, p. 6.

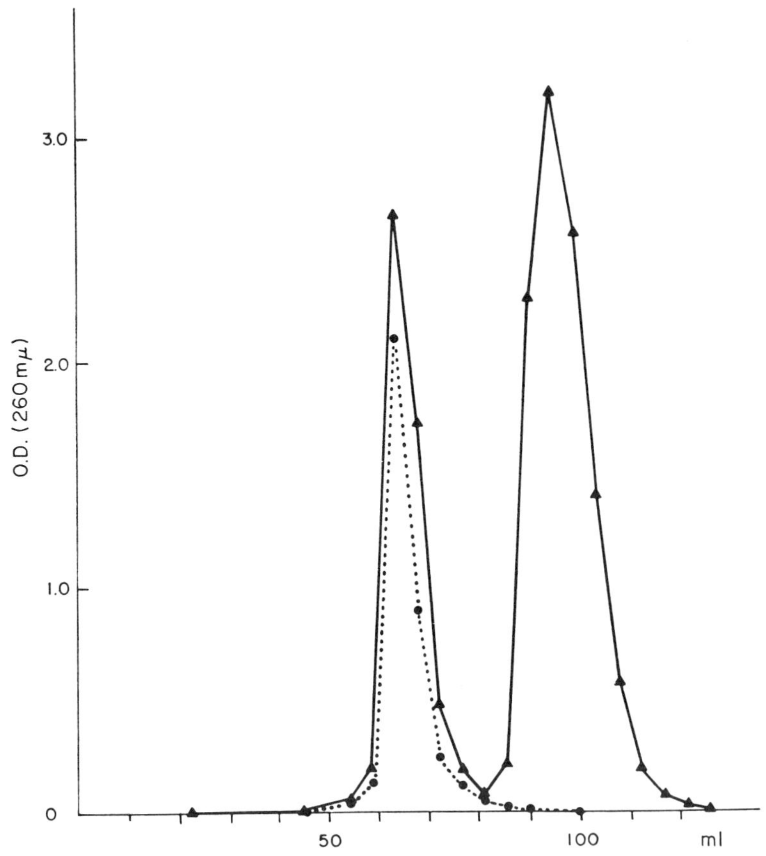


FIG. 1. DEAE chromatography of ribonuclease-treated soluble RNA (100 mg), partially esterified with 5 nonradioactive amino acids and leucine-$^{14}$C. Volume of sample applied: 2 ml. Volume of bed: 100 ml. Elution buffer: 0.02 *M* triethylammonium formate, pH 4.6. [Reprinted from R. Wolfenden, D. H. Rammler, and F. Lipmann, *Biochemistry* 3, 330 (1964). Copyright 1964 by the American Chemical Society and reprinted by permission of the copyright owner.]

to the walls are carefully washed onto the column. Elution is carried out with the same buffer, collecting 4-ml fractions and counting fractions to include that collected during the application of the sample.

The results of this procedure applied to 100 mg of sRNA from *Escherichia coli,* esterified with glycine, L-alanine, L-isoleucine, L-valine, L-phenylalanine, and L-leucine-$^{14}$C, are illustrated in Fig. 1. The first peak, which contained 25.5 optical density (OD) units at 260 mμ, appeared between 55 and 80 ml of eluent. The second, containing 45.5 OD units, was eluted between 80 and 120 ml of eluent. Both peaks were identical in spectrum with adenosine in the same buffer. The dotted line in Fig. 1 represents the distribution of leucine-$^{14}C$. As may be seen from the radioactivity distribution, the leucine was exclusively associated with the first peak, and ninhydrin analysis showed that this peak contained 1.7 micromoles of amino acid (1 mole per mole of adenosine) while the second contained none. That the first peak contained amino acid esters of adenosine was indicated by its failure to react with periodate and by paper electrophoresis.[1] After hydrolysis in dilute ammonia, the only ultraviolet-absorbing material was adenosine, identified by paper electrophoresis under the same conditions. The hydrolyzed material, as well as peak 2, gave only an ultraviolet spot corresponding to adenosine upon descending chromatography on Whatman No. 40 paper using saturated ammonium sulfate–1 *M* sodium acetate–isopropanol (80:18:2).[2]

For isolation, the amino acid ester formate salt peak may be lyophilized to dryness. Repetition of the freeze-drying (after addition of distilled water) is occasionally necessary to remove last traces of triethylammonium formate.

### Notes

1. Since the method depends on exclusion of the front-running material by the ion exchanger, and quite weak adsorption of the second peak, the ratio of sample volume to column volume is critical. In this procedure incomplete separation will result if the column bed volume does not exceed the sample volume by at least a factor of 20.

2. The different amino acid esters of adenosine do not emerge at exactly the same point. The inexact coincidence of the radioactive leucine peak with the first adenosine peak in Fig. 1 is due to the fact that the leucine esters move slightly more rapidly than the bulk of the amino acid esters present. In control experiments performed with sRNA esterified with a single radioactive amino acid, the specific radioactivity over the first adenosine peak remains constant, with a molar ratio of 1 between amino acid and adenosine. It has also been found that synthetic phenylalanyladenosine is slightly retarded, suggesting that esters of aromatic amino acids are weakly adsorbed. This retardation is insufficient to cause overlap with the free adenosine peak.

3. The above concentration of ribonuclease has been found sufficient to give complete release of aminoacyl adenosine and adenosine from soluble RNA under the present conditions. Some variability in the activity of ribonuclease preparations may be encountered, and it is advisable to check the conditions for quantitative release. The course of

release may be roughly followed by precipitating samples of sRNA esterified with a radioactive amino acid, after various intervals of treatment with ribonuclease, by addition of trichloroacetic acid or acid-ethanol.[4] Release is indicated by the appearance of radioactivity in the supernatant fluid.

4. Some preparations of enzymatically esterified *E. coli* sRNA, and virtually all preparations of sRNA from commercial yeast, yield considerable quantities of cytidine. In the present procedure, cytidine is completely separated in a peak (not shown) following the adenosine peak. Ultraviolet analysis and comparison of the two peaks permits direct comparison of the fraction of adenosine- and cytidine-terminal chains. The nucleoside peaks, easily identified by ultraviolet spectrum and paper chromatography, have never been found to be contaminated with nucleosides other than adenosine and cytidine, whereas the ester peak contains exclusively adenosine.

5. Ion-exclusion separation of this type has been carried out with Dowex 1, using dilute acetic acid as eluent.[7] Cation exchange, on Amberlite IRC-50 with acetic acid elution, has also been employed for the isolation of aminoacyl adenosine,[8] but requires considerably longer for elution of the esters than does ion exclusion. The original method of isolation, paper electrophoresis,[1] is useful for qualitative analysis of esters, although somewhat less convenient than column methods for quantitative recovery.

[7] C. S. McLaughlin and V. M. Ingram, *Biochemistry* **4**, 1442 (1965).
[8] J. Sonnenbichler, H. Feldman, and H. G. Zachau, *Z. Physiol. Chem.* **334**, 283 (1963).

# [8] Enzymic Modification of the C-C-A Terminus of tRNA

*By* MATHIAS SPRINZL and HANS STERNBACH

tRNA species with an altered C-C-A terminus have been used for investigation of the mechanism of aminoacylation and ribosomal protein biosynthesis, as well as for spectroscopic and X-ray crystallographic studies.[1] ATP(CTP):tRNA nucleotidyltransferase (EC 2.7.7.25), which catalyzes the incorporation of CMP and AMP into tRNA lacking the

[1] M. Sprinzl and F. Cramer, *Prog. Nucl. Acid Res. Mol. Biol.* **22** (1978). In press.



ISBN 0-12-504180-2

C-C-A part of its 3′ terminus,[2] can be used for preparation of modified tRNA species as it has been shown that some analogs of ATP and CTP are also substrates for this enzyme[3] (see Scheme 1).

An essential prerequisite for the preparation of uniformly modified tRNAs via incorporation of AMP and CMP analogs by ATP(CTP):tRNA nucleotidyltransferase into the 3′ end of tRNA is a tRNA species with a uniformly shortened 3′ terminus. Furthermore, a highly purified enzyme free of nuclease is required because long incubation times and high enzyme concentrations usually have to be applied,[3] owing to the higher $K_m$ values and lower reaction velocities, with the ATP and CTP analogs. It is essential that the ATP and CTP analogs be free of the natural substrates ATP and CTP. Described herein are procedures for the isolation of ATP(CTP):tRNA nucleotidyltransferase from yeast, preparation of $tRNA^{Phe}$-A, $tRNA^{Phe}$-A-C, and $tRNA^{Phe}$-A-C-C from yeast, enzymic synthesis of modified tRNAs, and their analysis.

## Assay for the Incorporation of Nucleotides into the 3′ Terminus of tRNA

The reaction mixture (100 μl) containing 100 m*M* Tris·HCl pH 9.0, 100 m*M* KCl, 10 m*M* $MgSO_4$, 1 m*M* dithiothreitol, 2.0 $A_{260}$ units of appropriate tRNA with a shortened 3′ end, 1.0 m*M* [$^{14}$C]ATP or [$^{14}$C]CTP, respectively (about 10.000 cpm/nmol), and 0.1 μg of ATP(CTP):tRNA nucleotidyltransferase is incubated at 32°. At appropriate time intervals, usually every minute, 10-μl aliquots are spotted onto Whatman 3 MM paper disks, which are then washed twice for 20 min with 5% aqueous trichloroacetic acid and finally with ethanol and ether. The tRNA-bound

$$\text{tRNA-N-C-C-A} \xrightarrow{\text{degradation}} \begin{cases}\text{tRNA-N} \\ \text{tRNA-N-C} \\ \text{tRNA-N-C-C}\end{cases}$$

$$\begin{matrix}\text{tRNA-N} & + 2\,\text{C*TP} + \text{ATP} & & \text{tRNA-N-C*-C*-A} + 3\,\text{PP}_i \\ \text{tRNA-N-C} & + \text{C*TP} \quad + \text{ATP} & \xrightarrow{\text{NTase}} & \text{tRNA-N-C-C*-A} + 2\,\text{PP}_i \\ \text{tRNA-N-C-C} & + \text{A*TP} & & \text{tRNA-N-C-C-A*} + \text{PP}_i\end{matrix}$$

SCHEME 1. NTase = ATP(CTP):tRNA nucleotidyltransferase; C* = cytidine or its analog; A* = adenosine or its analog.

[2] M. P. Deutscher, *Prog. Nucl. Acid Res. Mol. Biol.* **13,** 51 (1973).

[3] M. Sprinzl, H. Sternbach, F. von der Haar, and F. Cramer, *Eur. J. Biochem.* **81,** 579 (1977).

radioactivity is determined using a toluene scintillation fluid. The incorporation of ATP is tested using tRNA$^{Phe}$-A-C-C. The incorporation of CTP is assayed using tRNA$^{Phe}$-A or tRNA$^{Phe}$A-C in the absence and in the presence of ATP. During purification the enzyme assay is performed in an analogous way, but, instead of a tRNA$^{Phe}$ species, 2.0 $A_{260}$ units of crude baker's yeast tRNA and 0.03–30 μg of protein are used.

## Aminoacylation Assay for tRNA Species with a Partially Hydrolyzed 3′ End

The reaction mixture (0.1 ml) contains 150 m$M$ Tris·HCl, pH 7.6, 200 m$M$ KCl, 50 m$M$ $MgSO_4$, 5 m$M$ ATP, 0.02 m$M$ [$^{14}$C]phenylalanine, 1 m$M$ CTP, and 0.2 $A_{260}$ unit of tRNA$^{Phe}$. The mixture is preincubated with 0.5 μg of ATP(CTP):tRNA nucleotidyltransferase from yeast for 5 min at 37°. Aminoacylation is then started by the addition of 1 μg of phenylalanyl-tRNA synthetase (see volume LIX [19]); 10-μl aliquots are removed at appropriate times, and the acid-precipitable radioactivity is determined. By variations of this assay the amount of tRNAs with shortened 3′ end can be determined; i.e., tRNA$^{Phe}$-A-C-C-A can also be aminoacylated in the absence of ATP(CTP):tRNA nucleotidyltransferase. tRNA$^{Phe}$-A-C-C is aminoacylated only in the presence of the regenerating enzyme, but CTP is not needed for the reaction. tRNA$^{Phe}$-A and tRNA$^{Phe}$-A-C can be aminoacylated only in the presence of ATP, CTP, and ATP(CTP):tRNA nucleotidyltransferase. tRNA$^{Phe}$ with more than three nucleotides missing from the 3′ end cannot be regenerated and aminoacylated.

## Isolation of ATP(CTP):tRNA Nucleotidyltransferase

The first steps of purification are described by F. von der Haar in volume LIX[19]. Starting from 12 kg of baker's yeast, the given procedure is followed up to step h.

*Chromatography on CM-Sephadex C-50.* CM-Sephadex C-50 is equilibrated with 30 m$M$ potassium phosphate, pH 7.2, containing 1 m$M$ EDTA, 1 m$M$ dithiothreitol (DTT), and 0.01 m$M$ PMSF (buffer A) and poured into a 6 × 60 cm column. The column is rinsed with 2 liters of the same buffer. The dialyzate obtained in step g is diluted 1:1.5 with buffer A and applied to the column. The column is washed with buffer A containing 0.1 $M$ KCl until the absorbance of the eluent at 280 nm is lower than 0.1. The ATP(CTP):tRNA nucleotidyltransferase activity is eluted with buffer A containing 0.2 $M$ KCl. This eluent is diluted with 2 volumes of buffer A and again applied to a CM-Sephadex C-50 column (4 × 50 cm) equilibrated with buffer A containing 0.05 *M* KCl. A linear salt gradient of buffer A + 50 m*M* KCl to buffer A with 250 m*M* KCl (total volume 4 liters) is then applied. ATP(CTP):tRNA nucleotidyltransferase is eluted under these conditions at 0.18 *M* KCl. The fractions containing the highest enzyme activity are pooled and saturated with ammonium sulfate (440 g/liter). The precipitate is collected by centrifugation (30 min at 17000 *g*) and then dissolved in a minimum of buffer A. This solution is dialyzed overnight against the same buffer containing 170 g of ammonium sulfate per liter.

*Sepharose 4B Column.* The clear dialyzate from the preceding step is applied to a 3 × 30 cm column filled with Sepharose 4B and equilibrated with buffer A containing 390 g of ammonium sulfate per liter. The protein of the dialyzate precipitates on the top of the Sepharose column. The column is washed with about 250 ml of buffer A containing 390 g of ammonium sulfate per liter and then developed running a reversed concentration gradient from 390 g to 275 g of ammonium sulfate per liter in buffer A (total volume 1500 ml); 13-ml fractions are collected. ATP(CTP):tRNA nucleotidyltransferase is eluted at 51% (315 g/l) ammonium sulfate. In spite of the high ammonium sulfate concentration, enzyme activity is not depressed and can easily be tested. The active fractions are pooled and the protein is precipitated by addition of an equal volume of buffer A saturated with ammonium sulfate. The precipitate is collected by centrifugation (30 min at 17 000 *g*) and dissolved in a small amount of buffer A. At this stage tRNA nucleotidyltransferase is more than 90% pure, as can be shown by sodium dodecyl sulfate gel electrophoresis, and can be stored at −20° in the presence of 50% glycerol (v/v) for a month without measurable loss of activity.

*Affinity Elution.*[4] A pure and very highly active enzyme can be prepared by adsorption of the protein to a cation-exchange resin and specific elution of the ATP(CTP):tRNA nucleotidyltransferase with its substrate tRNA-N-C-C. The solution of tRNA nucleotidyltransferase from the preceding step is dialyzed overnight against buffer A containing 5% glycerol. The dialyzate is diluted with the same amount of buffer A containing 5% glycerol and applied to a CM-Sephadex C-50 column (1.5 × 10 cm) equilibrated with the same buffer. ATP(CTP):tRNA nucleotidyltransferase is eluted with a solution of tRNA-N-C-C (100 ml, 2 $A_{260}$ units/ml) in buffer A containing 5 m*M* $Mg^{2+}$. Generally yeast $tRNA^{Phe}$-A-C-C is used, but the elution can also be performed with tRNA-N-C-C from baker's

[4] F. von der Haar, this series, Vol. 34, p. 163.

yeast. The flow rate of the column during the elution of the enzyme should be not more than 30 ml/hr. The fractions containing the active complex of ATP(CTP):tRNA nucleotidyltransferase·tRNA-N-C-C are pooled. In order to dissociate the complex, the solution is passed through a DEAE-Sephadex A-25 column (1 × 5 cm) equilibrated with buffer A containing 50 m*M* KCl and 5% glycerol. tRNA-N-C-C binds quantitatively while the enzyme elutes and can be adsorbed on a CM-Sephadex C-50 column (1 × 5 cm) equilibrated with the same buffer. The ATP(CTP):tRNA nucleotidyltransferase is eluted from this column with a small volume of buffer C containing 0.5 *M* KCl. For storage at −20° an equal volume of glycerol is added to this enzyme solution. At this stage the ATP(CTP):tRNA nucleotidyltransferase is stable for a month, is free from proteases and nucleases, and can be used for incorporation of modified nucleotides into the 3′ end of tRNA. The individual steps of the purification procedure are summarized in Table I.

## Preparation of $tRNA^{Phe}$ from Yeast with Shortened 3′ Terminus

Crude tRNA from baker's yeast contains 85% tRNA-N-C-C and 15% tRNA-N-C-C-A.[3] $tRNA^{Phe}$ is isolated from this material by BD-cellulose

TABLE I
ISOLATION OF ATP(CTP): tRNA NUCLEOTIDYLTRANSFERASE FROM 12 KG OF BAKER'S YEAST

| Purification step | Volume (ml) | Protein (mg) | Total activity (kU) | Specific activity (U/mg) |
|---|---|---|---|---|
| Polymin B supernatant | 8600 | 296,800 | 1128 | 3.8 |
| Dialyzate after ammonium sulfate precipitation | 2300 | 88,000 | 1092 | 12.4 |
| CM-Sephadex C-50 elution | 700 | 6,960 | 814 | 117 |
| CM-Sephadex C-50 chromatography | 210 | 742 | 786 | 1,060 |
| Sepharose 4B column | 90 | 34.8 | 637 | 18,300 |
| Affinity elution | 4 | 11.4 | 526 | 46,200 |

chromatography[5] and is composed of a mixture of $tRNA^{Phe}$-A-C-C and $tRNA^{Phe}$-A-C-C-A (Table II). The separation of these two species is accomplished as follows:

[5] D. Schneider, R. Solfert, and F. von der Haar, *Hoppe Seyler's Z. Physiol. Chem.* **353**, 1330 (1972).

TABLE II
PROPERTIES OF SHORTENED tRNA$^{Phe}$ SPECIES AS ACCEPTORS FOR AMP, CMP, AND PHENYLALANINE

| | Incorporation into tRNA (nmol/$A^{260}$ unit) | | | |
|---|---|---|---|---|
| | | | [$^{14}$C]Phenylalanine | |
| tRNA | [$^{14}$C]CMP | [$^{14}$C]AMP | With NTase[b] | Without NTase[b] |
| tRNA$^{Phe}$[a] | 0.23 | 1.19 | 1.35 | 0.20 |
| tRNA$^{Phe}$-A-C-C | 0.018 | 1.49 | 1.50 | 0.02 |
| tRNA$^{Phe}$-A-C | 1.48 | 1.56 | 1.55 | 0 |
| tRNA$^{Phe}$-A | 2.95 | 1.58 | 1.55 | 0 |

[a] tRNA$^{Phe}$ isolated by chromatography on BD-cellulose.
[b] NTase = ATP(CTP):tRNA nucleotidyltransferase.

tRNA$^{Phe}$ (BD-cellulose fraction), 5300 $A_{260}$ units, dissolved in 100 ml of 20 m$M$ sodium acetate, pH 5.2, containing 10 m$M$ $MgCl_2$ is applied to a column of Sephadex A-25 (2.5 × 100 cm) equilibrated with the same buffer containing 500 m$M$ NaCl and then eluted with a 3000-ml linear gradient from 500 m$M$ to 650 m$M$ NaCl in the same buffer. Fractions of 20 ml are collected and assayed for enzymic aminoacylation with phenylalanine in the presence or in the absence of ATP(CTP):tRNA nucleotidyltransferase as described above. The same chromatography procedure is used for final purification of all shortened tRNAs$^{Phe}$. Appropriate fractions containing tRNA$^{Phe}$-A-C-C or tRNA$^{Phe}$-A-C-C-A are pooled, and the tRNA is isolated by alcohol precipitation and centrifugation. Finally, it is desalted by passage through a column (3.5 × 40 cm) of BioGel P-2 using water as the eluent. tRNA is freeze-dried and can be stored at −20° without loss of activity for over a year.

tRNA$^{Phe}$-A-C is prepared from tRNA$^{Phe}$-A-C-C by oxidation with sodium periodate, elimination of the terminal nucleoside, and alkaline phosphatase treatment as follows. tRNA$^{Phe}$-A-C-C, 300 $A_{260}$ units, is incubated in the dark with 60 ml of 0.8 m$M$ $NaIO_4$ in 50 m$M$ sodium acetate, pH 6.5, for 2 hr at room temperature. Excess periodate is then destroyed by the addition of glucose to a final concentration of 0.8 m$M$. After 30 min at room temperature the oxidized tRNA is isolated by ethanol precipitation and centrifugation. After desalting on a BioGel P-2 column as described above, the tRNA is dissolved in 1 ml of water; 1 ml of 500 m$M$ L-lysine hydrochloride, pH 9.0, is added. The mixture is incubated for 4 hr at 20° in the dark, the pH is then adjusted to 5.2, and the tRNA is isolated by alcohol precipitation and desalted by gel filtration. The yield is 2840 $A_{260}$ units of tRNA$^{Phe}$-A-Cp. To remove the 3′ terminal phosphate residue, the tRNA is dissolved in 50 ml of 100 m$M$ Tris·HCl buffer, pH 8.0, containing 10 m$M$ $MgCl_2$ and incubated with 0.9 mg of alkaline phosphatase for 1 hr at 37°. The pH is then adjusted to 5.2 by the addition of acetic acid, and the mixture is applied to a Sephadex A-25 column. Chromatography is performed under the conditions described above for isolation of tRNA$^{Phe}$-A-C-C. The fractions are tested for phenylalanine acceptance. The yield of tRNA$^{Phe}$-A-C is 2200 $A_{260}$ units.

tRNA$^{Phe}$-A is prepared by degradation of tRNA$^{Phe}$-A-C using the sodium periodate, lysine, and alkaline phosphatase treatment. Individual steps are performed in an analogous way to that described above for the preparation of tRNA$^{Phe}$-A-C. Starting from 2200 $A_{260}$ units of tRNA$^{Phe}$-A-C, 1600 $A_{260}$ units of tRNA$^{Phe}$-A are obtained.

Data on analysis of the 3′-terminal nucleoside of shortened tRNAs$^{Phe}$ are given in Table II. Incorporation of [$^{14}$C]CTP and [$^{14}$C]ATP, respectively, into various tRNA$^{Phe}$ species and their enzymic phenylalanylation are summarized in Table III.

The method described for preparation of shortened tRNAs can be applied also to the preparation of other tRNA species from yeast. tRNA species from *Escherichia coli* with missing 3′-terminal adenosine are obtained by treatment of tRNA-N-C-C-A with limiting amounts of snake venom phosphodiesterase at 0°.[6] This leads to a mixture of tRNAs with a 3′ end degraded to different degrees, which are then converted to tRNA-N-C-C by incorporation of CMP using ATP(CTP):tRNA nucleotidyltransferase in the absence of ATP. Final chromatographic purification of tRNA-N-C-C on Sephadex A-25, is, however, necessary for separation of the side products of the enzymic reactions, such as tRNA-N-C-C-C, which is formed by unnatural incorporation of additional CMP in the absence of ATP.[2]

However, the stepwise degradation of the 3′ end involving the sodium periodate reaction is limited to tRNA species in which only the terminal nucleoside is sensitive to this reagent. It was observed in several cases that some specific tRNAs from *E. coli* are irreversibly inactivated by sodium periodate.[7]

## Preparation of tRNA$^{Phe}$ with Modified 3′ Terminus

One hundred $A_{260}$ units of tRNA$^{Phe}$-A-C-C are incubated in a solution (3.3 ml) containing 100 m$M$ Tris·HCl pH 9.0, 100 m$M$ KCl, 10 m$M$ $MgCl_2$, 1 m$M$ dithiothreitol, and 1.0–5 m$M$ (depending on the $K_m$ of the

[6] M. Sprinzl, F. von der Haar, E. Schlimme, H. Sternbach, and F. Cramer, *Eur. J. Biochem.* **25,** 262 (1972).

[7] M. Sprinzl and F. Cramer, *Proc. Natl. Acad. Sci. U.S.A.* **72,** 3049 (1975).

substrate) nucleoside 5′-triphosphate with 50–100 units of ATP(CTP):tRNA nucleotidyltransferase for 1 hr at 32°. The pH of the mixture is adjusted to 5.0 by adding 2 *M* sodium acetate buffer, pH 4.5, and then an equal volume of water is added. The solution is then applied to a column of Sephadex A-25 (1 × 4 cm) equilibrated with 20 m*M* sodium acetate buffer, pH 5.2. The column was washed with the same buffer (20 ml) and then with a buffer containing 400 m*M* NaCl (100 ml). Under these conditions the excess nucleoside 5′-triphosphate is washed off. tRNA is finally eluted with the buffer containing 1.0 *M* NaCl and isolated by alcohol precipitation and centrifugation. After desalting on a BioGel P-2 column, the product was freeze-dried. Recovery of tRNA is 80–90%.

### Determination of the 3′ End Nucleoside of tRNA

One $A_{260}$ unit of tRNA is incubated in a 25-μl solution containing 100 m*M* Tris·HCl, pH 7.5, and 5 μg of pancreatic ribonuclease at 37° for 2 hr. In the cases where the terminal nucleoside has a purine as a 5′-neighbor, 1 $A_{260}$ unit tRNA is incubated with 2.5 units of T2 ribonuclease for 2 hr at 37° in 25 μl of 50 m*M* sodium acetate, pH 5.2. The 3′-terminal nucleoside is hydrolyzed by this treatment and can be easily separated from the remaining nucleotides and oligonucleotides by chromatography on Beckman M 71 cation-exchange resin. This chromatography is carried out at 50°. Column size is 0.6 × 40 cm; 0.4 *M* ammonium formate buffer, pH 4.15, is used as a eluent at a flow rate of 0.3 ml/min and at about 15 atm (1520 kPa) pressure. Samples of up to 50 μl are injected onto the column through a septum injector. The ultraviolet (UV) absorption of the eluate is monitored simultaneously at 254 and 280 nm by a Spectra-Physics dual-channel UV detector, Model 230 (Spectra-Physics, Santa Clara, California). The maximal sensitivity is 0.01 $A_{260}$ unit for the full scale. Absorption values are recorded every 4 sec for each wavelength with a Withof Transcomp twelve-channel point recorder 288 (Withof, Kassel, Germany). For qualitative determination of nucleosides, the elution volume of the nucleoside and the ratio of absorbance at 280 and 254 nm are compared with the elution volume and 280:254 ratio of authentic standards. Elution volumes for uridine, guanosine, adenosine, and cytidine are 3.6 ml, 9.1 ml, 22.9 ml, and 33.6 ml, respectively. The modified nucleosides elute at distinct volumes and their optical properties (280:254 ratio) can be used for safe identification. Nucleotides elute in the breakthrough volume. Quantitative determination of nucleosides was performed by graphical integration of the appropriate peaks with an accuracy of ±2% for 0.05 $A_{260}$ unit of analyzed nucleoside (Table III).

TABLE III
ANALYSIS OF THE 3'-END NUCLEOSIDE OF THE SHORTENED tRNA$^{Phe}$ SPECIES

| tRNA | 3'-End nucleoside present (%) | | | |
|---|---|---|---|---|
| | A-73 | C-74 | C-75 | A-76 |
| tRNA$^{Phe}$ [a] | 0 | 1 | 85 | 15 |
| tRNA$^{Phe}$-A-C-C | 0 | 0 | 100 | 0 |
| tRNA$^{Phe}$-A-C | 0 | 96 | 4 | 0 |
| tRNA$^{Phe}$-A | 97 | 1 | 3 | 0 |

[a] tRNA$^{Phe}$ isolated by chromatography on BD-cellulose.

Using this chromatographic method, the progress of incorporation of modified nucleotide into shortened tRNA could be also followed during the ATP(CTP):tRNA nucleotidyltransferase-catalyzed reaction. The modified nucleoside 5'-triphosphate, which for this assay does not have to be radioactively labeled, is incubated with shortened tRNA as given in the procedure for preparation of tRNA$^{Phe}$ with modified 3' terminus. At appropriate times, 30-$\mu$l samples corresponding to about 1 $A_{260}$ unit of tRNA are removed, pH is adjusted to 7.0, and 5 $\mu$g of pancreatic ribonuclease are added. The mixture is incubated at 37° for 2 hr and applied onto a Beckman M-71 column. The detected nucleoside must originate from the 3' end of tRNA. If an original 3'-end nucleoside, e.g., cytidine for tRNA$^{Phe}$-A-C-C, disappears in the course of incubation and a new nucleoside in the chromatogram is detected, ATP(CTP):tRNA nucleotidyltransferase-catalyzed incorporation into the 3' end of tRNA takes place.

Using this method, substrate properties of several ATP and CTP analogs for ATP(CTP):tRNA nucleotidyltransferase were tested.[2] Although the procedures in this communication are described for enzyme and tRNA$^{Phe}$ from yeast, similar methods were applied for the isolation and preparation of the components and modified tRNAs from *E. coli* or other sources.

## Materials

Crude tRNA from baker's yeast, Boehringer (Mannheim, Germany)
Baker's yeast (Reinzuchthefe), A. Asbeck, Presshefefabrik (Hamm, Germany)
[$^{14}$C]CTP (46 Ci/mol), [$^{14}$C]ATP (50 Ci/mol); phenylalanine (50 Ci/mol), Schwarz Bioresearch Inc. (Orangeburg)

BD-Cellulose, Boehringer (Mannheim, Germany)
BioGel P-2, 100–200 mesh, Bio-Rad Laboratories (Richmond, California)
CM Sephadex C-50, DEAE-Sephadex A-25, Sepharose 4B, Pharmacia Fine Chemicals (Uppsala, Sweden)
Phenylmethylsulfonylfluoride (PMSF), Merck (Darmstadt, Germany)
Pancreatic ribonuclease (EC 3.1.4.22), Boehringer (Mannheim, Germany)
Alkaline phosphatase (EC 3.1.3.1), Boehringer (Mannheim, Germany)
T2 ribonuclease (EC 3.1.4.29), Sankyo (Tokyo, Japan)

# [9] Methods for Determining the Extent of tRNA Aminoacylation *in Vivo* in Cultured Mammalian Cells

*By* IRENE L. ANDRULIS and STUART M. ARFIN

The central role of tRNA in the mechanism of protein synthesis, together with the multiplicity of tRNA species (isoaccepting tRNAs) and the degeneracy of the genetic code, have led to numerous suggestions that tRNAs may be involved as regulators of both translational and transcriptional events *in vivo*. Among the best-documented regulatory functions for tRNAs in prokaryotes are: repression control of some amino acid biosynthetic pathways[1]; amino acid transport[2]; and the synthesis of guanosine 3′-diphosphate 5′-diphosphate and guanosine 3′-diphosphate 5′-triphosphate.[3] tRNA appears to have some similar regulatory roles in mammalian cells.[4,5]

Much of the evidence for these tRNA-mediated regulatory phenomena is based on experiments in which the activity of one or more of the aminoacyl-tRNA synthetases was restricted *in vivo*. Under these conditions it is assumed that the concentration of a specific charged tRNA is decreased. Methods for determining the intracellular level of specific charged tRNAs have also proved to be useful in evaluating these proposed regulatory roles and in determining which form of tRNA, charged or uncharged, is the active form in regulation. The methods most fre-

[1] J. E. Brenchley and L. S. Williams, *Annu. Rev. Microbiol.* **29,** 251 (1975).

[2] S. C. Quay and D. L. Oxender, *J. Bacteriol.* **127,** 1225 (1976).

[3] M. Cashel, *Annu. Rev. Microbiol.* **29,** 301 (1975).

[4] S. M. Arfin, D. R. Simpson, C. S. Chiang, I. L. Andrulis, and G. W. Hatfield, *Proc. Natl. Acad. Sci. U.S.A.* **74,** 2367 (1977).

[5] P. A. Moore, D. W. Jayme, and D. L. Oxender, *J. Biol. Chem.* **252,** 7427 (1977).





ISBN 0-12-504180-2

quently employed for determining *in vivo* charging levels of tRNA are resistance to periodate oxidation of tRNA extracted from whole cells and aminoacylation with radioactive amino acids in intact cells.

## Periodate Oxidation

*Principle.* tRNA molecules that are not esterified with an amino acid are inactivated by oxidation with periodate. After destruction of the excess periodate, the sample and a control sample are reesterified with radioactive amino acid. The fraction of a particular tRNA esterified *in vivo* is estimated from the ratio of the acceptor activity of the periodate-treated tRNA to that of the untreated control. The chemical action of periodate on RNA has been reviewed previously in this series.[6]

*Reagents*

Buffer A: 50 m*M* sodium acetate/0.15 *M* NaCl, pH 4.5
Buffer B: 10 m*M* sodium acetate/10 m*M* $MgCl_2$/1 m*M* EDTA/15 m*M* β-mercaptoethanol, pH 4.5
$NaIO_4$, freshly prepared in buffer B. The concentration required will depend upon the amount of RNA to be oxidized.

*Procedure.* It is important to rapidly terminate cellular metabolic processes in order to prevent changes in the amount of charged tRNA due to differential effects on the rates of tRNA aminoacylation and utilization during the harvesting of cells. In bacteria this has been achieved by the addition of trichloroacetic acid directly to the cultures.[7,8] The high concentrations of serum proteins in most tissue culture media makes this impractical for cultured animal cells. For suspension cultures, we have found that the best method of rapidly stopping growth is to pour the cultures over an equal volume of crushed ice prepared from 0.1 *M* sodium acetate, pH 4.5. $7 \times 10^7$ cells are harvested by low speed centrifugation and resuspended in 6 ml of 50 m*M* sodium acetate/0.15 *M* NaCl, pH 4.5. An equal volume of phenol (saturated with 50 m*M* sodium acetate/0.15 *M* NaCl, pH 4.5) is added, and the mixture is vortexed for 1 min. After centrifugation, the aqueous layer is withdrawn and reextracted with an equal volume of buffer-saturated phenol. The RNA is precipitated from the aqueous phase by the addition of two volumes of ethanol. After at least 1 hr at −20°, the RNA is collected by centrifugation

[6] G. Schmidt, this series, Vol. 12B [116a].
[7] W. R. Folk and P. Berg, *J. Bacteriol.* **102,** 204 (1970).
[8] J. A. Lewis and B. N. Ames, *J. Mol. Biol.* **66,** 131 (1972).

and dissolved in 0.9 ml of 10 m$M$ sodium acetate/10 m$M$ $MgCl_2$/1 m$M$ EDTA/15 m$M$ mercaptoethanol, pH 4.5. The RNA is divided into two equal fractions, and the $A_{260}$ is determined.

The amount of periodate required to fully oxidize unesterified tRNA without damaging the total acceptance activity is best determined by preliminary experiments. In this laboratory it has been found that a 200-fold molar excess of $NaIO_4$ to RNA works well for $tRNA^{Asn}$, $tRNA^{Leu}$, and $tRNA^{His}$ from Chinese hamster ovary (CHO) cells. Sixty microliters of 20 m$M$ sodium periodate is added to one portion of the RNA and buffer B alone to the other. The samples are kept in the dark at 25° for 15 min, 0.1 ml of ethylene glycol is added to destroy any remaining periodate, and the samples are incubated for an additional 10 min before the RNA is precipitated with two volumes of ethanol. After at least 1 hr at −20°, the RNA is collected by centrifugation and dissolved in 1 ml of 50 m$M$ sodium acetate/0.15 $M$ NaCl, pH 4.5. The RNA is reprecipitated with two volumes of ethanol and washed twice more by dissolving it in 50 m$M$ sodium acetate/0.15 $M$ NaCl, pH 4.5, and reprecipitating with ethanol.

We have found that for $tRNA^{Asn}$, $tRNA^{Leu}$, and $tRNA^{His}$ of CHO cells, chemical deacylation prior to esterification with radioactive amino acid is unnecessary, since the cognate synthetases from hamster liver are able to catalyze a complete exchange between free radioactive amino acid and amino acid esterified to RNA. Similar findings have been made in *Salmonella typhimurium*.[8] Chemical deacylation by means of mild alkaline hydrolysis has been employed for tRNA from animal cells. Yang and Novelli[9] employed a 30-min incubation in 0.3 $M$ Tris·HCl, pH 8.0, to deacylate tRNA. Vaughan and Hansen[10] used a 24–30-hour incubation in 0.2 $M$ lysine buffer, pH 10, at 0° for stripping tRNA from HeLa cells, and Smith and McNamara[11] adjusted the pH of reticulocyte tRNA solutions to 9.5 by the addition of 1 $M$ LiOH and incubated at 37° for 30 min to strip amino acids from the tRNA.

The precipitated RNA is dissolved in a small volume of $H_2O$ and the $A_{260}$ is determined. Esterification of the tRNA of interest is carried out in an appropriate incubation mixture with excess enzyme.[9,12,13] A plateau of radioactivity incorporated into acid-insoluble material with time and proportionality between the amount of amino acid esterified and the

[9] W.-K. Yang and G. D. Novelli, this series, Vol. 20 [5].
[10] M. H. Vaughan and B. S. Hansen, *J. Biol. Chem.* **248,** 7087 (1973).
[11] D. W. E. Smith and A. L. McNamara, *J. Biol. Chem.* **249,** 1330 (1974).
[12] A. H. Mehler, this series, Vol. 20 [23].
[13] C. W. Hancher, R. L. Pearson, and A. D. Kelmers, this series, Vol. 20 [41].

amount of RNA added to the incubation mixture indicate that the reaction has gone to completion.

## Aminoacylation in Intact Cells

*Principle*. The intracellular steady-state concentration of a particular aminoacylated tRNA species depends upon its rate of aminoacylation and the frequency with which it is used in protein synthesis. The amount of aminoacylated tRNA present under steady-state conditions is determined by incubating cells with radioactive amino acid. The total amount of tRNA available for aminoacylation is determined in a parallel culture in which the utilization of tRNA for protein synthesis is inhibited by a short exposure to cycloheximide. The procedure described below is a modification of a number of earlier methods.[14-16]

*Procedure*. This procedure has been developed for cultured cells growing in suspension. Cells are maintained in exponential growth ($\sim 2$ to $3 \times 10^5$ cells/ml for CHO cells) in complete medium. During normal growth or after suitable expression time for mutant phenotypes,[14-16] $1 \times 10^7$ cells are collected by low speed centrifugation and concentrated in 1 ml of medium lacking the amino acid cognate to the tRNA of interest. After a 15-min incubation the radioactive amino acid is added to 1-5 $\mu$Ci/ml at the standard medium concentration. The labeled mixture is incubated for 7.5-60 min until a steady-state rate of incorporation of amino acid into aminoacyl-tRNA is achieved. Cycloheximide (200 $\mu$g/ml) is added to a duplicate sample during the last 5 min of incubation to obtain a value for fully esterified tRNA. The reaction is terminated by the addition of 9 ml of ice-cold 10 m*M* sodium acetate containing unlabeled excess (10 × the medium concentration) amino acid, pH 5.0. In the original procedures ice-cold phosphate-buffered saline was used. This gave spuriously high charging levels for some tRNAs because protein synthesis was stopped but the aminoacylation of tRNA continued to some extent.[15]

The cells are centrifuged and resuspended in 2 ml of ice cold 50 m*M* sodium acetate/0.15 *M* NaCl, pH 5.2. The total cellular RNA is extracted with phenol saturated with this buffer. The aqueous phase is further extracted with chloroform containing 1% isoamyl alcohol and divided

[14] L. H. Thompson, J. L. Harkins, and C. P. Stanners, *Proc. Natl. Acad. Sci. U.S.A.* **70,** 3094 (1973).

[15] L. H. Thompson, D. Lofgren, and G. Adair, *Cell* **11,** 157 (1977).

[16] I. L. Andrulis, C.S. Chiang, S. M. Arfin, T. A. Miner, and G.W. Hatfield, *J. Biol. Chem.* **253,** 58 (1978).

into two equal samples. The RNA in one sample is directly precipitated with cold 10% trichloroacetic acid. The second sample is treated with an equal volume of 0.2 *N* NaOH for 10 min to hydrolyze the amino acids from tRNA before precipitation with trichloroacetic acid. The precipitate from each sample is collected on glass-fiber disks and the radioactivity is determined. The amount of amino acid attached to tRNA *in vivo* is calculated by subtracting the radioactivity remaining in the NaOH-treated sample from the radioactivity precipitated with the total cellular RNA.

Compared to the periodate oxidation method, aminoacylation in intact cells has the advantage of being more rapid and requiring considerably fewer steps. However, the actual labeling is performed with cell suspensions 30–50 times more concentrated than exponentially growing cultures, and this may introduce complicating factors.

# [10] The Chemical Preparation of Acetylaminoacyl-tRNA*

*By* SARA RAPPOPORT and YEHUDA LAPIDOT

## General Principle

The problems involved in the preparation of acetylaminoacyl-tRNA are similar to those encountered in the synthesis of peptidyl-tRNA. First, the acetylating reagent must specifically acetylate the amino group of the amino acid attached to the tRNA and react with any of the different functional groups of the tRNA (such as the hydroxyl groups of the ribose moiety; the amino groups of cytosine, adenine, and guanine, or the phosphoric acid residues). Second, the acetylation reaction should be carried out under conditions that leave the biological properties of the tRNA unchanged. The lability of the ester bond between the amino acid and the tRNA requires special attention.

Haenni and Chapeville[1] reported on the acetylation of Phe-tRNA by using acetic anhydride as acetylating agent. When [$^{14}$C]acetic anhydride was used, and the acetylated Phe-tRNA was treated with Tris buffer at pH 9.5, the hydrolyzate gave four radioactive spots on paper electrophoresis. Only 50% of the radioactivity was found in the spot corresponding to acetylphenylalanine. It seems that acetic anhydride reacts not only with the amino group of the amino acid attached to the tRNA but also with some functional groups of the tRNA.

* This work was supported in part by a grant from the Stiftung Volkswagenwerk Germany, No. 111604.

[1] A. L. Haenni and F. Chapeville, *Biochim. Biophys. Acta* **114**, 135 (1966).





ISBN 0-12-504180-2

The method which will be described in detail below is based on the reaction between aminoacyl-tRNA and *N*-hydroxysuccinimide ester of acetic acid. The active ester (III) is prepared by allowing the acetic acid (I) to react with *N*-hydroxysuccinimide (II) in the presence of dicyclohexylcarbodiimide. The reaction is summarized in the following scheme:

$$\underset{\text{(I)}}{RCOOH} + \underset{\text{(II)}}{HON\begin{matrix} / \; CO - CH_2 \\ \backslash \; CO - CH_2 \end{matrix}} \xrightarrow{DCC} \underset{\text{(III)}}{RCOON\begin{matrix} / \; CO - CH_2 \\ \backslash \; CO - CH_2 \end{matrix}}$$

$$III + \underset{\text{(IV)}}{H_2N\overset{R'}{\overset{|}{C}}HCOO\text{-}tRNA} \rightarrow \underset{\text{(V)}}{RCONH\overset{R'}{\overset{|}{C}}HCOO\text{-}tRNA}$$

DCC = *N*, *N*-dicyclohexylcarbodiimide

This method was found to be a general one and has been used successfully with aliphatic carboxylic acids, e.g., formic, acetic, caprylic, lauric, and palmitic acids[2–5] as well as with aromatic compounds containing a free carboxylic groups.[6,7] It was found that when *N*-hydroxysuccinimide ester of [$^3$H]acetic acid was allowed to react with deacylated tRNA, the radioactivity associated with the tRNA was equivalent to one acetyl residue per 100 molecules of tRNA.

## Procedure

*Reagents*

[$^{14}$C]Aminoacyl-tRNA prepared as described elsewhere[8]
*N*-Hydroxysuccinimide purchased from Aldrich, Wisconsin
[$^3$H]Acetic acid purchased from the Radiochemical Centre, Amersham, England (batch No. 13).

*N-Hydroxysuccinimide Ester of Acetic Acid*

Glacial acetic acid (1.7 ml, 30 mmoles) was added to a solution of *N*-hydroxysuccinimide (3.45 g, 30 mmoles) in dry ethyl acetate (200 ml). A

[2] N. de Groot, Y. Lapidot, A. Panet, and Y. Wolman, *Biochem. Biophys. Res. Commun.* **25,** 17 (1966).
[3] Y. Lapidot, N. de Groot, and I. Fry-Shafrir, *Biochim. Biophys. Acta* **145,** 292 (1967).
[4] N. de Groot, I. Fry-Shafrir, and Y. Lapidot, *Eur. J. Biochem.* **8,** 571 (1969).
[5] Y. Lapidot, S. Rappoport, and Y. Wolman, *J. Lipid Res.* **8,** 142 (1967).
[6] I. Gillam, D. Blew, R. C. Warrington, M. von Tigerstrom, and G. M. Tener, *Biochemistry* **7,** 3459 (1968).
[7] P. Schofield, B. M. Hoffman, and A. Rich, *Biochemistry* **9,** 2525 (1970).
[8] E. Ziv, N. de Groot, and Y. Lapidot, *Biochim. Biophys. Acta* **213,** 115 (1970).

solution of dicyclohexylcarbodiimide (6.18 g, 30 mmoles) in dry ethyl acetate (10 ml) was then added, and after stirring for 15 minutes, the reaction mixture was kept at room temperature overnight. The dicyclohexyl urea, which precipitated out, was removed by filtration and the filtrate was evaporated to dryness under reduced pressure (under 40°). The residue (4.3 g) was recrystallized in the following way: water (40 ml) was added, and it was warmed in a boiling water bath for 2 minutes. The warm solution was filtered and kept at room temperature until crystallization of the product, and then at 4° for 15 hours. The crystalline *N*-hydroxysuccinimide ester of acetic acid was isolated by filtration and kept in a vacuum desiccator over $P_2O_5$. Yield 2.3 g (50%), m.p. 130°. The final product moved as a single spot on thin-layer chromatography (Fig. 1).

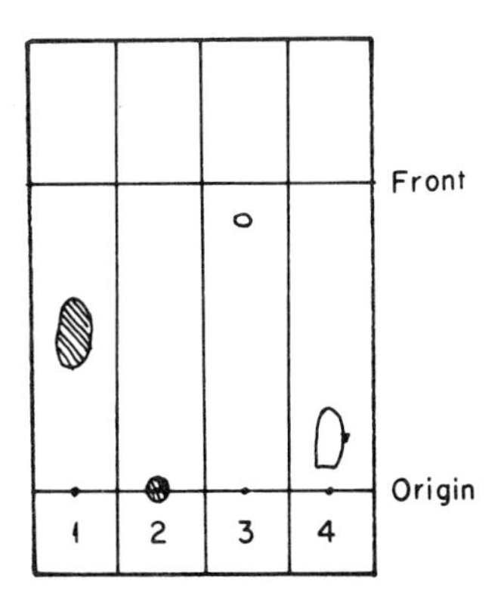


FIG. 1. Thin-layer chromatography on microchromatoplates coated with silica gel by dipping [J. Peifer, *J. Mikrochim. Acta*, p. 529 (1962)]. Solvent: chloroform. Indicators: open spots were made visible by charring with aqueous sulfuric acid 1:1 (v/v). Solid spots were made visible (red color) by applying a spray consisting of a mixture of 14% hydroxylamine solution in water (20 ml) and 14% NaOH (8.5 ml), followed after 2 minutes, by spraying with 5% $FeCl_3$ in 1.2 *N* HCl. 1, *N*-hydroxysuccinimide ester of acetic acid; 2, *N*-hydroxysuccinimide; 3, dicyclohexylcarbodiimide; 4, dicyclohexyl urea.

### *N-Hydroxysuccinimide Ester of [$^3$H]Acetic Acid*

[$^3$H]Acetic acid (3 mg, 50 μmoles, specific activity 500 mCi/mmole) together with glacial acetic acid (0.28 ml, 5 mmoles) were added to a solution of *N*-hydroxysuccinimide (575 mg, 5 mmoles) in dry ethyl acetate (25 ml). A solution of dicyclohexylcarbodiimide (1.03 g, 5 mmoles) in dry ethyl acetate (5 ml) was then added. After stirring for 15 minutes the reaction mixture was kept at room temperature overnight. The dicyclohexyl urea which precipitated out was removed by filtration, the filtrate was evaporated to dryness under reduced pressure, the residue was triturated with ether, filtered, and dried in a vacuum desiccator. The dry product

(330 mg, yield 44%) which had a specific activity of 5 mCi/mmole contained traces of *N*-hydroxysuccinimide but was used without further purification.

*N-Acetyl [$^{14}$C]Val-tRNA*

*Method A.* [$^{14}$C]Val-tRNA ($6 \times 10^5$ cpm, specific activity 260 mCi/mmole, 24 $A_{260\ nm}$ units) dissolved in 0.1 *M* acetate buffer pH 5.0 (0.1 ml) was added to a solution of *N*-hydroxysuccinimide ester of acetic acid (30 mg) in freshly distilled dimethyl sulfoxide (0.4 ml). The reaction mixture was kept at 0° for 2 hours, and the tRNA was precipitated by adding a solution of 10% dichloroacetic acid (0.5 ml) (final concentration 5% dichloroacetic acid). After 1 hour at 0°, the precipitate was isolated by centrifugation (12,000 rpm at 0° for 10 minutes), washed with cold absolute ethanol ($3 \times 3$ ml), and finally dried in a vacuum desiccator. The dry product was dissolved in 20 m*M* acetate buffer, pH 5.0 (0.5 ml) (any insoluble material was removed by centrifugation), and the clear solution was kept at −20°.

*Method B.* A solution of *N*-hydroxysuccinimide ester of acetic acid (30 mg) in dimethyl formamide (0.25 ml) was added to a solution of [$^{14}$C]Val-tRNA ($5 \times 10^5$ cpm, specific activity 260 mCi/mmole, 20 $A_{260\ nm}$ units) in 0.1 *M* triethanolamine hydrochloride, pH 4.3, containing 10 m*M* $MgCl_2$ (2.5 ml). The pH was adjusted to 8 by adding 0.1 *M* NaOH (about 1.3 ml). The reaction mixture was kept at 0° for 20 minutes, and the tRNA was precipitated by adding a solution of 50% dichloroacetic acid (0.46 ml) (final concentration 5% dichloroacetic acid). After 1 hour at 0°, the precipitate was isolated by centrifugation (12,000 rpm at 0° for 10 minutes), washed with cold absolute ethanol ($3 \times 3$ ml) and finally dried in a vacuum desiccator. The dry product was dissolved in 20 m*M* acetate buffer pH 5.0 (0.5 ml) (any insoluble material was removed by centrifugation) and the clear solution is kept at −20°.

# [11] The Synthesis of Oligopeptidyl-tRNA*

*By* YEHUDA LAPIDOT and SARA RAPPOPORT

In protein biosynthesis the peptide bond is formed by the reaction between the amino group of one aminoacyl-tRNA and the carboxyl group

* This work was supported in part by a grant from the Stiftung Volkswagenwerk Germany No. 111604.




ISBN 0-12-504180-2

of a second aminoacyl-tRNA or peptidyl-tRNA. The peptidyl-tRNA, which is an intermediate, remains bound to the ribosomes until the peptide chain of the protein is complete. Finally, the newly made protein is released while the tRNA is bound to the ribosomes.

Peptidyl-tRNA can be isolated from ribosomes active in protein synthesis. The peptidyl-tRNA so obtained is a mixture of different tRNA's to which peptides of different chain lengths and different amino acid composition are bound. By using synthetic homopolynucleotides or copolynucleotides with a known sequence, one can control the amino acid composition of the peptidyl-tRNA, but it is very difficult to control the peptide chain length.

In order to obtain chemically defined peptidyl-tRNA in sufficient quantities for biochemical and physical investigations, peptidyl-tRNA's have been prepared chemically, using aminoacyl-tRNA as the starting material.

## General Principle

The chemical synthesis of peptidyl-tRNA using aminoacyl-tRNA as starting material can be accomplished in three major steps: (1) preparation of a suitable N-blocked carboxyl activated amino acid or peptide; (2) condensing the N-blocked carboxyl activated amino acid (or peptide) with aminoacyl-tRNA; (3) removing the N-blocking group from the peptidyl-tRNA.

The chemical synthesis of peptidyl-tRNA involves two major problems. First, the N-blocked carboxyl activated amino acid (or peptide) should react specifically with the amino group of the amino acid attached to the tRNA. Second, conditions for the acylation reaction as well as for the removal of the protecting group from the N-blocked peptidyl-tRNA, which will leave the tRNA unchanged in its chemical and biological properties must be found.

Scheme I summarizes the chemical synthesis of peptidyl-tRNA. The amino group of the amino acid or peptide is blocked either by a monomethoxytriphenylmethyl group or by an *o*-nitrophenylsulfenyl group (III). The carboxyl group of the N-blocked amino acid or peptide is activated by converting it to the *N*-hydroxysuccinimide ester (V). After the acylation reaction takes place, the N-blocking group is removed from the N-blocked peptidyl-tRNA (VII) under mild conditions which leave the tRNA undamaged: The monomethoxytriphenylmethyl group is removed by treatment with 5% dichloroacetic acid and the *o*-nitrophenylsulfenyl group is removed by treatment with 0.25 *M* sodium thiosulfate.

The monomethoxytriphenylmethyl group was used successfully as

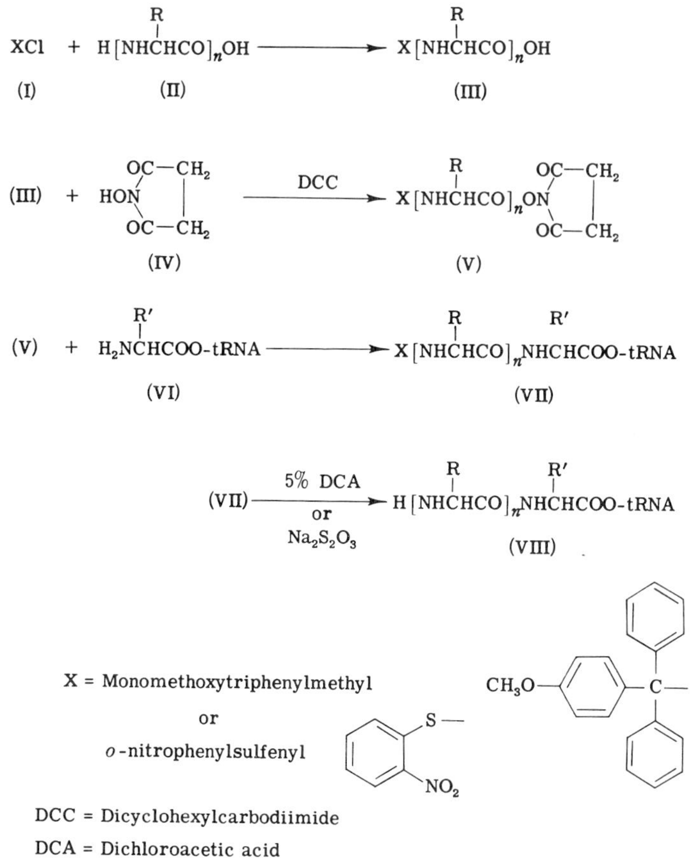

$$\underset{\text{(I)}}{XCl} + \underset{\text{(II)}}{H[NH\overset{R}{C}HCO]_nOH} \longrightarrow \underset{\text{(III)}}{X[NH\overset{R}{C}HCO]_nOH}$$

$$\text{(III)} + \underset{\text{(IV)}}{HON(COCH_2)_2} \xrightarrow{DCC} \underset{\text{(V)}}{X[NH\overset{R}{C}HCO]_nON(COCH_2)_2}$$

$$\text{(V)} + \underset{\text{(VI)}}{H_2N\overset{R'}{C}HCOO\text{-}tRNA} \longrightarrow \underset{\text{(VII)}}{X[NH\overset{R}{C}HCO]_nNH\overset{R'}{C}HCOO\text{-}tRNA}$$

$$\text{(VII)} \xrightarrow[\text{or } Na_2S_2O_3]{5\%\ DCA} \underset{\text{(VIII)}}{H[NH\overset{R}{C}HCO]_nNH\overset{R'}{C}HCOO\text{-}tRNA}$$

X = Monomethoxytriphenylmethyl ($CH_3O-C_6H_4-C(C_6H_5)_2-$)

or

*o*-nitrophenylsulfenyl ($o\text{-}NO_2C_6H_4S-$)

DCC = Dicyclohexylcarbodiimide

DCA = Dichloroacetic acid

Scheme I

N-blocking group in the preparation of glycine-[1–4] and alanine-[4] containing peptidyl-tRNA. However, for the preparation of phenylalanine-,[5] serine-,[6] tryptophan-,[6] tyrosine-,[6] or valine-[7] containing peptidyl-tRNA, the *o*-nitrophenylsulfenyl group was used as N-blocking group.

[1] Y. Lapidot, Y. Wolman, M. Weiss, R. Peled, and N. de Groot, *Isr. J. Chem.* **4,** 62p (1966).

[2] Y. Lapidot, N. de Groot, M. Weiss, R. Peled, and Y. Wolman, *Biochim. Biophys. Acta* **138,** 241 (1967).

[3] Y. Lapidot, N. de Groot, and S. Rappoport, *Biochim. Biophys. Acta* **182,** 105 (1969).

[4] Y. Lapidot, N. de Groot, S. Rappoport, and D. Elat, *Biochim. Biophys. Acta* **190,** 304 (1969).

[5] A. D. Hamburger, N. de Groot, and Y. Lapidot, *Biochim. Biophys. Acta* **213,** 115 (1970).

[6] M. Rubinstein, N. de Groot, and Y. Lapidot, *Biochim. Biophys. Acta* **209,** 183 (1970).

[7] Y. Lapidot, D. Elat, S. Rappoport, and N. de Groot, *Biochem. Biophys. Res. Commun.* **38,** 559 (1970).

A list of different peptidyl-tRNA's prepared according to the above procedure is given in Table I.

TABLE I
THE SYNTHESIS OF DIFFERENT PEPTIDYL-tRNA

| *N*-Hydroxysuccinimide ester of N-blocked amino acid or peptide | +Aminoacyl-tRNA →peptidyl-tRNA[a] | | Reference |
|---|---|---|---|
| $(Gly)_n$ | $(Gly)_n$-AA-tRNA | $n = 1$–3 | *b* |
| $(Ala)_n$ | $(Ala)_n$-AA-tRNA | $n = 1$–5 | *b* |
| $(Phe)_n$ | $(Phe)_n$-AA-tRNA | $n = 1$–2 | *c* |
| $(Phe)_n$-Gly | $(Phe)_n$-Gly-AA-tRNA | $n = 1$–3 | *c* |
| $(Ser)_n$ | $(Ser)_n$-AA-tRNA | $n = 1$–2 | *d* |
| $(Tyr)_n$-Gly | $(Tyr)_n$-Gly-AA-tRNA | $n = 1$–3 | *d* |
| Trp-Gly | Trp-Gly-AA-tRNA | | *d* |
| Val-Gly | Val-Gly-AA-tRNA | | *e* |

[a] AA = Phe, Val, Ala, Ser, Tyr, Leu.
[b] Y. Lapidot, N. de Groot, S. Rappoport, and D. Elat, *Biochim. Biophys. Acta* **190,** 304 (1969).
[c] A. D. Hamburger, N. de Groot, and Y. Lapidot, *Biochim. Biophys. Acta* **213,** 115 (1970).
[d] M. Rubinstein, N. de Groot, and Y. Lapidot, *Biochim. Biophys. Acta* **209,** 183 (1970).
[e] Y. Lapidot, D. Elat, S. Rappoport, and N. de Groot, *Biochem. Biophys. Res. Commun.* **38,** 559 (1970).

The preparation of peptidyl-tRNA containing a long peptide chain necessitates the synthesis of N-blocked carboxyl-activated long-chain peptide. As it is difficult to synthesize such peptides, the stepwise technique was used for the preparation of oligopeptidyl-tRNA. After each step, the acylated aminoacyl-tRNA was isolated by centrifugation, and after removal of the N-blocking group the peptidyl-tRNA was ready for the second acylation. Different oligopeptidyl-tRNA's prepared by the use of the stepwise synthesis are given in Table II. The stepwise synthesis is demonstrated by the preparation of $Gly_5$Val-tRNA (see Procedure).

## Procedure

*Reagents*

*Aminoacyl-tRNA*. Prepared as described elsewhere.[8]

*N-o-nitrophenylsulfenyl amino acid and peptides*. Prepared as described by Zernas *et al.*[9]

[8] E. Ziv, N. de Groot, and Y. Lapidot, *Biochim. Biophys. Acta* **213,** 115 (1970).
[9] L. Zervas, D. Borovas, and E. Gazis, *J. Amer. Chem. Soc.* **85,** 3660 (1963).

TABLE II
THE STEPWISE SYNTHESIS OF DIFFERENT OLIGOPEPTIDYL-tRNA

| N-Hydroxy-succinimide ester of N-blocked peptide | Substrate | Number of steps | Oligopeptidyl-tRNA | Reference |
|---|---|---|---|---|
| | | | *Peptapeptidyl-tRNA* | |
| $(Tyr)_3$-Gly | Phe-tRNA | 1 | $(Tyr)_3$-Gly-Phe-tRNA | *a* |
| $(Ala)_4$ | Ala-tRNA | 1 | $(Ala)_4$-Ala-tRNA | *b* |
| $(Gly)_2$ | $(Gly)_2$-Val-tRNA | 2 | $(Gly)_4$-Val-tRNA | *b* |
| $(Ser)_2$ | $(Ser)_2$-Phe-tRNA | 2 | $(Ser)_4$-Phe-tRNA | *a* |
| | | | *Hexapeptidyl-tRNA* | |
| $(Gly)_3$ | $(Gly)_2$-Val-tRNA | 2 | $(Gly)_5$-Val-tRNA | *b* |
| $(Ala)_2$ | $(Ala)_3$-Ala-tRNA | 2 | $(Ala)_5$-Ala-tRNA | *b* |
| $(Phe)_2$-Gly | Phe-Gly-Phe-tRNA | 2 | $(Phe)_2$-Gly-Phe-Gly-Phe-tRNA | *c* |
| | | | *Higher peptidyl-tRNA* | |
| $(Gly)_3$ | $(Gly)_3$-Phe-tRNA | 2 | $(Gly)_6$-Phe-tRNA | *b* |
| $(Ser)_2$ | $(Ser)_4$-Phe-tRNA | 3 | $(Ser)_6$-Phe-tRNA | *a* |
| $(Gly)_3$ | $(Gly)_4$-Phe-tRNA | 3 | $(Gly)_7$-Phe-tRNA | *b* |
| $(Ala)_4$ | $(Ala)_3$-Ala-tRNA | 2 | $(Ala)_7$-Ala-tRNA | *b* |
| $(Ala)_4$ | $(Ala)_4$-Phe-tRNA | 2 | $(Ala)_8$-Phe-tRNA | *b* |
| $(Gly)_2$ | $(Gly)_6$-Phe-tRNA | 3 | $(Gly)_8$-Phe-tRNA | *b* |
| $(Gly)_3$ | $(Gly)_6$-Phe-tRNA | 3 | $(Gly)_9$-Phe-tRNA | *b* |
| $(Ala)_3$ | $(Ala)_6$-Phe-tRNA | 3 | $(Ala)_9$-Phe-tRNA | *b* |
| $(Gly)_3$ | $(Gly)_9$-Phe-tRNA | 4 | $(Gly)_{12}$-Phe-tRNA | *b* |

[a] M. Rubinstein, N. de Groot, and Y. Lapidot, *Biochim. Biophys. Acta* **209,** 183 (1970).
[b] Y. Lapidot, N. de Groot, S. Rappoport, and D. Elat, *Biochim. Biophys. Acta* **190,** 304 (1969).
[c] A. D. Hamburger, N. de Groot, and Y. Lapidot, *Biochim. Biophys. Acta* **213,** 115 (1970).

*Monomethoxytriphenylmethyl Chloride.* The preparation is according to that of triphenylmethylchloride.[10] Dry magnesium turnings (15.5 g) are placed in a 1-liter three-necked flask fitted with a dropping funnel, mercury-sealed mechanical stirrer, and a double surface reflux condenser. A solution of dry bromobenzene (15 g, 10 ml) in sodium-dried ether (35 ml) is placed in the dropping funnel. The dropping funnel and the reflux condenser are provided with calcium chloride guard tubes in order to prevent the entrance of moisture into the reaction mixture. The bromobenzene

[10] A. I. Vogel, "Practical Organic Chemistry," p. 813. Longmans, Green and Co. London, 1957.

solution is added to the magnesium and warmed gently on a water bath until the reaction becomes vigorous. If no reaction ensues, add a small crystal of iodine to start the reaction; the use of iodine is generally unnecessary if the reagents and the apparatus are thoroughly dry. As soon as the reaction is moderately vigorous, the flask is immersed in a bath of cold water. Start the stirrer and add a solution of dry bromobenzene (75.5 g, 50.5 ml) in sodium-dried ether (200 ml) at such a rate as to cause vigorous refluxing (during about 1 hour); when all the bromobenzene solution has been introduced, the mixture is stirred for 20–30 minutes, until most (or all) of the magnesium has dissolved. To the resulting Grignard reagent (phenyl magnesium bromide) cooled in a cold water bath, a solution of dry ethyl anizate (45 g, 42 ml) in dry ether (100 ml) is added at such a rate that the mixture refluxes gently (about 1 hour). Then the mixture is refluxed for 1 hour on a water bath. Cool in a freezing mixture of ice and salt and pour it slowly, with constant stirring, into a mixture of crushed ice (750 g) and concentrated sulfuric acid (25 ml). Continue the stirring until all the solid dissolves; it may be necessary to add solid ammonium chloride (25 g) to facilitate the decomposition of the magnesium complex, and also a little more ether to dissolve all the product. When all the solids have passed into solution, the ether layer is separated and washed successively with water (100 ml), 5% sodium bicarbonate solution (100 ml), and water (100 ml). The ether is evaporated to dryness, and the evaporation is continued with an oil pump for at least 2 hours.

To mono-*p*-methoxytriphenylcarbinol (60 g) acetyl chloride (100 ml) is added dropwise. The mixture is refluxed for 3 hours with guard tube of calcium chloride in the reflux condenser. The resulting mixture is kept overnight at −20°. The solid product is collected by filtration, washed with about 10 ml of petroleum ether and put immediately in a dry vacuum desiccator.

*N-Mono-p-methoxytriphenylmethyl Gly-Gly.* Monomethoxytriphenylmethyl chloride (7.2 g, 23 mmoles) is added in portions with continuous stirring to a solution of Gly-Gly (2.64 g, 20 mmoles) in water (40 ml) containing diethylamine (4 ml, 40 mmoles) and isopropanol (80 ml). The addition is accomplished in 3 hours. The reaction mixture is stirred for additional 15 minutes, and water (120 ml) is added. The emulsion is centrifuged (10,000 rpm for 30 minutes at 0°) and the supernatant is acidified by adding dropwise a solution of acetic acid (1.2 ml, 20 mmoles) in water (10 ml). The mixture is immediately extracted with ethyl acetate (200 ml) and the two layers are separated. The ethyl acetate layer is washed several times with water (3 × 100 ml), then dried over anhydrous magnesium sulfate, and the solution is concentrated under reduced pressure to a volume of about 50 ml (the solution is filtered if a precipitation is formed). The clear solution is added dropwise with continuous stirring to petroleum ether (1.5 liters). The precipitate is collected by filtration and dried immediately in a vacuum desiccator for 30 minutes giving 5.2 g of dry product (yield 65%).

It is necessary to continue immediately with the next step.

*N-Hydroxysuccinimide Ester of N-Monomethoxytriphenylmethyl Gly-Gly.* To a solution of *N*-hydroxysuccinimide (1.495 g, 13 mmoles) and *N*-monomethoxytriphenylmethyl Gly-Gly (5.2 g, 13 mmoles) in dry ethyl acetate (200 ml), a solution of dicyclohexylcarbodiimide (2.73 g, 13 mmoles) in dry ethyl acetate (20 ml) is added. The reaction mixture is stirred for 30 minutes and then kept at room temperature overnight. The dicyclohexyl urea which precipitates out is removed by filtration. The filtrate is concentrated under reduced pressure to a volume of about 50 ml and then added dropwise with continuous stirring to petroleum ether (1.5 liters). The white precipitate is collected by filtration and stored in a vacuum desiccator over NaOH. Yield 4.9 g (69%).

*N-Hydroxysuccinimide Ester of N-o-Nitrophenylsulfenyl Phe-Gly.* To a solution of *N*-hydroxysuccinimide (2.185 g, 19 mmoles) and *N*-*o*-nitrophenylsulfenyl Phe-Gly (7.27 g, 19 mmoles) in dry tetrahydrofurane (50 ml), a solution of dicyclohexylcarbodiimide (3.99 g, 19 mmoles) in dry tetrahydrofurane (30 ml) is added. The reaction mixture was stirred for 30 minutes and then kept at room temperature overnight. The dicyclohexyl urea which precipitates out is removed by filtration, and the filtrate is evaporated to dryness under reduced pressure. The residue is dissolved in ethyl acetate (200 ml), washed with 5% $NaHCO_3$ (3 × 50 ml), then with water until the washing is neutral, and dried over unhydrous $MgSO_4$. The clear ethyl acetate is evaporated to dryness, and the residue is triturated with ether. The solid material is collected by filtration and stored in a vacuum desiccator over $P_2O_5$. Yield 7.95 g (89%).

*Gly-Gly-[$^{14}$C]Val-tRNA.* [$^{14}$C]Val-tRNA (3.6 × $10^5$ cpm specific activity 260 mCi/mmole, 10 $A_{260\,nm}$ units) dissolved in distilled water (0.05 ml) is added to a solution of freshly prepared *N*-hydroxysuccinimide ester of *N*-monomethoxytriphenylmethyl Gly-Gly (30 mg) in freshly distilled dimethyl sulfoxide (0.4 ml); 0.2 *M* triethanolamine sulfate buffer, pH 8.0 (0.05 ml), is added, and the clear solution is kept at 0° for 2 hours. The tRNA is precipitated by adding a solution of 10% dichloroacetic acid (0.5 ml) (final concentration 5% dichloroacetic acid). After 1 hour at 0°, the precipitate is isolated by centrifugation (12,000 rpm at 0° for 10 minutes), washed with cold absolute ethanol (3 × 3 ml), and finally dried in a vacuum desiccator. The dry product is crushed to powder with a glass rod. To the dry powder, 20 m*M* acetate buffer pH 5.0 (0.2 ml) is added, the mixture is mixed in a test tube mixer (Vortex) and kept at 60° for 5 minutes (in case the solution is not clear, the insoluble material is removed by centrifugation). The clear solution is stored at $-20°$. For analysis of the peptidyl-tRNA, an aliquot is treated with 0.2 *M* NaOH for 1 hour at 37° and the alkaline hydrolyzate was analyzed by high voltage paper electrophoresis together with the appropriate marker. In the preparation of Gly-Gly-[$^{14}$C]Val-tRNA, all the radioactivity moved as Gly-Gly-Val. In a sample of the peptidyl-tRNA which had not been treated with alkali, all the radioactivity remained at the origin.

*Gly$_5$[$^{14}$C]Val-tRNA.* Gly-Gly-[$^{14}$C]Val-tRNA ($7.2 \times 10^5$ cpm, 20 $A_{260\,nm}$ units) dissolved in distilled water (0.05 ml) is added to a solution of *N*-hydroxysuccinimide ester of *N*-monomethoxytriphenylmethyl Gly$_3$ (30 mg) in freshly distilled dimethyl sulfoxide (0.4 ml); 0.2 *M* triethanolamine sulfate buffer, pH 8.0 (0.05 ml), is added, and the clear solution is kept at 0° for 2 hours. The tRNA is precipitated by adding a solution of 10% dichloroacetic acid (0.5 ml) (final concentration 5% dichloroacetic acid). After 1 hour at 0°, the precipitate is isolated by centrifugation (12,000 rpm at 0° for 10 minutes), washed with cold absolute ethanol ($3 \times 3$ ml) and finally dried in a vacuum desiccator. The dry product is dissolved in 20 m*M* acetate buffer pH 5.0 (0.5 ml) and analyzed as described above. The clear solution is stored at $-20°$.

*Phe-Gly-[$^{14}$C]Val-tRNA.* [$^{14}$C]Val-tRNA ($6 \times 10^6$ cpm, specific activity 260 mCi/mmole, 40 $A_{260\,nm}$ units) dissolved in distilled water (0.05 ml) is added to a solution of *N*-hydroxysuccinimide ester of *o*-nitrophenylsulfenyl-Phe-Gly (30 mg) in freshly distilled dimethyl sulfoxide (0.4 ml); 0.2 *M* triethanolamine sulfate buffer, pH 8.0 (0.05 ml), is added and the clear solution is kept at 0° for 2 hours. The tRNA is precipitated by adding a solution of 10% dichloroacetic acid (0.5 ml) (final concentration 5% dichloroacetic acid). After 1 hour at 0°, the precipitate is isolated by centrifugation (12,000 rpm at 0° for 10 minutes), washed with cold dimethyl formamide ($3 \times 3$ ml) and then with cold absolute ethanol (3 ml), and finally dried in a vacuum desiccator. The dry product is dissolved in 0.1 *M* acetate buffer pH 5 (0.75 ml) containing 6 *M* urea and 1 *M* $Na_2S_2O_3$ (0.25 ml) was added. The reaction mixture is kept at 30°, after 2 hours 1 *M* NaCl (0.4 ml) and cold ethanol (2.8 ml) are added and after 3 hours at $-20°$ the tRNA is isolated by centrifugation. The oily precipitate was washed with cold 70% ethanol ($3 \times 3$ ml), then with absolute ethanol (3 ml) and finally dried in a vacuum desiccator. The dry product is dissolved in 20 m*M* acetate buffer pH 5.0 (0.5 ml) and analyzed as described above for the preparation of Gly-Gly-[$^{14}$C]Val-tRNA. The clear solution is stored at $-20°$.

# [12] Preparation of Aminoacyl-Oligonucleotides and Their Binding to Ribosomes

*By* Sidney Pestka

The binding of the terminal fragment of aminoacyl-tRNA to ribosomes enables one to distinguish among several events involving peptide bond formation (Fig. 1): the binding of the peptidyl- or aminoacyl-tRNA terminus (pCpCpA end) to ribosomes [steps (1) and (2), respectively] and the subsequent transfer of the nascent peptide to the adjacent aminoacyl-tRNA to form a peptide bond [peptidyl transfer reaction, step (3)]. In Fig. 2 is shown an illustration of the binding of phenylalanyl-oligonucleotide to ribosomes. Assay of the binding of the phenylalanyl-oligonucleotide to ribosomes has been used to study the effect of various antibiotics and other agents on the binding of the aminoacyl-end of tRNA to ribosomes. These studies help to elucidate the mechanism of action of many antibiotics and the nature of the ribosomal sites for binding this portion of the tRNA molecule. A simple procedure for the preparation of these aminoacyl-oligonucleotides and their binding to ribosomes is presented below. The procedure for their preparation is a modification of that reported previously by Herbert and Smith.[1]

## Preparation of Aminoacyl-Oligonucleotides

*Reagents*

Aminoacyl-tRNA[2] (labeled with a radioactive amino acid)
T1 ribonuclease (Cal Biochem)
Ammonium formate, 0.01, 0.3, 2.0, and 5.0 *M*, pH 5.0
EDTA 10 m*M*, adjusted to pH 5.0 with KOH
DEAE-Sephadex (A-25)

[1] E. Herbert and C. J. Smith, *J. Mol. Biol.* **28**, 281 (1967).
[2] Abbreviations used are as follows: Phe-oligonucleotide, phenylalanyl-oligonucleotide; Met-oligonucleotide, methionyl-oligonucleotide; AA-oligonucleotides, the 3′-terminal oligonucleotides containing a mixture of many amino acids (AA); [$^3$H]Phe-tRNA, unfractionated *E. coli* tRNA labeled with [$^3$H]phenylalanine and 19 nonradioactive amino acids; $tRNA^{Phe}$, phenylalanine-accepting species of tRNA; $tRNA^{fMet}$, the methionine-accepting species of tRNA, whose methionine can also accept a formyl group; one $A_{260}$ unit is the amount of material which in 1.0 ml would yield a value of 1.0 for the absorbance measured at 260 m$\mu$ in a cuvette with a path length of 1.0 cm at pH 7.0.



ISBN 0-12-504180-2

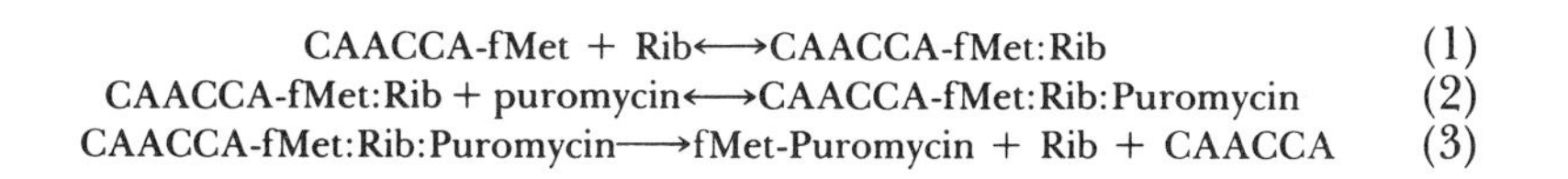

CAACCA-fMet + Rib⟷CAACCA-fMet:Rib (1)
CAACCA-fMet:Rib + puromycin⟷CAACCA-fMet:Rib:Puromycin (2)
CAACCA-fMet:Rib:Puromycin⟶fMet-Puromycin + Rib + CAACCA (3)

CAACCA-fMet + Puromycin⟶fMet-Puromycin + CAACCA

FIG. 1. Illustration of steps in puromycin reaction.

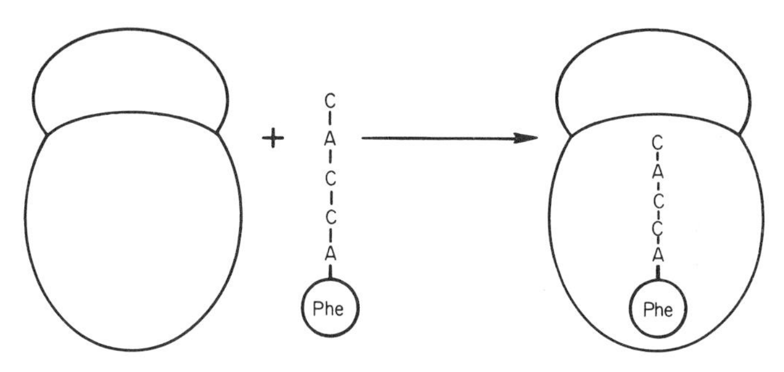


FIG. 2. Schematic illustration of the binding of phenylalanyl-oligonucleotide (CACCA-Phe) to ribosomes.

Digestion of aminoacyl-tRNA by T1 RNase was carried out at 37° for 30 minutes in reaction mixtures containing the following components: 10 m$M$ ammonium formate, pH 5.0; 2 m$M$ EDTA, pH 5.0; 80 $A_{260}$ units of aminoacyl-tRNA (about 4 mg); 100 units of T1 RNAse.

Immediately after incubation, the reaction is placed directly onto a small column of DEAE-Sephadex (A-25) equilibrated with 10 m$M$ ammonium formate, pH 5.0. The column should be previously washed with 5 ml of 5 $M$ ammonium formate. At least 2.3 mg tRNA can be applied per milliliter of the column. Free amino acids are eluted with 10 m$M$ ammonium formate with 3-ml fractions that can be directly counted in a scintillation fluor until little radioactivity is found in the eluted fractions. Next, the aminoacyl-oligonucleotides are eluted from the column with 0.3 $M$ ammonium formate, pH 5.0. Fractions of the 0.3 $M$ ammonium formate elution are collected, and portions are counted for determination of radioactivity. The aminoacyl-oligonucleotides are eluted until little further radioactivity appears. Any incompletely digested aminoacyl-oligonucleotides or undigested aminoacyl-tRNA are then eluted with 2 $M$ ammonium formate. A preparation of [$^3$H]phenylalanyl-oligonucleotide obtained from [$^3$H]Phe-tRNA is shown in Fig. 3. The 0.3 $M$ ammonium formate peak is pooled and subsequently lyophilized to remove the ammonium formate. About 90% of the radioactivity of [$^3$H]Phe-tRNA is recoverable as

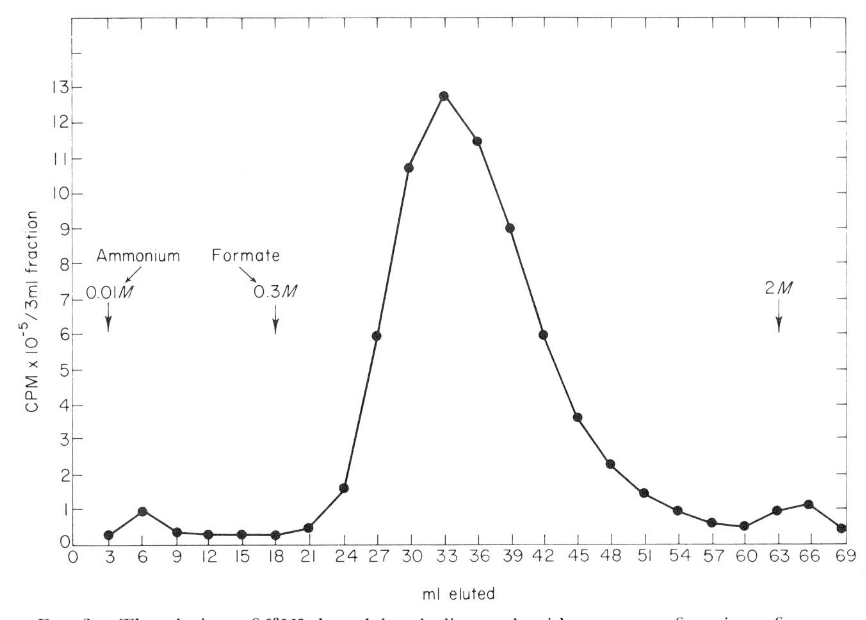


FIG. 3. The elution of [$^3$H]phenylalanyl-oligonucleotide as a step function of ammonium formate concentration. The 0.50 ml reaction mixture after T1 RNase digestion of [$^3$H]Phe-tRNA ($4.7 \times 10^6$ cpm; 185 $A_{260}$ units) was placed on a 4.5 ml column of DEAE-Sephadex (A-25) equilibrated with 10 m*M* ammonium formate, pH 5.0. The column was then eluted successively with 3-ml portions of 10 m*M*, 0.3 *M*, and 2 *M* ammonium formate (pH 5.0) as shown in the figure. Column procedures were performed at 24°.

[$^3$H]Phe-oligonucleotide. The residual phenylalanyl-oligonucleotide is stored at −20° in the lyophilized state.

The procedure is applicable to both unfractionated and fractionated tRNA preparations. With unfractionated tRNA preparations the procedure yields a mixture of aminoacyl-oligonucleotides. Some aminoacyl-oligonucleotides are more sensitive to T1 RNase digestion than others, so that lesser amounts of the T1 RNase may be required. It is thus advisable to run a small pilot study of digestion of the tRNA as a function of T1 RNase concentration when using a new aminoacyl-tRNA. This is especially true for purified tRNA fractions.

To obtain better resolution of aminoacyl-oligonucleotides from contaminating species, it is advisable to use a gradient of ammonium formate. The position of elution of a specific aminoacyl-oligonucleotide from the column depends on the net negative charge of the molecule. Thus, the smaller aminoacyl-oligonucleotides such as C-C-A(Ser) elute at 0.08 *M* ammonium formate whereas the large aminoacyl-oligonucleotide such as U-C-A-U-C-A-C-C-C-A-C-C-A(Val) at 0.8 *M* ammoni-

$T_1$ Ribonuclease Terminal Fragments from *E. coli* Aminoacyl-tRNA Species[a]

| tRNA Species | Aminoacyl-oligonucleotides | Ammonium Formate Molarity for Elution (*M*) | Reference[b] |
|---|---|---|---|
| Leu 1 | C-A-C-C-A(Leu) | 0.27 | 1 |
| Leu 2 | U-A-C-C-A(Leu) | 0.27 | 1,2 |
| Met f | C-A-A-C-C-A(Met) | 0.38 | 3 |
| Met m | C-C-A-C-C-A(Met) | 0.38 | 4 |
| Phe | C-A-C-C-A(Phe) | 0.28 | 5,6 |
| Ser | C-C-A(Ser) | 0.08 | 7 |
| Val 1 | U-C-A-U-C-A-C-C-C-A-C-C-A(Val) | 0.8 | 8 |
| Val 2 | C-A-C-C-A(Val) | 0.27 | 9 |
| Lys | A-C-C-C-A-C-C-A(Lys) | 0.43 | 10 |

[a]The sequences of aminoacyl-oligonucleotides which have been prepared are given in the table. The references refer to reports describing the sequences of the aminoacyl-oligonucleotides. The ammonium formate molarity (pH 5.0) is the concentration at which the peak of that aminoacyl-oligonucleotide fraction can be expected to appear using a gradient elution[11] at room temperature.

[b]Key to References.

[1]P. Berg, U. Lagerkvist, and M. Dieckmann, *J. Mol. Biol.*, **5**, 159 (1962).

[2]H. Ishikura, Y. Yamada, K. Ishii, and S. Nishimura, *J. Japanese Biochem. Soc.*, **41**, 418 (1969).

[3]S. K. Dube, K. A. Marcker, B. F. C. Clark, and S. Cory, *Nature,* **218**, 232 (1968).

[4]S. Cory, K. A. Marcker, S. K. Dube, and B. F. C. Clark, *Nature,* **220**, 1039 (1968).

[5]M. Uziel, and H. G. Gassen, *Biochemistry,* **8**, 1643 (1969).

[6]B. G. Barrell and F. Sanger, *FEBS Lett.*, **3**, 275 (1969).

[7]H. Maruyama and G. Cantoni, personal communication; H. Ishikura, personal communication.

[8]M. Yaniv and B. G. Barrell, *Nature,* **222**, 278 (1969).

[9]M. Yaniv and B. G. Barrell, personal communication.

[10]R. V. Case and A. H. Mehler, personal communication.

[11]S. Pestka, T. Hishizawa, and J. L. Lessard, *J. Biol. Chem.*, **245**, 6208 (1970).

um formate. The table lists several aminoacyl-oligonucleotides that have been prepared and their approximate elutions from the DEAE-Sephadex column at room temperature.

T1 RNase may contaminate the aminoacyl-oligonucleotide preparations. It does not contaminate C-C-A(Ser) and other aminoacyl-oligonucleotides eluting at 0.1 *M* ammonium formate or less. It does, however, elute with many of the aminoacyl-oligonucleotides eluting at 0.2 *M* ammonium formate and greater. Electrophoresis can be used to remove T1 ribonuclease from the aminoacyl-oligonucleotides if desirable. The small amounts of T1 ribonuclease contaminating the aminoacyl-oligonucleotide preparation does not generally interfere with binding of the aminoacyl-oligonucleotides to ribosomes. Nevertheless, it is desirable to remove T1 ribonuclease from these preparations. Electrophoresis is carried out under the following conditions: 0.5% pyridine and 5% acetic acid, pH 3.5, buffer; 2 hours with a voltage-gradient of 50 volts per centimeter; Whatman 3 MM paper. Under these conditions T1 ribonuclease is negatively charged and therefore will travel towards the anode whereas most of the aminoacyl-oligonucleotides are generally positively charged and will travel toward the cathode. The radioactive band of the aminoacyl-oligonucleotide is eluted from the paper, lyophilized and stored as indicated above.

## Binding of Aminoacyl-Oligonucleotides to Ribosomes

Binding of aminoacyl-oligonucleotides to ribosomes is determined in reaction mixtures containing the following components: 50 m$M$ Tris·acetate, pH 7.2; 0.4 $M$ KCl; 0.1 $M$ $NH_4Cl$; 40 m$M$ $MgCl_2$; 3–4 $A_{260}$ units of ribosomes washed four times in 1 $M$ ammonium chloride[3]; and varying amounts of aminoacyl-oligonucleotides. Test tubes are kept

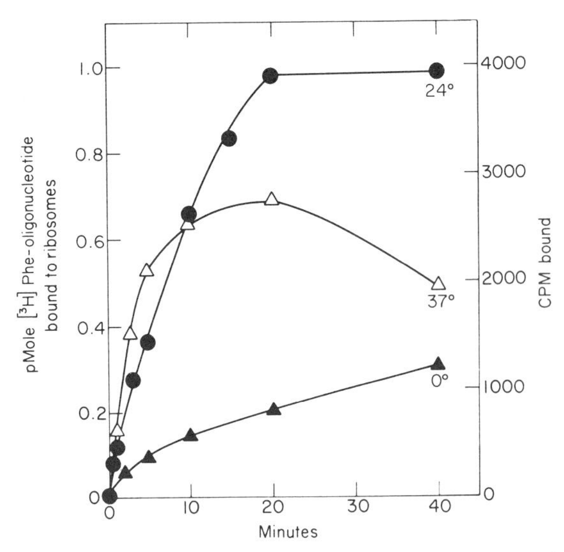


FIG. 4. Binding of Phe-oligonucleotide to ribosomes as a function of time at 0°, 24°, and 37°. Each 0.050-ml reaction mixture contained the following components: 50 m$M$ Tris·acetate, pH 7.2; 0.1 $M$ $NH_4Cl$; 50 m$M$ potassium acetate; 40 m$M$ magnesium acetate; 3.8 $A_{260}$ units of ribosomes; and 2.7 pmoles of [$^3$H]Phe-oligonucleotide (0.07 $A_{260}$ unit; 6300 mCi/mmole). Assays were performed as described in the text at the temperature and time indicated on this figure. ▲ 0°; ●, 24°; △, 37°.

[3] S. Pestka, *J. Biol. Chem.* **243**, 2810 (1968).

at 0° during addition of all components. Usually, [$^3$H]Phe-oligonucleotide or other aminoacyl-oligonucleotide is added last to initiate the binding of aminoacyl-oligonucleotide to ribosomes. Changing the order or addition of buffer, KCl, $NH_4Cl$, $MgCl_2$, ribosomes, and aminoacyl-oligonucleotide did not modify the results. Reactions were usually incubated at 24° for 10–20 minutes. Binding of the phenylalanyl-oligonucleotides to ribosomes was determined by filtering the reaction mixtures after incubation through cellulose nitrate filters.[4] Each tube and filter was then immediately washed three times with 3 ml portions of the buffer and salt concentration used in the reaction mixture. The binding of [$^3$H]Phe-oligonucleotide to ribosomes as a function of incubation time at several temperatures is shown in Fig. 4. For maximal binding of aminoacyl-oligonucleotides to ribosomes, 20% ethanol was present additionally in the reaction mixtures and wash solutions.

[4] M. Nirenberg and P. Leder, *Science* **145**, 1399 (1964).

# [13] Chemical Modifications of Amino Acids Esterified to tRNA (Other Than Acylations)

*By* S. Chousterman and F. Chapeville

Aminoacyl-tRNA's chemically modified on the amino acid moieties were first prepared to test the "adaptor hypothesis." They have proved useful also in the study of the mechanism of peptide chain initiation and elongation, as well as in the study of the specificity of various enzymes that use aminoacyl-tRNA's as substrates.

Several acylation procedures of the α-amino group of tRNA-bound amino acids are described elsewhere in this volume.

Four other types of modifications are considered here:

I. Reduction of Cys-tRNA$^{Cys}$[1] to Ala-tRNA$^{Cys}$

II. Deamination of aminoacyl-tRNA's to the corresponding α-hydroxy derivatives

[1] Abbreviations: Cys-tRNA$^{Cys}$, Ala-tRNA$^{Cys}$ or Cys($O_3$)H-tRNA$^{Cys}$ = cysteine, alanine, or cysteic acid esterified to tRNA$^{Cys}$.
Met-tRNA$^{Met}$ or Met(O)-tRNA$^{Met}$ = methionine or sulfoxymethionine esterified to tRNA$^{Met}$
Ser-tRNA$^{Ser}$ or Gxl-tRNA$^{Ser}$ = serine or glyoxylic acid esterified to tRNA$^{Ser}$
Thr-tRNA$^{Thr}$ Gxl-tRNA$^{Thr}$ = threonine or glyoxylic acid esterified to tRNA$^{Thr}$
Phe-tRNA$^{Phe}$ or Phl-tRNA$^{Phe}$ = phenylalanine or phenyllactic acid esterified to tRNA$^{Phe}$



ISBN 0-12-504180-2

III. Oxidation of Cys-tRNA$^{Cys}$ to Cys($O_3$)H-tRNA$^{Cys}$ and of Met-tRNA$^{Met}$ to Met(O)-tRNA$^{Met}$
IV. Oxidation of Ser-tRNA$^{Ser}$ and Thr-tRNA$^{Thr}$ to Gxl-tRNA$^{Ser}$ and Gxl-tRNA$^{Thr}$, respectively

## Type I. Reduction of Cys-tRNA$^{Cys}$ to Ala-tRNA$^{Cys}$ [2]

Raney nickel No. 28, suspended in water, is obtained from Raney Catalyst Company, Chattanooga, Tennessee, and stored at 0°C. The preparation is washed with water before use. Organic sulfur compounds are easily reduced by Raney nickel.[3,4] It acts at room temperature, and, in the case of cysteine, noncatalytically (cf. for example, Mozingo *et al.*[4]):

$$RCH_2SH + Ni(H)_x \rightarrow RCH_3 + H_2S + Ni$$

Generally, 1.5 ml of a solution containing 8.5 mg of [$^{14}$C]Cys-tRNA$^{Cys}$, 0.4 *M* acetate buffer, pH 5, 0.12 ml of saturated EDTA, and 170 mg of Raney nickel is shaken at room temperature for 30 minutes. After adding another 0.12 ml of saturated EDTA solution, the nickel is centrifuged off. To the clear supernatant, two volumes of cold ethanol are added, and the mixture is kept at −20° for 2 hours. After centrifugation, the precipitate is taken up in 0.5 ml of water and dialyzed against 0.5 m*M* cold acetate buffer, pH 5. The recovery of the tRNA after dialysis is about 65%.

After stripping of an aliquot at pH 11, in the presence of the carriers [$^{12}$C]Cys and [$^{12}$C]Ala, the amino acids released from the tRNA are analyzed by paper electrophoresis at pH 1.85 in 7.8% acetic acid and 2.5% formic acid (70 V/cm, 60 minutes). Up to 70% of the radioactive material is usually found as [$^{14}$C]Ala and the remaining 30% as [$^{14}$C]Cys.

## Type II. Deamination of Aminoacyl-tRNA's to the $\alpha$-Hydroxy Derivatives[5]

To 1 ml of [$^{14}$C]Phe-tRNA$^{Phe}$ (5–10 mg/ml), 1 ml of 1% acetic acid and 1 ml of sodium nitrite solution saturated at 0°C are added. The mixture is maintained for 2 hours at 20°, two volumes of ethanol are then added, and after centrifugation at 4°C, the tRNA is dissolved in 0.1 m*M* acetate buffer, pH 5, reprecipitated with ethanol, and redissolved in the same buffer.

[2] F. Chapeville, F. Lipmann, G. Von Ehrenstein, B. Weisblum, W. J. Ray, Jr., and S. Benzer, *Proc. Nat. Acad. Sci. U. S.* **48**, 1086 (1962).
[3] R. Schröter, *in* "Newer Methods of Preparative Organic Chemistry," p. 72. Wiley (Interscience) New York, 1948.
[4] R. Mozingo, D. E. Wolf, S. A. Harris, and K. Folkers, *J. Amer. Chem. Soc.* **65**, 1013 (1943).
[5] G. Hervé, and F. Chapeville, *J. Mol. Biol.* **13**, 757 (1965).

An aliquot of the material thus obtained is stripped at pH 11 and analyzed by paper electrophoresis in 0.5 *M* formic acid pH 2 (20 V/cm, 90 minutes) or by paper chromatography in *n*-butanol: acetic acid: water (78:5:17); the $R_f$ of phenylalanine and phenyllactic acid are 0.4 and 0.85, respectively.

When this procedure is used, total deamination of phenylalanine is observed. At least 90% of the radioactive material is found as phenyllactic acid, and less than 10% as an unidentified deaminated product, whose $R_f$ is higher than that of phenyllactic acid.

The tRNA obtained by stripping at pH 11 of Phl-$tRNA^{Phe}$ can be reacylated with phenylalanine to the extent of 20–30% of the level reached with tRNA not previously treated by nitrous acid.

The binding properties of [$^{14}$C]Phl-$tRNA^{Phe}$ in the presence of *E. coli* ribosomes and poly(U), as measured by the method of Nirenberg and Leder,[6] are similar to those observed with [$^{14}$C]Phe-$tRNA^{Phe}$.[7]

By the same procedure, different aminoacyl-tRNA's are quantitatively transformed into the corresponding $\alpha$-hydroxy derivatives, but the recharging capacity varies with the tRNA used.

### Type III. Oxidation of Cys-$tRNA^{Cys}$ to Cys($O_3$)H-$tRNA^{Cys}$ [8]

The mixture containing 3.5 mg of [$^{35}$S]Cys-$tRNA^{Cys}$, 0.25 mmole of sodium periodate, 0.5 mmole of sodium acetate pH 5 in 0.5 ml is maintained in the dark for 2 hours at 0°. To this mixture, 0.5 ml of 1 *M* KCl is added; 15 minutes later, the precipitation of potassium periodate is eliminated by centrifugation at 0°. The tRNA is precipitated in the presence of 0.2 mmole of potassium acetate pH 5 by two volumes of ethanol. The tRNA is then dissolved in 0.1 m*M* potassium acetate buffer, pH 5, and extensively dialyzed against the same buffer.

After stripping of an aliquot at pH 10, in the presence of [$^{12}$C]cysteine and [$^{12}$C]cysteic acid, the amino acids released from the tRNA are analyzed by paper electrophoresis. Usually, 60–80% of the radioactivity is associated with [$^{14}$C]cysteic acid, whereas the remaining radioactivity is found in [$^{14}$C]cysteine.

Higher concentrations of sodium periodate or prolonged treatment do not increase the yield of the oxidation.

The same procedure is used for the oxidation of Met-$tRNA^{Met}$ to Met(O)-$tRNA^{Met}$ with similar yields. Sulfoxymethionine is the only oxidation product, and no methionine sulfone is formed.

[6] M. Nirenberg and P. Leder, *Science* **145**, 1399 (1964).
[7] G. Hervé, Thèse, C.N.R.S. No. 417, Paris, 1966.
[8] S. Chousterman, G. Hervé, and F. Chapeville, *Bull. Soc. Chim. Biol.* **48**, 1295 (1966).

### Type IV. Oxidation of Ser-tRNA$^{Ser}$ and Thr-tRNA$^{Thr}$ to Gxl-tRNA$^{Ser}$ and Gxl-tRNA$^{Thr}$, Respectively[8]

It is known that $\beta$-hydroxy-$\alpha$-amino acids are easily oxidized by periodate; serine is oxidized to ammonia, formaldehyde and glyoxylic acid; and threonine, to ammonia, acetaldehyde, and glyoxylic acid.

The method used is the same as that described for oxidation of Cys-tRNA$^{Cys}$ to Cys($O_3$)H-tRNA$^{Cys}$.

An aliquot of the oxidized material thus obtained is removed, stripped at pH 8.5, and analyzed by paper electrophoresis in 0.1 *M* ammonium carbonate at pH 8.5 (20 V/cm, 45 minutes).

During the oxidation of [$^{14}$C]serine by periodate, one carbon atom is eliminated and, under the conditions used, is removed from the incubation mixture by the subsequent alcohol treatment and dialysis. Taking this into account, the yield of oxidation leading to [$^{14}$C]Gxl-tRNA$^{Ser}$ is usually about 90%.

The ester linkage between the glyoxylyl- and the tRNA moieties is very unstable to alkaline hydrolysis, and attention should be taken to maintain the compound at a pH value not above 5. The half-life time[8] of the Gxl-tRNA at pH 8.5 is less than 2 minutes, whereas that of Ser-tRNA$^{Ser}$ is 17 minutes and that of Thr-tRNA$^{Thr}$ is 38 minutes.

# [14] Use of *in Vitro* $^{32}$P Labeling in the Sequence Analysis of Nonradioactive tRNAs

*By* MELVIN SILBERKLANG, AMANDA M. GILLUM, and UTTAM L. RAJBHANDARY

The large quantity of a purified tRNA required for classical spectrophotometric sequencing procedures[1] precluded the study of those potentially interesting tRNAs that were available only in small quantity and could not be efficiently labeled with $^{32}$P *in vivo*. The development of *in vitro* $^{32}$P-labeling techniques has now permitted sequence analysis of nonradioactive tRNAs utilizing adaptations of methods developed by Sanger and co-workers for analysis of uniformly $^{32}$P-labeled RNAs.[2-4]

[1] R. W. Holley, *Prog. Nucl. Acid Res. Mol. Biol.* **8**, 37 (1968).

[2] F. Sanger, G. G. Brownlee and B. G. Barrell, *J. Mol. Biol.* **13**, 373 (1965).

[3] B. G. Barrell, *in* "Procedures in Nucleic Acid Research" (G. L. Cantoni and D. R. Davies, eds.), Vol. 2, p. 751. Harper & Row, New York, 1971.

[4] G. G. Brownlee, "Determination of Sequences in RNA." North-Holland Publ., Amsterdam, 1972.





ISBN 0-12-504180-2

This chapter describes a general method, which has made possible the sequence analysis of tRNAs from mitochondria, chloroplasts, and a wide variety of prokaryotic and eukaryotic organisms using in some cases as little as 20 μg of purified tRNA.[5–11]

Székely and Sanger[12,13] demonstrated that oligonucleotides with a free 5′-hydroxyl end which are present in digests of nonradioactive nucleic acids could be labeled to high specific activity using T4 polynucleotide kinase and [γ-$^{32}$P]ATP. We have used this approach for labeling with $^{32}$P the 5′ ends of both mononucleotides present in complete tRNA hydrolysates[14,15] and oligonucleotides present in T1 or pancreatic RNase digests of tRNA.[16,17] The 5′-$^{32}$P-labeled mononucleotides are identified by two-dimensional thin-layer chromatography and can, therefore, be used for analysis of the modified nucleotide content of the tRNA.[10,11,14] The 5′-$^{32}$P-labeled oligonucleotides are separated ("fingerprinted"[2]), recovered, and partially digested with snake venom phosphodiesterase. These partial digestion products are in turn separated,[3,4,18] and the sequence of the oligonucleotide in question is deduced from the characteristic mobility shifts[3,4] resulting from the successive removal of nucleotides from the 3′ end.[3,4,6,16] Using oligonucleotides present in tRNAs of known sequence as standards, we have also characterized the mobility shifts of many

[5] M. Simsek, U. L. RajBhandary, M. Boisnard, and G. Petrissant, *Nature (London)* **247**, 518 (1974).

[6] A. M. Gillum, N. Urquhart, M. Smith, and U. L. RajBhandary, *Cell* **6**, 395 (1975).

[7] A. M. Gillum, B. A. Roe, M. P. J. S. Anandaraj, and U. L. RajBhandary, *Cell* **6**, 407 (1975).

[8] S. H. Chang, C. K. Brum, M. Silberklang, U. L. RajBhandary, L. I. Hecker, and W. E. Barnett, *Cell* **9**, 717 (1976).

[9] A. M. Gillum, L. I. Hecker, M. Silberklang, S. D. Schwartzbach, U. L. RajBhandary, and W. E. Barnett, *Nucl. Acids Res.* **4**, 4109 (1977).

[10] J. E. Heckman, L. I. Hecker, S. D. Schwartzbach, W. E. Barnett, B. Baumstark, and U. L. RajBhandary, *Cell* **13**, 83 (1978).

[11] R. T. Walker and U. L. RajBhandary, *Nucl. Acids Res.* **5**, 57 (1978).

[12] M. Székely and F. Sanger, *J. Mol. Biol.* **43**, 607 (1969).

[13] M. Székely, *in* "Procedures in Nucleic Acid Research" (G. L. Cantoni and D. R. Davies, eds.), Vol. 2, p. 780. Harper & Row, New York, 1971.

[14] M. Silberklang, Ph.D. thesis, Massachusetts Institute of Technology, Cambridge (1976).

[15] M. Silberklang, A. Prochiantz, A.-L. Haenni, and U. L. RajBhandary, *Eur. J. Biochem.* **72**, 465 (1977).

[16] M. Simsek, J. Ziegenmeyer, J. Heckman, and U. L. RajBhandary, *Proc. Natl. Acad. Sci. U.S.A.* **70**, 1041 (1973).

[17] M. Simsek, G. Petrissant, and U. L. RajBhandary, *Proc. Natl. Acad. Sci. U.S.A.* **70**, 2600 (1973).

[18] F. Sanger, J. E. Donelson, A. R. Coulson, H. Kössel, and D. Fischer, *Proc. Natl. Acad. Sci. U.S.A.* **70**, 1209 (1973).

commonly occurring modified nucleotides, which has proved useful in recognizing these modified nucleotides within unknown sequences.[19]

To order the shorter oligonucleotides present in complete T1 or pancreatic RNase digests into a total tRNA sequence, larger (overlapping) oligonucleotide fragments may be obtained by specific chemical cleavage of the tRNA or by limited digestion with base-specific nucleases. The use of *in vitro* $^{32}$P-labeling allows sequence analysis of such fragments on a picomole scale. In addition, using 5′-$^{32}$P-labeled or 3′-$^{32}$P-labeled intact tRNA,[20] we have developed two methods for direct sequence analysis of large nucleotide stretches at either end of the molecule. One method involves partial digestion of the end-labeled tRNA with nuclease P1, a relatively nonspecific endonuclease from *Penicillium citrinum*,[21] followed by separation and mobility shift analysis of the digestion products.[20] The other method involves partial digestion of the end-labeled tRNA with base-specific nucleases followed by mapping of the base-specific cleavage sites by polyacrylamide gel electrophoresis.[22–24]

These basic steps in the sequence analysis of a nonradioactive tRNA are summarized in Fig. 1. In most instances, steps 1, 2, and 4 may already provide the information necessary for the derivation of the total sequence. In such cases, step 3, which requires relatively larger amounts of tRNA, can be dispensed with.

## Materials and General Methods

### *Enzymes*

Ribonucleases T1 and T2 were purchased from Sankyo Chemical Co., Ltd., Tokyo, Japan, through Calbiochem, Inc.

Pancreatic ribonuclease A and snake venom phosphodiesterase were obtained from Worthington Biochemical Corp. Snake venom phosphodiesterase was dissolved in 50 m*M* Tris·HCl, 5 m*M* potassium phosphate (pH 8.9) at 1 mg/ml and stored at 4°; when stored at this concentration there is no appreciable loss of activity for up to 2 months.

Ammonium sulfate suspensions of bacterial alkaline phosphatase, calf intestinal alkaline phosphatase, yeast hexokinase, yeast phosphoglycerate kinase, and rabbit muscle glyceraldehyde-3-phosphate dehydrogenase

[19] A. M. Gillum, Ph.D. thesis, Massachusetts Institute of Technology, Cambridge 1975.

[20] M. Silberklang, A. M. Gillum, and U. L. RajBhandary, *Nucl. Acids Res.* **4,** 4091 (1977).

[21] M. Fujimoto, A. Kuninaka, and H. Yoshino, *Agric. Biol. Chem.* **38,** 1555 (1974).

[22] H. Donis-Keller, A. Maxam, and W. Gilbert, *Nucl. Acids Res.* **4,** 2527 (1977).

[23] R. E. Lockard, B. Alzner-DeWeerd, J. E. Heckman, M. W. Tabor, J. MacGee, and U. L. RajBhandary, *Nucl. Acids Res.* **5,** 37 (1978).

[24] A. Simoncsits, G. G. Brownlee, R. S. Brown, J. R. Rubin, and H. Guilley, *Nature (London)* **269,** 833 (1977).

1. Analysis of Modified Nucleotide Content
   A. Complete digestion with T2-RNase
   B. 5′-$^{32}$P-labeling of nucleotides
   C. Identification of labeled nucleotides by two-dimensional thin-layer chromatography
2. Analysis of Oligonucleotides Present in T1 or Pancreatic RNase Digests
   A. Complete digestion with T1 or pancreatic RNase
   B. 5′-$^{32}$P-labeling of oligonucleotides
   C. Separation of 5′-$^{32}$P-labeled oligonucleotides by two-dimensional electrophoresis
   D. Sequencing of 5′-$^{32}$P-labeled oligonucleotides
      i. 5′-end group analysis
      ii. Partial digestion with snake venom phosphodiesterase and/or nuclease P1 and analysis by two-dimensional homochromatography (or electrophoresis on DEAE-cellulose paper)
3. Alignment of Oligonucleotides Present in 2 (above) by Isolation and Sequencing of Large Oligonucleotide Fragments
   A. i. Controlled digestion with T1 or pancreatic RNase, or
      ii. Specific chemical cleavage
   B. 5′-$^{32}$P-labeling of the large fragments
   C. Separation of the 5′-$^{32}$P large fragments by two-dimensional gel electrophoresis and their sequencing as in 4 (below)
4. Direct Sequence Analysis of $^{32}$P-End Labeled tRNAs or Large Fragments
   A. End group labeling of tRNAs
      i. 5′-end using polynucleotide kinase and [$\gamma$-$^{32}$P]ATP
      ii. 3′-end using tRNA nucleotidyltransferase and [$\alpha$-$^{32}$P]ATP
   B. Partial digestion of end-labeled tRNAs with nuclease P1 and analysis by two-dimensional homochromatography
   C. Partial digestion with base-specific nucleases and analysis by polyacrylamide gel electrophoresis

FIG. 1. Basic steps in the sequence analysis of a nonradioactive tRNA.

were obtained from Boehringer Mannheim Corp. and stored at 4°. Hexokinase could be diluted to approximately 2 units/ml in 50% glycerol, 5 m*M* sodium acetate (pH 5.0), and stored at −20°; no loss of activity is observed for up to 1 month.

Nuclease P1 from *Penicillium citrinum*[21] was purchased from Yamasa Shoyu Co., Ltd. (Tokyo, Japan). The lyophilized powder was dissolved in 50 m*M* Tris-maleate (pH 6.0) at 1 mg/ml and stored in aliquots at −20°. It was diluted as necessary in 50 m*M* ammonium acetate (pH 5.3); at a concentration of 2 $\mu$g/ml in this buffer, it is stable for 2 weeks at −20°.

T4 polynucleotide kinase[25] was prepared according to Panet *et al.*[26] or purchased from New England Biolabs, Inc. It is very useful to have this enzyme at a concentration of ≥1000 units/ml. Highly purified tRNA nucleotidyltransferase from *Escherichia coli* was a gift of Dr. D. Carré.

[25] C. C. Richardson, *Proc. Natl. Acad. Sci. U.S.A.* **54,** 158 (1965).
[26] A. Panet, J. H. van de Sande, P. C. Loewen, H. G. Khorana, A. J. Raae, J. R. Lillehaug, and K. Kleppe, *Biochemistry* **12,** 5045 (1973).

*Specialized Reagents and Supplies*

Triethylammonium bicarbonate (TEAB) was prepared by bubbling $CO_2$ gas filtered through glass wool into a suspension of 1150 ml of twice-distilled triethylamine (Eastman Organic Chemicals) in 2.5 liters of distilled water, kept near 0° in a salt–ice water bath, until the pH of the solution dropped to a value of 7.5 to 8. The TEAB concentration was determined by standardized titration with methyl orange indicator and was adjusted to 2.0 *M*. It was stored in opaque bottles at 4°, or, for long-term storage, at −20°.

[$\gamma$-$^{32}$P]ATP was prepared by the phosphate exchange procedure of Glynn and Chappell,[27] using carrier-free [$^{32}$P]orthophosphoric acid in water solution obtained from New England Nuclear Corp. The [$\gamma$-$^{32}$P]ATP (generally at 500–1000 Ci/mmol) was purified by DEAE-Sephadex A25 chromatography using a 0.1 to 1 *M* gradient of TEAB (pH 8.0) as eluent. After removal of TEAB by repeated evaporation, the ATP was neutralized with NaOH and stored at 10 $\mu M$ concentration in 50% ethanol, 5 m*M* Tris·HCl (pH 7.5) at −20°; it was stable up to 2 weeks.

[$\alpha$-$^{32}$P]ATP at ~300 Ci/mmol was purchased from New England Nuclear Corp.

Ribonuclease-free bovine serum albumin (BSA), Grade A, was from Calbiochem, Inc., and was stored in water at 1–10 mg/ml at −20°.

Nitrilotriacetic acid (NTA), Gold Label grade, was from Aldrich Chemical Co. It was suspended in water, titrated to pH 7 with NaOH, and stored at 50 m*M* or 100 m*M* at 4° or −20°.

Aniline hydrochloride (pH 4.5) was prepared by titration of aniline (Aldrich Chemical Co.; twice distilled) with concentrated HCl. It was stored in aliquots at a concentration of 0.3 *M* at −20°.

An Eppendorf Model 3200 microcentrifuge was used for centrifugation of small samples in polypropylene tubes. Polypropylene tubes in 1.5-ml, 0.7-ml, and 0.4-ml sizes with caps were purchased from Walter Sarstedt, Inc. (Princeton, New Jersey).

*Thin-Layer Chromatography*

Thin-layer partition chromatography is performed on glass-backed 100 $\mu$m cellulose plates from E. Merck (EM). Plates (20 × 20 cm) are either used directly from the box, or prerun in the intended chromatography solvent (volatile solvents only) and dried before use. Samples to be applied to thin-layer plates are generally adjusted to 0.5–2 $\mu$l volume; the area of application should be minimized. All thin-layer chromatog-

[27] I. M. Glynn and J. B. Chappell, *Biochem. J.* **90,** 147 (1964).

raphy (except homochromatography; see below) is run at room temperature. Where necessary (see below, *Modified Nucleoside Composition Analysis*), a 10 cm-long Whatman 3 MM paper wick may be attached to the top of a plate with a 1/8 inch-wide plastic binding clip cut to size (i.e., 20 cm).

Solvent systems used are: (a) isobutyric acid–concentrated $NH_4OH$–$H_2O$ (66/1/33, v/v/v), (b) *t*-butanol–concentrated HCl–$H_2O$ (70/15/15, v/v/v), (c) 0.1 *M* sodium phosphate, pH 6.8–ammonium sulfate–*n*-propanol (100/60/2, v/w/v). Solvent systems (a) and (b) are freshly prepared and stored no longer than 1 week in airtight bottles.

Ultraviolet (UV) absorbing material is detected under a 254 nm-emitting UV light; radioactive material is detected by autoradiography. Removal of areas containing radioactive material from thin-layer plates for liquid scintillation counting is according to the procedure of Turchinsky and Shershneva.[28] The spot to be removed is circumscribed, and a drop of solution containing 5–10% nitrocellulose in ethanol–acetone (1:1) is spread on it. After drying, the nitrocellulose–cellulose platelet is lifted from the plate and transferred to a vial for counting. Removal of material from cellulose plates for elution is performed by scraping the cellulose in the desired area from the surface of the plate with a flamed surgical blade; the scrapings are collected by suction into a conical plastic pipette tip (Eppendorf) plugged at the thin end with glass wool (this is essentially a homemade equivalent of a more sophisticated device described by Barrell[3]). Elution is by centrifugation of liquid through the packed cellulose into a 0.4-ml polypropylene test tube (Sarstedt). Elution of nucleotidic material from cellulose is with water, whereas elution from ion-exchange resins, such as DEAE-cellulose and polyethylenimine-(PEI)-impregnated cellulose is with 2 *M* TEAB.

*Homochromatography*

Homochromatography is performed on DEAE-cellulose 250 $\mu$m layer glass-backed plates purchased to order from Analtech, Inc.: 20 × 20 cm plates are of Machery-Nagel (MN) celluloses MN 300 HR–MN 300 DEAE (15/2), and 20 × 40 cm plates are of MN 300 HR–Avicel (EM)–MN 300 DEAE (10/5/2).

RNA hydrolyzates ("homomixes") for DEAE-cellulose thin-layer chromatography are prepared as follows. For 1 liter, 30 g of yeast RNA (Sigma type VI) and 420 g of urea (Baker Analyzed Reagent) are dissolved in 400 ml of $H_2O$ at 37°. The solution is brought to neutrality with 1–2 *N* KOH and then additional KOH is added to a further concentration of 10,

[28] M. F. Turchinsky and L. P. Shershneva, *Anal. Biochem.* **54**, 315 (1973).

25, 50, or 75 m$M$ (relative to the final total volume of 1 liter). The RNA solution is then incubated at 65° for 20–24 hr. After cooling to room temperature, the pH of the solution is reduced to 4.5–4.7 with acetic acid; the volume is then adjusted to 1 liter with water to give a final RNA concentration of 3% in 7 $M$ urea. The less KOH in the hydrolysis step, the greater the average chain length of the RNA in the respective solution, and the stronger its elution power. For long (20 × 40 cm) plates in particular, the best results are obtained if the homomix used is titrated down to pH 4.5 to provide additional buffering capacity for prolonged chromatographic runs. We have also tried the method of Jay *et al.*[29] for the preparation of homomixes and have found it satisfactory; however, best results are obtained, in our hands, when this homomix is used at pH 4.5–4.7.

Prior to chromatography, plates are topped with Whatman 3 MM paper wicks (10–15 cm long for 20 × 20 cm plates, 25–30 cm long for 20 × 40 cm plates) held in place by 1/8 inch-wide plastic binder clips (cut to 20 cm). For 20 × 40 cm plates, it is useful to also cover the bottom of the plate with a 1–2 cm wick to prevent the DEAE-cellulose from flaking off the bottom of the plate during prolonged chromatography. Each plate is heated to 65° in an oven, prerun at 65° to two-thirds its height in distilled water, and then transferred to a tank of the appropriate homomix; 20 × 20 cm plates are developed in glass chromatography tanks (Desaga type; Brinkmann). The 20 × 40 cm plates can be developed in stacked glass tanks, one inverted and positioned atop the other with a good silicone grease seal between; however, we have found it considerably more convenient to use custom-designed Plexiglas tanks built for us by Wilbur Scientific (Boston, Massachusetts). Plates are generally developed until the xylene cyanole blue marker dye is within 2–5 cm of the wick at the top of the plate; 20 × 40 cm plates may also be run farther for better resolution of longer oligonucleotides.

Two-dimensional homochromatography is essentially as described by Sanger and co-workers,[18] using electrophoresis on cellulose acetate or Cellogel strips at pH 3.5 (see below, *Standard Fingerprinting Procedure*) in the first dimension and DEAE-cellulose thin-layer chromatography in homomix in the second dimension. Transfer of material between the first and second dimension is according to Southern[30] (see below, *Alternative Fingerprinting Procedure*), using the xylene cyanole blue dye as a guide to select the appropriate region of the cellulose acetate strip for transfer (Table I).

[29] E. Jay, R. Bambara, R. Padmanabhan, and R. Wu, *Nucl. Acids Res.* **1**, 331 (1974).
[30] E. M. Southern, *Anal. Biochem.* **62**, 317 (1974).

TABLE I
RELATIVE ELECTROPHORETIC MOBILITIES OF THE FOUR NUCLEOSIDE 5′-PHOSPHATES

| | Electrophoretic mobility | | | |
|---|---|---|---|---|
| | DEAE-paper | | Cellulose acetate | |
| Nucleotide | pH 3.5, $R_{Blue}$[a] | pH 1.9, $R_{Blue}$[b] | pH 3.5, $R_{Blue}$[c] | pH 3.5, $R_{pU}$[d] |
| pA | 1.9 | 2.9 | 0.44 | 0.30 |
| pG | 1.6 | 2.2 | 1.2 | 0.80 |
| pC | 2.1 | 3.0 | 0.16 | 0.11 |
| pU | 2.6 | 2.7 | 1.5 | — |

[a] Mobility relative to xylene cyanole blue dye marker upon electrophoresis on DEAE-cellulose paper at 10°–15° in pH 3.5 buffer [0.5% pyridine, 5% acetic acid (v/v)].

[b] See footnote *a*; electrophoresis at 10°–15° in pH 1.9 buffer: 2.5% formic acid, 8.7% acetic acid (v/v).

[c] See footnote *a*; electrophoresis on cellulose acetate strip at 30° in pH 3.5 buffer under standard fingerprinting conditions (*General Methods*).

[d] The same data as in column 4[c] recalculated as mobility relative to pU, which has a charge at pH 3.5 of −1, to indicate the apparent charge observed under these experimental conditions for pA, pG, and pC [cf. C. P. D. Tu, E. Jay, C. P. Bahl, and R. Wu, *Anal. Biochem.* **74,** 73 (1976)].

## *Electrophoresis*

*Materials and Equipment.* Whatman DEAE-cellulose paper (DE 81), purchased in rolls, and Whatman 3 MM and 540 paper sheets are from Reeve Angel and Co. Cellulose acetate strips (3 × 57 cm) and PEI-impregnated cellulose (PEI-cellulose) thin-layer chromatography plates (glass-backed, without fluorescent indicator) are from Schleicher and Schuell, Inc. Cellulose acetate lots occasionally vary in quality; we have generally tested lots under our usual high-voltage electrophoresis conditions at pH 3.5 (see below) with the compound [5′-$^{32}$P]$(pdT)_{10}$, and purchased the best (minimum streaking) lot in bulk. Cellogel strips (2.5 × 55 cm) are obtained from Kalex, Inc., and stored in 40–50% methanol at 4°. We have used Gilson Instrument Co. high-voltage (5000 V) electrophoresis equipment for paper electrophoresis.

Electrophoresis grade acrylamide and *N,N′*-methylenebisacrylamide can be purchased from Bio-Rad laboratories. Ammonium persulfate and tetramethylethylenediamine (TEMED) are from Eastman Organic Chemicals. Ultrapure urea and sucrose are from Schwarz/Mann, Inc. Reagent grade formamide is deionized with Fisher Rexyn I-300 mixed-bed ion-exchange resin immediately prior to use.

One polyacrylamide slab gel apparatus we have used consists of two chambers of buffered electrolyte, with a double thickness of Whatman 3 MM paper wick connecting the top of the gel to the upper buffer chamber, as described by Akroyd.[31] Alternatively, we have used an apparatus allowing direct gel to buffer contact in the upper chamber, as described by Studier.[32] For moderate-voltage (100–400 V) gel electrophoresis, a Heathkit Model IP-17 power supply is satisfactory; for high-voltage (500–1000 V) gel electrophoresis, we have used a Wilbur Scientific (Boston, Massachusetts) power supply.

*Standard Fingerprinting Procedure.* Two-dimensional electrophoresis of oligonucleotides (5′-$^{32}$P-labeled) from T1 or pancreatic RNase digests of a tRNA is by the procedure of Sanger and co-workers,[2] as described in detail by Barrell,[3] with the following minor modifications in the first dimension: (1) the cellulose acetate strip is wetted in pH 3.5 buffer that contains 2m*M* EDTA in addition to 5% pyridinium acetate and 7 *M* urea; (2) electrophoresis is run in a tank at or above 20°, preferably prewarmed to 30°–35° (by running a small sheet of 3 MM paper wetted in 5% pyridinium acetate, pH 3.5, in the tank for approximately 15 min at maximum voltage, without cooling). If the total incubation mixture from a 5′-labeling reaction (see below, *5′-$^{32}$P-Labeling of Oligonucleotides*) are to be used for fractionation, Cellogel strips are preferred over the usual thinner cellulose acetate because of their greater capacity; the strips are soaked in the pH 3.5 buffer at 4° overnight prior to use, and the sample is lyophilized and applied, as usual, in 3–5 $\mu$l. The blotting procedure for transfer of oligonucleotides from cellulose acetate on DEAE-cellulose paper is essentially as described by Barrell,[3] except that three strips of dry 3 MM paper are placed beneath the DEAE-paper along the line of transfer to absorb penetrating water and facilitate the blotting process. The second-dimensional electrophoresis on DEAE-paper in 7% formic acid (v/v) or pH 1.9 buffer (2.5% formic acid, 8.7% acetic acid, v/v), and recovery of $^{32}$P-labeled oligonucleotides from excised pieces of DEAE-paper with 2 *M* TEAB are as described by Barrell.[3] TEAB is removed by 4-fold dilution with water followed by repeated lyophilization from water in 1.5-ml polypropylene test tubes; carrier yeast tRNA (5–50 $\mu$g) is added prior to lyophilization to minimize losses of radioactive material due to adsorption.

*Alternative Fingerprinting Procedure.* The first dimension is electrophoresis on cellulose acetate at pH 3.5 as described above; however, the electrophoresis time is halved so that the xylene cyanole blue tracking

[31] P. Akroyd, *Anal. Biochem.* **19**, 399 (1967).
[32] F. W. Studier, *J. Mol. Biol.* **79**, 237 (1973).

dye migrates only 8–9 cm. The second dimension is chromatography on a 20 × 20 cm, glass-backed PEI cellulose thin-layer plate[33] using freshly prepared 2 *M* pyridinium formate (pH 3.4)[34] as solvent (the plates are doubly prerun before use,[35] first in 2 *M* pyridinium formate, pH 2.2, and then, after drying, in distilled water; prerun plates are stored at 4°). A wick of 3 MM paper is clamped along the top of the plate so that chromatography can continue until the xylene cyanole tracking dye has migrated 10–15 cm up the plate. Transfer of oligonucleotides from the cellulose acetate strip to the PEI-cellulose thin-layer plate is according to Southern;[30] this method relies on capillary action rather than pressure to achieve the transfer and concentrates the oligonucleotides to a thin line along the second-dimensional origin. Oligonucleotides are removed from the plate by scraping and eluted from the ion-exchange resin with 2 *M* TEAB as described above (see above, *Thin-Layer Chromatography*).

*One-Dimensional Paper Electrophoresis.* Electrophoresis on DEAE-cellulose paper at pH 3.5 or pH 1.9 of partial enzymic digests of a 5′-$^{32}$P-labeled oligonucleotide can be used to deduce the sequence of the oligonucleotide[3] (see below, *Sequence Analysis of 5′-$^{32}$P-Labeled Oligonucleotides*). Aliquots taken out at timed intervals are applied in narrow 1-cm bands along an origin line on each of two sheets of Whatman DE 81 paper (one for the pH 3.5 and one for the pH 1.9 buffer systems[3]); the zero time point is spotted separately, followed by other timed aliquots, individually or in pools (10,000 cpm per track is suitable) selected so as best to represent the progress of the reaction. UV markers of the four nonradioactive nucleoside 5′-phosphates (2–4 $A_{260}$ units) are applied to those tracks containing 5′-$^{32}$P-labeled mononucleotides (longer periods of incubation). The marker dye mixture used for fingerprinting[3] is also used here and spotted at several positions along the origin line. Electrophoresis is in a cooled tank at a voltage selected so that current is limited to approximately 100 mA per sheet (to prevent heating) and is continued, by following the blue dye, until the nucleoside 5′-phosphate UV marker(s) is expected to be near the end of the sheet; the electrophoretic mobilities of the 4 nucleoside 5′-phosphates relative to the blue dye are given in Table I. The paper is dried in a fume hood; if necessary, to facilitate detection of UV markers the dried paper may be briefly exposed to ammonia vapors in a closed chamber.

*One-Dimensional Polyacrylamide Gel Electrophoresis.* Analysis and purification of intact tRNA, or large oligonucleotides produced by chem-

[33] B. E. Griffin, *FEBS Lett.* **15**, 165 (1971).

[34] E. M. Southern and A. R. Mitchell, *Biochem. J.* **123**, 613 (1971).

[35] G. Volckaert and W. Fiers, *Anal. Biochem.* **83**, 222 (1977).

ical cleavage at $m^7G$ (see below, *Preparation of Large Oligonucleotides*), is on 12% or 15% polyacrylamide (acrylamide: *N,N'*-methylenebisacrylamide = 20:1) slabs (20 × 20 × 0.2 cm) containing 7 *M* urea.[36] The gel buffer is 90 m*M* Tris borate, 4 m*M* EDTA (pH 8.3)[37]; no urea is used in the running buffer. Lyophilized samples of RNA are generally dissolved in 10–20 μl of 98% formamide (deionized) containing 20 m*M* sodium phosphate (pH 7.6), 20 m*M* EDTA, and 0.1% each of bromphenol blue and xylene cyanole FF; solutions are heated for 5 min at 55° prior to electrophoresis. Alternatively, the lyophilized RNA samples are dissolved in 10–20 μl of 90 m*M* Tris, 90 m*M* borate (pH 8.3), 20 m*M* EDTA, 7 *M* urea, 25% sucrose (w/v), and 0.1% each bromphenol blue and xylene cyanole FF, and heated for 1–2 min at 90°–100° prior to electrophoresis. To achieve optimal resolution, gels may be preelectrophoresed for at least 1 hr at 250 V at room temperature. Fresh running buffer is then introduced, samples are applied, and electrophoresis is continued at room temperature with a constant voltage gradient of 10–20 V/cm. When purifying tRNA after *in vitro* labeling, several strips of DEAE-paper are wedged under the gel slab in the anode buffer chamber to trap [$^{32}$P]ATP as it leaves the bottom of the gel.

Partial digests of 5'-$^{32}$P-labeled tRNAs with base-specific nucleases[22–24] (see below, *Direct Sequence Analysis of tRNAs*) are separated in adjacent lanes of a 20% polyacrylamide gel (20 × 40 × 0.15 cm) containing 7 *M* urea, as described.[38] Preelectrophoresis is at room temperature for 4 hr at 500 V (prior to loading samples), and then electrophoresis is continued at 1000 V.

For autoradiography, one glass plate is removed and the gel is covered with plastic wrap; DEAE-paper tabs are attached and labeled with radioactive ($^{35}$S) ink to act as autoradiographic markers, and the gel is exposed to X-ray film at −20° (or, with an intensifying screen, at −55° to −70°; see below, *Autoradiography*). It is helpful to place considerable weight (up to 50 pounds) atop the film holder during exposure to press the film flush against the gel and ensure optimal autoradiographic resolution.

*Two-Dimensional Polyacrylamide Gel Electrophoresis*. Oligonucleotides present in partial T1 RNase or partial pancreatic RNase digests of a tRNA (see below, *Preparation of Large Oligonucleotides*) are separated by two-dimensional gel electrophoresis. Our system[23] is adapted from that of De Wachter and Fiers.[39] Electrophoresis in the first dimension is

[36] G. R. Philipps, *Anal. Biochem.* **44,** 345 (1971).
[37] A. C. Peacock and C. W. Dingman, *Biochemistry* **6,** 1818 (1967).
[38] A. Maxam and W. Gilbert, *Proc. Natl. Acad. Sci. U.S.A.* **74,** 560 (1977).
[39] R. De Wachter and W. Fiers, *Anal. Biochem.* **49,** 184 (1972).

in a 10% polyacrylamide slab (20 × 40 × 0.2 cm) in citrate buffer and 7 *M* urea at pH 3.5; the second dimension is in a 20% polyacrylamide slab (20 × 40 × 0.2 cm) in Tris-borate buffer and 7 *M* urea at pH 8.3.

The first-dimensional gel solution contains 10% acrylamide, 0.5% bisacrylamide in 25 m*M* citrate, 7 *M* urea, pH 3.5, and is deaerated under reduced pressure and polymerized by addition of $FeSO_4$, ascorbic acid, and $H_2O_2$.[39] Preelectrophoresis is at 4° for 2 hr at 200 V; running buffer contains 25 m*M* citric acid and 4 m*M* EDTA adjusted to pH 3.5 with sodium hydroxide, and the slots (1 cm wide) are filled with 7 *M* urea in the same buffer. After preelectrophoresis, the slots are emptied and the slots and buffer trays are filled with fresh running buffer (without urea). Samples are applied in 25 m*M* citrate, 20 m*M* EDTA, 7 *M* urea, 25% sucrose, and 0.1% each bromphenol blue and xylene cyanole blue; electrophoresis is at 4° at 200 V. Electrophoresis is continued until the bromphenol blue tracking dye (the faster dye) has migrated approximately 25 cm.

After electrophoresis, the gel may be autoradiographed briefly at 4° to locate the radioactive material. A 1 cm-wide and 16 cm-long track (containing the radioactive material and the two dye markers) is then excised from the gel and placed symmetrically across the width of a 20 × 40 cm glass plate, 1 cm from the bottom of the plate; spacer bars are laid along the edges of the plate, and a second glass plate is placed on top and fastened to seal the gel strip for the pouring of the second-dimensional gel solution.

The gel solution (20% acrylamide, 1% bisacrylamide, 7 *M* urea, 90 m*M* Tris, 90 m*M* borate, 4 m*M* EDTA, pH 8.3) for the second dimension is poured in two stages. First, enough gel solution, containing excess catalyst (to ensure rapid polymerization), is poured to cover the upper edge of the first-dimensional gel strip. After polymerization of this bottom gel block, the remainder of the gel solution is poured and allowed to polymerize for about 4–6 hr. The running buffer for the second-dimensional electrophoresis slab is 90 m*M* Tris, 90 m*M* borate, 4 m*M* EDTA (pH 8.3). Electrophoresis is reversed in direction, being from bottom (−) to top (+), and is at room temperature at 300–400 V. Oligonucleotides resulting from T1 RNase or pancreatic RNase partial digestion of tRNA (see below, *Preparation of Large Oligonucleotides*) are generally 5–60 nucleotides in size and are separated by running the second dimension of the gel until the bromphenol blue dye has migrated about 30 cm.

*Recovery of RNA from Polyacrylamide.* Intact tRNA and large oligonucleotides (>20 nucleotides) are generally recovered by electrophoretic elution as described by Knecht and Busch.[40] Electrophoresis is in Pasteur pipettes with shortened tips that are plugged with 5% polyacrylamide in either Tris-borate gel buffer (without urea) or in 25 m*M* Tris·HCl, 1 m*M* EDTA (pH 8.0), using the corresponding buffer as running buffer in a tube-gel apparatus. After preelectrophoresis for 3–6 hr at room temperature at 1 mA per pipette, the pipettes are filled with fresh running buffer and dialysis bags, containing approximately 0.5 ml of running buffer, are attached. The excised bands are inserted into the pipettes and, if the RNA is already end-labeled, 10–50 $\mu$g of carrier yeast tRNA (free of ribonuclease) in 25% sucrose is layered onto the surface of the gel plug; the carrier tRNA improves the recovery of small quantities of $^{32}$P-labeled RNA. Electrophoresis is at room temperature for 6–12 hr at 1–2 mA/tube. Elution of radioactive RNA can be monitored with a Geiger counter. The eluted RNA is either freed of salts by dialysis and stored at −70° or precipitated with ethanol. Recovery is generally greater than 90%.

For recovery of smaller oligonucleotides (<20 nucleotides), excised gel pieces are ground in 2 ml (per 0.5-ml gel slice) of 0.2 *M* NaCl, 30 m*M* sodium citrate (pH 8) with a motor-driven glass–Teflon homogenizer chilled on ice.[41] The gel fragments are removed by centrifugation and reextracted; the combined supernatants, diluted 5-fold with water, are loaded onto a column of DEAE-cellulose equilibrated in 25–50 m*M* TEAB (pH 8); for small quantities of $^{32}$P-labeled oligonucleotides, recovery is improved by adding 25–50 $\mu$g $A_{260}$ of carrier yeast tRNA to the solution before loading. The column is washed with 25–40 m*M* TEAB, and $^{32}$P-labeled oligonucleotide is eluted with 2 *M* TEAB. The eluate is evaporated to dryness four or five times, and the residue containing the 5′-$^{32}$P-labeled oligonucleotide and carrier tRNA is dissolved in water and stored at −20° or −70°.

### *Autoradiography*

Autoradiography of paper or thin-layer chromatograms or of polyacrylamide gels is with either Kodak Royal X-O-Mat or Kodak No-Screen Medical X-ray film; the latter has been found to be about two times more sensitive to $^{32}$P than the former, and thus allows shorter exposure times. The sensitivity of Royal X-O-Mat film to $^{32}$P, on the other hand, may be enhanced up to 10-fold by using a Du Pont Cronex "Lightning Plus" intensifying screen and performing the autoradiography

[40] M. G. Knecht and H. Busch, *Life Sci.* **10,** II, 1297 (1971).
[41] R. T. Walker and U. L. RajBhandary, *Nucl. Acids Res.* **2,** 61 (1975).

at −55° to −70°.[42] Except where noted, the suggested quantities of $^{32}$P for various procedures described in this chapter assume exposure *without* an intensifying screen and should be reduced 5-fold if a screen is used.

Radioactive marker ink solution is made by dissolving xylene cyanole FF dye in sodium [$^{35}$S]sulfate in water.

## Identification of Modified Nucleosides in tRNAs

A relatively high content of posttranscriptional nucleoside modification is one of the distinguishing features of transfer RNA, and it is useful to begin the analysis of a tRNA species by identifying its modified nucleosides. Almost invariably, such determinations rely on a chemical or enzymic method for complete hydrolysis of the tRNA. Where 50–150 μg of the tRNA is available for this purpose, various analytical thin-layer[7,43,44] or column[45–47] chromatographic systems may be used and the individual nucleotides or nucleosides detected by UV absorbance. Where quantities of tRNA are limited, however, *in vitro* radiolabeling procedures provide efficient alternative means of detection. A $^{3}$H-labeling procedure for nucleoside composition analysis has been described by Randerath and Randerath[48]; we describe here a highly sensitive alternative procedure involving $^{32}$P labeling.[14,15]

*Preparation of $^{32}$P-Labeled tRNA Hydrolysate.* The procedure consists of the following steps: (a) complete digestion of the tRNA with T2 RNase; (b) phosphorylation of the nucleoside 3′-phosphates (Np) in the digest to [5′-$^{32}$P]Np using polynucleotide kinase and [$\gamma$-$^{32}$P]ATP; (c) elimination of excess [$\gamma$-$^{32}$P]ATP by phosphorylation of glucose to glucose-6-phosphate with yeast hexokinase; (d) deproteinization; (e) conversion of [5′-$^{32}$P]Np to [5′-$^{32}$P]N using the 3′-phosphatase activity present in nuclease P1; (f) identification of the resulting 5′-$^{32}$P-labeled mononucleotides by two-dimensional thin-layer chromatography in the presence of nonradioactive UV markers.

For the RNase digestion, the incubation mixture (10 μl) contains 0.1–0.5 μg of tRNA, T2 RNase (0.05 unit), and 10 m*M* ammonium acetate buffer (pH 4.5). Incubation is for 5 hr at 37°. The digest (or a fraction of

[42] R. Swanstrom and P. R. Shank, *Anal. Biochem.* **86**, 184 (1978).

[43] S. Nishimura, F. Harada, U. Narushima, and T. Seno, *Biochim. Biophys. Acta* **142**, 133 (1967).

[44] H. Rogg, R. Brambilla, G. Keith, and M. Staehelin, *Nucl. Acids Res.* **3**, 285 (1976).

[45] M. Uziel, C. Koh, and W. E. Cohn, *Anal. Biochem.* **25**, 77 (1968).

[46] R. P. Singhal, *Arch. Biochem. Biophys.* **152**, 800 (1972).

[47] G. C. Sen and H. P. Ghosh, *Anal. Biochem.* **58**, 578 (1974).

[48] K. Randerath and E. Randerath, *Methods Cancer Res.* **9**, 3 (1973).

it) is lyophilized and redissolved in 25 m$M$ Tris·HCl (pH 8), 10 m$M$ dithiothreitol, 10 $\mu$g of BSA per milliliter, 250 $\mu M$ [$\gamma$-$^{32}$P]ATP (specific activity adjusted to 20–100 Ci/mmol), and polynucleotide kinase (2 units) in a final volume of 10 $\mu$l. The mixture is made 10% overall in glycerol and incubated for 30 min at 37°. Excess [$\gamma$-$^{32}$P]ATP is eliminated by adding glucose to 2 m$M$ concentration and 0.008 unit of yeast hexokinase (this is four times the amount of hexokinase used in "fingerprinting"—see below). After 10 min at 37°, 2.5 nmol of nonradioactive ATP are added; this addition of ATP is repeated after 10 more minutes and then, after 10 final minutes of incubation at 37°, the reaction is stopped by cooling on ice.

Before treating with nuclease P1, the $^{32}$P-labeled tRNA hydrolysate is deproteinized by extracting twice, at room temperature, with equal volumes of chloroform–isoamyl alcohol (24:1). The pooled organic phases are back-extracted twice, each time with 10 $\mu$l of water, and the pooled aqueous phases are then extracted 6 times with diethyl ether. The aqueous solution is left for 15 min at 37° to drive off residual ether, lyophilized, and resuspended in 20 $\mu$l of water. Of this solution, 5–10 $\mu$l are incubated with 2 $\mu$g of nuclease P1 in 75 m$M$ ammonium acetate (pH 5.3) in a total volume of 10–15 $\mu$l; incubation is for 3 hr at 37°. The material is stored frozen until used.

*Two-Dimensional Thin-Layer Chromatography.* Aliquots (0.5–1 $\mu$l) of the $^{32}$P-labeled tRNA hydrolysates (either before or after treatment with nuclease P1) are mixed with 1 $\mu$l of nonradioactive 5′-mononucleotides (containing, per microliter, 0.05 $A_{260}$ unit each of the four major nucleotides, pC, pU, pA, and pG) and used for two-dimensional thin-layer chromatography. Chromatography is on 20 × 20 cm cellulose plates in solvent (a) in the first dimension and solvent (c) in the second dimension. To maximize resolution in solvent (a), we have found it useful to attach a Whatman 3 MM paper wick to the top of the plate and continue development until the solvent front has run approximately 0.5 cm beyond the top of the plate (about 12 hr running time at 20°). In addition, it is necessary to dry plates at room temperature in a fume hood for at least 24 hr, between first- and second-dimensional runs, to allow the first-dimensional solvent to evaporate completely. Chromatography in the second dimension is carried out until the solvent has run to about 1–2 cm from the top of the plate. Finally, plates are oven-dried at 65°, marked in the corners with radioactive ($^{35}$S) ink solution, and autoradiographed.

Identification of modified nucleotides present in a $^{32}$P-labeled tRNA hydrolysate can be carried out either at the level of [$^{32}$P]pNp[14,15] or at the level of [$^{32}$P]pN (i.e., after treatment with nuclease P1). Figure 2

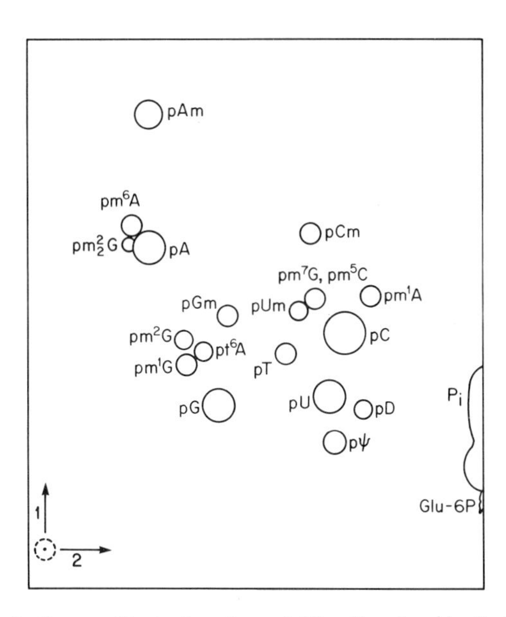


FIG. 2. Schematic diagram illustrating the mobility of nucleoside 5′-$^{32}$P-labeled monophosphates upon two-dimensional thin-layer chromatography on cellulose. First dimension, solvent (a); second dimension, solvent (c). $P_i$ is inorganic phosphate and Glu-6P is glucose 6-phosphate.

indicates schematically the relative location of the various 5′-mononucleotides in the two-dimensional thin-layer chromatography system; Table II lists the $R_f$ values of both the *pN and *pNp [48a] derivatives of the various modified nucleosides in these systems. Since solvent composition, temperature, etc., affect relative mobilities, the data in Table II and Fig. 2 should be used only as a guide; conclusive identification of a modified nucleotide should be based on chromatographic comigration with corresponding UV markers. Alternatively, where UV markers are not available, $^{32}$P-labeled nucleotides obtained by digestion of 5′-$^{32}$P-labeled oligonucleotides of known sequence may be used as markers; identifications can then be confirmed by addition of the marker to the $^{32}$P-labeled tRNA hydrolysate and analysis for comigration in the usual, and also in alternative, solvent systems (e.g., solvent b). Since UV markers of modified nucleotides are more readily available in the pN form than in the pNp form, it is usually more convenient to identify them as their 5′-mononucleotide derivatives. However, we recommend that analyses also be carried out at the pNp level whenever possible.

[48a] *p, 5′-$^{32}$P-labeled; *pN, 5′-$^{32}$P-labeled nucleotide; *pNp, nucleoside 3′, 5′-diphosphate in which the 5′-phosphate is $^{32}$P labeled.

TABLE II
THIN-LAYER CHROMATOGRAPHIC MOBILITIES OF MONONUCLEOTIDES

| Nucleoside (N) | $R_{pA}$ in Solvent (a)[a] pN | $R_{pAp}$ in Solvent (a)[a] pNp | $R_{pU}$ in Solvent (c)[a] pN | $R_{pUp}$ in Solvent (c)[a] pNp |
|---|---|---|---|---|
| A[d] | 1.00 | 1.00 (0.93)[b] | 0.34 | 0.39 (0.48) |
| $m^1A$ | 0.92 | 0.84 | 1.07 | 1.10 |
| $m^6A$ | 1.07 | 1.12 | 0.33 | 0.36 |
| $t^6A$[d] | ~0.64[c] | ~0.54 | ~0.47 | ~0.54 |
| Am | 1.25 | — | 0.33 | — |
| C | 0.83 | 0.68 | 1.00 | 1.00 (1.09) |
| $m^5C$ | 0.86 | — | 0.96 | — |
| Cm | 1.05 | — | 0.93 | — |
| U | 0.57 | 0.45 | 1.00 | 1.00 (1.09) |
| D | 0.53 | 0.40 | 1.07 | 1.03 |
| T | 0.71 | 0.59 | 0.85 | — |
| Ψ | 0.46 | 0.36 | 1.01 | — |
| Um | 0.86 | — | 0.88 | — |
| G | 0.50 | 0.41 | 0.63 | 0.70 |
| $m^2G$ | ~0.75 | ~0.80 | ~0.47 | — |
| $m^2_2G$[d] | 0.99 | 0.95 | ~0.34 | 0.44 |
| $m^7G$ | 0.90 | 0.77 | 0.93 | 0.95 |
| Gm | 0.86 | — | 0.58 | — |
| ATP | — | ~0.70 | — | ~0.43 |
| $P_i$ | — | 0.75 | At solvent front | |

[a] See *General Methods, Thin-Layer Chromatography* for solvent composition.
[b] Mobilities in parentheses are of the isomeric nucleoside-2′,5′-diphosphate.
[c] When approximate (~) mobilities are given, these nucleotides tend to form elongated spots.
[d] $pt^6A$ and $pm^1G$ are clearly distinguishable from one another by electrophoresis on DEAE-cellulose paper (pH 3.5). Mobilities relative to xylene cyanole blue dye ($R_{blue}$) are: $pt^6A$ = 1.02, $pm^1G$ = 1.87. They can also be separated by two-dimensional thin-layer chromatography (Fig. 2). Similarly, pA and $pm^2_2G$ can be separated either by electrophoresis (Fig. 9) or partly by two-dimensional thin-layer chromatography (Fig. 2).

It should be noted that, although the procedure described above can be used to identify modified nucleosides present in a tRNA, a quantitative estimation of nucleotide composition is not possible since polynucleotide kinase may not phosphorylate all the modified nucleotides quantitatively.[14] On the other hand, this procedure can be carried out as well on oligonucleotides as on tRNA and can, therefore, be used also for analysis of modified nucleotides present in oligonucleotides isolated from complete enzymic digests of tRNAs (see below, *Identification of Modified Nucleotides in 5′-$^{32}$P-Labeled Oligonucleotides*).

## 5′-End-Group Labeling and Separation of Oligonucleotides Present in Complete Enzymic Digests of tRNA

*In vitro* $^{32}$P labeling of oligonucleotides present in complete T1 or pancreatic RNase digests of a tRNA involves the following steps: (a) complete digestion of tRNA with T1 or pancreatic RNase and concomitant removal of phosphomonoester groups from the oligonucleotides by *E. coli* alkaline phosphatase; (b) inactivation of the phosphatase by heating in the presence of the chelating agent nitrilotriacetic acid (NTA)[16]; (c) phosphorylation of 5′-hydroxyl end groups of oligonucleotides with $^{32}$P using polynucleotide kinase and [$\gamma$-$^{32}$P]ATP; (d) elimination of excess [$\gamma$-$^{32}$P]ATP by phosphorylation of glucose to [$^{32}$P]glucose 6-phosphate with yeast hexokinase.[49] The resulting mixture of 5′-$^{32}$P-labeled oligonucleotides are separated by two-dimensional electrophoresis[2–4] or, in some cases, by an alternative two-dimensional system involving electrophoresis and ion-exchange thin-layer chromatography (see General Methods).

### *T1 or Pancreatic RNase Digestion and Dephosphorylation*

RNase digestion and dephosphorylation are carried out in the same reaction tube. Conditions are as described before,[16] except that the reaction has been scaled down to 10–25 $\mu$l and contains 0.5–5 $\mu$g of tRNA.[6] The reaction contains 50 m$M$ Tris·HCl (pH 8) and either pancreatic RNase (0.05 $\mu$g per microgram of tRNA) or T1 RNase (0.2 unit per microgram of tRNA); samples to be digested with T1 RNase are denatured by heating at 100° for 1 min and quick-cooling on ice before adding enzyme. After 3 hr of incubation at 37°, the dephosphorylation reaction is begun by the addition of *E. coli* alkaline phosphatase ($5 \times 10^{-4}$ unit per microgram of tRNA). The incubation is continued at 37° for 2 hr. The mixture is then cooled to room temperature and 0.1 volume of 50 m$M$ NTA is added. After 15–20 min, the mixture is heated to 100° for 90 sec to inactivate the phosphatase, then cooled, collected in the tip of the reaction tube by centrifugation, and stored at −20° until needed for *in vitro* labeling.

### *5′-$^{32}$P-Labeling of Oligonucleotides*

The *in vitro* phosphorylation reaction (10 $\mu$l) contains 0.5 $\mu$g of the tRNA, digested as described above, 10 m$M$ $MgCl_2$, 15 m$M$ $\beta$-mercap-

[49] R. Wu, *J. Mol. Biol.* **51**, 501 (1970).

toethanol, 1.6–1.8 nmol of [$\gamma$-$^{32}$P]ATP (100–500 Ci/mmol), and 2 units of T4 polynucleotide kinase (enough Tris·HCl is present in the tRNA digest and in the [$\gamma$-$^{32}$P]ATP to provide a final concentration of 15–100 m*M*). Incubation is at 37°. After 30 min glucose (20 nmol) and yeast hexokinase (0.002–0.003 unit) are added, and incubation is continued for 10 min; 2.5 nmol of nonradioactive ATP are then added and, after 10 min at 37° another 2.5 nmol of ATP. After a final period of incubation for 10 min at 37°, the reaction mixture is stored frozen until used for two-dimensional separation.

The material that results from the sequential operations described is a mixture of oligonucleotides bearing a 3′-hydroxyl group and a [5′-$^{32}$P]phosphomonoester group. Mononucleotides present in the original digest are mostly converted to nucleosides by alkaline phosphatase and are not substrates for subsequent enzymic phosphorylation by polynucleotide kinase.[25] However, some nucleoside-2′,3′-cyclic phosphate intermediates, which are resistant to alkaline phosphatase, may remain at the end of the digestion; these are subsequently phosphorylated at the 5′ end by polynucleotide kinase, and are, therefore, also present among the $^{32}$P-labeled oligonucleotides (see below, Fig. 3).[5]

Although the reaction conditions given have been designed to promote quantitative 5′-end-group labeling of most oligonucleotides, some of the oligonucleotides may not be quantitatively phosphorylated; these include oligonucleotides containing one or more modified nucleotides at the 5′ end or several G residues at or near the 5′ end; the exact degree of labeling varies somewhat from experiment to experiment. The yield of $^{32}$P in each oligonucleotide is therefore only a semiquantitative indication of actual molar ratios in the tRNA sequence (see below, Table III).

*Separation of 5′-$^{32}$P-Labeled Oligonucleotides*

The mixture of *in vitro* 5′-end group-labeled oligonucleotides from a T1 or pancreatic RNase digestion of a tRNA is usually separated by the standard two-dimensional electrophoresis fingerprinting procedure.[2] In general, the effects of nucleotide length and composition on the two-dimensional mobility of an oligonucleotide are the same for *in vitro* $^{32}$P-labeled oligonucleotides, which have an external phosphomonoester group at the 5′ end, as for uniformly $^{32}$P-labeled oligonucleotides, which have an external phosphomonoester group at the 3′ end.[13] Overall, therefore, the fingerprint patterns of a tRNA will be very similar whether prepared by *in vitro* labeling or from *in vivo* uniformly $^{32}$P-labeled tRNA. Some differences between the two types of fingerprint include the following. First, the *in vitro* labeled tRNA fingerprints lack spots corresponding to nucleoside 3′-monophosphates (present in fingerprints obtained from

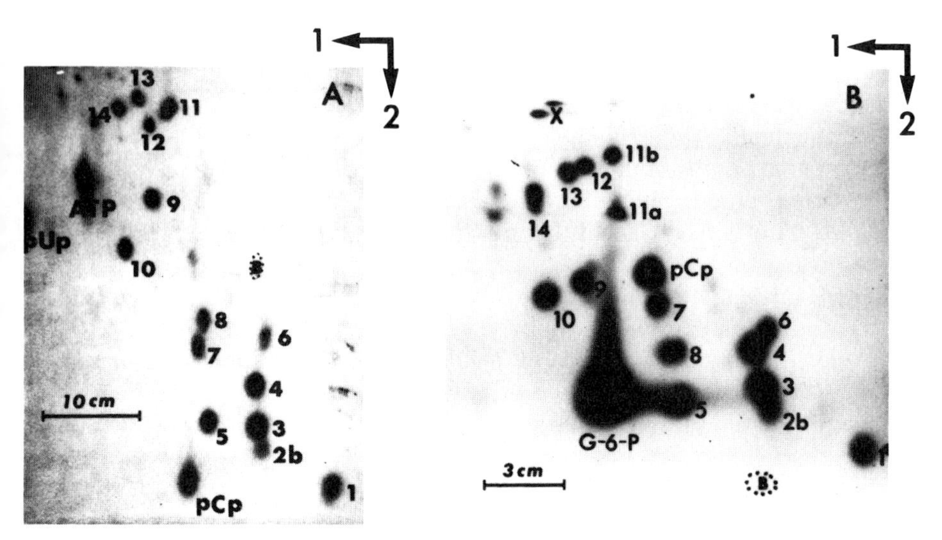


FIG. 3. Autoradiograms of 5′-$^{32}$P-labeled oligonucleotides obtained from pancreatic RNase digests of human placental tRNA$_i^{Met}$. (A) Standard fingerprinting procedure (DEAE-cellulose paper). [Reprinted with permission from *Cell* **6**, 407 (1975). Copyright © MIT.] (B) Alternative fingerprinting procedure (PEI-cellulose thin layer). Note difference in scale between the two fingerprints. Circled B is position of xylene cyanole blue dye. X is nonspecific radioactive material seen along the origin in all PEI-cellulose fingerprints of *in vitro*-labeled oligonucleotides. G-6-P, glucose 6-phosphate.

TABLE III

SEQUENCE AND MOLAR YIELD OF OLIGONUCLEOTIDES PRESENT IN FINGERPRINTS OF PANCREATIC RNASE DIGESTIONS OF HUMAN PLACENTA tRNA$_i^{Met}$

| Fingerprint spot | Sequence of oligonucleotide | Molar yield[a] found (theoretical) | Fingerprint spot | Sequence of oligonucleotide | Molar yield[a] found (theoretical) |
|---|---|---|---|---|---|
| 1 | pAC | 1.2 (1) | 9 | pGAU | 1.1 (1) |
| 2b | pm$^2$GC | 0.6 (1) | 10 | pGU | 1.1 (1) |
| 3 | pGC | 2.0 (2) | 11a | pAGAGm$^7$GD | 1.4 (2) |
| 4 | pAGC | 1.8 (2) | 11b | pGGAAGC | |
| 5 | pAU | 1.8 (2) | 12 | pGGGC | 0.7 (1) |
| 6 | pGm$^1$AAAC | 0.8 (1) | 13 | pAGAGU | 1.0 (1) |
| 7 | pm$^1$Gm$^2$GC | 0.8 (1) | 14 | pGGAU | 1.0 (1) |
| 8 | pt$^6$AAC | 1.0 (1) | | | |

[a] Based on Cerenkov counting of fingerprint spots (Fig. 3A) from two *in vitro* labeling reactions. Theoretical yield is predicted by the tRNA sequence [A. M. Gillum, B. A. Roe, M. P. J. S. Anandaraj, and U. L. RajBhandary, *Cell* **6**, 407 (1975)].

uniformly labeled tRNA), although they do contain some nucleoside 3′, 5′-diphosphates. Second, the relative amounts of $^{32}P$ in the individual oligonucleotide spots of an *in vitro* labeled tRNA fingerprint (and hence their densities on the autoradiogram) reflect the relative molar quantities of the oligonucleotides (with the exceptions noted above), rather than the relative number of total phosphate groups per oligonucleotide. Third, the two-dimensional mobilities of the oligonucleotides from the 5′ and 3′ termini of the tRNA are different in the two types of fingerprint. For instance, since the 3′-terminal sequence of all tRNAs is ...C-C-$A_{OH}$, the 3′-terminal oligonucleotide in a T1 RNase digest of uniformly labeled tRNA will have hydroxyl groups at both 5′ and 3′ ends, whereas all *in vitro* labeled oligonucleotides, including the 3′-terminal one, will contain a phosphate at the 5′ end.

Figure 3A shows a typical fingerprint of *in vitro* $^{32}P$-labeled oligonucleotides from a pancreatic RNase digest of human placenta initiator tRNA,[7] and Fig. 3B shows a fingerprint of the same *in vitro* labeled oligonucleotides using an alternative two-dimensional system (electrophoresis in the first dimension followed by PEI-cellulose thin-layer chromatography). Table III lists sequences of all the numbered oligonucleotides in Fig. 3A and their relative molar yields. It can be seen that the relative mobilities of several oligonucleotides differ between the two fingerprints in Fig. 3; in particular, the mixture of oligonucleotides present in spot 11 (Fig. 3A) resolves into two spots in the PEI-cellulose thin-layer fingerprint (Fig. 3B). The PEI-cellulose chromatography system is thus a useful alternative for resolving those oligonucleotides that may migrate close together (or streak[15]) on DEAE-paper electrophoresis. Occasionally, this system may also be used to resolve oligonucleotide mixtures recovered from a DEAE-paper fingerprint. In such cases, because the mixture consists of two compounds that migrate close together on cellulose acetate electrophoresis, it is useful also to extend the first-dimensional electrophoresis run for maximal resolution before transfer to the thin-layer plate.

Other alternative fingerprinting systems, such as two-dimensional homochromatography (see above, *General Methods*) may also be advantageous in certain cases.[50] Conceivably, if mainly the larger oligonucleotides resulting from T1 RNase or pancreatic RNase digestion of an RNA are of interest, fingerprinting by two-dimensional gel electrophoresis[51–53]

[50] H. J. Gross, H. Domdey, and H. L. Sanger, *Nucl. Acids Res.* **4,** 2021 (1977).

[51] M. A. Billeter, J. T. Parsons, and J. M. Coffin, *Proc. Natl. Acad. Sci U.S.A.* **71,** 3560 (1974).

[52] Y. F. Lee and E. Wimmer, *Nucl. Acids Res.* **3,** 1647 (1976).

[53] D. Frisby, *Nucl. Acids Res.* **4,** 2975 (1977).

(see above, *General Methods*) might be useful. This system is, in fact, used in our laboratory for separation of the large oligonucleotides resulting from *partial* T1 RNase or pancreatic RNase digestion of a tRNA (see below, *Preparation and Sequence Analysis of Large Oligonucleotides*).

## Sequence Analysis of 5′-$^{32}$P-Labeled Oligonucleotides

Sequence analysis of oligonucleotides recovered from fingerprints is in two stages, 5′-end-group analysis and complete sequence analysis.

### *5′-End-Group Analysis*

Two different methods are used for identification of the 5′-terminal nucleotide. The first involves treatment of the 5′-$^{32}$P-labeled oligonucleotide with T2 RNase and identification of the *pNp thus produced. The second method involves complete digestion of the 5′-end-labeled oligonucleotide with either snake venom phosphodiesterase or nuclease P1 and characterization of the *pN produced. Identification of *pNp and *pN is based on comigration with corresponding UV markers in several different thin-layer chromatography systems.

*Enzymic Digestion.* For end-group analysis, 5000 cpm of an oligonucleotide is used for complete digestion with each nuclease; if intensifying screens[42] are used during autoradiography, as little as 200–500 cpm will suffice. The T2 RNase reaction (5 μl) contains 0.1 unit of enzyme per microgram of carrier yeast tRNA in 20 m*M* ammonium acetate buffer (pH 4.5); incubation is for 3 hr at 37°. Incubation mixture (5 μl) with snake venom phosphodiesterase contains 1 μg of the enzyme per microgram of carrier yeast tRNA in 50 m*M* Tris·HCl, 5 m*M* potassium phosphate (pH 8.9) (the phosphate inhibits contaminating 5′-nucleotidase activity in the snake venom phosphodiesterase); incubation is for 2 hr at 37°. For digestion with nuclease P1, the reaction (5 μl) contains 1 μg of nuclease P1 per 1–10 μg of carrier yeast tRNA in 50 m*M* ammonium acetate buffer (pH 5.3) (this enzyme-to-substrate ratio is generally sufficient to cleave even highly resistant internucleotide bonds, such as those at sites of ribose methylation[54]); incubation is for 5 hr at 37°.

*Chromatography.* To each enzymic digest, 1 μl of a mixture of appropriate UV marker compounds (0.1 $A_{260}$ unit each) is added at the end of the incubation, and 2–3 μl of the solution are then applied directly onto each of two 20 × 20 cm cellulose thin-layer plates. If the samples are applied at 0.8–1 cm intervals, 16–20 samples (and hence 5′-end-group

[54] Y. Yamada and H. Ishikura, *Biochim. Biophys. Acta* **402,** 285 (1975).

analyses of all the $^{32}P$-labeled spots in a tRNA fingerprint) can be analyzed on the same plate. The plates are run at room temperature, one in solvent (a) and one in solvent (c), until the solvent front is about 1 cm from the top of the plate. They are then dried in a fume hood, marked with radioactive ($^{35}S$) ink solution, and autoradiographed.

*Identification.* Because T1 RNase cleaves an RNA chain after G residues and pancreatic RNase after C and U residues, the 5′-end groups of oligonucleotides present in T1 RNase digests must be A, U, or C (or a modified nucleoside), while the 5′-end groups of oligonucleotides present in pancreatic RNase digests must be A or G (or a modified nucleoside). $^{32}P$-labeled 5′-end groups that do not comigrate with any of the expected major nucleotide markers are prepared again and chromatographed with appropriately selected markers of modified nucleotides (generally, the 5′-monophosphate forms). The relative mobilities of many of the nucleotides commonly encountered in these analyses are listed in Table II. Some of the less common modified nucleotides, for which no UV markers are readily available, or dihydrouridine, which has no UV absorbance, must at first be tentatively identified by relative chromatographic mobility alone. Such tentative identifications may subsequently be confirmed by comparison of chromatographic mobilities with those of corresponding compounds derived from 5′-$^{32}P$-labeled oligonucleotides isolated from tRNAs of known sequence. In these cases, it is desirable to use thin-layer chromatography in additional solvent systems [such as solvent (b); see above, *General Methods*] and/or electrophoresis on Whatman 540 paper or DEAE-paper at pH 3.5 for further identification. Reference to tabulated or schematic chromatographic or electrophoretic mobilities[3,4,7,43,55] is also useful.

Occasionally, one fingerprint spot contains a mixture of oligonucleotides (e.g., sequence isomers), and may, therefore, yield two (or more) radioactive spots from the 5′ end. In such cases, the spots should be removed from the thin-layer plate as nitrocellulose platelets[28] (see above, *General Methods, Thin-Layer Chromatography*) and the radioactivity quantitated by liquid scintillation counting to determine the relative amount of each 5′ end in the mixture. Mixtures of oligonucleotides may sometimes be resolved by two-dimensional separation using electrophoresis on cellulose acetate at pH 3.5 followed by PEI-cellulose thin-layer chromatography (see above, *General Methods, Alternative Fingerprinting Procedure*).

[55] D. B. Dunn and R. H. Hall, *in* "Handbook of Biochemistry and Molecular Biology" (G. D. Fasman, ed.), 3rd Ed., Vol. 1, p. 65. Chemical Rubber Co. Press, Cleveland, Ohio, 1975.

End-group analysis may also be performed on larger 5′-$^{32}$P-labeled oligonucleotides or intact 5′-$^{32}$P-labeled tRNA.[41] Conditions for enzymic digestion are exactly as described above.

*Derivation of Nucleotide Sequence of 5′-$^{32}$P-Labeled Oligonucleotides by Mobility Shift Analysis*

The sequence of a 5′-$^{32}$P-labeled oligonucleotide can be determined by partial digestion of the oligonucleotide with snake venom phosphodiesterase and/or nuclease P1. By removing aliquots of the incubations at appropriately timed intervals, a range of 5′-$^{32}$P-labeled oligonucleotides is obtained that comprises a homologous partial digestion series containing every intermediate from the original oligonucleotide down to the 5′-terminal mononucleotide. These 5′-$^{32}$P-labeled oligonucleotides are separated by two-dimensional homochromatography and by one-dimensional electrophoresis on DEAE-cellulose paper, and the sequence of the oligonucleotide in question is derived from the characteristic mobility shifts between the successive intermediates present in the partial digest.

*Partial Digestion with Snake Venom Phosphodiesterase.* This is carried out at room temperature for 160 min in 50 m*M* Tris·HCl, 5 m*M* potassium phosphate (pH 8.9) with 1 μg of enzyme per 5 μg of carrier yeast tRNA; a typical reaction contains 1–2 × $10^5$ cpm of $^{32}$P-labeled oligonucleotide in a total volume of 50 μl. Aliquots are removed at 0, 2, 5, 10, 20, 40, 80, and 160 min and transferred to 0.4-ml polypropylene capped tubes containing an equal volume of 2 m*M* EDTA (pH 8); the aliquots are mixed briefly in a Vortex mixer, heated at 100° for 2 min to inactivate the enzyme, and stored at −20° until used.

*One-Dimensional Homochromatography Analysis.* Prior to analysis of the partial digests by two-dimensional homochromatography, the accumulation of partial degradation products in the aliquots is tested on small portions (0.5–2 μl) by one-dimensional homochromatography on 20 × 20 cm DEAE-cellulose thin-layer plates in 50 or 75 m*M* KOH-strength homomix. Based on the autoradiographic pattern obtained, appropriate aliquots are then pooled so as to best display the full range of partial digestion products for further analysis. The number of discrete partial digestion products in such a pattern gives a tentative indication of the size of the oligonucleotide, while the easily distinguishable purine and pyrimidine distances (see below, *Mobility-Shift Analysis by Two-Dimensional Homochromatography*) between successive intermediates in the series provides some preliminary indication of the nature of the sequence involved.

*Partial Digestion of 5′-$^{32}$P-Labeled Oligonucleotides with Nuclease P1.* In the sequencing of 5′-terminally labeled oligonucleotides, a special problem arises with oligonucleotide sequences that contain 3′- or internal modified nucleoside residues that block the 3′→5′ exonucleolytic progress of snake venom phosphodiesterase; this situation will be detected when the partial digest is analyzed by one-dimensional homochromatography. In such cases, the terminally $^{32}$P-labeled partial degradation products resulting from cleavage of phosphodiester bonds on the 5′ side of the modified nucleoside can be obtained in better yield by partial endonucleolytic digestion with nuclease P1. Since both snake venom phosphodiesterase and nuclease P1 cleave phosphodiester bonds to leave 3′-hydroxyl and 5′-phosphate ends, aliquots of partial digests produced by each of these enzymes can subsequently be combined to give an optimal representation of all the terminally $^{32}$P-labeled partial degradation products of an oligonucleotide. This is illustrated schematically in Fig. 4. The combined partial digest is then analyzed by two-dimensional homochromatography (see below, Fig. 7).

Partial digestion with nuclease P1 is in 50 m*M* ammonium acetate buffer (pH 5.3) at 20°, with 7.5 ng of enzyme per 50 μg of carrier RNA; a typical reaction contains 25,000–50,000 cpm of $^{32}$P-labeled oligonucleotide and 10–20 μg of carrier RNA in 20 μl. Aliquots (5 μl) are removed at 2, 5, 10, and 20 min, immediately made 5 m*M* in EDTA, and heated at 100° for 4 min. A small portion (0.5–1 μl) of each aliquot is analyzed for the extent of digestion by one-dimensional homochromatography as described above.

*Mobility-Shift Analysis by Two-Dimensional Homochromatography.* This method involves electrophoresis of selected, pooled aliquots of an oligonucleotide partial digest on cellulose acetate at pH 3.5 in the first dimension, followed by DEAE-cellulose thin-layer chromatography in homomix in the second dimension. Detailed descriptions of the two-dimensional mobility shift analysis of DNA sequences in this system have been given by Wu and colleagues[29,56,57]; we have found the mobility shift analysis of RNA sequences to be qualitatively very similar.

In the first dimension, the relative mobilities of two oligonucleotide intermediates in a homologous partial digestion series which differ by a single nucleotide will depend both on the p*K* of the nucleotide by which they differ and on the size and base composition of the sequence common to them. The formula derived by Tu *et al.*[57] can be used to predict such first-dimensional mobility shifts (approximately) from the apparent

[56] R. Bambara, E. Jay, and R. Wu, *Nucl. Acids Res.* **1**, 1503 (1974).
[57] C. P. D. Tu, E. Jay, C. P. Bahl, and R. Wu, *Anal. Biochem.* **74**, 73 (1976).

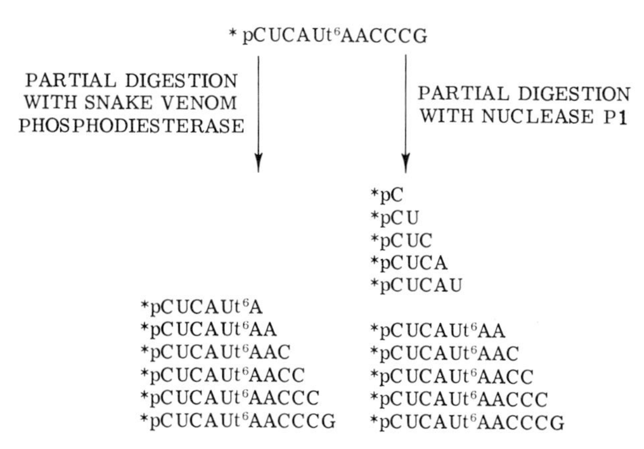


FIG. 4. Scheme for the use of nuclease P1 in conjunction with snake venom phosphodiesterase in the sequence analysis of 5′-³²P-labeled oligonucleotides containing modified bases that block the stepwise exonucleolytic progress of snake venom phosphodiesterase. Reprinted with permission from *Nucl. Acids Res.*, **4**, 4091 (1977).

charge observed for the 5′-mononucleotides under the conditions of electrophoresis (Table I, column 4).

The second dimension of this two-dimensional fractionation system resolves an oligonucleotide mixture on the basis of size, shorter oligonucleotides moving faster than longer ones. In addition, a difference of one purine residue between two successive intermediates in a series of partial degradation products causes a larger (by a factor of 1.5–2.5) mobility shift than a difference of one pyrimidine residue.[29] The overall result in two dimensions is that the stepwise removal of nucleotides in a homologous partial degradation series can be "read" as angular mobility shifts between successively shorter fragments (or, inversely, as the angular mobility shifts due to successive *additions* of nucleotides to the 5′-end mononucleotide). Thus, although U and G have a similar mobility shift in the first dimension, they can be distinguished from each other by their mobility shift in the second dimension (and likewise for C and A shifts). The mobility shifts of all four nucleotides are thus characteristic and identifiable.

In practice, it is simplest to deduce sequences by inspection, using the xylene cyanole blue dye marker as a reference point. A qualitative illustration of the relative direction of the two-dimensional mobility shifts caused by *adding* pA, pU, pC, or pG to an oligonucleotide running slower than, with, or faster than the blue dye (in the first dimension) is presented in Fig. 5. It should be noted that, whereas the *direction* of mobility shifts may be considered, empirically, to vary only with the first-dimensional position of the starting oligonucleotide relative to the blue dye, the *mag-*

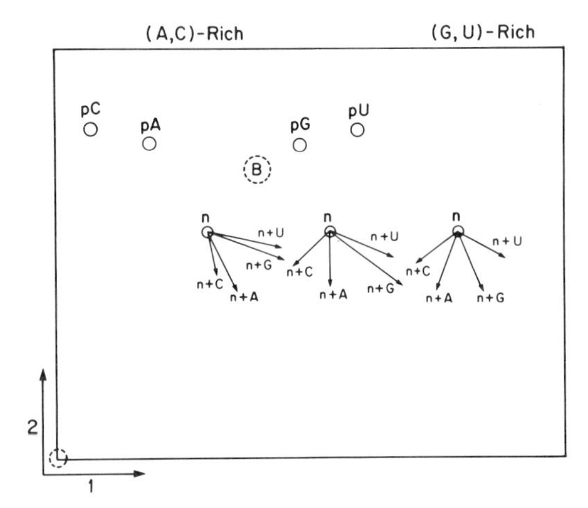


FIG. 5. Direction of angular mobility shifts in two-dimensional homochromatography. Mobility directions indicated are only approximate and are shown relative to the first-dimensional mobility of the oligonucleotide with respect to the xylene cyanole blue dye marker (circled B). Note that the vertical component of mobility shift is larger (1.5–2.5-fold) in the case of purine nucleotide additions than pyrimidine nucleotide additions. The overall magnitude of a mobility shift is partially also a function of oligonucleotide size, and decreases with increasing length of the oligonucleotide [i.e., with decreasing mobility in the second (vertical) dimension]. The positions of the 5′-mononucleotides relative to the xylene cyanole blue dye are also indicated.

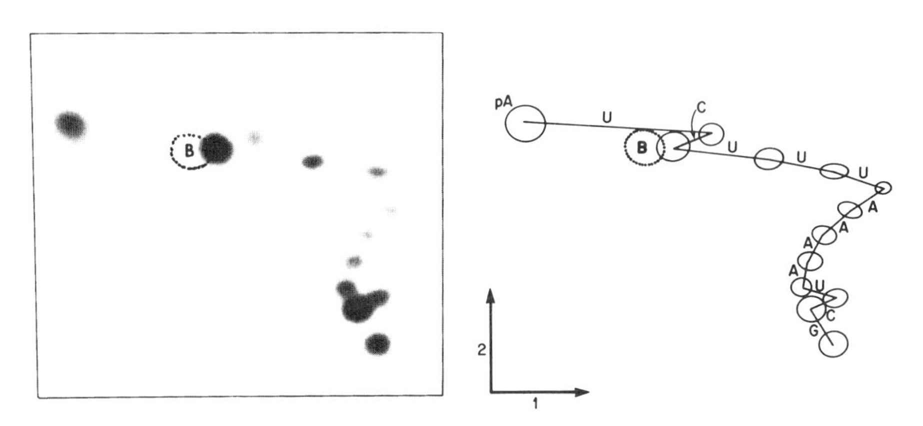


FIG. 6. (A) Autoradiogram of partial snake venom phosphodiesterase digest of [5′-$^{32}$P]-A-U-C-U-U-U-A-A-A-A-U-C-G analyzed by two-dimensional homochromatography. First dimension, electrophoresis on cellulose acetate at pH 3.5; second dimension, chromatography in 25 m*M* KOH-strength homomix. (B) Schematic diagram indicating identification of mobility shifts (cf. Fig. 5). Circled B is xylene cyanole blue dye marker.

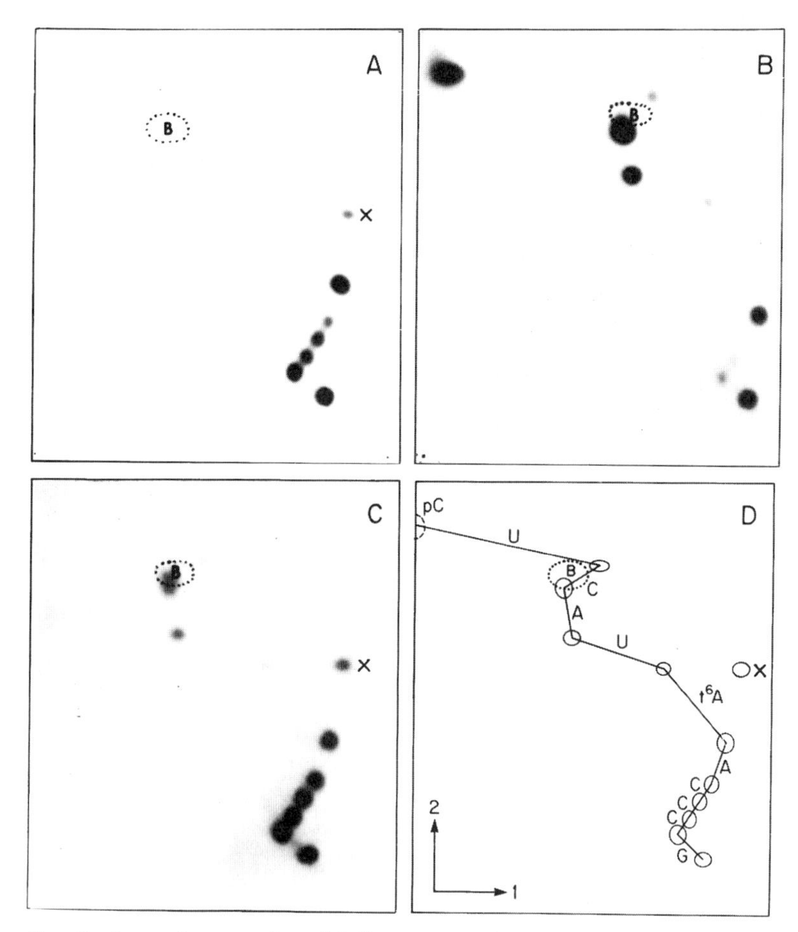


FIG. 7. Autoradiogram of partial digests of [5′-$^{32}$P]C-U-C-A-U-t$^6$A-A-C-C-C-G. First dimension, electrophoresis on cellulose acetate, pH 3.5; second dimension, homochromatography in 50 m*M* KOH-strength homomix. (A) Partial digestion with snake venom phosphodiesterase. (B) Partial digestion with nuclease P1. (C) Combined aliquots of digests indicated in (A) and (B). (D) Replica of C indicating identification of mobility shifts. Circled B is xylene cyanole blue dye. [The spot marked X, seen in A and C, is probably *pCUCp caused by contamination of the snake venom phosphodiesterase used by a specific endonuclease that cleaves within the sequence C-A (J. Heckman, personal communication).] Reprinted with permission from *Nucl. Acids Res.*, **4**, 4091 (1977).

*nitude* of mobility shifts is also a function of oligonucleotide size (i.e., of the *vertical* position of the oligonucleotide on the homochromatogram) and decreases with increasing oligonucleotide length.

An example of the application of these methods is presented in Fig.

6; the oligonucleotide in question, *pA-U-C-U-U-U-A-A-A-A-U-C-G,[57a] was encountered in the sequence analysis of a "tRNA-like" fragment from the 3′ end of turnip yellow mosaic virus RNA.[15] All the observed mobility shifts are unambiguous, and the entire sequence can be deduced directly (Fig. 6, right) by applying the criteria discussed above (*cf.* Fig. 5).

The use of pooled snake venom phosphodiesterase and nuclease P1 partial digests to sequence oligonucleotides that contain modified nucleosides at the 3′ end or internally (see above, Fig. 4) is illustrated in Figs. 7 and 8. The oligonucleotide in question in Fig. 7 is *pC-U-C-A-U-$t^6$A-A-C-C-C-G, present in a T1 RNase digest of *Neurospora crassa* $tRNA_i^{Met}$.[9] The snake venom phosphodiesterase digestion (Fig. 7A) is blocked at the $t^6$A residue. The nuclease P1 digest, on the other hand, being a nearly random endonucleolytic digest, accumulates the other intermediates except the one resulting from cleavage at the $t^6$A-A phosphodiester bond (Fig. 7B). The two digests are, therefore, complementary, and when pooled and analyzed together yield the complete sequence of the oligonucleotide (Fig. 7C,D). The nature of the unusual mobility shift displayed by $t^6$A was known from previous work on similar oligonucleotides present in T1 RNase digests of other eukaryotic initiator methionine tRNAs.[6,7,19] This and other unusual mobility shifts for certain modified nucleotides are discussed more fully below (*Identification of Modified Nucleotides in 5′-$^{32}$P-Labeled Oligonucleotides*). Figure 8 shows a similar example of the combined use of snake venom phosphodiesterase and nuclease P1 on the oligonucleotide *pA-A-$ms^2i^6$A-A-ψ-C-C-U-U-G.[8]

The most common complications in "reading" two-dimensional homochromatography patterns arise either from the presence of extraneous spots in the autoradiogram or from the presence of ambiguous or "unusual" mobility shifts.

Extraneous spots may either result from radioactive impurities present in the oligonucleotide as eluted from the fingerprint, or be generated during partial digestion by contaminating endonucleases (see, for example, spot X in Fig. 7). Impurities present in the oligonucleotide, whether consisting of other oligonucleotides not well resolved by the two-dimensional fingerprint separation or of nonnucleotidic $^{32}$P, may often be separated by prolonged electrophoresis of the oligonucleotide on cellulose acetate at pH 3.5, followed by thin-layer chromatography on PEI-cellulose (see above). Some sequence isomer oligonucleotides, however, may not resolve even by this procedure. If so, they are best sequenced by application of the DEAE-cellulose paper electrophoresis system de-

[57a] *p, 5′-$^{32}$P-labeled oligonucleotide.

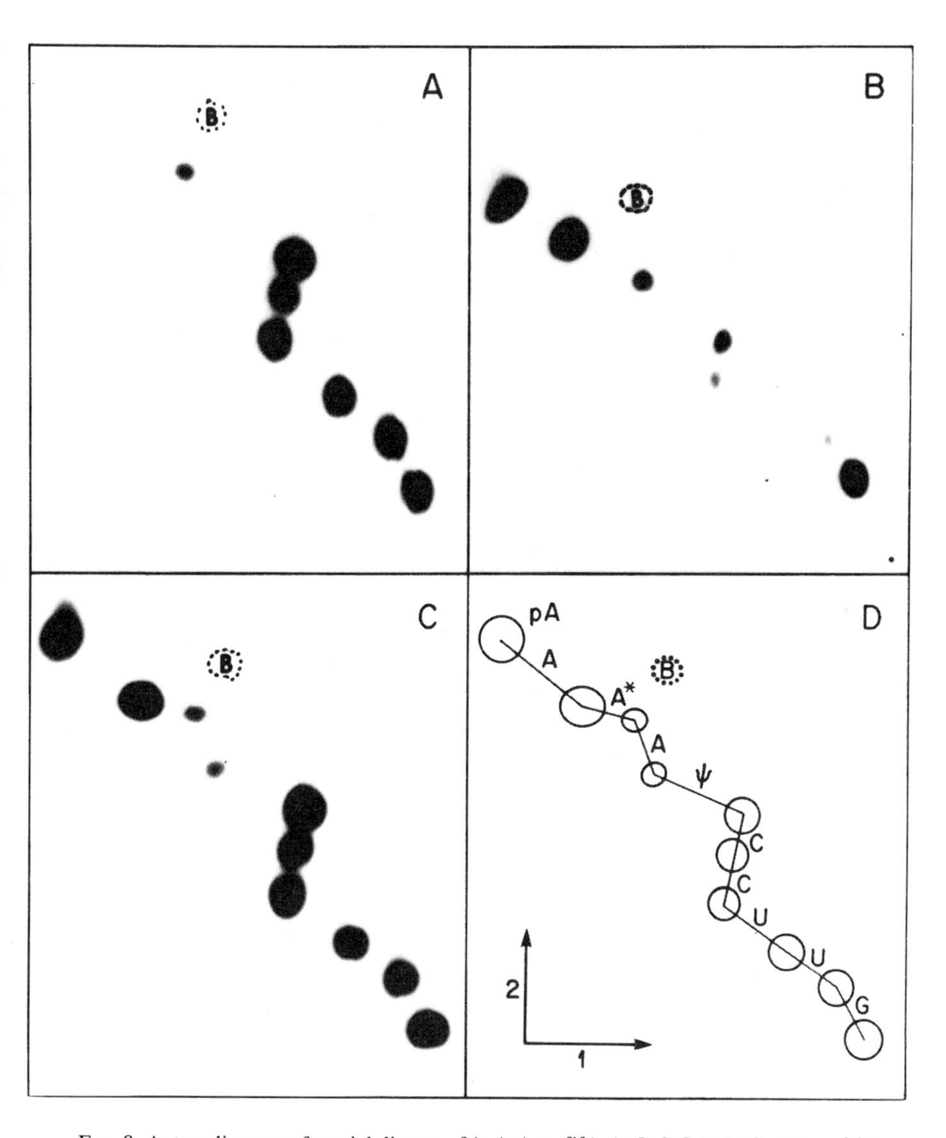


FIG. 8. Autoradiogram of partial digests of *pA-A-ms$^2$i$^6$A-A-Ψ-C-C-U-U-G. A*, ms$^2$i$^6$A. First dimension, electrophoresis on cellulose acetate, pH 3.5; second dimension, homochromatography in 50 m*M* KOH-strength homomix. (A) Partial digestion with snake venom phosphodiesterase. (B) Partial digestion with nuclease P1. (C) Combined aliquots of digests indicated in (A) and (B). (D) Replica of C indicating identification of mobility shifts. Circled B is xylene cyanole blue dye. Note the pronounced block toward the 3′-exonucleolytic progress of snake venom phosphodiesterase at Ψ and then at ms$^2$i$^6$A.

scribed below, in which partial digestion intermediates derived from short sequence isomers about 4–5 long can frequently be resolved; once the mixed isomer nature of the starting material is recognized, the two overlapping sequence patterns present on the homochromatogram can also be interpreted, confirming the sequence assignment. If the oligonucleotide being analyzed contains a sequence that is preferentially cut by an endonuclease commonly contaminating commercial snake venom phosphodiesterase, partial digestion may be repeated with nuclease P1; the pattern from the nuclease P1 digest should be free of extraneous spots (see, for example, Fig. 7B and legend).

Unusual mobility shifts in the two-dimensional homochromatography patterns of partial digests may indicate the presence of modified nucleosides; the detection and identification of these is discussed more fully below. It should be noted that the presence of many simple base methylations (except those that alter the charge of the bases, such as $m^1A$, $m^7G$, $m^3C$, etc.) and other modifications often cannot be detected by two-dimensional homochromatography; these are easily detectable, however, by DEAE-cellulose paper electrophoresis, which is uniquely sensitive to nucleoside modifications. It is, therefore, important to use both DEAE-cellulose paper electrophoresis and two-dimensional homochromatography for the analysis of partial digests of 5′-$^{32}$P-labeled oligonucleotides.

*Mobility-Shift Analysis by Electrophoresis on DEAE-Cellulose Paper.* Electrophoretic analysis on DEAE-cellulose paper (Whatman DE-81) of a series of 5′-$^{32}$P-labeled oligonucleotide partial-digestion products uses the pH 3.5 and pH 1.9 buffer systems described by Sanger and collaborators.[2–4] At either pH, the electrophoretic mobility of the shorter oligonucleotides of a homologous series is greater than that of the longer. Mobility shifts between successive homologs are characteristic of the specific nucleotide by which they differ, and these shifts, quantitated as $M$ values[2–4] (Table IV), can be used to derive the sequence of an oligonucleotide.

In general, we have found the most reliable method of sequence analysis, using DEAE-cellulose paper electrophoresis of partial digests, to be the use of 5′-$^{32}$P-labeled oligonucleotide standards of known sequence, or partial digests of these, as mobility markers, which are run alongside the unknown.[15,16] Appropriate standards may be selected based on the sequence assigned to the oligonucleotide in question by two-dimensional homochromatography. The variety of commercially available purified RNA species provides, through the use of fingerprinting, an extensive library of mono- and oligonucleotide reference standards for this purpose.

TABLE IV
RANGES OF *M* VALUES ON DEAE-PAPER ELECTROPHORESIS AT pH 3.5 OR 1.9[a]

| 3′-Terminal nucleotide removed | Value of *M* | |
|---|---|---|
| | pH 3.5 | pH 1.9 |
| pC | 0.6–1.2 | 0.05–0.3 |
| pA | 2.1–2.9 (4.1)[b] | 0.4–1.1 |
| pU | 1.7–1.9 (2.5)[b] | 1.5–2.5 |
| pG | 2.6–4.4 | 1.2–3.5 |

[a] G. G. Brownlee, "Determination of Sequences in RNA," North-Holland Publ., Amsterdam, 1972.

[b] Numbers in parentheses indicate deviations from the *M*-value ranges published by Sanger and collaborators,[a] which were observed occasionally in our work with 5′-terminally labeled oligonucleotides.

The major limitation of DEAE-paper electrophoresis for sequencing is that oligonucleotides longer than 5 or 6 generally do not resolve well enough from the origin to be analyzed. In principle, one might apply the 3′-$^{32}$P-labeling technique of Szeto and Söll[58] to circumvent this problem by sequencing the same oligonucleotide from its opposite (i.e., 3′) end and thereby extend this approach to the sequencing of most of the fragments usually present in pancreatic or T1 RNase digests of a tRNA.

## *Identification of Modified Nucleotides in 5′-$^{32}$P-Labeled Oligonucleotides*

When a modified nucleotide occurs at the 5′ end of an oligonucleotide its detection and identification are straightforward, because the unknown residue is $^{32}$P-labeled and can be identified directly upon 5′-end-group analysis, as described above (similarly, in principle, a modified nucleotide at the 3′ end of a fragment can be $^{32}$P-labeled using primer-dependent polynucleotide phosphorylase[58]). However, when a modified nucleotide is located at an internal position in an oligonucleotide, more indirect methods of analysis can be applied; in many cases, a modified nucleotide at the 3′ end of an oligonucleotide can also be identified indirectly.

The presence of a modified nucleotide in an oligonucleotide is indicated by several lines of evidence. First, the presence of a modified nucleotide (even a simple one such as m$^1$G, D, ψ) will result in an alteration of the position of the oligonucleotide in a fingerprint pattern (see, for instance, Fig. 3A, spots 3 and 2b, and Table II) compared to

[58] K. S. Szeto and D. Söll, *Nucl. Acids Res.* **1**, 171 (1974).

that of an oligonucleotide having the same sequence but no modification. Second, when partial digests of both unknown and marker oligonucleotide containing no modification are analyzed in parallel tracks on DEAE-cellulose paper electrophoresis, oligonucleotides containing the modified residue(s) will not comigrate with their unmodified counterparts. Third, because certain modified nucleotides, such as p$\psi$, pt$^6$A, pY, pms$^2$i$^6$A, and pNm are removed relatively slowly by snake venom phosphodiesterase, a block in the 3′-exonucleotytic action of this enzyme under typical partial digestion conditions is a strong indication of the presence of one of the above modified nucleotides at that position in the oligonucleotide.

Once the presence of a modified nucleotide is suspected, the modified nucleotide can be identified by a comparison of its migration and mobility shifts in a variety of systems with those of an appropriate oligonucleotide of known sequence that contains a characterized modified residue. A knowledge of the modified nucleotide composition of the tRNA is important for the judicious selection of marker oligonucleotides containing the modified nucleotide of interest for comparison. It is also extremely helpful to be aware of the types of effects various modified nucleotides have in the chromatographic and electrophoretic systems used in sequence analysis. Therefore, we have characterized the behavior of several of the common modified nucleotides that we have encountered and have summarized the results in Table V (also see below). If no available marker oligonucleotides are found to be identical to the unknown oligonucleotide, then the unknown modified nucleotide residue can be only tentatively identified, based on its effects on oligonucleotide mobility, its position within the total tRNA sequence, and so on.

Finally, the *in vitro* $^{32}$P-labeling method for modified nucleotide composition analysis described above also makes possible the analysis and identification of the modified nucleotide content of all the oligonucleotides present in a complete T1 and/or pancreatic RNase digest of a tRNA. Oligonucleotides present in such digest (50 $\mu$g of tRNA for each digest) can be separated by two-dimensional thin-layer chromatography.[59,60] These can then be eluted and used for modified nucleotide composition analysis. Although 0.05–0.5 $\mu$g of an oligonucleotide is sufficient for such analysis, complete digestion of tRNA with T1 and/or pancreatic RNase should be performed on 50 $\mu$g or so of the tRNA to facilitate detection of the oligonucleotides on the thin-layer chromatogram by their ultraviolet absorbance.

[59] M. Simsek, Ph.D. thesis, Massachusetts Institute of Technology, Cambridge, 1974.

[60] H. J. P. Schoemaker and P. R. Schimmel, *J. Mol. Biol.* **84**, 503 (1974).

*Effects of Hydrophobic Modifications on Properties of Oligonucleotides*

The most common posttranscriptional modification, methylation, generally causes some increase in oligonucleotide electrophoretic mobility on DEAE-cellulose paper in the pH 3.5, pH 1.9, or 7% formic acid buffer systems (methylations that result in a charged base, such as $m^1A$ or $m^7G$, are discussed separately below). There is also an increased mobility on PEI-cellulose thin-layer chromatography, and a very slight increased

TABLE V
EFFECT OF NUCLEOSIDE MODIFICATIONS ON OLIGONUCLEOTIDE MOBILITY[a]

| Modified nucleoside | Electrophoretic mobility on DEAE-paper[b] | | Two-dimensional homochromatography mobility shift |
|---|---|---|---|
| | Formic acid[c] | pH 3.5 | |
| $m^2G$ | >G | ≥G (Fig. 9) | G-like |
| $m^1G$ | >$m^2G$ > G | >$m^2G$ > G | G-like |
| $m^2_2G$ | ≫$m^1G$ > G | ≥$m^1G$ > G (Fig. 9) | — |
| $m^7G$ [d,e] | — | M value ≅ 0 | See Fig. 11B |
| Y [d–f] | — | — | See Fig. 10A |
| $m^6A$ | >A (also pH 1.9) | >A | A-like (Fig. 12) |
| $m^1A$ [d,f] | ≫A (also pH 1.9) | *M* value ≅ 0 | See Fig. 11A |
| $t^6A$ [d,e] | ≪A | ≪A | See Fig. 7 |
| $ms^2$-$i^6A$ [d,e] | — | — | See Fig. 10B |
| D | >U | >U | U-like (Fig. 11B) |
| T | >U | — | U-like |
| Ψ[e] | >U | <U | U-like (Figs. 8 and 10) |
| $s^4U$ [f] | — | — | — |
| $m^5C$ | >C | >C | — |
| Gm [g] | >G | >G | G-like |
| Cm [g] | >C | >C | C-like |

[a] Alteration of the mobility of an oligonucleotide as a result of substitution of a modified nucleoside for the corresponding unmodified nucleoside. Note: These alterations may not necessarily extrapolate to the relative mobility of the modified nucleoside 5′-monophosphate.

[b] Mobilities of oligonucleotides containing a modified nucleoside are given as: much greater than (≫), greater than (>), slightly greater than (≥), not resolving from (≅), less than (<), or much less than (≪) those of the corresponding oligonucleotides, which contain instead the indicated unmodified nucleoside.

[c] Second dimension of standard fingerprint analysis.[2–4]

[d] Discussion in text.

[e] 5′-Phosphodiester bond hydrolyzed slowly by snake venom phosphodiesterase.

[f] Degradation due to chemical instability may cause blurred or multiple patterns during analysis of partial digestion products.

[g] 3′-Phosphodiester bond hydrolyzed very slowly by snake venom phosphodiesterase.

mobility on DEAE-cellulose homochromatography. Conversely, electrophoretic mobility at pH 3.5 on cellulose acetate is somewhat retarded. These effects (summarized in Table V) are probably attributable to increased mass and hydrophobicity, as well as a slight increase in $pK_b$ of the residue after methylation.[55] The relative influence of a methylated residue must necessarily decrease as the size of the oligonucleotide increases. In general, the two-dimensional mobility shift alterations are too subtle for identification (typically a slight decrease in the vertical gap distance), but may be used to confirm the presence of a methylated residue that has already been detected and identified by mobility analysis of partial digests alongside the appropriate standard oligonucleotides by DEAE-cellulose paper electrophoresis. Although we have never encountered it, one instance where the presence of methylation could conceivably be overlooked is when the modified residue is located farther than 6 or 7 nucleotides from the 5′ end of the corresponding fragments produced both by T1 and pancreatic RNase, since unknown and standard oligonucleotides this large which differ only by a methyl residue are not expected to be resolved by DEAE-cellulose paper electrophoresis. In such a case, the modified nucleotide, predicted by nucleotide composition analysis, would not have been assigned a position in any oligonucleotide sequence. The longer oligonucleotides present in T1 and/or pancreatic RNase digests could then be 3′-end-labeled, as described by Szeto and Söll,[58] and analyzed by partial digestion to locate the modification.

An example of the effect of methylation on oligonucleotide mobilities is the resolution of $*pm^2GC$ (spot 2b) from *pGC (spot 3) in both fingerprinting procedures (Fig. 3A, B). Addition of a second methyl group generally results in a more profound influence on oligonucleotide mobility, as illustrated in Fig. 9. The snake venom phosphodiesterase digestion product of $*pm^2_2GC$ migrates ahead of those of $*pm^2GC$ and *pGC. In this case, the unknown oligonucleotide (B) is shown to be different from the standard oligonucleotide (C) and identical to the standard oligonucleotide (A).[6]

If modified nucleotide composition analysis indicates the occurrence of 2′-*O*-methylation(s) in a tRNA, the oligonucleotides containing such modifications may be identified by mobility-shift alterations similar to those described above, as well as by a strong block at the site of sugar methylation during snake venom phosphodiesterase partial digestion.[61]

The effects of D and $\psi$ relative to U are rather similar to those of methylation, but can be distinguished by comparison to standards. The effect of hydrophobic hypermodification, such as Y, $i^6A$ and $ms^2i^6A$, cannot be generalized and will depend on the properties of the side chain, p*K* of

[61] M. W. Gray and B. G. Lane, *Biochim. Biophys. Acta* **134,** 243 (1967).

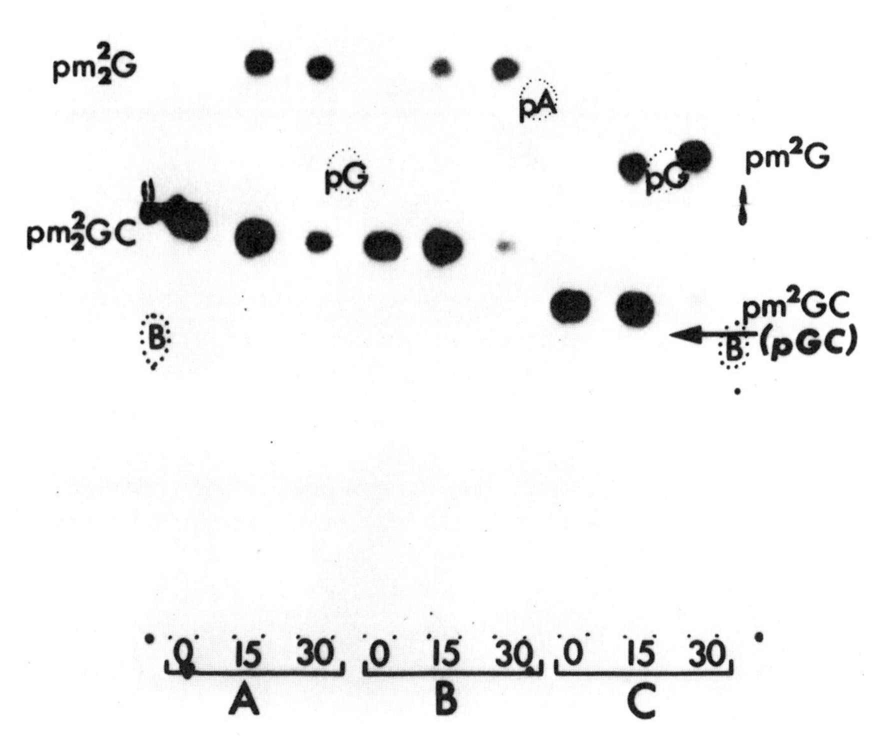


FIG. 9. Autoradiogram of partial snake venom phosphodiesterase digests of (A) *pm$_2^2$GC from yeast tRNA$_i^{Met}$; (B) *pm$_2^2$GC from salmon liver tRNA$_i^{Met}$; (C) *pm$^2$GC from rabbit liver tRNA$_i^{Met}$ as analyzed by electrophoresis on DEAE-cellulose paper at pH 3.5. Arrow indicates where unmodified *pGC would migrate. pA and pG, circled, indicate the location of added UV-absorbing markers of pA and pG, respectively. Circled B is position of xylene cyanole blue dye. Numbers at the origin indicate incubation times, in minutes, with snake venom phosphodiesterase. Reprinted with permission from *Cell* **6**, 395 (1975).

the modified base, and so on. A common effect of such modifications that we have seen is a substantial increase in the mobility of oligonucleotides during homochromatography on DEAE-cellulose plates caused by the presence of these hypermodified residues. Figure 10 shows the results obtained during the analysis of partial digests of *pGm-A-A-Y-A-ψ (Fig. 10A) and *pG-A-A-ms$^2$i$^6$A-A-ψ (Fig. 10B) by two-dimensional homochromatography. It can be noted that the presence of Y has increased the chromatographic mobility of *pGm-A-A-Y to such an extent that in the second dimension it has almost the mobility of its lower homolog, *pGm-A-A. Similarly, in Fig. 10B, it is found that the mobilities of *pG-

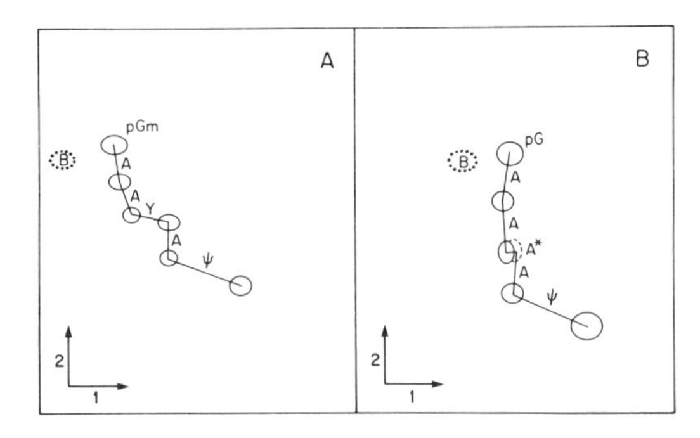


FIG. 10. Replicas of autoradiograms of combined partial snake venom phosphodiesterase and nuclease P 1 digests of (A) *pGm-A-A-Y-A-Ψ and (B) *pG-A-A-ms$^2$i$^6$A-A-Ψ as analyzed by two-dimensional homochromatography. First dimension, electrophoresis on cellulose acetate, pH 3.5; second dimension, homochromatography in 50 m*M* KOH-strength homomix. Circled B is xylene cyanole blue dye.

A-A and *pG-A-A-ms$^2$i$^6$A in the DEAE-cellulose chromatography dimension are virtually identical.

*Effects of Modifications Altering Electrostatic Charge on the Bases*

Certain posttranscriptional modifications cause large alterations in the p*K* of a nucleoside,[55] and alter the electrostatic charge on the ring, such as m$^1$A, m$^7$G, t$^6$A.

The aminoacryl side chain of the nucleoside t$^6$A may exist *in vivo* as a carboxylester.[62] However, the free carboxylic acid is the form isolated from tRNA, which gives the residue a net negative charge (p$K_a$ ~ 3).[55,62,63] Therefore, migration of a pt$^6$A-containing oligonucleotide relative to an unmodified oligonucleotide of the same size is increased during electrophoresis on cellulose acetate and dramatically retarded (owing to increased ionic affinity to the resin) during homochromatography or electrophoresis on DEAE-cellulose paper. For example, the location of *pt$^6$AAC (spot 8, Fig. 3) is greatly displaced in both fingerprint systems, compared to that of unmodified *pAAC (not present in Fig. 3), which would be just to the left and above *pAC (spot 1, Fig. 3). At pH 3.5, the mononucleotide pt$^6$A migrates slightly ahead of xylene cyanole blue on DEAE-cellulose paper, which is much slower than any unmodified mononucleotide and also than many dinucleotides. During two-dimensional

[62] R. H. Hall, "The Modified Nucleosides in Nucleic Acids." Columbia Univ. Press, New York, 1971.

[63] R. S. Cunningham and M. W. Gray, *Biochemistry* **13**, 543 (1974).

homochromatography, oligonucleotides containing t$^6$A are accelerated in the first dimension and slowed in the homochromatography dimension. Therefore, the mobility shift due to release of pt$^6$A from the 3′ end is an exaggerated shift upward and leftward (Fig. 7). The vertical gap between these intermediates is so large it might be mistaken for one due to a missing digestion product. The inhibition of snake venom phosphodiesterase digestion by the 5′-phosphodiester bond of t$^6$A is also shown in Fig. 7A.[63] Nuclease P1 cleaves the 5′-phosphodiester bond of t$^6$A much more readily than the 3′-phosphodiester bond (Fig. 7B), so that a combination of both partial digests contains every intermediate (Fig. 7C, D).[20]

The addition of pm$^1$A or pm$^7$G to an oligonucleotide actually results in either no change in net negative charge (owing to the cancelation of one phosphate negative charge by the positive charge on the base), or even a small net decrease in the total charge of the oligonucleotide, depending on the pH used for the analysis. Therefore, the mobility of an oligonucleotide containing either m$^1$A or m$^7$G on cellulose acetate (pH 3.5) is much slower than that of its unmodified counterpart. Conversely, during electrophoresis (pH $\leq$3.5) or homochromatography (pH 4.5–4.7) on DEAE-cellulose, affinity of the m$^1$A- or m$^7$G-containing oligonucleotide for the positively charged ion-exchange support is decreased, so that its mobility is faster than expected for an oligonucleotide of that size. Thus, during two-dimensional homochromatographic analysis of partial digests of *pGm$^1$AAAC (spot 6, Fig. 3), for example, the m$^1$A-containing oligonucleotides are retarded in the first dimension and advanced in the second dimension relative to the m$^6$A-containing isomers (Fig. 11A). Once pm$^1$A is removed from the 3′ end, the mobility of the remaining oligonucleotide is faster on cellulose acetate, yet slower on DEAE-cellulose, causing a highly unusual "reverse mobility shift" *downward* and rightward. A similar phenomenon is seen for removal of pm$^7$G from the 3′ end of an oligonucleotide (Fig. 11B). During electrophoresis on DEAE-cellulose paper, oligonucleotides containing m$^1$A or m$^7$G have such increased mobilities compared to their unmodified counterparts that upon removal of pm$^1$A or pm$^7$G the remaining oligonucleotide shows little or no mobility increase (i.e., *M* value 0). Thus, during analysis of *pGm$^1$AAAC, only three of the four snake venom phosphodiesterase digestion products are apparent at pH 3.5,[64] because *pGm$^1$A comigrates with *pG (cf. Fig. 11, ref. 64). Similarly, in the analysis of m$^7$G-containing oligonucleotides, such as *pAGAGm$^7$GD (spot 11a, Fig. 3), *pAGAGm$^7$G comigrates with *pAGAG at pH 3.5, i.e., *M* value = 0 for release of

[64] K. Ghosh, H. P. Ghosh, M. Simsek, and U. L. RajBhandary, *J. Biol. Chem.* **249,** 4720 (1974).

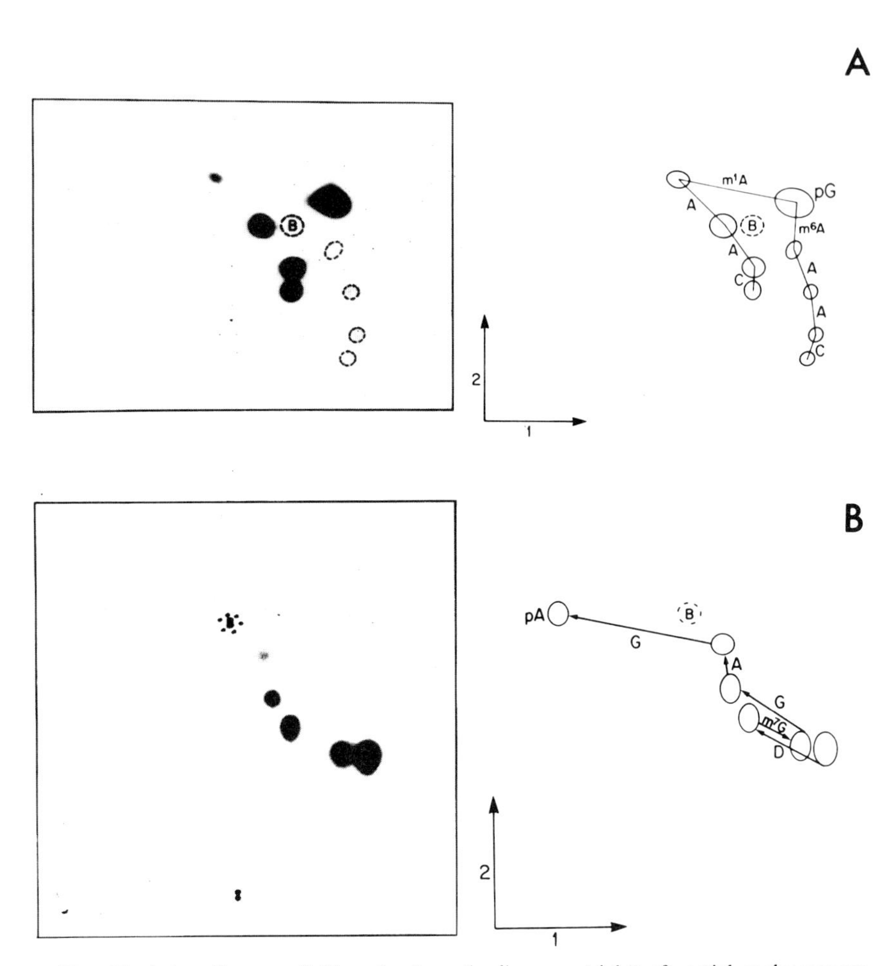


FIG. 11. Autoradiograms (left) and schematic diagrams (right) of partial snake venom phosphodiesterase digests of (A) *pG-m¹A-A-A-C (spot 6 of Fig. 3A) and (B) *pA-G-A-G-m⁷G-D (spot 11a of Fig. 3B) as analyzed by two-dimensional homochromatography. Circled B is xylene cyanole blue dye.

pm⁷G.[19] Fortunately, as shown above (Fig. 11A and B), analysis by two-dimensional homochromatography reveals the highly characteristic mobility shifts for these residues. However, in the special case where m⁷G or m¹A is at the 5′ end of an oligonucleotide, mobility shift analysis is difficult, because the 5′-end mono- and dinucleotides are not detected, or streak badly during homochromatography. On cellulose acetate electrophoresis at pH 3.5, pm⁷G barely migrates from the origin and pm¹A actually migrates toward the cathode ($R_{blue}$ = -0.28).[19] Since there is little or no effective negative charge on either nucleotide, it is not bound by DEAE-cellulose during the transfer step. The dinucleotides (sometimes even trinucleotides) may also be poorly retained by DEAE-cellulose during transfer, and therefore appear as smears after homochromatography. We have circumvented some of these problems by chemically converting these positively charged ring systems to a neutral form, as discussed below.

*Conversion of m¹A to m⁶A*

The nucleoside m¹A will convert readily to m⁶A at a rate proportional to hydroxide ion concentration.[65,66] Usually, during the isolation and analysis of oligonucleotides containing m¹A, some conversion to m⁶A has already occurred (e.g., Fig. 11A), and this can complicate the interpretation of mobility shifts. One approach to this problem is the complete conversion of an oligonucleotide containing m¹A to the m⁶A isomer, which allows the analysis of a single form (Fig. 12; cf. Fig. 11A), and also relieves the problems mentioned above when pm¹A is the 5′-end residue. Conditions used for quantitative conversion of m¹A to m⁶A as the free base, nucleoside,[66] or within short oligonucleotides[67] were found to be too harsh for use on longer oligonucleotides.[19] We describe here a milder procedure, which generally gives ≥50% rearrangement to m⁶A and less than 30% degradation of oligonucleotides up to 13 long.[19]

*Reaction Conditions.* The m¹A-containing oligonucleotide is obtained in the usual manner (see above, *General Methods*) from a fingerprint of 5′-end-labeled oligonucleotides. An aliquot of the material, containing, if possible, at least twice the amount of radioactive oligonucleotide required for subsequent analyses, is adjusted to contain 30–50 μg of carrier tRNA and lyophilized to dryness. The residue is dissolved in 15 μl of 50 m*M* $NH_4HCO_3$ (pH 9.0), sealed in a 20-μl capillary pipette, and immersed in an 85° water bath for 30–40 min. Contents are then transferred into a polypropylene tube. At this point, the material can be stored at −20° while an aliquot is analyzed for the extent of conversion to m⁶A (and extent of degradation).

If m¹A is the 5′-terminal residue, a routine end-group analysis by complete T2 RNase digestion and chromatography on a cellulose thin-layer using solvent (a) readily resolves pm¹Ap from *pm⁶Ap for subsequent quantitation (see Table II). Otherwise, an estimate of the extent of

[65] P. Brooks and P. D. Lawley, *J. Chem. Soc. (London)* (Part II), 539 (1960).

[66] J. B. Macon and R. Wolfenden, *Biochemistry* **7**, 3453 (1968).

[67] L. L. Spremulli, P. F. Agris, G. M. Brown, and U. L. RajBhandary, *Arch. Biochem. Biophys.* **162**, 22 (1974).

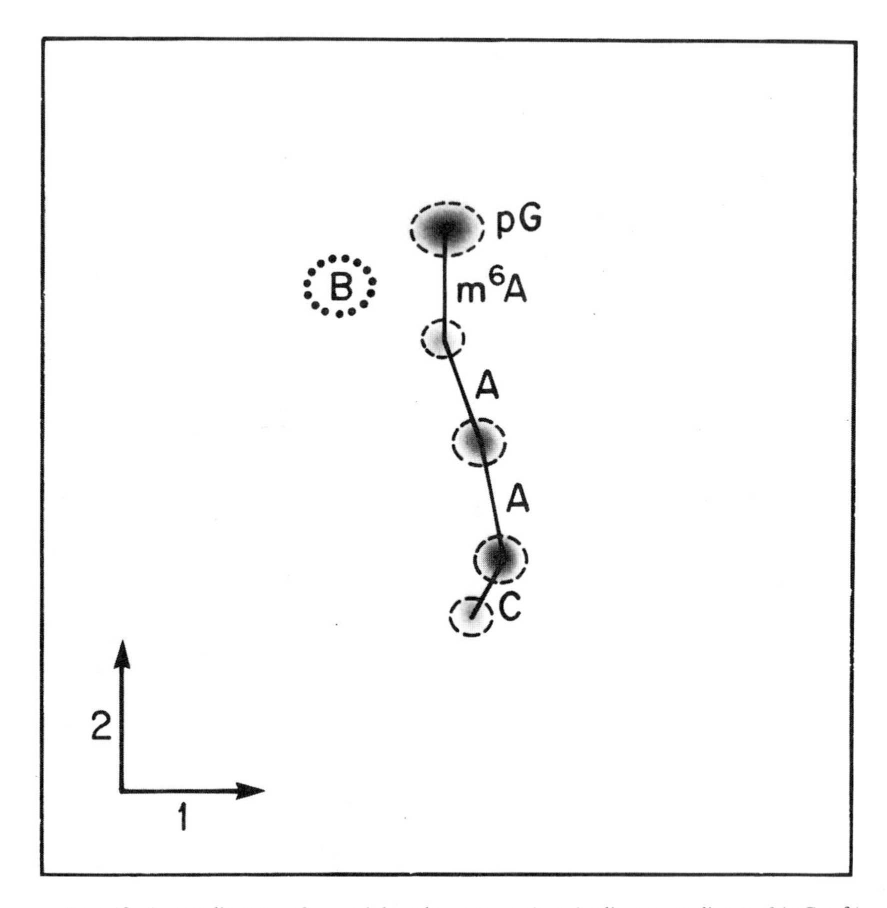


FIG. 12. Autoradiogram of a partial snake venom phosphodiesterase digest of *pG-$m^6A$-A-A-C as analyzed by two-dimensional homochromatography. Circled B indicates location of xylene cyanole blue dye. Note the pronounced difference between this pattern and the major track present in a similar digest of *pG-$m^1A$-A-A-C in Fig. 11A. The pattern here is, in fact, identical to the satellite track in Fig. 11A resulting from the partial conversion of $m^1A$ to $m^6A$ during isolation and analysis of the oligonucleotide.

degradation and of conversion to intact oligonucleotide containing $m^6A$ is obtained by one-dimensional homochromatography of the reaction mixture alongside a control sample containing unreacted starting material. The intact oligonucleotide containing $m^6A$ will be the only spot running more slowly than the original $m^1A$-containing oligonucleotide (if deemed necessary, at this point the reaction mixture may be thawed and incubated in a capillary at 85° for another 20–30 min).

The desired product must now be purified from unreacted and degraded material. A procedure we have found to be especially convenient for oligonucleotides up to 13 long is thin-layer chromatography on PEI-cellulose.[33,34] The sample is lyophilized to dryness, dissolved in 5 μl of water and applied as a narrow band about 1 cm wide on a glass-backed PEI-cellulose plate, alongside the tracking dye mixture. An aliquot of starting material is also applied for use as a marker to identify the desired product. A Whatman 3 MM paper wick is then attached to the top of the thin-layer plate (see above under *Homochromatography*). The plate is developed to half its height in distilled water at room temperature and then with 2 *M* pyridinium formate (pH 3.4),[34] (cf. above, *Alternative Fingerprinting Procedure*) until the xylene cyanole tracking dye has migrated 15 cm, or more, depending on the size of the oligonucleotide. The product is located by autoradiography, isolated from the resin, and used for further analysis.

*Application to Oligonucleotides Containing $m^7G$.* At neutral and alkaline pH values, the positively charged ring system of $m^7G$ can undergo a ring scission reaction to form 5-(*N*-methyl) formamido-6-ribosylaminocytosine,[62,68] which does not carry a net positive charge. Brum and Chang[69] have used the same reaction conditions to convert $m^7G$ in an oligonucleotide to the ring-opened form. After purification of the $m^7G$ ring-opened product, both the original and converted oligonucleotide were analyzed by partial digestion and two-dimensional homochromatography. The unusual problems associated with analysis of $m^7G$-containing oligonucleotides can thus be overcome using basically the same approach as that used toward oligonucleotides containing $m^1A$.

## Preparation and Sequence Analysis of Large Oligonucleotides

To complete a tRNA sequence, the oligonucleotides present in T1 RNase or pancreatic RNase digests must be ordered to give a unique total sequence. Toward this end, it is useful to prepare large specific oligonucleotide fragments of the tRNA and determine the arrangement within them of oligonucleotides whose sequences are already known. We describe here two methods for the preparation of large tRNA fragments, partial digestion under controlled conditions with T1 or pancreatic RNase and chemical cleavage at the $m^7G$ residue (present in many tRNA species); both methods are amenable to small-scale applications.

### *Controlled Digestion with T1 or Pancreatic RNase*

The reaction mixture (5–10 μl) contains 12–13 μg of tRNA and T1 RNase (≤5 units) or pancreatic RNase (0.01–0.015 μg) in 50 m*M*

[68] J. A. Haines, C. B. Reese, and A. R. Todd, *J. Chem. Soc. London* (Part IV), 5281 (1962).

[69] C. K. Brum and S. H. Chang, personal communication.

Tris·HC1 (pH 8.0); if mostly very large fragments, such as those produced by cleavage within the anticodon loop, are desired, the incubation mixture contains, in addition, 0.3 *M* KC1. Incubation is at 4° for 1 hr. The reaction is terminated by adding an equal volume of water and buffer-saturated phenol (10–20 μl), the aqueous layer is extracted 3–4 times more with phenol, and the pooled organic phases are back-extracted once with water (20 μl). The pooled aqueous phases are extracted six times with equal volumes of diethyl ether, and the mixture of oligonucleotides is precipitated by adding 2 volumes of ethanol and left overnight at −20°.

Dephosphorylation of the 3′-phosphate groups of the oligonucleotides thus generated is carried out on 5 μg of oligonucleotides in 25 m*M* Tris·HCl, pH 8.0, in the presence of 0.0025 unit of *E. coli* alkaline phosphatase and in a volume of 30 μl. Incubation is at 37° for 2 hr. Inactivation of phosphatase by adding NTA is as described (see above, *5′-$^{32}$P-Labeling of Oligonucleotides*).

*5′-$^{32}$P-Labeling of Large Oligonucleotides.* Conditions for this are essentially as described above under *5′-$^{32}$P-Labeling of Oligonucleotides*. Incubation mixture (10 μl) usually contained 0.5 μg of the mixture of large oligonucleotides and 50–100 μ*M* [γ-$^{32}$P]ATP at a specific activity of 500–1000 Ci/mmol. After incubation at 37° for 30 min, the mixture is lyophilized and used for two-dimensional gel electrophoresis.

*Separation and Sequencing of Large Oligonucleotides.* The mixtures of 5′-$^{32}$P-labeled oligonucleotides obtained above can be separated by two-dimensional gel electrophoresis (details above, see *General Methods*).[8,11,14] Figure 13 shows the pattern obtained from a partial pancreatic RNase digest of *Neurospora crassa* tRNA$^{Phe}$ after 5′-end group labeling and two-dimensional electrophoresis. Oligonucleotides more than 25 long migrate more slowly than the xylene cyanole marker dye in the second dimension. The 5′-$^{32}$P-labeled large oligonucleotides are recovered from the gel, and the sequence of 15–20 nucleotides from their 5′-proximal end is determined by partial digestion with nuclease P1 (see below, *Direct Sequence Analysis*).

*Chemical Cleavage of tRNA at the Site Occupied by m$^7$G*

The modified nucleoside m$^7$G has been detected in a variety of tRNAs, and has always been found at a unique position in the variable loop of most tRNAs that contain a small 5-membered variable loop (loop III). If modified nucleotide composition analysis indicates the presence of m$^7$G in a tRNA molecule, it is fortuitous for sequence analysis because the polynucleotide chain can be selectively cleaved at this residue, as

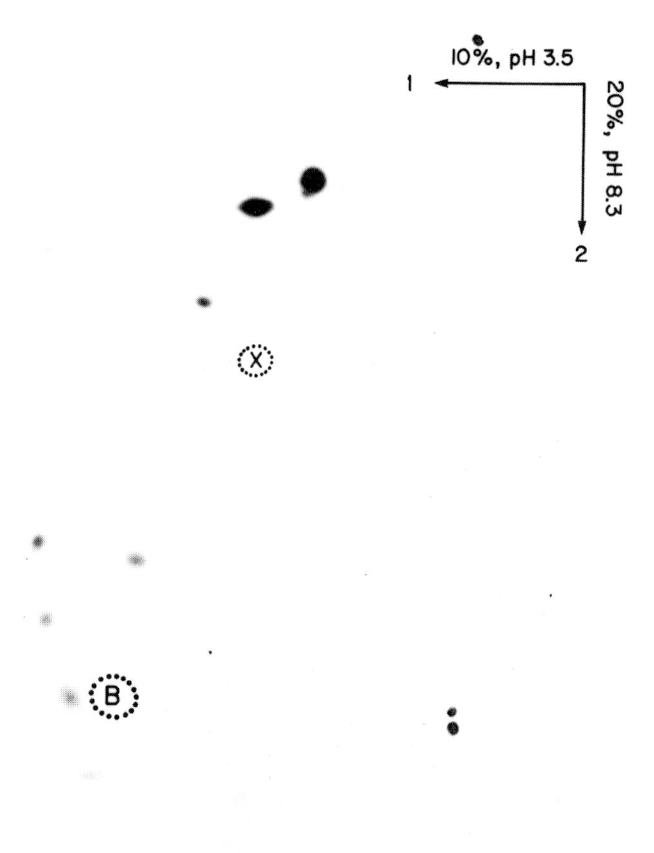


FIG. 13. Autoradiogram of two-dimensional gel separation of 5′-$^{32}$P-labeled oligonucleotides obtained upon partial pancreatic RNase digestion on *Neurospora crassa* $tRNA^{Phe}$ and subsequent labeling of oligonucleotide 5′ ends with $^{32}$P. Circled B and X indicate bromphenol blue and xylene cyanole marker dyes, respectively.

first demonstrated by Wintermeyer and Zachau.[70] Our procedure[17] is a modification of their original conditions.

The procedure involves brief treatment of the tRNA with alkali to open the imidazole ring of $m^7G$,[68] followed by treatment with aniline hydrochloride at pH 4.5 to cleave the glycosidic linkage and subsequently the phosphodiester bond adjacent to $m^7G$. The final products of the reaction are two large fragments: a 5′-terminal oligonucleotide 44–45

[70] W. Wintermeyer and H. G. Zachau, *FEBS Lett.* **11,** 160 (1970).

nucleotides long (depending upon the length of the intact tRNA) and carrying an uncharacterized remnant of the ribose moiety of the $m^7G$ residue; and a 3′-terminal oligonucleotide segment 30 nucleotides long, carrying a phosphate group at its 5′ end. These fragments are readily resolved from each other and from any intact tRNA starting material or nonspecific degradation products by polyacrylamide gel electrophoresis. Each fragment can then be analyzed[7,17] by complete digestion with T1 or pancreatic RNase and fingerprinting as described above for intact tRNA. Alternatively, if the amount of tRNA available is limited, the mixture of large oligonucleotides and any remaining intact tRNA can be labeled at their 5′ ends with $^{32}P$, as described above, and the sequence of 15–20 nucleotides from their 5′-proximal end determined by partial digestion with nuclease P1.

*Procedures.* The tRNA sample (at a concentration of 1–3 mg/ml) is extensively dialyzed against 0.1 m*M* EDTA. To 100 μl of the tRNA solution (large amounts of tRNA can be treated under identical conditions in proportionately large volumes) is added 100 μl of 100 m*M* NaOH (freshly prepared from a 10 *N* stock solution). The pH of the reaction should be >pH 12 (checked by spotting ~1 μl on pH indicator paper). After 15 min at room temperature, the pH of the solution is adjusted to ~5 by adding 2–3 μl of 5 *N* acetic acid, and 200 μl of 0.3 *M* aniline hydrochloride (pH 4.5) is added. Incubation is for 4 hr at 37° (the reaction mixture can be stored at −20° at this point). The solution is diluted 10-fold with water and is desalted by chromatography on DEAE-cellulose using 2 *M* TEAB (pH 8.0) as eluent.[41,59] After removal of TEAB by repeated evaporation, the large fragments are separated from each other and from any intact tRNA by polyacrylamide gel electrophoresis.[7,17]

## Direct Sequence Analysis of tRNAs

Recently, procedures have been developed for the analysis of RNA sequences directly by partial enzymic digestion of 5′-or 3′-end group $^{32}P$-labeled intact RNA.[20,22–24] To some degree, these methods are less precise than those discussed above for the sequence analysis of oligonucleotides; modified nucleotides, for example, are either not detected well or, if detected, cannot be identified specifically. However, once the sequence of oligonucleotides present in complete enzymic digests has been established, direct sequence analysis provides a rapid and highly sensitive method for ordering these shorter fragments of known sequence into a complete, unique, tRNA sequence.

### *5′-End-Group Labeling of tRNA with $^{32}P$*

The tRNA is first treated with phosphatase and then labeled with $^{32}$P at its 5′ end using polynucleotide kinase. The incubation mixture (10 $\mu$l) for the first step contains 0.5–2.5 $\mu$g of RNA, bacterial or calf intestinal alkaline phosphatase (0.005 unit), and 50 m*M* Tris·HCl (pH 8.0). Incubation is at 55° for 30 min. The phosphatase is inactivated by making the solution 5 m*M* in NTA and incubating for 20 min at room temperature and then for 2 min (bacterial alkaline phosphatase) or 4 min (calf intestinal alkaline phosphatase) at 100°. The phosphatase reaction may also be terminated by extraction with phenol–chloroform (1:1) (for details of extraction procedure, see below under *3′-End-Group Labeling of tRNA*). The incubation mixture (10 $\mu$l) for the second step contains 0.1–0.5 $\mu$g of the phosphatase-treated RNA, T4 polynucleotide kinase (2 units), 25–50 m*M* Tris·HCl (pH 8.0), 10 m*M* $MgCl_2$ (where necessary, additional $MgCl_2$ is added to balance the NTA concentration), 10 m*M* dithiothreitol (Calbiochem), 10 $\mu$g of BSA per milliliter, and 50–100 $\mu$*M* [$\gamma$-$^{32}$P]ATP at 500–1000 Ci/mmol (glycerol in the reaction mixture, resulting from that present in the polynucleotide kinase storage buffer,[25,26] may be adjusted to an optimal 10% concentration). Incubation is for 30 min at 37°. The reaction mixture is lyophilized and then subjected to polyacrylamide gel electrophoresis. An indication of the purity of material recovered from the polyacrylamide gel is obtained by 5′-end-group analysis (see above, *Sequence Analysis of 5′-$^{32}$P-Labeled Oligonucleotides*).

The efficiency of labeling at the 5′ end of an intact tRNA molecule is a function of three factors: (a) extent of enzymic dephosphorylation of the nonradioactive 5′-end phosphate group; (b) accessibility of the 5′-hydroxyl group to phosphorylation by polynucleotide kinase; and (c) degradative losses due to nuclease contamination during incubations. We routinely obtain yields of 25–70% end-labeled tRNA, varying mostly with tRNA species.

We have found calf intestinal alkaline phosphatase to have better activity at 55° than does bacterial alkaline phosphatase and generally prefer it for such applications as described here. However, commercial enzyme lots should be tested for nuclease activity using a known tRNA sample as control before use with valuable samples.

### *3′-End-Group Labeling of tRNA with $^{32}P$*

The tRNA is first treated with snake venom phosphodiesterase under mild conditions to remove part of the 3′-terminal C-C-A; it is then labeled with $^{32}$P at the 3′ end by using tRNA nucleotidyltransferase in the presence of [$\alpha$-$^{32}$P]ATP and nonradioactive CTP. The incubation mixture (10 $\mu$l) for the first step contains 5 $\mu$g of tRNA and 0.25 $\mu$g of snake venom phosphodiesterase in 50 m$M$ Tris·HCl (pH 8.0), 10 m$M$ $MgCl_2$; incubation is at room temperature for 10–15 min. The mixture is then extracted twice with equal volumes of phenol–chloroform (1:1). After two back-extractions of the pooled organic phases with water, the pooled aqueous phases are extracted six times with ether; the solution is left in an open tube for 15 min at 37° to drive off residual ether and is then lyophilized and redissolved in a small volume of water. For the second step, 0.5–2.5 $\mu$g of snake venom phosphodiesterase-treated tRNA is incubated (10 $\mu$l) with 1 $\mu$g of tRNA nucleotidyltransferase and 25–30 m$M$ Tris·HCl (pH 8.0), 10 m$M$ $MgCl_2$, 25–30 $\mu M$ [$\alpha$-$^{32}$P]ATP, 25–30 $\mu M$ CTP, and 8 m$M$ dithiothreitol, at 37° for 45 min. The reaction mixture is lyophilized and then subjected to polyacrylamide gel electrophoresis.

*Sequence Analysis by Partial Digestion of $^{32}$P-End-Labeled tRNAs with Nuclease P1*

*Procedures.* The use of nuclease P1 for partial digestion of end-group labeled RNA has been described above (*Sequence Analysis of 5′-$^{32}$P-Labeled Oligonucleotides*). When applied to intact 5′- or 3′-end-group $^{32}$P-labeled tRNA (or 5′-$^{32}$P-labeled large oligonucleotides), conditions are identical except that removal of aliquots from the digestion reaction at only two time points, 2 min and 5 min, is sufficient to give a good distribution of partial degradation products. After inactivation of the enzyme by heating (100°) in the presence of 5 m$M$ EDTA, the two time point aliquots are pooled for use in further analyses. One-dimensional analysis by homochromatography for extent of digestion is usually not necessary.

The pooled reaction aliquots are lyophilized, resuspended in a small volume of water, and stored at −20°; 2–3-$\mu$l portions are then used for two-dimensional homochromatography (10,000 cpm per analysis is sufficient if intensifying screens[42] are used during autoradiography). Analysis is on 20 × 40 cm plates in 10 m$M$ KOH-strength homomix. It is often useful to perform two analyses of each end-group-labeled sample, especially if the sequence is longer than 15 nucleotides. In one run, the first-dimensional electrophoresis on cellulose acetate is short (blue dye migration ≤10 cm), and the second-dimensional chromatography is as usual (blue dye migration within 3–4 cm of the wick at the top of the plate); this serves to determine the end-proximal sequence. In a second run, the cellulose acetate electrophoresis is prolonged (blue dye migration ~15 cm), and only the region of the cellulose acetate strip surrounding the intact material (i.e., region beyond the blue dye, which carries the most radioactivity as judged with a Geiger counter) is transferred to the DEAE-cellulose thin-layer plate; homochromatography on the DEAE-cellulose plate is also prolonged (blue dye migration well into wick). This latter run provides higher resolution of longer oligonucleotide intermediates.

A typical analysis of a 3′-$^{32}$P-labeled tRNA, that of *E. coli* $tRNA_{II}^{Tyr}$, is illustrated in Fig. 14. Since only the $^{32}$P-labeled products of the endonucleolytic nuclease P1 partial digest are visualized by autoradiography, the pattern of Fig. 14 represents a homologous series of successively longer oligonucleotides with a common $^{32}$P-labeled end group (in this case, [$^{32}$P]adenosine 5′-phosphate, the 3′-end group of the tRNA). The nucleotide sequences represented by the patterns can be determined by analysis of the two-dimensional mobility shifts between successive oligonucleotide spots[18,20,29,56,57] (see above, *Sequence Analysis of 5′-$^{32}$P-Labeled Oligonucleotides*), as illustrated in the schematic drawing. Sequences of over 20 nucleotides can frequently be determined from a single pattern, the actual number being limited mainly by the resolution of the two-dimensional separation.

As nuclease P1 is not entirely random in its selection of cleavage sites,[21] certain internucleotide bonds may be cut less frequently than others; this results in a final autoradiogram containing lighter and denser spots. Polypyrimidine clusters, especially oligo (C) stretches, in single-stranded conformations show up as a series of lighter spots (Fig. 7B); base-paired oligo (C) stretches are, however, cleaved more readily by nuclease P1 (Fig. 14). Similarly, the phosphodiester bond on the 3′ side of certain modified nucleosides is resistant to nuclease P1, and under the partial digestion conditions used, the corresponding oligonucleotide spot may be entirely missing from the autoradiogram (Figs. 7B and 8B). Also, modified nucleosides, even when they are not strongly inhibitory toward nuclease P1, may give rise to unusual mobility shifts between successive oligonucleotide spots in a homologous partial digestion series (see above, *Identification of Modified Nucleotides in 5′-$^{32}$P-Labeled Oligonucleotides*). The "gaps" and unusual mobility shifts encountered in two-dimensional homochromatographic patterns are rare, but when they occur are readily apparent. The sequence determination of such regions in a tRNA molecule must rely on prior knowledge of the modified nucleosides present and of the sequences of the oligonucleotides in complete T1 and pancreatic RNase digests of the tRNA and recognition of these sequences within the two-dimensional homochromatographic pattern of the end-group labeled tRNA or large (partial T1 or pancreatic RNase partial digestion product, see above) oligonucleotide. Partial digestion with nuclease P1 is thus a sequencing method particularly well suited to ordering shorter fragments of known sequence into a longer RNA sequence.

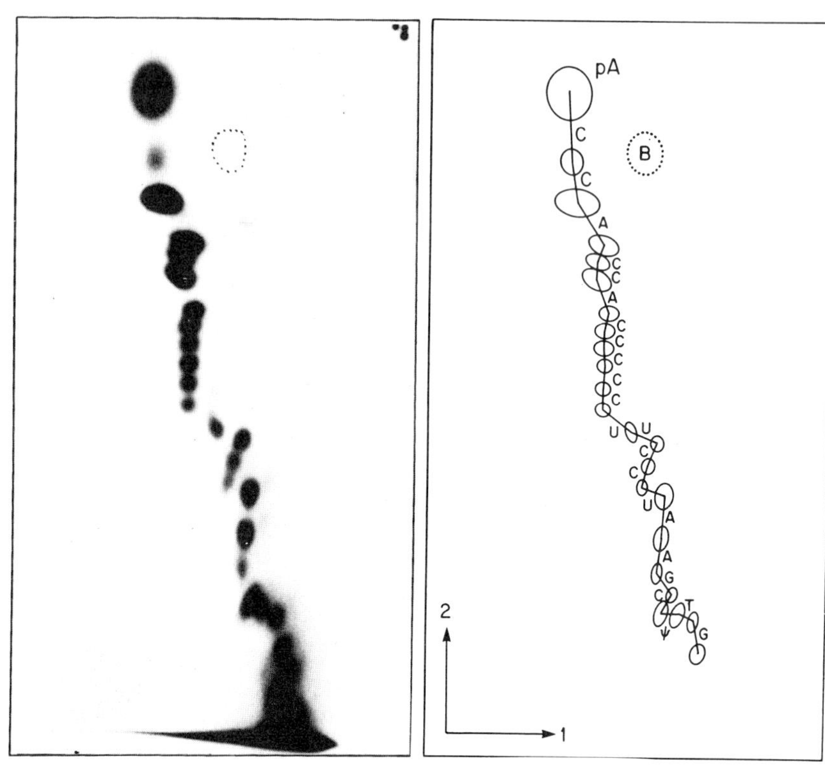


FIG. 14. Autoradiogram of a partial nuclease P1 digest of 3′-$^{32}$P-labeled *Escherichia coli* tRNA$^{Tyr}_{II}$; first dimension, electrophoresis on cellulose acetate, pH 3.5; second dimension, homochromatography in 10 m*M* KOH-strength homomix. Circled B is xylene cyanole blue dye. Note that the 3′-end group, pA, is diffuse and has moved rapidly in the second dimension. This is due to the presence of urea on the plate (from the cellulose acetate strip) during the water prerun; pA, being only slightly negatively charged at pH 3.5 (the pH upon transfer), is not adsorbed strongly onto the DEAE-cellulose. The first mobility shift, pA→pCpA, is, therefore, artifactually larger in the second dimension. Reprinted with permission from *Nucl. Acids Res.* **4**, 4091 (1977).

### *Sequence Analysis by Partial Digestion of 5′-$^{32}$P-Labeled tRNAs with Base-Specific Nucleases and Analysis by Polyacrylamide Gel Electrophoresis*

In principle, this method is similar to the Maxam and Gilbert method[38] for sequencing DNA. However, unlike the DNA sequencing method, which uses chemical treatment of $^{32}$P-end-labeled DNA to obtain partial cleavage at specific bases, enzymes of differing base specificity are used instead on 5′-$^{32}$P-labeled RNA.[22–24] The enzymes used are T1 RNase for cleavage at G residues, U2 RNase for cleavage at A residues, alkali or, in some cases, T2 RNase[23] for cleavage at all four residues, and an extracellular RNase from *B. cereus*[23,71] or, in some cases, pancreatic RNase for cleavage at pyrimidine residues.[24,72] The 5′-$^{32}$P-labeled partial digestion fragments are separated according to their size by polyacrylamide gel electrophoresis under denaturing conditions, and the relative location of G, A, and pyrimidine residues from the 5′ end of the RNA can then be determined. Methods that allow one to further distinguish among the pyrimidine residues C and U include two-dimensional polyacrylamide gel electrophoresis of partial alkali digests[23,71] or the use of an extracellular RNase from *Physarum polycephalum*, which cleaves all phosphodiester bonds except those involving CpN sequences[72–74] (H. Donis-Keller, personal communication).

*Procedure*. Except for partial digestion with the *B. cereus* nuclease, all other incubations are in 7 *M* urea, 20 m*M* sodium citrate, pH 5.0, 1 m*M* EDTA, 0.025% each xylene cyanole blue and bromphenol blue, and are at 50°. The incubation mixture (20 μl) contains 25,000–50,000 cpm of 5′-$^{32}$P-labeled tRNA and 20 μg of carrier tRNA. To ensure complete denaturation, the 5′-$^{32}$P-labeled tRNA is heated in the reaction cocktail at 100° for 90 sec and quickly chilled in ice prior to addition of enzymes. Appropriate nuclease: RNA ratios (units of RNase per microgram of RNA) for partial digestion are determined for each individual RNA preparation. These conditions are established by serial dilution of enzyme into reaction mixtures containing buffers, 7 *M* urea and 5′-$^{32}$P-labeled tRNA as described.[22] Two microliters of RNase T1 (Sankyo), 0.1 unit/μl, are added to the first of three test tubes of reaction mixture, followed by two successive 10-fold serial dilutions. RNase U2 (Sankyo), 1 unit/μl, is similarly diluted. These enzyme dilutions provide a range of digestion products from which two are usually selected for analysis.

Partial digestions with T2 RNase usually contain $10^{-2}$ and $10^{-3}$ unit of the enzyme per microgram of carrier tRNA.

Partial digestion of 5′-$^{32}$P-labeled tRNA with the pyrimidine-specific RNase from *B. cereus* is carried out using the same reaction cocktail used for T1, U2, and T2 RNase, except that the 7 *M* urea is omitted. The *B. cereus* enzyme when assayed by digestion of [$^{3}$H]poly(C) had an activity comparable to 4 units of pancreatic ribonuclease per milliliter. Partial digestions are usually carried out with 0.001–0.003 unit of enzyme

[71] E. C. Koper-Zwarthoff, R. E. Lockard, B. Alzner-DeWeerd, U. L. RajBhandary, and J. F. Bol, *Proc. Natl. Acad. Sci. U.S.A.* **74,** 5504 (1977).

[72] R. S. Brown, J. R. Rubin, D. Rhodes, H. Guilley, A. Simoncsits, and G. G. Brownlee, *Nucl. Acids Res.* **5,** 23 (1978).

[73] R. C. Gupta and K. Randerath, *Nucl. Acids Res.* **4,** 3441 (1977).

[74] R. Braun and K. Behrens, *Biochim. Biophys. Acta* **195,** 87 (1969).

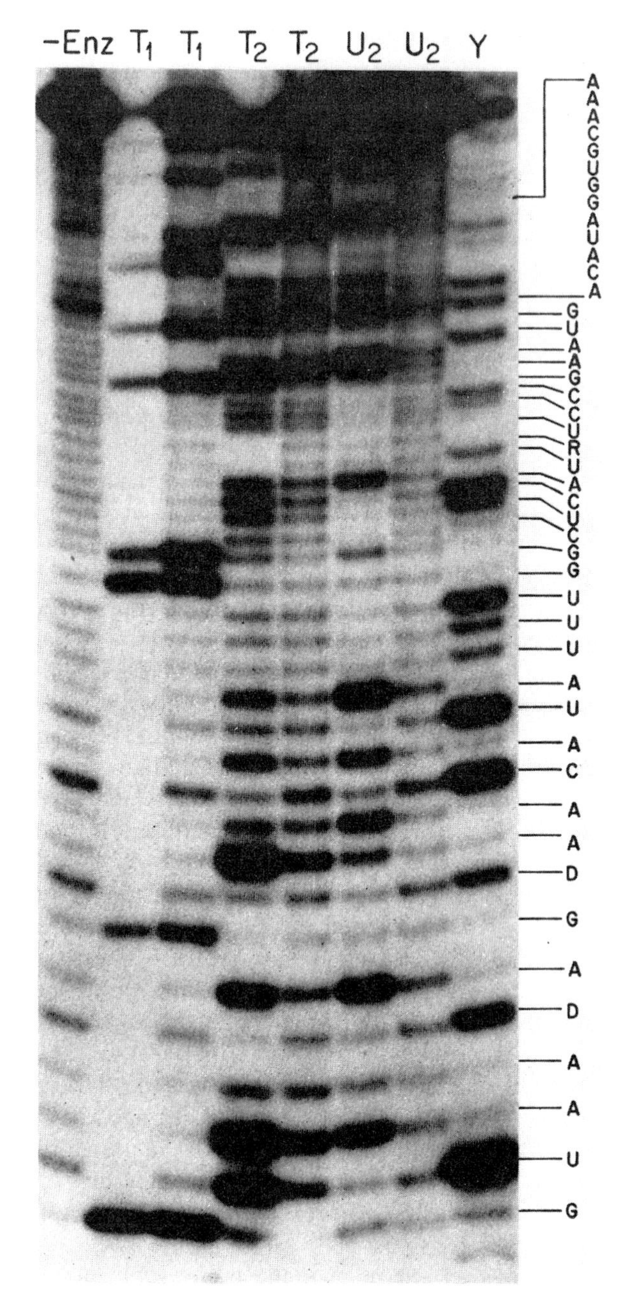


FIG. 15. Autoradiogram of partial digests of 5′-$^{32}$P-labeled *Neurospora crassa* mitochondrial initiator tRNA as analyzed by polyacrylamide gel electrophoresis [J. E. Heckman, L. I. Hecker, S. D. Schwartzbach, W. E. Barnett, B. Baumstark, and U. L. RajBhandary, *Cell* **13,** 83 (1978). Copyright © MIT.] Partial digestions contained the following amounts of enzyme per microgram of carrier tRNA and per microliter of incubation. From left to right: -Enz, no enzyme added; T1, 0.01 unit; T1, 0.001 unit; T2, 0.01 unit; T2, 0.001 unit; U2, 0.1 unit; U2, 0.01 unit; and Y (enzyme from *B. cereus*), 0.001 unit. In the sequence shown on the right, R is m$^1$G.

per microgram of carrier tRNA and at 50°. One half of the reaction mixture is removed after 10 min of incubation, and the reaction is terminated by addition of an equal volume of 10 *M* urea. After 20 min the remainder of the reaction is also terminated by addition of urea, as above. The two portions of the incubation mixture are combined and used for gel electrophoresis.

Figure 15 shows an example of the application of the above direct sequencing procedure to 5′-$^{32}P$-labeled *N. crassa* mitochondrial initiator methionine tRNA. The bands in the T1 RNase tracks provide spacings between the G residues, and those in the U2 RNase tracks provide spacings between the A residues. The bands in *B. cereus* nuclease track (Y) indicate the location of pyrimidine residues in this tRNA. Since partial digestion with the nuclease from *Physarum polycephalum* was not included in these studies, the results of this experiment alone do not allow one to distinguish among the pyrimidines. However, since the total sequence of all the oligonucleotides present in complete T1 RNase and pancreatic RNase digests of this tRNA were known, the results shown in Fig. 15 provided sufficient information to align most of these oligonucleotides into a unique sequence for this tRNA.[10]

Although this direct sequencing method is applicable to sequencing mRNAs and most other RNAs, use of this method alone may not be adequate for establishing the total sequence of tRNAs, since tRNAs contain many modified nucleotides—some of which are quite resistant to partial digestion with nucleases and others, such as 2′-*O*-methylated nucleotides, are almost totally resistant toward digestion by either nucleases or alkali. In addition, because of their stable secondary structure, stem regions of certain tRNAs may not be accessible to partial enzymic digestions even in 7 *M* urea and at 50°. Consequently, it will still be necessary to have in hand a knowledge of oligonucleotide sequences present in complete RNase digests of the tRNA. The usefulness of the direct sequencing methods described here for sequencing tRNA lies more in that they can provide most of the overlap information necessary for ordering these oligonucleotides into a unique sequence within a relatively short period of time and require less than a microgram of the tRNA.

## Acknowledgments

We are grateful to our colleagues Mehmet Simsek, Joyce Heckman, James Ziegenmeyer, Barbara Baumstark, Birgit Alzner-DeWeerd, Simon Chang, Raymond Lockard, and Jan Hack for participation in the development of methodology detailed in this paper and for allowing us to quote their unpublished results. This work was supported by Grants GM 17151 from NIH and NP-114 from the American Cancer Society. M. S. and A. M. G. were supported by NIH predoctoral fellowships.

# [15] A New Method for Attachment of Fluorescent Probes to tRNA [1]

*By* SCOTT A. REINES and LADONNE H. SCHULMAN

Fluorescent probes have been attached to tRNAs by modification of specific minor bases,[2-7] by coupling to the periodate-oxidized 3′ terminus,[8-11] through pyrophosphate linkage to the 5′ terminus,[12] by reaction with the primary amino group of aminoacyl-tRNA,[13,14] by replacement of the 3′-terminal adenosine with formycin,[15-17] and by modification of guanosine residues.[18] In the present report, we describe a new method for attachment of fluorescent dyes to cytidine residues in tRNA.

## Principle

Cytidine and uridine residues in single-stranded regions of nucleic acids are readily modified by addition of sodium bisulfite to the 5,6 double bond of the pyrimidine base.[19] Cytidine–bisulfite adducts undergo deamination by reaction with water and are converted to $N^4$-substituted

[1] This research was supported by grants from the National Institutes of Health (GM 16995) and the American Cancer Society (NP-19). L. H. S. is recipient of an American Cancer Society Faculty Research Award (FRA 129).
[2] W. Wintermeyer and H. G. Zachau, this series, Vol. 29, p. 667.
[3] C. H. Yang and D. Söll, *J. Biochem.* **73,** 1243 (1973).
[4] C. H. Yang and D. Söll, *Biochemistry* **13,** 3615 (1974).
[5] C. H. Yang and D. Söll, *Proc. Natl. Acad. Sci. U.S.A.* **71,** 2838 (1974).
[6] A. Pinguod, R. Kownatzki, and G. Maass, *Nucl. Acids Res.* **4,** 327 (1977).
[7] P. W. Schiller and A. N. Schechter, *Nucl. Acids Res.* **4,** 2161 (1977).
[8] J. E. Churchich, *Biochim. Biophys. Acta* **75,** 274 (1963).
[9] D. B. Millar and R. F. Steiner, *Biochemistry* **9,** 2289 (1966).
[10] K. Beardsley and C. R. Cantor, *Proc. Natl. Acad. Sci. U.S.A.* **65,** 39 (1970).
[11] S. A. Reines and C. R. Cantor, *Nucl. Acids Res.* **1,** 767 (1974).
[12] C. H. Yang and D. Söll, *Arch. Biochem. Biophys.* **155,** 70 (1973).
[13] D. C. Lynch and P. R. Schimmel, *Biochemistry* **13,** 1841 (1974).
[14] A. E. Johnson, R. H. Fairclough, and C. R. Cantor, *in* "Nucleic Acid–Protein Recognition" (H. J. Vogel, ed.), p. 469. Academic Press, New York, 1977.
[15] D. C. Ward, E. Reich, and L. Stryer, *J. Biol. Chem.* **244,** 1228 (1969).
[16] A. Maelicke, M. Sprinzl, F. von der Haar, T. A. Khwaja, and F. Cramer, *Eur. J. Biochem.* **43,** 617 (1974).
[17] S. M. Coutts, D. Riesner, R. Römer, C. R. Rabl, and G. Maass, *Biophys. Chem.* **3,** 275 (1975).
[18] L. M. Fink, S. Nishimura, and I. B. Weinstein, *Biochemistry* **9,** 496 (1970).
[19] For a review of bisulfite reactions with nucleic acids, see H. Hayatsu, *Prog. Nucl. Acid Res. Mol. Biol.* **16,** 75 (1976).



ISBN 0-12-504180-2

cytidine derivatives by transamination with an appropriate amine. We have found that cytidine undergoes a rapid reaction with the bifunctional amine carbohydrazide in the presence of bisulfite, leading to formation of a 4-carbohydrazidocytidine derivative.[20] This intermediate is reactive with a variety of amine-specific reagents. The procedure described below is used to attach the intensely fluorescent fluorescein moiety to tRNA by the scheme outlined in Fig. 1.

## Materials

*Escherichia coli* $tRNA^{fMet}$, purified as described before[21] to a specific activity of 1.9 nmol/$A_{260}$ unit
Yeast $tRNA^{Phe}$, specific activity 0.95 nmol/$A_{260}$ unit, from Boehringer Mannheim
Crude *E. coli* K12 tRNA, from General Biochemicals
Poly(C), from Miles Laboratories
Fluorescein isothiocyanate (96%) from Aldrich Chemical Co., used without further purification
Carbohydrazide, from Aldrich Chemical Co.
Sodium metabisulfite, grade I, from Sigma Chemical Co.
Sodium sulfite, from Fisher Scientific Co.
Sodium sulfite, $^{35}S$-labeled, under nitrogen, 50–200 mCi/mmol, from New England Nuclear Corp.
Dimethyl sulfoxide (DMSO), spectro grade, from Mallinckrodt

## Procedures

### *Modification of tRNAs and Poly(C) with Carbohydrazide in the Presence of Sodium Bisulfite*

A solution of 2 *M* sodium bisulfite, pH 6.0, 1 *M* carbohydrazide, and 10 m*M* $MgCl_2$ is prepared by dissolving 0.63 g of $Na_2SO_3$, 1.43 g of $Na_2S_2O_5$, and 0.90 g of carbohydrazide in 10 ml of 10 m*M* $MgCl_2$. An ethanol precipitate of RNA is dissolved in this solution to give a final concentration of 20 $A_{260}$/ml. The reaction mixture is incubated at 25° for a given amount of time and the reaction is essentially stopped by addition of 10 volumes of water. The sample is dialyzed overnight at 4° vs 1000 volumes of 0.15 *M* NaCl, 10 m*M* Tris·HCl, pH 7.0, and then for 3 hr at 4° vs the same volume of 50 m*M* NaCl, 10 m*M* Tris·HCl, pH 7.0. The sample is evaporated at room temperature to a concentration of 20 $A_{260}$/ml and precipitated by addition of 2 volumes of 95% ethanol.

[20] S. A. Reines and L. H. Schulman, in preparation.
[21] L. H. Schulman, *J. Mol. Biol.* **58,** 117 (1971).

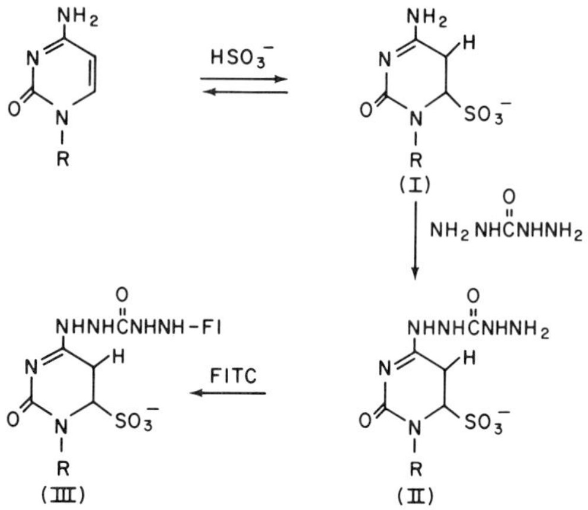

FIG. 1. Sequence of reactions leading to covalent attachment of fluorescein to cytosine derivatives in the presence of bisulfite and carbohydrazide.

A similar procedure is used for modification of poly(C), except that the carbohydrate concentration is reduced to 0.5 *M*. Modification of RNAs with [$^{35}$S]bisulfite is carried out as described above using [$^{35}$S]$Na_2SO_3$ instead of unlabeled sodium sulfite.

*Determination of the Yield of Carbohydrazide/Bisulfite Adduct II in Poly(C)*

Cytidine-bisulfite adducts (I) are unstable and rapidly revert to free cytidine following removal of excess bisulfite. The carbohydrazide-modified adduct (II) is stable for several days at 4°, pH 7, in the absence of free bisulfite. The yield of carbohydrazide/bisulfite adducts can therefore be determined by incorporation of radioactivity into poly(C) from [$^{35}$S]bisulfite in the presence of carbohydrazide after removal of excess reagents (Fig. 2). This value gives the number of groups in the polymer that can potentially be labeled with dye. There is little or no deamination of cytidine residues under the reaction conditions used.

Uridine–bisulfite adducts are stable at pH 6 and only slowly revert to uridine at neutral pH under the conditions described above. Since many tRNAs contain one or more exposed uridine residues in looped-out regions of the structure, the incorporation of $^{35}$S into tRNAs reflects the amount of uridine–bisulfite adduct formation plus the yield of adduct (II). Uridine adducts in tRNAs can be reversed after dye labeling (see Remarks).

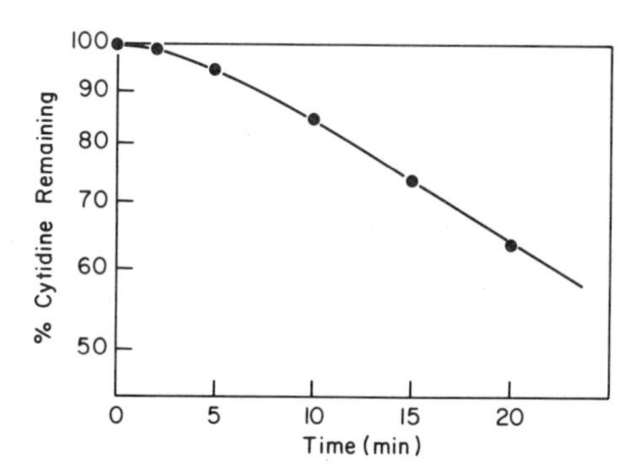


FIG. 2. Rate of modification of cytidine residues in poly(C) in the presence of 2 *M* [$^{35}$S]bisulfite, pH 6.0, and 0.5 *M* carbohydrazide at 25°. Formation of adduct (II) was determined by incorporation of $^{35}$S into poly(C) after removal of excess reagents as described in the text. The short lag period corresponds to the time required for formation of an equilibrium concentration of adduct (I).

### *Labeling of Carbohydrazide/Bisulfite-Modified tRNA and Poly(C) with FITC*[22]

Conditions for quantitative dye-labeling have been determined using carbohydrazide/bisulfite-modified poly(C) by correlating $^{35}$S incorporation with the yield of covalently attached fluorescein.

The ethanol precipitate of carbohydrazide/bisulfite-modified RNA is dissolved in 0.2 *M* Tris·HCl, pH 7.0. FITC is dissolved in DMSO just before use to give a concentration of 10 mg/ml. Equal volumes of the RNA and FITC solutions are mixed to give a final reaction mixture containing 20 $A_{260}$ per milliliter of RNA and 5 mg/ml of FITC in 50% DMSO, 0.1 *M* Tris pH 7.0. The solution is incubated in the dark at 37° for 2 hr, during which time the pH of the solution drops from 7.0 to 5.5 owing to hydrolysis of free FITC. A larger excess of dye should not be used, since the pH may drop below 5 and little or no labeling will occur. One-tenth volume of 4 *M* NaCl and 3 volumes of 95% ethanol are added to the reaction mixture, the solution is chilled at −20° for 10 min, and the precipitate is collected by centrifugation. The supernatant is discarded, and the RNA is reprecipitated four times from a solution containing 0.1 *M* Tris·HCl, pH 7.0, 0.5 *M* NaCl, and 5 m*M* $MgCl_2$ by addition of 3 volumes of ethanol. The precipitation procedure removes free FITC from the reaction mixture, as indicated by negligible absorption of the

[22] Abbreviations: FITC, fluorescein isothiocyanate; Fl-tRNA, fluorescein-labeled tRNA; Fl, fluoresceinthiocarbamyl-; DMSO, dimethyl sulfoxide.

final supernatant solution at 495 nm. The free dye is not completely removed by exhaustive dialysis.

Labeling of carbohydrazide/bisulfite-modified RNA is carried out in 50% DMSO in order to drive the reaction to completion within 2 hr by the addition of a large excess of FITC. The solubility of the dye is dependent on the final concentration of both DMSO and buffer in the reaction mixture. Labeling can also be carried out in 0.1 *M* Tris·HCl, pH 7.0 containing 10% DMSO. Under these conditions, the maximum concentration of FITC that can be used is 2 mg/ml and the rate of labeling is substantially reduced (Fig. 3). The reaction fails to go to completion because of hydrolysis of FITC during the incubation. In order to obtain quantitative labeling at low DMSO concentrations, the modified RNA is incubated in the dark at 37° for 6 hr, precipitated and treated with fresh FITC as before. After three 6-hr incubations at 37° in 0.1 *M* Tris·HCl, pH 7.0, containing 10% DMSO, 2 mg of FITC per milliliter, the labeling is complete (Fig. 3). If desired, labeling can be carried out at a lower pH using 0.5 *M* sodium acetate, pH 6.0, containing 50% DMSO and 5 mg of FITC per milliliter. Under these conditions the pH of the reaction is constant during the incubation and labeling is complete within 3 hr (Fig. 3).

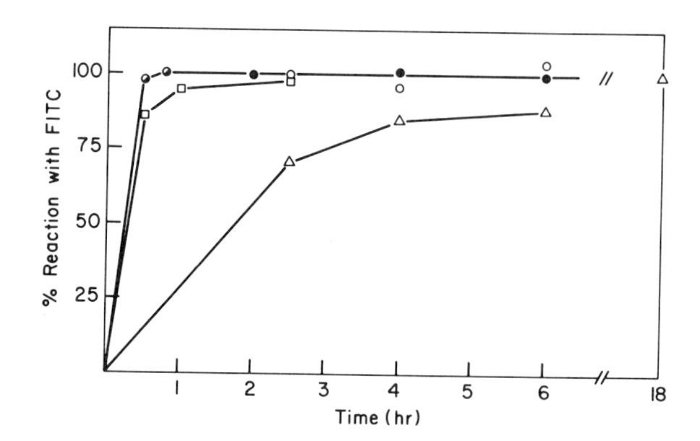


FIG. 3. Rate of reaction of fluorescein isothiocyanate (FITC) with carbohydrazide/bisulfite-modified RNAs. Poly(C) containing 0.1 mol of adduct (II) per mole of CMP: ○——○, in 0.1 *M* Tris·HCl, pH 7.0, 50% dimethyl sulfoxide (DMSO), 5 mg FITC/ml; □——□, in 0.5 *M* sodium acetate, pH 6.0, 50% DMSO, 5 mg FITC/ml; △——△, in 0.1 *M* Tris·HCl, pH 7.0, 10% DMSO, 2 mg of FITC/ml. *Escherichia coli* $tRNA^{fMet}$ containing 1.2 mol of adduct (II) per mole of tRNA: ●——●, in 0.1 *M* Tris·HCl, pH 7.0, 50% DMSO, 5 mg of FITC/ml.

*Calculation of Moles of Dye per Mole of RNA*

The absorption of RNA-bound fluorescein is lower than that of free dye and is dependent on the number of molecules of dye per molecule of RNA. The exact amount of fluorescein covalently bound to a given RNA is determined after hydrolysis of the sample to mononucleotides and release of free dye by treatment with 0.3 $N$ KOH at 37° for 18 hr. The extinction coefficient of fluorescein varies significantly with pH and dye concentration. After adjusting the solution to pH 7 and 0.1-0.5 $A_{495}$/ml, an $\epsilon_{495}$ of $5.03 \times 10^4\ M^{-1}\ \text{cm}^{-1}$ is used to calculate the yield of free dye.[23] The absorption of fluorescein at 260 nm is subtracted from the total $A_{260}$ in order to obtain the absorption of the hydrolyzed RNA alone.

At low levels of dye labeling, e.g., approximately 1 mol of dye per mole of tRNA, the average extinction coefficient of tRNA-bound fluorescein at 495 nm determined by the above procedure is $4.25 \times 10^4\ M^{-1}\ \text{cm}^{-1}$ in 0.1 $M$ Tris·HCl, pH 7.0, 5 m$M$ $MgCl_2$. An extinction coefficient of $5 \times 10^5\ M^{-1}\ \text{cm}^{-1}$ is used for intact tRNA at 260 nm in this buffer. Under these conditions the amount of dye per tRNA is calculated from the equation:

$$\text{Moles of fluorescein per mole of tRNA} = (A_{495}^{\text{Fl-tRNA}}/\epsilon_{495}^{\text{Fl-tRNA}})/(A_{260}^{\text{tRNA}}/\epsilon_{260}^{\text{tRNA}}) = (A_{495}^{\text{Fl-tRNA}}/A_{260}^{\text{tRNA}}) \times 11.8$$

where $A_{260}^{\text{tRNA}} = A_{260}^{\text{Fl-tRNA}} - 0.37 \times A_{495}^{\text{Fl-tRNA}}$. The ratio of experimentally observed $A_{495}/A_{260}$ is linear with concentration of Fl-tRNA up to 0.5 $A_{495}$/ml, and absorbance measurements are made after adjusting the concentration of tRNA to 0.1-0.5 $A_{495}$/ml.

*Yield of Dye per Mole of tRNA*

The yield of dye per mole of tRNA using the procedures described here depends on the rate of modification of a given tRNA with carbohydrazide/bisulfite. This rate is determined by the number and accessibility of cytidine residues in exposed regions of the structure. The yields of dye for several tRNAs under similar reaction conditions are compared in Table I. *E. coli* $\text{tRNA}^{\text{fMet}}$ contains six potentially reactive cytidine residues[24] that are modified at different rates. Yeast $\text{tRNA}^{\text{Phe}}$ contains only two exposed cytidines in the 3′-terminal CCA sequence[25] and re-

[23] R. P. Tengerdy and C.-A. Chang, *Anal. Biochem.* **16,** 377 (1966).
[24] J. P. Goddard and L. H. Schulman, *J. Biol. Chem.* **247,** 3864 (1972).
[25] D. Rhodes, *J. Mol. Biol.* **94,** 449 (1975).

TABLE I
EXTENT OF DYE LABELING FOLLOWING MODIFICATION OF tRNAs WITH CARBOHYDRAZIDE AND BISULFITE

| Sample | Reaction time[a] (min) | Fluorescein/mole tRNA |
|---|---|---|
| *Escherichia coli* $tRNA^{fMet}$ | 10 | 0.99 |
| | 20 | 1.74 |
| Yeast $tRNA^{Phe}$ | 30 | 1.57 |
| Crude *E. coli* tRNA | 10 | 1.02 |
| | 20 | 1.62 |

[a] Time of reaction at 25° in 2 *M* sodium bisulfite, pH 6.0, 1 *M* carbohydrazide, 10 m*M* $MgCl_2$.

quires a longer reaction time to achieve the same extent of dye labeling. An average *E. coli* tRNA is labeled with one dye per mole of tRNA following 10 min of reaction with carbohydrazide/bisulfite as described above.

## Optical Properties of Fluorescein-Labeled RNA

The absorption and fluorescence spectra of *E.coli* $tRNA^{fMet}$ labeled with 1.5 mol of fluorescein per mole of tRNA are illustrated in Fig. 4. An absorption maximum of 495 nm and fluorescence excitation and emission maxima of 490 nm and 525 nm have been observed at all dye concentrations examined. Increasing the amount of dye per mole of RNA results in a significant decrease in the extinction coefficient at 495 nm and in an increase in the $A_{470}:A_{490}$ ratio of Fl-RNA. In addition, a dramatic decrease in fluorescence intensity due to fluorescence quenching from dye-dye interactions is observed, as illustrated in Fig. 5 for fluorescein-labeled poly(C).

## Effect of Modifications on Amino Acid Acceptor Activity of tRNAs

The effect of the modification procedures described here on the amino acid acceptor activity of tRNAs depends on the sensitivity of the cognate aminoacyl-tRNA synthetases to structural alterations of exposed cytidine residues in the tRNAs. Such modifications are known to reduce the biological activity of *E. coli* $tRNA^{fMet}$.[26,27] We have found that carbohy-

[26] L. H. Schulman and J. P. Goddard, *J. Biol. Chem.* **248**, 1341 (1973).
[27] L. H. Schulman and H. Pelka, *Biochemistry* **16**, 4256 (1977).

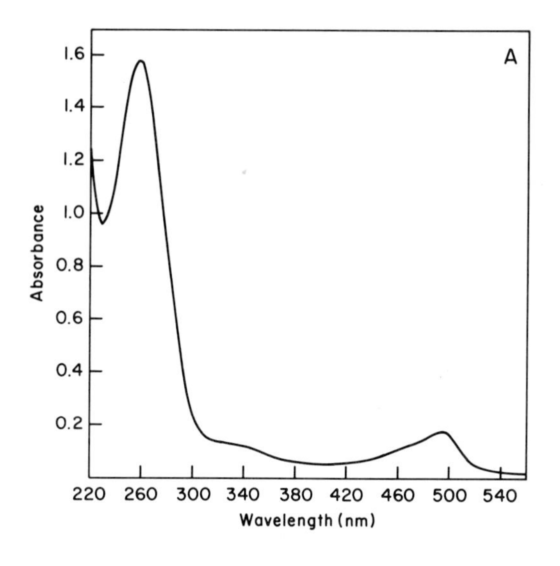


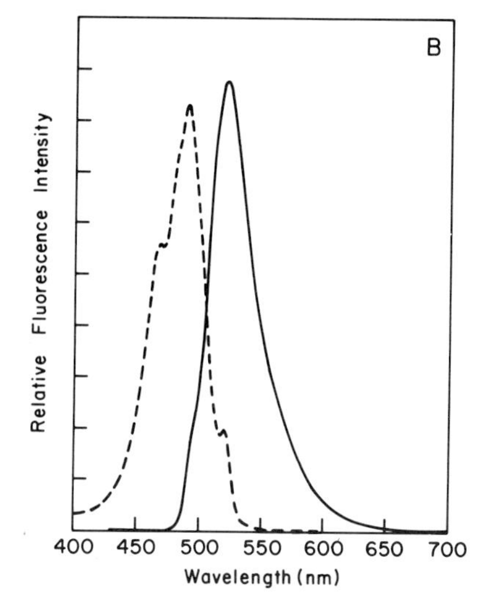


FIG. 4. Optical properties of fluorescein-labeled *Escherichia coli* $tRNA^{fMet}$. (A) Absorption spectrum of *E. coli* $tRNA^{fMet}$ containing 1.5 mol of fluorescein per mole of tRNA in 0.1 *M* Tris·HCl, pH 7.0, 5 m*M* $MgCl_2$. (B) - - -, Technical fluorescence excitation spectrum (emission at 525 nm); ——, emission spectrum (excitation at 490 nm) of the same sample.

drazide/bisulfite modification of an average of one cytidine per molecule of this tRNA reduces methionine acceptor activity to about 60% of that exhibited by the unmodified tRNA. Attachment of fluorescein has little

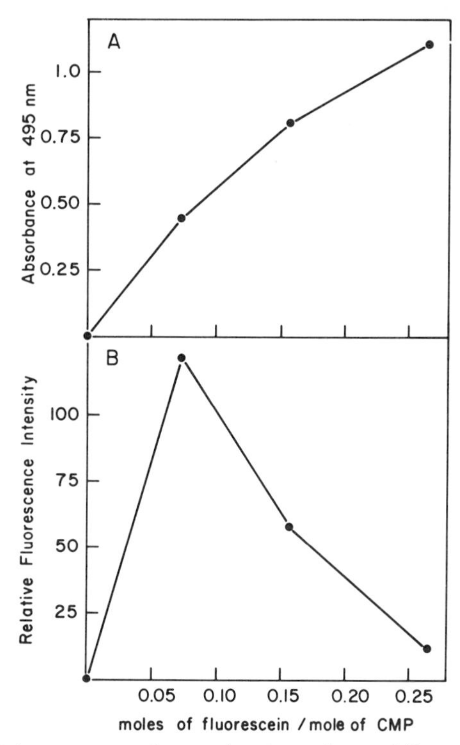


FIG. 5. Effect of dye concentration on the absorption and fluorescence properties of fluorescein-labeled poly(C). (A) Absorbance at 495 nm. (B) Fluorescence emission at 525 nm (excitation at 490 nm). The solvent is 0.1 $M$ Tris·HCl, pH 7.0, 5 m$M$ $MgCl_2$. Fluorescence units are arbitrary.

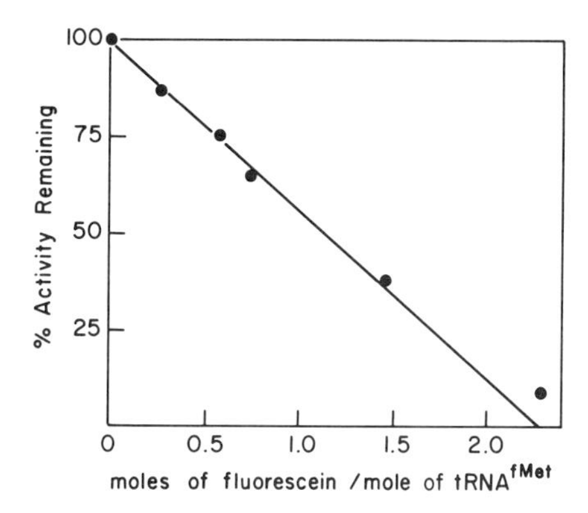


FIG. 6. Effect of the extent of fluorescein labeling of *Escherichia coli* $tRNA^{fMet}$ on methionine acceptor activity. Methionine acceptance was measured as described by L. H. Schulman, *J. Mol. Biol.* **58,** 117 (1971).

or no further effect on the activity, and methionine acceptance is a linear function of the number of dye molecules per molecule of tRNA (Fig. 6).

## Remarks

Carbohydrazide/bisulfite adduct (II) is less stable than the fluorescein-modified derivative (III). The presence of the unblocked primary amino group in adduct (II) allows this cytidine derivative to undergo intramolecular rearrangements that prevent subsequent reaction with FITC. These side reactions occur rapidly in the presence of acid. After incubation at 4° for 3 hr in acetate buffer, pH 3.5, only 10% of the reactive side chains in carbohydrazide/bisulfite-modified $tRNA^{fMet}$ remain available for reaction with FITC. A rapid loss of FITC-reactive groups is also observed above pH 9; however, adduct (II) is relatively stable at neutral pH. Carbohydrazide/bisulfite-modified poly(C) shows a 25% loss of reactivity with FITC following incubation at 25° in 10 m*M* Tris·HCl, pH 7.0, for 1 month. Slow rearrangement of adduct (II) also occurs when carbohydrazide/bisulfite-modified RNAs are stored as precipitates at −20°. It is therefore recommended that FITC labeling be carried out on freshly prepared samples of modified RNAs. It should also be noted that carbohydrazide/bisulfite-modified tRNAs are potentially capable of forming covalent cross-links to proteins by attack of the $\epsilon$-$NH_2$ groups of lysine residues at C-4 of the modified pyrimidine base, with displacement of the carbohydrazide side chain.

The fluorescein moiety of dye-labeled RNAs is stable during incubation of the modified RNAs at 37° for 6 hr at pH 5-8. A 10% loss of fluorescein is observed following incubation of dye-labeled poly(C) at 25° in 10 m*M* Tris·HCl, pH 7.0, for 2 months. Fluorescein-labeled $tRNA^{fMet}$ shows no loss of dye after storage as a precipitate for 6 months at −20°, and samples can probably be stored indefinitely in this manner.

The procedures used for carbohydrazide/bisulfite modification of tRNAs lead to formation of uridine–bisulfite adducts in regions of the structure that contain exposed uridine residues. Bisulfite addition to uridine occurs 3–10 times more slowly than formation of adduct (II). Uridine–bisulfite adducts in fluorescein-labeled tRNAs can be completely reversed to unmodified uridine by incubation of the dye-labeled tRNAs in 0.1 *M* Tris·HCl, pH 9.0, at 37° for 8 hr. These conditions result in a 20% release of fluorescein and a 40% release of free bisulfite from adduct (III).

Other types of amine-specific reagents can be used to attach a variety of fluorescent probes to carbohydrazide/bisulfite-modified tRNAs. For example, the carbohydrazide side chain of adduct (II) reacts rapidly with

*N*-hydroxysuccinimide esters (see this volume [16]), and we have covalently attached the naphthoxy moiety to tRNA using the activated ester of naphthoxyacetic acid.[28] It is expected that the *N*-hydroxysuccinimide esters of dansylglycine and *N*-methylanthranilic acid[7] can be used to attach these fluorescent probes to the modified tRNAs in a similar manner.

[28] L. H. Schulman, unpublished results.

# [16] Attachment of Cross-Linking Reagents to tRNA for Protein Affinity Labeling Studies[1]

*By* Asok K. Sarkar and LaDonne H. Schulman

A variety of protein affinity-labeling reagents have been attached to tRNAs by covalent linkage to the amino acid moiety of aminoacyl-tRNAs. These peptidyl-tRNA and aminoacyl-tRNA analogs have been used to probe the structure of tRNA binding sites on ribosomes[2] and to cross-link tRNAs to aminoacyl-tRNA synthetases.[3-10] A photoreactive group has also been attached to the periodate-oxidized 3′ terminus of tRNA.[11] Few methods are presently available for attachment of affinity labels to other regions of tRNA structure. Photolabile azido derivatives have been coupled to the 4-thiouridine residue in several *Escherichia coli*

[1] This research was supported by grants from the National Institutes of Health (GM 16995) and the American Cancer Society (NP-19). L. H. S. is recipient of an American Cancer Society Faculty Research Award (FRA 129).

[2] For a recent review, see A. E. Johnson, R. H. Fairclough, and C. R. Cantor, in "Nucleic Acid-Protein Recognition" (H. J. Vogel, ed.), p. 469. Academic Press, New York, 1977.

[3] C. J. Bruton and B. S. Hartley, *J. Mol. Biol.* **52,** 165 (1970).

[4] D. V. Santi and S. O. Cunnion, this series, Vol. 29, p. 695.

[5] D. V. Santi, W. Marchant, and M. Yarus, *Biochem. Biophys. Res. Commun.* **51,** 370 (1973).

[6] O. I. Lavrik and L. Z. Khutoryanskaya, *FEBS Lett.* **39,** 287 (1974).

[7] D. V. Santi and S. O. Cunnion, *Biochemistry* **13,** 481 (1974).

[8] P. Bartmann, T. Hanke, B. Hammer-Raber, and E. Holler, *Biochem. Biophys. Res. Commun.* **60,** 743 (1974).

[9] I. I. Gorshkova and O. I. Lavrik, *FEBS Lett.* **52,** 135 (1975).

[10] V. Z. Akhverdyan, L. L. Kisselev, D. G. Knorre, O. I. Lavrik, and G. A. Nevinsky, *J. Mol. Biol.* **113,** 475 (1977).

[11] R. Wetzel and D. Söll, *Nucl. Acids Res.* **4,** 1681 (1977).





ISBN 0-12-504180-2

tRNAs,[11-15] and a chemical affinity labeling group has been attached to modified cytidine residues in the 3′-terminal CCA sequence of yeast $tRNA^{Phe}$.[16] Described herein is a method for coupling a variety of protein affinity-labeling reagents to internal sites in tRNAs.

## Principle

Cytidine residues in exposed regions of tRNA structure are chemically modified in the presence of carbohydrazide and sodium bisulfite to give 4-carbohydrazidocytidine derivatives.[17] The primary amino group of the carbohydrazide side chain of the modified cytidine residues reacts with *N*-hydroxysuccinimide esters under mild conditions to yield the corresponding amides (Fig. 1). The procedures described below are used to couple several types of protein affinity labeling groups to tRNAs by this general method.

## Materials

*Escherichia coli* $tRNA_1^{fMet}$, purified as described before[18] to a specific activity of 1.9 nmol/$A_{260}$ unit.
Crude *E. coli* K12 tRNA, from General Biochemicals
Poly (C), from Miles Laboratories
Bromoacetic acid, from Aldrich Chemical Co.
Succinic acid, from Mallinckrodt
Dicyclohexylcarbodiimide, from Eastman Kodak Co.
*N*-Hydroxysuccinimide, from Eastman Kodak Co.
Dithiobis (succinimidyl propionate), from Pierce Chemical Co.
Carbohydrazide, from Aldrich Chemical Co.
Sodium metabisulfite, grade I, from Sigma Chemical Co.
Sodium sulfite, from Fisher Scientific Co.
Fluorescein isothiocyanate, 96%, from Aldrich Chemical Co., used without further purification
Dimethyl sulfoxide, spectro grade, from Mallinckrodt
*N,N*-Dimethylformamide, spectro grade, from Aldrich Chemical Co.

[12] I. Schwartz and J. Ofengand, *Proc. Natl. Acad. Sci. U.S.A.* **71,** 3951 (1974).
[13] I. Schwartz, E. Gorden, and J. Ofengand, *Biochemistry* **14,** 2907 (1975).
[14] V. G. Budker, D. G. Knorre, V. V. Kravchenko, O. I. Lavrik, G. A. Nevinsky, and N. M. Teplova, *FEBS Lett.* **49,** 159 (1974).
[15] I. I. Gorshkova, D. G. Knorre, O. I. Lavrik, and G. A. Nevinsky, *Nucl. Acids Res.* **3,** 1577 (1976).
[16] H. Sternbach, M. Sprinzl, J. B. Hobbs, and F. Cramer, *Eur. J. Biochem.* **67,** 215 (1976).
[17] S. A. Reines and L. H. Schulman, this volume [15].
[18] L. H. Schulman and H. Pelka, *J. Biol. Chem.* **252,** 814 (1977).

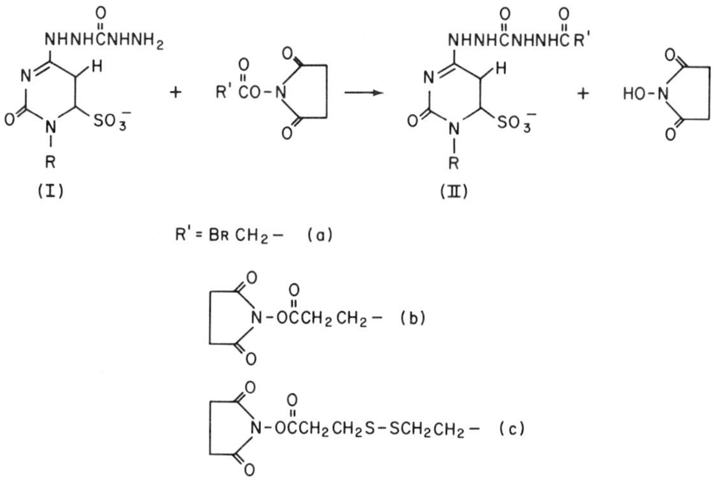

Fig. 1. Reaction leading to covalent attachment of affinity labeling reagents to carbohydrazide/bisulfite-modified cytosine derivatives.

Acetone, from Fisher Scientific Co., dried over calcium chloride and distilled
1,4-Dioxane, from Fisher Scientific Co., dried over calcium chloride and distilled
Isopropyl alcohol, from Fisher Scientific Co., dried over calcium chloride and distilled

## Procedures

### *N-Hydroxysuccinimide Ester of Bromoacetic Acid*

The *N*-hydroxysuccinimide ester of bromoacetic acid is prepared by the procedure of Santi and Cunnion.[7] The crystalline ester has a melting point of 115°–117°, gives the correct combustion analysis (C, H, N), migrates as a single spot on thin-layer chromatography (TLC) after visualization of the developed chromatogram with group-specific spray reagents,[19] and shows the infrared absorption bands ($\nu_{max}$ 1790 and 1815 $cm^{-1}$) characteristic of *N*-hydroxysuccinimide esters.[20] The product is

[19] S. Rappoport and Y. Lapidot, this series, Vol. 29, p. 685. *N*-Hydroxysuccinimide esters are visualized (violet spot) by spraying the chromatogram with a mixture of 14% $NH_2OH \cdot HCl$ in water (20 ml) and 14% NaOH (8.5 ml), followed by spraying with a solution of 5% $FeCl_3$ in 1.2 *N* HCl after 2 min.
[20] G. Fölsch, *Acta Chem. Scand.* **24,** 1115 (1969).

stored in a vacuum desiccator over $P_2O_5$ at $-20°$. Under these conditions, the ester undergoes noticeable decomposition to bromoacetic acid and *N*-hydroxysuccinimide in 2 weeks. These impurities can be removed by washing the solid with several small portions of dry isopropyl alcohol on a filter. The ester is quite unstable in aqueous buffers, being completely hydrolyzed in 1 hr at room temperature in 0.25 *M* sodium acetate, pH 6.0, containing 50% DMF.[21]

*Di-N-hydroxysuccinimide Ester of Succinic Acid*

Dicyclohexylcarbodiimide (4.12 g, 20 mmol) in 50 ml of dry 1,4-dioxane is added with stirring to a mixture of dry succinic acid (1.18 g, 10 mmol) and *N*-hydroxysuccinimide (2.3 g, 20 mmol) in 150 ml of dry dioxane. The solution is stirred for 6 hr at room temperature in a flask protected from moisture and then allowed to stand overnight at room temperature. The precipitated dicyclohexylurea is removed by filtration, and the filtrate is evaporated *in vacuo* at room temperature. The residue is recrystallized from dry acetone, giving a 50% yield of disuccinimidyl succinate (DSS),[21] m.p. 304°–305° (dec.). This procedure differs from the usual method[22] for preparation of *N*-hydroxysuccinimide esters in that a large excess of solvent is used in the reaction mixture in order to keep the desired product in solution, while allowing the dicyclohexylurea to precipitate. The purified diester has the correct combustion analysis (C, H, N), shows infrared absorption bands at 1745, 1790, and 1820 $cm^{-1}$, and gives a single spot ($R_f$ 0.70) after chromatography on silica gel plates in chloroform–methanol (80:20) and visualization with group-specific spray reagents.[19] The diester shows no decomposition when stored in a vacuum desiccator over $P_2O_5$ at $-20°$ for 6 months.

*Di-N-hydroxysuccinimide Ester of 3,3'-Dithiodipropionic Acid*

Dithiobis(succinimidyl propionate) (DTSP), prepared by the procedure of Lomant and Fairbanks,[23] is commercially available from Pierce Chemical Co. The commercial product has a melting point of 132°–134° and gives a single spot ($R_f$ 0.75) with *N*-hydroxysuccinimide ester-specific spray reagents[19] when chromatographed as described above. It contains a small amount of *N*-hydroxysuccinimide; however, this does not inter-

[21] Abbreviations: DSS, disuccinimidyl succinate; DTSP, dithiobis(succinimidyl propionate); DMSO, dimethyl sulfoxide; DMF, *N,N*-dimethylformamide; FITC, fluorescein isothiocyanate.

[22] G. W. Anderson, J. E. Zimmerman, and F. M. Callahan, *J. Am. Chem. Soc.* **86,** 1839 (1964).

[23] A. J. Lomant and G. Fairbanks, *J. Mol. Biol.* **104,** 243 (1976).

fere with the reaction of the diester with adduct (I) and the compound can be used without further purification. It is stable for at least 1 month when stored as described above.

*Preparation of Carbohydrazide/Bisulfite-Modified RNAs*

Poly(C) and tRNAs are modified with carbohydrazide and sodium bisulfite using the procedures described in this volume [15].

*Reaction of Carbohydrazide/Bisulfite-Modified RNAs with the N-Hydroxysuccinimide Ester of Bromoacetic Acid*

Succinimidyl bromoacetate is dissolved in DMF just before use. An ethanol precipitate of freshly prepared carbohydrazide/bisulfite-modified RNA is dissolved in 0.17 *M* sodium acetate, pH 6.0, and mixed with the ester solution to give a final reaction mixture containing 20 $A_{260}$/ml of RNA and a 200-fold molar excess of ester over adduct (I) in 0.1 *M* sodium acetate, pH 6.0, 40% DMF. The solution is incubated at room temperature for 1 hr. The RNA is precipitated by addition of two volumes of 95% ethanol. Excess ester is removed from the pellet by reprecipitating the RNA twice from 0.1 *M* sodium acetate, pH 6.0.

Attachment of α-bromoacetamide groups to RNAs decreases their solubility in aqueous buffers. Extensively modified samples, e.g., 15 nmol of adduct (IIa)/$A_{260}$, are redissolved with difficulty in 0.1 *M* Tris·HCl, pH 7.0, following ethanol precipitation from the reaction mixture.

The *N*-hydroxysuccinimide ester of bromoacetic acid is potentially capable of coupling to adduct (I) by alkylation of the carbohydrazide side chain; however, this reaction is one-tenth as fast as formation of amide (IIa) and represents a minor side reaction under the conditions described above.

*Reaction of Carbohydrazide/Bisulfite-Modified RNAs with DSS and DTSP*

The ester is dissolved in DMSO just before use. An ethanol precipitate of freshly prepared carbohydrazide/bisulfite-modified RNA is dissolved in 0.25 *M* sodium acetate, pH 6.0, and mixed with the ester solution to give a final reaction mixture containing 0.1 *M* sodium acetate, pH 6.0, and 60% DMSO. The solution is incubated at room temperature for a given amount of time and the RNA isolated as described above.

It is desirable to keep the concentration of RNA is the reaction mixture sufficiently high so that the product can be rapidly isolated by the ethanol precipitation procedure. It is also desirable to use a large excess of ester in order to complete the reaction within a short period of time. The maximum concentration of DSS that can be used is 2 mg/ml of reaction mixture, owing to its sparing solubility in aqueous buffers. Under these conditions, a 200-fold excess of ester is present when the concentration of carbohydrazide/bisulfite-modified RNA is adjusted to 30 nmol of adduct (I) per milliliter. DTSP is soluble in the reaction mixture at a maximum concentration of 4 mg/ml. Under these conditions, a 200-fold excess of DTSP is present when the RNA concentration is adjusted to 50 nmol of adduct (I) per milliliter.

RNAs containing high concentrations of adduct (I) can be essentially quantitatively labeled using per milliliter 2 mg of DSS or 4 mg of DTSP by reducing the concentration of RNA in the reaction mixture to maintain a 200-fold excess of ester. Dilute solutions of RNA require brief dialysis after incubation with the ester, and concentration of the sample by evaporation prior to ethanol precipitation of the product. Alternatively, a lower concentration of ester and a longer incubation time can be used to increase the amount of reaction of extensively modified RNAs at lower ratios of ester to adduct (I). It should be noted, however, that experimental conditions that increase the amount of time required for isolation of the RNAs after ester modification may lead to partial hydrolysis of the affinity labeling group on the RNA.

*Yield of Adduct (II)*

The extent of formation of adduct II is determined by measuring the decrease in labeling of carbohydrazide/bisulfite-modified RNAs with FITC.[17] Reaction of adduct (I) with *N*-hydroxysuccinimide esters blocks the primary amino group of the carbohydrazide side chain and prevents its reaction with the dye.

An additional procedure can be used to determine the yield of adduct (II) in tRNAs after treatment with DSS or DTSP. These adducts contain a reactive *N*-hydroxysuccinimide ester group. This group reacts rapidly with free carbohydrazide, yielding a derivative analogous to adduct (I), but containing an extended side chain. Such derivatives react with FITC in the same manner as adduct (I). The extent of formation of adduct (II) can therefore be determined by comparing the amount of FITC that can be covalently attached to the modified tRNA before and after treatment with free carbohydrazide. The freshly modified tRNA (20 $A_{260}$/ml) is incubated in 1 *M* carbohydrazide-HCl, pH 7.0, for 1 hr at room temperature; then the tRNA is precipitated by addition of two volumes of 95% ethanol. Traces of free carbohydrazide are removed from the pellet by reprecipitation from 0.2 *M* NaCl, 0.1 *M* Tris·HCl, pH 7.0. The amount of FITC that can be covalently attached to the product is determined by the procedure described in this volume [15]. This method gives values for the yield of adduct (II) that are within 10% of those obtained by directly measuring the amount of adduct (I) rendered resistant to FITC labeling after reaction with DSS or DTSP. The method is not suitable for measuring the yield of adduct (II) in poly(C) or in tRNAs containing modified cytidines in close proximity to each other in the structure, since partial intramolecular cross-linking of the side chains of (IIb) and (IIc) occurs by reaction of one molecule of carbohydrazide with two molecules of adduct (II) within the same polynucleotide. These cross-linked derivatives no longer contain a primary amino group and are therefore unreactive with FITC.

## Rate of Reaction of Carbohydrazide/Bisulfite-Modified RNAs with *N*-Hydroxysuccinimide Esters

Succinimidyl bromoacetate is the most reactive of the three *N*-hydroxysuccinimide esters used here. Quantitative ester labeling of carbohydrazide/bisulfite-modified poly(C) or tRNA is achieved within 1 hr under the conditions described above.

Carbohydrazide/bisulfite adducts (I) in single-stranded poly(C) also react rapidly with DSS and DTSP. The modification is almost complete within 15 min at room temperature using a 140-fold excess of ester over adduct (I) (Fig. 2). A 35-fold excess of ester results in 93% reaction in 30 min at room temperature; however, at lower ratios of ester: adduct (I), the reaction is much slower and fails to go to completion within 4 hr (Fig.

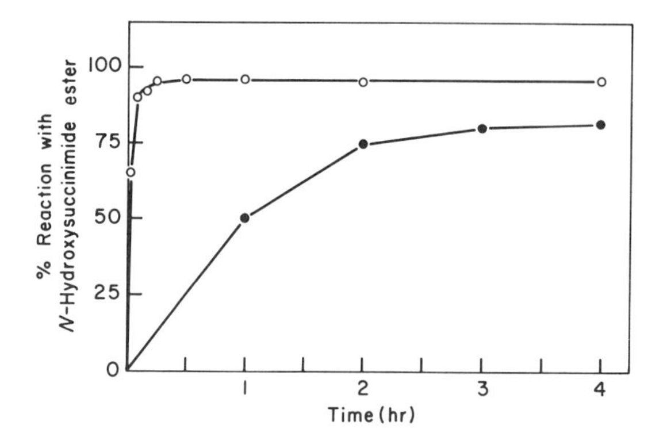


FIG. 2. Rate of reaction of carbohydrazide/bisulfite-modified poly (C) with disuccinimidyl succinate (DSS) at 25°. Poly (C) containing 0.1 mol of adduct (I) per mole of CMP in 0.1 *M* sodium acetate, pH 6.0, and 60% dimethyl sulfoxide ○—○, 140-fold molar excess of DSS/adduct (I); ●—●, 10-fold molar excess of DSS/adduct (I).

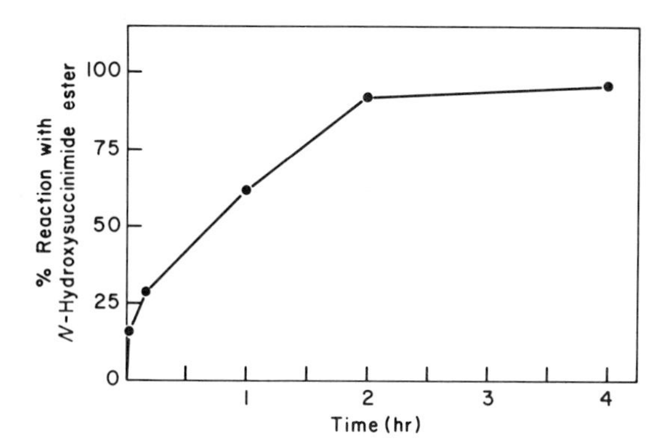


FIG. 3. Rate of reaction of carbohydrazide/bisulfite-modified *Escherichia coli* $tRNA_1^{fMet}$ with disuccinimidyl succinate (DSS) at 25°. The reaction mixture contains 20 $A_{260}$ per milliliter of *E. coli* $tRNA_1^{fMet}$ modified with 1.5 mol of adduct (I) per mole of tRNA and 2 mg of DSS per milliliter [100-fold molar excess of ester/adduct (I)] in 0.1 *M* sodium acetate, pH 6.0, and 60% dimethyl sulfoxide.

2). This is probably due to the slow hydrolysis of the ester during the incubation.

Reaction of adduct (I) with DSS or DTSP occurs more slowly with carbohydrazide/bisulfite-modified tRNAs than with modified poly(C). The kinetics of labeling of *E. coli* $tRNA_1^{fMet}$ containing 1.5 mol of adduct (I) per mole of tRNA in the presence of a 100-fold excess of DSS show an initial fast reaction, followed by a slower reaction (Fig. 3). This suggests that adducts in different environments of the tRNA structure react at different rates with the ester. Higher concentrations of ester drive the reaction to completion in a shorter time. Table I summarizes the extent of modification of adduct (I) in *E. coli* $tRNA_1^{fMet}$ in 1 hr at room temperature using different concentrations of *N*-hydroxysuccinimide esters.

## Stability of Adducts (IIb) and (IIc)

The hydroxysuccinimide ester group of adduct (II) derived by treatment of carbohydrazide/bisulfite-modified RNAs with DSS or DTSP is susceptible to hydrolysis by water. Such hydrolysis prevents subsequent use of the modified RNA in protein affinity-labeling experiments. These RNA derivatives should therefore be prepared just before use in cross-linking experiments.

The rate of hydrolysis of adducts (IIb) and (IIc) in different solvents can be followed by measuring the amount of FITC that can be covalently

TABLE I
EXTENT OF REACTION OF CARBOHYDRAZIDE/BISULFITE-MODIFIED *Escherichia coli* $tRNA_1^{fMet}$ WITH *N*-HYDROXYSUCCINIMIDE ESTERS

| *N*-Hydroxysuccinimide ester | $[Ester]_0/[adduct (I)]_0$ [a] | % Reaction[b] |
|---|---|---|
| Bromoacetate | 100 | 98 |
| | 200 | 100 |
| Succinate | 100 | 65 |
| | 200 | 97 |
| DTSP[c] | 100 | 78 |
| | 200 | 94 |

[a] The initial concentration of adduct (I) in *E. coli* $tRNA_1^{fMet}$ is 1.5 mol per mole of tRNA.

[b] Percentage of reaction in 1 hr at room temperature under the conditions described in the text.

[c] Dithiobis(succinimidyl propionate).

attached to the modified RNA following incubation of the sample with free carbohydrazide, as described in the preceding section. Figure 4 shows the rate of hydrolysis of adduct (IIb) in crude *E. coli* tRNA during incubation of the modified tRNA in 5 m*M* $MgCl_2$, 20 m*M* imidazole buffer, pH 7.0, at 25°. The half-life of the hydrolysis is 7 hr in this buffer and about 9 hr in 0.1 *M* sodium acetate, pH 6.0, at 25°. Adduct (IIc) is somewhat more unstable and is hydrolyzed with a half-life of 5 hr at 25°

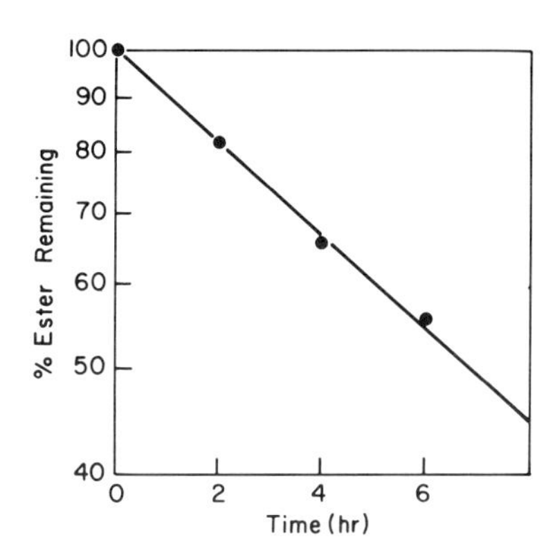


FIG. 4. Rate of hydrolysis of adduct (IIb). Crude *Escherichia coli* tRNA containing 0.9 mol of adduct (IIb) per mole of tRNA was incubated at 25° in 5 m*M* $MgCl_2$, 20 m*M* imidazole buffer, pH 7.0, for various times, and the amount of unhydrolyzed *N*-hydroxysuccinimide ester was determined as described in the text.

in 5 m*M* $MgCl_2$, 20 m*M* imidazole buffer, pH 7.0. Hydrolysis of the reactive ester groups is expected to occur significantly faster at higher pH's or at elevated temperatures. In addition, buffers containing primary amino groups must not be used as solvents for RNAs modified with these cross-linking groups. Modified tRNAs containing adducts (IIb) and (IIc) can be stored as ethanol precipitates at −20° for at least 1 week with no detectable destruction of the affinity-labeling group.

## Remarks

Polynucleotides and tRNAs are unreactive with *N*-hydroxysuccinimide esters,[24,25] except for a few species of tRNA that contain the minor base 3-(3-amino-3-carboxypropyl)uridine.[26–30] Modification of tRNAs with carbohydrazide and sodium bisulfite introduces reactive primary amino groups at the sites of exposed cytidine residues in the structure, providing a general method for attachment of cross-linking reagents and other probes to tRNAs by reaction with a variety of *N*-hydroxysuccinimide esters. The bromoacetamide group of adduct (IIa) is capable of reacting with cysteine and several other amino acids in proteins.[31–34] In addition, covalent coupling of bromoacetyl derivatives of aminoacyl-tRNAs to ribosomal 23 S RNA has been observed.[2,35,36] This suggests that tRNAs containing adduct (IIa) might undergo intramolecular alkylation under certain conditions. The *N*-hydroxysuccinimide ester groups of adducts (IIb) and (IIc) are expected to react rapidly with the $\epsilon$-$NH_2$ groups of protein lysine residues. Free DTSP has been shown to react completely with accessible lysine residues in hemoglobin within 2 min at 0° and pH 7.[23] Adduct (IIc) is also capable of forming cross-links to

[24] P. Schofield, B. M. Hoffman, and A. Rich, *Biochemistry* **9,** 2525 (1970).
[25] N. de Groot, Y. Lapidot, A. Panet, and Y. Wolman, *Biochem. Biophys. Res. Commun.* **25,** 17 (1966).
[26] S. Friedman, *Biochemistry* **11,** 3435 (1972).
[27] S. Friedman, *Nature (London), New Biol.* **244,** 18 (1973).
[28] U. Nauheimer and C. Hedgcoth, *Arch. Biochem. Biophys.* **160,** 631 (1974).
[29] M. Caron and M. Dugas, *Nucl. Acids Res.* **3,** 19 (1976).
[30] P. W. Schiller and A. N. Schechter, *Nucl. Acids Res.* **4,** 2161 (1977).
[31] M. E. Kirtley and D. E. Koshland, *J. Biol. Chem.* **245,** 276 (1970).
[32] P. Cuatrecasas, M. Wilchek, and C. B. Anfinsen, *J. Biol. Chem.* **244,** 4316 (1969).
[33] L. T. Lan and R. P. Carty, *Biochem. Biophys. Res. Commun.* **48,** 585 (1972).
[34] G. M. Hass and H. Neurath, *Biochemistry* **10,** 3541 (1971).
[35] J. B. Breitmeyer and H. F. Noller, *J. Mol. Biol.* **101,** 297 (1976).
[36] M. Pellegrini, H. Oen, and C. R. Cantor, *Proc. Natl. Acad. Sci. U.S.A.* **69,** 837 (1972).

proteins by exchange of the disulfide linkage with reactive sulfhydryl groups of cysteine residues.[23] Both types of cross-links can be quantitatively cleaved by reduction with dithioerythritol,[23] allowing reversible coupling of tRNAs to proteins.

In addition to the cross-linking groups described here, it should be possible to attach other types of affinity labeling reagents to tRNAs by reaction of carbohydrazide/bisulfite-modified tRNAs with appropriate *N*-hydroxysuccinimide esters. For example, reactive esters of photolabile compounds[37–40] could be coupled to tRNAs in this manner. The length of the chemically reactive side chain of adduct (II) could also be varied by using *N*-hydroxysuccinimide esters of other aliphatic carboxylic acids having different chain lengths.

It is desirable to use radioactive probes in affinity labeling experiments in order to facilitate detection of covalent cross-links and isolation of labeled peptides. Procedures have been described for preparation of radioactively labeled succinimidyl bromoacetate[16] and DTSP.[23] $^{14}$C-Labeled succinic acid is also commercially available for preparation of radioactive DSS.

[37] A. S. Girshovich, E. S. Bochkareva, V. M. Kramarov, and Y. A. Ovchinnikov, *FEBS Lett.* **45,** 213 (1974).

[38] N. Hsiung, S. A. Reines, and C. R. Cantor, *J. Mol. Biol.* **88,** 841 (1974).

[39] H. Hsiung and C. R. Cantor, *Nucl. Acids Res.* **1,** 1753 (1974).

[40] A. Barta, E. Kuechler, C. Branlant, J. Sriwidada, A. Krol, and J. P. Ebel, *FEBS Lett.* **56,** 170 (1975).

# [17] Enzymatic Synthesis of Trinucleoside Diphosphates of Known Sequence

*By* PHILIP LEDER

## Preparation

*Principle.* Synthesis involves the polynucleotide phosphorylase-catalyzed addition of nucleoside 5′-diphosphate to dinucleoside monophosphate under conditions favoring the production of trinucleoside diphosphate [Eq. (1)].[1]

$$\text{XpY} + \text{ppN} \xrightarrow[\text{Phosphorylase}]{\text{Polynucleotide}} \text{XpYpN} + \text{P}_i \tag{1}$$

The products of the reaction are isolated by column chromatography.

[1] P. Leder, M. F. Singer, and R. L. C. Brimacombe, *Biochemistry* **4,** 1961 (1965).





ISBN 0-12-504180-2

*Reagents*

1. Mixture I contains: 0.5 *M* Tris-HCl buffer, pH 9.0; 0.5 m*M* ethylenediaminetetraacetic acid; 0.1 *M* magnesium chloride
2. Nucleoside 5′-diphosphate: 0.1 *M* uridine 5-diphosphate; 0.1 *M* guanosine 5′-diphosphate; 0.05 *M* cytosine 5′-diphosphate; 0.04 *M* adenosine 5′-diphosphate.
3. 10 m*M* dinucleoside monophosphate (Miles Chemical Co., Elkhart, Indiana).[2]
4. Polynucleotide phosphorylase prepared according to Singer *et al.*;[3] about 50 units/mg.[4]
5. Alkaline phosphatase (*E. coli*), chromatographically purified, about 30 units/mg[6] (Worthington Biochemical Corp., Freehold, New Jersey).
6. 2 *M* triethylammonium bicarbonate (TEAB) pH 7.6, stock solution.[5]
7. Chromatographic solvents: I, N-propanol-water-ammonia, 55:35:10, v/v/v/; II, 40 g ammonium sulfate/100 ml 0.1 *M* sodium phosphate, pH 7.0.

*Procedure.* Preparative syntheses are conveniently carried out in 5-ml reaction mixtures which contain 0.5 ml mixture I; 0.5 ml nucleoside 5′-diphosphate solution; 3.0 ml dinucleoside monophosphate solution; 20 μg polynucleotide phosphorylase; $H_2O$ to volume. If complementary bases are involved, the reaction mixture is heated to 70° for 5 minutes and cooled to 37° prior to the addition of the enzyme. The reaction mixture is incubated for 12 hours or overnight. The reaction is stopped by the addition of 25 units of alkaline phosphatase and incubation is continued at 37° for 90 minutes. The reaction mixture is then heated to 90° for 5 minutes to reduce enzymatic activity and any precipitated protein is removed by low-speed centrifugation.

*Purification.* The reaction mixture is diluted 100-fold with 0.01 *M* TEAB and passed over a 2 × 20 cm DEAE-cellulose column previously equilibrated with this solution. Nucleoside is eluted from the column with a 200 ml 0.01 *M* TEAB wash. Oligonucleotides are eluted from the col-

[2] Dinucleoside monophosphates as purchased often contain impurities and prior to use should be characterized as previously noted. If impurities are detected, dinucleoside monophosphates may be purified in a manner identical to that detailed above for trinucleoside diphosphate; however, the linear TEAB gradient need only extend to 0.3 *M*.

[3] M. F. Singer and J. K. Guss, *J. Biol. Chem.* **237**, 182 (1962); M. F. Singer and B. M. O'Brien, *ibid.* **238**, 328 (1963).

[4] Units are determined by the phosphorolysis assay of Singer and Guss.[3] One unit is equivalent to the formation of 1 μmole of ADP from poly A in 15 minutes.

[5] TEAB (2 moles) is suspended in $H_2O$ at about 4° and volume is adjusted to one liter. The mixture is kept in an ice-water bath while $CO_2$ is bubbled through until all the TEAB goes into solution and the pH reaches approximately 7.5.

[6] A. Garen and C. Leventhal, *Biochim. Biophys. Acta* **38**, 470 (1960). Commercial enzyme, under conditions outlined above, does not exhibit significant phosphodiesterase activity.

umn with a 1000-ml linear TEAB gradient between 0.01 *M* and 0.35 *M* TEAB, both freshly diluted from stock solution. Fractions of 15 ml are collected and their optical density at 260 m$\mu$ is plotted to obtain an elution profile. The first major peak corresponds to unreacted dinucleoside monophosphate and each subsequent peak corresponds to oligonucleotides containing one additional nucleoside monophosphate residue. Generally trinucleoside diphosphate is the major product, corresponding to a utilization of from 5 to 40% of the dinucleoside monophosphate. Fractions are concentrated and TEAB is removed by repeated lyphilization or evaporation to dryness on a rotary evaporator. Oligonucleotides purified in this way are generally contaminated with residual phosphatase activity. Further purification can be achieved by applying approximately 500 $A_{260}$ units oligonucleotide per sheet Whatman 3MM paper and developing in descending solvent I for 20 hours. The dinucleoside monophosphate should be applied as a standard. The trinucleoside monophosphate will be the immediately trailing ultraviolet absorbing band. The band may be cut out and eluted from the paper with $H_2O$. These compounds are stable at room temperature for 3–6 months when lyphilized and may also be stored frozen in solution for 3–6 months.

*Characterization.* Three $A_{260}$ units of each oligonucleotide should migrate as a single spot on Whatman No. 1 paper when developed in solvent I for 24 hours or solvent II for 12–24 hours. $R_f$ values vary with respect to composition of the trinucleoside diphosphate.[1] Detailed base composition and sequence determination may be carried out according to Leder *et al.*[1]

# [18] Use of Purified Isoacceptor tRNAs for the Study of Codon-Anticodon Recognition *in Vitro* with Sequenced Natural Messenger RNA

*By* EMANUEL GOLDMAN and G. WESLEY HATFIELD

The broad outline of the genetic code was originally elucidated from two basic types of experiments: ribosome binding studies, in which defined trinucleotides or polynucleotides stimulated the binding to ribosomes of tRNA acylated with individual amino acids[1]; and protein syn-

[1] M. Nirenberg, T. Caskey, R. Marshall, R. Brimacombe, D. Kellogg, B. Doctor, D. Hatfield, J. Levin, F. Rottman, S. Pestka, M. Wilcox, and F. Anderson, *Cold Spring Harbor Symp. Quant. Biol.* **31,** 11 (1966).



ISBN 0-12-504180-2

thesis *in vitro* using synthetic polyribonucleotides as messenger RNA (mRNA) molecules.[2] Deviations from results predicted by Watson–Crick base pairing were largely accommodated by the wobble hypothesis, which provided certain rules for non-Watson–Crick pairing in the third position between anitcodon and codon.[3]

Attempts to apply these techniques to the specific study of individual purified tRNA isoaccepting species have been less satisfying. Ribosome binding studies have had to contend with very high backgrounds of nonspecific binding of tRNA to ribosomes, with only a small percent of increase in counts taken as a positive result. Protein synthesis *in vitro* with synthetic polyribonucleotides has to be performed at high magnesium concentrations, which eliminate the need for some protein synthesis factors and are detrimental to translational fidelity. Further, these techniques cannot determine whether there are effects of mRNA secondary/tertiary structure, or nearby primary sequences (reading context[4]), upon codon–anticodon interaction.

The large number of individual tRNA molecules which have been sequenced, combined with a few recently elaborated sequences of certain natural mRNA molecules and their corresponding proteins, now places us on the threshold of high-resolution *in vitro* analysis of codon–anticodon interaction under conditions closer to those *in vivo* than have previously been possible.

## Experimental Design

Individual, purified tRNA isoaccepting species are acylated with a radioactive amino acid and then added to an *in vitro* protein-synthesizing system, which is directed by a sequenced natural mRNA. The protein products are isolated, digested with protease (usually trypsin), and subjected to peptide mapping techniques. With the help of markers, specific peptides are purified and identified, and the amount of radioactivity transferred from different tRNA isoaccepting species is measured. Thus, it is possible to determine the efficiency of a codon–anticodon interaction at a known position in the message, by means of a specific isotope incorporated into a given amino acid residue at a known position in the polypeptide.

[2] H. G. Khorana, H. Büchi, H. Ghosh, N. Gupta, T. M. Jacob, H. Kössel, R. Morgan, S. A. Narang, E. Ohtsuka, and R. D. Wells, *Cold Spring Harbor Symp. Quant. Biol.* **31,** 39 (1966).

[3] F. H. C. Crick, *J. Mol. Biol.* **19,** 548 (1966).

[4] M. M. Fluck, W. Salser, and R. H. Epstein, *Mol. Gen. Genet.* **151,** 137 (1977).

tRNA-Dependent Systems *versus* Use of tRNAs as Tracers

The validity of this approach depends entirely upon the presumption that the radioactive amino acid that winds up in the final polypeptide was in fact transferred only from the individual tRNA isoaccepting species aminoacylated at the outset. It is possible, for example, that the amino acid attached to one tRNA species might be exchanged with a different tRNA isoaccepting species of the same family present in the extract, via the homologous aminoacyl-tRNA synthetase enzyme.

One approach that circumvents this objection is to prepare extracts from cells that are temperature sensitive in one of the aminoacyl-tRNA synthetases. Such extracts would have the property of being dependent on added aminoacylated tRNA for protein synthesis to occur, and no exchange of amino acid should then be possible. Such experiments have been elegantly performed by Mitra *et al.* for the valine family of tRNA isoaccepting species in *Escherichia coli.*[5] It was concluded on the basis of these experiments that valine codons were read operationally as a two-letter code by valyl-$tRNA^{Val}$ isoaccepting species *in vitro*.

However, subsequent work has indicated that these tRNA-dependent conditions foster misreading of the genetic code, since it was observed that such systems synthesize peptides downstream of codons that should not have been read in the absence of the aminoacylated tRNA upon which the system was dependent.[6]

The other approach to meeting the objection of amino acid exchange between tRNA isoaccepting species is simply to add a large excess of nonradioactive free amino acid to the (wild type) extract, and to use the tRNA isoaccepting species acylated with the homologous radioactive amino acid as a tracer. The following control experiment verifies the validity of this approach: to a standard *E. coli* protein synthesis reaction, acylated [$^3$H]Leu-$tRNA_3^{Leu}$ was added; one tube contained an excess of unacylated $tRNA_1^{Leu}$, and another, a vast excess of bulk *E. coli* tRNA. No exogenous mRNA was added. After incubation at 37° for 15 min, all RNA was extracted and both samples were chromatographed on reverse-phase chromatography 5 (RPC-5),[7] along with bulk [$^{14}$C]Leu-$tRNA^{Leu}$ marker. The results of this experiment demonstrate the virtual absence of transfer of [$^3$H]leucine from [$^3$H]Leu-$tRNA_3^{Leu}$ to $tRNA_1^{Leu}$ (Fig. 1A) or any other $tRNA^{Leu}$ isoaccepting species (Fig. 1B) under these experimental conditions.

[5] S. K. Mitra, F. Lustig, B. Åkesson, U. Lagerkvist, and L. Strid, *J. Biol. Chem.* **252,** 471 (1977).

[6] W. M. Holmes, G. W. Hatfield, and E. Goldman, *J. Biol. Chem.,* **253,** 3482 (1978).

[7] R. L. Pearson, J. F. Weiss, and A. D. Kelmers, *Biochim. Biophys. Acta* 228, 770 (1971).

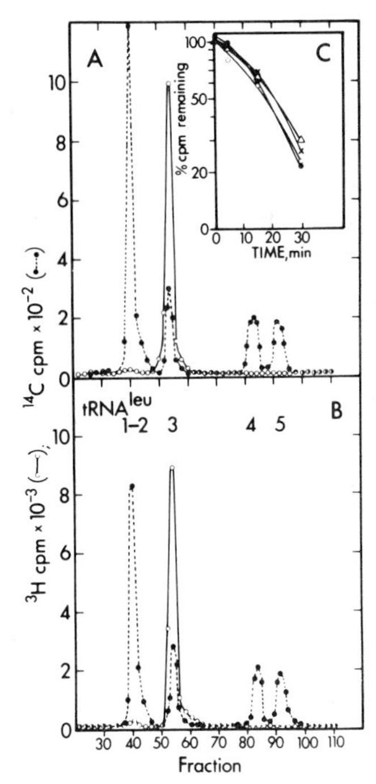


FIG. 1. Absence of exchange of [$^3$H]leucine between tRNA$^{Leu}$ isoaccepting species. (A) RPC-5 chromatogram of [$^3$H]Leu-tRNA$_3$$^{Leu}$ recovered from S30 with added tRNA$_1$$^{Leu}$. (B) RPC-5 chromatogram of $^3$H-Leu-tRNA$_3$$^{Leu}$ recovered from S30 with added bulk tRNA. ○—○, $^3$H cpm; ● - - - ●, $^{14}$C cpm., [$^3$H]Leu-tRNA$_3$$^{Leu}$, 50 pmol/ml, was mixed with 3.2 nmol/ml (0.1 mg/ml) tRNA$_1$$^{Leu}$ (A) or 128 nmol/ml (4 mg/ml) bulk tRNA (B), in a reaction mixture (0.1 ml) consisting of 50 m*M* *N*-2-hydroxyethylpiperazine-*N*-2-ethanesulfonic acid (HEPES), pH 7.0, 0.1 *M* $NH_4Cl$, 10 m*M* mercaptoethanol, 2 m*M* adenosine triphosphate (ATP), 4.2 m*M* phosphoenolpyruvate, 0.3 m*M* guanosine triphosphate (GTP), 30 mg/ml polyethylene glycol (MW 6000), 1 m*M* leucine, a mixture of all the other 19 amino acids at 0.1 m*M* for each amino acid, 35 $A_{260}$ units per milliliter of an *E. coli* K 12 crude extract (S30), and 5 m*M* Mg $(CH_3COO)_2$ (7 m*M* final concentration, including the contribution of the S30, which was the optimal magnesium concentration for protein synthesis with this extract). No exogenous mRNA was added to these samples, although a control sample did incorporate counts into protein directed by phage RNA, ensuring that the system was active for protein synthesis. After incubation at 37° for 15 min, RNA was extracted and chromatographed on RPC-5, with a [$^{14}$C]Leu-tRNA$^{Leu}$ marker prepared from bulk tRNA. Column fractions were dissolved in aquasol and counted in a Beckman LS 230 scintillation counter. (C) Deacylation of [$^3$H]Leu-tRNA$^{Leu}$ isoaccepting species in S30s. [$^3$H]Leu-tRNA$_1$$^{Leu}$ (●); [$^3$H]Leu-tRNA$_3$$^{Leu}$ (○); [$^3$H]Leu-tRNA$_4$$^{Leu}$ (△); [$^3$H]Leu-tRNA$_5$$^{Leu}$ (×). Individual purified tRNA$^{Leu}$ isoaccepting species [W. M. Holmes, R. E. Hurd, B. R. Reid, R. A. Rimmerman, and G. W. Hatfield, *Proc. Natl. Acad. Sci. U.S.A.* **72,** 1068 (1975)] were acylated with [$^3$H]leucine (47 Ci/mmol), with a theoretical acceptance of about 20%. Of each isoaccepting species, $1 \times 10^6$ cpm/ml were added to separate reaction mixtures as described above, without exogenous mRNA, and incubated at 37°. At various times, aliquots from each sample were removed, and the amount of radioactivity precipitable by 5% trichloroacetic acid (TCA) was determined by filtering on glass-fiber filters and counting in 4 g of Omnifluor per liter dissolved in toluene.

One additional concern is that different tRNA isoaccepting species might have different rates of deacylation, thereby changing the concentration of the tracer and undermining the validity of a comparison of relative incorporations with the different tRNA species. Therefore, it is advisable to measure the deacylation rates of each tRNA isoaccepting species under study. In the case of four Leu-$tRNA^{Leu}$ isoaccepting species of *E. coli*, the deacylation rates are all comparable, with a half-life of about 18 min (Fig. 1C).

### MS2 Specific Proteins

The remainder of this article describes methods for examining codon–anticodon interactions in MS2 RNA-directed protein synthesis reactions in extracts of *E. coli*. MS2 RNA codes for three proteins: coat, replicase, and A or maturation protein.[8] The latter is synthesized in very small amounts *in vitro* (as well as *in vivo*) in *E. coli* extracts,[9] and so is not useful for this analysis. However, since extracts of *Bacillus stearothermophilus* have been shown to efficiently synthesize the A-protein,[10] the possibility does exist for studying codon–anticodon interaction with this gene. The replicase protein is made *in vitro* in amounts sufficient to permit isolation and identification of specific peptides. However, the amino acid sequence of the replicase has not been independently elucidated and has been deduced from the nucleic acid sequence[11]; thus, the identification of replicase peptides is somewhat less solid. The coat protein, which is the primary product of the reaction, provides the best peptides for these kinds of studies. Unfortunately, a few codons are not represented in the coat,[12] or are located in positions that make it difficult to determine the codon-anticodon interaction. Most of the results thus far have come from coat protein peptides, though in our laboratory, we have determined codon-anticodon interactions in a few replicase peptides as well.

#### *Reagents*

Tris(hydroxymethyl)aminomethane (Tris), titrated with concentrated HCl, 1 *M*, pH 7.8; 1.6 *M*, pH 7.2; and 1 *M*, pH 7.4

[8] K. Weber and W. Konigsberg, *in* "RNA Phages" (N. Zinder, ed.), p. 51. Cold Spring Harbor Press, Cold Spring Harbor, New York, 1975.

[9] H. F. Lodish and H. D. Robertson, *J. Mol. Biol.* **45,** 9 (1969).

[10] H. F. Lodish, *Nature (London)* **226,** 705 (1970).

[11] W. Fiers, R. Contreras, F. Duerinck, G. Haegeman, D. Iserentant, J. Merregaert, W. Min Jou, F. Molemans, A. Raeymaekers, A. Van den Berghe, G. Volckaert, and M. Ysebaert, *Nature (London)* **260,** 500 (1976).

[12] W. Min Jou, G. Haegeman, M. Ysebaert, and W. Fiers, *Nature (London)* **237,** 82 (1972).

KCl, 2 *M*
$Mg(CH_3COO)_2$, 1 *M* and 0.1 *M*
Mercaptoethanol, 1 *M*
TMK buffer: 30 m*M* Tris, pH 7.8, 60 m*M* KCl, 10 m*M* $Mg(CH_3COO)_2$, 5 m*M* mercaptoethanol
Adenosine triphosphate (ATP), 50 m*M* (stored at −20°)
Phosphoenolpyruvate (P-enolpyruvate), 70 m*M* (−20°)
Guanosine triphosphate (GTP), 10 m*M* (−20°)
$KCH_3COO$, 1 *M*, pH 4.9
$NH_4CH_3COO$, 1 *M*
Concentrated broth: 32 g of tryptone, 20 g of yeast extract, 5 g of NaCl, 0.2 of NaOH, and 2 g of glucose per liter
$MgCl_2$, 1 *M*
Deoxyribonuclease (Worthington), 1 mg/ml (−20°)
Polyethylene glycol, MW 6,000 (Matheson, Coleman, and Bell), 50% w/w
Suspension medium: 1 g of tryptone, 2.5 g of NaCl per liter
Lysozyme (Sigma), 5 mg/ml (made fresh)
Ethylenediaminetetraacetic acid (EDTA), 0.5 *M*, titrated to pH 8.0 with 5 *M* NaOH
*N*-2-Hydroxyethylpiperazine-*N*-2-ethanesulfonic acid (HEPES), 1 *M* titrated to pH 7.2 with concentrated $NH_4OH$ (when diluted to 50 m*M*, the pH should be 7.0)
$NH_4Cl$, 2 *M*
Amino acids, each 0.1 *M*, dissolved or in suspension in $H_2O$ (−20°).
Amino acid mixtures: each of 19 amino acids at 5 m*M* (omitting the one used for radioactive labeling) (−20°)
Trichloroacetic acid (TCA), 10% w/v
Ribonuclease A (Miles), 5 mg/ml (−20°)
TPCK Trypsin (Worthington), 5 mg/ml (made fresh)
$NH_4HCO_3$, 1% w/v (made fresh)
pH 3.5 buffer: 5% w/v acetic acid, 0.5% w/v pyridine
Brilliant Cresyl Blue, saturated solution in pH 3.5 buffer
pH 1.9 buffer: 8% w/v acetic acid, 2% w/v formic acid
pH 4.7 buffer: 2% w/v acetic acid, 2% w/v pyridine
BPAW: 37.5% w/v butanol, 25% w/v pyridine, 7.5% w/v acetic acid, 30% $H_2O$
Radioactive ink: ordinary ink with any $^{14}C$ isotope added at about 0.5–1 μCi/ml

*Experimental Procedures*

*Strains.* Bacteriophage MS2 is obtained from W. Fiers, laboratory of Molecular Biology, University of Ghent, Belgium. Since the entire nucleotide sequence of MS2 was determined by Fiers and co-workers,[11–13] it is important to use phage from this laboratory in order to minimize the possibility of sequence variations in strains of MS2 that have been cultured for some length of time in other laboratories. Any wild-type *E. coli* strain is appropriate for the extract; we have used *E. coli* K12. It may be advisable to use stringent rather than relaxed *E. coli* because there is evidence that relaxed strains show misreading of the genetic code *in vivo* under amino acid starvation conditions.[14] If temperature-sensitive tRNA synthetase mutants are desired for the extracts, the *E. coli* genetic stock center, Yale University, New Haven, Connecticut, can supply the appropriate strains.

*Preparation of Extracts. E. coli* cells are grown in rich broth to mid-log phase (about $5 \times 10^8$ cells/ml). The culture is poured on ice, harvested, and washed with TMK buffer (30 m*M* Tris·HCl, pH 7.8, 60 m*M* KCl, 10 m*M* Mg $(CH_3COO)_2$, 5 m*M* mercaptoethanol); 3 ml of TMK buffer are added per gram of wet weight of cells, and the suspension is passed with a moderate flow rate through a French pressure cell at 3000 psi. The extract is centrifuged at 12,000 *g* for 10 min, and the supernatant is clarified at 30,000 *g* for 20 min. The supernatant (S30) is incubated at 30° for 1 hr with 2.5 m*M* adenosine triphosphate (ATP), 3.5 m*M* phosphoenolpyruvate (P-enolpyruvate), and 0.5 m*M* guanosine triphosphate (GTP), followed by dialysis (4 hr or overnight) against TMK buffer at 4°, and then another dialysis (overnight or 4 hr) against TMK buffer with 1 g of glutathione per liter replacing the mercaptoethanol. The extract (200–300 $A_{260}$ units/ml) is quick-frozen in aliquots in a Dry-Ice ethanol bath and stored at −70°.

*Preparation of Aminoacylated tRNA Isoaccepting Species*. Purified individual tRNA isoaccepting species are isolated by chromatography on BD-cellulose,[15] DEAE-Sephadex,[16,17] and Sepharose using reverse salt gradients, as described.[18,19] Aminoacylation reactions (0.4 ml) contain

[13] W. Fiers, R. Contreras, F. Duerinck, G. Haegeman, J. Merregaert, W. Min Jou, A. Raeymakers, G. Volckaert, M. Ysebaert, J. Van de Kerckhove, F. Nolf, and M. Van Montagu, *Nature (London)* **256**, 273 (1975).

[14] P. H. O'Farrell, *Cell* **14**, 545 (1978).

[15] I. Gillam, S. Millward, D. Blew, M. von Tigerstrom, E. Wimmer, and G. M. Tener, *Biochemistry* **6**, 3043 (1967).

[16] Y. Kawade, T. Okamoto, and Y. Yamamoto, *Biochem. Biophys. Res. Commun.* **10**, 200 (1963).

[17] J. D. Cherayil and R. M. Bock, *Biochemistry* **4**, 1174 (1965).

[18] W. M. Holmes, R. E. Hurd, B. R. Reid, R. A. Rimmerman, and G. W. Hatfield, *Proc. Natl. Acad. Sci. U.S.A.* **72**, 1068 (1975).

[19] G. W. Hatfield, Vol. LIX [15].

0.16 $M$ Tris·HCl, pH 7.2, 10 m$M$ mercaptoethanol, 10 m$M$ Mg $(CH_3COO)_2$, 10 m$M$ KCl, 5 m$M$ ATP, 7 m$M$ P-enolpyruvate, 10 mg/ml of an *E. coli* extract passed over a DEAE column,[20] and the appropriate radioactive amino acid at maximum specific acitivity. The reaction mixture is distributed to tubes with 0.4 mg of purified individual tRNA isoaccepting species, or 4 mg of total *E. coli* tRNA (Plenum), and incubated for 15 min at 37°. Samples are made 0.2 $M$ in $KCH_3COO$, pH 4.9, and extracted with an equal volume of 1:1 $H_2O$-saturated redistilled phenol:$CHCl_3$. RNA is precipitated with 2–3 volumes of ethanol at −20° for 1 hr or more, centrifuged, washed with ethanol and ether, dried by blowing $N_2$ gas over the pellet, dissolved in 0.5 ml of a solution of 1 m$M$ $NH_4CH_3COO$, 10 $\mu M$ $Mg(CH_3COO)_2$, and 1 m$M$ mercaptoethanol, and stored at −70°. The overall concentration of individual tRNA isoaccepting species is about 25 nmol/ml (1 $A_{260}$ unit = 0.05 mg = 1.6 nmol). Percent acceptance of the amino acid by tRNA should be calculated to ascertain the concentration of aminoacylated tRNA. These preparations are diluted 10-fold in protein synthesis reactions. Different isoaccepting species in a family should be aminoacylated with the same reaction mixture to ensure that the concentration of counts per minute will be comparable from one species to the next.

*Preparation of MS2 RNA.* The following procedures are somewhat modified from those previously published.[21] An *E. coli* Hfr strain, 10–15 liters, is grown in a fermentor (or with other forced aeration) at 37° in concentrated broth[22] (32 g of tryptone, 20 g of yeast extract, 5 g of NaCl, 0.2 g of NaOH, and 2 g of glucose per liter) to 4 × $10^9$ cells/ml ($OD_{550}$ = 4). $MgCl_2$ or $CaCl_2$ is added to 2 m$M$, and the culture is infected with bacteriophage MS2 at a multiplicity of 1–2. After 3–4 hr, or when the culture has lysed, add 1 mg of DNase, 30 g of NaCl, and 100 g of polyethylene glycol (MW 6000) per liter.[23,24] Stir into solution with a magnetic stirrer. Let sediment overnight at 4°. Aspirate or siphon off as much of the supernatant as possible without disturbing the sediment; spin the remainder at 5000 $g$ for 10 min. Discard the supernatant (contains about 1% of the phage yield). Take up the pellet in 300–400 ml of suspension media (1 g of tryptone, 2.5 g of NaCl per liter); stir into uniform suspension, and distribute into 250 ml Nalgene screw-cap centrifuge

[20] K. Muench and P. Berg, *in* "Procedures in Nucleic Acid Research" (G. L. Cantoni and D. R. Davies, eds.), p. 375. Harper & Row, New York, 1966.

[21] E. Goldman and H. F. Lodish, *J. Virol.* **8,** 417 (1971).

[22] H. F. Lodish, *J. Mol. Biol.* **50,** 689 (1970).

[23] R. Leberman, *Virology* **30,** 341 (1966).

[24] K. R. Yamamoto, B. M. Alberts, R. Benziger, L. Lawhorne, and G. Treiber, *Virology* **40,** 734 (1970).

buckets. Add about 20 ml of $CHCl_3$ per bucket, screw caps on tightly, and shake vigorously by hand. Centrifuge at 10,000 *g* for 10 min; save the supernatant. The pellet will contain $CHCl_3$, covered by cell debris and polyethylene glycol. Add 50 ml of suspension media to pellets, shake, and centrifuge again; pool supernatants. Repeat extraction once more. About 85-90% of the phage are in the initial extraction; some 7-10% in the second; 1-2% in the third. To the pooled supernatants (about 500 ml), again add 30 g of NaCl and 100 g of polyethylene glycol per liter. Stir into solution, let stand in the cold 2 hr or more, then centrifuge at 15,000 *g*. Take up pellet in 50 ml of suspension medium and extract with 5 ml of $CHCl_3$. Repeat extraction twice more with 10 ml of suspension medium. To the pooled supernatants (70-80 ml), add 0.625 g of CsCl per milliliter. Centrifuge at 44,000 rpm for 18 hr at 4° in a Beckman Ti 60 rotor, and collect and pool visible phage bands. Dilute 1:4 with 0.1 *M* Tris, pH 7.4, and pellet phage at 49,000 rpm for 90 min at 4° in a Beckman Ti 60 rotor. Take up phage pellets overnight in 0.1 *M* Tris, pH 7.4, at 4°. To extract RNA, take a few milliliters of phage solution and add $KCH_3COO$, pH 4.9, to 0.2 *M*. Extract with an equal volume of 1:1 $H_2O$-saturated redistilled phenol:$CHCl_3$. RNA is precipitated with 2-3 volumes of ethanol at −20°, centrifuged, washed with ethanol and ether, dried, and dissolved in $H_2O$ at a final concentration of 300 $A_{260}$ units/ml. This preparation is diluted 15- to 20-fold in protein synthesis reactions. Up to 900 mg of pure MS2 RNA have been obtained from one 10-liter fermentor batch.

For smaller MS2-infected cultures (1-2 liters), it is easier to amend the procedure as follows: make certain that the culture is lysed at the outset by adding 50 $\mu$g of lysozyme per milliliter, 1 ml of $CHCl_3$, and 10 m*M* EDTA, then centrifuge away the cell debris at 5000 *g* for 5 min. Phage are precipitated out of the supernatant with 0.5 *M* NaCl and 10% w/v polyethylene glycol, as above. The sediment is taken up directly in 0.1 *M* Tris, pH 7.4, and 0.625 g of CsCl are added per milliliter. Centrifugation is for 18 hr at 35,000 rpm in a Beckman SW 41 rotor at 4°. The phage band (if not readily visible, it can usually be seen under ultraviolet light) is collected, diluted 4-fold with 0.1 *M* Tris, pH 7.4, and centrifuged at 49,000 rpm for 2 hr in a Beckman Ti 50 rotor at 4°. The phage pellet is again dissolved in 0.1 *M* Tris, pH 7.4, and stored at 4°. Extraction of RNA is as described above.

*Cell-Free Protein Synthesis.*[9, 25] Each milliliter of reaction mixture contains: 0.2-0.3 ml of an *E. coli* S30 extract (200-300 $A_{260}$ units/ml), for a final concentration of 40-60 $A_{260}$ units/ml; 0.05 ml of 1 *M* HEPES, pH

[25] E. Goldman and H. F. Lodish, *J. Mol. Biol.* **67**, 35 (1972).

7.0 (final concentration 50 m*M*); 0.05 ml of 2 *M* $NH_4Cl$ (final, 0.1 *M*); 0.01 ml of 1 *M* mercaptoethanol (final, 10 m*M*); 0.04 ml of 50 m*M* ATP (final, 2 m*M*); 0.06 ml of 70 m*M* P-enolpyruvate (final, 4.2 m*M*); 0.03 ml of 10 m*M* GTP (final, 0.3 m*M*); 0.065 ml of 50% w/w polyethylene glycol, MW 6000 (final, 30 mg/ml) (this reagent is optional; it appears to increase amounts of protein synthesis 2 to 3-fold without affecting the products or changing the background[26]); 0.05–0.07 ml of 0.1 *M* $Mg(CH_3COO)_2$ (final, 5–7 m*M*) (magnesium optimum should be determined for each extract); 0.02 ml of a 5 m*M* mixture of 19 amino acids (final, 0.1 m *M* for each); 0.01 ml of a 0.1 *M* solution of the amino acid homologous to the isotope used to aminoacylate tRNA (final, 1 m*M*). This mixture is distributed to individual tubes on ice containing, per 0.5 ml reaction, either 0.03 ml of 300 $A_{260}$ units/ml MS2 RNA (final, 0.9 mg/ml, which is just saturating[27]) or 0.03 ml of $H_2O$ (blank); 0.05 ml tRNA isoaccepting species aminoacylated with the appropriate radioactive amino acid (see above, section on Preparation of Aminoacylated tRNA Isoaccepting Species); if incorporation of free radioactive amino acid is desired (for example, as a marker), use an amino acid mixture omitting the desired amino acid, and add the radioactive amino acid at a maximum specific activity that allows a final concentration of 5–10 $\mu M$ (usually around 2–4 $\mu$Ci per 0.5 ml reaction for $^{14}C$-labeled amino acids).

Samples are incubated at 37° for 10- 15 min and chilled on ice. Portions (0.025 ml) of each sample are removed to determine the initial incorporation, as follows: 0.5 ml of 0.1 *M* NaOH is added, and the samples are incubated at 37° for 10 min. Then 0.5 ml of 10% trichloroacetic acid (TCA) is added, and the samples are allowed to sit on ice for 15 min before they are filtered through glass-fiber filters and counted in a scintillation counter in 4 g of Omnifluor (New England Nuclear) per liter dissolved in toluene. The remainder of the sample is subjected to peptide mapping, described below.

*Analysis of Peptides.*[25, 28] Although the coat protein can be purified prior to peptide mapping,[5] this is probably not necessary in all cases since many coat peptides are present in large amounts and are readily identifiable from a mixture of peptides. Also, processing of all the products synthesized *in vitro* allows the possibility of identifying replicase peptides as well.

After incubation at 37°, and removal of a portion to determine initial incorporation, samples are treated with 40 m *M* EDTA, pH 8.0, and 0.1

[26] E. Goldman, Ph.D. thesis, p. 147. Massachusetts Institute of Technology, Cambridge, 1972.

[27] E. Goldman and H. F. Lodish, *Biochem. Biophys. Res. Commun.* **64**, 663 (1975).

[28] H. F. Lodish, *Nature (London)* **220**, 345 (1968).

mg of ribonuclease A (Miles) per milliliter at 37° for 10 min, precipitated in the cold, and washed three times with 5% TCA, lyophilized twice, digested with 0.5 mg/ml TPCK trypsin (Worthington) at 37° for 3-4 hr in a fresh solution of 1% $NH_4HCO_3$, lyophilized twice more, and subjected to electrophoresis on Whatman 3 MM paper at pH 3.5 (5% acetic acid, 0.5% pyridine) for 2 hr at 4 kV in a Varsol-cooled tank. A saturated solution of the visible dye marker Brilliant Cresyl Blue, in pH 3.5 buffer, serves to monitor the electrophoresis and can be used as a visible divider when spotted between adjacent samples. The dried electrophoretogram is exposed to Kodak RP Royal X-Omat X-ray film for 3-6 days, and the autoradiogram is aligned with the electrophoretogram. Writing with radioactive ink on the electrophoretogram prior to exposure permits aligning the paper with the film. Peptide bands are excised, sewn onto fresh sheets of Whatman 3 MM, subjected to another electrophoresis at pH 1.9 (8% acetic acid, 2% formic acid), and autoradiographed. Some peptides may be sufficiently pure at this point for excision and quantitation in the scintillation counter. Other peptides may require an additional dimension for purification, which can be electrophoresis at pH 4.7 (2% acetic acid, 2% pyridine), or chromatography in BPAW (37.5% butanol, 25% pyridine, 7.5% acetic acid, 30% $H_2O$).

Once the peptide is pure, the spots from individual samples are cut out and quantitated in the scintillation counter. Identity of peptides can be determined by the presence or the absence of radioactive marker amino acids run in parallel tracks. For coat protein peptides, it is also possible to add carrier coat protein at the outset; coat peptides are then visualized with ninhydrin, and amino acid compositions are determined directly.[5] As an example of identifying peptides by the presence or absence of radioactive markers, Table I lists all the leucine-containing tryptic peptides in the three MS2 proteins, showing what markers are present or absent in each peptide. Identification of a peptide is then determined by comparing the marker patterns obtained in a given peptide with the marker patterns expected from the chart. The presence of amino acids that appear infrequently (Trp, Tyr, Phe, His, Met, Cys) facilitates preliminary identification of certain peptides.

*Example*[29]

MS2 RNA-directed reactions containing either [$^{14}$C]Leu-$tRNA_1^{Leu}$ (Figure 2, line 1), [$^{14}$C]Leu-$tRNA_3^{Leu}$ (line 2), [$^{14}$C]Leu-$tRNA_4^{Leu}$ (line 3), [$^{14}$C]Leu-$tRNA_5^{Leu}$ (line 4), or various marker radioactive amino acids (lines 5–14, 16), or [$^{14}$C]leucine with no MS2 RNA (blank; line 15), were

[29] E. Goldman, W. M. Holmes, and G. W. Hatfield, submitted to *J. Mol. Biol.*

TABLE I
MARKER PATTERNS OF LEUCINE CONTAINING TRYPTIC PEPTIDES OF MS2 PROTEINS[a]

| Peptide | Position | No. of residues | Leucine codon(s) | Arg | Lys | Trp | Tyr | Phe | His | Met | Asn | Asp | Glu | Pro | Val | Ile | Ala | Thr | Ser | Gly | Gln | Cys |
|---|---|---|---|---|---|---|---|---|---|---|---|---|---|---|---|---|---|---|---|---|---|---|
| Coat T2 | 67 | 17 | CUU | + | − | + | − | − | − | − | − | − | + | + | + | − | + | + | − | + | + | − |
| T10 | 84 | 23 | UUA,CUA CUU | − | + | − | + | + | − | + | + | + | + | + | + | + | + | + | + | − | − | + |
| T11 | 1 | 38 | CUC | + | − | + | − | + | − | − | + | + | + | + | + | + | + | + | + | + | + | − |
| T6 | 107 | 7 | CUC,CUA | − | + | − | − | − | − | + | − | − | − | − | − | − | + | − | − | + | + | − |
| Replicase | | | | | | | | | | | | | | | | | | | | | | |
| 1 | 7 | 10 | UUA,CUU | + | − | − | − | + | − | − | + | + | − | + | − | + | − | − | + | − | − | + |
| 2 | 49 | 6 | UUG,CUC | − | + | − | − | − | − | − | − | + | + | − | − | − | − | + | − | − | − | − |
| 3 | 55 | 13 | UUA | + | − | − | − | − | + | − | + | + | + | + | − | − | + | + | + | + | − | − |
| 4 | 69 | 6 | UUA | − | + | − | − | − | − | − | − | − | − | − | − | + | + | + | − | − | − | − |
| 5 | 88 | 7 | UUA | − | + | − | − | + | + | − | − | + | + | − | − | − | − | − | − | + | − | − |
| 6 | 95 | 14 | UUG,UUA | + | − | + | − | − | − | − | − | + | − | + | + | + | − | + | + | − | + | − |
| 7 | 109 | 11 | CUU,CUC(2) | + | − | − | + | − | − | − | + | − | − | − | − | + | − | − | + | + | − | − |
| 8 | 120 | 18 | UUG | − | + | − | − | + | + | + | + | − | − | + | − | − | + | + | + | + | + | + |
| 9 | 138 | 8 | UUG | − | + | − | + | − | − | − | − | + | − | + | − | − | + | − | − | − | + | − |
| 10 | 160 | 6 | CUA,UUG | + | − | − | − | − | − | − | − | − | − | − | + | − | + | − | − | − | − | − |
| 11 | 186 | 12 | CUC | − | + | − | − | + | − | − | + | − | − | + | + | − | − | + | − | + | − | − |
| 12 | 208 | 10 | CUC | − | + | − | + | − | − | + | + | + | + | + | − | − | − | − | − | − | + | − |
| 13 | 229 | 14 | CUG | + | − | − | − | − | − | − | + | + | − | − | + | + | − | − | + | + | + | − |
| 14 | 243 | 26 | CUG,CUU UUA | + | − | − | − | − | − | − | − | + | − | − | + | + | + | + | + | + | + | − |
| 15 | 269 | 16 | CUG,CUA CUC(2) | + | − | + | + | + | − | − | − | + | + | + | + | − | − | − | + | − | − | − |

(continued)

TABLE I—(*Continued*)

| Peptide | Position | No. of residues | Leucine codon(s) | Arg | Lys | Trp | Tyr | Phe | His | Met | Asn | Asp | Glu | Pro | Val | Ile | Ala | Thr | Ser | Gly | Gln | Cys |
|---|---|---|---|---|---|---|---|---|---|---|---|---|---|---|---|---|---|---|---|---|---|---|
| 16 | 299 | 25 | CUA(2) | − | + | + | − | + | − | + | + | − | + | − | + | + | + | + | + | + | − | − |
| 17 | 352 | 11 | CUA,CUU | − | + | − | + | + | − | − | − | − | + | − | + | − | + | − | − | + | − | − |
| 18 | 363 | 4 | CUU | + | − | − | − | − | − | − | + | − | − | + | − | − | − | − | − | − | − | − |
| 19 | 368 | 8 | CUC | + | − | − | − | + | − | − | − | − | − | − | + | − | − | + | + | + | − | − |
| 20 | 396 | 14 | UUA,CUC CUG(2) | + | − | − | − | + | − | + | + | + | − | + | + | + | + | − | − | − | − | − |
| 21 | 424 | 3 | CUC | − | + | − | + | − | − | − | − | − | − | − | − | − | − | − | − | − | − | − |
| 22 | 469 | 6 | CUG,CUC | + | − | − | − | − | − | − | − | + | − | − | − | − | + | + | − | − | − | − |
| 23 | 523 | 22 | CUA | + | − | + | − | + | − | − | − | − | + | + | + | − | + | + | + | + | + | + |
| 24 | 410 | 2 | CUA | + | − | − | − | − | − | − | − | − | − | − | − | − | − | − | − | − | − | − |
|  | 75 | 2 | CUA | + | − | − | − | − | − | − | − | − | − | − | − | − | − | − | − | − | − | − |
| 25 | 227 | 2 | CUC | − | + | − | − | − | − | − | − | − | − | − | − | − | − | − | − | − | − | − |
| 26 | 480 | 3 | CUU | + | − | − | − | − | − | − | − | − | − | − | − | − | + | − | − | − | − | − |
|  | 157 | 3 | CUG | + | − | − | − | − | − | − | − | − | − | − | − | − | + | − | − | − | − | − |
| 27 | 17 | 32 | CUU,CUC UUA | + | − | − | + | + | + | − | − | + | + | + | + | + | + | + | + | + | + | − |
| 28 | 431 | 33 | CUC(2) | − | + | − | + | + | − | + | − | + | − | + | + | − | + | + | + | + | + | − |

| | | | | | | | | | | | | | | | | | | | | | | |
|---|---|---|---|---|---|---|---|---|---|---|---|---|---|---|---|---|---|---|---|---|---|---|
| Maturation | | | | | | | | | | | | | | | | | | | | | | |
| 1 | 3 | 7 | CUC | + | − | − | − | + | − | − | − | + | − | − | − | − | + | + | + | − | − | − |
| 2 | 20 | 15 | UUA | − | + | − | + | + | − | − | + | + | + | − | + | − | + | + | + | + | − | − |
| 3 | 63 | 14 | UUA | − | + | − | + | − | − | − | − | + | − | − | + | + | + | + | + | + | + | − |
| 4 | 90 | 10 | CUC | + | − | − | − | + | − | − | − | + | − | − | − | − | + | − | + | + | − | − |
| 5 | 100 | 24 | CUC,UUG | + | − | − | + | + | − | − | + | − | + | + | + | + | + | + | + | − | + | + |
| 6 | 124 | 8 | CUG | − | + | − | − | − | − | − | − | − | + | − | + | − | + | + | + | − | + | − |
| 7 | 132 | 16 | CUU,UUA | + | − | − | − | + | − | − | + | − | + | − | + | − | + | + | − | + | + | − |
| 8 | 148 | 15 | CUC(2) | − | + | − | − | − | − | − | − | − | − | − | + | + | + | + | + | − | + | − |
| 9 | 174 | 4 | CUC | + | − | − | − | − | − | − | − | − | − | − | − | − | + | − | − | − | + | − |
| 10 | 178 | 8 | CUU,CUA | + | − | − | + | − | − | − | + | + | + | − | − | − | + | − | − | − | − | − |
| 11 | 196 | 24 | CUA,CUU UUA,UUG(2) | − | + | + | + | + | − | + | − | + | + | + | − | + | + | + | + | + | + | − |
| 12 | 220 | 10 | CUU(2) | + | − | − | − | + | + | + | − | − | + | + | + | − | − | − | − | − | + | − |
| 13 | 240 | 4 | UUA | + | − | − | − | − | − | − | − | + | − | − | − | − | − | − | − | + | − | − |
| 14 | 244 | 16 | CUG | + | − | − | + | + | − | − | + | − | − | + | − | + | + | + | + | − | + | + |
| 15 | 272 | 19 | UUG(3) CUA(2) | − | + | + | − | − | − | − | + | − | + | + | + | + | + | − | + | + | − | − |
| 16 | 373 | 16 | UUA(3) | + | − | − | − | + | + | + | − | + | − | + | + | + | + | + | + | − | − | − |
| 17 | 391 | 3 | CUC | + | − | − | − | − | − | − | − | − | − | − | − | − | − | − | + | − | − | − |
| 18 | 291 | 55 | CUU,CUA CUC(2) | + | − | + | + | + | + | + | + | + | + | + | + | + | + | + | + | + | − | + |

[a] Derived from published sequence data: see text footnotes 11–13, 31, 38. Numbering of coat protein peptides is after Konigsberg *et al.*[30] Numbering of peptides in the replicase and maturation (A) protein is arbitrary. Position refers to the amino terminal residue of the peptide. Under "leucine codons," numbers in parentheses indicate the number of times a leucine codon appears in a peptide, if more than once; − indicates the absence, + the presence, of one or more residues of the respective amino acid in the peptide.

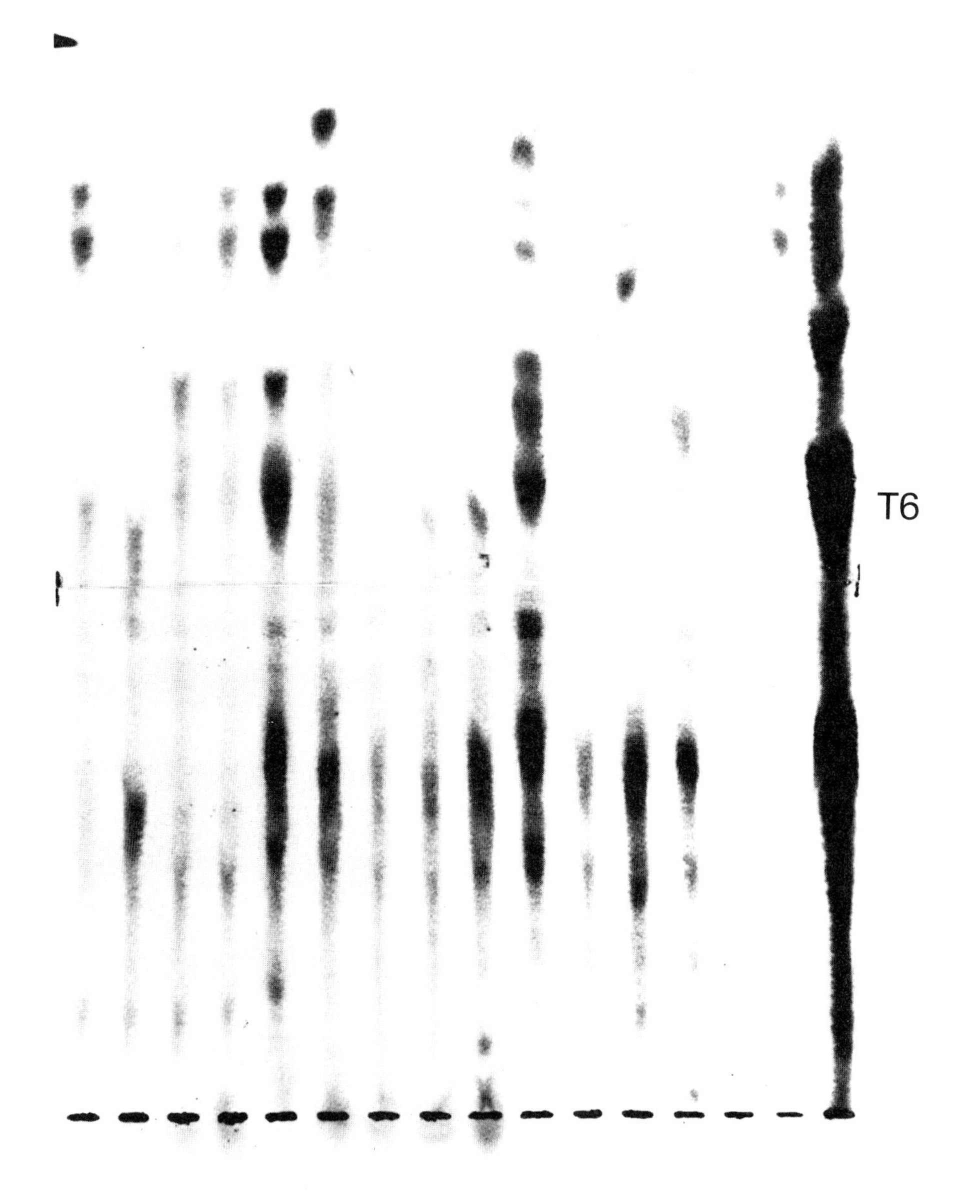
T6
1
2
3
4
5
6
7
8
9
10
11
12
13
14
15
16

TABLE II

TRANSFER OF [$^{14}$C]LEUCINE FROM [$^{14}$C]LEU-tRNA$^{Leu}$ ISOACCEPTING SPECIES INTO THE T6 PEPTIDE OF THE MS2 COAT PROTEIN[a,b]

| Line | Isotope(s) | Specific activity (mCi/ mmol) | Amount added ($\mu$Ci) | Initial total incorporation (cpm × $10^{-4}$) | Cpm recovered in peptide[b] |
|---|---|---|---|---|---|
| 1 | [$^{14}$C]Leu-tRNA$_1$$^{Leu}$ | 312 | 0.17 | 4.3 | 737 |
| 2 | [$^{14}$C]Leu-tRNA$_3$$^{Leu}$ | 312 | 0.15 | 4.1 | 564 |
| 3 | [$^{14}$C]Leu-tRNA$_4$$^{Leu}$ | 312 | 0.17 | 3.4 | 57 |
| 4 | [$^{14}$C]Leu-tRNA$_5$$^{Leu}$ | 312 | 0.15 | 3.5 | 49 |
| 5 | [$^{14}$C]Leucine | 312 | 1 | 12.7 | 1943 |
| 6 | [$^{14}$C]Arginine | 292 | 2 | 11.9 | 46 |
| | [$^{3}$H]Isoleucine | 25800 | 20 | 31.4 | 32 |
| 7 | [$^{14}$C]Aspartic acid | 193 | 2 | 2.5 | 30 |
| | [$^{3}$H]Valine | 1260 | 20 | 19.1 | 20 |
| 8 | [$^{14}$C]Glutamic acid | 223 | 2 | 3.0 | 25 |
| | [$^{3}$H]Tyrosine | 18000 | 20 | 44.0 | 35 |
| 9 | [$^{14}$C]Proline | 260 | 2 | 6.5 | 40 |
| | [$^{3}$H]Tryptophan | 4250 | 10 | 19.9 | 80 |
| 10 | [$^{14}$C]Lysine | 318 | 2 | 22.0 | 1323 |
| 11 | [$^{14}$C]Asparagine | 174 | 2 | 2.7 | 41 |
| 12 | [$^{14}$C]Phenylalanine | 460 | 1 | 10.8 | 45 |
| 13 | [$^{14}$C]Alanine | 156 | 2 | 4.8 | 362 |
| 14 | [$^{14}$C]Histidine | 270 | 1.5 | 1.3 | 76 |
| 15 | [$^{14}$C]Leucine blank (no added MS2 RNA) | 312 | 1 | 2.5 | 80 |
| 16 | [$^{35}$S]Methionine | 83000 | 50 | 205.9 | 26236 |

[a] Sequence: $_{107}$Ala-Met-Gln-Gly-Leu-Leu-Lys. (CUC CUA printed above Leu-Leu)

[b] The regions of paper corresponding to each sample at the location of the T6 peptide in Fig. 3C were cut out and counted in the scintillation counter. Sequence is from published data: see text footnotes 12 and 31.

FIG. 2. Autoradiogram of tryptic digest of MS2 RNA-directed proteins synthesized *in vitro* and labeled with various radioactive amino acids or [$^{14}$C]Leu-tRNA$^{Leu}$ isoaccepting species. Individual purified tRNA$^{Leu}$ isoaccepting species had been acylated with [$^{14}$C]leucine, with a theoretical acceptance of about 40%. The final concentration of each tRNA$^{Leu}$ isoaccepting species was about 2.7 nmol/ml; therefore the final concentration of acylated [$^{14}$C]Leu-tRNA$^{Leu}$ was about 1.1 nmol/ml. Isotopes were obtained from New England Nuclear or SchwarzMann. Samples are identified in Table II. Reaction volumes were 0.25 ml for each except lines 1–4, which were 0.5 ml. Experimental details are described in the text.

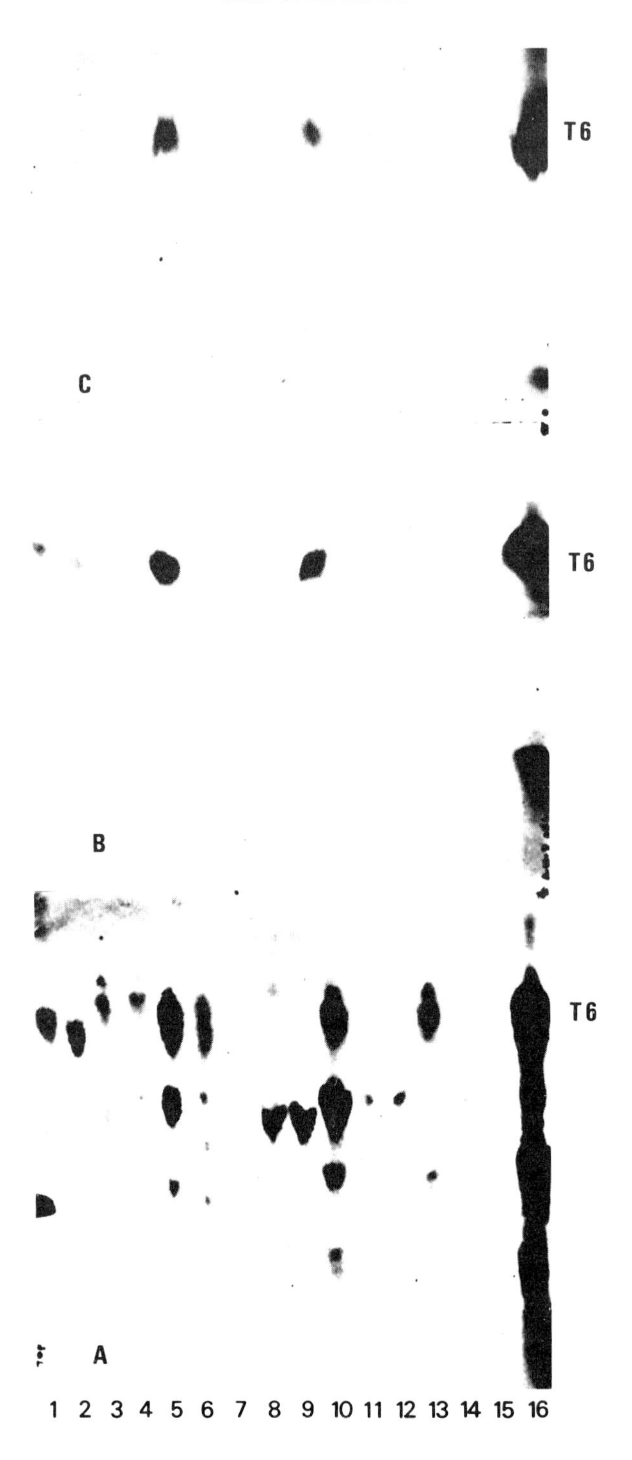

T6
C
T6
B
T6
A
1 2 3 4 5 6 7 8 9 10 11 12 13 14 15 16

digested with trypsin, subjected to electrophoresis at pH 3.5, and autoradiographed. The region of paper between 24 and 28 cm from the origin was thought likely to contain the T6 peptide of the coat protein,[30,31] because of the large quantities of [$^{14}$C]leucine (line 5), [$^{14}$C]lysine (line 10), and [$^{35}$S]methionine (line 16) present in this region. The only other methionine in the coat protein is in the insoluble T10 peptide (see Table I). This region of paper was excised, sewn onto a fresh sheet of Whatman 3 MM paper, subjected to electrophoresis at pH 4.7, and again autoradiographed (Fig. 3A). It was evident that the T6 peptide was not yet sufficiently pure, since some [$^{14}$C]arginine (line 6) comigrated with the other markers. Hence, the region of paper between 12 and 18 cm from the origin of this second dimension was again excised and subjected to electrophoresis at pH 1.9 (Fig. 3B). At this point, the T6 peptide appeared to be pure in the autoradiogram, but to be absolutely certain, the peptide band was again excised and subjected to descending chromatography in BPAW (Fig. 3C). The autoradiogram was aligned with the chromatogram, and portions of the paper corresponding to each sample at the same migration distance as the peptide spots were cut out and counted in the scintillation counter. The counts obtained for each sample (Table II) unambiguously identify this to be the pure T6 peptide, which contains two leucines in tandem encoded by CUC, CUA. $tRNA_1^{Leu}$ and $tRNA_3^{Leu}$ were both about equally efficient in inserting [$^{14}$C]leucine into these residues, while $tRNA_4^{Leu}$ and $tRNA_5^{Leu}$ failed to insert leucine, though these latter two species did transfer [$^{14}$C]leucine into protein in the initial reaction.

Since the sequenced anticodon of $tRNA_1^{Leu}$ is CAG[32] (corresponding to a CUG codon), the fact that this species inserts leucine into a peptide containing CUC and CUA codons means that unorthodox wobble is occurring, since neither C-C nor A-C wobble was deemed likely by the original wobble hypothesis. Further, though the sequenced anticodon of $tRNA_3^{Leu}$ (originally called $tRNA_2^{Leu}$)[33] is GAG[34] (corresponding to a

[30] W. Konigsberg, K. Weber, G. Notani, and N. Zinder, *J. Biol. Chem.* **241,** 2579 (1966).

[31] J. Lin, C. Tsung, and H. Fraenkel-Conrat, *J. Mol. Biol.* **24,** 1 (1967).

[32] S. Dube, K. Marcker, and A. Yudelovich, *FEBS Lett.* **9,** 168 (1970).

[33] W. M. Holmes, E. Goldman, T. A. Miner, and G. W. Hatfield, *Proc. Natl. Acad. Sci. U.S.A.* **74,** 1393 (1977).

[34] H. U. Blank and D. Söll, *Biochem. Biophys. Res. Commun.* **43,** 1192 (1971).

FIG. 3. Peptide mapping the T6 peptide. Experimental details are described in the text. (A) The region of paper marked T6 in Fig. 2 was excised, sewn onto a new piece of Whatman 3 MM paper, subjected to electrophoresis at pH 4.7, and autoradiographed. (B) Third dimension of T6 peptide: electrophoresis at pH 1.9. (C) Fourth dimension of T6 peptide: descending chromatography in BPAW (37.5% w/v butanol, 25% w/v pyridine, 7.5% w/v acetic acid, 30% $H_2O$).

CUC codon), this species is not favored over $tRNA_1^{Leu}$ despite the fact that a CUC-encoded leucine appears in this peptide.

By contrast, other experiments have shown that $tRNA_3^{Leu}$ *is* favored over $tRNA_1^{Leu}$ for the CUU codon in the T2 peptide of the coat,[29] indicating that the theoretically favored U-G wobble is more successful than the unorthodox U-C wobble occurring with $tRNA_1^{Leu}$ at this site.

### Final Remarks

A great deal more needs to be done to sort out all the details of codon–anticodon interaction; the kind of approach outlined in this article seems to hold the promise of more accurately defining these interactions than has previously been possible. The early results obtained thus far have already shown that wobble theory will have to be revised.

With the availability of other sequenced natural mRNAs, the principles of this approach can be extended, for example, to eukaryotic systems translating sequenced rabbit $\beta$-globin RNA.[35–37]

### Acknowledgments

This work was supported by grants from the National Science Foundation (PCM 75-23482), the American Cancer Society (VC 219 D), and the National Institutes of Health (GM 24330). Emanuel Goldman is a Lievre Senior Fellow (D303) of the California Division–American Cancer Society. G. Wesley Hatfield was the recipient of a USPHS Career Development Award (GM 70530). Special acknowledgment is made also to Dr. W. Michael Holmes, who collaborated and contributed in the development of this methodology.

[35] F. E. Baralle, *Cell* **10,** 549 (1977).
[36] N. J. Proudfoot, *Cell* **10,** 559 (1977).
[37] A. Estratiadis, F. L. Kafatos, and T. Maniatis, *Cell* **10,** 571 (1977).
[38] J. S. Van de Kerckhove and M. Van Montagu, *J. Biol. Chem.* **252,** 7773 (1977).

# [19] Rabbit Liver tRNA Nucleotidyltransferase

*By* MURRAY P. DEUTSCHER

All tRNA molecules examined to date contain the identical trinucleotide sequence, -C-C-A, at their 3′ terminus. These residues appear to be required for the biological activity of the tRNA molecule. At the present time it is not known whether this common sequence of nucleotides is synthesized on each tRNA molecule during transcription, or added during the maturation of the tRNA precursor.[1,2] Nevertheless, enzymes that

[1] S. Altman and J. D. Smith, *Nature (London) New Biol.* **233,** 35 (1971).
[2] V. Daniel, S. Sarid, and U. Z. Littauer, *Science* **167,** 1682 (1970).



ISBN 0-12-504180-2

incorporate nucleotides into the terminal positions of tRNA have been known for a number of years, and have been detected in a variety of sources.[3–7] A single protein is responsible for incorporation of all three nucleotide residues.[8] Procedures for the purification of the enzyme from *Escherichia coli*,[9–12] yeast,[13,14] and rabbit liver[15] have been described. The procedure described below yields a preparation of the enzyme from rabbit liver which is homogeneous and about 30,000-fold purified.

## Assay Method

*Principle.* The assay for tRNA nucleotidyltransferase measures the incorporation of radioactive, acid-soluble ATP or CTP into an acid-insoluble product. tRNA molecules lacking all, or part, of their terminal sequences can be used as substrates as follows:

$$\text{tRNA-C-C} + \text{ATP}^* \leftrightharpoons \text{tRNA-C-C-A}^* + \text{PP}_i \quad (1)$$

$$\text{tRNA-C} + \text{CTP}^* \leftrightharpoons \text{tRNA-C-C}^* + \text{PP}_i \quad (2)$$

$$\text{tRNA-N} + 2\text{CTP}^* \leftrightharpoons \text{tRNA-C}^*\text{-C}^* + \text{PP}_i \quad (3)$$

Generally, reaction (1) is utilized throughout purification, and only that assay will be described. Other enzymatic reactions that might incorporate ATP into an acid-insoluble product can be assessed by leaving out the tRNA substrate. Even in crude extracts, such contaminating activities contribute less than 5% of the acid-precipitable counts.

*Reagents*

Glycine-NaOH buffer, 1 *M*, pH 9.4[16]
$MgCl_2$, 0.1 *M*
[$^{14}$C]- or [$^{32}$P]ATP, 10 m*M* (about $10^3$ cpm/nmole)
tRNA-C-C, 10 mg/ml[17]

[3] L. I. Hecht, P. C. Zamecnik, M. L. Stephenson, and J. F. Scott, *J. Biol. Chem.* **233,** 954 (1958).
[4] J. Preiss, M. Dieckmann, and P. Berg, *J. Biol. Chem.* **236,** 1748 (1961).
[5] P. Lebowitz, P. L. Ipata, M. Makman, H. H. Richards, and G. L. Cantoni, *Biochemistry* **5,** 3617 (1966).
[6] V. Daniel and U. Z. Littauer, *J. Biol. Chem.* **238,** 2102 (1963).
[7] J. L. Starr and D. A. Goldthwait, *J. Biol. Chem.* **238,** 682 (1963).
[8] M. P. Deutscher, *J. Biol. Chem.* **245,** 4225 (1970).
[9] J. P. Miller and G. R. Philipps, *J. Biol. Chem.* **246,** 1274 (1971).
[10] D. S. Carré, S. Litvak, and F. Chapeville, *Biochim. Biophys. Acta* **224,** 371 (1970).
[11] A. N. Best and G. D. Novelli, *Arch. Biochem. Biophys.* **142,** 527 (1971).
[12] H. J. Gross, F. R. Duerinck, and W. C. Fiers, *Eur. J. Biochem.* **17,** 116 (1970).
[13] R. W. Morris and E. Herbert, *Biochemistry* **9,** 4828 (1970).
[14] H. Sternbach, F. von der Haar, E. Schlimme, E. Gaertner, and F. Cramer, *Eur. J. Biochem.* **22,** 166 (1971).
[15] M. P. Deutscher, *J. Biol. Chem.* **247,** 450 (1972).
[16] pH was determined on a 1 *M* solution at 37°.
[17] Concentration determined assuming a 1 mg/ml solution in $H_2O$ would have an $A_{260}$ of 21.4.

Trichloroacetic acid, 10% and 2.5%, containing 20 m$M$ sodium pyrophosphate

*Preparation of tRNA-C-C.* This substrate is generally prepared by periodate oxidation of intact tRNA. However, one can also use commercial baker's yeast tRNA[18] which contains about 50–60% tRNA-C-C. At saturating concentrations, the latter tRNA is as efficient a substrate as material prepared by periodate oxidation. The periodate oxidation procedure employed is a modification of the method of Neu and Heppel.[19] All solutions are autoclaved to destroy RNase. To 18 mg of tRNA in 1 ml of $H_2O$ are added 0.8 ml of 1 $M$ lysine, pH 8.9 and 0.4 ml of 0.1 $M$ sodium metaperiodate. The reaction mixture is heated at 60° for 15 minutes. Ethylene glycol, 10 μl, is added, and the solution is incubated for 20 minutes at 30° to destroy excess periodate. Alkaline phosphatase[20] (15 μl, 6 units) and 1 ml of 0.1 $M$ HCl are added; the mixture is again heated at 60° for 15 minutes. The cooled sample is placed on a Sephadex G-25 column (1.5 × 28 cm) containing 0.6 g of silicic acid on top (silicic acid adsorbs the alkaline phosphatase). The column is eluted with 10 m$M$ Tris·HCl, pH 7.3, 0.5 $M$ NaCl, 1 m$M$ magnesium acetate, and the tRNA is separated from low molecular weight contaminants. Fractions containing tRNA are combined, precipitated with 2.5 volumes of ethanol, and left overnight at −20°. The precipitate is washed with ethanol, dried, and dissolved in $H_2O$.

*Procedure.* The standard assay mixture contains: glycine-NaOH, 0.01 ml; $MgCl_2$, 0.01 ml; radioactive ATP, 0.01 ml; tRNA-C-C, 0.015 ml; enzyme; and $H_2O$ to a final volume of 0.20 ml. An amount of enzyme is added which leads to the incorporation of 0.2–1.0 nmole of AMP. Incorporation of nucleotides under standard conditions is linear with time and enzyme concentration until about one-third of the tRNA substrate is converted to product. After incubation at 37° for the appropriate time (usually 5 minutes), the reaction is terminated by placing the mixture in ice and adding 3 ml of cold 10% trichloroacetic acid–pyrophosphate solution. The assay tubes are kept in ice for 10 minutes to allow precipitation of the tRNA. The contents are then filtered through Whatman GF/C filters, washed with six portions (3 ml each) of 2.5% trichloroacetic-pyrophosphate and with one portion of 5 ml of ethanol:ether (1:1). The filters are placed at the bottom of scintillation vials and dried under an infrared lamp, and the radioactivity is determined in a scintillation counter with 5 ml of toluene-based scintillation solution.

Blank values (either no enzyme or zero time) obtained with the washing

[18] Obtained from Schwarz/Mann.

[19] H. C. Neu and L. A. Heppel, *J. Biol. Chem.* **239,** 2927 (1964).

[20] Obtained from Worthington, grade BAPF.

procedure described are within 20 cpm of the machine background and are subtracted from each assay. If required, the enzyme is diluted for assays in 10 m$M$ glycine buffer containing 1 mg/ml of tRNA-C-C. The presence of the tRNA is necessary to prevent loss of enzyme activity upon dilution. The ATP concentration (0.5 m$M$) used in the routine assay is considerably below the $K_m$ for this substrate, and optimal incorporation is not obtained until about 5 m$M$ ATP is used. The use of suboptimal triphosphate concentrations does not affect the accuracy of the assay, but conserves radioactive material. If higher ATP concentrations are used, the $MgCl_2$ concentration must also be increased.

*Definition of Unit and Specific Activity.* One unit of tRNA nucleotidyltransferase is defined as the incorporation into tRNA-C-C of 1 μmole of AMP per hour at 37° under standard conditions. Specific activity is expressed as units per milligram of protein.

## Purification Procedure

Buffer A: 10 m$M$ Tris·HCl, pH 7.3; 10 m$M$ $MgCl_2$; 0.1 m$M$ EDTA; 0.1 m$M$ dithiothreitol; 5% glycerol

Buffer B: 10 m$M$ potassium phosphate, pH 7.5; 1 m$M$ $MgCl_2$; 0.1 m$M$ EDTA; 0.1 m$M$ dithiothreitol; 5% glycerol

All steps in the purification are carried out in a cold room at 4°.

*Step 1. Preparation of Homogenate and Centrifugation.* Generally, 675 g of rabbit liver (1.5 pounds) are taken through steps 1 and 2 at a time. Frozen rabbit livers[21] are partially thawed, cut into small pieces, and homogenized in 75-g portions in a Sorvall Omni-mixer (400 ml cup, setting 4) for 30 seconds with 5 volumes of buffer A (375 ml). The combined homogenate (about 4000 ml) is centrifuged for 15 minutes at 9000 rpm in the GS-3 rotor of a Sorvall refrigerated centrifuge in order to remove cell debris, nuclei, and mitochondria. The supernatant liquid is poured through glass wool to remove fatty material, and the pellet is reextracted with 2 volumes of buffer A (1350 ml) by homogenizing for 10 seconds. After centrifuging as above, the two supernatant fractions are combined (4700 ml) for the next step. The second extraction releases an additional 20–30% of the activity.

*Step 2. Ammonium Sulfate Fractionation.* Ammonium sulfate, 1471 g, is added to the combined 15,000 $g$ supernatant over a period of 30 minutes. After stirring for an additional 30 minutes, the mixture is centrifuged at 9000 rpm for 15 minutes as above, and the pellet is discarded. To the clear supernatant liquid is added an additional 827 g of ammonium sulfate, with

[21] Obtained from Pel-Freez Biologicals, Rogers, Arkansas.

stirring and centrifugation as before. This pellet is dissolved in 225 ml of buffer A and dialyzed for 18–24 hours (with one change) against 9 liters of buffer A lacking both glycerol and $MgCl_2$. The solution after dialysis (about 600 ml) is made 5% in glycerol, and the pH is lowered to 6.6, if necessary. After dialysis the specific conductivity of the solution should be below 0.0025 $ohm^{-1}$ $cm^{-1}$ to ensure removal of all the ammonium sulfate. Protein that precipitates during dialysis is removed by centrifugation.

*Step 3. Alumina* $C_\gamma$ *Fractionation.* Two batches of the dialyzed ammonium sulfate fraction (about 1200 ml) are usually combined for this step. One hundred ninety milliliters of an alumina $C_\gamma$ suspension[22] (55 mg/ml) are added and stirred for 30 minutes, followed by centrifugation at 9000 rpm for 15 minutes as before. The pellets are washed twice for 30 minutes with 1200 ml each of buffer B containing 5 m*M* potassium phosphate, pH 7.5. The enzyme is eluted by three 30-minute washes of 300 ml each with the same buffer containing 25 m*M* potassium phosphate. The three enzyme eluates are combined for further purification. The alumina $C_\gamma$ fraction loses almost no activity when frozen at −20° for at least 3 months.

*Step 4. DEAE-Cellulose Chromatography.* For this step, four preparations of enzyme from step 3 are combined (about 3600 ml) and concentrated to about 700 ml by ultrafiltration with an Amicon PM-30 Diaflo membrane. The concentrated material is then diluted with buffer B lacking phosphate to lower the specific conductivity to 0.0013 $ohm^{-1}$ $cm^{-1}$. The sample (about 1200 ml) is added to a column of DEAE-cellulose (7.5 × 18 cm) previously equilibrated with buffer B, and washed with this buffer until the $A_{280}$ in the effluent is below 0.2. The activity is eluted with a gradient containing 3 liters of buffer B and 3 liters of buffer B with 70 m*M* potassium phosphate, pH 7.5. The enzyme is eluted as a broad peak, and the fractions with highest specific activities are combined. Side fractions are saved for subsequent purifications. Alternatively, the enzyme may be eluted in a batchwise manner with buffer B containing 30 m*M* potassium phosphate, pH 7.5. In this case the specific activity is only about one-half that of enzyme prepared by gradient elution. However, the activity is more concentrated, and the yield is higher. The lower specific activity in the batchwise procedure does not present a problem since after the next step this enzyme is identical to that prepared using a gradient. The DEAE-cellulose step also serves to remove the RNA present as judged by the increase of the $A_{280}:A_{260}$ ratio to 1.7. Enzyme purified through the DEAE-cellulose step has been stable for at least 6 months when frozen at −20°.

*Step 5. Hydroxyapatite Chromatography.* The active fractions from the

[22] Obtained from Sigma Chemical Co.

previous step are concentrated to about 100 ml by ultrafiltration with a PM-10 Diaflo membrane,[23] and made 50 m$M$ in potassium phosphate, pH 7.5. The sample is applied to a hydroxyapatite column (3.5 × 3 cm) and washed with buffer B containing 50 m$M$ potassium phosphate until the $A_{280}$ is close to base-line levels. The enzyme is eluted in 10 ml fractions at a flow rate of 25 ml/hr with a linear gradient of 1050 ml going from 50 m$M$ to 250 m$M$ potassium phosphate, pH 7.5, in buffer B. Fractions with specific activities greater than 15 are combined, and the side fractions are saved for subsequent purifications. Hydroxyapatite fractions have been stored frozen at −20° for 6 months with no loss in activity.

*Step 6. Sephadex G-100 Chromatography.* The sample from hydroxyapatite is concentrated to 10 ml with a PM-10 Diaflo membrane and applied to a Sephadex G-100 column (2.5 × 84 cm) equilibrated with buffer B. The column is washed with the same buffer at a flow rate of about 20 ml per hour, and 5-ml fractions are collected. The enzyme is eluted as a symmetrical peak between fractions 35 and 45. Tubes with specific activities greater than 50 are combined for further purification. Enzyme at this stage of purification has been stored frozen at −20° for 2 months with essentially no loss in activity.

*Step 7. Phosphocellulose Chromatography.* The combined fractions from step 6 are added to a 1.5 × 8 cm phosphocellulose column previously equilibrated with buffer B lacking $MgCl_2$. The column is washed with the same buffer until the $A_{220}$[24] is at baseline, and the enzyme is eluted with a 300-ml linear gradient going from 0 to 0.2 $M$ KCl in the starting buffer. Fractions of 5 ml are collected at a flow rate of about 30 ml per hour. The enzyme is eluted in two peaks near the end of the gradient. Each peak is combined separately and stored or concentrated as described below.

The purification procedure described here (summarized in the table) has been reproducible in consistently leading to homogeneous enzyme. However, the final yield and specific activities of intermediate steps have varied somewhat depending on the size of the preparation. Generally, yields and specific activities are higher with smaller preparations.

## Properties of the Purified Enzyme

*Purity.* Both active fractions isolated on phosphocellulose are highly purified, and the major fraction appears homogeneous by several criteria: the specific activity across phosphocellulose peak II is constant; a single band is detected by gel electrophoresis at pH 8.9 and by electrophoresis in

[23] After the DEAE-cellulose step, considerable activity came through the PM-30 membrane.

[24] Absorbance is measured at 220 nm because of the extremely low $A_{280}$ values.

SUMMARY OF PURIFICATION OF tRNA NUCLEOTIDYLTRANSFERASE

| Step | Total activity (units)[a] | Specific activity (units/mg) | Relative purification (−fold) | Yield (%) |
|---|---|---|---|---|
| 1a. Homogenate | 16,000 | 0.017 | — | 100 |
| 1b. 15,000 *g* supernatant | 16,000 | 0.024 | 1.4 | 100 |
| 2. Ammonium sulfate | 11,000 | 0.072 | 4.2 | 69 |
| 3. Alumina C$\gamma$ | 6,700 | 0.550 | 32 | 42 |
| 4. DEAE-cellulose | 3,500 | 5.05 | 297 | 22 |
| 5. Hydroxyapatite | 2,400 | 28.3 | 1660 | 15 |
| 6. Sephadex G-100 | 1,800 | 100 | 5880 | 11 |
| 7. Phosphocellulose | | | | |
| Peak I | 520 | ~600[b] | ~35,000 | 3 |
| Peak II | 990 | ~550[b] | ~32,000 | 6 |

[a] Based on 12 pounds of rabbit liver.

[b] These values are approximate owing to the difficulty of determining protein concentrations adequately at this stage of purification.

sodium dodecyl sulfate; and only a single species is observed upon equilibrium ultracentrifugation. On the basis of acrylamide gel electrophoresis, peak I appears to be about 80–90% pure. Both purified enzymes can incorporate AMP and CMP into tRNA.

*Contamination by Other Enzymes.* tRNA nucleotidyltransferase at a specific activity of about 100 units/mg is devoid of alkaline phosphatase, phosphodiesterase, RNase, aminoacyl-tRNA synthetases, ATPase, CTPase, and CMP kinase. A small amount of AMP kinase, which catalyzes the formation of 60 nmoles of ADP per minute per milligram of protein, can be detected.

*Stability.* Phosphocellulose peak I is somewhat more labile than peak II. However, both enzymes retained more than 75% of their activity for at least 1 year when stored at −20° in 50% glycerol or when frozen with 0.2 mg of commercial yeast tRNA per milliliter. Individual fractions from the phosphocellulose chromatography, at concentrations as low as 30 $\mu$g/ml, have been stored at −20° for 15 months with no loss in activity.

Considerable difficulty has been encountered in trying to concentrate the pure enzymes. Large amounts of activity are lost when the pure enzyme is concentrated by ultrafiltration with Diaflo membranes or collodion bags, as well as by dialysis against dry Sephadex G-200. These problems do not occur with less pure enzyme. The pure protein has been successfully concentrated with no loss in activity by adsorption and elution from small columns of hydroxyapatite (prepared in Pasteur pipettes).

The pure proteins are relatively sensitive to heat, losing 50% of their activity in 5 minutes at 45°, and essentially all their activity at 50°. At the latter temperature substrates do not protect against inactivation, but heating in the presence of 0.15 $M$ potassium phosphate, pH 7.5, instead of 10 m$M$ buffer stabilizes the enzymes such that only about 40% of the activities are lost.

*Sedimentation Coefficient and Molecular Weight.* The sedimentation coefficients of phosphocellulose peaks I and II are 4.05 and 3.95, respectively, at protein concentrations of 0.15 mg/ml. The molecular weights of the two proteins, based on sedimentation equilibrium, are 49,000 and 48,000. Essentially identical values are obtained at protein concentrations between 0.06 and 0.40 mg/ml. The molecular weights of the two phosphocellulose peaks on SDS-acrylamide gel electrophoresis are also about 45,000 to 49,000. Also, the molecular weight of enzyme purified through step 6, based on the elution position of its activity on a Sephadex G-100 column, is 44,000. Prior to the DEAE-cellulose step the molecular weight of the enzyme, determined on Sephadex G-100, is 80,000. This change in molecular weight upon chromatography on DEAE-cellulose is attributed to removal of one molecule of bound tRNA, although conclusive evidence for this point has not been obtained.

*Chemical Composition and Effect of Sulfhydryl- or Reducing Agents.* The amino acid composition of the two phosphocellulose peaks are very similar except that peak I contains 7 half-cystine residues and peak II only 4. Both enzymes contain cystine, but the amounts are too small to be determined reliably. No unidentified components are seen on the chromatogram after acid hydrolysis.

Both enzymes are relatively resistant to the sulfhydryl reagents *p*-mercuribenzoate, mersalyl, and 5,5′-dithiobis(2-nitrobenzoic acid), although peak I, which contains more half-cystines, is also more sensitive to these compounds. Exposure of peak I to the reagents, present at 1 m$M$ concentration, for 3 hours leads to a loss of only 40–60% of the AMP-incorporating activity. Under the same conditions, peak II loses at most only 20% of its activity. The two proteins are somewhat more sensitive to 10 $\mu M$ $Hg^{2+}$, with peak I losing about 80% of its activity in 3 hours, and peak II about 40%.

Phosphocellulose peaks I and II are both resistant to mercaptoethanol at concentrations up to 25 m$M$. In fact, at lower concentrations the reducing agent appears to stimulate AMP incorporation. In contrast, the activity of both enzymes is inhibited about 70% by 10 m$M$ dithiothreitol, although this inhibition does not increase further at higher concentrations of the reagent.

The ultraviolet spectra indicate that less than one residue of nucleotide material could be present per molecule of either protein. Both enzymes have typical protein spectra with maxima at 278 nm. The $A_{280}:A_{260}$ ratios have been between 1.72 and 1.81 in different preparations. Tentative extinction coefficients, $E_{280}^{0.1\%}$ of 0.80 to 0.86 have been assigned to each protein.

## Catalytic Properties

*Specific Activities and Turnover Number.* Each of the phosphocellulose fractions has a specific activity of about 550 μmoles of AMP incorporated per hour per milligram of protein under standard assay conditions. In the presence of saturating concentrations of ATP and tRNA-C-C, the specific activity is closer to 2500. The latter value corresponds to a turnover number of close to 2000 molecules of AMP incorporated per minute per molecule of enzyme.

*pH Optima.* The pH optima for both enzymes in 50 m*M* glycine-NaOH buffer varies from pH 9.3 to 10.0 depending on the nucleotide and enzyme under study. At pH 8.8, the rates of AMP and CMP incorporation in Tris·HCl buffer are similar to those in glycine. Nucleotide incorporation in Tris buffer, pH 7.1 is 20–30% of maximum, and is about 50% lower in potassium phosphate of the same pH.

*Cation Requirements.* The divalent cation requirement can be satisfied by $Mg^{2+}$, $Mn^{2+}$, or $Co^{2+}$ present at 10 m*M*. At this concentration, $Zn^{2+}$, $Ca^{2+}$, $Ba^{2+}$, $Cu^{2+}$, and $Ni^{2+}$ are inactive. Optimal nucleotide incorporation is obtained at 10–25 m*M* $Mg^{2+}$ (the optimal cation concentration varies with the level of triphosphate and tRNA present; for this experiment ATP is present at 5 m*M* and CTP at 2 m*M*). Optimal AMP incorporation is found at about 4 m*M* $Mn^{2+}$ and 7.5–10 m*M* $Co^{2+}$; optimal CMP incorporation is obtained at 0.5–1 m*M* $Mn^{2+}$ and 5 m*M* $Co^{2+}$. At their respective maxima, $Mn^{2+}$ is about 30% as effective and $Co^{2+}$ about 15% as effective as $Mg^{2+}$ for AMP incorporation; for CMP these values are about 40–50% for $Mn^{2+}$ and 20–30% for $Co^{2+}$. In the case of $Mn^{2+}$, the optimal cation concentration is lower than the level of triphosphate present and higher concentrations strongly inhibit the reaction. In addition, care must be exercised in the use of $Mn^{2+}$ at the alkaline pH of the assay since a time-dependent, nonenzymatic precipitation of radioactive triphosphates has been found to occur at higher levels of this cation.

*Specificity for Nucleoside Triphosphates.* Purified liver tRNA nucleotidyltransferase can incorporate AMP, CMP, or UMP into tRNA.[25] The ratio of these various incorporating activities are constant throughout purification, although they are somewhat different in the two purified phosphocellulose peaks. UTP acts as an analog of CTP, and can be incorporated into either of the two positions in the tRNA terminus normally occupied by CMP. However, incorporation into both positions occurs

[25] M. P. Deutscher, *J. Biol. Chem.* **247**, 459 (1972).

extremely slowly.[26–28] The apparent $K_m$ values for nucleoside triphosphates are essentially identical for both purified enzymes: about 2 m$M$ for ATP, 50 m$M$ for UTP, and 0.004 or 0.4 m$M$ for CTP. The latter substrate leads to biphasic double reciprocal plots from which the two $K_m$ values have been calculated.[25]

Other nucleotide analogs, such as bromo CTP[29]; iodo CTP[30]; tubercidin, toyocamycin, sangivamycin[31] or formycin triphosphate[32]; and adenosine 5′-*O*-(1-thiotriphosphate)[33] are substrates for various tRNA nucleotidyltransferases. In contrast, GTP[20]; dATP[25]; 2-aminopurine triphosphate or 2,6 diaminopurine triphosphate[34] are inactive.

*Nucleic Acid Specificity.* Rabbit liver tRNA nucleotidyltransferase displays no species specificity. tRNA-C-C, tRNA-C, and tRNA-N from liver, yeast, or *E. coli* are equally active as substrates. In addition, all tRNA molecules in a mixed population are active as acceptors. Reactions with 5 S RNA, rRNA, and modified tRNA's, such as tRNA-C-A, tRNA-C-U, and tRNA-C-C-C, also occur, although generally at much slower rates. Synthetic homopolymers and DNA are inactive as substrates. Apparent $K_m$ values for the normal tRNA substrates vary between 4 and 12 $\mu M$.

*Effect of Other Reagents.* AMP incorporation is inhibited by $(NH_4)_2SO_4$, KCl, or NaCl, with 50% inhibition attained at about 0.2 ionic strength. In contrast CMP incorporation can be stimulated as much as 2-fold at similar salt concentrations. Similarly, the polyamines spermine, spermidine, putrescine, and cadaverine inhibit AMP incorporation, but stimulate CMP incorporation. The polyamines, however, cannot satisfy the requirement for a divalent cation when present alone. The chelating agents, *o*-quinolinol, $\alpha,\alpha'$-dipyridyl, EGTA, EDTA, and *o*-phenanthroline have no effect on AMP or CMP incorporation at concentrations up to 1 m$M$. Similarly, rifampicin at concentrations as high as 150 $\mu$g/ml has no effect on enzyme activity.

[26] M. P. Deutscher, *J. Biol. Chem.* **247,** 469 (1972).
[27] H. G. Klemperer and E. S. Canellakis, *Biochim. Biophys. Acta* **129,** 157 (1966).
[28] A. Fernandez-Sorensen, D. D. Anthony, and D. A. Goldthwait, *J. Biol. Chem.* **241,** 5019 (1966).
[29] R. L. Soffer, S. Uretsky, L. Altwerger, and G. Acs, *Biochem. Biophys. Res. Commun.* **24,** 376 (1966).
[30] M. Sprinzl, F. von der Haar, E. Schlimme, H. Sternbach, and F. Cramer, *Eur. J. Biochem.* **25,** 262 (1972).
[31] S. C. Uretsky, G. Acs, E. Reich, M. Mori, and L. Altwerger, *J. Biol. Chem.* **243,** 306 (1968).
[32] D. C. Ward, A. Cerami, E. Reich, G. Acs, and L. Altwerger, *J. Biol. Chem.* **244,** 3243 (1969).
[33] E. Schlimme, F. von der Haar, F. Eckstein, and F. Cramer, *Eur. J. Biochem.* **14,** 351 (1970).
[34] D. C. Ward, E. Reich, and L. Stryer, *J. Biol. Chem.* **244,** 1228 (1969).

*Poly(C) Polymerase*. Purified preparations of rabbit liver tRNA nucleotidyltransferase contain a poly(C) polymerase activity as an integral part of the protein.[26] This activity can attach long sequences of CMP residues to any intact or partially degraded tRNA molecule or to rRNA. The rate of this activity under the usual assay conditions is only about 1% of the normal CMP incorporation and is generally not detected. However, in experiments in which high levels of enzyme or long periods of incubation are used (such as for the synthesis of tRNA's with labeled terminal CMP residues), this activity can interfere. The poly(C) polymerase can be completely inhibited by 0.25 *M* KCl, whereas normal CMP incorporation is stimulated by this concentration of salt.

*Subcellular Localization*. In rat liver, tRNA nucleotidyltransferase is located predominantly in the high-speed supernatant fraction. However, about one-third of the total cellular activity is associated with the mitochondrial fraction, and this activity is present in the mitochondrial matrix.[35] Microsomes and nuclei are essentially devoid of activity. It is not known whether the mitochondrial enzyme is a distinct protein.

[35] S. K. Mukerji and M. P. Deutscher, *J. Biol. Chem.* **247**, 481 (1972).

# [20] Bacterial tRNA Methyltransferases

*By* Robert Greenberg and Bernard S. Dudock

tRNA methyltransferases from bacteria[1] and eukaryotes[2] have previously been described in this series, and a general review of these enzymes has recently been published.[3] Within the past few years a number of methyltransferases have been highly purified[4-8] and several new techniques have been developed, e.g., affinity chromatography, affinity elution chromatography, and isoelectric focusing, which are now readily available for use in the isolation of these enzymes. Several of these methods have been used in our laboratory for the purification of 5-methyluridine methyltransferase and uridine 5-oxyacetic acid methylester methyltransferase from *Escherichia coli*.

## Assay of Enzymes

*Principle*. The tRNA methyltransferase is incubated in the presence of a substrate tRNA and a radioactively labeled methyl donor, usually *S*-



ISBN 0-12-504180-2

adenosyl-L-[*methyl*-$^{3}$H]methionine. After incubation, the extent of methylation is determined by measuring the amount of radioactivity that has been incorporated into acid-insoluble material.

*Reagents.* A typical tRNA methyltransferase assay contains the enzyme being studied, *S*-adenosyl-L-[*methyl*-$^{3}$H]methionine (5–15 Ci/mmol) and a suitable tRNA substrate. In general, the tRNA should be from a source different from that of the methyltransferase since endogenous tRNA would usually be expected already to contain the modification produced by this enzyme. Exceptions to this rule, which enable a normal homologous system to be studied, are the methylester methyltransferases, one of which will be discussed later.

Optimal methylation conditions vary considerably depending upon the specific reaction being studied. In addition to the methyltransferase, tRNA, and *S*-adenosylmethionine (Ado-Met), the major assay components or conditions that must be optimized include the nature of the buffer and its specific pH, the ionic strength, and the presence or absence of salts, polyamines, EDTA, and $Mg^{2+}$. Optimal reaction conditions for 5-methyluridine methyltransferase are: 0.5–1.0 μg of wheat germ $tRNA_1^{Gly}$, 10–20 μ*M* *S*-adenosyl-L-[*methyl*-$^{3}$H]methionine (sp. act. 5–15 Ci/mmol), 100 m*M* HEPES, pH 8.4, 1 m*M* $Na_2EDTA$, 75 m*M* ammonium acetate, 20 m*M* spermidine, and 1–5 μl of enzyme extract. The total reaction volume is 20 μl.

For uridine 5-oxyacetic acid methylester methyltransferase the optimal reaction conditions are: 0.5–1.0 μg of a suitable tRNA substrate such as *E. coli* $tRNA_1^{Ala}$, 10–20 μ*M* *S*-adenosyl-L-[*methyl*-$^{3}$H]methionine (sp. act 5–15 Ci/mmol), 100 m*M* HEPES, pH 8.4, 2 m*M* $Na_2EDTA$, and 1–5 μl of enzyme extract. The total reaction volume is 20 μl.

*Procedure.* After a 30–120 min incubation at 30°, the reaction is stopped by the addition of 0.5–1.0 ml of ice cold 10% trichloroacetic acid (TCA). The precipitated reaction mixtures are kept at 0°–4° for 10 min to ensure complete precipitation of the RNA. The samples are then filtered through 2.4 cm glass-fiber filters (Whatman GF/A) and washed five times

[1] J. Hurwitz and M. Gold, this series, Vol. 12B, p. 480.

[2] S. Kerr, this series, Vol. 29, p. 716.

[3] F. Nau, *Biochimie* **58,** 629 (1976).

[4] Y. Taya and S. Nishimura, *Biochem. Biophys. Res. Commun.* **51,** 1062 (1973).

[5] H. J. Aschhoff, H. Elten, H. H. Arnold, G. Mahal, W. Kersten, and H. Kersten, *Nucl. Acids Res.* **3,** 3109 (1976).

[6] J. M. Glick and P. S. Leboy, *J. Biol. Chem.* **252,** 4790 (1977).

[7] J. M. Glick, V. M. Averyhart, and P. S. Leboy, *Biochim. Biophys. Acta,* **518,** 158 (1978).

[8] H. Wierzbicka, H. Jakubowski, and J. Pawelkiewicz, *Nucl. Acids Res.* **2,** 101 (1975).

with 2 ml each of cold 2% TCA. Filters are dried under a heat lamp and counted by liquid scintillation spectroscopy in Omnifluor-toluene. It should be noted that occasional lots of filters bind abnormally large amounts of Ado-Met resulting in unacceptably high background values. Prior to use, therefore, new lots of filters should be tested for Ado-Met binding by filtering precipitated, unincubated assays lacking tRNA.

Identification of the reaction products is readily accomplished. After incubation the reaction is brought to 0.5 ml with 0.01 *M* Tris·HCl, pH 7.6, 1.0 m*M* EDTA, 0.1 *M* NaCl, and 7 $A_{260}$ units of carrier tRNA are added. The mixture is extracted three times with 0.5 ml of buffer-saturated phenol (Eastman crystalline). The RNA is then recovered by ETOH precipitation and digested to mononucleotides with either RNase T2[9] or KOH;[10] the products are separated for identification by two-dimensional thin-layer chromatography on cellulose plates (E. Merck) as published.[11]

The reference nucleotides observed under UV light are carefully marked with a No. 1 pencil on the cellulose plates. The radioactivity on the chromatogram is then detected by fluorography.[12] The plate is rapidly and evenly coated with a solution of 7% PPO (New England Nuclear) in diethyl ether. The ether is allowed to evaporate, then the corners are marked with $^{14}$C-labeled ink (Schwarz/Mann) so that the plate can be realigned with the fluorogram after development. Flashed[13] X-ray film (Kodak XR-5) is exposed to the chromatogram at −70°. The exposure time needed varies with approximate guidelines as follows: 300,000 dpm of tritium-labeled 5-methyluridylic acid can readily be seen in 8 hr, whereas a level of 15,000 dpm requires 72 hr of exposure. After the film has been developed, the radioactive spots on the chromatogram are marked, cut out, and eluted by shaking in 1 ml of 2 *M* $NH_4OH$ for 1 hr. An aliquot of the eluent is counted in Aquasol (New England Nuclear).

The use of fluorography facilitates these studies in that it greatly increases the sensitivity of detection of the radioactive areas and also eliminates the need for gridding and counting the entire plate. This procedure has been further enhanced by the use of new, faster X-ray film and the development of techniques for further sensitizing this film by flashing.[13] In addition, the technique has been aided by the availability of tritiated Ado-Met of relatively high specific activity (5–15 Ci/mmol), which enables several hundred thousand counts to be readily incorporated into 1–2 μg of tRNA. For even greater sensitivity, tritiated Ado-

[9] K. B. Marcu, R. E. Mignery, and B. S. Dudock, *Biochemistry* **16**, 797 (1977).

[10] G. Katz and B. S. Dudock, *J. Biol. Chem.* **244**, 3062 (1969).

[11] F. Kimura-Harada, M. Saneyoshi, and S. Nishimura, *FEBS Lett.* **13**, 335 (1971).

[12] K. Randerath and E. Randerath, *Methods Cancer Res.* **9**, 3 (1973).

[13] R. A. Laskey and A. D. Mills, *Eur. J. Biochem.* **56**, 335 (1975).

Met with specific activity as high as 70 Ci/mmol is also available (New England Nuclear). The use of tritium-labeled Ado-Met of specific activity much higher than that available with $^{14}C$-labeled Ado-Met allows the methylation reaction to be performed in a total volume of only 20 $\mu$l, thus conserving, by at least 10-fold, tRNA, enzyme, and Ado-Met.

As discussed above, the specific reaction conditions should be optimized for each individual methyltransferase with respect to such parameters as incubation temperature, buffer, pH, and concentrations of Ado-Met, RNA, EDTA, $Mg^{2+}$, $NH_4^{2+}$ $NH_4^+$, and polyamines. Over a fairly wide concentration range, polyamines, e.g., spermidine, spermine, and putrescine, frequently enhance methyltransferase activity.[14–18] In the absence of $NH_4^+$ ions but polyamines, $Mg^{2+}$ ions often stimulate methylation, but frequently tend to inhibit the reaction in the presence of optimal concentrations of these cations.[14,17]

The study of a particular tRNA methyltransferase is greatly facilitated by the use of a pure species of tRNA as the methyl acceptor, as the methylation product does not then have to be analyzed at each stage of the purification or optimization study. For example, 5-methyluridine methyltransferase can be specifically assayed, even in a crude *E. coli* enzyme extract, using pure wheat germ $tRNA_1^{Gly}$ as a substrate, since with this pure tRNA, 5-methyluridine (ribothymidine) is the only methylation product. Sometimes, however, even a pure tRNA may still be a substrate for more than one methyltransferase in a crude extract. In this case a partial purification must be achieved before the desired methyltransferase can be studied without analysis of the reaction product at each stage. If one does optimize for a particular methylation reaction in a system in which several products are being synthesized, the observed optimal conditions may merely be those conditions that most tend to reduce competing methylations. In such cases, as the enzyme becomes more highly purified optimal assay conditions may significantly change.

## Purification of tRNA Methyltransferases from *E. coli*

*Growth of Cells. E. coli* MRE 600 (3/4 log phase) used in these studies was either freshly grown in 1.3% Tryptone (Difco) and 0.7% NaCl at 37° with rapid aeration, or was purchased commercially (Grain Processing, Muscatine, Iowa). Commercially grown cells have been found to give

[14] P. S. Leboy, *Biochemistry* **9**, 1577 (1970).
[15] P. S. Leboy and J. M. Glick, *Biochim. Biophys. Acta* **435**, 30 (1976).
[16] C. S. Salas, C. J. Cummins, and O. Z. Sellinger, *Neurochem. Res.* **1**, 369 (1976).
[17] P. S. Leboy, *FEBS Lett.* **16**, 117 (1971).
[18] A. E. Pegg, *Biochim. Biophys. Acta* **232**, 630 (1971).

enzymic activity equal to that of freshly grown cells. Cells are stored either at −20° or −70° prior to use.

*Isolation of tRNA.* Crude *E. coli* tRNA[19] and wheat germ tRNA[20] were isolated according to published procedures. Pure *E. coli* $tRNA_1^{Ala}$[21] and pure wheat germ $tRNA_1^{Gly}$[9] and wheat germ $tRNA_2^{Gly}$[22] were purified as described. At the present time a variety of highly purified tRNAs are available commercially (Boehringer Mannheim; Miles; Research Plus, Denville, New Jersey).

*Preparation of Crude Extract.* All procedures in the enzyme purification are carried out at 0°–5°. All buffers are made from stock solutions that have been filtered through Millipore type HA 0.45 $\mu$m filters and autoclaved. All assays are performed in 2.5-ml disposable plastic culture tubes (Walter Sarstedt, Princeton, New Jersey), and components are added with an adjustable volume Pipetman (Gilson).

*E. coli* MRE 600 (frozen cell paste), 465 gm, is thawed in 465 ml of ice-cold breaking buffer consisting of 10 m*M* Tris·HCl, pH 7.6, 10 m*M* $MgCl_2$, 0.5 m*M* dithiothreitol (DTT), and 10% (v/v) glycerol. The cells are broken in a French pressure cell at 12,000 psi. When freshly grown cells are used, it is necessary to reduce the viscosity at this point either by dilution with an equal volume of breaking buffer or by DNase treatment. In the latter case, the cell extract is incubated at 4° in the presence of several micrograms of RNase-free DNase I (Worthington Biochemical Corp.) until the extract can be readily pipetted with a Pasteur pipette. This procedure is not necessary when frozen cells are used because the extract is considerably less viscous. The extract is centrifuged at 10,000 rpm (16,000 *g*) for 40 min in a Sorvall GSA rotor. The supernatant is then centrifuged at 55,000 rpm (215,000 *g*) for 2 hr in a Beckman type 60 Ti rotor. The clear, golden supernatant (S215) has a volume of approximately 455 ml, an $A_{280}$ of 190/ml and an $A_{260}$ of 300/ml. It may be frozen at this point if desired, and is stable at −70° for at least 6 months.

*DEAE-Cellulose Chromatography.* The S215 supernatant is brought to 0.22 *M* KCl by the addition of 2.2 *M* KCl. The extract is then loaded onto a 2-liter DEAE-cellulose column (40 × 8 cm) previously equilibrated with DEAE-cellulose column buffer consisting of 10 m*M* Tris·HCl, pH 7.6, 10 m*M* $MgCl_2$, 10% (v/v) glycerol, 0.5 m*M* DTT, and 0.22 *M* KCl.

[19] B. Roe, K. Marcu, and B. Dudock, *Biochim. Biophys. Acta* **319,** 25 (1973).

[20] B. S. Dudock, G. Katz, E. K. Taylor, and R. W. Holley, *Proc. Natl. Acad. Sci. U.S.A.* **62,** 941 (1969).

[21] R. J. Williams, W. Nagel, B. Roe, and B. Dudock, *Biochem. Biophys. Res. Commun.* **60,** 1215 (1974).

[22] K. Marcu, D. Marcu, and B. Dudock, *Nucl. Acids Res.* **5,** 1075 (1978).

The DEAE-cellulose (No. 70, Standard, Schleicher and Schuell, Keene, New Hampshire) is prepared by successive treatments with 0.1 $M$ NaOH, $H_2O$, 0.2 $M$ $CH_3COOH$, $H_2O$, and then 50 m$M$ Tris·HCl, pH 7.6, until the pH returns to 7.6 as previously described.[23] The enzyme is eluted from the column at a flow rate of 1.2 liters/hr using DEAE-cellulose column buffer. Ten-milliliter fractions are collected, and the absorbance at 260 and 280 nm is monitored. Fractions in the initial $A_{280}$ peak having an $A_{280}:A_{260}$ ratio greater than 0.8 are pooled. The pooled sample has a volume of approximately 925 ml and has an $A_{280}$ of 16/ml and an $A_{260}$ of 13/ml.

*Ammonium Sulfate Fractionation.* The pooled peak from the DEAE-cellulose column is then subjected to $(NH_4)_2SO_4$ fractionation. Solid $(NH_4)_2SO_4$ (Schwarz/Mann, Ultrapure) is added until the desired concentration is reached, and the extract is stirred at 5° until all the $(NH_4)_2SO_4$ is in solution. The solution is then stirred for an additional 30 min and centrifuged at 10,000 rpm (16,000 $g$) for 30 min in a Sorvall GSA rotor. Additional $(NH_4)_2SO_4$ is then added to the supernatant to bring it to the next desired concentration, and the solution is stirred and centrifuged as before. Fractions of 0–15%, 15–30%, 30–45%, 45–60%, and 60–85% of saturation are collected. The resulting $(NH_4)_2SO_4$ pellets from each fraction are each dissolved in 2–10 ml of dialysis buffer (10 m$M$ Tris·HCl, pH 7.6, 10% (v/v) glycerol, 0.5 m$M$ DTT), depending on the size of the pellet. The resuspended pellets are dialyzed against several changes of at least 100 times the sample volume of dialysis buffer. Dialysis is continued until the addition of several drops of 1 $M$ $BaCl_2$ to a 1-ml aliquot of the dialysis buffer fails to produce a white precipitate of $BaSO_4$.

Almost 90% of the 5-methyluridine methyltransferase is present in the 45–60% $(NH_4)_2SO_4$ fraction which has an $A_{280}$ of 42/ml, an $A_{260}$ of 26/ml, and a volume of 100 ml.

A methyltransferase activity independent of added tRNA is found in the 30–45% $(NH_4)_2SO_4$ fraction. This methyltransferase activity, which is probably protein methyltransferase(s), contains little or no 5-methyluridine methyltransferase activity and has not been further studied.

*Phosphocellulose Affinity-Elution Chromatography.* Affinity elution chromatography has been used by von der Haar[24] for the purification of aminoacyl-tRNA synthetases. This procedure has been modified in our laboratory for the purification of 5-methyluridine methyltransferase.

Whatman P-11 phosphocellulose is prepared as described by Bur-

[23] B. Roe, M. Sirover, and B. Dudock, *Biochemistry* **12**, 4146 (1973).
[24] F. von der Haar, this series, Vol. 34, p. 163.

gess.[25] The 45–60% $(NH_4)_2SO_4$ fraction, 100 ml is loaded onto a 250-ml phosphocellulose column (19 × 4 cm) previously equilibrated with 30 m*M* potassium phosphate, pH 7.2, 1 m*M* EDTA, 10% (v/v) glycerol, and 0.5 m*M* DTT (PC-A buffer). The column is washed with 50 ml of PC-A buffer. Most of the proteins void the column under these conditions. Another $A_{280}$ absorbance peak is then eluted with 300 ml of PC-A buffer containing 0.2 *M* KCl. This is slightly less salt than is required to elute the 5-methyluridine methyltransferase. The column is then reequilibrated with 500 ml of PC-A buffer. The 5-methyluridine methyltransferase is then affinity-eluted from the column with 500 ml of PC-A buffer containing 0.1 $A_{260}$ units of wheat germ $tRNA_1^{Gly}$ or wheat germ $tRNA_2^{Gly}$ per milliliter. Both of these tRNAs completely lack 5-methyluridine and are substrates for *E. coli* 5-methyluridine methyltransferase.[9,22] The phosphocellulose column has a running time of 8 hr. The column fractions

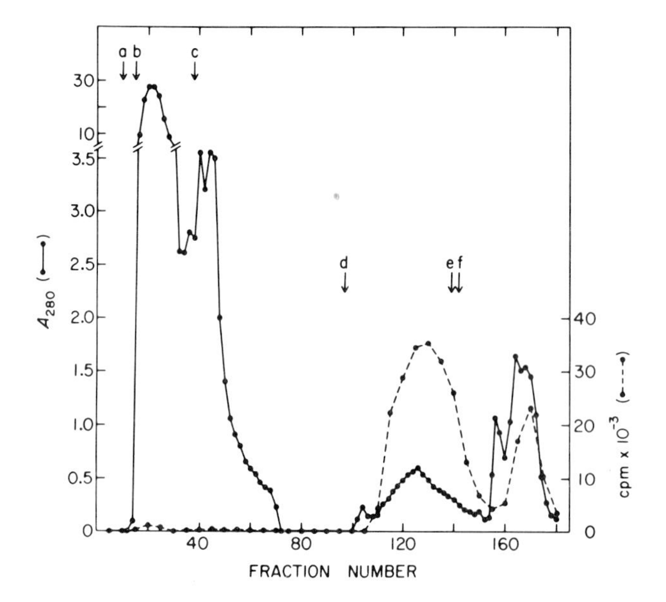


FIG. 1. Phosphocellulose affinity-elution chromatography. The 45–60% $(NH_4)_2$ $SO_4$ fraction of the pooled enzyme from the initial DEAE-cellulose column was applied to a phosphocellulose column as described in the text. Arrows indicate the start of each buffer as follows: (a) PC-A buffer; (b) PC-A buffer containing 0.2 KCl; (c) PC-A buffer; (d) PC-A buffer containing wheat germ $tRNA_2^{Gly}$; (e) PC-A buffer; (f) PC-A buffer containing 0.5 *M* KCl. —, $A_{280}$; - - -, 5-methyluridine methyltransferase activity. Ten-milliliter fractions were collected at a rate of 120 ml per hour and 4-$\mu$l aliquots of each were assayed for 30 min as described.

[25] R. R. Burgess, *J. Biol. Chem.* **244**, 6160 (1969).

(10 ml) are assayed as discussed above, and the affinity-eluted methyltransferase peak is pooled. The column chromatographic profile of this purification step is shown in Fig. 1.

In the affinity-elution procedure described above, the buffer containing tRNA removes only about two-thirds of the 5-methyluridine methyltransferase activity and results, following the DEAE cellulose gradient column described below, in a 45- to 50-fold purification. Further refinement of the affinity elution procedure is in progress in an attempt to enhance the recovery while retaining the same level of purification.

*DEAE-Cellulose Chromatography.* The affinity-eluted enzyme (370 ml) is then applied to a 50-ml DEAE-cellulose column (53 × 1.1 cm) previously equilibrated with 10 m*M* Tris·HCl, pH 7.6, 10 m*M* $MgCl_2$, 10% (v/v) glycerol, and 0.5 m*M* DTT. The column is washed with 10 ml of this buffer, and the 5-methyluridine methyltransferase is eluted with a 200-ml linear gradient of 0–0.28 *M* KCl in 10 m*M* Tris·HCl, pH 7.6, 10 m*M* $MgCl_2$, 10% (v/v) glycerol, and 0.5 m*M* DTT (Fig. 2). Under these conditions the tRNA is quantitatively retained. The column effluent, collected in 1-ml fractions, is assayed, the activity peak pooled, and the resulting 25-ml sample brought to 85% of $(NH_4)_2SO_4$ saturation and centrifuged at 10,000 rpm (12,000 *g*) in a Sorvall SS-34 rotor for 30 min.

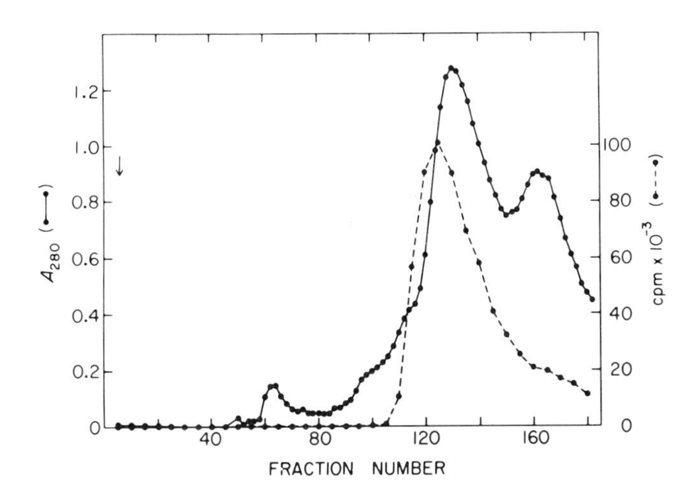


FIG. 2. DEAE-cellulose chromatography. 5-Methyluridine methyltransferase purified by phosphocellulose affinity elution chromatography was applied to a DEAE-cellulose column as described in the text. No $A_{280}$ absorbing material eluted from the column during application of the sample. The arrow indicates the start of the 0–0.28 *M* KCl gradient. ——, $A_{280}$; - - -, 5-methyluridine methyltransferase activity. One milliliter fractions were collected at a rate of 80 ml/hr, and 4-$\mu$l aliquots of each were assayed for 30 min as described.

*Hydroxyapatite Chromatography.* Hydroxyapatite (BioGel HTP, Bio-Rad Laboratories) is mixed in a ratio of 4 g of HTP/1 g of Whatman powdered cellulose and defined five times prior to use.[23] The 0–85% $(NH_4)_2SO_4$ pellet from the DEAE-cellulose column which follows the phosphocellulose affinity elution step is dialyzed into a solution of 10 m*M* potassium phosphate, pH 7.0, 2 m*M* EDTA, 20% (v/v) glycerol, and 0.5 m*M* DTT (HA-A buffer). The resulting 3-ml sample has an $A_{280}$ of 6/ml and an $A_{260}$ of 3.6/ml. One milliliter is diluted with HA-A buffer to a concentration of 1 $A_{280}$ unit/ml and is applied to a 10-ml hydroxyapatite column (29 × 0.7 cm) that was previously equilibrated with HA-A buffer. The column is washed with 10 ml of HA-A buffer and 10 ml of buffer containing 50 m*M* potassium phosphate, pH 7.0, 2 m*M* EDTA, 20%- (v/v) glycerol, and 0.5 m*M* DTT. The enzyme is eluted with a linear gradient of 50 m*M* potassium phosphate, pH 7.0 (start buffer) to 0.1 *M* potassium phosphate, pH 7.0 (limit buffer), both solutions containing 2 m*M* EDTA, 20% (v/v) glycerol, and 0.5 m*M* DTT. The total volume of the gradient is 40 ml. The column is washed with two additional column volumes of

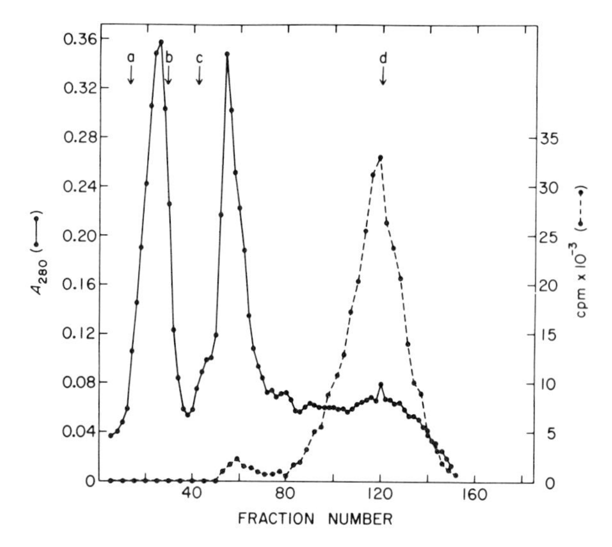


FIG. 3. Hydroxyapatite chromatography. 5-Methyluridine methyltransferase purified by phosphocellulose affinity elution chromatography followed by DEAE cellulose chromatography was applied to a hydroxyapatite column as described in the text. Arrows indicate the start of each buffer as follows: (a) 0.01 *M* potassium phosphate buffer; (b) 0.05 *M* potassium phosphate buffer; (c) 0.05–0.1 *M* potassium phosphate gradient; (d) 0.1 *M* potassium phosphate buffer. The solid line represents $A_{280}$ and the dashed line represents 5-methyluridine methyltransferase activity. 0.5 ml fractions were collected at a rate of 20 ml/hr and 4 $\mu$l aliquots of each were assayed for 30 minutes as described.

limit buffer. The column effluent is collected in 0.5-ml fractions and assayed for enzymic activity (Fig. 3). The pooled enzyme peak is concentrated by precipitation with $(NH_4)_2SO_4$ (85% of saturation) and dialyzed into a solution of 10 m$M$ Tris · HCl, pH 7.6, 10% (v/v) glycerol, and 0.5 m$M$ DTT. The enzyme is distributed in 50- or 100-μl quantities into prefrozen plastic tubes and is stable at −70° for at least 6 months. The purification of 5-methyluridine methyltransferase through the hydroxyapatite step is summarized in Table I.

TABLE I
PURIFICATION OF 5-METHYLURIDINE METHYLTRANSFERASE FROM *E. coli* MRE 600

| Enzyme fraction | Total protein[a] (mg) | Enzyme units[b] | Specific activity (enzyme units/mg) |
|---|---|---|---|
| S215 | 11,830 | 14,608[c] | 1.23 |
| 45–60% $(NH_4)_2SO_4$ fraction | 1,740 | 49,326 | 28.3 |
| Affinity-eluted DEAE-cellulose enzyme | 5.68 | 7,595 | 1337.2 |
| Hydroxyapatite enzyme | 0.32[d] | 204[d] | 638.2 |

[a] Protein concentration is determined by the $A_{280}/A_{260}$ method of Warburg and Christian [O. Warburg and W. Christian, *Biochem. Z.* **310,** 384 (1941)] using the table by Layne (this series, Vol. 3, p. 447).

[b] An enzyme unit is the amount of enzyme that catalyzes 1 pmol of methylation per minute.

[c] The apparently low number of enzyme units in the S215 extract is due to tRNA methylation inhibitors (such as RNA-independent methyltransferases), which are removed by $(NH_4)_2SO_4$ fractionation.

[d] This yield was obtained when one-third of the protein, i.e., 1.9 mg of the 5.68 mg obtained from the affinity-eluted DEAE cellulose chromatographed enzyme is applied to a 10 ml hydroxyapatite column.

*Isoelectric Focusing.* Recently, work has begun in our laboratory using Ampholine isoelectric focusing for the purification of 5-methyluridine methyltransferase. A 0 to 60% (w/v) glycerol gradient containing 1% Ampholine solution, pH 4–8, and 0.1 m$M$ DTT is built in an LKB Ampholine column No. LKB8100-1. The column is prerun to form the pH gradient as described.[26,27] The concentrated enzyme sample is di-

[26] LKB Instruction Manual I-8100-EO4.

[27] P. G. Righetti and J. W. Drysdale, "Isoelectric Focusing. Laboratory Techniques in Biochemistry and Molecular Biology" (T. S. Work and E. Work, eds.), p. 337. North-Holland/American Elsevier, Amsterdam/New York, 1976.

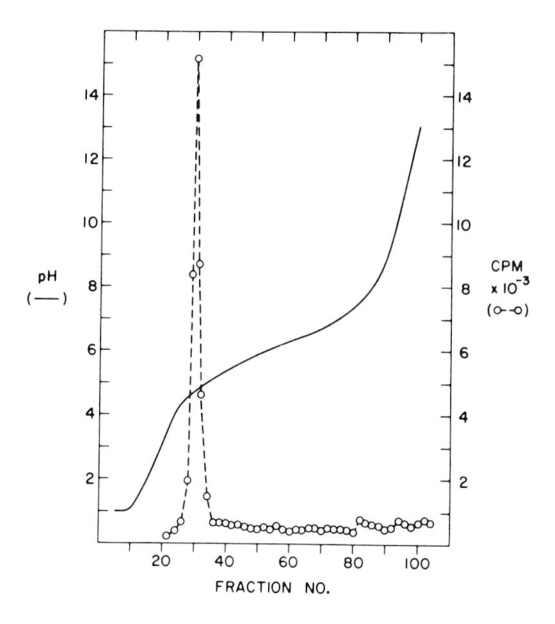


FIG. 4. Isoelectric focusing. 5-Methyluridine methyltransferase, 800 μl purified by hydroxyapatite chromatography, was applied to the column and focused as described in the text. ○ - - - ○, 5-Methyluridine methyltransferase activity; ——, pH gradient. Fractions contained 1 ml each, and 4-μl aliquots were assayed for 2 hr as described.

alyzed into a solution of 10 m*M* Tris·HCl, pH 7.6, 10% (v/v) glycerol, and 0.5 m*M* DTT and is layered inside the gradient at approximately the position at which it is expected to be focused. If the p*I* is not known, it may be layered in the center of the column. The enzyme is then focused for 12 hr at 1000 V, and the gradient is eluted from the column. Fractions are then assayed directly as described above. The resolution obtained is shown in Fig. 4. More recently a pH gradient of 4–6 has been found to give even better resolution.

*Phosphocellulose Chromatography.* An alternative method of methyltransferase purification on phosphocellulose has been published by Glick and Leboy.[6] A modification of this procedure has been used in our laboratory for the partial purification of both 5-methyluridine methyltransferase and uridine 5-oxyacetic acid methylester methyltransferase. Pellets from a 0 to 60% $(NH_4)_2SO_4$ fraction from the first DEAE-cellulose column are resuspended and dialyzed into a solution of 0.5 m*M* EDTA, 0.5 m*M* DTT, and 10% (v/v) glycerol (PC-G buffer) containing 0.1 m*M* potassium phosphate, pH 6.5. The dialyzed sample (19 ml) has an $A_{260}$

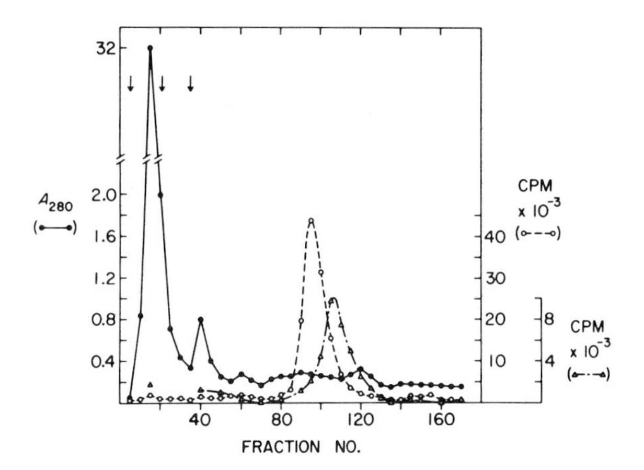


FIG. 5. Phosphocellulose gradient chromatography. The 0 to 60% $(NH_4)_2SO_4$ fraction of the pooled enzyme from the initial DEAE-cellulose column was applied to a phosphocellulose column as described in the text. ●——●, $A_{280}$; ○ - - - ○, 5-methyluridine methyltransferase activity; △ —·— △, uridine 5-oxyacetic acid methylester methyltransferase activity. Arrows indicate the start of the 0.1 *M* potassium phosphate buffer, the 0.15 *M* potassium phosphate buffer, and the 0.15 to 0.35 *M* potassium phosphate gradient, respectively. Fractions contained 5 ml each, and 4-μl aliquots were assayed for 30 min as described in the text.

of 38/ml and $A_{280}$ of 64/ml. The material is then loaded onto a 100-ml phosphocellulose column (35 × 1.9 cm) previously equilibrated with PC-G buffer containing 0.1 *M* potassium phosphate, pH 6.5. The column is washed with 80 ml of this buffer followed by 80 ml of PC-G buffer containing 0.15 *M* potassium phosphate, pH 6.5; the enzyme is then eluted with a linear gradient of 0.15 *M* potassium phosphate, pH 6.5, to 0.35 *M* potassium phosphate, pH 7.2, both in PC-G buffer. Fractions of 5 ml are collected at a flow rate of about 1 ml/min and assayed. The activity peak is pooled, and solid $(NH_4)_2SO_4$ is added to 85% of saturation. The pelleted precipitate is dialyzed into dialysis buffer (10 m*M* Tris·HCl, pH 7.6, 10% (v/v) glycerol, and 0.5 m*M* DTT), distributed in 50- or 100-μl quantities into prefrozen tubes, and stored at −70°. The chromatographic profile for this purification is shown in Fig. 5.

As previously stated, this procedure is used in our laboratory for the partial purification of uridine 5-oxyacetic acid methylester methyltransferase. This enzyme catalyzes the formation of uridine 5-oxyacetic acid methylester on several *E. coli* tRNAs including $tRNA_1^{Ala}$, $tRNA_1^{Ser}$, $tRNA^{Pro}_{major}$, and $tRNA^{Pro}_{minor}$ (unpublished results of B. D.). This methylation product is quite labile to both mild acid and alkaline treatment. Indeed, only through gentle extraction of the tRNA is the hydrolysis of this modification avoided. Thus, this modification can be readily removed from the tRNA, and the methylation reaction, as well as its effects, can be studied in a totally homologous system. Unlike the uridine 5-oxyacetic acid methylester methyltransferase, which can be assayed in a normal homologous system, tRNA methyltransferases are generally assayed with heterologous tRNA, or undermethylated homologous tRNA. The latter is usually obtained by drug treatment[28-33] or through the use of methyl-deficient tRNA mutants.[34-36]

## Properties

*Specificity.* tRNA methyltransferases show a high degree of substrate specificity. A given methyltransferase modifies only a specific nucleotide at a specific site in a particular environment. Indeed, these enzymes require the tRNA to be in a specific spatial configuration for methylation to take place[37,38]; mere possession of the "correct" nucleotide sequence appears to be insufficient to ensure proper methylation. Because of this high level of substate specificity shown by tRNA methyltransferases, 5-methyluridine methyltransferase has been used in our laboratory as a probe for detecting tRNA-like moieties in viral RNA.[39]

As previously stated, Ado-Met is most often found to be the methyl donor in tRNA methylation reactions. More recent studies, however, indicate that some organisms, such as *Streptococcus faecalis, Bacillus subtilis* and probably *B. cereus* use tetrahydrofolate derivatives as the source of the methyl group in the formation of 5-methyluridine.[40-42]

In the study of tRNA methyltransferases *in vitro, S*-adenosylhomo-

[28] E. Wainfan, J. S. Tscherne, F. A. Maschio, and M. E. Balis, *Cancer Res.* **37,** 865 (1977).
[29] S. J. Kerr, *Cancer Res.* **35,** 2969 (1975).
[30] E. Wainfan, M. L. Moller, F. A. Maschio, and M. E. Balis, *Cancer Res.* **35,** 2830 (1975).
[31] E. Wainfan, J. Chu, and G. B. Chheda, *Biochem. Pathol.* **24,** 83 (1975).
[32] P. Jank and H. J. Gross, *Nucl. Acids Res.* **1,** 1259 (1974).
[33] J. S. Tscherne and E. Wainfan, *Nucl. Acids Res.* **5,** 451 (1978).
[34] N. Biezunski, D. Giveon, and U. Z. Littauer, *Biochim. Biophys. Acta* **199,** 382 (1970).
[35] G. R. Björk and F. C. Neidhardt, *J. Bacteriol.* **124,** 99 (1975).
[36] M. G. Marinus, N. R. Morris, D. Söll, and T. C. Kwong, *J. Bacteriol.* **122,** 257 (1975).
[37] L. P. Shershneva, T. V. Venkstern, and A. A. Bayev, *Biochim. Biophys. Acta* **294,** 250 (1973).
[38] Y. Kuchino, T. Seno, and S. Nishimura, *Biochem. Biophys. Res. Commun.* **43,** 476 (1971).
[39] K. Marcu and B. Dudock, *Biochem. Biophys. Res. Commun.* **62,** 798 (1975).
[40] A. S. Delk, J. M. Romeo, D. P. Nagle, Jr., and J. C. Rabinowitz, *J. Biol. Chem.* **251,** 7649 (1976).
[41] H. Kersten, L. Sandig, H. H. Arnold, *FEBS Lett.* **55,** 57 (1975).
[42] W. Schmidt, H. H. Arnold, and H. Kersten, *Nucl. Acids Res.* **2,** 1043 (1975).

cysteine (Ado-Hcy) is an important inhibitor and can interfere with the methylation reaction. Different methyltransferases show quite varied sensitivities to this by-product of the methylation reaction. For example, a recent study of three tRNA methyltransferases from rat liver found that they differ more than 20-fold in their sensitivities to Ado-Hcy.[43] Both 5-methyluridine methyltransferase and uridine-5 oxyacetic acid methylester methyltransferase show 50% inhibition of the rate of methylation at an Ado-Hcy concentration of 2 $\mu M$ (Lesiewicz and Dudock[44] and our unpublished results). Since Ado-Hcy is an end product of methylation, the study of a particular methyltransferase may be significantly complicated by the presence of competing methyltransferases. The use of a pure tRNA that is a specific substrate for the particular methyltransferase to be studied is often a way to overcome this difficulty.

*Precautions with S-Adenosyl-L-methionine*. It is important to handle the methyl donor, Ado-Met, with considerable care, as it is quite labile under some conditions. It should be stored as an acidic solution, frozen, preferably at −70° or colder, and should not be allowed to warm to a temperature above 0° prior to use in the assay. Ado-Met rapidly degrades in neutral and alkaline environments[45] and therefore should not be mixed with the other assay components until immediately before incubation.

[43] J. M. Glick, S. Ross, and P. S. Leboy, *Nucl. Acids Res.* **2**, 1639 (1975).
[44] J. Lesiewicz and B. Dudock, *Fed. Proc., Fed. Am. Soc. Exp. Biol.* **36**, 705 (1977).
[45] L. W. Parks and F. Schlenk, *J. Biol. Chem.* **230**, 295 (1958).

# [21] Assay for AA-tRNA Recognition by the EFTu-GTP Complex of *Escherichia coli*

*By* JAMES OFENGAND

## Principle

Recognition of AA-tRNA (but not tRNA) by the EFTu-GTP complex of *Escherichia coli* and by the EFI-GTP complex of other organisms is central to the process of protein synthesis since this complex is an essential intermediate in the process by which all AA-tRNA's become attached to their proper site on the ribosome (the A site). The prerequisite formation of this intermediate also provides a way to screen out unwanted tRNA's such as deacylated tRNA or *N*-acylated AA-tRNA, which might otherwise block the protein-synthesizing machinery. Either of these two species



ISBN 0-12-504180-2

would cause premature termination of peptide chains if bound to the A site during peptide chain elongation.

On the other hand, the EFTu-GTP complex does not, and indeed must not, discriminate against any normal AA-tRNA despite their wide variation in such parameters as size, shape, and number of modified bases. The recognition system involves more than the nature of the 3′-end of the tRNA since at least two "denatured" forms of AA-tRNA, the stable denatured form of Leu-tRNA of yeast and the initiator Met-tRNA$_i^{Met}$ of most species, are known not to interact with the EFTu-GTP complex, or to do so only weakly.[1–5] In this sense, EFTu-GTP recognition of AA-tRNA structure is the reverse of AA-tRNA synthetase recognition of tRNA, since while the latter must be able to distinguish those features unique to a given tRNA or set of tRNA's, the former must be able to filter out these differences and instead focus on the common features shared by all tRNA molecules.

A number of modifications of AA-tRNA's have been studied in view of the considerations mentioned above, in an attempt to map out the regions of the tRNA that are essential to interaction with the Tu-GTP complex. This article describes a simple technique for measuring this interaction, based on earlier observations[6,7] that while EFTu and EFTu-GTP complexes adsorb to nitrocellulose membrane filters, the ternary complex of AA-tRNA-EFTu-GTP does not. This effect is the basis of the present assay. Several aspects of this assay are also covered in volume XXX, Part F [22]. A second method is also described for use when it is not certain that the modified AA-tRNA-EFTu-GTP ternary complex, if formed, would pass through the filter. This possibility arises since a molecular explanation for the change in adsorption properties upon forming a ternary complex is not available. In this competition method, the test AA-tRNA competes with a standard AA-tRNA for a limiting amount of EFTu-GTP.

## Materials

[$^3$H]GTP, approximately 1 Ci/mmole
EFTu-GDP, see article [22], Volume XXX, Part F

[1] C. Jerez, A. Sandoval, J. Allende, C. Henes, and J. Ofengand, *Biochemistry* **8,** 3006 (1969).
[2] C.-M. Chen, *Fed. Proc., Fed. Amer. Soc. Exp. Biol.* (Abstract) 1283 (1971).
[3] Y. Ono, A. Skoultchi, A. Klein, and P. Lengyel, *Nature (London)* **220,** 1304 (1968).
[4] D. Richter and F. Lipmann, *Nature (London)* **227,** 1212 (1970).
[5] D. Richter, F. Lipmann, A. Tarragó, and J. E. Allende, *Proc. Nat. Acad. Sci. U.S.* **68,** 1805 (1971).
[6] J. M. Ravel, R. L. Shorey, and W. Shive, *Biochem. Biophys. Res. Commun.* **29,** 68 (1967).
[7] J. Gordon, *Proc. Nat. Acad. Sci. U.S.* **59,** 179 (1968).

Aminoacyl-tRNA
Pyruvate kinase (Boehringer), crystalline, in 2 *M* $(NH_4)_2SO_4$
Phosphoenol pyruvate, trisodium
Wash buffer: 10 m*M* $MgCl_2$, 10 m*M* $NH_4Cl$, 10 m*M* Tris·HCl, pH 7.4
Cellulose nitrate filters, Millipore HAWP, 0.45 μ, 2.5 cm, or equivalent

### Procedure

*Direct Assay.* EFTu-GDP and any contaminating GDP in the commercial radioactive GTP preparation is first converted to EFTu-GTP and GTP, respectively, by preliminary incubation at 37° for 10 minutes in a 100–150 μl reaction mixture containing 7.5 μmoles of Tris·HCl, pH 7.4, 7.5 μmoles of $NH_4Cl$, 1.5 μmole of $MgCl_2$, 0.75 μmole of dithiothreitol, 1 μmole of phosphoenolpyruvate, 15 μg of pyruvate kinase, 1.5 nmole of [$^3$H]GTP, and 15–50 units[8] of EFTu-GDP. This preliminary incubation also ensures complete equilibration of the added radioactive GTP with the protein-bound GTP (see article [22], Volume XXX, Part F). The reaction mixtures are then placed at 0°, AA-tRNA is added to make the final volume 150 μl, and incubation is continued for 5 minutes at 0°. At the end of the reaction, a 100-μl aliquot is directly pipetted onto a Millipore filter prewetted with wash buffer and immediately filtered. The filter is washed 3 times with 2 ml of cold wash buffer, dissolved in 10 ml of Bray's solution, and counted. The amount of ternary complex formed is computed as the difference in filter-bound radioactive GDP in the absence and in the presence of AA-tRNA. An illustration of the use of the direct assay is given in Fig. 1.

*Competition Assay.* Using the standard assay conditions described above and a limiting amount of EFTu-GTP, a mixture of the test AA-tRNA and standard AA-tRNA (preferably a control AA-tRNA of the same type) is added and allowed to react. The amount of ternary complex is calculated as above. Any *decrease* in the amount of ternary complex as compared to the control AA-tRNA alone is taken to be due to formation of a ternary complex between EFTu-GTP and the test AA-tRNA, which is not filterable. An example of this assay is given in Fig. 2.

Alternatively, if the amino acid group of the AA-tRNA is sufficiently radioactive, nonfilterable AA-tRNA-EFTu-GTP can be detected directly since uncomplexed AA-tRNA will pass through the filter.

### Remarks

*EFTu-GTP Complex.* The units of EFTu as measured by GTP binding may not be a true measure of complexes functional for AA-tRNA inter-

[8] One unit is that amount of protein which causes 1 pmole of [$^3$H]GDP to be retained on a Millipore filter after incubation with $10^{-5}$ *M* GDP as described above for GTP.

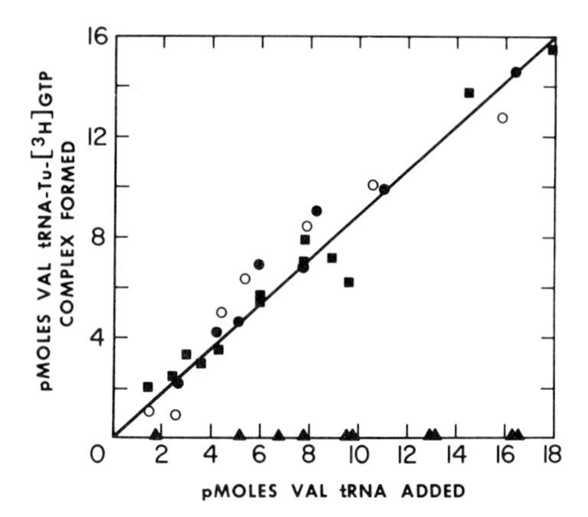


FIG. 1. Ability of valyl-tRNA and fragments to form a ternary complex with EFTu-GTP [M. Krauskopf, C.-M. Chen, and J. Ofengand, *J. Biol. Chem.* **247,** 842 (1972)]. The formation of Val-tRNA-Tu-GTP complex was measured by the direct filter assay as described in the section on procedure. In this composite of three experiments, 100% [$^3$H]GTP retained on the filter ranged from 34 to 52 pmoles. ■, untreated Val-tRNA; ●, Val-tRNA (cleaved at the anticodon); ▲, Val-tRNA(3′) (3′ half-molecule esterified with valine); ○, Val-tRNA(3′ + 5′) prepared by reannealing Val-tRNA(3′) with the 5′ half of the molecule.

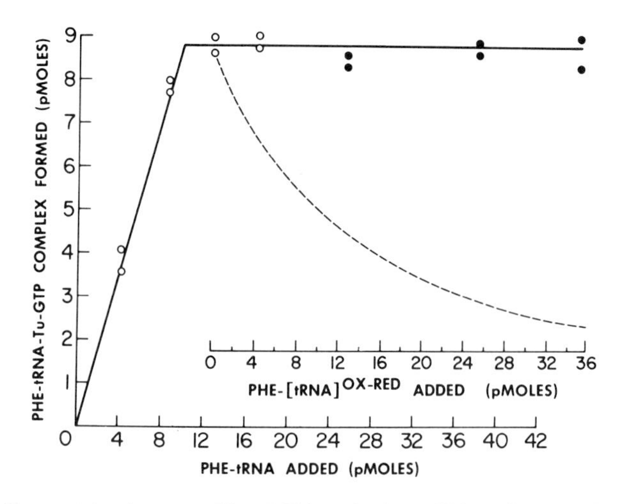


FIG. 2. Competition between Phe-tRNA and Phe-tRNA$^{ox\text{-}red}$ for binding to EFTu-GTP [J. Ofengand and C.-M. Chen, *J. Biol. Chem.* **247,** 2049 (1972)]. In the absence of Phe-tRNA, 14.5 pmoles of EFTu-GTP were bound. (○), Phe-tRNA; (●), Phe-tRNA$^{ox\text{-}red}$ in the presence of 12.8 pmoles of Phe-tRNA. The theoretical line (- - - - - - -) illustrates the expected behavior for equal competition by both Phe-tRNA's for the limiting amount of EFTu-GTP assuming that Phe-tRNA-EFTu-GTP is filterable and Phe-tRNA$^{ox\text{-}red}$-EFTu-GTP is not filterable.

MODIFIED AA-tRNA'S TESTED FOR EFTu-GTP RECOGNITION

| Modified AA-tRNA | Modification region | Result | Reference |
|---|---|---|---|
| Deacylated | 3′ End | Inactive | *a* |
| *N*-Acetyl AA-tRNA | 3′ End | Inactive | *a* |
| α-Hydroxyacyl-tRNA (EFI – wheat embryo) | 3′ End | Inactive | *a* |
| Phenylacetyl tRNA | 3′ End | Weakly active | *b* |
| Phe-tRNA$^{ox-red}$ | 3′ End | Inactive | *c* |
| Val-tRNA (phosphodiester chain broken) | Anticodon | Active | *d* |
| Phe-tRNA (phosphodiester chain broken) | Anticodon | Active | *e* |
| Arg-tRNA (cyanoethylation of anticodon inosine) | Anticodon | Active | *f* |
| Phe-tRNA$^{yeast}$ (Y base removed by acid) | Anticodon | Active | *g* |
| Val-tRNA ($^4Srd_8$ – $C_{13}$ cross-linked) | hU arm | Active | *d* |
| Val-tRNA(3′) (5′ half of molecule removed) | General | Inactive | *d* |
| Leu-tRNA (denatured) | General | Inactive or weakly active | *a, h* |
| Met-tRNA$_f^{Met}$ | General | Inactive | *i* |
| Met-tRNA$_i^{Met}$ | General | Weakly active | *j, k* |
| Trp-tRNA (denatured) | General | Active | *l* |
| Fluorouracil-substituted Val-tRNA | General | Active | *m* |
| Phe-tRNA (CCCA) | 3′ end | Weakly active | *n* |

[a] C. Jerez, A. Sandoval, J. Allende, C. Henes, and J. Ofengand, *Biochemistry* **8,** 3006 (1969).
[b] S. Fahnestock, H. Weissbach, and A. Rich, *Biochim. Biophys. Acta* **269,** 62 (1972).
[c] J. Ofengand and C.-M. Chen, *J. Biol. Chem.* **247,** 2049 (1972).
[d] M. Krauskopf, C.-M. Chen, and J. Ofengand, *J. Biol. Chem.* **247,** 842 (1972).
[e] M. N. Thang, M. Springer, D. C. Thang, and M. Grunberg-Manago, *FEBS* (*Fed. Eur. Biochem. Soc.*) *Lett.* **17,** 221 (1971).
[f] C.-M. Chen, M. Krauskopf, J. Hachmann, H. Weissbach, and J. Ofengand, *Plant Physiol.* **46,** S-30 (1970).
[g] K. Ghosh and H. P. Ghosh, *Biochem. Biophys. Res. Commun.* **40,** 135 (1970).
[h] C.-M. Chen, *Fed. Proc., Fed. Amer. Soc. Exp. Biol.* **30** (Abstract) 1283 (1971).
[i] Y. Ono, A. Skoultchi, A. Klein, and P. Lengyel, *Nature* (*London*) **220,** 1304 (1968).
[j] D. Richter and F. Lipmann, *Nature* (*London*) **227,** 1212 (1970).
[k] D. Richter, F. Lipmann, A. Tarragó, and J. E. Allende, *Proc. Nat. Acad. Sci. U.S.* **68,** 1805 (1971).
[l] J. Ofengand and J. Bierbaum, unpublished results.
[m] J. Ofengand and J. Horowitz, unpublished results.
[n] M. N. Thang, L. Dondon, D. C. Thang, and B. Rether, *FEBS Letters* (*Fed. Eur. Biochem. Soc.*) **26,** 145 (1972).

action since when EFTu ages, it apparently loses its ability to bind AA-tRNA before the ability to bind GTP. This is detected by the failure to complex all the EFTu-GTP capable of adsorption to a filter even with a severalfold excess of AA-tRNA (see for example, Fig. 2 and article [22], Volume XXX, Part F). However, since the amount of ternary complex formed is calculated as the difference between the EFTu-GTP bound in the presence and in the absence of AA-tTNA, this increased "blank" value does not normally create a problem. In the assay as described in the Procedure, the GTP concentration of $10^{-5}$ $M$ is high enough to maintain 97% of the added EFTu in the form of a GTP complex, calculated from the reported $K_d$ of $3 \times 10^{-7}$ $M$ (see article [22], Volume XXX, Part F). Even if much lower concentrations, such as 0.5 $\mu M$, were used so that only 63% of the EFTu were bound to GTP, the only effect would be to decrease the expected stoichiometry of 1.0 for AA-tRNA bound to AA-tRNA added by the constant factor $[\text{GTP}]/(K_d + [\text{GTP}])$. This is so because the dissociation constant for the ternary complex is <1 n$M$ (see below), so that the addition of AA-tRNA acts essentially to remove a stoichiometric quantity of EFTu-GTP from the system.

*Apparent Failure to Form a Ternary Complex.* The complete failure to detect ternary complex formation, as illustrated in Fig. 1 for Val-tRNA(3′), actually means only that the dissociation constant is high relative to the control tRNA. For example, one can calculate that for this experiment, a $K_d > 4$ $\mu M$ would not have been reliably detected. We estimate that the error involved in determining the amount of EFTu-GTP bound to the filter to be 5% of that bound in the absence of AA-tRNA, so that decreases of less than 5% are not reliably measurable. This limit and the concentration of test tRNA added determine the minimum value for $K_d$. Thus when no detectable decrease in EFTu-GTP bound to the filter is observed, $K_d$ may be said to be greater than twenty times the concentration of AA-tRNA added.

*Apparent Failure to Inhibit Ternary Complex Formation.* The $K_d$ for unmodified AA-tRNA is probably <1 n$M$ (see article [22], Volume XXX, Part F). Estimation of this value is based on the experimental observations of Figure 2, and other similar experiments which show that at least 90% of the AA-tRNA added is bound in ternary complex. Taking the ratio (AA-tRNA):(AA-tRNA-EFTu-GTP) to be 0.1 or less on this basis, and estimating the minimum equilibrium concentration of free EFTu-GTP functional for complex formation as 10 n$M$, the $K_d$ value of <1 n$M$ follows from the relation, $K_d$ = (EFTu-GTP)(AA-tRNA):(AA-tRNA-EFTu-GTP).

It is important to recognize, therefore, that modifications that have relatively small effects on the affinity of AA-tRNA for the EFTu-GTP complex may not be detected by this technique in view of the very low $K_d$ for normal AA-tRNA. Thus a modification which changes $K_d$ by 10-fold may not show any effect at all if $K_d$ for the modified AA-tRNA is still below the detectability limits of the assay conditions used.

*tRNA Modifications Tested.* A compilation of the various modifications of AA-tRNA which have been specifically investigated in terms of their effect on complex formation with EFTu-GTP is given in the table.

## [22] Mapping the Structure of Specific Protein–Transfer RNA Complexes by a Tritium Labeling Method

*By* PAUL R. SCHIMMEL and HUBERT J. P. SCHOEMAKER

In recent years there has been a growing interest in the molecular architecture of complexes formed between two or more macromolecules. Many such macromolecular complexes are found in protein synthesis, such as complexes of transfer RNAs with their aminoacyl tRNA synthetases and with the various protein synthesis factors, as well as the large ribosomal complex involving many proteins bound with ribosomal RNAs. Protein-nucleic acid complexes are also found in many other situations, such as the association of repressors with their specific sites on genes, in the structure of chromatin, and in virus particles.

In attempting to determine the mechanism of specific protein-nucleic acid interactions, the sheer size of the reacting partners presents a major experimental obstacle. For example, techniques such as X-ray diffraction and high-resolution nuclear magnetic resonance (NMR), which are useful for elucidating the structures of smaller complexes, have great technical difficulties as well as problems in data interpretation when applied to complexes involving two or more macromolecules. As a result, much effort has been directed at developing alternative methods for determining architectural features of protein-nucleic acid complexes.

In developing any new technique, a prime concern is that the method itself does not substantially perturb the complex under investigation and thereby give meaningless results. For example, any technique that uses unusual solution conditions or harsh reagents is undesirable because it is likely to disturb the often delicate protein-nucleic acid complexes under investigation. In addition, large amounts of material are often not available so that a useful method must be able to analyze relatively small quantities (milligram amounts or less). It is a major challenge to develop



ISBN 0-12-504180-2

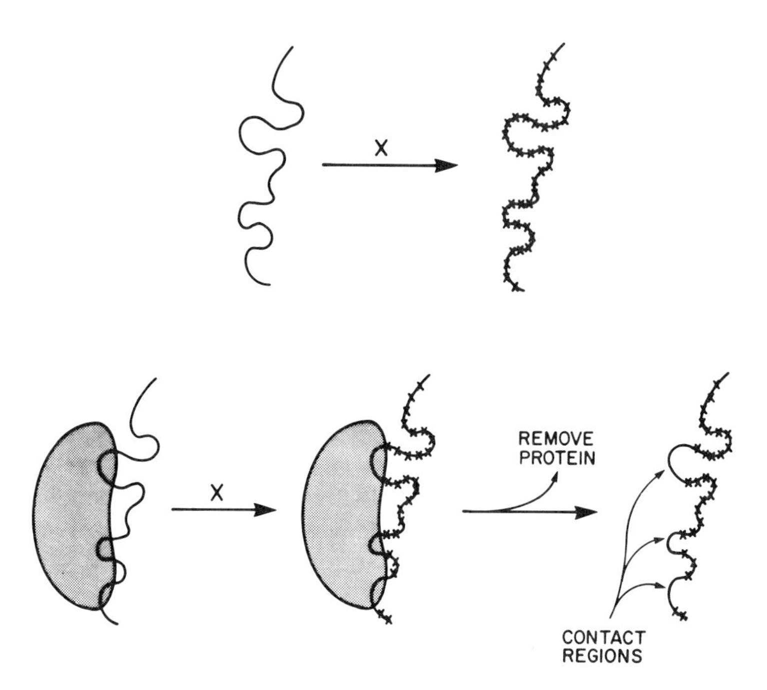


FIG. 1. Illustration of an experiment in which a nucleic acid (wavy line) is allowed to react with a reagent X in the presence and in the absence of bound protein. Adapted with permission, from P. R. Schimmel, *Acc. Chem. Res.* **10,** 411 (1977). Copyright by the American Chemical Society.

techniques that circumvent these difficulties and at the same time yield a substantial amount of structural information.

In considering this problem, we envisioned a general procedure that is outlined in Fig. 1.[1] The wavy line designates a nucleic acid and the globular particle is a protein. It is imagined that all parts of the nucleic acid can react with a reagent X, which neither disturbs the nucleic acid structure nor interferes with protein-nucleic acid complex formation. When protein is added, the reagent X cannot react with those parts of nucleic acid that are in contact with or shielded by the protein. By determining the reagent labeling pattern in the presence and in the absence of bound protein, one can determine which parts of the nucleic acid make contact with the protein. This is a big step forward in deducing the architectural features of a protein-nucleic acid complex.

The main problem is finding a reagent X and solution conditions that fulfill the criteria mentioned above. We found that an attractive reaction

[1] P. R. Schimmel, *Acc. Chem. Res.* **10,** 411 (1977).

is the H-8 exchange reaction of purine nucleotides.[2–13] This reaction may be followed by carrying out the exchange reaction in deuterated or tritiated water according to the following scheme, illustrated for an adenine nucleotide

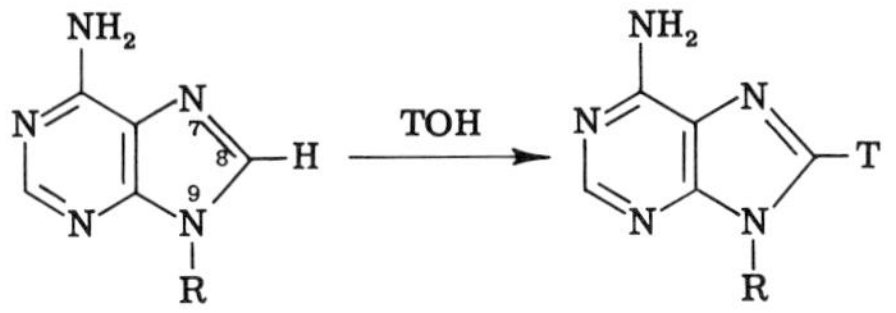

This slow reaction can be observed directly by NMR[2,3] or Raman spectroscopy.[4] The exchange occurs with all the purine bases in the nucleic acid chain.

Since tritium is an innocuous substitute for hydrogen, and since the exchange reaction can be carried out under physiological conditions,[11–13] it is clear that tritium labeling of purines is an ideal and gentle way to map the structure of protein–nucleic acid complexes. This is done by determining the labeling rates of the various bases in a nucleic acid in the presence and in the absence of the bound protein. Since tritium is an atomic-size probe, it provides a level of structural discrimination greater than that achievable with a more bulky reagent.

In the discussion below, the general features of the tritium labeling reaction as applied to protein–nucleic acid complexes are discussed. A specific application is illustrated with the complex between an aminoacyl-tRNA synthetase and its cognate tRNA.

[2] M. P. Schweizer, S. I. Chan, G. K. Helmkamp, and P. O. P. Ts'o, *J. Am. Chem. Soc.* **86,** 696 (1964).

[3] F. J. Bullock and O. Jardetzky, *J. Org. Chem.* **29,** 1988 (1964).

[4] K. R. Shelton and J. M. Clark, Jr., *Biochemistry* **6,** 2735 (1967).

[5] D. G. Search, *Biochim. Biophys. Acta* **166,** 360 (1968).

[6] F. Doppler-Bernardi and G. Felsenfeld, *Biopolymers* **8,** 733 (1969).

[7] R. N. Maslova, E. A. Lesnik, and Y. M. Varshavskii, *Mol. Biol.* **3,** 728 (1969).

[8] M. Maeda, M. Saneyoshi, and Y. Kawazoe, *Chem. Pharm. Bull.* **19,** 1641 (1971).

[9] M. Tomasz, J. Olson, and C. M. Mercado, *Biochemistry* **11,** 1235 (1972).

[10] E. A. Lesnik, R. N. Maslova, T. G. Samsonidze, and Y. M. Varshavskii, *FEBS Lett.* **33,** 7 (1973).

[11] R. C. Gamble, H. J. P. Schoemaker, E. Jekowsky, and P. R. Schimmel, *Biochemistry* **15,** 2791 (1976).

[12] H. J. P. Schoemaker, R. C. Gamble, G. P. Budzik, and P. R. Schimmel, *Biochemstry* **15,** 2800 (1976).

[13] H. J. P. Schoemaker and P. R. Schimmel, *J. Biol. Chem.* **251,** 6823 (1976).

General Considerations

At the outset it is useful to consider the general features of the H-8 exchange reaction and their implications for studying topological features of protein-nucleic acid complexes.

*Mechanism of H-8 Exchange.* Several reports indicate that the H-8 exchange proceeds through an ylid mechanism.[9,14,15] The reaction is envisioned to require first protonation at N-7; this is followed by removal of the proton from C-8 to give the ylid. The reaction is completed by reprotonation at C-8 and dissociation of the proton from N-7. In aqueous solution, for both adenosine and guanosine the reaction is pH independent in the range of pH 5-7.[8,9] It is clear from the proposed mechanism that there are several ways in which binding of a protein could perturb the H-8 exchange rate of a purine nucleotide. In addition to shielding the nucleotide from the solvent and thereby inhibiting exchange, alterations in the acidity of N-7 or of C-8 will also effect the exchange rate.

*Rate Constants for H-8 Exchange are Small.* The exchange reaction is first order in the purine nucleotide concentration[8,9,14-17] so that the tritium labeling may be written as

$$-d\Delta P^*/dt = k\Delta P^* \tag{1}$$

where $\Delta P^*$ is the deviation of the labeled purine concentration ($P^*$) from its final value ($\bar{P}^*$) achieved at isotopic equilibrium, and $k$ is the first order-rate constant. In accordance with Eq. (1) isotopic equilibrium is approached in a simple exponential fashion given by

$$\Delta P^* = (\Delta P^*)_0 e^{-kt} \tag{2}$$

where $(\Delta P^*)_0$ is the value of $\Delta P^*$ at $t = 0$.

The rate constants for H-8 exchange are very small. At 37° around neutral pH the rate constant for 5′-AMP is 0.083% $h^{-1}$ while for 5′-GMP it is 0.17% $h^{-1}$.[11] Rate constants for the 3-nucleotides and for the nucleosides are similar to those for the 5′-nucleotides.[18]

Although the rate constants for exchange are extremely small, tritium labeling is readily studied at 37° by using water with a sufficiently high

[14] J. A. Elvidge, J. R. Jones, C. O'Brien, E. A. Evans, and H. C. Sheppard, *J. Chem. Soc., Perkin Trans.* **2,** 2138 (1973).

[15] J. A. Elvidge, J. R. Jones, C. O'Brien, E. A. Evans, and H. C. Sheppard, *J. Chem. Soc., Perkin Trans.* **2,** 174 (1974).

[16] J. Livramento and G. J. Thomas, Jr., *J. Am. Chem. Soc.* **96,** 6529 (1974).

[17] J. A. Elvidge, J. R. Jones, and C. O'Brien, *Chem. Commun.* p. 394 (1971).

[18] R. C. Gamble, Ph.D. thesis, Massachusetts Institute of Technology, Cambridge, 1975.

specific activity. For example, consider a 1-ml solution of 0.1 m*M* AMP dissolved in tritiated water having a specific activity of 1 Ci/ml (even higher specific activities are available from New England Nuclear). In 1 hr at 37°, about 460 cpm are incorporated into the 8 position of AMP, assuming a 28% counting efficiency (which is a common efficiency in our experiments). In the same situation, approximately twice as much is incorporated into GMP. With a typical background of about 35 cpm, these levels, and even lower levels, are easy to detect by scintillation counting.

*The Small Rate Constants Are a Crucial Advantage.* The small rate constants associated with H-8 exchange give studies of this reaction a crucial advantage over the powerful techniques that monitor exchange of more rapidly exchanging sites, such as hydrogens on hydrogen-bonded amino groups.[19,20] The low exchange rates mean that once tritium has been incorporated into nucleic acid material at the 8 positions of the purine nucleotide units, this material may be worked up and analyzed without fear of losing a significant amount of the radioactivity that was originally incorporated. For example, if 400 cpm are incorporated into an A residue at a specific position in the sequence of a nucleic acid polymer, time-consuming nuclease digestions and chromatographic separations must be performed in order to isolate and determine the specific activity of this particular nucleotide unit within the polymer. These procedures may take 24 hr. Using the rate constant for H-8 exchange for AMP at 37°, it is easy to calculate that only $0.083 \times 10^{-2} \times 24 \times 400 \cong 8$ cpm are lost during this time. The actual amount lost can be significantly less because separations are generally done at room temperature or below, where the exchange rate is considerably less (see below). This type of analysis of specific sites is not possible with relatively fast exchanging hydrogens, where existing techniques do not permit the measurement of exchange rates of distinct sites throughout a nucleic acid chain. Instead, only the overall exchange of the entire polymer is measured.

The small values for the H-8 exchange rate constants also mean that it is clearly desirable to study tritium exchange-in as opposed to tritium exchange-out. When studying exchange-in, one starts with unlabeled nucleotide units and incorporates a few hundred or more cpm. If, on the other hand, one starts with [8-$^3$H]nucleotide in nonradioactive $H_2O$, the percentage change in the cpm of the labeled nucleotides in a period of a

[19] S. W. Englander, *Biochemistry* **2,** 798 (1963).

[20] S. W. Englander, N. W. Downer, and H. Teitelbaum, *Annu. Rev. Biochem.* **41,** 903 (1972).

few hours is insignificant, making accurate detection of the exchange reaction extremely difficult.

*H-8 Exchange Is Strongly Temperature Dependent.* The H-8 exchange reaction has an activation energy of about 22 kcal/mol for both nucleotides.[11] This means that for a 10° change in temperature the tritium labeling rate changes about 3-fold. This has certain practical ramifications. First of all, to extend tritium labeling measurements to temperatures substantially below 37° necessitates what generally are inconveniently long incubation times. However, this also has an advantage. As mentioned above, the work-up involved in determining the specific activity of a particular nucleotide unit in a nucleic acid chain may take many hours. Since these procedures are usually done at room temperature or lower, the significantly smaller rate constants at lower temperatures diminishes further the problem of the back-exchange reaction.

A second consideration is that there are certain situations in which it is desirable to do the exchange reaction at temperatures substantially above 37°. For example, because the tritium labeling rate is sensitive to microenvironment (see below), it is possible to study the melting of nucleic acid helices by the tritium-labeling method.[11,21] Since these transitions frequently occur above 60°, tritium labeling can be done in situations where the exchange rate is an order of magnitude or more greater than it is at 37°. This permits shorter incubation times. Along the same lines, it is sometimes advantageous to study the properties of protein-nucleic acid complexes from thermophilic organisms, where complexes may be stable up to 60° or more.

*Microenvironment Strongly Affects H-8 Exchange Rates.* To be useful in studying protein-nucleic acid complexes, it is crucial that the H-8 exchange rate be sensitive to microenvironment, in order that nucleotide units close to or embedded in a protein matrix have altered tritium-labeling rates. This criteria was shown to be fulfilled in a variety of different ways.

One of the most interesting observations is that the labeling rates of purines within nucleic acids are sensitive to the conformation of the nucleic acid itself. This is illustrated in Table I for several nucleic acids. The table gives values of $R$ at different temperatures for various nucleic acids. The parameter $R$ is defined as the ratio of the rate constant for H-8 exchange for a free purine mononucleotide to that of the corresponding nucleotide within the nucleic acid polymer. (A residues in a polymer are compared to AMP, and G residues to GMP.) Thus, a value of $R = 10$ for

[21] R. C. Gamble and P. R. Schimmel, *Proc. Natl. Acad. Sci. U.S.A.* **71**, 1356 (1974).

TABLE I
VALUES OF TRITIUM LABELING PARAMETER $R$ FOR VARIOUS NUCLEIC ACIDS AT DIFFERENT TEMPERATURES AT ABOUT pH 6.5[a]

| Poly(A) | Poly(A):poly(U) | Viral RNA | tRNA | DNA | Temperature (°C) |
|---|---|---|---|---|---|
| 1.05 | 0.97 | — | 1.0 | 0.97 | 90-100 |
| 1.6 | 4.8 | — | 4.5 | 1.6 | 60 |
| 3.1 | 17 | 36(A)<br>44(G) | 7.9 | 2.6 | 37 |

[a] Adapted from R. C. Gamble, H. J. P. Schoemaker, E. Jekowsky, and P. R. Schimmel, *Biochemistry* **15**, 2791 (1976).

an A residue in a polymer means that the A residue within the polymer exchanges at one-tenth the rate of free AMP. The table shows that at high temperatures, where the polymers are completely melted out, the $R$ values are each about 1.0 (in the case of tRNA and DNA, the $R$ values refer to an average for the A's and G's). In contrast, when the temperature is lowered, the $R$ values increase substantially, showing that the labeling rates are markedly retarded by formation of ordered nucleic acid structures. In the case of single-stranded poly(A), which is about 55–65% stacked at 37°,[22,23] the $R$ value rises to 3.1 at this temperature. But in the synthetic poly(A):poly(U) duplex, which is completely helical at 37° according to hyperchromicity measurements, the labeling rate is retarded 17-fold. An even greater effect is seen in the viral RNA duplex. In the case of DNA, the duplex shows only a 2.6-fold effect.

These data show that not only are the labeling rates perturbed substantially when the nucleic acid goes into a helical conformation, but the magnitude of the effect is significantly dependent upon the helix itself. The differences between the DNA and the viral RNA helix might be ascribed to the different geometries of the RNA and DNA helices, which in turn provide different microenvironments for the bases.[24] Other explanations can also be considered.[11] But the important point is that the labeling rate itself is sensitive to microenvironment, as many studies have shown.[7,9–13,21] This suggests that the reaction can have useful applications.

For the purpose of exploring protein-nucleic acid complexes, the

[22] M. Leng and G. Felsenfeld, *J. Mol. Biol.* **15**, 455 (1966).
[23] J. Brahms, A. M. Michelson, and K. E. van Holde, *J. Mol. Biol.* **15**, 467 (1966).
[24] S. Arnott, S. D. Dover, and A. J. Wonacott, *Acta Crystallogr., Sect. B* **25**, 2192 (1969).

Table II
VALUES OF $R$ FOR FREE AND BOUND ADENINE NUCLEOTIDES AT 37°[a]

| Protein and nucleotide | $R$ |
|---|---|
| ATP | 1.0 |
| Serum albumin + ATP | 1.3 |
| Ile-tRNA synthetase + ATP | >3.0 |
| Ile-tRNA synthetase + ATP + Ile | 5.3 |

[a] Adapted from H. J. P. Schoemaker and P. R. Schimmel, *J. Biol. Chem.* **251,** 6823 (1976).

simplest preliminary experiment is to check the effect of a bound protein on the labeling rate of a mononucleotide. A straightforward experiment is to examine the effect of an aminoacyl tRNA synthetase on the labeling rate of the substrate ATP. The results of this kind of experiment with Ile-tRNA synthetase are summarized in Table II. In the presence of a nonspecific protein, such as serum albumin, the $R$ value of ATP is only slightly changed. This could be due to some nonspecific binding of the nucleotide to serum albumin. But in the presence of Ile-tRNA synthetase, $R$ rises to a value greater than 3.0, showing a significant retardation in the exchange rate. Because the $K_m$ for ATP is relatively high,[25-27] it was not convenient to use synthetase concentrations that saturate the nucleotide; this is the reason why only a lower bound to $R$ is reported in Table II. However, the binding of the nucleotide can be greatly enhanced by adding isoleucine, so that the more tightly bound aminoacyl adenylate is formed.[28] Measurements of this species show that $R$ rises to 5.3, indicating a substantial effect of the protein on the bound nucleotide.

The experiments cited in Tables I and II leave little question that tritium labeling at the C-8 position is sensitive to microenvironment and offers promise for studying in detail the contact points on the nucleic acid in a protein-nucleic acid complex. This expectation has been fulfilled, as illustrated below for a particular system.

[25] F. X. Cole and P. R. Schimmel, *Biochemistry* **9,** 480 (1970).

[26] E. Holler, E. L. Bennett, and M. Calvin, *Biochem. Biophys. Res. Commun.* **45,** 409 (1971).

[27] M. R. McNeil and P. R. Schimmel, *Arch. Biochem. Biophys.* **152,** 175 (1972).

[28] E. Holler, P. Rainey, A. Orme, E. L. Bennett, and M. Calvin, *Biochemistry* **12,** 1150 (1973).

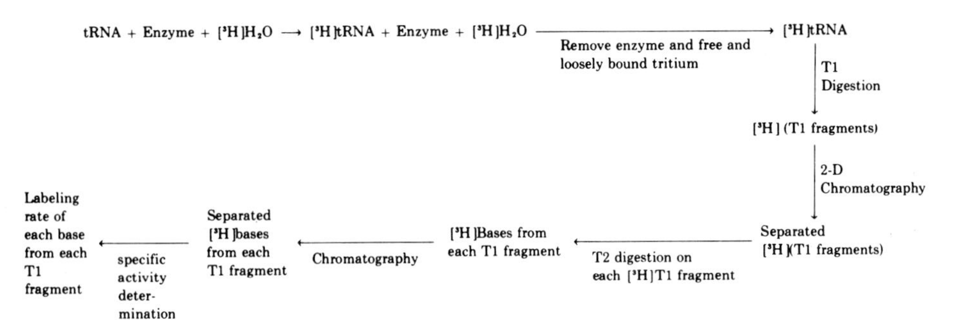


SCHEME 1. General protocol for obtaining labeling rates of specific sites. Adapted from H. J. P. Schoemaker and P. R. Schimmel, *J. Biol. Chem.* **251**, 6823 (1976).

## Mapping Protein–Nucleic Acid Complexes

*General Protocol.* The main objective is to determine the tritium labeling rates of free and bound nucleotide units distributed throughout a nucleic acid structure. In the discussion below, an illustration is given for protein-tRNA complexes, but the general procedures and ideas are applicable to virtually any protein-nucleic acid system.

The general protocol is outlined in Scheme 1. The nucleic acid and protein are first incubated in [$^3$H]$H_2O$ at 37° for several hours. After the incubation, the protein and free and loosely bound tritium are removed by one of several procedures (see below). The labeled nucleic acid is then subjected to a T1 ribonuclease digestion. This nuclease cleaves only after G residues and generates a characteristic series of fragments known as T1 fragments (see below). The fragments are separated by two-dimensional chromatography. Each of these fragments contains 1 G (at the 3′ terminus of each fragment) and may or may not contain one or more A residues. To determine the individual labeling rates of A's and G's within the T1 fragments, each fragment is subjected to a T2 digestion; this enzyme cleaves after every base. The bases from the T2 digestion on each T1 fragment are then separated by chromatography and the specific activities determined on the individual A's and G's. In this way the labeling rates of the purines from each T1 fragment are obtained.

The purpose of the nuclease digestions in the protocol of Scheme 1 is to enable isolation of specific purine units from the nucleic acid chain. This can be seen more clearly by considering the cloverleaf structure of tRNA$^{\text{Ile}}$[29] given in Fig. 2. In this figure, a lowercase number designates every fifth base from the 5′ end of the chain. Dotted lines enclose T1

[29] M. Yarus and B. G Barrell, *Biochem. Biophys. Res. Commun.* **43**, 729 (1971).

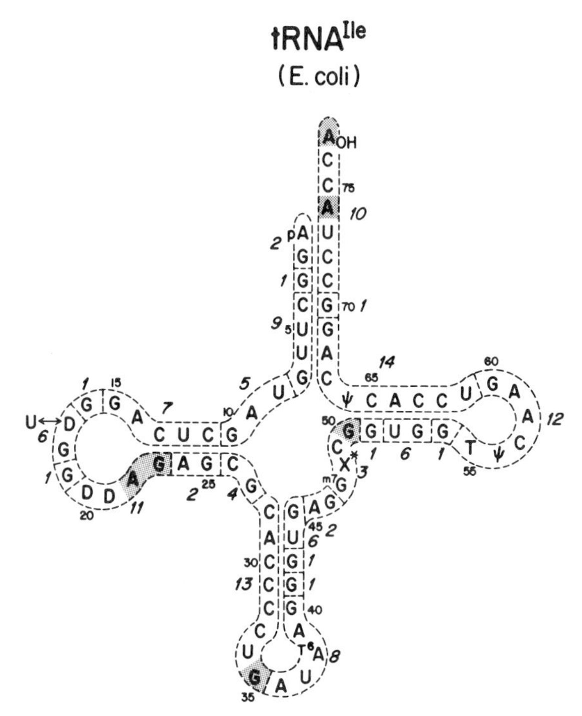


FIG. 2. Sequence and cloverleaf structure of tRNA$^{\mathrm{Ile}}$ [M. Yarus and B. G. Barrell, *Biochem. Biophys. Res. Commun.* **43,** 729 (1971)]. Dotted outlines enclose T1 fragments that are numbered in accordance with their positions on a chromatogram. Lowercase numbers indicate every 5th base from the 5′ end. Shaded bases are those perturbed in their tritium labeling by bound tRNA$^{\mathrm{Ile}}$. Adapted from H. J. P. Schoemaker and P. R. Schimmel, *J. Biol. Chem.* **251,** 6823 (1976).

fragments that are numbered with uppercase numbers. The numbering assignment corresponds to positions of these fragments on a two-dimensional chromatogram that is illustrated and discussed below. It is clear from this figure that by isolating the T1 fragments, and consequently performing a T2 digestion it is possible to obtain labeling rates on a large number of the specific A's and G's in a chain. For example, by determining the labeling rate of the A and the G in fragment 5, specific information on the 9 and 10 position in the nucleic acid is obtained. On the other hand, this combination of nucleases does not resolve every site in the chain; some fragments are redundant (such as fragment 2), and some contain more than one A unit (such as fragment 10). In these cases the digestions mix together residues from two or more distinct locations in the chain. However, this complication can be overcome by using a different combination of nucleases, in addition to the T1-T2 combination, or other special procedures.[11]

*Amounts of Materials Required.* As mentioned above, the unperturbed tritium labeling rate of AMP is such that with a 28% counting efficiency a 1 ml, 0.1 m*M* solution in tritiated water of specific activity 1 Ci/ml incorporates about 460 cpm in 1 hr at 37° ; this is generally well above background levels (35 cpm). In the case of protein-nucleic acid complexes, allowance must be made for the retarded labeling rates of the A's and G's in the nucleic acid structure, and for the further alterations in labeling rates that occur at specific sites when these are embedded in a protein. Compensations for these effects can be made by using water with a somewhat higher specific activity (2-3 Ci/ml) and by increasing the incubation time to several hours. In this way it is possible to obtain a sufficiently accurate (± 15%) measure of the labeling rates of individual bases in a nucleic acid chain.[13]

*Incubation Conditions and Work-up.* The incubation conditions are illustrated for the complex of Ile-tRNA synthetase and $tRNA^{Ile}$. The 400 $\mu$l reaction mixture contains 50 m*M* sodium cacodylate (pH 6.5), 10 m*M* $MgCl_2$, 0.1 m*M* dithiothreitol, 15% glycerol, 200 $\mu M$ $tRNA^{Ile}$ and a roughly 1.5-fold molar excess of enzyme, in tritiated water having a specific activity of 2–3 Ci/ml. These concentrations of enzyme and tRNA are well above the dissociation constant of the complex.[30] Incubations are carried out at 37° for 5–15 hr.

In order to analyze the labeled nucleic acid at the end of the incubation, it is necessary to separate it from the enzyme and subsequently subject it to digestions with nucleases (see Scheme 1). To achieve separation from enzyme, and from free and loosely bound tritium, three different procedures may be used. In one of these the reaction mixture is diluted to 1.5 ml with $H_2O$ and subsequently extracted with an equal amount of phenol. Back-extractions with several volumes of ether remove the residual phenol and the aqueous phase containing the tRNA is lyophilized to dryness. The tRNA is then successively lyophilized 5 additional times from 100-$\mu$l volumes. This procedure removes free and lossely bound tritium. After this, the tRNA is dissolved in 500 $\mu$l of water and dialyzed for about 12 hr against approximately 1 liter of $H_2O$ (with three changes).

As a second alternative procedure, the phenol-extracted tRNA is dialyzed for about 24 hr against approximately 1 liter of $H_2O$ (with six changes). No lyophilizations are performed.

[30] S. S. M. Lam and P. R. Schimmel, *Biochemistry* **14,** 2775 (1975).

Finally, as a third alternative, the tRNA and protein are precipitated from the incubation solution by 2.5 volumes of ethanol. The precipitate is collected by centrifugation and then partially redissolved in 1.5 ml of $H_2O$. After extracting the solution with an equal volume of phenol, the tRNA is worked up as described for the first approach.

The three different work-up procedures give equivalent results.

*Nuclease Digestion and Chromatographic Separations.* The isolated tRNA (about 1 mg or 20 $A_{260}$ units) is lyophilized to dryness and then dissolved in a 200-$\mu$l reaction mixture containing 100-200 units of T1 RNase (Calbiochem; units are defined by Takahashi[31]) and 60-80 $\mu$g of bacterial alkaline phosphatase (Worthington) in 10 m$M$ $NH_4HCO_3$ (pH 7.5). (Bacterial alkaline phosphatase is included in the digestion because removal of the 5′-terminal phosphates from the T1 fragments facilitates chromatographic separations.) Digestion is carried out for 5 hr at 37°, and the reaction is then lyophilized. The residue is taken up in aout 50 $\mu$l of $H_2O$.

This solution is then spotted onto several 20 × 20 cm cellulose thin-layer plates (0.1 mm, Brinkmann). About 1.5 $A_{260}$ units are spotted onto each plate, so that with 20 $A_{260}$ units about 12-14 plates are used. Chromatography is done first in solvent 1 (55% 1-propanol, 35% $H_2O$, and 10% concentrated $NH_4OH$) and then at a right angle in solvent II (66% isobutyric acid, 33% $H_2O$, and 1% concentrated $NH_4OH$). The separated oligonucleotides are visualized under ultraviolet light.

Figure 3 shows the two-dimensional chromatogram of $tRNA^{Ile}$. The numbers on the chromatogram correspond to those of the T1 fragments in Fig. 1. This chromatographic system separates oligonucleotides approximately according to size in the first dimension and according to base composition in the second dimension. It is seen that good separation of each fragment is achieved. The identity of each fragment is established by standard methods.[32]

The spots of cellulose corresponding to each oligonucleotide are scraped, and scrapings for corresponding oligomers are pooled and suspended in 1 ml of 10 m$M$ $NH_4OAc$, pH 4.5. The cellulose is removed by centrifugation; concentrations of oligomers can be determined by $A_{260}$ measurements, using appropriate extinction coefficients.[17] The yield from the elution is about 40-95%, depending on the size of the oligomer (larger ones are eluted less efficiently).

For the purpose of T2 digestions, the solutions containing the oligo-

[31] K. Takahashi, *J. Biochem. (Tokyo)* **49**, 1 (1961).

[32] G. P. Budzik, S. S. M. Lam, H. J. P. Schoemaker, and P. R. Schimmel, *J. Biol. Chem.* **250**, 4433 (1975).

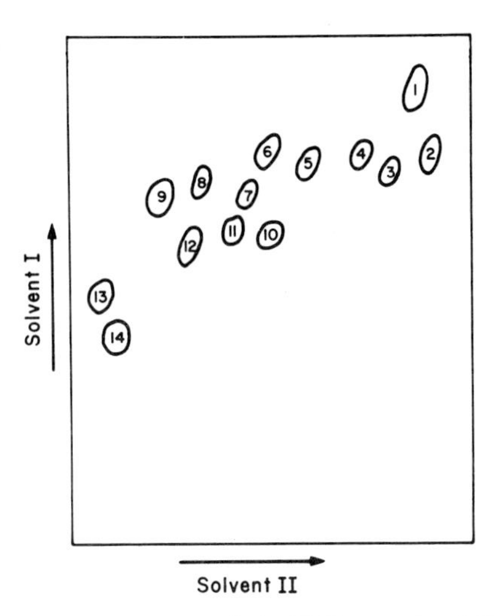


FIG. 3. Two-dimensional chromatogram of the T1 fragments of $tRNA^{Ile}$. Each number corresponds to one of the T1 fragments given in Fig. 2. Adapted from G. P. Budzik, S. S. M. Lam, H. J. P. Shoemaker, and P. R. Schimmel, *J. Biol. Chem.* **250,** 4433 (1975).

mers are lyophilized to dryness and then each is taken up in 100 $\mu$l of 50 m*M* $NH_4OAc$ (pH 4.5) containing 2 units (defined by Uchida[33]) of T2 RNase (Calbiochem) per $A_{260}$ unit of nucleic acid oligomer. Digestion proceeds for 5 hr at 37° and is followed by lyophilization. After several lyophilizations to remove the $NH_4OAc$, the residue of each digestion is taken up in a 100-$\mu$l reaction mixture of 10 m*M* $NH_4HCO_3$ (pH 8) and 3-5 $\mu$g of bacterial alkaline phosphatase per $A_{260}$ unit of nucleotide material. The reaction mixture is incubated for 5 hr at 37°, lyophilized to dryness, taken up in a few microliters of $H_2O$, and spotted on a cellulose thin-layer plate developed in two dimensions with solvents I and II. [It is important that all salt ($NH_4OAc$) be removed by the lyophilization, since too much salt causes streaking on the thin-layer chromatogram. Thus, it may be necessary to repeat the lyophilization before proceeding with the chromatography.]

After chromatography the individual mononucleosides are located by ultraviolet light and eluted from the plates as described above. Concentrations are determined by absorbance measurements at 260 nm, and radioactivities are determined by suspending the samples in 3-6 ml of Aquasol (New England Nuclear) and counting by scintillation. In this way, specific activities for each of the purine bases in each T1 fragment are obtained.

[33] T. Uchida, *J. Biochem. (Tokyo)* **60,** 115 (1966).

Experiments in the absence of protein are carried out in a similar fashion.

As mentioned above, after T1 ribonuclease digestion and chromatography the yield of an oligonucleotide is about 40–95%. The subsequent T2 ribonuclease and alkaline phosphatase digestions, followed by chromatography and elution, give each base in about 80% yield. Thus, for the entire procedure, the overall yield for a given base in a tRNA chain is approximately 30-75%.

As a reference point, and for the purpose of calculating $R$ values (see above), it is desirable to obtain labeling rates for AMP and GMP. This can be done in the manner described above, except that no nuclease digestions are required. After tritiation, the labeled nucleotide is removed from free and loosely bound tritium by the lyophilizations and, additionally or alternatively, dialysis (see above). It is then chromatographed (without phosphatase treatment) and analyzed as described above for the mononucleosides obtained from digesting labeled tRNA.

*Accuracy of Results.* In the example cited above, about 1 mg, or 40 nmol, of tRNA was subjected to tritium labeling. At the end of the entire procedure each base should be obtained in approximately 30–75% or greater yield. Thus, commonly, 20 nmol of each specific purine in the tRNA are obtained. These amounts are easy to determine accurately since 20 nmol of A or G in a 1-ml solution should give an absorbance reading at 260 nm of about 0.31 (A) or 0.13 (G).

Assuming a 28% counting efficiency, an A residue that is unperturbed and labels like free AMP should give a measured incorporation of about 4.6 cpm/nmol per hour in a solution with a $[^3H]H_2O$ specific activity of 1 Ci/ml. For a 10-hr incubation with a $[^3H]H_2O$ specific activity of 2 Ci/ml the unperturbed unit should give about 92 cpm/nmol, or about 1800 cpm for 20 nmol. However, in a folded RNA chain, such as tRNA at 37°, labeling rates may be reduced from 2- to 20-fold from that of the unperturbed values.[11,12] The reduction in labeling rate can be even greater when a unit is bound tightly to a protein (see below). Thus, if an A residue is perturbed about 10-fold (i.e., $R = 10$), about 200 cpm (above background) will be detected in that A residue in the above-described experiment. For G residues, for which the incorporation rate is a little over 2-fold faster than for A residues, the observed counts are correspondingly higher.

Thus, incorporation of 200–400 cpm (above background) into a specific nucleotide unit is commonly achieved. With a background of 35 cpm, and with accurate concentration determinations made possible by the ample $A_{260}$ material, the reproducibility of an experimental result on a particular nucleotide is generally better than ± 15%.[11,13] Even in situations where a base is so highly perturbed that only 50–100 cpm (above background) are incorporated, the fairly constant background level enables reasonably accurate results to be obtained.

In situations where the incorporation is unusually low owing to marked perturbation of a residue, greater amounts of nucleic acid can be used or enhanced incorporation may be achieved by extending the incubation time and raising somewhat the specific activity of [$^3$H]$H_2O$. Prolonged incubations increase the chance for denaturation of the material under study; this must be checked by performing assays at the end of the incubation to be certain that no significant deterioration has taken place.

*Presentation of Results.* A convenient way to present tritium-labeling data is by a histogram, or bar graph. This is done in Fig. 4, which summarizes data on $tRNA^{Ile}$ in the presence and in the absence of Ile-tRNA synthetase. The abscissa designates specific purine nucleotides that can be identified in Fig. 2, and the ordinate gives values of $R$. The dark, narrow bars correspond to labeling rates for the free tRNA, and the light, shaded bars refer to the bound tRNA. The bar graph is separated into two sections according to residues that are unperturbed and perturbed, respectively, by the bound enzyme. This type of representation of the data gives a quick view of the most important findings.

*General Features of the Data.* In the absence of bound enzyme, the $R$ values fluctuate considerably according to their position in the sequence. For example, the retardation is only a little more than 2-fold for the terminal adenosine (A77), but is close to 20 for G69. In this tRNA, and in two other specific tRNAs variations of these kinds have been shown to correlate closely with secondary and tertiary structural features of tRNA.[11,12] Thus, the tritium labeling rates are remarkably sensitive to the microenvironments at the different positions along the chain.

In the presence of enzyme, six bases are significantly perturbed by the bound protein. In all cases, the labeling is retarded by the protein. These bases are indicated by shading in Fig. 2.

A number of the purines have been omitted from the bar graph in Fig. 4. The omitted purines correspond to those that occur in redundant fragments [such as fragment 2, which is 3-fold redundant (see Fig. 2)] or which occur in those fragments that have more than one copy of an A residue (such as fragment 12). An exception is fragment 10, for which special procedures were used to separate A74 from A77.[11] The data for the purines that are omitted from Fig. 4 could also be included in the bar graph, but interpretation of these sites is ambiguous because the observed labeling rates cannot be ascribed to a single base. However, for the

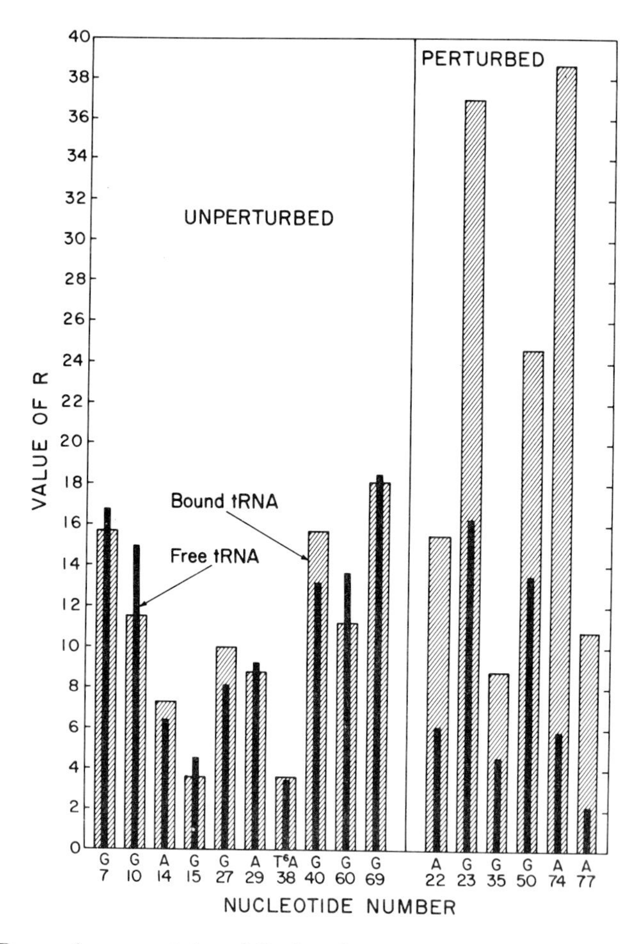


FIG. 4. Bar graph representation of $R$ values for specific bases of $tRNA^{Ile}$ in the presence and in the absence of bound Ile-tRNA synthetase at pH 6.5, 37°. Nucleotides are identified by kind (A or G) and numerical position in the sequence. See Fig. 2 for exact locations of bases in the structure. Adapted from H. J. P. Schoemaker and P. R. Schimmel, *J. Biol. Chem.* **251**, 6823 (1976).

example of $tRNA^{Ile}$ given in Fig. 4, none of the "ambiguous" purines show an altered labeling rate due to bound synthetase. Of course, with a redundant fragment, for example, one cannot rule out the possibility that an effect in one specific position in the tRNA is offset by an opposite effect in another position, where both positions fall in the same redundant fragment. As mentioned above, these ambiguities can be clarified by using a different set of nucleases or other special procedures.

## Discussion

*Validity of the Results.* An important question is whether the retarded labeling rates caused by bound protein are due to the close proximity of parts of the bound protein to the affected sites or to secondary effects. For example, a conformational distortion of the tRNA by the bound protein might affect the labeling rates in certain positions without the protein actually coming close to the affected sites.

To answer this question, it is helpful to examine more closely which bases are affected by the enzyme in the example of Fig. 4. The biggest effects occur at A22, G23, A74, and A77. In another type of study, using photochemical crosslinking to identify enzyme–tRNA contact sites in this same complex, these four bases have been shown to be close to the surface of the enzyme.[32] In the case of G35, another of the affected sites (see Fig. 4), three other experimental approaches have indicated that this base is close to the surface of the enzyme.[34–36] Thus, five out of the six sites identified in the tritium-labeling studies have been implicated in other investigations as enzyme–tRNA contact or close proximity points. With the exception of G50, for which no other comparative data are available that either implicate or do not implicate this base as close to the surface of the enzyme, the weight of the combined evidence is that the tritium labeling method identified parts of the nucleic acids that are close to the surface of the enzyme. Thus, the locations of the perturbed bases on the three-dimensional tRNA structure give a topological picture of how the enzyme comes into contact with the tRNA.[13]

*Interpretation of Altered Labeling Rates.* With an ylid mechanism, there are several ways in which a nucleotide unit within a nucleic acid chain can have an altered labeling rate as a result of being bound to or close to a protein moiety. For example, the bound protein may alter the p*K*s of N-7 and C-8; the protein may facilitate exchange by acid–base catalysis; or the protein may increase or decrease the exposure to solvent of the C-8 side of the purine ring.

Altogether labeling rates of specific bases in the native structures of three tRNA species, in addition to the effect of bound synthetase on the labeling rates of bases in $tRNA^{Ile}$, have been examined.[11–13] In all cases where structure formation (such as the native tRNA conformation) or complex formation perturbs the labeling rate well outside of experimental error, the perturbation produces a *reduction* in the tritium incorporation.

[34] P. R. Schimmel, O. C. Uhlenbeck, J. B. Lewis, L. A. Dickson, E. W. Eldred, and A. A. Schreier, *Biochemistry* **11,** 642 (1972).

[35] L. A. Dickson and P. R. Schimmel, *Arch. Biochem. Biophys.* **67,** 638 (1975).

[36] H. J. P. Schoemaker and P. R. Schimmel, *Biochemistry,* **16,** 5454 (1977).

It is plausible to expect that the variety of microenvironments for specific bases produced by the tRNA secondary and tertiary structure, and by the binding of the protein, could produce both accelerated and retarded exchange rates. Since only retarded rates are generally observed, it is possible that in all the cases studied thus far the major effect of structure formation or of the bound protein is on the accessibility of the exchanging site to the solvent.

*Advantages of the Method.* There are at least five major advantages to the tritium-labeling method.

First, an atomic-sized probe is used. Because of its size, the probe is not limited to exposed portions of the nucleic acid (such as single-stranded regions), but can penetrate even the most deeply buried parts of the structure. In addition, the small tritium atom is an innocuous substitute for hydrogen so that introduction of the probe does not significantly distort the structure of the nucleic acids under investigation.

A second advantage is that purines distributed throughout the entire nucleic acid structure can be probed. With other types of probes, such as kethoxal (which reacts with exposed guanosine residues),[37] very limited parts of the nucleic acid structure can be investigated. The removal of this limitation is a major attraction for the tritium-labeling approach.

Third, the conditions used to achieve labeling are gentle, i.e., physiological conditions. Thus, denaturation of delicate materials can be kept to a minimum. Moreover, it is possible to add stabilizing additives, such as glycerol, without affecting the labeling rates. For example, in the system illustrated above, 15% glycerol was used in the incubation mixture; this amount of glycerol produced no effect on the labeling rates.

Fourth, relatively small quantities of materials are required. For example, reliable results can be produced with 40 nmol of nucleic acid (see above). With longer incubation times and higher specific activities of $[^3H]H_2O$, even smaller quantities might be used.

Finally, nucleic acid labeled in its purine H-8 positions is stable with respect to its tritium content. This is a great advantage when further analyses and manipulations are to be done.

*Limitations.* At this point the main limitation is in providing a molecular interpretation of the altered labeling rates. Although data obtained thus far suggest that shielding of the exchanging site from the surrounding solvent plays a major role in producing retarded labeling rates, definitive data are still lacking. Thus, it would be helpful to have studies on model systems in which purine units are embedded into different types of known structures in which solvent accessibility and other factors are varied in

[37] M. Litt, *Biochemistry* **8**, 3249 (1969).

a systematic fashion. Such studies would establish a more rational foundation for the interpretation of results.

A second cautionary consideration is that a purine unit may be very close to a bound protein segment, yet not experience an altered labeling rate. This could happen, for example, if a protein side chain interacts primarily with a phosphate group instead of the base itself. This possibility seems likely in view of data showing that when tRNA phosphate groups are coated with a bulky polyamine, instead of $Mg^{2+}$, labeling rates appear to be about the same as they are for the $Mg^{2+}$-stabilized form.[11] Thus, if a base does not have an altered labeling rate due to the bound protein, one cannot conclude that the base is not close to the surface of the enzyme.

*Future Directions.* As mentioned above, a key goal for the future is to obtain good data on model systems so that the interpretation of labeling rates can be put on a better foundation. However, even without the advantage of these studies, there are many systems that invite investigation by the tritium labeling approach.

First of all, as already shown with transfer RNAs, the method is useful for mapping secondary and tertiary structural features. Thus, nucleic acids of an unknown structure, such as 5 S RNA, interesting sections of ribosomal RNAs and messenger RNAs, regions of particular structural interest in DNA, etc., may all be explored. In these cases, the method may be used to distinguish those units that participate in structure formation from those that do not. Thus, single-stranded, unpaired regions might readily be identified and thereby severely limit the number of structural models that can be considered.

There is also a rich diversity of protein-nucleic acid systems that can be investigated. Much more work can be done on synthetase-tRNA complexes and on complexes of aminoacyl tRNAs with a protein synthesis factor, such as EF-Tu from *E. coli*. The method may also be used to identify contact sites in ribosomal protein-ribosomal RNA complexes, repressor-DNA complexes, histone-DNA complexes, etc. Of course, for all these systems, and for those mentioned above, it will be necessary to develop the appropriate nuclease digestions and chromatographic separations to achieve resolution of individual bases.

In conclusion, the many attractive features of the tritium labeling method combined with its proved usefulness in studying tRNA structure in solution and in mapping the structure of a synthetase-tRNA complex, give considerable encouragement for pursuing investigations with this approach in a wide variety of systems.[38]

[38] The tritium labeling work in our laboratory has been supported by Grant No. GM 15539 from the National Institutes of Health.

# Section II

# Aminoacyl-tRNA Synthetases

# [23] Aminoacyl-tRNA Synthetase Complex from Rat Liver

*By* MURRAY P. DEUTSCHER

Protein biosynthesis is a complex process requiring the interaction of a variety of enzymes and nucleic acids. Although a great deal is known about the enzymatic mechanisms leading to the synthesis of completed proteins, relatively little information has accumulated about the structural organization of the numerous components involved in the process. It is known that a large fraction of the active ribosomes in higher organisms are associated with membranes of the endoplasmic reticulum.[1,2] However, other components such as tRNA, aminoacyl-tRNA synthetases, and transfer factors are generally thought to exist free in the cytoplasm, diffusing to their sites of action when needed. In contrast to such ideas, reports have appeared suggesting that these presumed "soluble" components may actually exist in more organized structures within the cell. Thus, tRNA has been found associated with proteins,[3] with microsomes,[4] and with 10–30 S particles[5]; aminoacyl-tRNA synthetases have been isolated with microsomal and membrane fractions,[6–8] with ribosomes,[9,10] or in a particulate complex[9–12]; and transfer factors have been found to exist in a particulate form.[13,14] These results have prompted the suggestion that the whole protein synthetic apparatus may exist as a highly organized structure which is disrupted upon opening of the cells.[11]

Aminoacyl-tRNA synthetases purified from a variety of sources generally have molecular weights between 70,000 and 200,000. The procedure

[1] G. E. Palade and P. Siekevitz, *J. Biophys. Biochem. Cytol.* **2,** 171 (1956).
[2] P. N. Campbell, *FEBS* (*Fed. Eur. Biochem. Soc.*) *Lett.* **7,** 1 (1970).
[3] E. L. Hess, A. M. Herranen, and S. E. Lagg, *J. Biol. Chem.* **236,** 3020 (1961).
[4] S. H. Wilson and R. V. Quincey, *J. Biol. Chem.* **244,** 1092 (1969).
[5] A. E. Hampel, A. G. Saponara, R. A. Walters, and M. D. Enger, *Biochim. Biophys. Acta* **269,** 428 (1972).
[6] G. D. Hunter, P. Brookes, A. R. Crathorn, and J. A. V. Butler, *Biochem. J.* **73,** 369 (1959).
[7] B. Nisman, M. L. Hirsch, and A. M. Bernard, *Ann. Inst. Pasteur* **95,** 615 (1958).
[8] S. J. Norton, M. D. Key, and S. W. Scholes, *Arch. Biochem. Biophys.* **109,** 7 (1965).
[9] W. K. Roberts and W. H. Coleman, *Biochem. Biophys. Res. Commun.* **46,** 206 (1972).
[10] J. D. Irvin and B. Hardesty, *Biochemistry* **11,** 1915 (1972).
[11] A. K. Bandyopadhyay and M. P. Deutscher, *J. Mol. Biol.* **60,** 113 (1971).
[12] C. Vennegoor and H. Bloemendal, *Eur. J. Biochem.* **26,** 462 (1972).
[13] E. Shelton, E. L. Kuff, E. S. Maxwell, and J. T. Harrington, *J. Cell. Biol.* **45,** 1 (1970).
[14] O. Henriksen and M. Smulsen, *Fed. Proc., Fed. Amer. Soc. Exp. Biol.* **31,** 865 (Abstract) (1972).



ISBN 0-12-504180-2

described below leads to the isolation of a high molecular weight complex (1 to 2 $\times$ $10^6$) containing aminoacyl-tRNA synthetases and most of the cellular tRNA. Under certain circumstances, not completely understood, only partial complexes are isolated. Apparently, some of the less tightly bound synthetases are easily removed from the complex.

## Assay Method

*Principle.* The assay measures the conversion of radioactive, acid-soluble amino acids into an acid-insoluble product due to their attachment to tRNA.[15]

*Reagents*

Tris-acetate buffer, 1 *M*, pH 7.0[16]
$MgCl_2$, 0.1 *M*
ATP, 0.1 *M*
EDTA, 20 m*M*
Bovine serum albumin, 10 mg/ml
[$^{14}$C]Amino acid, 2.5 m*M* (8–30 cpm/pmole)
Rabbit liver tRNA, 10–15 mg/ml
Trichloroacetic acid, 10% and 2.5%, containing 20 m*M* sodium pyrophosphate

*Preparation of Rabbit Liver tRNA.* This procedure is a modification of one published earlier.[17] The reagents required are:

Tris·HCl, 0.1 *M*, pH 7.5—EDTA, 4 m*M*–NaCl, 1 *M*
Liquefied phenol[18] 90%, containing 0.1% *O*-quinolinol
Ethanol, 95%
Sodium acetate, 0.3 *M*, pH 7.0
Isopropanol

The buffers are autoclaved prior to use.

Partially thawed rabbit liver,[19] 100 g, is disrupted in a Waring Blendor for 90 seconds with 150 ml of the Tris–EDTA–NaCl solution and 150 ml of the phenol solution. The mixture is centrifuged at 13,000 rpm for 15 minutes in the GSA rotor of a Sorvall RC-2B centrifuge. The upper aqueous layer is poured off and reextracted with a half-volume of the phenol solution by shaking for 5 minutes. After centrifuging as before, the aqueous layer is removed with a syringe, combined with the first aqueous layer, and the

[15] See M. P. Stulberg and G. D. Novelli, this series, Vol. 5, p. 703.
[16] Measured on a 1 *M* solution at 37°.
[17] M. P. Deutscher, *J. Biol. Chem.* **242**, 1123 (1967).
[18] Obtained from Fischer Scientific Co.
[19] Frozen livers from fasted rabbits were obtained from Pel-Freez Biologicals.

combined fraction precipitated with 3 volumes of ethanol.[20] The precipitate (containing mainly tRNA, DNA, and glycogen) is collected by centrifugation for 10 minutes at 10,000 rpm and dissolved in 150 ml of 0.3 *M* Na acetate, pH 7.0, at room temperature. Isopropanol, 80 ml, is added to the solution dropwise, with stirring at room temperature, and the suspension is centrifuged at 20° for 10 minutes at 13,000 rpm. To the clear supernatant solution is added another 70 ml of isopropanol as above, and the suspension is centrifuged. The tRNA pellet is dried in a stream of air, redissolved in 15 ml of 0.3 *M* Na acetate, pH 7.0, and the isopropanol fractionation is repeated with 8-ml and 7-ml portions of isopropanol as before. The final tRNA pellet is dissolved in $H_2O$. The tRNA isolated by this procedure has an $A_{260}$:$A_{280}$ of 1.8–2.0 and contains no protein, high molecular weight RNA, DNA, or UV-absorbing low molecular weight contaminants. About 10% of the RNA present is 5 *S* RNA. The yield of tRNA (excluding 5 *S* RNA) is about 40 mg per 100 g of liver. In practice, the procedure is conveniently performed on 1 kg of liver.

*Assay Procedure.* The standard assay mixture contains: Tris-acetate, 25 μl; $MgCl_2$, 5 μl; ATP, 5 μl; EDTA, 1 μl; serum albumin, 1 μl; amino acid 10 μl; tRNA, 15 μl; enzyme; and $H_2O$ to 100 μl. In practice, a cocktail is made containing all the components except amino acid, tRNA, and enzyme, such that the addition of 40 μl includes all the reagents at their correct final concentrations. Synthetase activity for each amino acid is determined separately. KCl, 5 μmoles, is also added to the reaction mixture for the determination of tyrosyl-tRNA synthetase. The reaction is initiated by the addition of enzyme and terminated after incubation at 37° by 3 ml of cold 10% trichloroacetic acid solution. After 10 minutes in ice, the precipitate is collected on Whatman GF/C filters, washed with six 3-ml portions of the 2.5% trichloroacetic acid solution, and once with one 5-ml portion of ethanol:ether (1:1). The filters are dried under an infrared lamp, and the radioactivity determined in a scintillation counter with 5 ml of toluene-based scintillation fluid.

Owing to the large number of enzymes present in the complex, no attempt has been made to optimize the assay conditions for each synthetase. Furthermore, assays of crude extracts probably lead to minimal values for synthetase activities due to interference by contaminants, such as unlabeled amino acids, $Mg^{2+}$, ATPases, proteases, and RNases.

## Purification Procedure

Homogenizing buffer: 0.35 *M* sucrose; 0.1 *M* Tris·HCl, pH 7.0; 10 m*M* $MgCl_2$. All operations are carried out in a cold room at 4°.

*Step 1. Homogenization and Differential Centrifugation.* Livers obtained

[20] Higher yields of tRNA can be obtained if the phenol layers are reextracted with buffer.

from 150–200 *g* male Sprague-Dawley rats are washed with cold buffer, minced, and homogenized in about 5-g portions with 2 volumes of buffer. Homogenization consists of 10 strokes with a loose-fitting pestle and 3 strokes with a tight-fitting one in an all-glass Dounce homogenizer. The homogenate is centrifuged at 27,000 *g* for 15 minutes to remove cell debris, nuclei, and mitochondria. The supernatant fluid is poured through glass wool to remove fatty material and diluted with an equal volume of homogenizing buffer. This mixture is centrifuged for 90 minutes at 105,000 *g* in a 40 rotor in a Spinco Model L2-65B preparative ultracentrifuge to remove the microsomal fraction. It is important that the latter centrifugation be carried out for only 90 minutes, since the region immediately above the tightly packed microsomal pellet is greatly enriched for aminoacyl tRNA synthetases, and longer periods of centrifugation lead to pelleting of this material.

The complete 105,000 *g* supernatant fraction is included for the next step; the practice of removing only the top two-thirds or three-quarters of the supernatant should be avoided since a large percentage of the activity would be lost by such a procedure. Activity in the high-speed supernatant fraction may be concentrated by centrifugation at 105,000 *g* for 15 hours since greater than 90% of the synthetase activities pellet under these conditions.

*Step 2. Chromatography on Sephadex G-200.* The aminoacyl-tRNA synthetase complex can be partially purified by chromatography on Sephadex G-200, as shown in Fig. 1. An aliquot of the high-speed supernatant fraction is applied to the column and eluted with the homogenizing buffer. Fractions of 5 ml are collected at a flow rate of about 15 ml/hr, and the $A_{280}$, $A_{260}$, protein (determined by the Lowry method[21]), RNA (determined by the orcinol method[22]) and synthetase activities are measured. All the activities elute in the void volume of the column, whereas only about 25% of the protein is found in this region. About 70% of the RNA in the high-speed supernatant is also associated with the synthetase activities.

*Step 3. Chromatography on Sepharose 6B.* The active fractions from the Sephadex column are concentrated about 5-fold either with Diaflo PM-30 membranes[23] or by centrifugation for 15 hours in order to sediment the complex. The concentrated sample is then applied to a column of Sepharose 6B, and eluted with the homogenizing buffer at a flow rate of about 5 ml/hr. As shown in Fig. 2, all the activities elute as a single peak associated

[21] O. H. Lowry, N. J. Rosebrough, A. L. Farr, and R. J. Randall, *J. Biol. Chem.* **193**, 265 (1951).

[22] See W. C. Schneider, this series, Vol. 3, p. 680.

[23] Obtained from Amicon Corp., Lexington, Massachusetts.

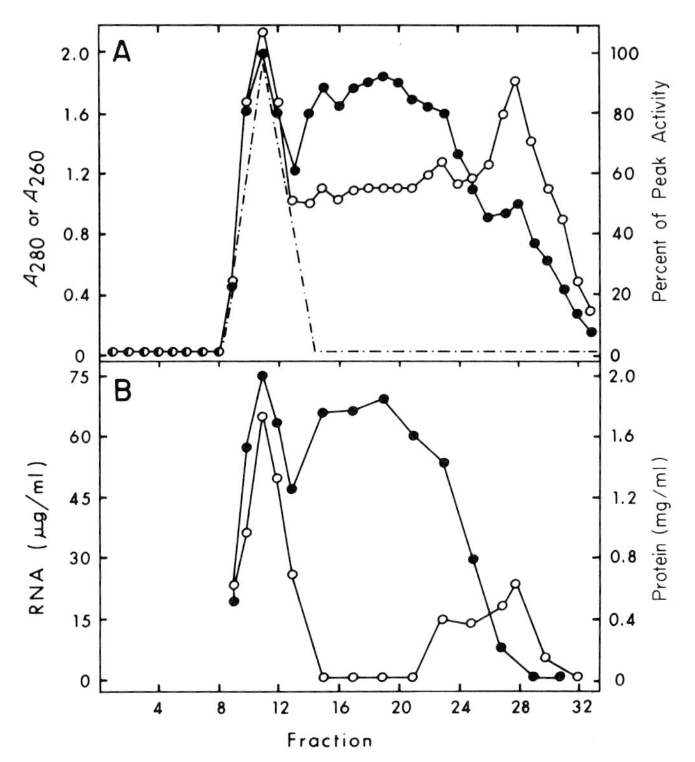


FIG. 1. Chromatography of the high-speed supernatant fraction on Sephadex G-200. The 105,000 *g* supernatant (4.5 ml) is applied to a 1.5 × 80 cm column of Sephadex G-200 and eluted with homogenizing buffer. Fractions of 5 ml are collected at a flow rate of about 15 ml per hour. Dextran Blue peaks in tube 11. (A) $A_{280}$ (●), $A_{260}$ (○), and synthetase activity (– · – ·) profiles. (B) Protein (●) and RNA (○) profiles.

with the trailing edge of the second protein peak. All the RNA in the sample is also present in the peak of synthetase activities.

Generally, 80–120% of the aminoacyl-tRNA synthetase activities are recovered in the Sephadex chromatography procedure, leading to approximately a 4-fold purification of the complex. On Sepharose, the recoveries for different synthetases varies from 40 to 70%, so although about two-thirds of the inactive protein is removed, the purification is only 2-fold at most. Thus, the total purification achieved is about 8-fold. Since mammalian aminoacyl-tRNA synthetases generally must be purified about 500-fold to achieve homogeneity, it would be expected that a complex of twenty synthetases would require about a 25-fold purification. Attempts to purify the complex further while maintaining its structure have been unsuccessful. For example, chromatography on DEAE-cellulose removes

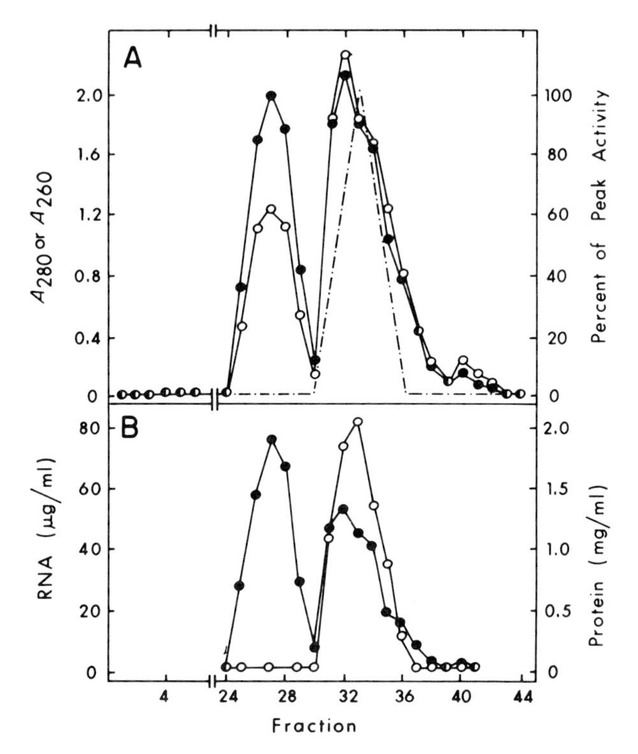


FIG. 2. Chromatography of concentrated Sephadex fractions on Sepharose 6B. Samples (4.5 ml) of the concentrated, active fractions from the Sephadex column are applied to a 1.5 × 80 cm column of Sepharose 6B and eluted with homogenizing buffer. Fractions of 2.4 ml are collected at a flow rate of about 5 ml per hour. (A) $A_{280}$ (●), $A_{260}$ (○), and synthetase activity (– · – ·) profiles. (B) Protein (●) and RNA (○) profiles.

the RNA present in the complex and leads to its disruption with the concomitant separation of many individual synthetases. Similarly, exposure to high salt concentrations, such as in an ammonium sulfate fractionation, also leads to partial disruption of the complex.

## Properties of the Complex

*Stability.* The aminoacyl-tRNA synthetase complex is extremely unstable and is disrupted by a variety of treatments. Thus, more vigorous homogenization of liver using a motor-driven homogenizer leads to dissociation of the complex. The effect of excess homogenization is thought not to be on the complex directly, since it is too small, but is probably due to disruption of other structures, such as lysosomes, which could release enzymes that act on components of the complex. The aminoacyl-tRNA synthetase complex is also destroyed by the process of freezing and thawing.

In addition, maintaining the complex in ice leads to aggregation and precipitation of synthetase activities after a few days of storage. Finally, the complex is dissociated by treatment with lipid solvents, high salt concentrations, EDTA and DEAE-cellulose.

*Chemical Composition.* The RNA associated with the complex is about 90% low molecular weight RNA. This material has amino acid acceptor activity and chromatographs on Sephadex with tRNA. However, it is not known whether all the low-molecular weight RNA is, in fact, tRNA. About 10% of the RNA is high molecular weight and is probably due to a small contamination by ribosomal subunits. The RNA in the purified complex amounts to 5–10% of the protein present.

The purified aminoacyl-tRNA synthetase complex also contains lipid which can be removed by treatment with acetone, chloroform:methanol (2:1), or ethanol:ether (3:1). Thin-layer chromatography and direct chemical analyses indicate that the lipid is almost exclusively cholesterol esters. The amount of phospholipid present is negligible. About 7 or 8 different fatty acids are present in the cholesterol esters. The weight of lipid material amounts to about 20–25% of the protein present in the complex.[24]

In addition to protein, RNA, and lipid, the complex may also contain a structural divalent cation since in the presence of 1 m*M* EDTA the complex is partially dissociated.

Based on its elution position on Sepharose 6B the aminoacyl-tRNA synthetase complex has a molecular weight between 1 and $2 \times 10^6$. Such a molecular weight is too low for a complex containing all 20 synthetases. Undoubtedly, this procedure leads to a spectrum of different complexes containing different groups of synthetases with the average molecular weight of 1 to $2 \times 10^6$. The extreme fragility of the complex probably leads to its partial breakage. In fact, in many instances, even greater breakage of the complex has been observed, with some synthetases still excluded from Sephadex G-200, and others included. However, even the included enzymes have molecular weights greater than $2 \times 10^5$, which is larger than that found for purified synthetases. The variables involved in this lability are not understood, but they may explain the reports in which only certain synthetases are found in a complex.[10,12] It appears that some of the synthetases are much more tightly bound in the complex than others.

*Catalytic Activity.* The activity of aminoacyl-tRNA synthetases in the complex are considerably higher than those in a free state. Disruption of the complex by any of the methods described above also leads to considerable loss of activity for these enzymes. Despite the fact that some ribosomes may still be present in the partially purified complex, all the activity is due to the formation of aminoacyl-tRNA, not to the synthesis of protein.

[24] A. K. Bandyopadhyay and M. P. Deutscher, *J. Mol. Biol.* **74**, 257 (1973).

# [24] Kinetic Techniques for the Investigation of Amino Acid : tRNA Ligases (Aminoacyl-tRNA Synthetases, Amino Acid Activating Enzymes)[1]

*By* ELIZABETH ANN EIGNER and ROBERT B. LOFTFIELD

In 1955 Hoagland[2] reported the existence of enzymes that "activated" amino acids while cleaving ATP into AMP and $PP_i$. Shortly thereafter, transfer ribonucleic acid was discovered and its role in protein synthesis was correctly adumbrated.[3] By 1959 it was clear that some or all of the "activating enzymes" were responsible also for the attachment of the amino acid to the cognate tRNA.[4] Each of these enzymes is, in more precise terminology, an amino acid:tRNA ligase (AMP) or an aminoacyl-tRNA synthetase.

As pointed out by Mehler in an earlier volume of this series,[5] many factors make study of this group of enzymes exceptionally difficult. There are at least twenty enzymes (at least one for each of the natural amino acids) in every cell that is synthesizing protein, and these appear to differ in such fundamental properties as molecular weight, subunit organization, cation requirements, etc. Moreover, it is likely that still another twenty enzymes are to be found in the same organism in other organelles like mitochondria. In the same tissue compartment we usually find more than one tRNA serving as an amino acid specific cosubstrate. Even if this multiplicity of enzymes and tRNA's from a single tissue did not discourage generalizations, other tissues from the same creature and all other creatures have their unique complement of ligases and tRNA's. It is certain that enzymes specific for a particular amino acid differ in heat stability, degree of polymerization, $K_m$'s, etc., depending on the biological source.

By analogy with other carboxylic acid activation reactions, it has been assumed that amino acids combine with enzyme and ATP to yield enzyme-bound aminoacyl adenylate and inorganic pyrophosphate. If $[^{32}P]PP_i$ is present in the reaction mixture, the radioactive pyrophosphate can react

[1] The preparation of this chapter has been supported, in part, by American Cancer Society Grant No. BC-11B and by United States Public Health Grant CA-08000.

[2] M. B. Hoagland, *Biochim. Biophys. Acta* **16,** 288 (1955).

[3] M. B. Hoagland, P. C. Zamecnik, and M. L. Stephenson, *Biochim. Biophys. Acta* **24,** 215 (1957).

[4] J. Preiss, P. Berg, J. Ofengand, F. H. Bergmann, and M. Dieckmann, *Proc. Nat. Acad. Sci. U.S.* **45,** 319 (1959).

[5] A. H. Mehler, this series, Vol. 20, p. 203.





ISBN 0-12-504180-2

with the Enz·AA ∼ AMP, reversing steps 5, then 4, then 3 to release [$^{32}$P]ATP (Fig. 1). Alternatively, if a nucleophilic reagent like hydroxylamine were present, either Enz·AA·ATP or Enz·AA ∼ AMP could react to form amino acid hydroxamate and AMP (steps 19 or 20). When it was recognized that tRNA was also a substrate for these enzymes, the immediate assumption was that tRNA bound to Enz·AA ∼ AMP (step 6) and then reacted (step 7) to form Enz·AA-tRNA, which then dissociated (step 8) yielding free AA-tRNA.

Such a formulation with a highly ordered sequence of reaction is supported by the following line of evidence: (a) Several of these enzymes form stable complexes with ATP (step 3), the isolated Enz·ATP complex being able to react with AA and tRNA to form AA-tRNA (steps 4–8).[6–9] Similar complexes of the type Enz·AA have not been reported. (b) Numerous stable complexes of the type Enz·AA ∼ AMP have been prepared by incubation of AA, ATP, enzyme and magnesium (steps 3–5) followed by isolation usually using Sephadex. The preparation of these has been reviewed by Allende and Allende.[10] (c) The Enz·AA ∼ AMP complexes react with cognate tRNA to give fair to good yields of AA-tRNA (steps 6 and 7) or with $PP_i$ to regenerate ATP (reverse steps 5, 4, and 3).[11] (d) In some cases the overall rate of esterification of tRNA is determined by the rate of dissociation of the AA-tRNA from the enzyme (step 8).[12,13] (e) In most cases Enz·AA ∼ AMP reacts with hydroxylamine to form the amino acid hydroxamate (step 20). In a few cases, it appears that Enz·AA ∼ AMP and Enz·AA-tRNA are unreactive and only free AA-tRNA can react with hydroxylamine (step 16).[14]

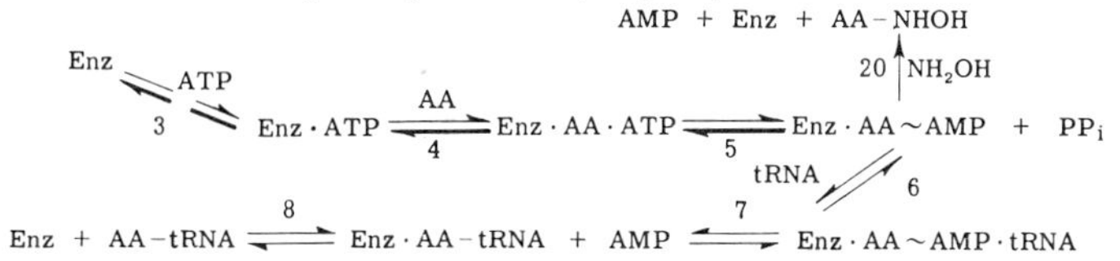

FIG. 1. A simple and widely accepted scheme describing the function of amino acid: tRNA ligases in catalyzing ATP: $PP_i$ exchange, hydroxamate formation and tRNA aminoacylation.

[6] C. C. Allende, H. Chaimovich, M. Gatica, and J. E. Allende, *J. Biol. Chem.* **245,** 93 (1970).

[7] A. V. Parin, E. P. Savelyev, and L. L. Kisselev, *FEBS* (*Fed. Eur. Biochem. Soc.*) *Lett.* **9,** 163 (1970).

[8] P. Rouget and F. Chapeville, *Eur. J. Biochem.* **4,** 310 (1968).

[9] S. A. Berry and M. Grunberg-Manago, *Biochim. Biophys. Acta* **217,** 83 (1970).

[10] J. E. Allende and C. C. Allende, this series, Vol. 20, p. 210.

[11] A. N. Baldwin and P. Berg, *J. Biol. Chem.* **241,** 839 (1966).

[12] M. Yarus and P. Berg, *J. Mol. Biol.* **42,** 171 (1969).

[13] E. W. Eldred and P. R. Schimmel, *Biochemistry* **11,** 17 (1972).

[14] D. I. Hirsh, *J. Biol. Chem.* **243,** 5731 (1968).

Although this sequence can be demonstrated to proceed step by step in the test tube, and although it may be the correct formulation for the attachment of some amino acids to tRNA *in vivo* in some organisms, there is now certainty that such a simple sequence does not obtain in all cases. This is perhaps not surprising since there are a total of three quite different substrates, and three products as well as the enzyme. The presence of any one of the substrates might be expected to exert allosteric effects on the enzyme so as to modify the binding or activity of other substrates. The following observations must be considered in interpreting any kinetic study:

1. To date all ligases specific for arginine,[15–18] glutamic acid, and glutamine[19–21] fail to carry out the ATP:$PP_i$ exchange except in the presence of intact cognate tRNA. At the very least this requires that, for these enzymes, the complex Enz·ATP·AA·tRNA must be formed. Perhaps the presence of tRNA suffices to "activate" the enzyme permitting the execution of step 14 (Fig. 2). The resulting Enz·AA ~ AMP·tRNA could now react with [$^{32}$P]$PP_i$ to reverse steps 14, 13 and 12, or 11, or 22, 4 and 3, or 2 to release [$^{32}$P]ATP from the enzyme. However, since, in each instance, the maximum rate of ATP:$PP_i$ exchange is approximately equal to the rate of aminoacylation of tRNA, and since receptor tRNA modified slightly so as not to be esterifiable is incompetent to catalyze the ATP:$PP_i$ exchange, it seems likely that Enz·AA-tRNA must first be formed before $PP_i$ and AMP can combine by way of steps 7 and 14 or by step 15.[18]

2. In the case of the tyrosine[22] and phenylalanine[23] enzymes, kinetic analysis of the effects of inhibitors on ATP:$PP_i$ exchange shows that the addition of AA and ATP to the enzyme to form Enz·ATP·AA is completely random (steps 1 and 2, and their reversal contribute equally with steps 3 and 4).

3. Although it is sometimes possible to isolate a remarkably stable Enz·ATP complex, various kinds of equilibrium studies show that amino acid binds with the enzyme in the absence of ATP.[24–26]

[15] S. Mitra and A. H. Mehler, *J. Biol. Chem.* **241,** 5161 (1966).
[16] S. Mitra and A. H. Mehler, *J. Biol. Chem.* **242,** 5490 (1967).
[17] S. Mitra and C. Smith, *Biochim. Biophys. Acta* **190,** 222 (1969).
[18] R. Parfait and H. Grosjean, *Eur. J. Biochem.* **30,** 242 (1972).
[19] J. Ravel, S. F. Wang, C. Heinemeyer, and W. Shive, *J. Biol. Chem.* **240,** 432 (1965).
[20] L. W. Lee, J. Ravel, and W. Shive, *Arch. Biochem. Biophys.* **121,** 614 (1967).
[21] M. P. Deutscher, *J. Biol. Chem.* **242,** 1132 (1967).
[22] D. V. Santi and V. A. Peña, *FEBS (Fed. Eur. Biochem. Soc.) Lett.* **13,** 157 (1971).
[23] D. V. Santi, P. V. Danenberg, and P. Satterly, *Biochemistry* **10,** 4804 (1971).
[24] C. J. Bruton and B. S. Hartley, *J. Mol. Biol.* **52,** 165 (1970).
[25] R. B. Loftfield and A. Pastuszyn, *Biophys. Soc. Abstr.* (1970).
[26] F. X. Cole and P. R. Schimmel, *Biochemistry* **9,** 480 (1970).

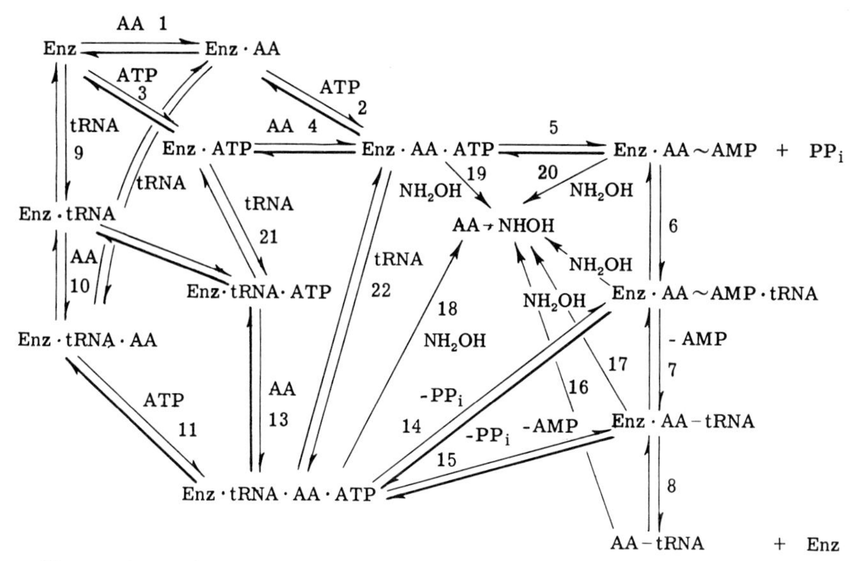


Fig. 2. A reaction scheme showing the multiple routes by which [$^{32}$P]PP$_i$, can be incorporated into ATP (heavy arrows), by which hydroxamates can be formed and by which AA-tRNA can be formed. In one system or another evidence exists for almost every one of the postulated 23 reactions.

4. All the ligases form stable complexes (Enz·tRNA) with the cognate tRNA regardless of the presence or absence of ATP.[10] Given the ambient concentrations of tRNA and enzyme in the cell and the low $K_D$ of such complexes, the dominant species in the cell will be Enz·tRNA, not Enz.[27] The catalytic properties of Enz·tRNA and of Enz are generally different.[23,28]

5. In at least two cases formation of hydroxamate has been shown to be by reaction of hydroxylamine with AA-tRNA (step 16).[14,29]

6. In some cases spermine (substituting for $Mg^{2+}$) catalyzes the formation of AA-tRNA under conditions where step 5 cannot be demonstrated. In such cases, ATP:PP$_i$ exchange probably occurs only by way of steps 14 or 15.[30,31]

7. The effect of monofunctional bases (to inhibit ATP:PP$_i$ exchange but to stimulate hydroxamate formation and tRNA esterification) is not consistent with an obligatory passage through Enz·AA ~ AMP.[32]

[27] K. B. Jacobson, *J. Cell. Physiol.* **74,** Suppl. 1, 99 (1969).
[28] R. B. Loftfield and E. A. Eigner, *J. Biol. Chem.* **240,** PC 1482 (1965).
[29] O. Favorova and L. Kisselev, *FEBS* (*Fed. Eur. Biochem. Soc.*) *Lett.* **6,** 65 (1970).
[30] I. Igarashi, K. Matzuzaki, and Y. Takeda, *Biochim. Biophys. Acta* **254,** 91 (1971).
[31] A. Pastuszyn and R. B. Loftfield, *Biochem. Biophys. Res. Commun.* **47,** 775 (1972).
[32] R. B. Loftfield and E. A. Eigner, *J. Biol. Chem.* **244,** 1746 (1969).

All the above points as well as several others have been brought forward in recent reviews.[5,33,34] The purpose in mentioning them in this chapter is to emphasize the complexity of these systems and the consequent ambiguity in interpretation of kinetic data. For instance, in one case or another it can be demonstrated that amino acid hydroxamates are produced by steps 16, 18, 19, and 20, and *a priori* there is no reason for excluding contributions from steps 17 and 21 in untested cases. The exchange of $[^{32}P]PP_i$ into ATP can occur by way of any combination of those reactions marked by heavy arrows; the exchange has been demonstrated, in various cases, to use at least four of the possible paths. Similarly, generation of AMP from ATP may occur as a result of the operation of steps 7, 15, 18, 19, 20, or 21. Although the production of AA-tRNA must certainly result from dissociation of a complex like Enz·AA-tRNA, this dissociation may be affected by the concentrations of amino acid, ATP, and tRNA,[11,12] by ionic strength,[35–37] by pH[38] and by temperature. Moreover, at least two of the possible pathways (through Enz·AA·tRNA and through Enz·AA $\sim$ AMP) are demonstrable.

Put succinctly, we cannot predict beforehand what step or series of steps are being measured when we use any one of the four kinetic assays described in this chapter. In every case additional studies are necessary before these kinetic techniques can be interpreted unambiguously.

In addition to the kinetic methods described here and by Santi *et al.*,[39] the following techniques have been especially useful in studying the kinetic properties of these enzymes. (1) Single-pass kinetics in which stoichiometric amounts of enzymes are used[13,40,41] and the rates of appearance or disappearance of Enz(AA $\sim$ AMP) or AA-tRNA are measured independently of the original activation reaction; (2) kinetic determinations of the isolated step of association or dissociation of the Enz·tRNA complex or of the Enz·AA-tRNA complex using radioactive tRNA or AA-tRNA and the membrane filter technique that permits isolation of only the undissociated complex[12,37]; (3) equilibrium and nonequilibrium dialysis tech-

[33] R. B. Loftfield, "Protein Synthesis" (McConkey, ed.), Vol. 1, p. 1, Marcel Dekker, New York, 1971.
[34] R. B. Loftfield, *Progr. Nucl. Acid Res. Mol. Biol.* **12**, 87 (1972).
[35] R. B. Loftfield and E. A. Eigner, *J. Biol. Chem.* **242**, 5355 (1967).
[36] R. Taglang, J. P. Waller, N. Befort, and F. Fasiolo, *Eur. J. Biochem.* **12**, 550 (1970).
[37] M. Yarus, *Biochemistry* **11**, 2050 (1972).
[38] M. Yaniv and F. Gros, *J. Mol. Biol.* **44**, 17 (1969).
[39] D. V. Santi, R. W. Webster, Jr., and W. W. Cleland, this volume [25].
[40] E. W. Eldred and P. R. Schimmel, *J. Biol. Chem.* **247**, 2961 (1972).
[41] M. Yarus and S. Rashbaum, *Biochemistry* **11**, 2043 (1972).

niques[24,31] which permit determination of the binding of one or another substrate to enzyme in the absence of chemical reaction; and (4) possibly the most powerful tool, fluorescence analysis.[42–45] Measurement of changes in the fluorescence of tryptophan residues in the enzyme or of Y-base residues of tRNA or of artificially included fluorescent dyes permits determination of the actual rates of several association and dissociation steps under conditions near physiological. This is especially important since most single-pass kinetic approaches are necessarily conducted at low temperature and low pH in order to slow the reactions sufficiently to be measurable.

## ATP:$PP_i$ Exchange

When the enzyme, ATP, and amino acid are mixed in the absence of any nucleophile such as $NH_2OH$ or the natural substrate tRNA, but in the presence of [$^{32}$P]$PP_i$, the reactions shown by the heavy arrows in Figs. 1 and 2 usually occur rapidly, the [$^{32}$P]$PP_i$ is converted into enzyme-bound ATP, and the [$^{32}$P]-ATP dissociates from the enzyme.

*Reagents*

ATP, 0.1 *M* magnesium salt (pH 7.4)
$Na_2H_2P_2O_7 \cdot 10H_2O$, 0.1 *M*, pH 7.4, containing 0.1 *M* $MgCl_2$
2-Amino-2-hydroxymethylpropane-1,3-diol (Tris), 1.0 *M*, pH 7.4 at 25°
[$^{32}$P]$PP_i$ prepared by pyrolysis of [$^{32}$P]-$Na_2HPO_4$ followed by anion exchange column purification,[46] or obtained from New England Nuclear. $Na_4{}^{32}P_2O_7$, 100–1000 mCi/mmole
Albumin, 10 mg/ml
[$^{12}$C]Amino acid, 10 m*M*
Purified ligase
$MgCl_2$, 0.1 *M*
Anion exchange paper, Reeve-Angel SB-2

The reaction conditions are essentially the same as those of Berg[47] except that we use anion exchange paper to separate ATP from pyrophos-

[42] C. Hélène, F. Brun, and M. Yaniv, *Biochem. Biophys. Res. Commun.* **37,** 393 (1969).
[43] C. J. Bruton, Ph.D. Thesis, Univ. of Cambridge, 1969.
[44] R. Riegler, E. Cronvall, R. Hirsch, V. Pachmann, and H. G. Zachau, *FEBS* (*Fed. Eur. Biochem. Soc.*) *Lett.* **11,** 320 (1970).
[45] E. Holler and M. Calvin, *Biochemistry* **11,** 3741 (1972).
[46] R. B. Loftfield and E. A. Eigner, *Biochim. Biophys. Acta* **130,** 426 (1966).
[47] P. Berg, *J. Biol. Chem.* **222,** 991 (1956).

phate. We find that adsorption of ATP on charcoal followed by elution with $NH_4OH$ is less reliable. A modification of the charcoal method[48] where the charcoal (and ATP) is collected on glass filters appears to be more successful.[23]

The method as described here yields three time points for each reaction, and allows seven reactions to be run simultaneously.

Into a single test tube, kept in ice, are pipetted 40 $\mu$l of 0.1 $M$ ATP, 40 $\mu$l of 0.1 $M$ $PP_i$, 80 $\mu$l of Tris, sufficient $[^{32}P]PP_i$ to give about $10^6$ cpm, (the half-life of $^{32}P$ is 12 days. Since the specific activity of $[^{32}P]PP_i$ is high, the molarity of $PP_i$ is essentially that of the cold $PP_i$ used), 8 $\mu$l of 0.1 $M$ $MgCl_2$, an appropriate amount of enzyme, and water to bring the volume up to 400 $\mu$l. After mixing thoroughly, 50 $\mu$l of this mixture is pipetted into each of seven 2-ml test tubes, also kept in ice. Various amounts of $^{12}C$-labeled amino acid are added to each tube if the $K_m$ of the amino acid is to be determined. Water is added to each tube to bring the volume to 100 $\mu$l. If the concentration of some component other than amino acid is to be varied (ATP, $PP_i$, salt, tRNA, inhibitor), it is added last, and the amino acid is included in the first mixture. The final 0.1 ml reaction mixture contains 5 m$M$ ATP, 5 m$M$ $[^{32}P]PP_i$, 0.1 $M$ Tris, 50 $\mu$g of albumin, 1 m$M$ $MgCl_2$ in excess of ATP and $PP_i$, and appropriate enzyme and amino acid (if crude enzyme preparations are used, 10 m$M$ KF is added to inhibit pyrophosphatase activity).

The reaction is started by placing the tubes in a 25° water bath at 1-minute intervals, and three aliquots of 25 $\mu$l are taken from each at times up to 30 minutes. Controls without enzymes are also run. The reaction proceeds at a slow rate at 0°, producing some conversion of $[^{32}P]PP_i$ to ATP. This gives a high "zero" intercept which is subtracted from later time points to give the correct rate at 25°. Alternatively, the reaction can be initiated by adding the enzyme to the reagents already at 25°, but this introduces possible pipetting errors in enzyme concentrations.

The aliquots are pipetted onto an origin line one inch from the end of a 6-inch strip of anion exchange paper and held in steam briefly to destroy the enzyme. (Reeve-Angel SB-2 paper is soaked in 0.10 $M$ $Na_2H_2P_2O_7$ at pH 8.0 for 15 minutes. The paper is washed once in water, dried in air, and then cut into 6-inch by 1-inch strips.) Twenty or more 6-inch strips can be suspended by paper clip hooks from a glass rod, sample end down, and separated by Teflon spacers. The bottom edges of the strips are dipped into a tray of 0.10 $M$ $Na_2H_2P_2O_7$ (pH 8.0). The ascending solution carries the $[^{32}P]PP_i$ while leaving the $[^{32}P]ATP$ near the origin. The elution is

[48] R. Calendar and P. Berg, *Biochemistry* **5**, 1681 (1966).

first carried out in a closed tank and chromatography is continued until the solvent reaches the end of the strip (about one hour). The cover of the tank is then removed and elution is continued for another hour while evaporation takes place, allowing the pyrophosphate to be moved farther upward since pyrophosphate does not move quite as fast as the solvent front. The ATP can be located by using a strip counter initially or otherwise by cutting out about 1.5 inches forward from the origin line and counting in a vial. Since the pyrophosphate is somewhat spread out, the total amount of radioactivity in each reaction vessel is best estimated by direct counting of a 10-$\mu$l aliquot from one reaction tube on a piece of the anion exchange paper. It is convenient to count the paper by rolling it into a nonoverlapping cylinder and placing it in a small (10 by 50 mm) test tube containing scintillation solution (0.3% PPO, 0.03% POPOP in toluene) and then counting it in a liquid scintillation counter.

The method is very sensitive because of the high specific activity of the pyrophosphate. Using multiple time points, it is possible to detect and determine with confidence as little as 0.01% conversion of $PP_i$ to ATP; with lower concentrations of $PP_i$ an exchange rate of 0.1 pmole per 25-$\mu$l aliquot per 25 minutes can be measured.[31] About 0.2 to 0.4% of the initial $^{32}P$ activity is found at the origin with ATP even when enzyme is absent, possibly due to polyphosphate contamination of the $[^{32}P]PP_i$. Typical experimental data have been presented.[31,46]

In the pyrophosphate exchange assay, the amino acid is regenerated at the same rate that it is activated, so that there is no significant change in its concentration. However, the pyrophosphate and ATP do change in specific activity as the reaction proceeds to equilibrium. It is usually desirable to work with less than 10% of the pyrophosphate converted to ATP. If a small trace of endogenous amino acid remains associated with the enzyme during its purification, it may catalyze exchange in the absence of added amino acid.

Even though most of the amino acid activating enzymes carry out this exchange, there are several reports of enzymes that do not catalyze this reaction except in the presence of tRNA. Mitra and Mehler[15] have shown that the arginine enzyme is incapable of carrying out the exchange except in the presence of RNA capable of accepting arginine. In this case, tRNA serves as an activator of the enzyme. Deutscher[21] observed a low pyrophosphate exchange under the usual conditions for the glutamate enzyme which could be stimulated by either high concentrations of glutamate or tRNA, indicating that aminoacyl tRNA may be an intermediate in the ATP : $PP_i$ exchange. Periodate-treated tRNA that has no acceptor activity for glutamate does not stimulate the exchange. It is clear that the absence of pyrophosphate exchange does not indicate that the amino acid activating enzymes are absent, but in general, ATP:$PP_i$ exchange is a good method for the detection of the ligases.

As with other ATP reactions, $Mg^{2+}$ is generally required. There are numerous reports of the successful substitution of other divalent cations, such as $Ca^{2+}$, $Cd^{2+}$, or $Mn^{2+}$ for $Mg^{2+}$, but such substitution apparently depends on the particular enzyme and the other reaction conditions.[41] The optimal $Mg^{2+}$ concentration also varies with the system being examined. As described here the assay mixture contains about 10% excess of $Mg^{2+}$ over the combined concentrations of ATP and $PP_i$. Under these conditions there is a maximal concentration of the monomagnesium complexes of both ATP and $PP_i$, which Cole and Schimmel[26,49] conclude are the functional species in the isoleucine:tRNA ligase system.

The ATP:$PP_i$ exchange assay is very sensitive to a variety of bases, such as imidazole, hydroxylamine, *o*-phenanthroline, $PP_i$, $P_i$, and Tris.[32] Some such materials may contaminate the enzyme preparation, others may be added deliberately but in varying amounts to adjust the pH. Comparison of rates at different pH's requires, for example, extensive kinetic studies with several kinds of buffer in order to exclude or recognize catalytic or inhibitory properties of the buffer combinations.

Changes in ionic strength or dielectric constant affect the rate of ATP:$PP_i$ exchange somewhat,[23] but not nearly as much as the tRNA aminoacylation assay.

The ATP:$PP_i$ exchange assay has been widely and profitably used to measure $K_i$'s for a variety of amino acid derivatives or ATP analogs that do not catalyze the ATP:$PP_i$ exchange. However, in many cases a near homolog (such as valine for isoleucine or dATP for ATP) will participate in the exchange. $K_m$'s and $V_{max}$'s for the homolog can be determined, but $K_i$'s for the homolog are difficult or impossible to obtain.

## The Hydroxamate Assay

The reaction of the activated carboxyl group of the amino acid with hydroxylamine to form a colored complex with ferric ion was first used by Hoagland.[2] Since then, the use of very pure radioactive amino acids of high specific activity followed by separation of hydroxamate from amino acid either by ion exchange paper[50,51] or ion exchange column[52,53] has made the

[49] F. X. Cole and P. R. Schimmel, *Biochemistry* **9,** 3143 (1970).
[50] R. B. Loftfield and E. A. Eigner, *J. Amer. Chem. Soc.* **81,** 4753 (1959).
[51] R. B. Loftfield and E. A. Eigner, *Biochim. Biophys. Acta* **72,** 372 (1963).
[52] W. H. Elliott and G. Coleman, *Biochim. Biophys. Acta* **57,** 236 (1962).
[53] C. Bublitz, *Biochim. Biophys. Acta* **128,** 165 (1966).

reaction much more sensitive. It is useful for studying a single ligase in a crude extract when other ligases are present since only the [$^{14}$C]hydroxamate of a given amino acid is measured. It has also been a valuable technique in measuring competition between two amino acids, both of which are substrates for a single enzyme. The ion exchange paper method is described here.

*Reagents*

ATP, magnesium salt, 0.1 *M*, pH 7.4
Hydroxylamine; 25 *M*, adjusted to pH 7.0 with HCl
1-[$^{14}$C]-labeled amino acid; 20–30 mCi/mmole
Purified ligase
Bovine albumin, 10 mg/ml
Cation exchange paper, Reeve-Angel SA-2 or Whatman IRP/69/M

Hydroxylamine is prepared by adding a solution of 40.0 g of NaOH in 350 ml methanol to a solution of 69.5 g of hydroxylamine·HCl in 350 ml of methanol. As reported by Lecher and Hofmann,[54] the yield of hydroxylamine is highest if the mixture never becomes alkaline. The NaCl is largely removed by filtration of the cooled mixture, methanol is removed *in vacuo* in a rotary still and the hydroxylamine is distilled *in vacuo*. Although the boiling flask is held at 60°, there is usually considerable decomposition as evidenced by gas evolution and an unstable pressure and boiling point. About 2 g of salt remains in the boiling flask. A second distillation proceeds without decomposition yielding a product of b.p. 34–36° (under 8 mm Hg), m.p. 25–30°. The yield of material is between 65% and 75%, usually about 25–30 *M* as determined by titration with HCl to pH 2.0. After neutralization with 0.1 equiv of concentrated HCl, 1-ml portions are frozen at −15°. Under these conditions the hydroxylamine is indefinitely stable.

If the variable substrate is the amino acid, a solution of all the other components is prepared at 0°. This includes 100 μl of 0.1 *M* ATP ($Mg^{2+}$ salt), 100 μl of $NH_2OH$, 20 μl of albumin solution, an appropriate amount of enzyme, and further additions of inhibitors, catalysts, etc., made up to a final volume of 750 μl with water. Aliquots of 75 μl are pipetted into each of seven to nine 8 × 35 mm test tubes which are then transferred to a water bath at 25° at 1-minute intervals. After 2 minutes' equilibration at 25°, appropriate amounts of [$^{14}$C]-labeled amino acid solution and of water to make 100 μl are added (again at 1-minute intervals).

After the first reaction has progressed some 10 minutes, a 25-μl aliquot

[54] H. Lecher and J. Hofmann, *Chem. Ber.* **55**, 912 (1922).

is withdrawn and spotted evenly on to the origin line (25 mm from one end) of a 22 mm × 114 mm cation exchange paper. The paper is immediately held in a jet of steam for 20 seconds to destroy the enzyme and then left to dry at room temperature. Each minute an aliquot is taken from another tube in the order in which the reactions were initiated. Thus three observations are made on every reaction at 10, 20, and 30 minutes. The strips of ion exchange paper, separated by Teflon spacers, are eluted by dipping the bottom (origin) end into a dish of 50 m$M$ $Na_2HPO_4$ (pH 7.0) inside a tank (a 30 cm × 30 cm × 50 cm glass fish aquarium is ideal). This prevents drafts and evaporation and allows the ascending solvent front to run smoothly. With neutral amino acids, the amino acid moves almost with the solvent front while the cationic amino acid hydroxamate is unmoved at the origin line. Chromatography is stopped when the buffer comes to within 1 cm of the top, the strips are removed and allowed to dry in air and then in a 60° oven. The final drying is necessary to reduce high background (chemiphosphorescence?) in the liquid scintillation counter. The radioactive areas may be located and assayed in a strip counter. However, it is usually more convenient to count in a liquid scintillation counter. For this, the hydroxamate area (6 mm before the origin to 25 mm beyond) and the top 42 mm of the strip (containing the free amino acid) are cut out and assayed for radioactivity.

Each strip is rolled into a cylinder and put inside a 10 × 50 mm test tube which in turn is supported inside a liquid scintillation vial. The usual toluene counting fluid is added to the small tube. If the paper is not formed into a cylinder, the counts are found to vary slightly as the orientation of the paper changes with respect to the counter phototubes. The type of cation exchange paper also determines the efficiency of counting. The Reeve-Angel SA-1, which is less colored, gives very high counting efficiencies, but it is no longer available. SA-2 paper has a little more color, and thus quenches the radiation somewhat. Whatman IRP/69/M paper runs more slowly, requiring an hour rather than a half hour for separation. However, the separation is equally good, and this can be used instead of SA-2 paper. This technique is useful for the reaction of all the neutral amino acids while other conditions have been worked out for the separation of basic or acidic amino acids from their hydroxamates.[54a]

The purity of the hydroxylamine can be quite important. Hydroxylamine prepared only from mixing methanolic NaOH with methanolic hydroxylamine hydrochloride, followed by filtration and concentration always contains substantial amounts of sodium chloride and some heavier

[54a] A. F. Lamkin and R. B. Hurlburt, *Biochim. Biophys. Acta* **272,** 321 (1972).

cations. Two vacuum distillations of such hydroxylamine yield a product whose $NH_2OH$ concentration is between 20 and 30 *M* and whose melting point is about 30°.[46]

The separation of amino acid from hydroxamate is moderately sensitive to the concentration of the eluting buffer. At low salt concentrations the amino acid remains near the origin. At higher salt concentrations, the hydroxamate moves from the origin. A concentration of 50 m*M* and a pH near neutrality permits the amino acid to move essentially with the solvent front and does not move the hydroxamate. Care should be taken that the eluting buffer does not evaporate and thus become more concentrated in salt. There is one "artifact" in this method since a very small amount of the amino acid seems to be bound to the ion exchange paper and remains at the origin instead of moving with the solvent front. With valine, leucine, or isoleucine, this percentage is in the range of 0.1% to 0.2% and is quite constant.[51] It appears that a stable, perhaps covalent, bond is formed between a small fraction of the amino acid and the ion-exchange resin. This is reduced if the strip is neither heated nor dried during the amino acid application; but since the fraction is so reproducible, and since several time points are always taken, its presence does not affect the determination or rates. (A single time point could give erroneous results because of this factor.) [$^{14}$C]Phenylalanine leaves somewhat more at the origin, and 1% of [$^{14}$C]threonine stays with the hydroxamate.[14] Clearly, the conditions for each amino acid have to be worked out carefully. In working with crude extracts, there is sometimes so much protein and other solid material present when the sample is applied to the paper that the solvent cannot make its way past this band, so it is necessary to keep the amount of protein down to a level that permits the solvent front to go through the origin smoothly.

Counting efficiency is about 30% (when the samples are counted in a liquid scintillation counter); however, it is necessary to study the counting conditions for the amino acid and the hydroxamates separately, since their $\beta$-radiation is attenuated to different degrees by the paper.[51] It is convenient to select counting conditions where the two compounds are counted with equal efficiency.

The rate of hydroxamate formation varies in many cases according to the square of hydroxylamine concentration followed by a much decreasing rate at still higher concentrations. Some amino acids appear to form the hydroxamate under no conditions and in other cases the maximum rate is found anywhere from 1.0 to 4.0 *M* hydroxylamine. Clearly it is necessary to find for each ligase the optimal hydroxylamine concentration and to control the concentration carefully. A typical example showing both the sensitivity of rate to hydroxylamine concentration and the constancy of rate during 20 minutes even at 10.0 $M$ hydroxylamine has been published.[55]

The hydroxamate assay consumes the amino acid. Therefore, the concentration will be changing, and for this reason the enzyme, substrate, and inhibitor concentrations should be chosen such that conversion to hydroxamate is less than 20% in 20 or 30 minutes. Using high specific activity amino acids of around 20 mCi/mmole and scintillation counting, it is possible to measure accurately the formation of 1.0 pmole of hydroxamate in 25 $\mu$l of sample.

The rate of hydroxamate formation is generally much slower than that of pyrophosphate exchange, and there is great variability in the enzymes with respect to their rate of amino acid hydroxamate formation and their aminoacyl-tRNA formation.[56] Tyrosyl-, phenylalanyl-, tryptophanyl-, and methionyl-tRNA synthetases catalyze hydroxamate formation at rates within 20% of the aminoacyl-tRNA formation. The enzymes that activate aliphatic amino acids rapidly also form hydroxamates rapidly. With several other enzymes the rates of hydroxamate formation progressively decrease, while glutaminyl- and threonyl-tRNA synthetases do not catalyze the hydroxamate formation at all. Threonine hydroxamate is formed only when tRNA is present from which it is believed that threonyl tRNA is an intermediate.[14] Valyl-, isoleucyl-, and leucyl-tRNA also react nonenzymatically with hydroxylamine to form hydroxamate, though the ester is not a necessary intermediate (our unpublished results), this reaction being slow compared to direct formation of hydroxamates.

A different effect is observed when tRNA is present while hydroxamate is formed by the valine and isoleucine enzymes.[28] tRNA increases the specificity of the enzyme for its correct substrate in the case of the isoleucine enzyme, thus lowering the $K_m$ for isoleucine and raising the $K_m$ for valine. This appears to be an allosteric effect which makes the enzyme more discriminating in its choice of substrates.

One product of the reaction, pyrophosphate, has been shown to be a very potent inhibitor of this reaction.[57] If the kinetics of this inhibition are to be studied, potassium fluoride should be added to the reaction mixture to prevent the action of pyrophosphatase since the small amount of pyrophosphate could be destroyed by a little enzyme and the inhibition would be overcome with time. The addition of potassium fluoride causes a constant extent of inhibition. The kinetics of this inhibition have shown that 1 molecule of pyrophosphate inhibits one molecule of enzyme.[32] (With

[55] R. B. Loftfield and E. A. Eigner, *Biochemistry* **7,** 1100 (1968).

[56] D. I. Hirsh and F. Lipmann, *J. Biol. Chem.* **243,** 5724 (1968).

[57] E. W. Davie, V. Koningsberger, and F. Lipmann, *Arch. Biochem. Biophys.* **65,** 21 (1956).

the valine enzyme, 1 m*M* pyrophosphate causes 70% inhibition of valine hydroxamate formation.)

Although magnesium is generally used in this reaction, George and Meister[58] have found that $Co^{2+}$, $Cd^{2+}$, $Mn^{2+}$, and $Mg^{2+}$ all catalyze hydroxamate formation by the *E. coli* valine enzyme.

### tRNA Esterification Assay

This method involves the collection of the tRNA labeled with [$^{14}$C]amino acid on Millipore filters as described by Scott.[59] It has been modified slightly to give best conditions for doing kinetic studies on purified ligases. This assay has the advantage of being almost always totally specific for a particular enzyme, amino acid, and tRNA. As such it is amenable to precise study of the inhibitory qualities of close structural analogs of ATP and of the amino acid or tRNA in question.

*Reagents*

- ATP, 0.1 *M* $Mg^{2+}$ salt, pH 7.6, Sigma
- 2-Amino-2-hydroxymethyl propane-1,3-diol (Tris), 1.0 *M*, pH 7.4 (25°)
- tRNA, 10 mg/ml $H_2O$ Schwarz
- [$^{14}$C]Amino acid, 20–40 mCi/mmole, New England Nuclear
- $MgCl_2$, 0.1 *M*
- Purified ligase
- Trichloroacetic acid (TCA), 5% (w/v)

As described in the other assays, about 600 μl of a solution containing appropriate concentrations of all of the nonvarying components (enzyme, ATP, tRNA, AA, salt or buffer, general bases, inhibitors) is prepared at 0°. Aliquots of 100 μl are transferred to a series of tubes, still at 0°; water and the variable component are added to a total volume of 175 μl. Typical final concentrations are 10 m*M* ATP ($Mg^{2+}$ salt), 0.1 *M* Tris, 2 mg/ml albumin, 0.1 m*M* amino acid, 2 mg/ml tRNA, and 2 m*M* $MgCl_2$. The tubes are immersed at intervals of 0.5 or 1 minute into a thermostatted bath. From each reaction vessel, a 50-μl aliquot is withdrawn after incubation periods of 2–5 minutes each. Three such sequential aliquots should be taken.

Each aliquot is removed from the test tube at a precise time after its incubation was begun and is expressed into 5% TCA, thus stopping the

[58] H. George and A. Meister, *Biochim. Biophys. Acta* **132,** 165 (1967).
[59] J. F. Scott, this series, Vol. 12, p. 173.

reaction and precipitating both tRNA and protein. After standing for 10 minutes, the mixture is stirred and poured onto a membrane filter (Millipore filter type AA WG02400 in a Millipore Pyrex microanalysis filter holder, No. XX1002500) and filtered under vacuum. The filter is washed once with 5 ml of cold 5% TCA and then the filter is transferred to a larger sintered-glass support and rinsed with an additional 10 ml of cold 5% TCA. The edges of the Millipore filter are especially carefully washed in this procedure. After drying on the filter, the sample is removed and placed in a Tracer-lab holder, type E-7B, and mounted for counting in a Micromil end-window low-background gas-flow counter. A final heating in a 60° oven (for about half an hour) is necessary to remove the TCA. If this is not done, the TCA will corrode the very thin end window on the gas-flow counter.

As described, the method measures the initial rate of reaction of amino acid with tRNA. The time intervals selected are generally 2- to 5-minute intervals, and reaction conditions are chosen so that no more than 30% of the tRNA is esterified. It is frequently desirable to determine total acceptor capacity of the tRNA in addition. In this case, one final aliquot is allowed to run for 30–60 minutes when the reaction is assumed to be complete.

This reaction does proceed at measurable velocity at 0° and slow or erratic treatment of the assay mixtures may lead to confusing data or incorrect conclusions. Some workers may find it more convenient to initiate the reaction by addition of tRNA after all the other components have been warmed to final temperature. If the enzymes are relatively pure, they contribute little protein and do not yield a significant amount of precipitate, making the assay technique equally convenient with any amount between 5 and 200 μg tRNA per 50-μl aliquot. The method is sensitive since the amino acids can be obtained in very high specific activity, in the range of 30 mCi/mmole, for the carboxyl-labeled [$^{14}$C]-labeled amino acids; and the thin-end window, low-background counters have a background of 5 cpm or less. This makes it possible to detect less than 1 pmole of formed aminoacyl-tRNA per aliquot in 20 minutes.

This assay is the most nearly physiological and, correspondingly, of greatest interest. It is more specific than either the ATP:$PP_i$ exchange, the hydroxamate forming or the adenylate-generating assays. If the salt concentration, the temperature, the pH, and the substrate concentrations are in a more or less physiological range and if the receptor tRNA is derived from the same source as the ligase, one natural protein amino acid is apparently never erroneously activated and attached to an incorrect tRNA molecule. Occasionally a closely related nonprotein amino acid (such as thiosine for lysine[60] or canavanine for arginine[16]) will be activated and esterified to tRNA. In contrast, in the other assay techniques, it is the rule that all near homologs of the correct amino acid will be active, usually with $K_m$'s ten to a thousand times larger than that of the natural substrate and with $V_{max}$'s that may vary from ten times greater to one hundred times smaller than the natural substrate.[46]

However, the tRNA esterification assay is subject to a large number of potential pitfalls. Even when tRNA and enzyme are derived from a single species, it must be remembered that the tRNA frequently contains several isoaccepting species, not all of which are equally active toward the ligase. Thus two $tRNA^{Val}_{E. coli}$ have $K_D$'s toward the *E. coli* valine ligase that differ by a factor of ten.[61] In other cases more than one enzyme specific for a single amino acid is present in a single tissue. These multiple enzymes may have entirely different activities toward the several isoaccepting tRNA's.[62] For convenience, the investigator may choose to employ the commercially available tRNA of yeast, *E. coli*, or liver or he may choose some other tRNA not from the same source as his enzyme. As a rough generalization, most bacterial tRNA's can be aminoacylated by heterologous bacterial enzymes, most yeast tRNA's and ligases interact with most plant or animal tRNA's and ligases.[63] However, there are numerous examples of failure to be aminoacylated, of very specific competitive inhibition by heterologous tRNA,[64] of greater activity of the heterologous pair than of homologous[65] and, finally, of erroneous aminoacylation.[66] Thus an *E. coli* enzyme specific for valine will cause the formation of Val-$tRNA^{Ile}_{yeast}$[40] particularly as reaction conditions become less physiological (high temperature, presence of organic solvents, etc.). Unless there is a specific interest in the kinetics of mischarging tRNA or in determining interspecific similarities and differences in tRNA's, every effort should be made to use homologous tRNA and enzyme. In some instances, interpretable kinetic data will be obtainable only with a purified isoacceptor tRNA receptor.

Commercial preparations of tRNA have frequently been found to be contaminated with some sort of ribonuclease and may require additional purification. It is essential that the terminal C-C-A groups of the tRNA be intact; if crude enzymes are used, these nucleotides may become hydrolyzed. Therefore, in crude enzyme preparations, CTP, phosphoenol

[60] R. Stern and A. H. Mehler, *Biochem. Z.* **342,** 400 (1970).
[61] M. Yaniv and F. Gros, *J. Mol. Biol.* **44,** 1 (1969).
[62] W. E. Barnett and J. L. Epler, *Proc. Nat. Acad. Sci. U.S.* **55,** 184 (1966).
[63] G. A. Tomlinson and J. J. R. Campbell, *Biochim. Biophys. Acta* **123,** 337 (1966).
[64] R. B. Loftfield and E. A. Eigner, *Acta Chem. Scand.* **17,** S 117 (1963).
[65] R. B. Loftfield, E. A. Eigner, and J. Nobel, *Biol. Bull.* **135,** 181 (1968).
[66] K. B. Jacobson, *Progr. Nucl. Acid Res. Mol. Biol.* **11,** 461 (1971).

pyruvate and phosphoenol pyruvate kinase are frequently helpful in maintaining the integrity of the RNA. With purified enzymes, this is unnecessary, and in fact undesirable as the CTP inhibits the esterification.

There are substantial conversions to [$^{14}$C]TCA-insoluble material already at the earliest time intervals. Partly this is because of a real though slow esterification occurring at 0°. Moreover most [$^{14}$C]-labeled amino acids suffer from radiolytic decomposition which yields radioactive products that collect on the Millipore filter. This is especially a problem with the aromatic amino acids. Rates rather than single time points are essential even if the purest [$^{14}$C]-labeled amino acids are used, and even if these are repurified prior to use.

Much more than the other assay methods, the esterification is very sensitive to reaction conditions. The esterification proceeds one hundred and fifty times faster at 25° than at 0°. The rate of esterification is depressed about 10-fold or more as the pH is decreased by 1.0 unit. (In this connection it should be noted that *a Tris buffer that is pH 7.6 at 25° is pH 8.2 at 0° and 7.3 at 37°.*) The choice of buffer introduces additional problems.[67] At constant pH and temperature, increasing concentrations of cacodylate accelerate or inhibit the aminoacylation reaction depending on the enzyme. Tris, imidazole, and other nitrogenous buffers accelerate the aminoacylation as their concentrations are increased, possibly because they participate in the reaction as a general base. In our hands, Veronal (diethyl barbiturate) buffers appear to be neither stimulatory nor inhibitory. However, the observed rates of esterification are lower than with Tris. Other newer buffer systems (HEPES, Bicine, etc.) have not been examined in detail.

Ionic strength or dielectric constant effects are another source of error. The presence of 0.2 *M* NaCl may slow the aminoacylation by 90%. Since a late step in the preparation of both tRNA or enzyme may involve salt-gradient chromatography or salt precipitation, it is possible inadvertently to introduce slightly different amounts of salt into the several reaction vessels.[68] Not only is the rate decreased by salt and increased by organic solvents, but the extent of the reaction is also reported to be less in high salt media[69,70] and much greater (and less specific[66]) in media of low dielectric constant. Numerous investigators have reported that the final extent of aminoacylation depends upon the amount of enzyme.

The accidental or inevitable presence of other materials may further complicate the kinetic analysis. Pyrophosphate is a very potent noncompetitive inhibitor of the homologous reaction with a $K_i$ in the range of

[67] H. Beikirch, F. von der Haar, and F. Cramer, *Eur. J. Biochem.* **26,** 182 (1972).
[68] I. B. Rubin, A. D. Kelmers, and G. Goldstein, *Anal. Biochem.* **20,** 533 (1967).
[69] A. Peterkofsky, S. J. Gee, and C. Jesensky, *Biochemistry* **5,** 2789 (1966).
[70] D. W. E. Smith, *J. Biol. Chem.* **244,** 896 (1969).

0.1 m$M$.[32] In the heterologous aminoacylation reactions, the $K_i$ for $PP_i$ is 1 n$M$,[71] in the range of the amount of $PP_i$ that would be produced by the cleavage of ATP; this can be destroyed with pyrophosphatase. Phosphate, citrate and a number of other anions, which might be present in an enzyme preparation, are generally inhibitory. General bases like Tris and imidazole are stimulatory as noted above, however the product aminoacyl-tRNA is subject to general base catalyzed hydrolysis so that the amount of aminoacyl-tRNA found after some minutes may be less than was actually formed in the absence of the base. This base-catalyzed hydrolysis is not serious with the branched side chain amino acids, valine and isoleucine, but it is a complicating factor with other amino acids, such as alanine and leucine.

### AMP Production

On occasion, especially if there is no net accumulation of amino acid hydroxamate or of aminoacyl-tRNA, it is useful to determine the rate of production of AMP and $PP_i$. The following method employs the use of thin-layer chromatography (TLC) described by Randerath and Randerath[72] and is essentially their method for nucleotide separation.

*Reagents*

ATP, 0.1 $M$, pH 7.4, $Mg^{2+}$ salt
Tris, 1.0 $M$, pH 7.4, containing 10 m$M$ $MgCl_2$
Albumin, 10 mg/ml
[$^{12}$C]-labeled amino acid, 1.0 m$M$
Adenosine-[8-$^3$H]5′-triphosphate, 7 Ci/mmole, in 50% ETOH, Schwarz BioResearch
Purified ligase
$NH_2OH$, 25–30 $M$
Polyethylenimine TLC plastic plate

The final reaction mixtures contain, in a total volume of 30 $\mu$l, 0.7 m$M$ ATP, 0.1 $M$ Tris, 0.1 m$M$ amino acid (isoleucine), 4 $\mu$g of albumin, 1.7 m$M$ $MgCl_2$, 300,000 cpm of [$^3$H]ATP, tRNA, $NH_2OH$, and other inhibitors or catalysts. The reagents, with the exception of that being varied, are combined in a small test tube in ice; after thorough mixing, aliquots of 25 $\mu$l are placed into each of 4 or 5 reaction tubes, also in ice. Zero to 5 $\mu$l of the variable reagent and water to make a total of 30 $\mu$l are added to 0°. The reaction is started by bringing the tubes to 25°, and aliquots of 5 $\mu$l are taken at 0, 5, 10, and 15 minutes and placed on the origin line of a poly-

[71] F. J. Kull, P. O. Ritter, and K. B. Jacobson, *Biochemistry* **8**, 3015 (1969).
[72] K. Randerath and E. Randerath, *J. Chromatogr.* **16**, 111 (1965).

ethylenimine TLC sheet. The reaction stops as soon as the aliquots are placed on the TLC medium. In order to locate the spots visually later, 1 μl of a solution 10 m*M* in ADP and AMP is also put on each of the zero-time spots. (Occasionally very high concentrations of salt or hydroxylamine cause the nucleotides to run at slightly different rates; visualization of the three nucleotides is then essential.) The thin-layer ion exchange sheet (polyethylenimine medium on a plastic support) is first eluted with 1.0 *M* LiCl outside the developing tank until the solvent is ¼ inch past the origin line. The TLC sheet is then transferred to a developing tank, and elution is continued with 1 *M* LiCl until the solvent is about 1 inch from the end of the plate. This takes about 1.5 hours. The AMP, ADP, and ATP separate cleanly, and, after drying can be located under ultraviolet light. The nucleotide-containing areas are cut from the plastic support, and assayed for radioactivity by putting each spot in a liquid scintillation vial, adding 5 ml of toluene counting fluid, and counting for tritium in a liquid scintillation counter. The plastic support does not interfere with the counting, although it is probably advisable to put the sample into the counting vial with the plastic support down. One commercial [$^3$H]ATP solution used was found to contain 1.1% of ADP and 0.4% of AMP. The conversion of ATP to AMP is only about as fast as hydroxamate formation or tRNA esterification; in general one-tenth or one-hundredth as fast as ATP:$PP_i$ exchange. It is desirable to reduce the ATP concentration to less than millimolar in order to maximize the fraction of ATP converted to AMP. If the enzymes are not entirely pure, there may be myokinases present which cause the formation of ADP and confuse the results. The presence of such myokinases may be detected by the addition of cold AMP, which stimulates the formation of ADP. The method can detect the formation of about 10 pmoles of AMP formed in 5 μl in 15 minutes.

The assay is responsive to any ATPases and to all kinds of ATP (AMP) ligases. It is necessary to establish either that there is no AMP formation in the absence of the chosen amino acid or that this background formation of AMP is adequately constant to permit subtraction.

# [25] Kinetics of Aminoacyl-tRNA[1] Synthetases Catalyzed ATP-$PP_i$ Exchange[2]

*By* DANIEL V. SANTI, ROBERT W. WEBSTER, JR., and W. W. CLELAND

Most of the aminoacyl-tRNA synthetases catalyze an amino acid-dependent ATP-$PP_i$ exchange in the absence of their cognate tRNA's which probably proceeds through an enzyme-bound aminoacyladenylate (AA-AMP) as depicted in Eq. (1).[3]

$$E + AA + ATP\cdot Mg \rightleftharpoons E\cdot AA\text{-}AMP + PP_i\cdot Mg \tag{1}$$

Unlike most two-substrate reactions, the liberated product ($PP_i$) and a modified enzyme form (E·AA-AMP) are generated in equimolar amounts, and the reaction reaches chemical equilibrium after a single turnover. Consequently, the rate expressions differ from those usually encountered for two-substrate reactions, as do predicted slope-intercept and inhibition patterns. The purpose of this article is to provide concise kinetic descriptions of the simplest cases of the ATP-$PP_i$ exchange reaction, and general methods of ascertaining the sequence of the interactions of amino acid and ATP with the enzyme. It is our intent that the methods presented, consisting of kinetic studies and studies of inhibitors, be sufficiently simple that the nonkineticist may determine order of substrate addition by procedures of inspection. Consideration is not given to the aminoacylation of tRNA, which may be analyzed by previously described methods,[4] the effects of tRNA on ATP-$PP_i$ exchange, or the effect of metal ions.

## Recommended Assay and Kinetic Procedures

The standard assay generally contains amino acid (6–10 times $K_m$), $ATP\cdot Mg^{2+}$ (6–10 times $K_m$), $MgCl_2$ (optimal concentration), $[^{32}P]PP_i\cdot Mg^{2+}$ (2 m$M$; ca. $10^5$ cpm/μmole), appropriate buffer, β-mercaptoethanol or dithiothreitol if necessary for enzyme stabilization, and a limiting amount of enzyme. Precautions should be taken to ensure that the exchange is

[1] Abbreviations: tRNA, transfer ribonucleic acid; E, enzyme; AA, amino acid; ATP, adenosine 5′-triphosphate; $PP_i$, inorganic pyrophosphate; ATP·Mg and $PP_i$·Mg, the magnesium complexes of ATP and $PP_i$, respectively.

[2] This work was supported by U.S. Public Health Service Grant CA-14266 from the National Cancer Institute.

[3] G. D. Novelli, *Annu. Rev. Biochem.* **30,** 449 (1967).

[4] W. W. Cleland, *in* "The Enzymes" (P. D. Boyer, ed.), Vol. 2, p. 1. Academic Press, New York, 1970.



ISBN 0-12-504180-2

linear with respect to time and proportional to enzyme concentration over the concentration range of the substrate used. The free magnesium concentration for optimal activity varies among the synthetases and should be determined for each enzyme; the level chosen should, if possible, fall on a plateau so that small variations have little effect on rates. ATP and $PP_i$ are varied as their magnesium complexes to ensure a constant free magnesium concentration. The same holds true for the pH used, with pH 8.0 offering the convenience of known stability constants for $ATP \cdot Mg^{2+}$ and $PP_i \cdot Mg^{2+}$ complexes.[5] Substrates and inhibitors are varied over the concentration ranges suggested by Cleland[6] for optimal spacing between points and statistical considerations. Controls usually omit enzyme and reactions are initiated by the addition of enzyme or $[^{32}P]PP_i$. Aliquots are removed at 5-, 10-, and 15-minute intervals to ensure initial rates, the ATP absorbed on charcoal, collected on glass filters,[7] and counted on a planchet counter. ATP-$PP_i$ exchange rates are calculated by standard equations,[8] and double reciprocal plots[9] are constructed. The validity of these kinetic methods for ATP-$PP_i$ exchange has been discussed,[10] and weighted least-squares programs for required data analysis have been reported.[6,11]

## Rate Equations

The binding of amino acid and ATP may proceed by an ordered addition (mechanisms I and II) or by a random sequence (mechanism III) where X represents all forms of the central complex ($E \cdot AA \cdot ATP \rightleftharpoons E \cdot AA \cdot AMP \cdot PP_i$) and F is the enzyme·aminoacyladenylate complex or a similar enzyme form.

I. Ordered: ATP binds first

$$E + ATP \underset{k_{-1}}{\overset{k_1}{\rightleftharpoons}} E \cdot ATP \underset{k_{-2}}{\overset{k_2[AA]}{\rightleftharpoons}} X \underset{k_{-3}}{\overset{k_3}{\rightleftharpoons}} F + PP_i$$

II. Ordered: AA binds first

$$E + AA \underset{k_{-1}}{\overset{k_1}{\rightleftharpoons}} E \cdot AA \underset{k_{-2}}{\overset{k_2[ATP]}{\rightleftharpoons}} X \underset{k_{-3}}{\overset{k_3}{\rightleftharpoons}} F + PP_i$$

[5] F. X. Cole and P. R. Schimmel, *Biochemistry* **9,** 3143 (1970).
[6] W. W. Cleland, *Advan. Enzymol. Relat. Areas Mol. Biol.* **29,** 1 (1967).
[7] R. Calendar and P. Berg, *Biochemistry* **5,** 1690 (1966).
[8] M. P. Stulberg and G. D. Novelli, this series, Vol. 5, p. 703.
[9] H. Lineweaver and D. Burk, *J. Amer. Chem. Soc.* **56,** 658 (1934).
[10] F. X. Cole and P. R. Schimmel, *Biochemistry* **9,** 480 (1970).
[11] G. N. Wilkinson, *Biochem. J.* **80,** 324 (1961).

III. Random

$$\begin{array}{l} E + AA \underset{k_{-1}}{\overset{k_1}{\rightleftharpoons}} E\cdot AA \underset{k_{-3}}{\overset{k_3[ATP]}{\rightleftharpoons}} \\ \qquad\qquad\qquad\qquad X \underset{k_{-5}}{\overset{k_5}{\rightleftharpoons}} F + PP_i \\ E + ATP \underset{k_{-2}}{\overset{k_2}{\rightleftharpoons}} E\cdot ATP \underset{k_{-4}}{\overset{k_4[AA]}{\rightleftharpoons}} \end{array}$$

The empirical forms of the rate equations corresponding to mechanisms I–III are given in reciprocal form in Eqs. (2)–(4).

*Mechanism I*

$$\frac{1}{v} = \frac{1}{\varphi_1}(1 + \varphi_2[AA])\left(1 + \frac{\varphi_3}{[AA]} + \frac{\varphi_4}{[PP_i]} + \frac{\varphi_5}{[AA][ATP]}\right) \qquad (2)$$

*Mechanism II*

$$\frac{1}{v} = \frac{1}{\varphi_1}\left(1 + \frac{\varphi_2}{[ATP]} + \frac{\varphi_3}{[PP_i]} + \frac{\varphi_4}{[AA][ATP]}\right) \qquad (3)$$

*Mechanism III*

$$\frac{1}{v} = \frac{1}{\varphi_1}\left(1 + \frac{\varphi_2}{[ATP]} + \frac{\varphi_3}{[AA]} + \frac{\varphi_4}{[PP_i]} + \frac{\varphi_5}{[AA][ATP]}\right) \qquad (4)$$

Expansions of these have been derived by the method of Cole and Schimmel[10] and by graphical methods.[12,13] For the purposes described here the empirical expressions suffice and the reader is referred to the aforementioned reports for complete equations.[14]

## Analysis of Kinetic Data

The methods consist of analysis of experimental slope and intercept effects when one substrate is varied in the presence of fixed changing levels of the other. The terms containing the variable substrate (S) in the denominators of Eqs. (2)–(4) are combined and expressed as:

[12] D. V. Santi, P. V. Danenberg, and P. Satterly, *Biochemistry* **10,** 4804 (1971).

[13] H. Cedar and J. H. Schwartz, *J. Biol. Chem.* **244,** 4122 (1969).

[14] It is noted that derivation of the complete rate equation for mechanism III gives a $\varphi_1$ term which is a complex function of [AA]. Such dependence on [AA] is not observed if the assumption is made that the interconversion of central complexes is rate determining (rapid equilibrium condition[15]). Regardless, Cleland[4] has pointed out that such dependence would not be experimentally detectable in the nonrapid equilibrium mechanism, and, for the purposes described here, the $\varphi_1$ term for the random mechanism may be regarded as independent of [AA].

[15] W. W. Cleland, *Biochim. Biophys. Acta* **67,** 104 (1963).

$$\frac{1}{v} = (\text{slope}) \frac{1}{\text{S}} + \text{intercept} \tag{5}$$

and subsequent analysis is performed by the general methods described by Cleland.[4] Eqs. (2)–(4) are presented below in this form.

*Mechanism I*

Varying AA

$$\frac{1}{v} = \frac{1}{\varphi_1}\left(\varphi_3 + \frac{\varphi_5}{[\text{ATP}]}\right)\frac{1}{[\text{AA}]} + \frac{1}{\varphi_1}\left(1 + \varphi_2\varphi_3 + \frac{\varphi_4}{[\text{PP}_\text{i}]} + \frac{\varphi_2\varphi_5}{[\text{ATP}]}\right) + \frac{\varphi_2[\text{AA}]}{\varphi_1}\left(1 + \frac{\varphi_4}{[\text{PP}_\text{i}]}\right) \tag{6}$$

Equation (6) predicts that substrate inhibition by amino acid will be encountered at high [AA]. Double reciprocal plots at differing ATP levels may curve upward at high [AA], but in the lower concentration range should extrapolate to intersect to the left of the vertical axis. Replots of slopes and intercepts of the extrapolated lines vs. 1/[ATP] are linear with nonzero vertical intercepts.

Varying ATP

$$\frac{1}{v} = \frac{1}{\varphi_1}\left(\varphi_2\varphi_5 + \frac{\varphi_5}{[\text{AA}]}\right)\frac{1}{[\text{ATP}]} + \frac{1}{\varphi_1}\left(1 + \frac{\varphi_3}{[\text{AA}]} + \varphi_2\varphi_3 + \frac{\varphi_4}{[\text{PP}_\text{i}]}\right) + \frac{\varphi_4[\text{AA}]}{\varphi_1}\left(1 + \frac{\varphi_4}{[\text{PP}_\text{i}]}\right) \tag{7}$$

High levels of [AA] will cause uncompetitive substrate inhibition; that is, the slopes decrease normally to a limiting value as [AA] is raised, but the intercepts decrease initially and then rise linearly as the level of AA is increased. At lower levels of [AA] lines will intersect to the left of the vertical axis, while at very high [AA] the patterns will approach a parallel one. The slope replot vs. 1/[AA] is normal, but the intercept replot is hyperbolic (i.e., shows substrate inhibitor).

If rate constants for ATP release are much greater than the maximum velocity for exchange (i.e., rapid equilibrium case), the $\varphi_2$ [AA] term becomes small such that it is not observed. As a consequence, (1) substrate inhibition by AA is not observed, (2) lines for varying 1/[AA] will intersect on the vertical axis and those for varying 1/[ATP] will intersect to the left of the axis, and (3) all slope and intercept replots will be linear. In contrast with the nonrapid equilibrium mechanism, replots of slopes vs. 1/[AA] obtained when ATP is the variable substrate will pass through the origin. In addition, inhibition patterns will differ from the nonrapid equilibrium mechanism (see below) and may be used as diagnostic aids. Although the rapid equilibrium ordered mechanism may appear unlikely, Rouget and Chapeville[16] have reported inhibition patterns which suggest that the leucine activating enzyme may proceed by this mechanism.

*Mechanism II*

Varying AA

$$\frac{1}{v} = \frac{1}{\varphi_1}\left(\frac{\varphi_4}{[\mathrm{ATP}]}\right)\frac{1}{[\mathrm{AA}]} + \frac{1}{\varphi_1}\left(1 + \frac{\varphi_2}{[\mathrm{ATP}]} + \frac{\varphi_3}{[\mathrm{PP_i}]}\right) \tag{8}$$

Varying ATP

$$\frac{1}{v} = \frac{1}{\varphi_1}\left(\varphi_2 + \frac{\varphi_4}{[\mathrm{AA}]}\right)\frac{1}{[\mathrm{ATP}]} + \frac{1}{\varphi_1}\left(1 + \frac{\varphi_3}{[\mathrm{PP_i}]}\right) \tag{9}$$

For varying AA, lines will intersect to the left of the vertical axis, but, in contrast to mechanisms I (nonrapid equilibrium) and III, a replot of slopes vs. 1/[ATP] passes through the origin. Primary plots for varying ATP will intersect on the vertical axis and slope replots are linear with a nonzero intercept.

*Mechanism III*

Varying AA

$$\frac{1}{v} = \frac{1}{\varphi_1}\left(\varphi_3 + \frac{\varphi_5}{[\mathrm{ATP}]}\right)\frac{1}{[\mathrm{AA}]} + \frac{1}{\varphi_1}\left(1 + \frac{\varphi_2}{[\mathrm{ATP}]} + \frac{\varphi_4}{[\mathrm{PP_i}]}\right) \tag{10}$$

Varying ATP

$$\frac{1}{v} = \frac{1}{\varphi_1}\left(\varphi_2 + \frac{\varphi_5}{[\mathrm{AA}]}\right)\frac{1}{[\mathrm{ATP}]} + \frac{1}{\varphi_1}\left(1 + \frac{\varphi_3}{[\mathrm{AA}]} + \frac{\varphi_4}{[\mathrm{PP_i}]}\right) \tag{11}$$

Lines will intersect to the left of the vertical axis when either substrate is varied; the random mechanism may be distinguished from mechanism I (nonrapid equilibrium) by linear intercept replots, the lack of substrate inhibition by high levels of AA, and by the use of competitive inhibitors.

These simple mechanisms do not include the formation of dead-end complexes by pyrophosphate with any enzyme form. However, since $[\mathrm{PP_i}]$ is held constant, this type of behavior would not alter the qualitative aspect of the diagnostic methods outlined above. The method of analysis for pyrophosphate inhibition and equations have been presented for the random mechanism.[12]

The patterns described above which are obtained when one substrate is

[16] P. Rouget and F. Chapeville, *Eur. J. Biochem.* **4,** 305 (1968).

TABLE I
PREDICTED EXCHANGE PATTERNS[a] WHEN ONE SUBSTRATE IS VARIED AT DIFFERENT LEVELS OF THE OTHER[b]

| Mechanism | Vary AA | Vary ATP |
|---|---|---|
| Ia. Nonrapid equilibrium | Primary plots show substrate inhibition by AA, but extrapolated pattern is intersecting | Intersecting with uncompetitive substrate inhibition by AA (intercept replot is hyperbolic, slope replot is linear) |
| Ib. Rapid equilibrium | Equilibrium ordered | Intersecting, slope replot passes through origin |
| II. | Intersecting, slope replot passes through origin | Equilibrium ordered |
| III. | Intersecting | Intersecting |

[a] Patterns are defined as follows: *intersecting,* intersection point of primary reciprocal plots is to the left of the vertical axis; *equilibrium ordered,* intersection point is on the vertical axis. Unless otherwise specified, replot of slopes or intercepts (in the intersecting pattern) versus the reciprocal of the nonvaried substrate is linear with a finite slope and vertical intercept.

[b] These patterns are described for conditions where $[PP_i]$ is held constant.

varied in the presence of differing concentrations of the other are summarized in Table I.

## Inhibition Studies

The kinetic patterns obtained in the presence of several levels of an inhibitor are, in themselves, diagnostic for the order of substrate addition and should be used as an adjunct to the kinetic studies described. The empirical rate equations may be modified to include dead-end inhibition by substrate analogs by simply multiplying the expression(s) in the empirical rate equations corresponding to the enzyme form(s) with which the inhibitor binds by a factor of $(1 + [I]/K_i)$.[17] For purposes of identification, the various enzyme forms corresponding to the terms of Eqs. (2)–(4) are presented in Table II; the resulting equations are identical to those derived by standard methods. The inhibition equations are then arranged in the form of Eq. (5), and interpretation of slope and intercept effects is performed in the same manner as previously discussed for the exchange experiments in the absence of inhibitors. Inhibitors may cause a change in the slopes (competitive), slopes and intercepts (noncompetitive), or intercepts (uncompetitive) of double reciprocal plots. A summary of the patterns which mechanisms I–III would produce in the presence of amino acid and

[17] W. W. Cleland, *Biochim. Biophys. Acta* **67,** 188 (1963).

TABLE II
CORRELATION OF ENZYME FORMS AND EMPIRICAL EXPRESSIONS

| Enzyme form | Mechanism I | Mechanism II | Mechanism III |
|---|---|---|---|
| Free enzyme | $\frac{\varphi_5}{[AA][ATP]}$ | $\frac{\varphi_4}{[AA][ATP]}$ | $\frac{\varphi_5}{[AA][ATP]}$ |
| E·AA | — | $\frac{\varphi_2}{[ATP]}$ | $\frac{\varphi_2}{[ATP]}$ |
| E·ATP | $\frac{\varphi_3}{[AA]}$ | — | $\frac{\varphi_3}{[AA]}$ |
| X (central complex) | 1 | 1 | 1 |
| F (E·AA-AMP) | $\frac{\varphi_4}{[PP_i]}$ | $\frac{\varphi_3}{[PP_i]}$ | $\frac{\varphi_4}{[PP_i]}$ |

ATP analogs is given in Table III. It is noted that each of the mechanisms shows a unique set of inhibition patterns, which is sufficient to establish the order of addition.

*Choice of Inhibitors.* Based on a number of reported inhibition studies of the activating enzymes,[6,18–22] it is possible to select with a high degree

TABLE III
INHIBITION PATTERNS IN THE PRESENCE OF SUBSTRATE ANALOGS

| Mechanism | Analogs of AA, varying substrate AA | Analogs of AA, varying substrate ATP | Analogs of ATP, varying substrate AA | Analogs of ATP, varying substrate ATP |
|---|---|---|---|---|
| Ia. Ordered, ATP first[a] | N[b] | U | N | C |
| Ib. Ordered, ATP first, rapid equilibrium | C | U | C | C |
| II. Ordered, AA first | C | C | U | C |
| III. Random | C | N | N | C |

[a] It is noted that in this case, the $(1 + \varphi_2[AA])$ term in Eq. (2) gives rise to a $(1 + [I]/K_i)$ term in the intercept of the reciprocal equation when either substrate is varied. In other cases, the number of $(1 + [I]/K_i)$ terms in the inhibition equation is equal to the number of enzyme forms with which the inhibitor may bind.

[b] C = competitive, U = uncompetitive, and N = noncompetitive.

[18] D. V. Santi and P. V. Danenberg, *Biochemistry* **10,** 4813 (1971).

[19] D. V. Santi, P. V. Danenberg, and K. A. Montgomery, *Biochemistry* **10,** 4821 (1971).

[20] T. S. Papas and A. H. Mehler, *J. Biol. Chem.* **245,** 1588 (1970).

[21] S. L. Owens and F. E. Bell, *J. Biol. Chem.* **245,** 5515 (1970).

[22] I. D. Baturina, N. V. Gnutchev, R. M. Khomutov, and L. L. Kisselev, *FEBS* (*Fed. Eur. Biochem. Soc.*) *Lett.* **22,** 235 (1972).

of certainty competitive inhibitors of the substrates that are suitable for kinetic studies. Substitution at the α-carboxyl position of the amino acid appears to be a general modification which may be made without losses in affinity or change in mode of binding. Analogs of this type may be easily prepared or obtained commercially and include those in which the α-carboxylate is replaced by $—CH_2OH$, $—CONH_2$, $—CH_3$, or —H. For competitive inhibitors of ATP, we suggest the use of α,β-methylene ATP, adenosine, $N^6$-hydroxyethyl-ATP, CrATP,[23,24] or dATP; it is noted that the last compound may exhibit substrate properties in certain cases.[19,25] Compounds that are unsuitable for determining order of binding are those that must occupy both AA and ATP sites for binding, e.g., aminoalkyl-adenylates.[26] Such inhibitors will probably bind only to free enzyme and would give competitive inhibition vs. both substrates for all orders of substrate addition except the nonrapid equilibrium mechanism I. In the latter case, where substrate inhibition by AA is observed, aminoalkyl-adenylates will show noncompetitive inhibition with respect to AA.

[23] M. L. De Pamphilis and W. W. Cleland, *Fed. Proc., Fed. Amer. Soc. Exp. Biol.* **30,** 1132 (1971).

[24] W. W. Cleland, K. D. Danenberg, and M. I. Schimerlik, *Fed. Proc., Fed. Amer. Soc. Exp. Biol.* **31,** 850 (1972).

[25] S. K. Mitra and A. H. Mehler, *Eur. J. Biochem.* **9,** 79 (1969).

[26] D. Cassio, F. Lemoine, J.-P. Waller, and R. A. Boisannas, *Biochemistry* **10,** 4804 (1971).

# [26] Applications of Kinetic Methods to Aminoacyl-tRNA Synthetases

*By* CHRISTIAN F. MIDELFORT and ALAN H. MEHLER

Contemporary kinetics as developed and popularized by Cleland[1] have several objectives: to contribute to an understanding of mechanism of enzyme action, to determine catalytic constants and to appreciate the roles of individual enzymes in biology. Although relatively few kinetic studies have been carried out with enzymes that catalyze reactions of the complexity of those catalyzed by aminoacyl-tRNA synthetases (3 substrates and 3 products), such studies are becoming more prominent as interest increases in the question of mechanisms. A parallel series of studies involves the use of modified substrates, amino acid analogs, nucleosides and nucleotides, and especially altered tRNA. The biological properties of

[1] W. W. Cleland, *Annu. Rev. Biochem.* **36,** 77 (1967).



ISBN 0-12-504180-2

tRNA have been examined with tRNA containing bases modified by chemical treatment, with heterologous tRNA, with genetically modified tRNA, and with fragments of tRNA isolated after enzymatic or chemical cleavage. For full appreciation of the consequences of modification of the structures of substrates, careful kinetic analysis is required. As Cleland has pointed out, multisubstrate reactions can have mechanisms that alter fundamentally the interpretation of Michaelis constants. It is, therefore, especially important in studying these complex systems to include careful kinetic analyses.

In an accompanying article Santi *et al.*[2] have described the analysis of the pyrophosphate-ATP exchange reaction, which is usually considered to describe the partial reaction in which an amino acid is activated, and for which equations have been developed to determine the order in which substrates bind to the enzyme. For the overall reaction it has generally been assumed that the addition of tRNA follows the release of pyrophosphate so that one of the following Ping-Pong mechanisms applies:

Mechanism 1, ordered

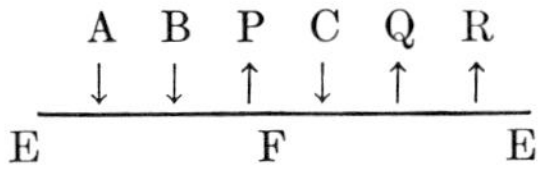

$$\begin{array}{ccccccc} & A & B & P & C & Q & R \\ & \downarrow & \downarrow & \uparrow & \downarrow & \uparrow & \uparrow \\ \hline E & & & F & & & E \end{array}$$

Mechanism 2, rapid equilibrium, random

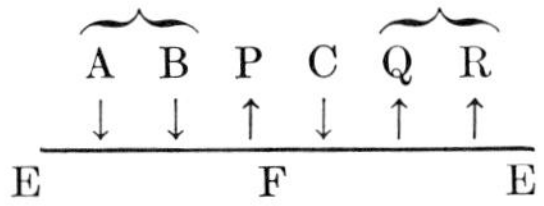

$$\begin{array}{ccccccc} & \overbrace{A \quad B} & & P & C & \overbrace{Q \quad R} & \\ & \downarrow \quad \downarrow & & \uparrow & \downarrow & \uparrow \quad \uparrow & \\ \hline E & & & F & & & E \end{array}$$

In these mechanisms, A and B represent ATP and amino acid, which might react in either order, P represents pyrophosphate, C represents tRNA, and P and Q represent AMP and aminoacyl-tRNA, again in unknown order. These mechanisms have been favored because of the ability of synthetic aminoacyladenylates to react as intermediates,[3] because of the activity of enzyme complexes formed by reaction with ATP and amino acid,[4] and because of the general lack of dependence of the pyrophosphate exchange reaction on the presence or absence of tRNA.

Recently an alternative mechanism has been favored by several workers. Loftfield[5] has noted that in several cases the rates of reaction of isolated enzyme complexes are less than the rates of the overall reaction; he there-

[2] D. V. Santi, R. W. Webster, Jr., and W. W. Cleland, this volume [25].

[3] P. Berg, *Annu. Rev. Biochem.* **30,** 293 (1961).

[4] A. N. Baldwin and P. P. Berg, *J. Biol. Chem.* **241,** 831 (1966).

[5] R. B. Loftfield, *Progr. Nucl. Acid Res. Mol. Biol.* **12,** 87 (1972).

fore proposed that the complex may be the product of a side reaction and that the overall reaction may be concerted. Support for this idea has come from reports of reaction conditions that permit esterification of tRNA but not exchange of pyrophosphate—namely, the substitution of spermine for magnesium ions.

The finding that three of twenty activating enzymes of *E. coli*[6] and probably many other organisms[7] carry out the exchange of pyrophosphate only in the presence of tRNA has been interpreted as further support for a concerted mechanism.[5]

The concerted mechanism, designated sequential by Cleland, may involve either an ordered or random addition of substrate:

Mechanism 3, ordered

$$\begin{array}{l} \quad A \;\; B \;\; C \qquad\qquad\qquad\qquad\qquad P \;\; Q \;\; R \\ \quad \downarrow \;\; \downarrow \;\; \downarrow \qquad\qquad\qquad\qquad\qquad \uparrow \;\; \uparrow \;\; \uparrow \\ \overline{E \qquad\qquad EABC \rightleftharpoons EPQR \qquad\qquad E} \end{array}$$

Mechanism 4, rapid equilibrium, random

$$\begin{array}{l} \quad \overbrace{A \;\; B \;\; C} \qquad\qquad\qquad\qquad\qquad \overbrace{P \;\; Q \;\; R} \\ \quad \downarrow \;\; \downarrow \;\; \downarrow \qquad\qquad\qquad\qquad\qquad \uparrow \;\; \uparrow \;\; \uparrow \\ \overline{E \qquad\qquad EABC \rightleftharpoons EPQR \qquad\qquad E} \end{array}$$

### Steady State Analyses

Steady state-initial velocity kinetics are a useful tool in the study of enzyme mechanisms. Although they cannot be used to prove a particular mechanism, they can often exclude certain possibilities. For example, let us consider the general mechanism involving the concerted formation of the aminoacyl ester bond and hydrolysis of the $\alpha,\beta$-phosphate anhydride bond of ATP as proposed by Loftfield.[5] This model necessarily involves a sequential mechanism, in which all substrates add to the enzyme before a product is released. The finding of a Ping-Pong mechanism, in which the first product, PP, is released before the third substrate is added, would exclude the concerted mechanism for the particular aminoacyl-tRNA synthetase. Such Ping-Pong patterns have been found in the case of mammalian threonine synthetase[8] and the *E. coli* proline synthetase.[9] However, the

[6] J. M. Ravel, S. F. Wang, C. Heinemeyer, and W. Shive, *J. Biol. Chem.* **240,** 432 (1965); S. K. Mitra and A. H. Mehler, *J. Biol. Chem.* **242,** 5490 (1967).

[7] M. P. Deutscher, *J. Biol. Chem.* **242,** 1932 (1967); L. W. Lee, J. M. Ravel, and W. Shive, *Arch. Biochem. Biophys.* **121,** 614 (1967).

[8] C. C. Allende, M. Chaimovich, M. Gatica, and J. E. Allende, *J. Biol. Chem.* **245,** 93 (1970).

[9] T. S. Papas and A. H. Mehler, *J. Biol. Chem.* **246,** 5924 (1971).

finding of a sequential kinetic pattern, as in the case of the *E. coli* arginine synthetase,[10] does not provide proof for a concerted mechanism since one or more distinct intermediate species may be formed during the conversion E:ABC E:PQR without changing the kinetic pattern.

Initial velocity equations have been developed by Cleland[11] by the method of King and Altman.[12] For the aminoacyl-tRNA synthetases the pertinent equations are given below:

*Sequential, ordered*

$$\frac{1}{v} = \frac{1}{v_1}\left[\left(1 + \frac{K\text{qP}}{K\text{iq}K\text{ip}} + \frac{K\text{rQ}}{K\text{ir}K\text{iq}}\right) + K\text{a}\left(1 + \frac{\text{R}}{K\text{ir}}\right)\left(\frac{1}{\text{A}}\right) + K\text{b}\left(\frac{1}{\text{B}}\right) \right.$$
$$+ K\text{c}\left(1 + \frac{K\text{qP}}{K\text{iq}K\text{p}}\right)\left(\frac{1}{\text{C}}\right) + K\text{ia}K\text{b}\left(1 + \frac{\text{R}}{K\text{ir}}\right)\left(\frac{1}{\text{AB}}\right)$$
$$+ K\text{ib}K\text{c}\left(1 + \frac{K\text{qP}}{K\text{p}K\text{iq}}\right)\frac{1}{\text{BC}}$$
$$\left. + K\text{ia}K\text{ib}K\text{c}\left(1 + \frac{K\text{qP}}{K\text{iq}K\text{p}} + \frac{R}{K\text{ir}}\right)\frac{1}{\text{ABC}}\right] \quad (1)$$

*Ping-Pong, ordered*

$$\frac{1}{v} = \frac{1}{V_1}\left(1 + \frac{K\text{rQ}}{K\text{ir}k\text{iq}}\right) + K\text{a}\left(1 + \frac{\text{R}}{K\text{ir}}\right)\left(\frac{1}{\text{A}}\right) + K\text{b}\left(\frac{1}{\text{B}}\right)$$
$$+ K\text{c}\left(1 + \frac{K\text{qP}}{K\text{iq}K\text{ip}} + \frac{K\text{rQ}}{K\text{ir}K\text{q}}\right)\left(\frac{1}{\text{C}}\right)$$
$$+ K\text{ia}k\text{b}\left(1 + \frac{\text{R}}{K\text{ir}}\right)\frac{1}{\text{AB}} + \frac{K\text{ib}K\text{c}K\text{qP}}{K\text{iq}K\text{p}}\left(\frac{1}{\text{AB}}\right)$$
$$+ K\text{ia}K\text{ib}K\text{c}\cdot\frac{K\text{qP}}{K\text{iq}K\text{p}}\left(\frac{1}{\text{ABC}}\right) \quad (2)$$

These equations are tested experimentally by varying one substrate ($S_1$) systematically at various changing fixed concentration of one other substrate (S), product (P), or dead end inhibitors.[1] The above equations are rearranged so that all forms containing (1/S) are grouped together yielding an equation of the form $1/v = K_1(1/\text{S}) + K_2$. $K_1$ is the slope term, and $K_2$ is the vertical intercept term. If the changing fixed substrate, product, or inhibitor appears in the $K_1$ term, it changes the slope only, and competitive kinetics are seen; if such a compound appears in the $K_2$ term only, the intercept, but not the slope, will change and uncompetitive kinetics are seen: if such a compound is found in both $K_1$ and $K_2$ terms, both

[10] T. S. Papas and A. Peterkofsky, *Biochemistry* **11** (1972).
[11] W. W. Cleland, *Biochim. Biophys. Acta* **67**, 104, 173, 188 (1963).
[12] E. L. King and C. Altman, *J. Phys. Chem.* **60**, 1375 (1956).

slope and intercept change and a noncompetitive pattern results. Regrouping of the above two equations in the form $1/v = K_1(1/\mathrm{A}) + K_2$ shows that (1/C) appears only in the $K_2$ term for the Ping-Pong case, but it is found in both $K_1$ and $K_2$ terms for the sequential case, resulting in uncompetitive and noncompetitive kinetics, respectively. However, if 1/B is saturating ($K\mathrm{ib}/\mathrm{B} \ll 1$) the slope term in 1/C for sequential kinetics is made very small also, and apparent uncompetitive kinetics will result. Therefore, the third substrate must be kept nonsaturating to differentiate the two mechanisms. Myers *et al.*[13] studied the leucine, serine, and valine enzymes of *E. coli* with purified species of tRNA; they noted that their finding of parallel uncompetitive kinetics when varying 1/(AA) against (1/tRNA) did not necessarily prove a Ping-Pong mechanism since they used only saturating concentration of ATP. These authors could have distinguished sequential from Ping-Pong by observing the change in their kinetic patterns caused by the presence of a small concentration of $PP_i$ [see Eqs. (1) and (2)]. Parallel lines would have been converted to intersecting ones for the Ping-Pong case only.

The order of product release can be determined from these two equations. In the sequential mechanism, the second product, Q, is uncompetitive toward all three substrates; the first product, P, is noncompetitive toward all three substrates; and the third product, R, is competitive toward A and noncompetitive toward B and C. In the Ping-Pong mechanism, P is competitive toward C and noncompetitive toward A and B; Q is noncompetitive toward C and uncompetitive toward A and B; and R is competitive toward A, noncompetitive toward B, and uncompetitive toward C.

The equation for random mechanisms cannot be analyzed so simply. Only rapid-equilibrium random mechanisms yield straight lines in $1/v$ vs. 1/S plots. The same patterns as above are obtained for (1/A) or (1/B) vs. (1/C) and a discrimination between sequential and Ping-Pong mechanisms is still possible.

*Dead-End Inhibition*

The use of dead-end inhibitors, which combine with only one enzyme form (E, EA, F, EP, etc.), offers a way to differentiate between ordered and rapid equilibrium-random mechanisms; also a distinction as to which substrate adds first in an ordered mechanism can be made. In this case one substrate (1/S) is varied at changing fixed concentrations of the inhibitor (I). The assumption that I combines with only one enzyme form is verified by replotting slope or intercept vs. (I). Linear replots should be obtained.

[13] G. Myers, H. V. Blank, and D. Söll, *J. Biol. Chem.* **246**, 4955 (1971).

The logic as developed by Cleland[11] is as follows: Whenever the enzyme form with which the variable substrate combines is reversibly connected in the reaction sequence with the enzyme form with which the inhibitors combines, an intersecting (competitive or noncompetitive) pattern is obtained. Whenever these two enzyme forms are not reversibly connected, a parallel (uncompetitive) pattern results. Irreversibility is produced by: (1) release of product at zero concentration; (2) saturation by a substrate which adds between the two enzyme forms in the reaction sequence; or (3) the inhibitor combining with an enzyme form that follows the addition of the variable substrate in the reaction sequence. Competitive inhibition is caused by the substrate and inhibitor combining with the same enzyme form. This last point is the most important one for discriminating which substrate is A and which is B in the ordered mechanisms. An inhibitor, B′, which combines with the same enzyme form as B, will be an uncompetitive inhibitor for A, but an analog of A, A′, will give noncompetitive inhibition when B is varied. In a random mechanism, B′ will be noncompetitive for A, and A′ noncompetitive for B.

Replots of dead-end inhibition patterns can be very informative if enough combinations of inhibitor and substrate are tested. A replot of intercept or slope vs. (I) can yield a value of, for example, $(K\text{ia}K\text{ib})/(K\text{iq})$. If enough numerical values for combinations of kinetic constants are obtained, absolute values for each constant can be derived by simultaneous equations. For each inhibitor, certain terms in the denominator of the rate equation are multiplied by $(1 + \text{I}/K_\text{I})$. These are the terms obtained for that enzyme form (numerator in $\text{E}/\text{E}_\text{t}$ = numerator/sum of all numerators when summing up all the enzyme forms to equal $\text{E}_\text{total}$ $(\text{E}_\text{t})$).

Terms in Eqs. (1) and (2) multiplied by $(1 + \text{I}/K_\text{I})$ are tabulated.

| I | Ping-Pong | Sequential |
|---|---|---|
| A′ | P/ABC, 1/A, 1/AB | 1/ABC, P/ABC, 1/A, 1/AB |
| B′ | 1/B, P/BC | 1/B, 1/B, P/BC |
| C′ | 1/C, Q/C | 1/C, P/C |
| P′ | (same as C′) | Constant (I), Q |
| Q′ | R/A, R/AB, constant (I) | R/A, R/AB, R/ABC, constant (I) |
| R′ | Same as A′ | Same as A′ |

*Exchange at Equilibrium*

The method of isotope exchange at equilibrium has been applied to a number of Uni Bi, Bi Uni, and Bi Bi reactions involving up to two substrates and two products [for example, see Eq. (1)]. The general principles developed for the simpler system can also be applied to the more complicated reaction of aminoacyl tRNA esterification. Recently, Papas and Peterkofsky[10] applied the method to the *E. coli* arginine synthetase and were able to confirm the conclusion derived from initial velocity studies of a random sequential (Ter Ter) mechanism.

A theoretical quantitative treatment was first developed by Boyer,[14] and more recently it was refined by Yagil and Hoberman,[15] by Cole and Schimmel,[16] and by Cleland.[1] The last method forms the basis of the equations used by Cedar and Schwartz[17] and by Santi *et al.*[2] in this volume. The method of Yagil and Hoberman was used to derive the equations given here.

Briefly, the method involves first establishing chemical equilibrium between nonradioactive reactants and products (either by adding enough enzyme or by adding both substrates and products) and then adding a trace of labeled substrate (or product) and measuring the rate of incorporation into one product (or reactant). The approach to radioactive equilibrium is first order, and the rate constant for the decay of A* into P is $V_{exch}(A + P)/AP$, where A* and P are the substrate–product pair exchanging label.

Equations relating this $V_{exch}$ to reactant concentrations and rate and equilibrium constants are obtained as follows:

Exchange between A and P occurs through two discrete steps[18]:

$$A + E \underset{k_1}{\overset{V_1}{\leftrightarrow}} AE \underset{k_2}{\overset{V_2}{\leftrightarrow}} E + P \qquad V_1 = k_{-1}AE$$

$$V_2 = k_2AE \tag{3}$$

$$\begin{aligned} 1/V_{exch} &= 1/V_1 + 1/V_2 \\ &= 1/k_{-1}AE + 1/k_2AE \\ &= 1/(AE)\left[\frac{1}{k_{-1}} + \frac{1}{k_2}\right] \end{aligned} \tag{4}$$

Since the reaction is at chemical equilibrium, all enzyme forms (E, EA) are also at equilibrium and $k_{-1}(AE) = k_1(A)(E)$. (AE) in the above equation is unknown but can be related to $E_t$ by using equilibrium constants:

[14] P. D. Boyer, *Arch. Biochem. Biophys.* **83,** 381 (1959).

[15] G. Yagil and H. Hoberman, *Biochemistry* **8,** 352 (1969).

[16] F. Cole and P. R. Schimmel, *Biochemistry* **9,** 480 (1970).

[17] H. Cedar and J. H. Schwartz, *J. Biol. Chem.* **244,** 4122 (1969).

[18] The exchange rate of a multistep process is related to the rates of a series of sequential steps by $1/V_E = 1/V_1 + 1/V_2 + \cdots$, while the exchange rate of a process occurring through several parallel steps, as in a random reaction mechanism, is related to the rate of each step by $V_E = V_1 + V_2 + V_3 + \cdots$.

$$(E_t) = (E) + (AE) = (AE)(1 + K_1)/(A) \tag{5}$$

$$1/V_{exch} = \left(\frac{1}{E_t}\right) \cdot \left(\frac{1}{k_{-1}} + \frac{1}{k_2}\right) \cdot [1 + K_1/(A)] \tag{6}$$

A term for P does not appear, but since chemical equilibrium must be maintained, (P) must be changed in proportion to (A). In more complicated mechanisms (ordered only), the last product released does not appear in the rate equation and can be used to maintain chemical equilibrium while a single substrate or product is varied.

In the table are listed the various equations obtained for certain three-substrate three-product reactions. Only sequential Ter Ter and Bi Uni Uni Bi Ping-Pong are considered, since these seem the only likely possibilities for the AA-tRNA synthetases. Each equation consists of a $k_{app}$ term containing rate constants and a distribution-of-enzyme-forms term containing equilibrium constants. If the simplest substrate-product pair is chosen ($B^* \rightarrow P$ for ordered Ping-Pong or $C \rightarrow P$ for ordered sequential), the $k_{app}$ contains no reactant concentration terms, and the equilibrium constants for all reactants except R can be determined. For instance, at saturating B ($B/K_B \gg 1$) the slope of $1/V$ versus $1/A$ gives $K_1$ for Ping-Pong and the slope of $1/V$ versus $1/C$ gives $K_3$ for sequential.

Sequential and Ping-Pong mechanisms are easily distinguished. The rate $A \leftrightarrow P$ exchange is zero at zero C and increases to a plateau at high C for sequential, whereas the rate is maximal at zero C and increases to zero at high C for Ping-Pong. We attempted such a qualitative distinction using the valine synthetase from *E. coli*. An exchange between tRNA* and Val-tRNA ($A^* \leftrightarrow P$ with symbols for substrates and products reversed) was measured by the method described by Papas and Peterkofsky.[10] The effect of increasing C ($PP_i$ in this case) and R (ATP) in constant ratio would be to increase the exchange rate from zero to a finite value for a sequential, and vice versa for a Ping-Pong, mechanism. The experimental results together with two theoretical curves are shown in Fig. 1. The maximum exchange rate can be seen at zero $PP_i$ concentration, and it only decreases at high pyrophosphate. Using the equations from the table for ordered Ping-Pong and ordered random, the theoretical curves were calculated. The results agree quite well with an ordered Ping-Pong mechanism, assuming the equilibrium constants shown in the table. An alternative mechanism, which seems more likely, is that the release of products and substrates is random[19,20] and formation and reaction of the adenylate is

[19] E. W. Eldred and P. R. Schimmel, *Biochemistry* **11**, 17 (1972).

[20] E. Holler and M. Calvin, *Biochemistry* **11**, 3741 (1972).

$$1/V_{\text{exch}} = \frac{1}{\text{E}_t} \cdot \frac{1}{k_{\text{app}}} \cdot \text{(Distribution of Enzyme Forms)}$$

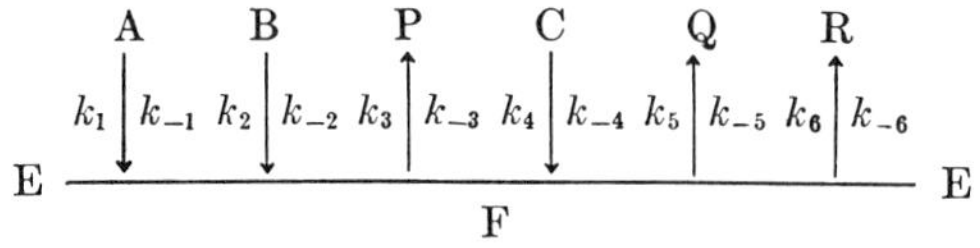


| | | Distribution of Enzyme Forms |
|---|---|---|
| *Ping-Pong Ordered* | $1/k_{\text{app}}$ | $\left[1 + K_2/\text{B} + \frac{K_1K_2}{\text{AB}} + \frac{K_3}{\text{P}} + \frac{K_3\text{C}}{\text{P}K_4} + \frac{K_3}{\text{P}} \cdot \frac{\text{C}}{K_4}\frac{K_5}{\text{Q}}\right]$ |
| B* ↔ P | $1/k_{-2} + 1/k_3$ | |
| A* ↔ P | $1/k_{-2} + 1/k_3 + \frac{\text{B}}{k_{-1}K_2}$ | |
| B* ↔ Q | $1/k_{-2} + 1/k_3 + \frac{K_4}{\text{C}}\frac{\text{P}}{K_3} \cdot \left[\frac{1}{k_{-4}} + \frac{1}{k_5}\right]$ | |
| A* ↔ Q | $1/k_{-2} + 1/k_3 + \text{B}/k_{-1}K_2 + \frac{K_4}{\text{C}} \cdot \frac{\text{P}}{K_3}\left[\frac{1}{k_{-4}} + \frac{1}{k_5}\right]$ | |
| A* ↔ R | $1/k_{-2} + 1/k_3 + \text{B}/k_{-1}K_2 + \frac{K_4}{\text{C}} \cdot \frac{\text{P}}{K_3}\left[1/k_{-4} + 1/k_5 + \frac{\text{Q}}{K_5} \cdot \frac{1}{k_6}\right]$ | |
| B ↔ R | $1/k_{-2} + 1/k_3 + \frac{K_4}{\text{C}} \cdot \frac{\text{P}}{K_5}\left[\frac{1}{k_{-4}} + \frac{1}{k_5} + \frac{\text{Q}}{K_5} \cdot \frac{1}{k_6}\right]$ | |
| *Sequential Ordered* | $1/k_{\text{app}}$ | Distribution of Enzyme Forms |
| C ↔ P | $\frac{+1}{k_{-3}} + \frac{1}{k_4}$ | $\left[1 + \frac{K_3}{\text{C}} + \frac{K_2K_3}{\text{BC}} + \frac{K_1K_2K_3}{\text{ABC}} + \frac{K_4}{\text{P}} + \frac{K_4K_5}{\text{PQ}}\right]$ |
| B ↔ P | $1/k_{-3} + 1/k_4 + \frac{\text{C}}{K_3k_{-2}}$ | |
| A ↔ P | $1/k_{-3} + 1/K_4 + \frac{\text{C}}{K_3k_{-2}} + \frac{\text{BC}}{k_{-1}K_2K_3}$ | |
| B ↔ Q | $1/k_{-3} + 1/k_4 + \frac{\text{C}}{k_{-2}K_3} + \frac{\text{P}}{k_5K_4}$ | |
| A ↔ Q | $1/k_{-3} + 1/k_4 + \frac{\text{C}}{k_{-2}K_3} + \frac{\text{BC}}{k_{-1}K_2K_3} + \frac{\text{P}}{k_5K_4}$ | |
| A ↔ R | $1/k_{-3} + 1/k_4 + \frac{\text{C}}{k_{-3}K_3} + \frac{\text{BC}}{k_2K_2K_3} + \frac{\text{P}}{k_5K_4} + \frac{\text{PQ}}{k_6K_4K}$ | |

*Random Mechanisms*

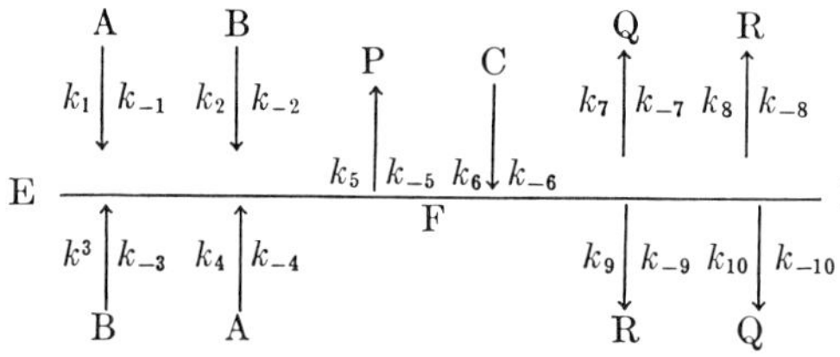


$$K_{\text{AB}} = \frac{(\text{EA})(\text{B})}{(\text{EAB})}$$

$$K_{\text{ABC}} = \frac{(\text{EAB})(\text{C})}{(\text{EABC})}$$

$$K_{\text{EA}} = \frac{(\text{EB})(\text{A})}{(\text{EAB})}$$

(*Continued*)

(*Continued*)

*Ping-Pong* $(1/k_{app})$ — *Distribution of Enzyme Forms*

$$A \leftrightarrow P \quad \frac{k_{-1}(k_{-2} + k_4 + k_5) + k_2B(k_{-4} + k_5)}{k_{-4}k_5(k_{-1} + k_2B) + k_{-1}k_{-2}k_5}$$

$$1 + \frac{K_{AB}}{B} + \frac{K_{BA}}{A} + \frac{K_AK_{AB}}{AB} + \frac{K_P}{P} + \frac{K_PC}{PK_6}\left(1 \frac{K_QR}{R} + \frac{K_{RQ}}{Q}\right)$$

$$A \leftrightarrow Q \quad \frac{k_{-1}(k_{-2} + k_{-4} + k_5) + k_2B(k_{-4} + k_5)}{k_{-4}k_5(k_{-1} + k_2B) + k_{-1}k_{-2}k_5} + \frac{K_6}{C} \cdot \frac{P}{K_5} \cdot \left[\frac{k_{-9}P(k_7 + k_{-6}) + K_{10}(k_7 + k_8 + k_{-6})}{k_{-6}(k_7k_{-9}(P) + k_7k_{10} + k_9k_{10})}\right]$$

*Sequential* — *Distribution of Enzyme Forms*

$$\left[1 + \frac{K_{ABC}}{C} + \frac{K_{AB}K_{ABC}}{BC} + \frac{K_AK_{AB}K_{ABC}}{ABC} + \frac{K_{BA}K_{ABC}}{AC} + \frac{K_{ACB}}{B} + \frac{K_{CA}K_{ACB}}{AB} + \frac{K_{BCA}}{A} + \frac{K_{QRP}}{P} + \frac{K_{PRQ}}{Q} + \frac{K_{PQR}}{R} + \frac{K_{QRP}K_{RQ}}{PQ} + \frac{K_{QRP}K_{QR}}{PR} + \frac{K_{PR}K_{PRQ}}{RQ}\right]$$

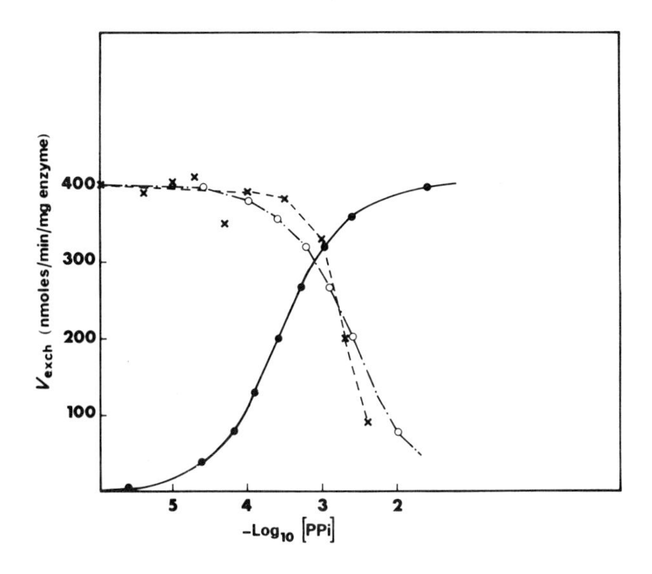


FIG. 1. Exchange of tRNA-valyl-tRNA at equilibrium. [$^{14}$C]tRNA labeled in the 3′-terminal adenosine was added to reaction mixtures at equilibrium with a fixed amount of enzyme.

$$K_{eq} = \frac{[\text{Val-tRNA}][\text{PP}_i][\text{AMP}]}{[\text{tRNA}][\text{ATP}][\text{Val}]}$$

was calculated to be 0.25. The theoretical curves were calculated for $k_{app} = 2.8 \times 10^{-4}\ M$ and $K_{tRNA} = 5.5 \times 10^{-7}\ M$. ——, Sequential ordered; – – –, experimental; – · – ·, Ping-Pong ordered.

fast compared to tRNA and AA-tRNA release.[21] The fast exchange at zero pyrophosphate is due to the rapid formation of Enz·AA-adenylate and the inhibition at high ($PP_i$) is due to pyrophosphorolysis of adenylate even when tRNA is present on the enzyme. Loftfield[5] has predicted that a sequential pathway would predominate since the stable enzyme form (E·AA-adenylate) leading to Ping-Pong kinetics would not be formed in the presence of tRNA. The inhibition at high $PP_i$ could be dead-end inhibition as described by Cole and Schimmel[16] or it could be due to the addition of $PP_i$ to an enzyme form which lies between the addition of tRNA* and AA-tRNA. This last type of inhibition is given by terms in numerators of the $[1/k_{app}]$ terms of the ordered mechanisms in the table. Although substrate inhibition at high substrate concentrations is characteristic of an ordered mechanism in which the exchanging substrate adds to the enzyme before the inhibiting substrate, the phenomenon is subject to other interpretations. Substrate inhibition can be caused by the substrate combining in a dead-end complex with a form of the enzyme other than the one with which it reacts productively. In addition, failure to find substrate inhibition may mean that high enough substrate was not used and the substrate terms in $k_{app}$ were still too small relative to the other rate constants to cause inhibition. Both Cedar and Schwartz[17] and Papas and Peterkofsky[10] searched in vain for substrate inhibition and concluded that they were dealing with random substrate addition. Perhaps the concentrations were too low for the substrate terms in $1/k_{app}$ to cause inhibition. The use of competitive inhibitors (substrate and product analogs) offers a surer way to diagnose ordered and random mechanisms.

*Dead-End Inhibitors*

As outlined by Santi *et al.*,[2] dead-end inhibitors provide a simple way to distinguish between ordered and random addition of substrates and to decide which substrate adds first in an ordered mechanism. The presence of an inhibitor in an ordered mechanism causes one term in the sum-of-enzyme-forms to be multiplied by $(1 + I/K_I)$. In considering, for example, an ordered Ping-Pong mechanism, the presence of an A′ changes the 1/ABC coefficient by $(1 + I/K_I)$; B′ changes the 1/BC coefficient, and C′ changes only the 1/C coefficient. When a substrate is varied (along with product R) at several fixed concentrations of an unreactive analog of one of the other substrates, either competitive or uncompetitive inhibition for an ordered mechanism or noncompetitive inhibition for a random mechanism will be seen. This is illustrated in the following scheme:

[21] M. Yarus and P. Berg, *J. Mol. Biol.* **28**, 479 (1967).

$$\begin{array}{c} \quad A(\text{or } A') \quad B(\text{or } B') \quad P \quad Q \\ \quad \downarrow \qquad\quad \downarrow \qquad \uparrow \quad \uparrow \\ \overline{E \qquad\qquad\qquad EAB \quad\; X} \end{array}$$

B′ gives uncompetitive inhibition when A/Q is the varied substrate–product pair, but A′ gives competitive inhibition when B/Q is the varied pair. In a random sequence of substrate addition, both A′ and B′ are noncompetitive for (B/Q) and (A/Q), respectively. However, an inhibitor, such as valinol-AMP, an analog of the type used by Santi *et al.*[22] and by Rouget and Chapeville,[23] which can bind only to the free enzyme (E), not to EA or EB, will be a competitive inhibitor toward both A/Q and B/Q in either a random or a sequential mechanism.

## Experimental

Methods for measuring ATP ↔ $PP_i$, ATP ↔ AMP, AA ↔ AA-tRNA exchanges can be found in this volume [25]. We briefly present the method for measuring the AMP-dependent-tRNA ↔ AA tRNA exchange. This reaction, when carried out in the absence of ATP, $PP_i$, and amino acid, provides a means for measuring E:AA-AMP formation from the product side.

*Reagents.* tRNA must be purified to near homogeneity. In our experiments, 80% pure valine-tRNA prepared according to Gillam and Tener[24] was used successfully. tRNA, with the 3′-OH-terminal adenosine labeled, is prepared in a 3-step process: Purified tRNA (300 $A_{260}$ units) is incubated with a small amount of phosphodiesterase 1 (from snake venom) in $10^{-3}$ *M* $MgCl_2$, 0.1 *M* Tris, pH 8.0 at 20°.[25] When all the amino acid acceptance activity has been lost, the reaction mixture is applied to a 0.3 × 1 cm DEAE-cellulose ($Cl^-$) column equilibrated with 0.3 *M* NaCl and 10 m*M* sodium acetate buffer (pH 4.5). The diesterase is eluted under these conditions. The tRNA is eluted with 1.0 *M* NaCl in a volume of 1.5–2 ml. Salt is removed by precipitation of the tRNA with 2 volumes of ethanol. The dissolved tRNA is incubated with 0.1 m*M* CTP, 0.1 m*M* [$^{14}$C]ATP (44 mCi/mmole), and 25–50 units of CCA enzyme.[26] Reaction is complete within 20 minutes. The tRNA is reisolated on DEAE-cellulose as above. The final specific radioactivity of the tRNA is about 100,000 dpm/$OD_{260}$ unit [(~1100 pmoles of AMP incorporated)/$OD_{260}$ unit]. Valine could be

[22] D. V. Santi, P. V. Danenberg, and P. Satterly, *Biochemistry* **10**, 4804 (1971).

[23] P. Rouget and F. Chapeville, *Eur. J. Biochem.* **4**, 305 (1968).

[24] I. C. Gillam and G. M. Tener, this series, Vol. 20, p. 55.

[25] J. P. Miller, M. E. Hirst-Bruns, and G. P. Phillips, *Biochim. Biophys. Acta* **217**, 176 (1970).

[26] J. Hurwitz and J. J. Furth, *in* "Procedures in Nucleic Acid Research" (G. L. Cantoni and R. D. Davies, eds.), Harper, New York, 1966.

esterified to 85% of this radioactive material. [$^{12}$C]Valyl-tRNA$^{Val}$ is prepared by incubating 1–2 mg of tRNA$^{Val}$ with 50 esterification units of valyl-tRNA synthetase[27] under standard assay conditions and is purified by column chromatography on DEAE-cellulose as above. $^{3}$H-labeled amino acid (1000 cpm/nmole) can be used to check the esterification and the hydrolysis of the AA-tRNA with time.

*[$^{14}$C]tRNA—AA-tRNA Exchange.* The reaction mixture contains in 0.05 ml volume at 37°: 1 m$M$ AMP, 0.2 $OD_{260}$ unit of [$^{14}$C]tRNA$^{Val}$ (about 13,000 cpm), 0.2 $OD_{260}$ unit of Val-tRNA$^{Val}$, 72 m$M$ $NH_4Cl$, 1 m$M$ mercaptoethanol, 0.1 mg of bovine serum albumin per milliliter and enough valyl-tRNA synthetase to exchange 10–20% of the radioactive label into Val-tRNA in 10 minutes. The reaction is terminated by adding 10 $\mu$l of 0.5 mg/ml pancreatic ribonuclease to each reaction mixture and incubating at room temperature for 5–10 minutes. The 3′-OH-adenosine and valyl-adenosine are liberated with half-times of the order of seconds under these conditions. The valyl-adenosine can be stabilized by adding 10 $\mu$l of 1 $M$ acetic acid. Each sample is applied to Whatman 3 MM paper and electrophoresed in pH 3.6 pyridine–acetate buffer (5% acetic acid, 0.5% pyridine) for 25 minutes at 2500 V. The papers are dried, and the radioactive spots are located by autoradiography overnight, or by cutting thin strips of the paper for liquid scintillation counting. The free and esterified nucleosides are well separated with valyl-adenosine having a mobility of twice that of adenosine (2–3 cm). The radioactivities are determined by cutting out the spots and counting in Bray's solution.

## Partial Reactions

Conventional kinetic analyses are complicated by the necessity of measuring complex functions. Several approaches have been developed for measuring binding constants and velocity constants independently. Binding of substrates and other ligands to aminoacyl-tRNA synthetases has been studied by three methods: protection against thermal denaturation, modification of fluorescence of the enzyme, and equilibrium dialysis. Velocity constants of partial reactions have been studied also by three methods: rapid binding of enzyme to nitrocellulose filters, transfer of amino acid from the enzyme–aminoacyl adenylate complex and stopped-flow measurements with a fluorescence reporter group.

### *Binding Constants*

Chuang and Bell[28] have developed equations to express a protection constant $\pi$ in terms of the rate constants of inactivation of an enzyme in

[27] M. Yaniv and F. Gros, *J. Mol. Biol.* **44,** 1 (1969).

[28] H. Y. K. Chuang and F. E. Bell, *Arch. Biochem. Biophys.* **152,** 502 (1972).

the presence and in the absence of ligand. Of the several types of graphic analysis considered, the Lineweaver-Burk type and the Eedie-Hofstee type give linear plots. For the former, a plot of $1/(k_0 - k)$ vs. $1/x$ and for the latter, a plot of $k_0 - k$ vs. $(k_0 - k)/x$, where $k_0$ and $k$ are the first-order rate constants of inactivation in the absence of ligand and in the presence of a concentration of ligand $x$, respectively. The corresponding forms of the equation relating these values are:

$$\frac{1}{k_0 - k} = \frac{\pi}{k_0 - k_\infty} \cdot \frac{1}{x} + \frac{1}{k_0 - k_\infty} \tag{7}$$

and

$$k_0 - k = \frac{k_0 - k}{x} \pi + k_0 - k_\infty \tag{8}$$

In these equations $k_\infty$ is the constant in the presence of saturating concentrations of ligand.

This method has been used by Chuang and Bell to measure the binding to the valine-activating enzyme of *E. coli* of analogs of valine that are inhibitors as well as substrates and to show a dependence of the binding constant on the presence of Mg-ATP. A less rigorous application of this approach was used by Mitra *et al.*[29] to study the contribution of residues at the 3′ terminus of $tRNA^{Arg}$ to the arginine-activating enzyme of *E. coli*.

Equilibrium dialysis has not been widely applied to aminoacyl-tRNA synthetases, probably because of the relatively large quantities of enzyme required. Blanquet *et al.*[30] have carried out studies on the binding of L-methionine and ATP with the methionine-activating enzyme of *E. coli* and obtained binding constants similar to the corresponding Michaelis constants, in both the presence and the absence of $Mg^{2+}$.

Measurements of binding by determination of changes in fluorescence depend upon the chance that such changes occur is a result of association of substrates with the enzyme. In the case of the valyl-tRNA synthetase of *E. coli*, Hélène *et al.*[31] found this approach to be sensitive and to permit measurements to be made on the binding of all substrates. This method has also been used with the methionine enzyme.[31]

### *Velocity Constants*

Yarus and Berg[32] have described conditions under which the isoleucine-activating enzyme of *E. coli* is effectively bound to nitrocellulose filters and have used this binding to measure the association of ligands with the

[29] S. K. Mitra, K. Chakraburtty, and A. H. Mehler, *J. Mol. Biol.* **49,** 139 (1970).
[30] S. Blanquet, G. Fayat, J. P. Waller, and M. Iwatsubo, *Eur. J. Biochem.* **24,** 461 (1972).
[31] C. Hélène, F. Brun, and M. Yaniv, *J. Mol. Biol.* **58,** 349 (1971).
[32] M. Yarus and P. Berg, *J. Mol. Biol.* **42,** 171 (1969).

enzyme. When the enzyme-tRNA complex was studied at reduced temperature (17°) and low pH (5.5), the rate of dissociation could be determined.[21] This was done by preparing complexes of enzyme and labeled tRNA in relatively high equal concentrations, where the equilibrium leaves little unbound tRNA, and measuring the liberation of tRNA after rapid dilution (×1000). Aliquots of the diluted sample are filtered at intervals and counted. An alternative method consists of diluting labeled complex with a large excess of unlabeled tRNA and measuring the rate at which label is lost from the form that adsorbs on the filter. This method provided the first evidence that dissociation of the aminoacyl-tRNA might be rate-limiting in the esterification reaction and that the release of this product can be accelerated by the binding of amino acid. From the measurements of the binding constant and rate of dissociation, the rate of association of enzyme and tRNA can be calculated. Since the binding constant does not change with the presence of isoleucine, the amino acid apparently increases equally the rates of association and dissociation.

The rate of transfer of amino acid from the complex enzyme–aminoacyl adenylate to tRNA was measured by Eldred and Schimmel.[19] They isolated the complex of the isoleucine enzyme with labeled isoleucyl adenylate as described by Baldwin and Berg[4] and measured the rate of transfer as conversion of acid-soluble label to acid-precipitable material.[33] By carrying out the reaction at low temperature (3°) and low pH (6.0), they were able to measure a reaction somewhat faster than the overall rate of esterification under similar conditions. Since the rate of formation of the complex from free amino acid and ATP (as measured by the exchange of pyrophosphate) is about an order of magnitude still faster, it appeared that neither catalytic step was rate-limiting. The conclusion of Yarus and Berg,[21] that dissociation of aminoacyl-tRNA is rate-limiting, was supported by studies with exogenous tritium-labeled isoleucyl-AMP. With this substrate added to a complex of enzyme with [$^{14}$C]isoleucyl-AMP, rapid transfer of $^{14}$C to tRNA was accompanied by a slower transfer of $^{3}$H. A similar experiment, using free enzyme and exogenous isoleucyl adenylate, gave a biphasic curve with an initial "burst" measuring the catalytic transfer and a second phase measuring dissociation of isoleucyl-tRNA.

In these experiments, the presence of excess aminoacyl-AMP increased the rate of dissociation of isoleucyl-tRNA as free isoleucine had been shown to do previously.[21] If the acceleration of dissociation is a part of the normal mechanism of action, free enzyme does not participate in the catalytic process, and the sequence of events should be diagrammed:

[33] The substrate intermediate that is highly bound to the native form of the enzyme is released as an acid-soluble material when the enzyme is denatured.

AA ↓ ATP ↓ $PP_i$ ↑ AA · tRNA ↑ tRNA ↓ AMP ↑

E · AA · tRNA ——— E · AA · AMP ——— E · AA · tRNA

The alteration of the environment of tryptophan residues or other fluorescing groups in enzymes on binding of ligands, including substrates, is an unpredictable event. Therefore, endogenous fluorescence cannot be relied upon as a general basis for measuring interactions of enzymes with substrates. An alternative method for performing similar studies was introduced by Holler *et al.*,[34] who found that 2-*p*-toluidinylnaphthalene-6 sulfonate has only slight fluorescence in aqueous solution but fluoresces strongly in the presence of isoleucyl-tRNA synthetase. The fluorescence is quenched by the addition of substrates. Holler and Calvin[20] have used the change of fluorescence produced by mixing substrate with enzyme containing the "reporter group" to measure several rates of reaction in a stopped-flow apparatus.

Independent measurements showed that the reporter does not interfere with the reactions of the enzyme and that it does not compete with ATP or isoleucine (tRNA results have not yet been published). Using an apparatus that can measure changes in milliseconds, rates were measured for the interaction of the enzyme with isoleucine, with ATP in the absence of $Mg^{2+}$, and with the two substrates under conditions of reaction. With this system the quenching of fluorescence when isoleucine was added to the enzyme showed two rates; ATP reacted faster, and in the presence of $Mg^{2+}$ reacted too fast for the rate to be measured; ATP plus isoleucine caused a slow quenching that is compatible with the formation of aminoacyl adenylate. The reaction with pyrophosphate appeared to cause a change in fluorescence, but the change was too fast to be measured. The measured values permit the calculation of equilibrium constants and rates of reverse reactions.

The slow second step seen with isoleucine in the absence of ATP, where there can be no catalytic reaction, is defined as an isomerization step and may be caused by a change in the conformation of the enzyme. This slow step complicates the kinetic analysis. The finding that ATP reacts rapidly with an enzyme already saturated with isoleucine indicates that ATP can act as a second substrate, but the more rapid reaction of ATP with free enzyme suggests that it usually is the first substrate and that the mechanism is properly designated as random. The isomerization step has been suggested by Holler and Calvin[20] as the reaction responsible for the stimulation by isoleucine of release of aminoacyl-tRNA from the enzyme.[19,21]

[34] E. Holler, E. L. Bennett, and M. Calvin, *Biochem. Biophys. Res. Commun.* **45,** 409 (1971).

# Section III

# Polypeptide Chain Initiation

# [27] Purification of Protein Synthesis Initiation Factors IF-1, IF-2, and IF-3 from *Escherichia coli*[1]

*By* JOHN W. B. HERSHEY, JOAN YANOV, and JOHN L. FAKUNDING

Initiation of protein synthesis in *Escherichia coli* is promoted by three protein factors. These are found associated with ribosomes but can be separated from them by centrifugation in buffer containing high salt. The initiation factors, called IF-1, IF-2, and IF-3, have been studied extensively during the past 10 years, and the broad features of their role in initiation of protein synthesis have been elucidated. We describe a simple and rapid method for the purification of the three factors in high yield. The degree of purity is carefully monitored, and factors greater than 98% pure are obtained. The three factors stimulate protein synthesis or various partial reactions of the initiation process with phage R-17 RNA as messenger RNA. Their physical characterization is described in detail elsewhere.[2]

*Buffers*

Buffer A: 10 m*M* potassium phosphate, pH 7.5; 0.1 m*M* EDTA; 7 m*M* 2-mercaptoethanol; 5% (v/v) glycerol; and various concentrations of KCl as indicated: buffer A-100 means buffer A containing 100 m*M* KCl; buffer A-500 means buffer A containing 500 m*M* KCl

Buffer B: 10 m*M* Tris · HCl, pH 7.4; 0.1 m*M* magnesium acetate; 7 m*M* 2-mercaptoethanol; 10% (v/v) glycerol; 6 *M* urea; and various amounts of $NH_4Cl$ indicated as in buffer A

*Reagents*

Urea, ultrapure, from Schwarz/Mann
GTP, from Calbiochem
2-Mercaptoethanol, from British Drug House
Dithiothreitol, from Pierce Chemical Co.
DNase, RNase-free, from Worthington
Glass beads, Superbrite 100, from 3M Company
Phosphocellulose, P-11, from Whatman
DEAE-Sephadex A-50 (3.5 meq/g) and Sephadex G-50, from Pharmacia Fine Chemicals
Toluene scintillation fluid, prepared by mixing 4 g of 2,5-diphenyloxazole and 0.05 g of 1,4-bis[2-(5-phenyloxazolyl)]benzene in 1 liter of toluene

[1] Supported by a grant from the American Cancer Society, NP-70.

[2] J. W. B. Hershey, J. Yanov, K. Johnston, and J. L. Fakunding, *Arch. Biochem. Biophys.* **182,** 626 (1977).




ISBN 0-12-504180-2

*Biological Materials*

AUG is prepared according to the method of Sundararajan and Thach.[3] f[$^{14}$C]Met-tRNA is prepared from unfractionated *E. coli* B tRNA (Schwarz/Mann) and [$^{14}$C]methionine (Schwarz/Mann, specific activity 265 Ci/mol) according to the method of Hershey and Thach.[4] R-17 RNA is prepared by phenol extraction of R-17 phage essentially as described by Gesteland and Spahr[5]; 70 S ribosomes are prepared and washed twice in high-salt buffer[6]; 30 S subunits are prepared by a modification[7] of a procedure described by Noll.[8]

*Escherichia coli* strain MRE600 is grown in a New Brunswick Model F-130 fermentor in 100-liter batches in a buffered enriched medium containing per liter: 11.6 g of yeast extract (Difco), 42.0 g of $KH_2PO_4$, 14 g of KOH, and 10 g of glucose, final pH 7.0. Each batch is inoculated with one liter of stationary-phase cells grown in the same medium overnight. Growth is carried out at 37° with maximum stirring (780 rpm) and aeration (10 ft$^3$/min); foaming is suppressed by the addition of 50 ml of Antifoam A (Sigma). Growth is stopped in late log phase ($A_{540}$ = 6.5 in a Gilford 2400S spectrophotometer) by emptying the culture into 50 liters of crushed ice to bring the temperature to 0°–4° in less than 1 min. The chilled cells are harvested quickly by centrifugation in a Sharples continuous-flow centrifuge, washed once by resuspension in 20 m*M* Tris · HCl, pH 7.5, 10 m*M* Mg acetate, and 0.5 m*M* EDTA, and pelleted in liter bottles at 8000 rpm in a Lourdes centrifuge. The cells, about 1 kg wet weight per 100-liter batch, are stored at −70° in 100-g amounts.

Cell lysates are prepared by combining 500 g of frozen cells with 1250 g of glass beads and 500 ml of buffer (10 m*M* Tris · HCl, pH 7.4; 10 m*M* Mg acetate) in a 1-gal Waring blender and mixing at high speed for a total of 15 min. The temperature is maintained at 0°–5° by the periodic addition of powdered Dry Ice. DNase (150 $\mu$g) is added, and the beads are removed by centrifugation at 700 rpm in a Sorvall GSA rotor. The beads are resuspended in 300 ml of buffer and centrifuged again. The supernatants from the two centrifugations are combined and centrifuged for 20 min at 16,000 rpm in the Sorvall SS-34 rotor. The clarified cell lysate (615 ml) is carefully removed from the pellet and held on ice for further fractionation as described below.

[3] T. A. Sundararajan and R. E. Thach, *J. Mol. Biol.* **19**, 74 (1966).
[4] J. W. B. Hershey and R. E. Thach, *Proc. Natl. Acad. Sci. U.S.A.* **57**, 759 (1967).
[5] R. F. Gesteland and P. F. Spahr, *Biochem. Biophys. Res. Commun.* **41**, 1267 (1970).
[6] J. W. B. Hershey, E. Remold-O'Donnell, D. Kolakofsky, K. F. Dewey, and R. E. Thach, this series, Vol. 20, p. 235.
[7] J. L. Fakunding and J. W. B. Hershey, *J. Biol. Chem.* **248**, 4206 (1973).
[8] L. M. Noll, Ph.D. Dissertation, Northwestern University, Evanston, Illinois, 1972.

*Procedures for Column Chromatography*

Phosphocellulose and DEAE-Sephadex are treated as directed by the manufacturers and then equilibrated with buffer A-100. To the settled resins are added 2 volumes of 0.1% bovine serum albumin in buffer A-0, the suspension is stirred an hour and the resin is filtered, washed thoroughly with buffer A-1000 to remove the BSA, and equilibrated with buffer A-100. Phosphocellulose columns are packed tightly under 200 cm of hydrostatic pressure, and DEAE-Sephadex A-50 columns are prepared at less than 40 cm hydrostatic pressure. Details of column size and elution conditions are given in the text. The salt concentrations in elutant fractions are determined with a conductivity meter (Radiometer), and absorbance at 280 nm is measured with a spectrophotometer (Gilford 2400).

*Assay Procedures*

Initiation factor-dependent binding of fMet-tRNA to 70 S ribosomes is used routinely to assay factor activity. Each assay mixture of 40 $\mu$l contains: 20 m*M* Tris·HCl, pH 7.4; 100 m*M* $NH_4Cl$ plus KCl; 5–7 m*M* Mg acetate; 1 m*M* dithiothreitol; 1 m*M* GTP; 0.6 $A_{260}$ unit of washed 70 S ribosomes; and 10 $\mu$g of unfractionated tRNA charged with 16 pmol of [$^{14}$C]methionine. For the IF-1 activity assay, 20 $\mu M$ AUG and 0.5–1.5 $\mu$g of IF-2 are included. For IF-2, 20 $\mu M$ AUG and 0.1–0.2 $\mu$g of IF-1 are added. For IF-3, 15 $\mu$g of R-17 RNA, 1.0–1.5 $\mu$g of IF-2 and 0.15–0.2 $\mu$g of IF-1 are added. The volume of factor assayed is usually 2–5 $\mu$l. The reaction is started by the addition of a mixture containing the GTP, AUG and f[$^{14}$C]Met-tRNA to the other components, and is incubated 5 min at 30°. The mixture is diluted with 1 ml of cold buffer (10 m*M* Tris·HCl, pH 7.5; 50 m*M* $NH_4Cl$; 10 m*M* Mg acetate) and filtered immediately through glass fiber filters (Whatman GF/C, 2.4 cm). After washing and drying, the filters are placed in 5 ml of scintillation fluid and counted in a Beckman LS-200 scintillation counter at 90% efficiency.

## Purification of Initiation Factors

*Preparation of Crude Factors.* One kilogram of *E. coli* MRE600 cells is lysed as described above and fractionated as follows. The lysate (1230 ml) is divided into 24 equal portions and layered over 25-ml cushions containing 10 m*M* Tris·HCl, pH 7.4; 20 m*M* $NH_4Cl$; 10 m*M* Mg acetate; 7 m*M* 2-mercaptoethanol, and 15% (v/v) glycerol in Beckman polycarbonate tubes. The solutions are centrifuged at 35,000 rpm for 10 hr at 4° in a Beckman type 35 rotor. The upper 80% of each supernatant is removed and saved as the S100 fraction; the next 10% is discarded. The lowest 10% of the supernatant and the pellet are suspended in buffer and brought to 20 m$M$ Tris·HCl, pH 7.4; 500 m$M$ $NH_4Cl$; 20 m$M$ Mg acetate; and 7 m$M$ 2-mercaptoethanol in a final volume of 1000 ml. The solution is stirred for 1 hr and then layered over 30-ml cushions of the same buffer containing 20% glycerol. The ribosome solutions are centrifuged for 14 hr as described above. The supernatant is decanted and saved as the 0.5 $M$ salt wash. The pellets are resuspended in buffer and brought to 20 m$M$ Tris·HCl, pH 7.4; 1000 m$M$ $NH_4Cl$; 40 m$M$ Mg acetate, and 7 m$M$ 2-mercaptoethanol in a final volume of 1000 ml. This solution is centrifuged through a glycerol cushion as before to provide the 1 $M$ salt wash and 20–30 g of pelleted ribosomes. As each wash is obtained, it is immediately subjected to ammonium sulfate fractionation (see below). Best result are obtained when delays are minimized; the elapsed time from cell lysis to the completion of the ammonium sulfate fractionation steps (below) should not be greater than 55 hr.

*Separation of IF-2 from IF-1 and IF-3 by Ammonium Sulfate Fractionation.* The 0.5 $M$ and 1.0 $M$ salt washes are separately brought to 45% saturation by slow (1 hr) addition of saturated ammonium sulfate solution at 4°. After an additional hour of stirring, the precipitates are collected by centrifugation, and the supernatants are brought to 80% saturation by the slow addition of solid ammonium sulfate. The protein precipitate formed at 45% saturation from the 0.5 $M$ salt wash contains most of the IF-2; that from the 1.0 $M$ salt wash contains less than 15% of the IF-2 and is discarded, since it also contains proteins that are difficult to separate from IF-2 by the procedures described below. The precipitate formed at 80% saturation from the 0.5 $M$ salt wash contains most of the IF-1 and IF-3; that from the 1.0 $M$ wash contains appreciable quantities of IF-3 and is combined with the other. The crude initiation factors can be stored for several weeks as frozen (−70°) ammonium sulfate pastes with little or no loss of activity.

*Purification of IF-2.* The precipitate formed at 45% saturated ammonium sulfate of the 0.5 $M$ wash is dissolved in about 100 ml of buffer A-0 and dialyzed overnight against 2 × 4 liters of buffer A-100. The dialyzed solution is diluted to 200 ml, and precipitated protein is removed by centrifugation. The soluble protein (780 mg) is adsorbed on a phosphocellulose column (1.5 × 95 cm) equilibrated with buffer A-100 and is eluted with a 2-liter linear gradient of KCl, 100 to 500 m$M$ in buffer A. One fraction (20 ml) is collected per hour. IF-2 is detected by its stimulation of fMet-tRNA binding to 70 S ribosomes in the presence of AUG and by SDS–polyacrylamide gel electrophoresis. The protein concentration and activity profiles are similar to those described elsewhere.[9] The IF-2 activity is

[9] J. L. Fakunding, J. A. Traugh, R. R. Traut, J. W. B. Hershey, this series, Vol. 30, p. 24.

eluted from 320–400 m*M* KCl, and active fractions are combined to yield 84 mg of protein in 440 ml.

The pooled fractions from the phosphocellulose column are diluted in a Teflon-coated beaker with 1 liter of buffer A-0 to bring the KCl concentration to about 100 m*M*. The solution is adsorbed on a DEAE-Sephadex column (2.0 × 60 cm) equilibrated with buffer A-100, and the protein is eluted with a 2-liter linear gradient of KCl, 150 to 450 m*M* in buffer A. Fractions (20 ml) are collected at the rate of one per hour and are assayed for IF-2 activity and analyzed by SDS–polyacrylamide gel electrophoresis. The IF-2 activity coincides with two distinct protein peaks. The first, which is eluted from 290 to 310 m*M* KCl, contains the larger molecular weight form, IF-2a. The second is eluted from 320 to 350 m*M* KCl and contains the smaller form, IF-2b. IF-2a is the predominant form (29 mg) and is usually about 90% pure at this stage of the purification. IF-2b varies between 15 and 30% of total IF-2, and is usually 70–80% pure.

IF-2a is obtained nearly homogeneous by rechromatography on a phosphocellulose column (0.9 × 30 cm). The material above is diluted to 100 m*M* KCl with buffer A-0, adsorbed to the column, and eluted with a 200-ml linear gradient of KCl, 150 to 500 m*M* in buffer A. Pure, active fractions are combined and stored frozen at −70°. Analysis of the protein by polyacrylamide gel electrophoresis in SDS or urea indicates that the IF-2a is 98% pure (see Fig. 2).

*Separation of IF-1 from IF-3.* The protein precipitating between 45% and 80% ammonium sulfate saturation from both the 0.5 and 1.0 *M* salt washes is combined and dissolved in buffer A-0. The solution is dialyzed overnight against 2 × 4 liters of buffer A-100, brought to 250 ml, and clarified by centrifugation. The soluble protein (5.1 g) is applied to a phosphocellulose column (1.5 × 100 cm) equilibrated with buffer A-100 and is eluted with a 2-liter linear gradient of KCl, 100 to 1000 m*M* in buffer A. One fraction (20 ml) is collected per hour. IF-1 is eluted in a sharp peak from 490 to 530 m*M* KCl (99 mg of protein); IF-3 is eluted in a broad peak from 680 to 810 m*M* KCl (46 mg of protein). Analysis of the pooled fractions by SDS–polyacrylamide gel electrophoresis shows that the IF-1 is quite impure and that the IF-3 is about 33% pure.

*Purification of IF-1.* In order to concentrate the impure IF-1 fraction above, the solution is diluted with buffer A-0 to bring the KCl concentration to 100 m*M* and is adsorbed onto a small phosphocellulose column (bed volume, 8 ml). The protein is eluted with buffer A-1000, and the fractions containing the bulk of the protein are pooled (volume, 7 ml). The concentrated IF-1 is divided in half, and each half is applied separately to a Sephadex G-50 column (1.5 × 100 cm) equilibrated with buffer A-200. The column is washed with the same buffer at 10 ml/hr, and fractions (about 2 ml) are analyzed by measuring absorbance at 280 nm and by SDS–polyacrylamide gel electrophoresis. IF-1 is eluted later than the bulk of the protein (9 mg of protein from both columns), and is either homogeneous or greater than 90% pure at this stage. If homogeneous IF-1 is not obtained, the remaining impurities are removed by chromatography on a small phosphocellulose column (0.9 × 25 cm). After this step, IF-1 is pure by the criterion of polyacrylamide gel electrophoresis in SDS or urea (see Fig. 2).

*Purification of IF-3.* The IF-3 fraction (33% pure) from the phosphocellulose column above is purified further by phosphocellulose column chromatography in 6 *M* urea essentially according to the procedure of Lee-Huang and Ochoa.[10] IF-3 (46 mg) is brought to 6 *M* urea with solid urea and is dialyzed overnight against 2 × 1 liters of buffer B-30. The sample is adsorbed onto a column (0.9 × 30 cm) of phosphocellulose equilibrated with the same buffer, and the protein is eluted with a 400-ml linear gradient of $NH_4Cl$, 100 to 500 m*M* in buffer B. Fractions of 4 ml are collected every 30 min, and 2-μl aliquots are analyzed directly for IF-3 activity. The IF-3 activity is eluted from 180 to 260 m*M* $NH_4Cl$. Based on analysis by SDS–polyacrylamide gel electrophoresis, the purest active fractions are combined to yield 3 mg of IF-3 that is more than 95% pure. Less-pure fractions (6 mg) are also pooled and are chromatographed a second time on phosphocellulose in urea to remove contaminating proteins. Although the IF-3 fractions in urea buffer can be assayed directly if small volumes (e.g., 2 μl) are used, it is ultimately necessary to remove the urea by a procedure that does not cause a large loss of protein. Combined fractions in urea are diluted with 2 volumes of water, and the protein is adsorbed on a small phosphocellulose column (1–2 ml bed volume). The column is washed with 20 ml of buffer A-100 to remove urea, and the protein is eluted with buffer A-800. This procedure serves not only to remove urea, but also to concentrate the IF-3 solution. Dilute solutions of IF-1 and IF-2 are concentrated in the same way.

*Yields and Purity.* The fractionation scheme for the three initiation factors is shown diagrammatically in Fig. 1. The amounts of protein[11] and factor activities obtained from a typical preparation are shown in Table I. The extract from 1 kg of cells routinely provides 4–7 mg of IF-1, 15–25 mg of IF-2a, and 5–15 mg of IF-3. Based on activity measurements, these values represent 30–50% yields of each factor. Efficient recovery is possible because only 2 or 3 chromatographic procedures are required to obtain nearly homogeneous proteins. Other elements that contribute to the high yields are: (a) minimal delay between lysing the cells and beginning the first

[10] S. Lee-Huang and S. Ochoa, *Arch. Biochem. Biophys.* **156,** 84 (1973).
[11] O. H. Lowry, N. J. Rosebrough, A. L. Farr, and R. J. Randall, *J. Biol. Chem.* **93,** 265 (1951).

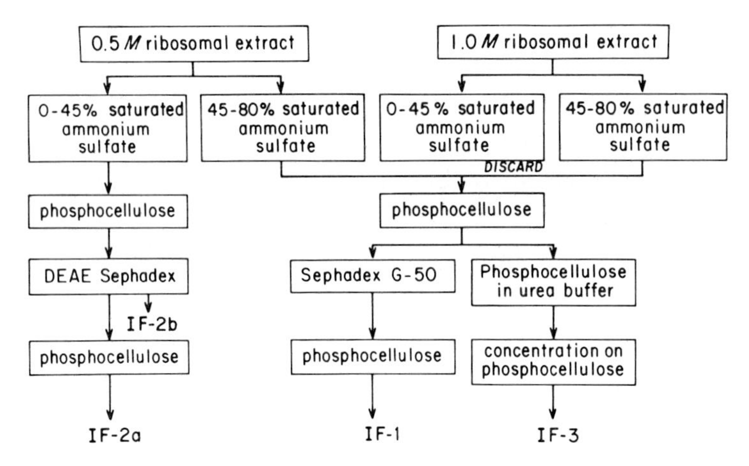


FIG. 1. Fractionation scheme for IF-1, IF-2, and IF-3.

TABLE I
SUMMARY OF PURIFICATION OF IF-1, IF-2, and IF-3[a]

| Purification step | Protein (mg) | Activity (units) | Specific activity (units/mg) | Yield (%) |
|---|---|---|---|---|
| High-salt washes: 0.5 *M* | 7020 | — | — | — |
| 1.0 *M* | 2214 | — | — | — |
| IF-2: Ammonium sulfate, 0–45% | 780 | 97 | 0.12 | 100 |
| Phosphocellulose | 84 | 78 | 0.93 | 80 |
| DEAE-cellulose | 29 | 48 | 1.65 | 50 |
| Phosphocellulose | 15.8 | 34 | 2.15 | 35 |
| IF-1: Ammonium sulfate, 45–80% | 5125 | — | — | — |
| Phosphocellulose | 99 | 61 | 0.62 | 100 |
| Sephadex G-50 | 9 | 36 | 4.0 | 59 |
| Phosphocellulose | 4.7 | 27 | 5.7 | 44 |
| IF-3: Ammonium sulfate, 45–80% | 5125 | 312 | 0.06 | 100 |
| Phosphocellulose | 46 | 207 | 4.50 | 67 |
| Phosphocellulose, urea | 7.1 | 100 | 14.1 | 32 |

[a] The values are derived from a typical preparation involving 1 kg of cells. Protein is determined by the method of O. H. Lowry, N. J. Rosebrough, A. L. Farr, and R. J. Randall, *J. Biol. Chem.* **93,** 265 (1951). Activity is determined by measuring fMet-tRNA binding to 70 S ribosomes in the presence of A-U-G (for IF-1 and IF-2) or R-17 RNA (for IF-3), as described in the text. One activity unit is defined as the ability to stimulate the ribosomal binding of 1 nmol of f[$^{14}$C]Met-tRNA. All determinations are made in the region of linear response of the factor. Since the absolute values for activity also depend on the quality of the ribosomes and other assay ingredients, and thus may vary from time to time, all values for a given factor are determined in the same experiment.

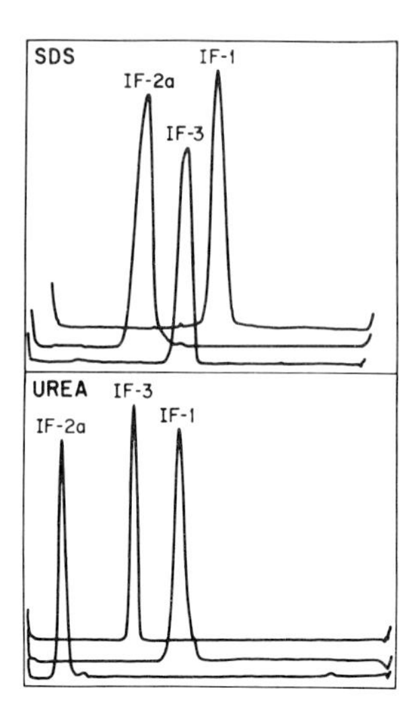


FIG. 2. Densitometric tracings of stained polyacrylamide gels. The upper panel shows polyacrylamide gels run in sodium dodecyl sulfate (SDS) buffer as described by T. A. Bickle and R. R. Traut [*J. Biol. Chem.* **246,** 6828 (1971)]. The acrylamide concentrations in the gels are 5% (IF-2a), 10% (IF-3), and 15% (IF-1). The lower panel shows scans of polyacrylamide gels containing 7.5% acrylamide and 0.2% bisacrylamide run in 8 *M* urea buffer at pH 4.4 as described by P. S. Leboy, E. C. Cox, and J. G. Flaks [*Proc. Natl. Acad. Sci. U.S.A.* **52,** 1367 (1964)] except that the stacking gel is omitted. For each gel, 4–8 $\mu$g of the most purified pooled fractions of each initiation factor are added; electrophoretic migration occurs from left to right. The gels are stained with Coomassie Brilliant Blue, destained, and scanned in a Gilford 2400 spectrophotometer with a linear transport attachment. The IF-3 sample is treated with iodoacetamide prior to analysis in the urea gel system; such treatment is necessary in order to prevent the formation of IF-3 dimers during electrophoresis.

chromatographic columns; (b) the use of siliconized or Teflon-coated vessels whenever possible; (c) the treatment of chromatography resins with protein (e.g., bovine serum albumin or lysozyme) prior to use; (d) the avoidance of dialysis at later stages of the preparations (dialysis membranes bind all three factors); (e) the avoidance of dilute solutions of purified factors; and (f) the use of small phosphocellulose columns to concentrate factor preparations. Most loss of activity appears to be due to the loss of protein on vessel surfaces. This is readily apparent when radioactively labeled factors are employed. Most of the items above are addressed to this problem.

The highly purified initiation factors are analyzed by polyacrylamide gel electrophoresis in buffers containing either SDS or urea. Densitometric tracings of the stained gels are shown in Fig. 2.[12,13] The purity of each factor

[12] T. A. Bickle and R. R. Traut, *J. Biol. Chem.* **246,** 6828 (1971).

[13] P. S. Leboy, E. C. Cox, and J. G. Flaks, *Proc. Natl. Acad. Sci. U.S.A.* **52,** 1367 (1964).

is calculated from the area under the peaks. IF-1 and IF-3 have no detectable contaminants; IF-2a contains about 2% of IF-2b. Thus the relatively simple purification procedures described here provide essentially homogeneous IF-1, IF-2a, and IF-3 in high yield.

## Properties

The physical and biological properties of IF-1, IF-2a and IF-3 are described in detail elsewhere[2]; a brief summary is given here. The molecular weights of the initiation factors are determined by SDS–polyacrylamide gel electrophoresis according to the method of Osborn and Weber[14]: for IF-1, 8700; for IF-2a, 118,000; and for IF-3, 22,500. The smaller form of IF-2, IF-2b, has a molecular weight of 90,000. Only one molecular weight form for IF-1 or IF-3 is detected. All three initiation factors are required for maximal responses in assays involving fMet-tRNA binding to 70 S ribosomes with AUG or R-17 RNA, formylmethionylpuromycin synthesis with R-17 RNA, or protein synthesis with R-17 RNA.

[14] K. Weber and M. Osborn, *J. Biol. Chem.* **244,** 4406 (1969).

# [28] Protein Synthesis Initiation Factors from Rabbit Reticulocytes: Purification, Characterization, and Radiochemical Labeling[1]

*By* ROB BENNE, MARIANNE L. BROWN-LUEDI, and JOHN W. B. HERSHEY

Initiation of protein synthesis in mammalian cells proceeds by a complex process whereby the assembly of the ribosome, mRNA, and methionyl-$tRNA_f$ into an 80 S initiation complex is promoted by proteins called initiation factors. In order to study the mechanisms of initiation and translational control, it is important to identify all the initiation factors and obtain them in a purified form. We describe here procedures for the preparation of eight highly purified initiation factors from rabbit reticulocytes: eIF-1,[2] eIF-2, eIF-3, eIF-4A, eIF-4B, eIF-4C, eIF-4D, and eIF-5. The procedures are based on methods previously described for the individual

[1] Supported by a grant from the U.S. Public Health Service, GM-22135.

[2] The initiation factor nomenclature used here is that proposed at the International Symposium on Protein Synthesis, Bethesda, Maryland in 1976 and described by W. F. Anderson, L. Bosch, W. E. Cohn, H. Lodish, W. C. Merrick, H. Weissbach, H. G. Wittmann, and I. G. Wool, *FEBS Lett.* **76,** 1 (1977).





ISBN 0-12-504180-2

initiation factors.[3-6] Comparable initiation factors have been purified by Staehelin and co-workers[7,8] and by the Anderson–Merrick group.[9-13] The eight factors are characterized physically and functionally, and an *in vitro* method is described for labeling the proteins with radioactivity without altering their biological activities.

*Buffers*

Buffer A: 20 m*M* Tris·HCl, pH 7.6; 7 m*M* 2-mercaptoethanol; 5% v/v glycerol; and various concentrations of KCl as indicated: buffer A-100 means buffer A containing 100 m*M* KCl; buffer A-500 means buffer A containing 500 m*M* KCl

Buffer B: 20 m*M* potassium phosphate, pH 7.2; 7 m*M* 2-mercaptoethanol; 5% glycerol; and varying amounts of KCl, designated as in buffer A

*Reagents*

2-Mercaptoethanol, from British Drug House

Tris (Trizma base), puromycin, creatine phosphokinase, and sodium borohydride, from Sigma Chemical Co.

Diphenyloxazole and 1,4-bis[2-(5-phenyloxazolyl)]benzene (POPOP), from Amersham/Searle

Urea (ultrapure), [$^3$H]methionine, and [$^3$H]leucine, from Schwarz/Mann

[$^{14}$C]Formaldehyde, from New England Nuclear Corp.

GTP, ATP, creatine phosphate, and Aquacide II-A, from Calbiochem

Dithioerythritol, from Pierce Chemical Co.

Acrylamide and bisacrylamide, from Eastman

Phosphocellulose (P-11) and DEAE-cellulose (DE-32), from Whatman

Sephadex G-75, G-100, and G-200, from Pharmacia Fine Chemicals

Toluene scintillation fluid, prepared by mixing 4 g of diphenyloxazole and 0.05 g of POPOP in 1 liter of toluene

[3] R. Benne and J. W. B. Hershey, *Proc. Natl. Acad. Sci. U.S.A.* **73,** 3005 (1976).

[4] R. Benne, C. Wong, M. Luedi, and J. W. B. Hershey, *J. Biol. Chem.* **251,** 7675 (1976).

[5] R. Benne, M. Luedi, and J. W. B. Hershey, *J. Biol. Chem.* **252,** 5798 (1977).

[6] R. Benne, M. L. Brown-Luedi, and J. W. B. Hershey, *J, Biol. Chem.* **253,** 3070 (1978).

[7] T. Staehelin, H. Trachsel, B. Erni, A. Boschetti, and M. H. Schreier, *Proc. FEBS Meet., 10th* 309 (1975).

[8] M. H. Schreier, B. Erni, and T. Staehelin, *J. Mol. Biol.* **116,** 727 (1977).

[9] B. Safer, W. F. Anderson, and W. C. Merrick, *J. Biol. Chem.* **250,** 9067 (1975).

[10] W. C. Merrick, W. M. Kemper, and W. F. Anderson, *Proc. Natl. Acad. Sci. U.S.A.* **72,** 5556 (1975).

[11] P. M. Prichard and W. F. Anderson, this series, Vol. 30, p. 136.

[12] B. Safer, S. L. Adams, W. M. Kemper, K. W. Berry, M. Lloyd, and W. C. Merrick, *Proc. Natl. Acad. Sci. U.S.A.* **73,** 2584 (1976).

[13] W. M. Kemper, K. W. Berry, and W. C. Merrick, *J. Biol. Chem.* **251,** 5551 (1976).

*Biological Materials*. Ribosomal subunits and pH 5 enzyme fractions are prepared from the livers of Sprague–Dawley rats according to the procedures of Falvey and Staehelin.[14] Unfractionated tRNA is isolated from rabbit liver[15] and the initiator tRNA is specifically charged with [$^3$H]methionine (specific activity, 3.7 Ci/mmol) by using an *Escherichia coli* synthetase preparation according to the method of Gupta[16] as modified by Stanley.[17] Globin 9 S mRNA is purified by the procedure of Staehelin[8] and iodinated essentially as described by Getz.[18] AUG is synthesized by the method of Sundararajan and Thach.[19] Crude fractions of reticulocyte initiation factors for assays are obtained by precipitation of the high-salt ribosomal wash[20] with ammonium sulfate (see text below): fraction A (0–40% saturation), fraction B (40–50%), and fraction C (50–70%). Precipitates are dissolved in and dialyzed against buffer A-100 and stored frozen at −70° in small aliquots.

*Preparation of Chromatography Columns*

Phosphocellulose, DEAE-cellulose, and Sephadex G-75, G-100, and G-200 are first treated as directed by the manufacturers. The settled resins are then treated with 2 volumes of 0.1% bovine serum albumin in buffer A-100 by stirring for 1 hr, filtered, and washed thoroughly with buffer A-1000 to remove protein. Columns of phosphocellulose equilibrated with buffer B-100 and DEAE-cellulose equilibrated in buffer A-100 are packed tightly under 150 cm of hydrostatic pressure. Sephadex columns are equilibrated in their appropriate buffer and packed at less than 40 cm hydrostatic pressure. After use, the columns are routinely washed with buffer A-1000 or buffer B-1000 containing 0.01% sodium azide and stored at 4°. They are re-used numerous times after equilibration with a suitable buffer.

## Initiation Factor Assays

### *Assay A: Synthesis of Globin*

*Principle*. eIF-1, eIF-3, eIF-4A, and eIF-4B are assayed in a cell-free protein synthesis system dependent on exogenous globin 9 S mRNA and

[14] A. K. Falvey and T. Staehelin, *J. Mol. Biol.* **53,** 1 (1970).
[15] H. Rogg, W. Wehrli, and M. Staehelin, *Biochim. Biophys. Acta* **195,** 13 (1969).
[16] K. K. Bose, N. K. Chatterjee, and N. K. Gupta, this series, Vol. 29, p. 522.
[17] W. M. Stanley, Jr., this series, Vol. 29, p. 530.
[18] M. J. Getz, *Biochim. Biophys. Acta* **287,** 485 (1972).
[19] T. A. Sundararajan and R. E. Thach, *J. Mol. Biol.* **19,** 74 (1966).
[20] M. H. Schreier and T. Staehelin, *J. Mol. Biol.* **73,** 329 (1973).

initiation factors. The assay is the one developed by Schreier and Staehelin[20] and utilizes: high salt-washed ribosomal subunits from rat liver; a pH 5 enzyme fraction from rat liver as the source of elongation factors, termination factors, and aminoacyl tRNA synthetases; additional tRNA from rabbit liver; and purified globin 9 S mRNA from rabbit reticulocytes. The major products are complete globin $\alpha$ and $\beta$ chains; however, the incorporation of radioactive amino acids into a hot trichloroacetic acid precipitate is measured here. The assay is suitable for determining eIF-2, eIF-4C, and eIF-5 activities as well.

*Procedure.* Each assay mixture of 100 $\mu$l contains: 20 m$M$ Tris·HCl, pH 7.6; 4 m$M$ Mg acetate; 70–80 m$M$ KCl; 10 m$M$ 2-mercaptoethanol; 30 $\mu M$ each of 19 nonradioactive amino acids and [$^3$H]leucine, specific activity 500 Ci/mol; 12 m$M$ creatine phosphate; 0.5 m$M$ ATP; 0.2 m$M$ GTP; 0.4 unit of creatine phosphokinase; 0.10 $A_{260}$ unit of 40 S ribosomal subunits; 0.25 $A_{260}$ unit of 60 S ribosomal subunits, 0.05 $A_{260}$ unit of globin mRNA; 5 $\mu$l of pH 5 enzyme fraction; 0.3 $A_{260}$ unit of rabbit liver tRNA, and initiation factors as follows:

eIF-1 assay: 20 $\mu$g of fraction A, 20 $\mu$g of fraction B, and 1.2 $\mu$g of purified eIF-4A

eIF-3 assay: 20 $\mu$g of fraction B, 20 $\mu$g of fraction C, and 1.2 $\mu$g of purified eIF-4B

eIF-4A assay: 20 $\mu$g of fraction A, 20 $\mu$g of fraction B, and 1.2 $\mu$g of partially purified eIF-1 from phosphocellulose (step 14)

eIF-4B assay: 20 $\mu$g of fraction B, 20 $\mu$g of fraction C, and 5 $\mu$g of purified eIF-3

The mixtures are incubated at 37° for 30 min, brought to 5% in trichloroacetic acid, and heated at 90° for 15 min. The precipitates are filtered through glass fiber filters (Whatman GF/C), and the filters are washed with 5% trichloroacetic acid, dried, and counted in 5 ml of scintillation fluid.

### *Assay B: Synthesis of Methionylpuromycin*

*Principle.* eIF-4C, eIF-4D, and eIF-5 are assayed in this totally purified model system for the synthesis of the first peptide bond. The assay has been described in detail by Anderson and co-workers.[21] As employed here it utilizes: high salt-washed ribosomal subunits from rat liver; AUG as template (although globin 9 S mRNA may be used instead[21,22]); [$^3$H]Met-tRNA$_f$; puromycin; and purified initiation factors. The pH 5 enzyme fraction used in assay A above is not required, so possible contamination of

[21] R. S. Crystal, N. A. Elson, and W. F. Anderson, this series, Vol. 30, p. 101.

[22] R. Benne and J. W. B. Hershey, *J. Biol. Chem.* **253**, 3078 (1978).

initiation factors due to this fraction is avoided. eIF-2 and eIF-3 may also be assayed by this procedure.

*Procedure.* Each assay mixture of 50 μl contains: 20 m*M* Tris · HCl, pH 7.6; 2 m*M* Mg acetate; 60–80 m*M* KCl; 10 m*M* 2-mercaptoethanol; 10 pmol of [$^3$H]Met-tRNA$_f$ (specific activity, 3.7 Ci/mmol); 0.8 m*M* GTP; 1 m*M* puromycin; 0.065 $A_{260}$ unit of AUG; 0.14 $A_{260}$ unit of 40 S ribosomal subunits; 0.35 $A_{260}$ unit of 60 S ribosomal subunits; and purified initiation factors as follows:

eIF-4C assay: 3 μg of eIF-2, 5 μg of eIF-3, 0.5 μg of eIF-5, and 0.25 μg of eIF-4D

eIF-4D assay: 5 μg of eIF-2, 10 μg of eIF-3, 0.5 μg of eIF-5, and 0.2 μg of eIF-4C

eIF-5 assay: 5 μg of eIF-2, 10 μg of eIF-3, 0.07 μg of eIF-4C, and 0.25 μg of eIF-4D

The samples are incubated for 20 min at 37°; 0.5 ml of 0.2 *M* potassium phosphate, pH 8.0, is added, and the methionylpuromycin is extracted with 1 ml of ethyl acetate as described by Leder and Bursztyn.[23]

### *Assay C: Ternary Complex Formation with eIF-2, Met-tRNA$_f$ and GTP*

*Principle.* eIF-2 is assayed in the absence of ribosomes and other initiation factors by measuring the formation of a complex of eIF-2, [$^3$H]-Met-tRNA$_f$, and GTP. The assay was developed by Gupta and co-workers[24] and exploits the fact that [$^3$H]Met-tRNA$_f$ adsorbs to nitrocellulose filters only when complexed with protein.

*Procedure.* Each assay mixture contains in 50 μl: 20 m*M* Tris · HCl, pH 7.6; 1.5 m*M* Mg acetate; 100 m*M* KCl; 7 m*M* 2-mercaptoethanol; 20 pmol of [$^3$H]Met-tRNA$_f$ (specific activity, 3.7 Ci/mmol); 0.8 m*M* GTP; and the fraction containing eIF-2 (range 1–5 μg). The mixtures are incubated at 37° for 10 min, diluted with 1 ml of cold buffer A-100, and filtered immediately through nitrocellulose filters (Millipore HAWP 024.00). The filters are rapidly washed twice with 1-ml portions of cold buffer A-100, dried, and counted in 5 ml of scintillation fluid.

## Purification of Initiation Factors

### *Overview*

Crude initiation factors are prepared according to procedures developed by Schreier and Staehelin.[20] Reticulocytes are routinely obtained

[23] P. Leder and M. Bursztyn, *Biochem. Biophys. Res. Commun.* **25**, 233 (1966).

[24] N. K. Gupta, C. L. Woodley, Y. C. Chen, and K. K. Bose, *J. Biol. Chem.* **248**, 4500 (1973).

from 30 anemic 4 to 5-lb rabbits and are lysed immediately. Total ribosomes in the lysates are sedimented, resuspended in buffer containing 0.5 $M$ KCl, and sedimented again. The resulting supernatant, or high-salt ribosomal wash, is the source of initiation factors. Methods for the purification of eight initiation factors are described in detail in this text. The procedures involve ammonium sulfate fractionation, fractionation on the basis of factor mass, and chromatography on ion-exchange columns. A flow diagram of the overall purification scheme, together with the column or gradient profiles, is shown in Fig. 1 and may serve as a guide for the text below. The amounts of protein and specific activities of factors are given in Table I for the various purification steps. All operations are performed at 0–4°.

## *Separation of eIF-3 and eIF-4B from the Other Initiation Factors*

*Step 1. Ammonium Sulfate Fractionation.* The high-salt ribosomal wash (concentration 20–25 $A_{280}$ units/ml) is brought to 40% saturation at 4° by the slow addition of solid ammonium sulfate with stirring, and the

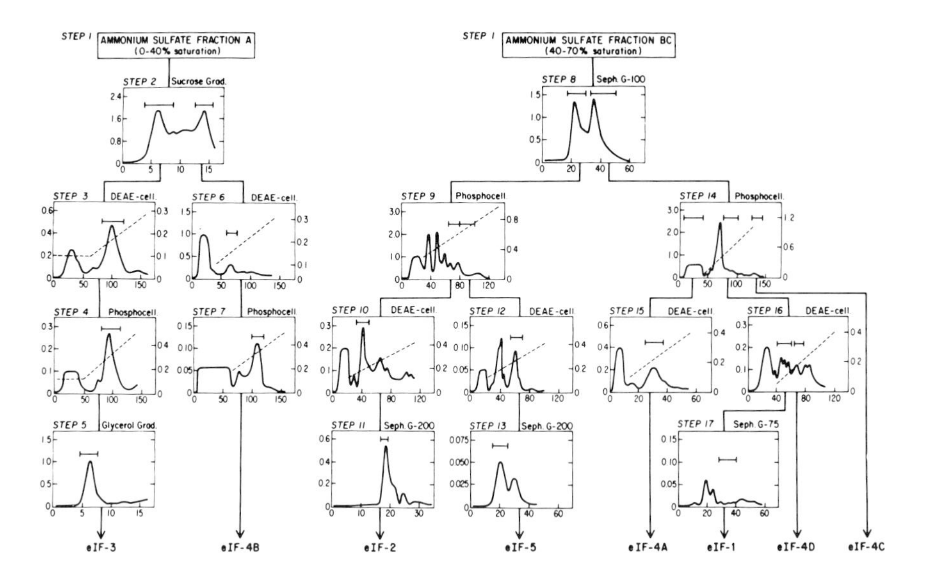


FIG. 1. Flow scheme for purification of initiation factors. The scheme shows the interrelationships of the various purification steps, which are described in detail in the text and in Table I. The panel for each step shows the $A_{280}$ profile (solid line; scale indicated along the left ordinate); the concentration of KCl in the eluting buffer (dashed line; scale in $M$ indicated along the right ordinate) and fraction numbers (indicated along the abscissa at bottom). The horizontal brackets indicate the pooled fractions containing initiation factor activity, as identified in the text.

TABLE I
PURIFICATION OF INITIATION FACTORS[a]

| Factor | Purification step | Protein ($A_{280\ nm}$ units) | Activity (units) | Specific activity | Recovery (%) |
|---|---|---|---|---|---|
| eIF-1 | Step 1 (Fraction BC) | 375 | — | — | — |
| | Step 8 (Seph. G-100) | 190 | 141 | 0.7 | 100 |
| | Step 14 (Phosphocell.) | 75 | 112 | 1.5 | 79 |
| | Step 16 (DEAE-cell.) | 1.3 | 8.0 | 6.2 | 6 |
| | Step 17 (Seph. G-75) | 0.2 | 9.5 | 47.6 | 7 |
| eIF-2 | Step 1 (Fraction BC) | 375 | — | — | — |
| | Step 8 (Seph. G-100) | 169 | 7.5 | 0.04 | 100 |
| | Step 9 (Phosphocell.) | 13.2 | 5.9 | 0.45 | 79 |
| | Step 10 (DEAE-cell.) | 5.1 | 5.0 | 0.98 | 67 |
| | Step 11 (Seph. G-200) | 2.6 | 4.1 | 1.58 | 52 |
| eIF-3 | Step 1 (Fraction A) | 310 | 40 | 0.13 | 100 |
| | Step 2 (sucrose grad.) | 135 | 68 | 0.50 | 170 |
| | Step 3 (DEAE-cell.) | 52.1 | 75 | 1.44 | 188 |
| | Step 4 (Phosphocell.) | 42.8 | 65 | 1.52 | 163 |
| | Step 5 (glycerol gradient) | 39.5 | 53 | 1.34 | 133 |
| eIF-4A | Step 1 (Fraction BC) | 375 | — | — | — |
| | Step 8 (Seph. G-100) | 190 | 36 | 0.19 | 100 |
| | Step 14 (Phosphocell.) | 66 | 23 | 0.35 | 64 |
| | Step 15 (DEAE-cell.) | 1.6 | 11 | 6.9 | 31 |
| eIF-4B | Step 1 (Fraction A) | 620 | — | — | — |
| | Step 2 (sucrose grad.) | 120 | 52.0 | 0.43 | 100 |
| | Step 6 (DEAE-cell.) | 18.8 | 37.5 | 2.00 | 72 |
| | Step 7 (Phosphocell.) | 2.4 | 18.7 | 7.80 | 36 |
| eIF-4C | Step 1 (Fraction BC) | 375 | — | — | — |
| | Step 8 (Seph. G-100) | 190 | — | — | — |
| | Step 14 (Phosphocell.) | 0.3 | 1.6 | 5.3 | — |
| eIF-4D | Step 1 (Fraction BC) | 375 | — | — | — |
| | Step 8 (Seph. G-100) | 190 | — | — | — |
| | Step 14 (Phosphocell.) | 75 | 1.8 | 0.02 | 100 |
| | Step 16 (DEAE-cell.) | 0.5 | 0.4 | 0.80 | 22 |
| eIF-5 | Step 1 (Fraction BC) | 375 | — | — | — |
| | Step 8 (Seph. G-100) | 169 | 53.5 | 0.32 | 100 |
| | Step 9 (Phosphocell.) | 21.4 | 44.3 | 2.1 | 83 |
| | Step 12 (DEAE-cell.) | 4.6 | 11.5 | 2.5 | 21 |
| | Step 13 (Seph. G-200) | 0.7 | 2.0 | 2.9 | 4 |

[a] The values are obtained from the procedures described in the text by using crude initiation factors from reticulocytes of the following number of rabbits: for eIF-3, 90 rabbits; for eIF-4B, 180 rabbits; for all other initiation factors, 120 rabbits. An $A_{280\ nm}$ unit is the amount of protein in 1 ml that gives an $A_{280\ nm}$ value of 1 in a 1-cm cell. An activity unit is the amount of initiation factor that stimulates the incorporation of 1 nmol of [$^3$H]leucine into protein (Assay A), the synthesis of 1 nmol of methionylpuromycin (Assay B), or the formation of 1 nmol of ternary complex (Assay C).

suspension is stirred for 1 hr at 4°. The precipitate (fraction A, 0–40% saturation) is removed by centrifugation and contains eIF-3 and eIF-4B. The supernatant is brought to 70% saturation as above, and the precipitate (fraction BC, 40–70% saturation) is pelleted and contains eIF-1, eIF-2, eIF-4A, eIF-4C, eIF-4D, and eIF-5.

*Fraction A: Separation of eIF-3 and eIF-4B*

*Step 2. Sucrose-Gradient Centrifugation.* Fraction A is dissolved in and dialyzed against buffer A-100 to give a concentration of about 30 $A_{280}$ units/ml and is clarified by centrifugation for 10 min at 10,000 rpm in a Sorvall SS 34 rotor. Aliquots of 2 ml are layered over 15 to 30% (w/v) sucrose gradients in buffer A-100, and the gradients are centrifuged for 24 hr at 25,000 rpm in a Beckman SW 27 rotor. The contents are fractionated, protein concentration is determined by measuring absorbance at 280 nm, and the fractions are assayed for eIF-3 and eIF-4B activities. As shown in Fig. 1, eIF-3 activity is found associated with fast-sedimenting material near the middle of the gradient (fractions 4–8), while eIF-4B activity is found near the top of the gradient (fractions 13–16). Active fractions for eIF-3 and for eIF-4B are pooled and further fractionated as described below.

*Purification of eIF-3*

*Step 3. DEAE-Cellulose.* The material from the sucrose gradients (135 $A_{280}$ units) is applied to a column of DEAE-cellulose (Whatman DE-32; 2 × 25 cm) equilibrated with buffer A-100. The adsorbed protein is eluted with a 300-ml linear gradient of KCl, 100 to 300 m*M* in buffer A. Protein concentration is determined by measuring absorption at 280 nm, and the fractions are assayed for eIF-3 activity. As shown in Fig. 1, eIF-3 activity is eluted at about 150 m*M* KCl and is associated with the major protein peak. Fractions 80–120 are combined to yield 52 $A_{280}$ units of protein.

*Step 4. Phosphocellulose.* The active fractions from DEAE-cellulose chromatography are dialyzed against buffer B-100 and applied to a column of phosphocellulose (Whatman P-11; 0.9 × 25 cm) equilibrated in buffer B-100. The adsorbed protein is eluted with a 200-ml linear gradient of KCl, 100 to 500 m*M* in buffer B. Protein concentration is measured by absorbance at 280 nm, and eIF-3 activity is assayed. As shown in Fig. 1, eIF-3 is eluted at 300 m*M* KCl as a sharp protein peak and active fractions (80–112) are combined. Chromatography on phosphocellulose removes only a few impurities from the eIF-3 complex, and the step may be omitted if a slightly less pure preparation is acceptable.

*Step 5. Glycerol Gradient Centrifugation.* The eIF-3 from the preceding step is concentrated to about 2 mg/ml by ultrafiltration (Amicon PM-10 filter) or by treatment in a dialysis bag with Aquacide. Aliquots containing about 2 mg of protein are layered over 10–30% (v/v) glycerol gradients in buffer A-100 and are centrifuged for 16 hr at 40,000 rpm in a Beckman SW 41 rotor. The gradients are fractionated, and protein concentration and eIF-3 activity are determined as above. The eIF-3 activity corresponds closely with the bulk of the fast-sedimenting protein (Fig. 1). Active fractions (5–8) are combined, pooled with those of other gradients, concentrated by ultrafiltration to about 1 mg/ml, and stored frozen in small aliquots at −70°. The procedures yield about 20 mg of eIF-3 (per 90 rabbits), which is purified about 11-fold from ammonium sulfate fraction A. The purified preparation analyzed by sodium dodecyl sulfate–polyacrylamide gel electrophoresis (Fig. 2) contains at least 9 polypeptides.

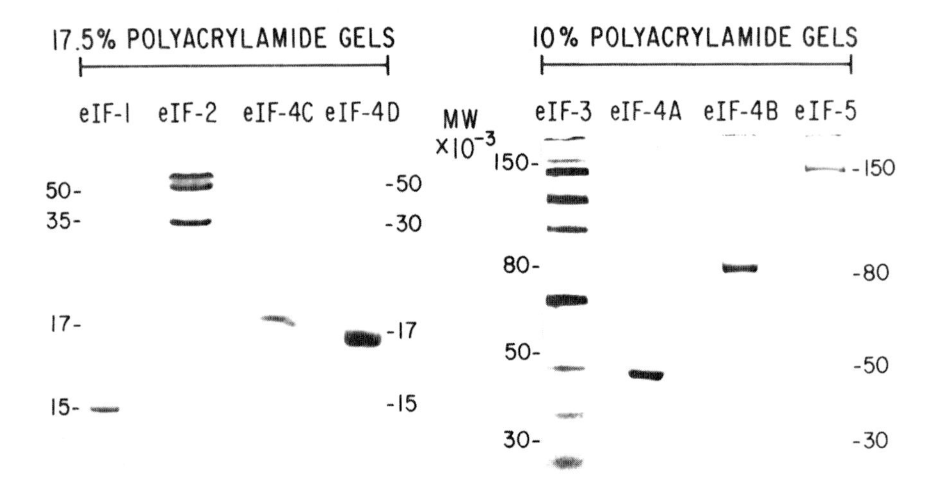


FIG. 2. Analysis of purified initiation factors by sodium dodecylsulfate–polyacrylamide gel electrophoresis. Purified preparations of all initiation factors except eIF-3 are analyzed by electrophoresis in 10 and 17% acrylamide slab gels (12 × 10 × 0.2 cm) according to the method of U. K. Laemmli [*Nature* (*London*) **227,** 680 (1970)]; eIF-3 is analyzed on a 10% polyacrylamide tube gel (10 × 0.5 cm) as described by K. Weber and M. Osborn [*J. Biol. Chem.* **244,** 4406 (1969)]. Electrophoresis of the slab gels was for 5 hr at 90 V; the gels were stained with Coomassie Brilliant Blue and photographed. The amounts of protein analyzed are: eIF-1, 2.0 μg; eIF-2, 12.5 μg; eIF-3, 25 μg; eIF-4A, 2.5 μg; eIF-4B, 2.5 μg; eIF-4C, 2.5 μg; eIF-4D, 4.0 μg; and eIF-5, 2.5 μg.

*Purification of eIF-4B*

*Step 6. DEAE-Cellulose.* Fractions containing eIF-4B from the sucrose gradients above (step 2; about 120 $A_{280}$ units, from combined preparations involving 180 rabbits) are added to a column of DEAE-cellulose (2 × 25 cm) equilibrated in buffer A-100. Adsorbed protein is eluted with a 300-ml linear gradient of KCl, 100–300 m*M* in buffer A; 3-ml fractions are collected every 15 min. Protein concentration is determined by absorbance at 280 nm, and eIF-4B activity is assayed. eIF-4B is eluted at about 160 m*M* KCl (Fig. 1) and active fractions (46–72) are combined to yield about 18$A_{280}$ units of protein.

*Step 7. Phosphocellulose.* The active fractions from the DEAE-cellulose column are diluted with 1 volume of buffer B-0 to reduce the salt concentration and are added to a column of phosphocellulose (1.5 × 20 cm) equilibrated in buffer B-100. The adsorbed protein is eluted with a 300-ml gradient of KCl, 100 to 500 m*M* in buffer B; fractions of 1.5 ml are collected every 15 min, and protein concentration and eIF-4B activity are assayed. eIF-4B activity is eluted at about 320 m*M* KCl (Fig. 1) and corresponds to the major protein peak. Active fractions (100–120) are combined and concentrated to 2–4 ml by ultrafiltration (Amicon PM-10 filter). The procedures yield about 2.4 mg (per 180 rabbits) of eIF-4B that is 70–85% pure. Analysis of an eIF-4B preparation by sodium dodecyl sulfate–polyacrylamide gel electrophoresis[25] is shown in Fig. 2.

*Fraction BC: Separation of eIF-2 and eIF-5 from eIF-1, eIF-4A, eIF-4C, and eIF-4D*

*Step 8. Sephadex G-100 Chromatography.* The crude initiation factor proteins precipitating with ammonium sulfate between 40 and 70% saturation (fraction BC, about 375 $A_{280}$ units from 120 rabbits) are dissolved in about 10 ml of buffer A-100 and clarified by centrifugation for 10 min at 10,000 rpm in a Sorvall SS 34 rotor. The solution is divided into two equal parts, and each is fractionated without prior dialysis on a column of Sephadex G-100 (2 × 60 cm) equilibrated and eluted with buffer A-100. Fractions of 3.1 ml are collected every 40 min and are assayed for protein concentration and for eIF-1, eIF-2, and eIF-4A activities. Two major protein peaks are obtained (Fig. 1) and are pooled separately: the first, due to large proteins, contains eIF-2 and eIF-5 (fractions 19–33); the second, due primarily to globin, contains eIF-4A and the three low-molecular-weight factors, eIF-1, eIF-4C, and eIF-4D (fractions 35–50). The fractions to be pooled in the high-molecular-weight region are determined from the

[25] U. K. Laemmli, *Nature (London)* **227**, 680 (1970).

eIF-2 activity profile; eIF-5 is not routinely assayed, but follows the eIF-2 activity closely. The fractions to be included in the second, lower-molecular-weight pool are chosen on the basis of the eIF-1 and eIF-4A activity profiles and on analysis of the fractions by sodium dodecyl sulfate–polyacrylamide gel electrophoresis, which readily enables identification of eIF-4A and the unresolved low-molecular-weight factors. Although eIF-4A is largely separated from the eIF-1 activity, no attempt is made at this time to prepare separate pools, because eIF-4A is readily resolved from the other factors in the next step by chromatography on phosphocellulose and because substantial amounts of eIF-4C are sometimes found in the fractions containing eIF-4A.

*Purification of eIF-2*

*Step 9. Phosphocellulose.* The combined fractions from the Sephadex G-100 column (Step 8) which contain eIF-2 and eIF-5 (about 170 $A_{280}$ units protein) are applied to a column of phosphocellulose (1.5 × 20 cm) equilibrated in buffer B-300. Adsorbed protein is eluted with a 250-ml linear gradient of KCl, 300 to 1000 m$M$ in buffer B. Fractions of 3 ml are collected every 20 minutes, protein concentration is measured by absorbance at 280 nm, and the fractions are assayed for eIF-2 and eIF-5 activity. As shown in Fig. 1, eIF-2 activity is eluted at about 580 m$M$ KCl (fractions 70–79) while eIF-5 activity follows at about 690 m$M$ KCl (fractions 80–98) and is pooled for further purification as described later in the text.

*Step 10. DEAE-Cellulose.* The pooled eIF-2 fractions from the phosphocellulose column, containing about 13 $A_{280}$ units of protein, are dialyzed against buffer A-100 and are applied to a column of DEAE-cellulose (0.9 × 27 cm) equilibrated with buffer A-100. The adsorbed protein is eluted with a 160-ml linear gradient of KCl, 100 to 400 m$M$ in buffer A; 2-ml fractions are collected every 20 min. Absorbance at 280 nm and eIF-2 activity are determined as before. eIF-2 activity is eluted at about 160 m$M$ KCl (Fig. 1) and corresponds to a sharp peak of protein (fractions 32–50). The eIF-2 preparation at this stage is 60–85% pure.

*Step 11. Sephadex G-200.* eIF-2 may be further purified by molecular sieve chromatography. The fractions containing eIF-2 from DEAE-cellulose (about 5 $A_{280}$ units of protein) are concentrated by precipitation with 80% saturated ammonium sulfate at 4°. The precipitate is dissolved in 2 ml of buffer A-300, and the solution is passed through a column of Sephadex G-200 (1.7 × 60 cm) equilibrated in the same buffer. Fractions of 2 ml are collected every 30 min, and protein concentration and eIF-2 activity are determined (Fig. 1). eIF-2 is eluted as a sharp peak (fractions 16–19) ahead of contaminating proteins to yield about 2.6 $A_{280}$ units of protein, 85–90% pure. The preparation is concentrated by ammonium sulfate precipitation as described above and dialyzed against buffer A-100. Analysis of eIF-2 by sodium dodecyl sulfate/polyacrylamide gel electrophoresis[25] (Fig. 2) shows three major polypeptides.

*Purification of eIF-5*

*Step 12. DEAE-Cellulose.* Fractions containing eIF-5 activity are obtained from the phosphocellulose column (step 9) described above for the purification of eIF-2. The pooled fractions containing about 21 $A_{280}$ units of protein are dialyzed against buffer A-100 and applied to a column of DEAE-cellulose (0.9 × 27 cm) equilibrated with the same buffer. Adsorbed protein is eluted with a 150-ml linear gradient of KCl, 100 to 400 m*M* in buffer A; 2-ml fractions are collected every 20 min, and protein concentration and eIF-5 activity are determined. As shown in Fig. 1, eIF-5 activity is eluted at about 250 m*M* KCl and is associated with a major protein peak. Active fractions (52–75) are pooled and concentrated by precipitation with 70% saturated ammonium sulfate, followed by dialysis against buffer A-300. The preparation is 30–50% pure at this stage.

*Step 13. Sephadex G-200.* The concentrated eIF-5 preparation (about 5 $A_{280}$ units) is further purified by passage through a column of Sephadex G-200 (1.5 × 60 cm) equilibrated with buffer A-300. Fractions of 2 ml are collected every 30 min, and protein concentration and eIF-5 activity are determined (Fig. 1). eIF-5 activity is eluted ahead of contaminating proteins. Active fractions (16–22) are combined and concentrated by ultrafiltration (Amicon PM-10 filters) to 2 ml to yield about 0.7 $A_{280}$ unit of protein, 75–80% pure. Analysis of eIF-5 by sodium dodecyl sulfate–polyacrylamide gel electrophoresis is shown in Fig. 2.

*Purification of eIF-4A*

*Step 14. Phosphocellulose.* Fractionation of fraction BC by Sephadex G-100 as described above (step 8) results in a preparation containing eIF-4A and the low-molecular-weight initiation factors eIF-1, eIF-4C, and eIF-4D. The combined fractions (about 190 $A_{280}$ units protein) are applied directly to a column of phosphocellulose (1.5 × 20 cm) equilibrated in buffer B-100. eIF-4A does not adsorb to the phosphocellulose under these conditions and is separated from the other three factors. The nonadsorbed proteins (Fig. 1, fractions 5–45) are pooled for the further purification of eIF-4A. Elution of adsorbed eIF-1, eIF-4C, and eIF-4D is described later.

*Step 15. DEAE-Cellulose.* The nonadsorbed proteins (66 $A_{280}$ units) from the phosphocellulose column above (step 14) are immediately applied to a column of DEAE-cellulose (0.9 × 27 cm) equilibrated in buffer A-100. Adsorbed protein is eluted with a 150-ml linear gradient of KCl, 100–400 m$M$ in buffer A; 2-ml fractions are collected every 20 min, and absorbance at 280 nm and eIF-4A activity are determined. As shown in Fig. 1, eIF-4A is eluted at 210 m$M$ KCl corresponding with the major protein peak. Active fractions (25–32) are combined and concentrated by precipitation with ammonium sulfate at 90% saturation, or by treatment of the solution in a dialysis bag with Aquacide, followed by dialysis against buffer A-100. The yield is 1.6 $A_{280}$ units of protein that is 75–85% pure as determined by sodium dodecyl sulfate/polyacrylamide gel electrophoresis (Fig. 2).

*Purification of eIF-4C*

*Step 14. Phosphocellulose.* eIF-4C is adsorbed to the phosphocellulose column (step 14) described for the purification of eIF-4A above. The adsorbed proteins are eluted with a 250-ml linear gradient of KCl, 100–1000 m$M$ in buffer B, followed with buffer B-1200. Fractions of 3 ml are collected every 20 min, and protein concentration is determined by measuring absorbance at 280 nm (Fig. 1). To detect eIF-4C, aliquots of late-eluting fractions (115–140) are dialyzed against buffer A-100, assayed for eIF-4C activity, and analyzed by sodium dodecyl sulfate–polyacrylamide gel electrophoresis. eIF-4C is eluted at 900–1000 m$M$ KCl or with the first fractions containing 1200 m$M$ KCl. Active fractions are pooled, concentrated in a dialysis bag with Aquacide, and dialyzed against buffer A-100 to yield about 0.3 $A_{280}$ unit of protein that is about 90% pure (Fig. 2).

*Purification of eIF-4D*

*Step 14. Phosphocellulose.* eIF-4D and eIF-1 are also adsorbed to the phosphocellulose column (step 14) first described for the purification of eIF-4A. Their elution is described above for the purification of eIF-4C. Both eIF-4D and eIF-1 activities are assayed. The activity profiles, shown in detail elsewhere,[6] indicate that eIF-4D is eluted at about 420 m$M$ KCl as a sharp peak that overlaps eIF-1, eluting more broadly at about 500 m$M$ KCl. No attempt is made to separate eIF-4D and eIF-1 at this stage, and fractions containing either or both activities (fractions 72–100) are combined.

*Step 16. DEAE-Cellulose.* The combined fractions containing eIF-4D and eIF-1 activities from the phosphocellulose column (step 14) are dialyzed against buffer A-50 and applied to a column of DEAE-cellulose (0.9 × 27 cm) equilibrated with the same buffer. Adsorbed protein is eluted with a 150-ml linear gradient of KCl, 50–400 m$M$ in buffer A. Fractions of 2 ml are collected every 20 min; protein is determined by absorbance at 280 nm, and eIF-1 and eIF-4D activities are assayed. eIF-1 activity is eluted as a broad peak from about 60–160 m*M* KCl (Fig. 1), and active fractions (34–60) are combined for further purification as described below. eIF-4D activity is eluted as a sharp peak at about 200 m*M* KCl and corresponds to a protein peak in the elution profile. The active fractions (62–75) containing eIF-4D are combined, concentrated by ultrafiltration (Amicon PM-10 filter) or Aquacide, and dialyzed against buffer A-100. The yield of eIF-4D is about 0.5 mg of protein, and the purity is 80–85% (Fig. 2).

*Purification of eIF-1*

*Step 17. Sephadex G-75.* Fractions containing eIF-1 from the DEAE-cellulose column described above for the purification of eIF-4D (step 16) are pooled, and the protein is concentrated by precipitation with solid ammonium sulfate to 90% saturation followed by dissolving the precipitate

TABLE II
PROPERTIES OF THE PURIFIED INITIATION FACTORS[a]

| Factor | Number of polypeptides | | Molecular weights | KCl concentration eluted from DEAE-cellulose (m*M*) | Phosphocellulose (m*M*) | Purity (%) | Yield (mg/100 rabbits) |
|---|---|---|---|---|---|---|---|
| eIF-1 | 1 | | 15,000 | 100 | 500 | 80–90 | 0.2 |
| eIF-2 | 3 | | 150,000 | 160 | 580 | 90 | 3.4 |
| | | #1 | 57,000 | | | | |
| | | #2 | 52,000 | | | | |
| | | #3 | 36,000 | | | | |
| eIF-3 | 9 | | 700,000 | 150 | 300 | 90 | 20–30 |
| | | #1 | 140,000 | | | | |
| | | #2 | 120,000 | | | | |
| | | #3 | 110,000 | | | | |
| | | #4 | 69,000 | | | | |
| | | #5 | 47,000 | | | | |
| | | #6 | 45,000 | | | | |
| | | #7 | 37,000 | | | | |
| | | #8 | 31,000 | | | | |
| | | #9 | 28,000 | | | | |
| eIF-4A | 1 | | 49,000 | 210 | 0–100 | 80–90 | 1.3 |
| eIF-4B | 1 | | 80,000 | 160 | 300 | 80–85 | 1.3 |
| eIF-4C | 1 | | 17,500 | — | 900–1200 | 85–90 | 0.25 |
| eIF-4D | 1 | | 16,500 | 200 | 450 | 80–90 | 0.4 |
| eIF-5 | 1 | | 150,000 | 250 | 690 | 75–80 | 0.6 |

[a] The values cited are derived from the text or from material published elsewhere (see text footnotes 3–6).

in 1 ml of buffer A-500. The solution (1.3 $A_{280}$ units of protein) is applied to a column of Sephadex G-75 (1.5 × 50 cm) equilibrated with buffer A-500. Protein is eluted with buffer A-500, and 2-ml fractions are collected every 30 min. Protein concentration is determined by absorbance at 280 nm, and eIF-1 activity is assayed. eIF-1 activity is eluted following the bulk of the protein, but does not correspond to a distinct protein peak (Fig. 1). Active fractions (28–38) are combined and are concentrated by treatment of the solution in a dialysis bag with Aquacide followed by dialysis against buffer A-100. The yield of eIF-1 is about 0.2 mg of protein with a purity of 75–85%.

### *General Discussion*

The 17 steps described above and shown in Fig. 1 lead to the purification of 8 initiation factors. The amounts of protein and specific activities at each step are given in Table I. Each of the preparations is analyzed by sodium dodecyl sulfate–polyacrylamide gel electrophoresis (Fig. 2). The purities, yields, and salt concentrations at which they are eluted from ion exchange columns are summarized in Table II, together with the number of different polypeptides per factor and their molecular weights. All the factors are stored frozen at −70° in small aliquots, generally at concentrations between 0.1 and 2 mg/ml. Each is stable for at least a year at −70° and can be frozen and thawed several times without significant loss of biological activity.

## Characterization of Initiation Factors

### *Physical Properties*

When analyzed by one-dimensional sodium dodecyl sulfate–polyacrylamide gel electrophoresis (Fig. 2), preparations of eIF-1, eIF-4A, eIF-4B, eIF-4C, eIF-4D, and eIF-5 show a single major polypeptide component. The respective molecular weights, determined by the method of Weber and Osborn,[26] are given in Table II. The preparation of eIF-2 contains three different polypeptides, which appear to be present in the factor complex in one copy each.[4] There are nine different major polypeptides in eIF-3, each present in one copy except for polypeptides 4 and 9, which are present in two copies per complex.[3] The nine eIF-3 components are poorly resolved in one-dimensional gels but are more clearly shown on a two-dimensional urea–sodium dodecyl sulfate–polyacrylamide gel as shown in Fig. 3. Thus the 8 initiation factors are comprised of 18 polypeptides, whose molecular weights total about 1,200,000.

[26] K. Weber and M. Osborn, *J. Biol. Chem.* **244,** 4406 (1969).

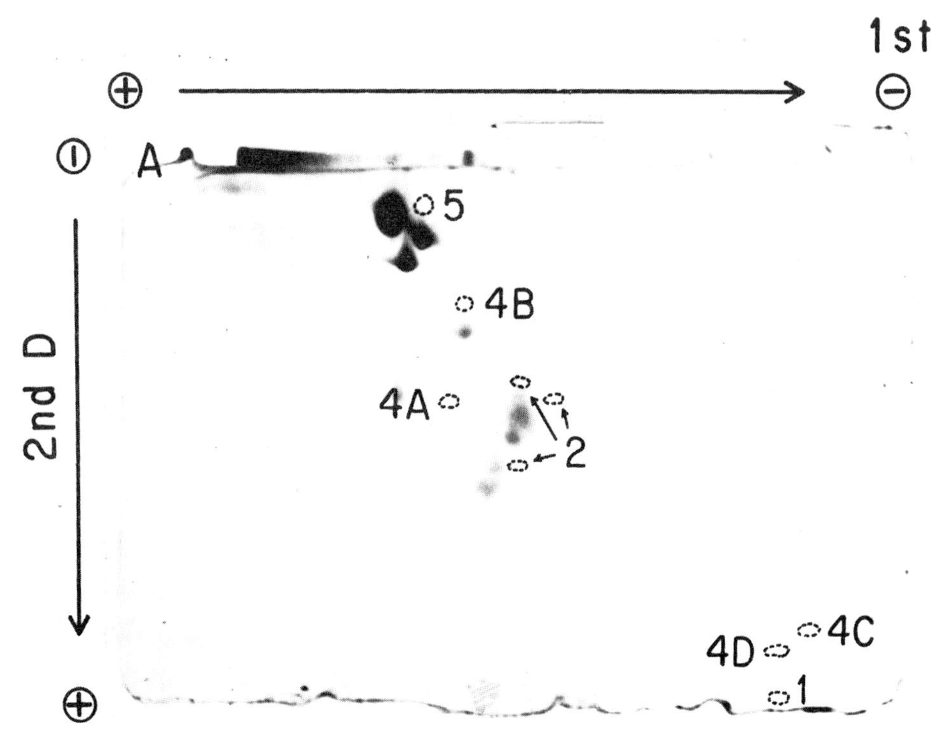


FIG. 3. Analysis of initiation factors by two-dimensional polyacrylamide gel electrophoresis. The gel system, described in detail elsewhere (see text footnotes 3–6), involves electrophoresis in urea buffer at pH 4.5 in the first dimension and electrophoresis in sodium dodecyl sulfate buffer in the second dimension; eIF-3 (50 $\mu$g) and the following radioactive initiation factors were analyzed by coelectrophoresis in different experiments: 0.5 $\mu$g of $^{14}CH_3$-eIF-1 (3000 cpm); 1 $\mu$g of $^{14}CH_3$-eIF-2 (25,000 cpm); 0.3 $\mu$g of $^{14}CH_3$-eIF-4A (3000 cpm); 0.3 $\mu$g of $^{14}CH_3$-eIF-4B (2250 cpm); 0.1 $\mu$g of $^{14}CH_3$-eIF-4C (3000 cpm); 0.3 $\mu$g of $^{14}CH_3$-eIF-4D (9600 cpm); and 0.25 $\mu$g of $^{14}CH_3$-eIF-5 (3750 cpm). The figure shows the pattern of the stained eIF-3 polypeptides whereas the dashed circles indicate the positions to which the labeled initiation factors migrated, as determined by the various autoradiograms.

Each of the factor polypeptides is distinctly different from the others, as demonstrated by coelectrophoresis of the various initiation factors in the two-dimensional gel system used for eIF-3. The stained pattern of eIF-3 components together with the positions to which the other initiation factors migrate on coelectrophoresis is shown in Fig. 3. The relative positions of the other 7 factors (labeled with radioactivity; see below) were determined by autoradiography as described in detail elsewhere.[3-6] The results show that none of the 18 polypeptides is identical to another. Comparison of initiation factors and ribosomal proteins (extracted from 40 S and 60 S

TABLE III
COMPARISON OF METHODS FOR DETERMINING THE PROTEIN CONCENTRATION OF INITIATION FACTORS[a]

| | Method of determination[b] | | | |
|---|---|---|---|---|
| Initiation factor | Bradford[27] (mg/ml) | Schaffner and Weissmann[28] (mg/ml) | Lowry *et al.*[29] (mg/ml) | Absorbance, 280 nm |
| eIF-1 | 0.031 | 0.032 | — | 0.037 |
| eIF-2 | 4.3 | 4.1 | 4.5 | 3.6 |
| eIF-3 | 0.46 | 0.38 | 0.42 | 0.43 |
| eIF-4A | 0.19 | 0.15 | 0.16 | 0.16 |
| eIF-4B | 0.036 | 0.039 | — | 0.04 |
| eIF-4C | 0.11 | 0.092 | — | 0.075 |
| eIF-4D | 0.30 | 0.22 | 0.23 | 0.20 |
| eIF-5 | 0.39 | — | — | 0.42 |

[a] Different initiation factor preparations were tested for protein concentration by using bovine serum albumin as a standard. The results are compared.
[b] Superscript numbers refer to text footnotes.

ribosomal subunits previously washed in 0.5 *M* KCl buffer) by coelectrophoresis indicates that none of the factor polypeptides are present in such ribosomes.[3-6]

An accurate, sensitive method for the determination of the protein concentration of pure initiation factor preparations is required. Four methods are compared in Table III: (1) absorbance at 280 nm; (2) a change in the absorbance of Coomassie Brilliant Blue, as described by Bradford[27]; (3) the staining by Coomassie Brilliant Blue of precipitated protein, as described by Schaffner and Weissmann[28]; and (4) the method of Lowry *et al.*[29] There is strikingly good agreement between the methods for all the purified factor preparations tested. Since the methods of Bradford and Schaffner–Weissmann are suitable for very dilute samples and are not influenced by other components in the buffer, they are especially appropriate for determining initiation factor protein concentrations.

### *Functional Properties*

Three assays are used to detect initiation factor activities during the purification procedures: Assay A, synthesis of globin chains, for eIF-1,

[27] M. M. Bradford, *Anal. Biochem.* **72,** 248 (1976).
[28] W. Schaffner and C. Weissmann, *Anal. Biochem.* **56,** 502 (1973).
[29] O. H. Lowry, N. J. Rosebrough, A. L. Farr, and R. L. Randall, *J. Biol. Chem.* **193,** 265 (1951).

eIF-3, eIF-4A, and eIF-4B; Assay B, synthesis of methionylpuromycin, for eIF-4C, eIF-4D, and eIF-5; and Assay C, ternary complex formation, for eIF-2. The effects of adding varying amounts of the initiation factor in their respective assays are shown in Fig. 4. Greater than 3-fold stimulation is seen at saturating concentrations for eIF-2, eIF-3, eIF-4A, eIF-4B, eIF-4D, and eIF-5; for eIF-1 and eIF-4C stimulation was somewhat less than 2-fold. In general, the molar amount of initiation factor required for maximal response approximates the molar amount of ribosomes in the assay.

A detailed analysis of the functional roles played by the eight initiation factors at various steps in the pathway of assembly of the 80 S initiation complex has been described elsewhere.[22] A list of the different reactions studied and the effects of the factors are given in Table IV. These results are briefly summarized as follows: eIF-2 first forms a ternary complex with Met-tRNA$_f$ and GTP. Binding of the ternary complex to the 40 S ribosomal subunit is enhanced by eIF-3 and eIF-4C. mRNA binding to the 40 S ribosomal subunit follows and is promoted by eIF-1, eIF-4A, and eIF-4B in addition to the factors required for maximal Met-tRNA$_f$ binding. eIF-5 stimulates junction of the 40 S initiation complex with the 60 S ribosomal subunit, while eIF-4D promotes the reaction of completed 80 S initiation complexes with puromycin. This tentative pathway of assembly of the 80 S initiation complex is shown in Fig. 5.

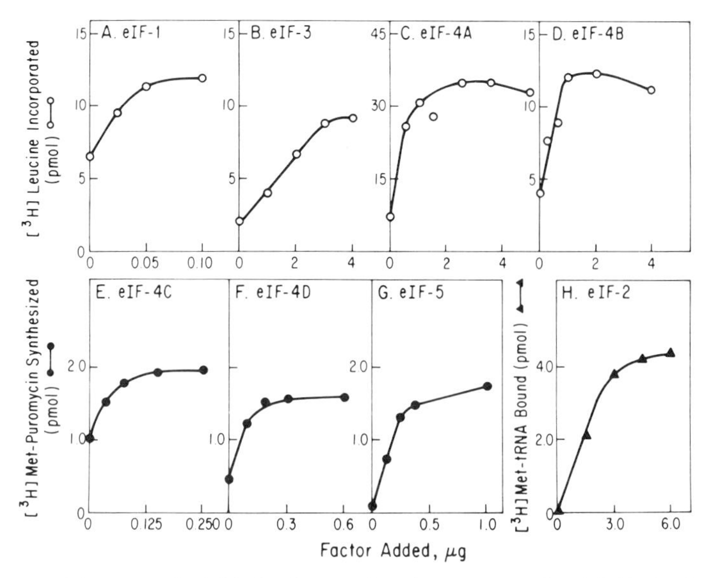


FIG. 4. Activity assays for initiation factors. Varying amounts of each initiation factor as indicated in the figure is added to either Assay A, B, or C as described in the text.

TABLE IV
FUNCTIONAL PROPERTIES OF PURIFIED INITIATION FACTORS[a]

| Initiation factor | Assay number | | | | | | | |
|---|---|---|---|---|---|---|---|---|
| | 1 | 2 | 3 | 4 | 5 | 6 | 7 | 8 |
| eIF-1 | − | − | ± | ± | ± | − | + | ± |
| eIF-2 | ++ | ++ | ++ | ++ | ++ | ++ | ++ | ++ |
| eIF-3 | − | ++ | ++ | ++ | ++ | ++ | ++ | ++ |
| eIF-4A | − | − | + | ± | + | − | ± | ++ |
| eIF-4B | − | − | ± | ± | ± | − | ± | ++ |
| eIF-4C | − | ± | ± | ± | ± | ± | ± | ± |
| eIF-4D | − | − | − | − | − | + | + | − |
| eIF-5 | − | − | − | ++ | + | ++ | ++ | ++ |

[a] The results are taken from a paper by Benne and Hershey (see text footnote 22), where the effects on various assay systems were measured by omitting one purified factor at a time. The effects of omission listed in the table are: ++, 0–25% of complete system; +, 25–50%; ±, 50–85%; −, little or no change. The following assays were used:

Assay 1: ternary complex formation with factor, GTP and Met-tRNA$_f$ (same as Assay C).

Assay 2: sucrose gradient analysis of [$^3$H]Met-tRNA$_f$ binding to 40 S subunits, with globin 9 S mRNA.

Assay 3: sucrose gradient analysis of globin 9 S [$^{125}$I]mRNA binding to 40 S subunits.

Assay 4: sucrose gradient analysis of [$^3$H]Met-tRNA$_f$ binding to 80 S ribosomes, with globin 9 S mRNA.

Assay 5: sucrose gradient analysis of globin 9 S [$^{125}$I]mRNA binding to 80 S ribosomes.

Assay 6: methionylpuromycin synthesis, with AUG template (same as Assay B).

Assay 7: methionylpuromycin synthesis, with globin 9 S mRNA.

Assay 8: globin synthesis with globin 9 S mRNA (same as Assay A).

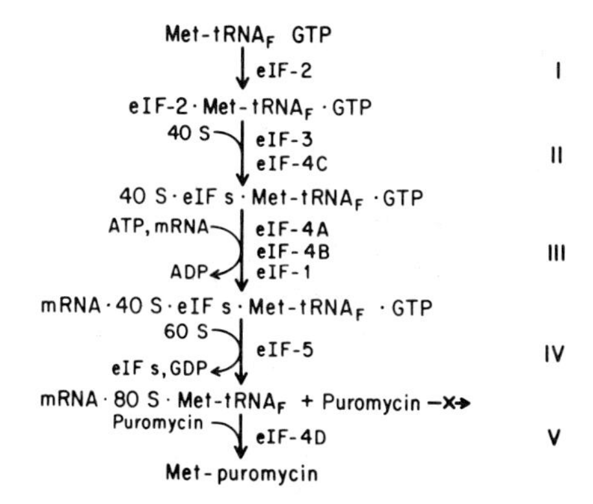


FIG. 5. A tentative pathway for assembly of initiation complexes. The reaction sequence is explained in the text and in detail elsewhere [R. Benne and J. W. B. Hershey, *J. Biol. Chem.* **253,** 3078 (1978)].

TABLE V
RADIOACTIVELY LABELED INITIATION FACTORS

| Initiation factor | Amount treated ($\mu$g) | Specific activity of product (cpm/$\mu$g) | Methyl groups attached per molecule factor | Radio-chemical purity (%) | Specific biological activity |
|---|---|---|---|---|---|
| eIF-1 | 50 | 6,000 | 1 | 90 | 0.40 |
| eIF-2 | 100 | 25,000 | 38 | 90 | 1.00 |
| eIF-3 | 1000 | 6,000 | 54 | 90 | 1.00 |
| eIF-4A | 100 | 10,000 | 5 | 90 | 0.80 |
| eIF-4B | 100 | 7,500 | 5 | 80 | 0.80 |
| eIF-4C | 50 | 23,000 | 4 | 95 | 1.00 |
| eIF-4D | 50 | 32,000 | 5 | 85 | 0.86 |
| eIF-5 | 50 | 15,000 | 22 | 75 | 0.22 |
| eIF-5 (mild) | 50 | 4,000 | 5 | 75 | 0.73 |

## Radioactive Labeling of Initiation Factors

Radioactive initiation factors may be usefully employed for studying their binding to ribosomes or other macromolecules. Reductive methylation with [$^{14}$C]formaldehyde and sodium borohydride is a mild *in vitro* method[30,31] for labeling proteins, and has been used to radiochemically label the eight highly purified initiation factors.[3-6] Nearly all the labeled initiation factors retain biological activity.

*Procedure.* The method of Ottesen and Svensson[31] is employed. About 50–100 $\mu$g of each initiation factor are dialyzed for 4 hr against buffer B-100 minus the 2-mercaptoethanol. The range of protein concentrations used is 0.2–0.5 mg/ml, although even more concentrated solutions allow more efficient labeling. The pH is brought to 9.0 by the addition of 0.1 volume of 1.0 *M* potassium borate, pH 11.4, and the solution is chilled to 0°. [$^{14}$C]-Formaldehyde (4 nmol per microgram of protein; specific activity 44 Ci/mol) is added, the solution is held for 30 sec, and then a solution of sodium borohydride in 1 m*M* NaOH is added 10 times at 3-min intervals (0.2 nmol per microgram of protein each time). Reagent concentrations are chosen so that dilution of the reaction mixture is minimized. The mixtures finally are dialyzed extensively against buffer A-100 to remove unreacted radioactive material.

The results for the reductive methylation of the 8 initiation factors are given in Table V. For each factor, the specific radioactivity and number of methyl groups attached per molecule of factor are reported. The

[30] R. Means and R. E. Feeney, *Biochemistry* **7**, 2192 (1968).
[31] M. Ottesen and B. J. Svensson, *C. R. Trav. Lab. Carlsberg* **38**, 445 (1971).

radiochemical purity is measured by electrophoresis in sodium dodecyl sulfate–polyacrylamide gels as described in Fig. 2, followed by slicing the gel and counting the radioactivity in the slices. The specific biological activity is reported as the fraction of activity measured compared with the nonlabeled initiation factor. Assays A, B, and C are employed with limiting amounts of the factor being tested. Little or no change in the functional activities of eIF-2, eIF-3, eIF-4A, eIF-4B, eIF-4C, and eIF-4D is observed following the labeling procedure. In contrast, the activities of both eIF-1 and eIF-5 are severely altered. When eIF-1 and eIF-5 are treated with only 25% of the [$^{14}$C]formaldehyde and sodium borohydride routinely used, significant labeling of eIF-5 and retention of biological activity is obtained. However, eIF-1 is only poorly labeled under the milder conditions.

A second method for labeling eIF-2, eIF-3, and eIF-4B without apparently altering their functional properties is to treat the factors with [$\gamma$-$^{32}$P]-ATP and a cyclic AMP-independent protein kinase obtained from rabbit reticulocytes or erythrocytes, as described in detail elsewhere.[32,33]

[32] O. G. Issinger, R. Benne, R. R. Traut, and J. W. B. Hershey, *J. Biol. Chem.* **252,** 6471 (1976).

[33] R. Benne, R. R. Traut, and J. W. B. Hershey, *Proc. Natl. Acad. Sci. U.S.A.* **75,** 108 (1978).

# [29] Initiation Factors for Protein Synthesis from Wheat Germ

*By* Benjamin V. Treadwell, Ljubica Mauser, and William G. Robinson

Factors involved in the formation of initiation complexes for protein synthesis have been isolated from a number of eukaryotic tissues.[1-10] In virtually every procedure crude ribosomes are prepared and extracted with

[1] M. H. Schreier, B. Erni, and T. Staehelin, *J. Mol. Biol.* **116,** 727 (1977).

[2] C. L. Woodley, Y. C. Chen, and N. K. Gupta, this series, Vol. 30, p. 141.

[3] D. Picciano and W. F. Anderson, this series, Vol. 30, p. 171.

[4] G. Kramer, A. B. Henderson, P. Pinphanichakarn, M. H. Wallis, and B. Hardesty, *Proc. Natl. Acad. Sci. U.S.A.* **74,** 1445 (1977).

[5] R. S. Ranu and I. G. Wool, *J. Biol. Chem.* **251,** 1926 (1976).

[6] G. E. Blair, H. H. M. Dahl, E. Truelsen, and J. C. Lelong, *Nature (London)* **265,** 651 (1977).

[7] J. F. Hejtmancik and J. P. Comstock, *Biochemistry* **15,** 3804 (1976).

[8] L. L. Spremulli, B. J. Walthall, S. R. Lax, and J. M. Ravel, *Arch. Biochem. Biophys.* **178,** 565 (1977).

[9] M. Giesen, R. Roman, S. N. Seal, and A. Marcus, *J. Biol. Chem.* **251,** 6057 (1976).

[10] H. A. Thompson, I. Sadnik, J. Scheinbuks, and K. Moldave, *Biochemistry* **16,** 2221 (1977).





ISBN 0-12-504180-2

buffered solutions of 0.5–0.7 *M* potassium chloride. In wheat germ, on the other hand, all the factors that we have attempted to detect have been present in the high-speed supernatant. These include eIF-2,[11] a presumptive eIF-3, and eIF-5; dissociation factor activity is also present in the supernatant. From the standpoint of preparative procedures, therefore, wheat germ offers a number of advantages. It is inexpensive, readily available in large quantity, and does not require prior isolation of ribosomes. In addition, the supernatant also contains eIF-M1, which catalyzes the GTP-independent binding of Met-tRNA or AcPhe-tRNA to 40 S ribosomes in the presence of AUG or poly(U), respectively. The physiological function of this factor is not known, but since it behaves similarly to eIF-2 on ion-exchange chromatography, the purification procedure for eIF-2 can be used to prepare both eIF-2 and eIF-M1. Wheat germ eIF-M1, eIF-2, and dissociation factor are functionally interchangeable with factors from other tissues; eIF-3 and eIF-5 have not yet been tested in this respect.

The function of eIF-2 is to form a complex with GTP and Met-$tRNA_i$ that is subsequently bound to 40 S ribosomes. An additional factor that is required for the transfer of Met-$tRNA_i$ to the 40 S subunit has been reported[4,12] to occur in reticulocytes, but its presence in wheat germ has not been demonstrated. Met-$tRNA_i$ is transferred to 40 S ribosomes in the presence of GTP, highly purified wheat germ eIF-2, and magnesium ions. Measurement of the amount of either the eIF-2·GTP·Met-$tRNA_i$ complex or the 40 S complex in reaction mixtures can serve as the basis of an eIF-2 assay. The ribosomal complex assay, however, can be used to advantage only with eIF-2 preparations that are free of eIF-M1 since, under the conditions of the assay, eIF-M1 catalyzes the destruction of the 40 S complex formed with eIF-2. A third but more cumbersome assay is the restoration by eIF-2 of protein synthesis in a wheat germ system inhibited by antibody to eIF-2.

Wheat germ supernatant also contains a protein that seems to be analogous to the cofactor for ternary complex formation in reticulocytes reported by Gupta.[13] At the protein concentration ordinarily used in the eIF-2 assay, addition of the cofactor (Co-eIF-2) has no effect on activity. When amounts of eIF-2 are used that give barely detectable amounts of ternary complex, addition of Co-eIF-2 stimulates complex formation about 3-fold. The Co-eIF-2 preparation alone has no eIF-2 activity.

[11] The nomenclature for the eukaryotic initiation factors conforms to the system proposed at an International Symposium on Protein Synthesis in October 1976 at the National Institutes of Health in Bethesda, Maryland.

[12] A. Majumdar, R. Roy, A. Das, A. Dasgupta, and N. K. Gupta, *Biochem. Biophys. Res. Commun.* **78,** 161 (1977).

[13] S. H. Reynolds, A. Dasgupta, S. Palmieri, A. Majumdar, and N. K. Gupta, *Arch. Biochem. Biophys.* **184,** 328 (1977).

Eukaryotic ribosomal subunits are often separated by centrifugation in sucrose gradients containing high concentrations of both potassium chloride (up to 1 *M*) and magnesium acetate (up to 12 m*M*). Wheat germ subunits prepared by these procedures have very little activity in binding assays and are inactive in translating globin mRNA in a reconstituted protein synthesizing system. The small subunit is damaged more severely by the high salt than the 60 S subunit. Highly active wheat germ subunits can be prepared, however, with sucrose gradients at low concentrations of potassium chloride and magnesium acetate as described by Weeks.[14] Subunits prepared in this way recombine spontaneously, translate natural mRNAs, and are fully active in binding assays.

## Assays

### *eIF-M1 Assay*

The activity of eIF-M1 preparations is determined by measuring the amount of [$^{14}$C]aminoacyl-tRNA retained on Millipore filters under appropriate conditions.[15] Two aminoacyl-tRNAs, fMet-tRNA and AcPhe-tRNA (both prepared with *E. coli* tRNA; but eukaryotic aminoacyl-tRNAs are also active) can be used; since the assay conditions are different for each of them, they are described separately.

The complex consisting of fMet-tRNA, AUG, and 40 S subunit forms readily at 0°, but decomposes rapidly as the temperature is raised. In order to obtain consistent results, therefore, all reagents and transfer pipettes are kept cold while the complex is being collected and washed on the filter. Reaction mixtures contain 90 m*M* Tris · HCl, pH 7.5, 150 m*M* KCl, 4 m*M* $Mg(Ac)_2$, 1.5 m*M* dithiothreitol, 0.3 $A_{260}$ unit of 40 S ribosomal subunits from wheat germ, 0.05 $A_{260}$ unit of AUG, 15 pmol of f[$^{14}$C]Met-tRNA (400 cpm/pmol), and eIF-M1. The final volume is 60 $\mu$l. [$^{14}$C]Met-tRNA can be substituted for f[$^{14}$C]Met-tRNA. After the samples have been maintained at 0° for 20 min, the reaction is terminated by adding 2 ml of ice-cold buffer containing 60 m*M* KCl, 50 m*M* Tris · HCl, pH 7.8, and 10 m*M* $Mg(Ac)_2$. The dilute mixture is immediately filtered through a Millipore filter, which is subsequently washed with three 2-ml portions of diluting buffer. Filters are placed in scintillation vials, dried under a heat lamp, suspended in a solution of Omnifluor in toluene, and counted in a Beckman scintillation counter.

When AcPhe-tRNA is used as substrate, the resulting complex is more stable, and for this reason is more convenient as a routine assay. The

[14] D. P. Weeks, D. P. S. Verma, S. N. Seal, and A. Marcus, *Nature (London)* **236,** 167 (1972).
[15] M. Zasloff and S. Ochoa, this series, Vol. 30, p. 197.

optimum magnesium ion concentration for this assay is 7 m$M$. Sixty-microliter reaction mixtures contain 90 m$M$ Tris, pH 7.5, 120 m$M$ KCl, 7 m$M$ $Mg(Ac)_2$, 2 m$M$ dithiothreitol, 0.3 $A_{260}$ unit of 40 S subunits from wheat germ, 1.0 $A_{260}$ unit of poly(U), 15 pmol of Ac[$^{14}$C]Phe-tRNA (800cpm/pmol), and eIF-M1. Incubation is for 20 min at 25°, after which the samples are diluted, filtered, washed, dried, and counted as described for the fMet-tRNA binding assay.

A unit of eIF-M1 is defined as the amount that catalyzes the binding of 1 pmol substrate in 20 min in the standard assay. Specific activity is defined as units per milligram of protein.

### *eIF-2 Assay*

The assay for eIF-2 is based upon the retention of a ternary complex containing Met-tRNA$_i$, eIF-2, and GTP on Millipore filters. A reaction mixture containing 20 m$M$ HEPES, pH 7.56, 0.2 m$M$ GTP, 100 m$M$ KCl, 2 m$M$ 2-mercaptoethanol, 3 pmol of wheat germ [$^{35}$S]Met-tRNA, and eIF-2 in a volume of 50 $\mu$l is incubated at 25° for 10 min. GTP is omitted from controls. The reaction is terminated by adding 2 ml of ice-cold buffer containing 20 m$M$ HEPES, pH 7.56, 100 m$M$ KCl, and 2 m$M$ 2-mercaptoethanol. The diluted reaction mixture is filtered on a Millipore filter, which is washed twice (2 ml each time) with the same buffer used to terminate the reaction. The Millipore filters are then dried and counted in a liquid scintillation counter as described in the procedure for the eIF-M1 assay. In all eIF-2 assays, the methionyl derivative of wheat germ tRNA was used. A unit of eIF-2 is the amount that binds 1 pmol of Met-tRNA$_i$ in 10 min. Specific activity is units per milligram of protein.

### *Co-eIF-2 Assay*

When the amount of Met-tRNA$_i$ bound in the standard eIF-2 assay is plotted against increasing concentrations of eIF-2, a lag in Met-tRNA$_i$ binding at low eIF-2 concentration is observed.[16] After the lag, Met-tRNA$_i$ binding is linear with respect to increasing eIF-2 concentrations. For the Co-eIF-2 assay, an eIF-2 concentration at the lowest part of the linear portion of the binding curve must be chosen; Met-tRNA$_i$ binding is not stimulated by Co-eIF-2 when higher eIF-2 concentrations are used. Except for the addition of Co-eIF-2, the components of this assay are identical to those of the standard eIF-2 assay. Three tubes are used for each assay. The

[16] B. V. Treadwell and W. G. Robinson, *Biochem. Biophys. Res. Commun.* **65**, 176 (1975).

first is a complete standard eIF-2 assay mixture containing eIF-2, but no Co-eIF-2. The second tube contains Co-eIF-2, but no eIF-2. The third contains both eIF-2 and Co-eIF-2. Incubation, termination of the reaction, filtration on Millipore filters, and counting are the same as for the standard eIF-2 assay. The sum of the amount of Met-tRNA$_i$ bound in the first and second tubes subtracted from the amount bound in the third is the stimulation of binding due to Co-eIF-2. The stimulation is usually 2- to 3-fold. A unit of Co-eIF-2 is the amount which stimulates the binding of 1 pmol of Met-tRNA$_i$ in the standard assay. Specific activity is defined as units per milligram of protein.

*eIF-2-Dependent Binding of Met-tRNA$_i$ to 40 S Ribosomes*

The addition of magnesium salts to the standard eIF-2 assay causes a marked decrease in the amount of ternary complex captured on Millipore filters. At 5 m*M* Mg(Ac)$_2$ very little complex is retained, but when 40 S subunits and AUG are also present in the incubation mixture, the amount of Met-tRNA$_i$ retained by filters is equal to the amount bound by eIF-2 alone in the absence of magnesium. In this sense there is a stoichiometric binding of ternary complex to ribosomes. The composition of the 50-$\mu$l reaction mixture is 20 m*M* HEPES, pH 7.56, 100 m*M* KCl, 2 m*M* 2-mercaptoethanol, 5 m*M* Mg(Ac)$_2$, 5 $\mu M$ GTP, 0.05 $A_{260}$ unit of AUG, 0.3 $A_{260}$ unit of wheat germ 40 S subunits, 3 pmol of wheat germ [$^{35}$S]Met-tRNA$_i$ and eIF-2. The eIF-2 must be free of eIF-M1. The ribosomal subunits are omitted from the blank. Samples are incubated for 10 min at 25° and the reaction is terminated by adding 2 ml of cold buffer containing 20 m*M* HEPES, pH 7.56, 100 m*M* KCl, 2 m*M* 2-mercaptoethanol, and 5 m*M* Mg(Ac)$_2$. Filtration, washing, and counting are the same as described for the standard eIF-2 assay.

## Preparation of Wheat Germ Ribosomal Subunits

*Materials*

Raw wheat germ: fully active ribosomal subunits can be prepared from samples of raw wheat germ that do not yield an active S30 protein-synthesizing system. The failure to synthesize protein is usually due to a defect in the supernatant. Brands of wheat germ used to prepare active translation systems are high in lipid, which complicates somewhat the preparation of ribosomes. Wheat germ suitable for ribosome preparation can be purchased from Dixie Portland Co., Arkansas City, Kansas.

Buffer A: 90 m*M* KCl, 4 m*M* Mg(Ac)$_2$, 2 m*M* $CaCl_2$, 6 m*M* $KHCO_3$

Buffer B: 40 m$M$ KCl, 0.33 m$M$ $Mg(Ac)_2$, 1 m$M$ 2-mercaptoethanol, 10 m$M$ Tris·HCl, pH 7.8
Buffer C: 40 m$M$ KCl, 80 $\mu M$ $Mg(Ac)_2$, 3 m$M$ 2-mercaptoethanol, 2 m$M$ Tris·HCl, pH 7.8
Buffer D: 40 m$M$ KCl, 8 m$M$ $Mg(Ac)_2$, 3 m$M$ 2-mercaptoethanol, 2 m$M$ Tris·HCl, pH 7.8
Buffer E: 80 m$M$ KCl, 8 m$M$ $Mg(Ac)_2$, 8 m$M$ 2-mercaptoethanol, 4 m$M$ Tris·HCl, pH 7.8
$Mg(Ac)_2$, 1 $M$
Tris·HCl, 1 $M$, pH 7.8
Washed sea sand

*Procedure.* Fifteen grams of wheat germ are ground for 10 min in a mortar with an equal volume of sand. The grinding and all subsequent steps are done at 0°–4°. The dry mixture, after grinding, is suspended in 150 ml of buffer A, and about 20 ml of the same buffer are used for rinsing the mortar when its contents are transferred to a beaker. Sand and other insoluble material are removed by centrifuging the combined suspension and rinses at 15,000 rpm in a Sorvall SS-34 rotor for 12 min. After this and all subsequent centrifugations, a yellow lipid layer is scooped from the top of the solutions, which are then poured though cheese cloth. The volume of the supernatant is measured (about 140 ml) and 0.1 volume of 0.1 $M$ $Mg(Ac)_2$ (about 14 ml) and 0.1 volume of 1 $M$ Tris·HCl, pH 7.8 (about 14 ml) are added. A precipitate that forms is removed by centrifuging the extract at 15,000 rpm in a Sorvall SS-34 rotor for 15 min. The upper seven-eighths of the supernatant is retained and transferred to tubes in which it is centrifuged in a Beckman 60 Ti rotor at 38,000 rpm for 2 hr. The supernatant is discarded, and the centrifuge tubes containing the ribosomal pellets are allowed to stand in ice for an hour. During this time a small amount of debris is released from the surface of the clear pellets of 80 S ribosomes. A minimal volume of buffer B is used to gently rinse the pellets before they are suspended in a total volume of 3 ml of buffer B. The concentration of 80 S ribosomes should be 600–650 $A_{260}$ units/ml.

To prepare 40 S and 60 S ribosomal subunits, 0.5 ml of the 80 S ribosome preparation is immediately layered onto each of three 50-ml 15–30% (w/v) sucrose gradients in buffer C and centrifuged for 14 hr at 21,000 rpm in a Beckman SW 25.2 rotor. Fractions (0.7–0.8 ml) from each gradient are collected, and their absorbancy at 260 nm is determined and plotted against tube number. By extrapolating the leading arm of the 40 S peak and the trailing arm of the 60 S peak, 40 S and 60 S fractions are chosen that give minimal cross-contamination. Fractions containing 40 S subunits from the three gradients are combined, and the magnesium ion concentration is raised to 8 mM by adding the calculated amount of 1 $M$ $Mg(Ac)_2$. The preparation is diluted to 25 ml with buffer D. An identical procedure is followed with the fractions containing 60 S subunits. The two 25-ml samples are centrifuged overnight at 35,000 rpm in a Beckman 60 Ti rotor in order to concentrate the subunits. The clear, colorless pellets are gently rinsed with buffer E and carefully suspended in a volume of about 0.2 ml of the same buffer. The concentrated suspension is centrifuged at 4000 rpm for 20 min in the Sorvall SS-34 rotor, and the precipitate is discarded. The supernatant is diluted with buffer E so that the 40 S subunits have a concentration of 200 $A_{260}$ units/ml, and the 60 S 400 $A_{260}$ units/ml. Each sample is diluted with 1 volume of glycerol and stored at −20°. The subunits remain fully active for several months, but upon very prolonged storage activity in Met-$tRNA_i$ binding experiments begins to decline. As determined in a poly(U) translational assay, the 40 S subunits are free of 60 S contamination, but the 60 S subunits have about 10% contamination with the smaller subunit.

## Purification of eIF-2 and EIF-M1

*Materials*

Wheat germ. The wheat germ used for isolation of ribosomal subunits is satisfactory also for this preparation.
Acetone
DEAE-Cellulose (Whatman DE-23)
DEAE-Sephadex A-50 (Pharmacia)
CM-Sephadex C-50 (Pharmacia)
Hydroxyapatite (Clarkson Hypatite C)
Potassium phosphate buffer, 1 *M*, pH 7.0
Buffer A: 10 m*M* Tris·HCl, pH 7.5, 10 m*M* 2-mercaptoethanol, 0.1 m*M* EDTA, and varying concentrations of KCl indicated by a number in parentheses, thus A (100) is buffer A containing 100 m*M* KCl.
Buffer B: 20 m*M* HEPES–KOH, pH 7.56, 20 m*M* 2-mercaptoethanol, 0.1 m*M* EDTA, and varying concentrations of KCl indicated by a number in parentheses, thus B (100) is buffer B containing 100 m*M* KCl.
Buffer C: 50 m*M* Tris·HCl, pH 7.5, 10 m*M* 2-mercaptoethanol, 0.1 m*M* EDTA, 100 m*M* KCl, 50% glycerol (v/v)

*Procedure.* A summary of the purification procedure is given in Table I. All operations are carried out at 4° unless otherwise stated. In chromatographic procedures the presence of UV-absorbing material in eluates is continuously monitored with an LKB Uvicord apparatus. Dialysis of samples against buffer C (contains 50% glycerol) before storage in steps 3–5

TABLE I
PURIFICATION OF eIF-2 AND eIF-M1

| Purification step | Volume (ml) | Protein (mg) | eIF-M1[a] (pmol/mg/-20 min) | eIF-2[b] (pmol/mg/-10 min |
|---|---|---|---|---|
| DEAE-cellulose, ammonium sulfate | 250 | 20,000 | — | — |
| CM-Sephadex C-50 | 15 | 84 | 9,000 | 70 |
| DEAE-Sephadex A-50 | 2 | 18 | 19,200 | 148 |
| Hydroxyapatite | | | | |
| Fractions 4 to 10 | 2 | 4.1 | — | 300 |
| Fractions 22 to 30 | 2 | 2.0 | 37,000 | — |

[a] The standard AcPhe-tRNA binding assay was used to determine eIF-M1 activity.
[b] The ternary complex formation with GTP and [$^{35}$S]Met-tRNA$_i$ was used to assay eIF-2.

results in a three-fourths reduction in volume. This can be a useful technique for concentrating protein solutions.

*Step 1. Preparation of Acetone Powder.* One-hundred-gram portions of commercial wheat germ are homogenized in a Waring blender for 45 sec with 500 ml of acetone that has been chilled to $-15°$. The homogenate is quickly filtered by suction on a large Büchner funnel, and the filtrate is washed in rapid succession with 100 ml of acetone and 200 ml of ether at $-15°$. The light gray powder is spread in a thin layer on paper and allowed to dry at room temperature until the odors of acetone and ether can no longer be detected. Each 100 g of wheat germ yields approximately 80 g of acetone powder. No loss of activity is observed after prolonged storage of the acetone powder at $-15°$.

*Step 2. Extraction, DEAE-Cellulose Treatment, and Ammonium Sulfate Fractionation.* Three hundred and twenty grams of wheat germ acetone powder are suspended in 1600 ml of buffer A (100), and the mixture is stirred for 2.5 hr. The suspension is centrifuged at 15,000 *g* for 20 min, and the supernatant is poured into a beaker containing 700 ml of settled DEAE-cellulose (DE-23), which has been equilibrated with buffer A (100). The DEAE-cellulose suspension is stirred and filtered with suction through Whatman No. 1 filter paper on a Büchner funnel. The packed DEAE-cellulose is suspended in about 400 ml of buffer A (100) and again filtered. The filter cake can be used as starting material for the preparation of Co-eIF-2. The combined filtrates are brought to 40% saturation with ammonium sulfate and centrifuged at 15,000 *g* for 20 min. The resulting precipitate is discarded, and the supernatant is adjusted to 65% saturation with solid ammonium sulfate. The protein pellet obtained after centrifugation at 15,000 *g* for 20 min is dissolved in 135 ml of buffer A (100), and the protein solution is dialyzed against 4 liters of the same buffer overnight.

*Step 3. Chromatography on CM-Sephadex C-50.* The success of chromatography on CM-Sephadex C-50 is somewhat dependent upon the procedure by which the column is prepared. CM-Sephadex C-50 (33 g) is allowed to swell in an excess of 0.5 *M* KCl for at least 3 days at 4°. The supernatant is decanted and the remaining volume of settled ion exchanger (about 800 ml) is diluted 5-fold with cold distilled water. The suspension is briefly stirred and allowed to settle; the supernatant is decanted. The settled Sephadex is resuspended in buffer A (100) and allowed to settle; the supernatant is decanted, and the thick Sephadex suspension is used to fill a 5 × 60 cm column. The column is immediately attached to a reservoir containing buffer A (100) at a pressure head of about 1 meter. After 350–400 ml of buffer has flowed through the column it is sufficiently packed for application of the sample.

The dialyzed preparation from the preceding step (25 g of protein in 300 ml) is applied to the column, and washing with buffer A (200) is begun immediately. Washing is continued until the absorbance of the effluent reaches a low steady level; this is usually after about 1800 ml of buffer have passed through the column. The factors eIF-2 and eIF-M1 are eluted with buffer A (300). Both these factors are in a single UV-absorbing peak after about 500 ml of buffer has passed through the column. The active fractions are pooled and concentrated to a volume of 15–20 ml with an Amicon ultrafiltration apparatus fitted with a PM-10 membrane. The concentrated solution is dialyzed against buffer C before storage at −15°, and before it is used in the next step it is dialyzed for 2–3 hr against buffer B (50).

*Step 4. Chromatography on DEAE-Sephadex A-50.* The protein solution from the preceding step is applied to a DEAE-Sephadex A-50 column (2.5 × 40 cm), which has been prepared similarly to the preparation of cation exchanger in the preceding section, except that it is equilibrated with buffer B (50). The column is washed with buffer B (100) after the sample has been applied until no protein can be detected in the eluate. The protein that elutes with buffer B (200) is collected and concentrated by ultrafiltration, as in the preceding step, to 4 ml. The preparation is dialyzed against buffer C for storage at −15° and against buffer A (100) for 2–3 hr before use in the next step.

*Step 5. Hydroxyapatite Chromatography.* The sample from the preceding step is applied to a hydroxyapatite column (1.5 × 11 cm) equilibrated with buffer A (100). The column is then washed with 100 m*M* potassium phosphate, pH 7.0, containing 10 m*M* 2-mercaptoethanol until no UV-absorbing material can be detected eluting from the column. The column is then developed with a linear gradient (100 × 100 ml) from 100 m*M* potassium phosphate, pH 7.0, to 500 m*M* potassium phosphate, pH 7.0 (both buffer solutions contain 10 m*M* 2-mercaptoethanol), and 4-ml fractions are collected. The peak of eIF-2 activity elutes from the column at 280 m$M$ potassium phosphate and the peak of eIF-M1 activity at 400 m$M$ potassium phosphate. Fractions containing eIF-M1 activity but no eIF-2 activity are pooled, and fractions with eIF-2 activity but no eIF-M1 are also pooled. Both these fractions are concentrated to 2 ml by Amicon ultrafiltration with a PM-10 membrane and dialyzed against buffer C before use or storage.

## *Comments*

*eIF-2.* The eIF-2 fractions chosen from the hydroxyapatite chromatography are free of eIF-M1, but they represent a little less than half of the total eIF-2 in the eluate. The remaining eIF-2 is present in intermediate fractions contaminated with eIF-M1. Rechromatography of these fractions has not been of practical value, but the fractions can still be used in experiments in which the presence of eIF-M1 is not an important factor. Upon storage at $-15°$ eIF-2 gradually loses activity, but when frozen in small aliquots in buffer C at $-80°$, full activity is retained for at least 6 months.

The elution profiles of eIF-2 and eIF-M1 on Sephadex G-200 are identical; they both have estimated molecular weights by this technique of 145,000. Upon electrophoresis in sodium dodecyl sulfate on polyacrylamide gels, three major bands are obtained from eIF-2 with molecular weights of 37,000, 41,000, and 57,000. Wheat germ eIF-2, therefore, is assumed to be like eIF-2 from reticulocytes in consisting of three different polypeptide chains. Ternary complex formation with wheat germ eIF-2 is inhibited by *N*-ethylmaleimide and *p*-chloromercuribenzoate. It is also inhibited by EDTA. The probable reason for this is that complex formation requires a low concentration of magnesium,[17] which is normally supplied as an impurity in other components of the standard reaction mixture. There is no difference in activity at 25° and 37° in the usual 10-min assay.

The concentration of GTP (0.2 m$M$) routinely used in the ternary complex assay is a large excess over the amount actually needed. With the amount of eIF-2 and Met-tRNA$_i$ employed in the standard assay, maximal complex formation occurs at a GTP concentration of about 0.5 $\mu M$. GTP concentration becomes an important consideration in regard to the binding of Met-tRNA$_i$ to 40 S ribosomes because some inhibition of binding appears to occur at the high concentration used for ternary complex formation. For this reason a lower GTP concentration (5 $\mu M$) is used for formation of the 40 S complex. Under the conditions of the standard assays, Met-tRNA$_i$ bound in both the ternary and 40 S complexes does not exchange with free Met-tRNA$_i$.

[17] L. M. Cashion, G. L. Dettman, and W. M. Stanley, Jr., this series, Vol. 30, p. 153.

Experimental conditions for Met-tRNA$_i$·AUG·40 S complex formation with eIF-2 are similar to those for binding Met-tRNA$_i$ to the small subunits with eIF-M1. A major difference in the two systems, however, is the temperature at which the assay is performed. The complex produced in the presence of eIF-M1 rapidly decomposes at temperatures above 0°. The decomposition is catalyzed by eIF-M1 itself. On the other hand, the eIF-2 complex is stable at 25°, but is formed very slowly at 0°. Addition of eIF-M1 at 25° to a Met-tRNA$_i$·AUG·40 S complex formed with eIF-2 catalyzes the destruction of the complex. For this reason Met-tRNA$_i$·AUG·40 S complex formation with eIF-2 can be detected only if the eIF-2 preparation is free of eIF-M1. In the systems described here eIF-2 acts stoichiometrically; it does not turn over.

In the absence of AUG, a Met-tRNA$_i$·40 S complex is formed with eIF-2. However, up to 5 times more complex is formed when the trinucleotide is present than when it is absent. The reason for this observation is that AUG stabilizes the ribosomal complex. Both types of complex, with and without AUG, are sufficiently stable to survive analytical sucrose density gradients. The complex formed in the absence of AUG is not destroyed by eIF-M1.

*eIF-M1*. The physical and biological properties of what germ eIF-M1 are similar to those of eIF-M1 isolated from other tissues.[15,18,19] Full activity of the purified factor is retained upon storage in 50% glycerol (buffer C) at −15° for several weeks, but after this time activity begins to decline because of the oxidation of sulfhydryl groups. In the early stages of inactivation, dialysis of the sample against buffer C containing fresh 2-mercaptoethanol will completely restore activity. Samples frozen at −80° retain activity for longer times.

As estimated by gel filtration on Sephadex G-200, the molecular weight of eIF-M1 is 145,000. The value obtained by electrophoresis in sodium dodecyl sulfate on polyacrylamide gels is 41,000, which could be an indication that the native protein is a tetramer. The factor acts catalytically, and, as determined for a molecular weight of 145,000, it has a turnover number estimated roughly to be about 10.

Although the isolation procedure described here uses a wheat germ acetone powder extract as starting material, the same procedure can be used to purify eIF-M1 from a 100,000 *g* wheat germ supernatant. In addition, wheat germ ribosomes also contain eIF-M1 that can be extracted with 700 m*M* KCl and also purified by this same procedure. The supernatant and ribosomal eIF-M1s are indistinguishable. Highly purified eIF-M1 binds to 40 S wheat germ ribosomes. One method of demonstrating this is to mix 40

[18] E. Gasior and K. Moldave, *J. Mol. Biol.* **66,** 391 (1972).

[19] W. C. Merrick and W. F. Anderson, *J. Biol. Chem.* **250,** 1197 (1975).

S ribosomes and eIF-M1, and separate the ribosome-eIF-M1 complex from free eIF-M1. The 40 S ribosome·eIF-M1 complex is active in binding aminoacyl-tRNA. Binding of eIF-M1 to ribosomes is the most likely explanation for the fact that 40 S ribosomes, prepared by the low-salt procedure described here, frequently give high blanks in the eIF-M1 assay. As the ribosome preparations age, the blank values decrease.

## Preparation of Co-eIF-2

*Materials*

Wheat germ, ion-exchange chromatography materials, and buffers A and C are the same as described for the preparation of eIF-2 and eIF-M1.

*Procedure.* The starting material for the preparation of Co-eIF-2 is the filtered DEAE-cellulose from step 2 in the purification of eIF-2 and eIF-M1. All operations are carried out at 4°. Table II is a summary of the purification procedure.

TABLE II
PURIFICATION OF CO-EIF-2

| Purification step | Volume (ml) | Protein (mg) | Co-eIF-2 (pmol/mg/10 min) |
|---|---|---|---|
| DEAE-cellulose, ammonium sulfate | 57 | 1900 | — |
| DEAE-cellulose | 2.5 | 69 | 9 |
| CM-Sephadex C-50 | 0.7 | 6.1 | 60 |

*Step 1. Elution of Co-eIF-2 and Ammonium Sulfate Precipitation.* The moist DEAE-cellulose filter cake from step 2 of the eIF-2–eIF-M1 preparative procedure is suspended in 55 ml of buffer A (400). The suspension is occasionally stirred over a period of 30 min and filtered with gentle suction through Whatman No. 1 filter paper on a Büchner funnel. Solid ammonium sulfate is added to the clear yellow filtrate to 80% saturation, and the precipitated protein is removed by centrifugation at 15,000 rpm in the SS-34 Sorvall rotor. The solid pellet, which contains the Co-eIF-2, can be stored at −15° for several months without loss of activity. Before assay or use in the next step, it is dissolved in approximately 25 ml of buffer A (50) and dialyzed overnight against the same buffer. A precipitate which forms at this low salt concentration is removed by centrifugation and discarded.

*Step 2. Chromatography on DEAE-Cellulose.* The protein solution from the preceding step is applied to a DEAE-cellulose column (2.5 × 35 cm) equilibrated in buffer A (50), and the column is washed with the same buffer. The major protein peak that washes from the column is retained and concentrated to approximately 20 ml by ultrafiltration using an Amicon cell with a PM-10 membrane. If the sample is to be stored at $-15°$, it should first be dialyzed against buffer C. Before use in the next step it is dialyzed against buffer A (100).

*Step 3. Chromatography on CM-Sephadex C-50.* The Co-eIF-2 preparation is applied to a CM-Sephadex C-50 column ($2.5 \times 30$ cm) equilibrated with buffer A (100). The column is washed successively with buffer A (100) and A (200). Each time the washing is continued until the amount of protein in the column effluent is close to zero. The protein fraction that elutes from the column with buffer A (100) contains some Co-eIF-2, but the bulk of the activity is in the A (200) eluate. This sample is concentrated to 4 ml by ultrafiltration as in step 2 and dialyzed against buffer C for storage at $-15°$.

### *Comments*

An unusual aspect of the purification procedure is that Co-eIF-2 in the crude extract binds to DEAE-cellulose, but in a later step it is not retained on the same anion exchanger even at a lower salt concentration. A possible reason for this is that Co-eIF-2 may be complexed with another protein in the crude extract, and it is the complex that binds to DEAE-cellulose. A subsequent manipulation, such as elution from the DEAE-cellulose with 400 m*M* potassium chloride, precipitation with ammonium sulfate, or dialysis against 50 m*M* KCl, could disrupt the complex, thereby altering the behavior of the cofactor. A distinct advantage of this preparative procedure is that Co-eIF-2 is readily separated from eIF-2. The buffer A (400) eluate is virtually free of eIF-2. Occasionally some preparations have a high GTP-independent binding of Met-$tRNA_i$ which is probably due to nonspecific binding.

A 2- to 3-fold stimulation of Met-$tRNA_i$ binding to eIF-2 by Co-eIF-2 is routinely obtained although a maximum of about 5-fold has been observed in a few cases. Addition of Co-eIF-2 to the standard reaction mixture for binding Met-$tRNA_i$ to 40 S ribosomes does not cause an increase in the amount of 40 S complex formed. The mechanism of action of Co-eIF-2 is not known, but it would be attractive to assume that it forms a complex with eIF-2 and that this complex is the species which combines with GTP and Met-$tRNA_i$. If this hypothesis is correct it should be possible to show a complete dependence upon Co-eIF-2 by very highly purified eIF-2. In the reticulocyte system[20] it has been shown that purified, but not crude, eIF-2 is stimulated by the cofactor.

[20] A. Dasgupta, A. Majumdar, A. D. George, and N. K. Gupta, *Biochem. Biophys. Res. Commun.* **71**, 1234 (1976).

# [30] Assay of the Binding of the Ternary Complex Met-tRNA$_f$·eIF-2·GTP to the 40 S Ribosomal Subunit by Sucrose Gradient and CsCl Gradient Analysis[1]

*By* EDGAR C. HENSHAW

An early step in polypeptide chain initiation is the binding of the ternary complex Met-tRNA$_f$·eIF-2·GTP to the native 40 S subunit.[2-15] This reaction can be assayed *in vitro* indirectly, using a Millipore filter assay,[16] or by direct visualization of the complexes on sucrose or CsCl gradients.[13] In the gradient techniques described here, [$^{35}$S]Met-tRNA$_f$, eIF-2, GTP (with or without a GTP-regenerating system), and 40 S subunits are incubated at 28°, chilled, and analyzed on gradients. KCl-washed 40 S subunits are used rather than native subunits, as the latter are poor substrates for ternary binding,[13,17] presumably because of other factors associated with them, such as Met-tRNA$_f$ deacylase.[18-20]

[1] Supported by NIH Grant CA-21663 and CA-11198.

[2] R. G. Crystal, D. A. Shafritz, P. M. Prichard, and W. F. Anderson, *Proc. Natl. Acad. Sci. U.S.A.* **68,** 1810 (1971).

[3] A. B. Burgess and B. Mach, *Nature (London), New Biol.* **233,** 209 (1971).

[4] C. Darnbrough, T. Hunt, and R. J. Jackson, *Biochem. Biophys. Res. Commun.* **48,** 1556 (1972).

[5] Y. C. Chen, C. L. Woodley, K. K. Bose, and N. K. Gupta, *Biochem. Biophys. Res. Commun.* **48,** 1 (1972).

[6] G. L. Dettman and W. M. Stanley, Jr., *Biochim. Biophys. Acta* **287,** 124 (1972).

[7] C. Baglioni, *Biochim. Biophys. Acta* **287,** 189 (1972).

[8] D. P. Weeks, D. P. S. Verma, S. N. Seal, and A. Marcus, *Nature (London)* **236,** 167 (1972).

[9] M. H. Schreier and T. Staehelin, *Nature (London), New Biol.* **242,** 35 (1973).

[10] D. H. Levin, D. Kyner, and G. Acs, *Proc. Natl. Acad. Sci. U.S.A.* **70,** 41 (1973).

[11] K. Balkow, S. Mizuno, and M. Rabinovitz, *Biochem. Biophys. Res. Commun.* **54,** 315 (1973).

[12] T. Ahern, J. Sampson, and J. E. Kay, *Nature (London)* **248,** 519 (1974).

[13] K. E. Smith and E. C. Henshaw, *Biochemistry* **14,** 1060 (1975).

[14] V. M. Pain and E. C. Henshaw, *Eur. J. Biochem.* **57,** 335 (1975).

[15] S. L. Adams, B. Safer, W. F. Anderson, and W. C. Merrick, *J. Biol. Chem.* **250,** 9083 (1975).

[16] N. K. Gupta, B. Chatterjee, and A. Majumdar, *Biochem. Biophys. Res. Commun.* **65,** 797 (1975).

[17] I. Sadnik, F. Herrera, J. McCuiston, H. A. Thompson, and K. Moldave, *Biochemistry* **14,** 5328 (1975).

[18] J. McCuiston, R. Parker, and K. Moldave, *Arch. Biochem. Biophys.* **172,** 387 (1976).

[19] J. Morrisey and B. Hardesty, *Arch. Biochem. Biophys.* **152,** 385 (1972).

[20] N. K. Gupta and R. J. Aerni, *Biochem. Biophys. Res. Commun.* **51,** 907 (1973).





ISBN 0-12-504180-2

Components and Solutions

Buffer A: triethanolamine–HCl, 20 m*M*; KCl, 25 m*M*; Mg acetate, 2 m*M*; EDTA, 0.1 m*M*; dithiothreitol, 0.5 m*M*, pH 7.0 (20°)

KCl-washed 40 S subunits. Subunits prepared by a number of techniques have been used successfully. We have used subunits released by 0.5 *M* KCl from monomers of Ehrlich cells treated with NaF,[21] and subunits made from rabbit reticulocytes by the method of Schreier and Staehelin.[22] Trachsel *et al.* have used subunits prepared by the latter method from mouse liver.[23]

[$^{35}$S]Met-tRNA$_f$ prepared by charging stripped rabbit or rat liver tRNA (Grand Island Biological) with [$^{35}$S]methionine by the method of Takeishi *et al.*,[24] slightly modified[25]. A 1 ml incubation contains 100 m*M* Na cacodylate, pH 7.0 (20°), 10 m*M* KCl, 2 m*M* ATP, 250 μCi of [$^{35}$S]methionine (about 0.5–1.5 μ*M* at specific activity of 150–500 Ci/mmol), 4 mg of tRNA, and 0.18 mg of *E. coli* mixed aminoacyl-tRNA synthetases. After incubation at 37° for 10 min, the reaction is chilled, followed by the addition of 50 μl of 2 *M* Na acetate (pH 4.35) and 2 ml of phenol saturated with water. The mixture is shaken for 5 min at 4°; the emulsion is broken by centrifugation; and the aqueous phase is removed. The phenol is reextracted with 0.8 ml of 50 m*M* Na acetate (pH 5.0), 5 m*M* Mg acetate. The combined aqueous phases are dialyzed for 6 hr against 0.5 M NaCl, 50 m*M* Na acetate (pH 5.0) and then for 17 hr against 20 m*M* Na acetate (pH 5.0).

Aminoacyl-tRNA synthetase prepared from *E. coli* by the method of RajBhandary and Ghosh[26] or obtained commercially (Grand Island Biological Co.). We have not found it necessary to separate the Met-tRNA$_f$ from the bulk of the (deacylated) tRNA. The specific activity of the [$^{35}$S]methionine is usually between 100 and 500 Ci/mmol, and the specific activity of the [$^{35}$S]Met-tRNA$_f$ is assumed to be the same. With a 1-ml incubation, yield is about 1.8 ml of dialyzed Met-tRNA$_f$ solution, usually containing 400,000–800,000 dpm per 5 μl.

eIF-2 at any stage of purification, from a crude ribosomal KCl wash to

[21] E. C. Henshaw, D. G. Guiney, and C. A. Hirsch, *J. Biol. Chem.* **248,** 4367 (1973).

[22] M. H. Schreier and T. Staehelin, *J. Mol. Biol.* **73,** 329 (1973).

[23] H. Trachsel, B. Erni, M. H. Schreier, and T. Staehelin, *J. Mol. Biol.* **116,** 755 (1977).

[24] K. Takeishi, T. Ukita, and S. Nishimura, *J. Biol. Chem.* **243,** 5761 (1968).

[25] N. K. Gupta, N. K. Chatterjee, C. L. Woodley, and K. K. Bose, *J. Biol. Chem.* **246,** 7460 (1971).

[26] U. L. RajBhandary and H. P. Ghosh, *J. Biol. Chem.* **244,** 1104 (1969).

the homogeneous, purified factor.[27-30] It is likely that in the crude preparations components in addition to eIF-2 affect the reaction.

### Incubation

The incubations, 1.0 ml for CsCl gradient analysis and 0.5 ml for sucrose gradient analysis, contain 50 m*M* triethanolamine-HCl, pH 7.2 (20°), 100 m*M* KCl, 2 m*M* Mg acetate, 1 m*M* dithiothreitol, 0.2 m*M* GTP, and 40 $\mu$g of poly(A,U,G) per milliliter. Both incubations, 1.0 ml and 0.5 ml contain 0.6 $A_{260}$ unit of KCl-washed 40 S ribosomal subunits (about 40 pmol), 1.5 $\mu$Ci of [$^{35}$S]Met-tRNA$_f$(3–15 pmol), and eIF-2 preparation. If purified eIF-2 is used, 0.1 $\mu$g (about 0.7 pmol) is sufficient. For a crude ribosomal KCl wash, 50–200 $\mu$g of protein are used. Phosphoenolpyruvate, 3.2 m*M*, and pyruvate kinase, 40 $\mu$g/ml, may be included. The [$^{35}$S]Met-tRNA$_f$, neutralized immediately before use, is added last, and the assay is incubated at 28° for 15 min.

### Gradient Analysis

*Sucrose Gradient Analysis*. The 0.5-ml incubations are terminated by chilling on ice, 0.06 ml of 37° (w/w) formaldehyde (pH 7.0) is added, and the mixture is layered directly on 12-ml linear 20 to 40% sucrose gradients made up in buffer A. Centrifugation is at 41,000 rpm for 5 hr at 3° in the SW 41 rotor. Absorbance is monitored, and fractions are collected on ice. One drop of 0.5% bovine serum albumin is added to each fraction as carrier, followed by 5 volumes of cold 5% TCA. The precipitate is collected and washed on Whatman GF/C glass fiber filters, which are dried and counted in a scintillation counter. Since the Met-tRNA$_f$ is unstable, fractions should be kept cold and processing should be done expeditiously.

Alternatively, since ribosomal subunits bind to membrane filters, the fractions can be filtered directly through cellulose nitrate-cellulose acetate filters (type HAWP, Millipore Filter Corporation, Bedford, Massachusetts 01730). Recovery has been excellent, and the procedure is rapid.

*CsCl Gradient Analysis*. The 1.0 ml incubations are terminated by the addition of 6 ml of cold buffer containing 10 m*M* morpholinosulfonic acid–KOH, pH 7.0 (20°), 25 m*M* KCl, 2 m*M* Mg acetate, and 4% (w/v) formaldehyde. The 7-ml fixed preparation is layered on 5 ml of CsCl solution, density 1.51 g/cm$^3$ as described elsewhere.[31] Gradients are moni-

[27] B. Safer, W. F. Anderson, and W. C. Merrick, *J. Biol. Chem.* **250,** 9067 (1975).
[28] M. H. Schreier, B. Erni, and T. Staehelin, *J. Mol. Biol.* **116,** 727 (1977).
[29] L. M. Cashion and W. M. Stanley, Jr., *Biochim. Biophys. Acta* **324,** 410 (1973).
[30] A. Barrieux and M. G. Rosenfeld, *J. Biol. Chem.* **252,** 3843 (1977).
[31] E. C. Henshaw, this series, Vol. 59 [33], p. 410–421.

tored for absorbance and density, and cold TCA-precipitable radioactivity is measured as described above for sucrose gradients.

## Comments

1. The major advantage of the sucrose gradient technique is speed. The CsCl technique gives added information concerning the density of the subunit complexes formed. For the CsCl technique fixation with formaldehyde is necessary to prevent dissociation of the bound complexes by high CsCl concentrations. Isotope is lost from the labeled subunit complex in a time-dependent manner, at a rate of about 10% per 16 hr.[13] This is presumably due to hydrolysis of [$^{35}$S]Met-tRNA$_f$. We also "fix" the preparations before analysis on sucrose gradients, but even so there is usually better recovery of isotope in the CsCl technique (Fig. 1A,B).

2. The salt-washed 40 S subunits tend to form dimers, which appear as a separate peak on sucrose gradient analyses (Fig. 1A). These are dissociated by higher KCl concentrations, or by proteins contained in crude eIF-2 preparations, which bind to them.

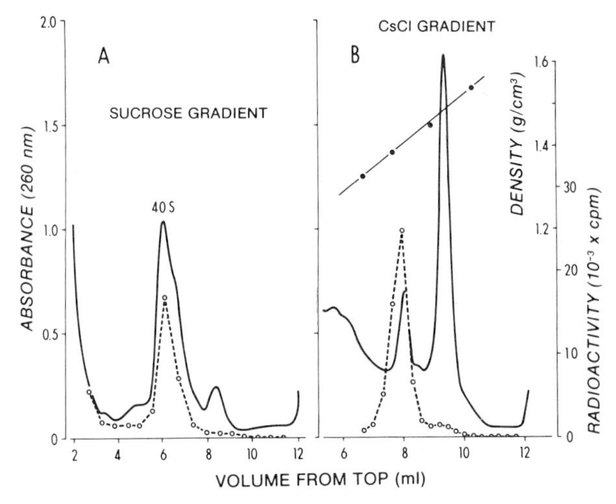


FIG. 1. Sucrose gradient and CsCl gradient analyses of the same reaction mixture. Incubations (0.5 ml for sucrose gradient and 1.0 ml for CsCl gradient) containing 0.6 $A_{260}$ unit of 40 S ribosomal subunits, 8.5 pmol of [$^{35}$S]Met-tRNA$_f$ (210 Ci/mmol), and 85 $\mu$g of the eIF-2 preparation were fixed and analyzed as described. The eIF-2 preparation was obtained from an Ehrlich ascites tumor cell ribosomal KCl wash by ammonium sulfate fractionation (40–70% saturation) and elution from DEAE-cellulose with 0.1–0.3 *M* KCl, as described by M. H. Schreier and T. Staehelin [*J. Mol. Biol.* **73,** 329 (1973)] and it contained eIF-3. (A) Sucrose gradient analysis. (B) CsCl gradient analysis.

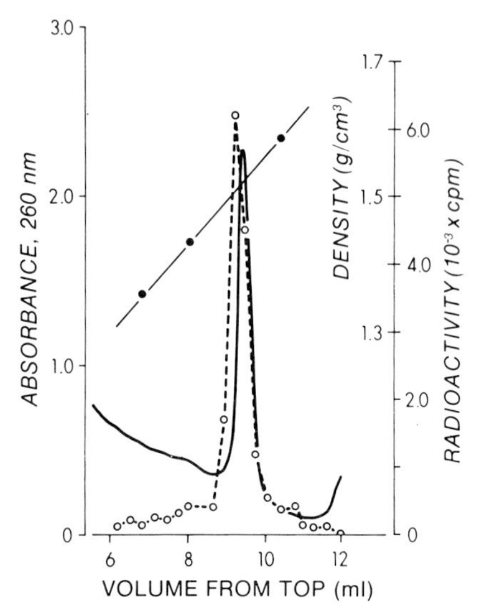


FIG. 2. CsCl gradient analysis of a binding reaction using purified eIF-2. An incubation containing 0.5 $A_{260}$ unit of 40 S subunits, 3.3 pmol of [$^{35}$S]Met-tRNA$_f$ (100 Ci/mmol), 0.8 $\mu$g of purified eIF-2, and phosphoenolpyruvate plus pyruvate kinase, was fixed and analyzed as described. The eIF-2 was purified from an Ehrlich cell ribosomal KCl wash through ammonium sulfate, phosphocellulose, and two DEAE-cellulose steps.

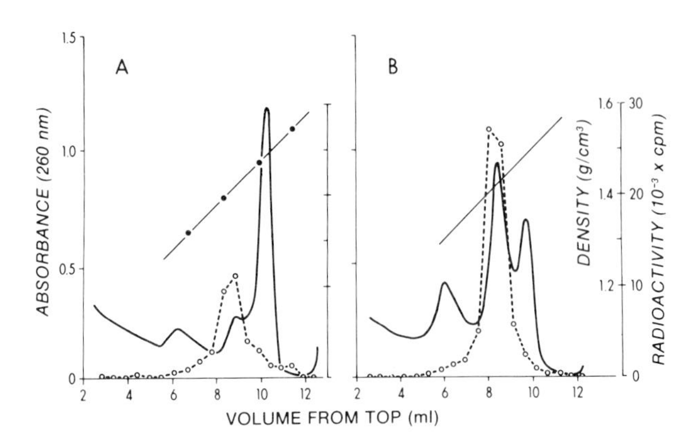


FIG. 3. CsCl gradient analyses of incubations containing different amounts of a crude eIF-2 preparation. Incubations containing 0.6 $A_{260}$ 40 S subunit, 10 pmol of [$^{35}$S]Met-tRNA$_f$ (170 Ci/mmol), and a crude ribosomal KCl wash prepared from Ehrlich cells by the method of D. A. Shafritz and W. F. Anderson [*J. Biol. Chem.* **245,** 5553 (1970)] were fixed and analyzed as described. (A) 45 $\mu$g of KCl wash. (B) 180 $\mu$g of KCl wash.

3. The buoyant density of salt-washed 40 S subunits is about 1.51 g/cm$^3$, and when highly purified eIF-2 is used, the binding of the ternary complex to the subunit adds sufficient protein to lower its buoyant density only slightly (Fig. 2). When a crude KCl wash is used as the source of eIF-2, other proteins in the wash bind to the 40 S subunits and lower their density. These proteins appear to be present in roughly the amounts found on native 40 S subunits, so that two predominant subunit populations are seen, of buoyant densities 1.49 and 1.40 g/cm$^3$. The shift to density 1.40 g/cm$^3$ is presumably due in part to the binding of the very large initiation factor eIF-3.[8] If increased amounts of crude KCl wash are added, more of the subunits are converted to 1.40 g/cm$^3$ (Fig. 3, A and B). The ternary complex binds preferentially to the subunits with the large amount of added protein (Fig. 1B). In the assay described here, [$^{35}$S]Met-tRNA$_f$ of very high specific activity is used, and 40 S subunits are in large molar excess. The formation of bound complexes increases linearly with added Met-tRNA$_f$. Binding is also proportional to the amount of eIF-2 added (Fig. 3, A and B).

4. Although RNA template is not required for 40 S-ternary complex formation, we add poly(A U G) because stability of the complex may be enhanced.[32,33]

5. Inclusion of phosphoenol pyruvate and pyruvate kinase (Henshaw, unpublished) or ATP and nucleoside diphosphate kinase[34] as a GTP generating system improves complex formation 3-fold or more under the conditions described. This is at least in part because commercial GTP often contains 3–10% GDP. However, introduction of only partially purified kinase preparations might complicate certain types of experiments.

[32] K. E. Smith, A. C. Richards, and H. R. V. Arnstein, *Eur. J. Biochem.* **62,** 243 (1976).
[33] B. Chatterjee, A. Dasgupta, S. Palmieri, and N. K. Gupta, *J. Biol. Chem.* **251,** 6379 (1976).
[34] G. M. Walton and G. N. Gill, *Biochim. Biophys. Acta* **390,** 231 (1975).

# [31] Prokaryotic Ribosome Binding Sites

*By* JOAN ARGETSINGER STEITZ

Takanami, Yan, and Jukes[1] first demonstrated that ribosomes could protect associated segments of natural messenger RNA (mRNA) from ribonuclease digestion in 1965. A few years later, three other groups adapted their basic procedure to isolate protein synthesis initiator regions

[1] M. Takanami, Y. Yan, and T. H. Jukes, *J. Mol. Biol.* **12,** 761 (1965).



ISBN 0-12-504180-2

from the genomes of the small RNA-containing coliphages.[2-4] Sequence analysis of these messenger fragments showed that, in *in vitro* reactions containing all the components required for polypeptide chain initiation, *E. coli* ribosomes are quite selective; they interact with the mRNA almost exclusively at the beginnings of the three phage cistrons, where they protect regions of about 35 nucleotides from nuclease digestion.

Since then, the same technique has been utilized to isolate protein synthesis initiator regions from a variety of coliphage- and *E. coli*-specified mRNAs. Ribosomes and initiation factors from both *E. coli* and other bacteria[5] have been used. Although not all protein synthesis initiator regions can be detected by ribosome protection, so far there is no documented case where *E. coli* ribosomes bind to *native E. coli* or coliphage mRNA at a site that does not direct polypeptide chain initiation *in vivo* and/or *in vitro*. On the other hand, spurious binding can arise when using modified mRNAs—including fragmented,[6] thermally or chemically unfolded,[7] or mutated molecules.[8] Likewise, mixed *in vitro* systems containing components from several different bacterial species can yield protection of sites unable to initiate polypeptide chains.[5]

Other uses of prokaryotic ribosome binding sites have included: (1) identification of a protein synthesis "initiator region" in a single-stranded DNA[9]; (2) mapping of initiator regions on single mRNA molecules[10-12]; (3) analysis of the relative contributions to initiation specificity made by initiation factors and other ribosomal components[5,7,13-15]; (4) determination of the position of the ribosome relative to the mRNA at various stages of the

[2] J. A. Steitz, *Nature (London)* **224,** 957 (1969).

[3] J. Hindley and D. H. Staples, *Nature (London)* **224,** 964 (1969).

[4] S. L. Gupta, J. Chen, L. Schaefer, P. Lengyel, and S. M. Weissman, *Biochem. Biophys. Res. Commun.* **39,** 883 (1970).

[5] J. A. Steitz, *J. Mol. Biol.* **73,** 1 (1973).

[6] J. A. Steitz, *Proc. Natl. Acad. Sci. U.S.A.* **70,** 2605 (1973).

[7] H. Berissi, Y. Groner, and M. Revel, *Nature (London), New Biol.* **234,** 44 (1971).

[8] J. J. Dunn, E. Buzash-Pollert, and F. W. Studier, *Proc. Natl. Acad. Sci. U.S.A.* **75,** 2741 (1978).

[9] H. D. Robertson, B. G. Barrell, H. L. Weith, and J. E. Donelson, *Nature (London), New Biol.* **241,** 38 (1973).

[10] P. G. N. Jeppesen, J. A. Steitz, R. F. Gesteland, and P. F. Spahr, *Nature (London)* **226,** 230 (1970).

[11] D. H. Staples, J. Hindley, M. A. Billeter, and C. Weissmann, *Nature (London), New Biol.* **234,** 202 (1971).

[12] J. A. Steitz and R. A. Bryan, *J. Mol. Biol.* **114,** 527 (1977).

[13] J. A. Steitz, S. K. Dube, and P. S. Rudland, *Nature (London)* **226,** 824 (1970).

[14] M. Yoshida and P. S. Rudland, *J. Mol. Biol.* **68,** 465 (1972).

[15] J. A. Steitz, A. J. Wahba, M. Laughrea, and P. B. Moore, *Nucl. Acids Res.* **4,** 1 (1977).

translation cycle[16-18]; and (5) probing the mechanism of initiation.[19] Most recently, ribosome protection methods for obtaining initiator regions from mRNAs in eukaryotic translation systems have been developed.[20,21]

This article describes the general experimental procedures for isolating ribosome-protected regions from *in vivo* $^{32}P$-labeled messenger RNAs using initiation systems containing bacterial ribosomes. Techniques for utilizing *in vitro* synthesized mRNA or complete cell-free translation systems blocked at initiation are detailed in the following article by Model and Robertson.[22] Then, characterization of initiation sites bound by eukaryotic ribosomes is covered by Kozak and Shatkin.[23] A list of prokaryotic ribosome-binding site sequences (complete as of June, 1977) has recently been presented and discussed elsewhere.[24]

## Strategy

In order to isolate a ribosome binding site from a $^{32}P$-labeled mRNA in a form suitable for further analysis, the experimentalist always wishes to obtain the highest possible yield of a specific region(s) from the lowest number of input RNA counts. To achieve these goals, I have found that the following guidelines apply.

1. In the initiation reaction, use the highest possible concentrations of all components. This means ribosomes at about 200 $A_{260}$ units/ml (about 3 $\mu M$) and relative molar ratios of 3:1:3 for ribosomes: mRNA: fmet-tRNA (assuming 5% of total *E. coli* tRNA to be $tRNA_f^{Met}$).

2. Use the most active ribosomes possible. This generally means ribosomes prepared by washing in low salt (buffer A below) so that initiation factors are retained, rather than having to be added as a separate component to the initiation reaction.

3. Once initiation complexes have been formed, carry out all steps leading to isolation of the protected mRNA fragment(s) in the simplest possible manner. That is, do not isolate the 70 S complexes first, but trim with nuclease directly in the initiation reaction. Next, fractionate the

[16] E. Kuechler and A. Rich, *Nature (London)* **225,** 920 (1970).

[17] E. Kuechler, *Nature (London), New Biol.* **234,** 216 (1971).

[18] S. L. Gupta, J. Waterson, M. L. Sopori, S. M. Weissman, and P. Lengyel, *Biochemistry* **10,** 4410 (1971).

[19] J. A. Steitz and K. Jakes, *Proc. Natl. Acad. Sci. U.S.A.* **72,** 4734 (1975).

[20] S. Legon, H. D. Robertson, and W. Prensky, *J. Mol. Biol.* **106,** 23 (1976).

[21] M. Kozak and A. J. Shatkin, *J. Biol. Chem.* **251,** 4259 (1976).

[22] P. Model and H. D. Robertson, Vol. LX [28].

[23] M. Kozak and A. J. Shatkin, this volume [32].

[24] J. A. Steitz, *in* "Biological Regulation and Development" (R. Goldberger, ed.). Plenum, New York, 1978.

ribosome-bound from the released mRNA fragments by the fastest method that provides adequate resolution (i.e., on sucrose gradients). Finally, isolate the mRNA from the pooled 70 S ribosome peak by direct phenol extraction. Proceed with analysis.

Of course, all the above rules can be (and have been) successfully broken. To accomplish certain goals, it is necessary to do so. In fact, the ribosome binding system is surprisingly flexible and can easily tolerate at least 10-fold variations in the relative concentrations of ribosomes, mRNA, tRNA, and initiation factors. (Only the efficiency of 70 S initiation complex formation will thereby be affected, and the time required to achieve maximal mRNA binding may increase.) Even changes in $Mg^{2+}$ concentration (from 5 m*M* to 12 m*M* have been observed *not* to alter the specificity of ribosome protection of RNA bacteriophage messengers. At Mg concentrations lower than 5 m*M*, yields of 70 S initiation complexes are reduced.

The one variable, however, which does seem to be crucial to successful ribosome binding site isolation, is the amount of nuclease used to trim the preformed 70 S initiation complex. Too much nuclease causes degradation of the protected fragments during isolation. Too little may yield mRNA fragments too long to be analyzed easily.

## The Basic Binding System

### *Buffers*

A. Low-salt ribosome buffer[25]: 10 m*M* Tris·HCl (pH 7.4); 10 m*M* $MgCl_2$; 60 m*M* $NH_4Cl$; 6 m*M* β-mercaptoethanol
B. High-salt ribosome buffer[25]: 10 m*M* Tris·HCl (pH 7.4); 10 m*M* $MgCl_2$; 2.0 *M* $NH_4Cl$;6 m*M* β-mercaptoethanol
C. 5X Charging buffer[26]: 40 ml of $H_2O$; 10 ml of 1 *M* Tris·HCl (pH 8.5); 4 ml of 2 *M* KCl; 3.0 ml of 1 *M* Mg acetate
D. 10X Initiation buffer: 1 *M* Tris·HCl (pH 7.4); 0.5 *M* $NH_4Cl$; 50 m*M* Mg acetate
E. Sucrose gradient buffer[27]: 50 m*M* Tris·HCl (pH 7.8); 50 m*M* $NH_4Cl$; 5 m*M* Mg acetate

### *Ribosomes*

Ribosomes from *E. coli* or other bacteria are prepared basically as described by Anderson *et al.*[25] Cells are grown to mid-log phase, quick-

[25] J. S. Anderson, M. S. Bretscher, B. F. C. Clark, and K. A. Marcker, *Nature (London)* **215**, 490 (1967).

[26] B. F. C. Clark and K. A. Marcker, *J. Mol. Biol.* **17**, 394 (1966).

[27] M. Kondo, G. Eggerston, J. Eisenstadt, and P. Lengyel, *Nature (London)* **220**, 368 (1968).

cooled, pelleted, and frozen. All subsequent steps are carried out at 4°–6°. Using a mortar and pestle, 10 g of frozen cells are ground with 20 g of alumina to a complete paste (10–15 min). Then 20 ml of buffer A plus 20 $\mu$l of DNase (Worthington, at 5 mg/ml) are added. The slurry is spun in the Sorvall SS 34 rotor for 20 min at 16,000 rpm to eliminate unbroken cells and alumina. Next, the supernatant is centrifuged in the Spinco 50 Ti rotor for 2.5 hr at 50,000 rpm or in the 65 Ti rotor for 90 min at 55,000 rpm. This high speed supernatant is saved (see 4 below). The ribosome pellet is resuspended in 2–4 ml of buffer A per tube by homogenization with a Dounce homogenizer. After clarification (10 min at 12,000 rpm in the Sorvall SS 34 rotor) the ribosomes are repelleted at high speed as above.

If low salt-washed ribosomes are desired, the above clarification and high-speed pelleting steps are repeated once more. The ribosomes are finally resuspended in buffer A at 500–2000 $A_{260\ nm}$ units/ml, clarified and stored in small aliquots at −70°. They appear to be stable for years.

If high salt-washed ribosomes lacking initiation factors are desired, the ribosomal pellet from the second high-speed spin is resuspended in 2 ml of buffer B per tube by stirring with a glass rod (lumps will remain). After standing overnight at 4°, the volume is made up to 10 ml per tube with buffer B. The ribosomes are homogenized, clarified as above, and pelleted by spinning at 65,000 rpm for 2 hr or at 50,000 rpm for 3 hr. The pellet is resuspended in buffer A, clarified, and stored as described for low salt-washed ribosomes.

Crude initiation factors are recovered by adding 0.52 g/ml solid $(NH_4)_2SO_4$ to the buffer B high-speed supernatant and stirring for 0.5 hr in the cold. Precipitated protein is pelleted by centrifugation for 30 min at 15,000 rpm in the Sorvall SS 34 rotor and resuspended in 0.5 ml per 10 g of cells in 20 m*M* Tris·HCl (pH 7.8), 2 m*M* Mg acetate, 0.18 *M* $NH_4Cl$. After overnight dialysis against several liters of the same buffer, the initiation factor fraction is titrated for optimal fMet-tRNA binding activity[25] using AUG or RNA bacteriophage RNA and high salt-washed ribosomes. Crude initiation factors are stored in small aliquots at −70°.

Alternatively, equally active ribosomes can be prepared by other methods; see Gold and Schweiger[28] or Model and Robertson.[22] Purified initiation factors[29] can also be utilized, but again must be titrated against high salt-washed ribosomes to determine the concentration that will yield optimal 70 S initiation complex formation.

[28] L. M. Gold and M. Schweiger, this series, Vol. 20, p. 537.

[29] A. J. Wahba and M. J. Miller, this series, Vol. 30, p. 3.

*Messenger RNA*

$^{32}P$-labeled mRNA at specific activities of 1 to 5 × $10^6$ cpm/μg can easily be prepared from the small RNA bacteriophages. *Escherichia coli* strain S26 (for R17, MS2, or other group I phage) or M27 (specific for Qβ) is grown at 37° to 2 × $10^8$ cells/ml in 100 ml of 1% Difco Bacto-peptone, 0.5% NaCl, 0.1% glucose, 0.1 *M* Tris·HCl (pH 7.5), 2 m*M* $CaCl_2$. The cells are infected with phage at a multiplicity of 10, and within 5 min $^{32}PO_4$ is added at 0.1–0.2 mCi/ml. After vigorous aeration for 3–5 hr, a few drops of chloroform are added followed by 33 g $(NH_4)_2SO_4$ per 100 ml. Titers of about 5 × $10^{11}$ R17/ml and 1 × $10^{11}$ Qβ/ml are routinely obtained. All subsequent steps are carried out at 4°–6°. The phage are recovered after overnight precipitation by pelleting at 13,000 rpm for 30 min in the Sorvall SS 34 rotor. The pellets must be well resuspended in 10 ml SSC [0.15 *M* NaCl, 15 m*M* Na citrate (pH 7.0)] per 100 ml of lysate by occasional vigorous vortexing over a period of 2–3 hr. After clarification by centrifugation at 13,000 rpm for 10 min in the SS 34 rotor, the phage are pelleted from the supernatant at 37,000 rpm for 2 hr in the Spinco 50 Ti rotor. One milliliter of SSC is added, and the phage are allowed to resuspend overnight. Next 2.6 g of CsCl are added to the phage in a 5-ml cellulose nitrate tube for the SW 50.1 Spinco rotor, and the volume is made up to 4.5 ml with SSC. The phage are banded by spinning at 37,000 rpm for a minimum of 16 hr, and the visible phage band is removed. After dilution to 10 ml with SSC, the purified phage are pelleted for 4–6 hr at 37,000–40,000 rpm in the 50 Ti rotor and resuspended in 1 ml of SSC. Then the phage solution is made 1% in SDS, the RNA is extracted with phenol (saturated with SSC), and ethanol-precipitated. To remove all SSC, the RNA is finally reprecipitated 2 times from 0.2 *M* Tris. The yield from 100 ml of lysate is usually 20–50 $A_{260\ nm}$ units of pure RNA. To avoid autoradiolysis, the RNA is stored at about 3–5 $A_{260\ nm}$/ml in 30% ethanol in water at −20°. Nonetheless it becomes severely fragmented within a week or so (see Steitz[30]).

Relatively stable mRNA species labeled *in vivo* with $^{32}P$ can also be isolated from several other types of phage-infected cells. Methods for obtaining T7 early mRNAs suitable for use in ribosome binding studies are described fully by Kramer *et al.*[31] and Steitz and Bryan.[12] Note here, however, that unlabeled carrier RNA phage RNA must be added to the T7 mRNA in order to obtain specific ribosome-protected regions[12]; see also Model and Robertson[22] on this point.

[30] J. A. Steitz, *Nature (London), New Biol.* **236,** 71 (1972).

[31] R. A. Kramer, M. Rosenberg, and J. A. Steitz, *J. Mol. Biol.* **89,** 767 (1974).

*Transfer RNA*

Whereas Model and Robertson[22] describe the use of purified fMet-$tRNA_f^{Met}$ in their initiation reactions, I have always used unfractionated *E. coli* tRNAs. Stripped tRNAs are prepared in the conventional manner[32] with the addition of a DEAE-cellulose chromatography step.[33] They are then charged with all 20 amino acids and formylated in the same reaction mixture. It contains, per ml: buffer C (5X charging buffer), 200 $\mu$l; $H_2O$, 390 $\mu$l; tRNA at 25 mg/ml, 100 $\mu$l, ATP (50 m$M$ in 20 m$M$ Tris·HCl, pH 7.5), 30 $\mu$l; phosphoenol pyruvate (0.5 $M$), 20 $\mu$l; $\beta$-mercaptoethanol (0.14 $M$), 1 $\mu$l; L-methionine (5 m$M$), 5 $\mu$l; 19 amino acids (−Met, each at 5 m$M$), 5 $\mu$l; supernatant [saved from the first high speed ribosome pelleting (Section 2 above) and dialyzed against 20 m$M$ Tris·HCl (pH 7.5), 2 m$M$ $\beta$-mercaptoethanol before storage in aliquots at −70°, 250 $\mu$l.

After 2 min of incubation at 37°, 50 $\mu$l of formyltetrahydrofolate solution are added. $N^{10}$-Formylfolate is synthesized from folic acid as described by Jones *et al.*[34] and is reduced by bubbling $H_2$ for 2 hr through 1 ml of 20 m$M$ phosphate buffer (pH 7.0) containing 200 mg of formylfolate and 100 mg of 5% rhodium on alumina; the formyltetrahydrofolate solution can be stored frozen for months.

The charging formylation reaction is incubated for another 13 min at 37°. The tRNA is then phenol extracted and run over a 5-ml column of Sephadex G-50 (medium) in 20 m$M$ Na acetate to remove triphosphates. After concentration by ethanol precipitation, it is stored at −20° in 20 m$M$ Tris·HCl (pH 7.5). Usually a charging reaction containing isotopically labeled methionine is run in parallel to check the charging/formylation.

*Formation of 70 S Initiation Complexes*

Initiation can be performed in 15 $\mu$l (or more) depending on whether the ribosome binding sites are being isolated for analytical or preparative reasons. In a typical 50-$\mu$l reaction, components are added to a small glass test tube in the following order: Buffer D (10X initiation buffer), 5 $\mu$l; 5 m$M$ GTP in 10 m$M$ Tris·HCl (pH 7.5) = 0.2 m$M$ final concentration, 2 $\mu$l; charged formylated tRNA at 200 $A_{260}$ nm units/ml in 20 m$M$ Tris·HCl (pH 7.5) = total of 3 $A_{260}$ nm units, 15 $\mu$l; ribosomes at 2000 $A_{260}$ nm units/ml in buffer A = total of 10 $A_{260}$ nm units, 5 $\mu$l; mRNA (2 $A_{260}$ nm units), which has been ethanol precipitated in a siliconized tube, pelleted, dried, sus-

[32] G. Zubay, *in* "Procedure in Nucleic Acid Research" (G. L. Cantoni and D. R. Davies, eds.), Vol. 1, p. 455. Harper, New York, 1966.

[33] R. W. Holley, J. Apgar, B. P. Doctor, J. Farrow, M. A. Marini, and S. H. Merrill, *J. Biol. Chem.* **236,** 200 (1961).

[34] K. M. Jones, J. R. Guest, and D. D. Woods, *Biochem. J.* **79,** 566 (1961).

pended in 0.1 m*M* EDTA and heated at 37° for 5–10 min before addition to the initiation reaction. (This step increases the efficiency without altering the specificity of ribosome binding.)

After gentle agitation (do *not* vortex), the reaction mixture is incubated at 37°–38° for 7–10 min with occasional shaking. At the above concentration of components, the formation of 70 S initiation complexes is complete in 2–5 min.

*Special Notes:*

1. If initiation complexes are to be formed at temperatures higher than 42°, $NH_4$ cacodylate (pH 7.3) must be substituted for Tris · HCl.
2. If separate initiation factors are being used, they should be added after the ribosomes.
3. If the total volume of the reaction mixture becomes a problem, the mRNA pellet may be resuspended in the charged formylated tRNA.

*Nuclease Trimming*

After initiation complexes are formed, the reaction tube is plunged into ice and pancreatic or T1 ribonuclease is added from a concentrated enzyme stock (made up in 10 m*M* Tris · HCl (pH 7.5), 2 m*M* EDTA). 3 μl Pancreatic RNase (3 μl at 0.1 mg/ml) brings the final concentration in a 50-μl initiation reaction to 6 μl/ml; or 3 μl of T1 RNase at 1.0 mg/ml yields 60 μg/ml final concentration. After gentle mixing the reaction tube is incubated at room temperature (18°–23°) for 15 min. Simply because of its enzymic specificity, T1 RNase yields larger protected fragments than pancreatic ribonuclease.

*Sucrose Gradient Fractionation*

The nuclease-treated reaction mixture is next cooled to 0° on ice, and ice-cold buffer E (sucrose gradient buffer) is added to bring the volume to 150–200 μl per 5-ml gradient. Dilution is not essential; when isolating ribosome binding sites on a preparative scale, 170 μl of nuclease-treated reaction mixture may be loaded directly. Hence, as many as 30 $A_{260}$ units of ribosomes (or as few as 2–3 $A_{260}$ units) can be successfully fractionated on one 5-ml sucrose gradient.

Sucrose gradients are linear 5 to 20% (w/v) gradients in buffer E, which is sometimes made containing 10 m*M* $Mg^{2+}$. These are prepared in cellulose nitrate tubes for the SW 50.1 Spinco rotor and cooled to 4° before use. The nuclease-treated initiation reactions are fractionated by spinning at 4° for 105 min at 39,000 rpm. Use of higher centrifugation speeds can lead to pressure-induced dissociation of the 70 S initiation complexes (and consequent loss of ribosome-protected material).

After centrifugation, the bottom of the gradient tube is pierced with a 21-gauge (1 inch) disposable hypodermic needle, and 3-drop fractions are collected directly through the needle into siliconized glass tubes or scintillation vial inserts kept cool on ice. About 25 fractions are obtained. (They may be stored overnight at $-20°$ at this point.) The radioactivity profile is assayed either by counting small aliquots in Bray's solution or by direct Cerenkov counting of the entire sample in a scintillation counter. Yields of protected mRNA in the 70 S region of the gradient can be as high as 80% of the theoretical value, assuming the average ribosome binding site to be about 30 nucleotides long.

*Preparation of Sites for Sequence Analysis*

Fractions containing the trimmed 70 S initiation complexes, as designated in Fig. 1, are pooled into a siliconized 15- or 30-ml Corex centrifuge tube. The material may be processed further in one of three ways. Methods

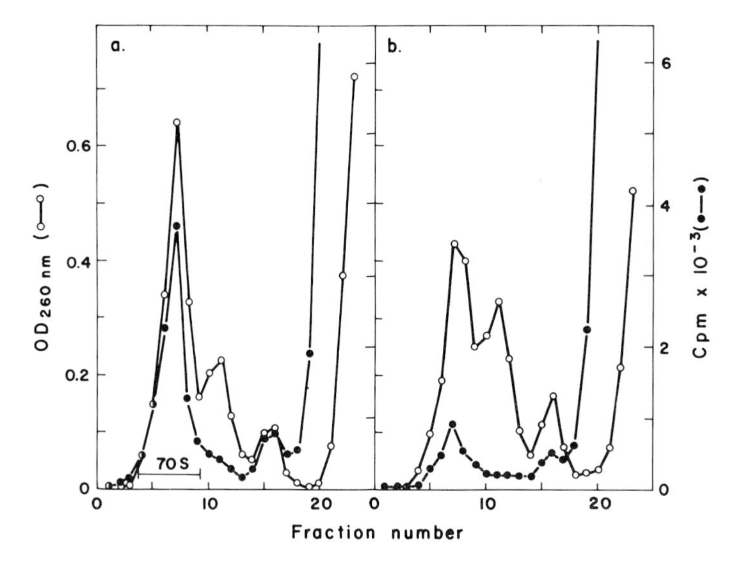


FIG. 1. Fractionation of ribosome binding sites from $^{32}$P-labeled Q$\beta$ RNA on sucrose gradients. All methods are described in the text. Reaction mixture in (a) contained charged formylated tRNA, and reaction mixture in (b), uncharged tRNA. Complexes were trimmed with pancreatic ribonuclease. Centrifugation was from right to left. A small aliquot of each fraction was counted. In (a), the 70 S fractions pooled to extract mRNA fragments for analysis are indicated. Note the small radioactivity peak at 30 S; when subjected to fingerprint analysis, this material always proves to be identical to the 70 S ribosome-protected RNA [J. A. Steitz, unpublished observations; S. Legon, P. Model, and H. D. Robertson, *Proc. Natl. Acad. Sci. U.S.A.* **74,** 2692 (1977)].

1 and 2 are suitable only if the total amount of RNA present can be tolerated in subsequent analytical procedures (i.e., no greater than 2 $A_{260}$ nm units per fingerprint). On the other hand, if the pooled fractions contain less than 1 $A_{260}$ unit of total RNA, cold carrier RNA must be added to a level of 1 $A_{260}$ unit/ml before proceeding with method 1 or 2.

*Method 1.* Tris·HCl (pH 7.5) at 0.2 *M* and 2.5 volumes of ethanol are added, and the RNA (and protein) is precipitated at −20° for 2–5 hr. (Longer precipitation may result in undesired nuclease degradation of the ribosome-protected fragments.) After spinning for 10 min at 13,000 cpm, or 25 min at 8000 rpm in the SS 34 Sorvall rotor, the pellet is resuspended in 1 ml of 0.2 *M* Tris·HCl (pH 7.5), 0.1% SDS. 1 ml of SSC-saturated phenol is added, and the mixture is vortexed for 2 min before separating the phases by centrifugation. A second phenol extraction may be performed. Finally, the isolated RNA is precipitated with 2.5 volumes of ethanol and reprecipitated from 1.0 ml of 0.2 *M* Tris·HCl (pH 7.5) before analysis.

*Method 2.* The pooled sucrose gradient fractions may be phenol extracted directly by addition of Tris·HCl (pH 7.5) to 0.2 *M*, SDS to 0.1%, and an equal volume of phenol. Two ethanol precipitations should follow.

*Method 3.* If the mRNA was of low specific activity or if, for other reasons, the ribosome binding sites must be freed of most of the cold RNA derived from the 70 S region of the gradient, the following technique can be used. The pooled 70 S fractions are made 8 *M* in urea and applied directly to a 0.5–1 ml DEAE-cellulose (Whatman DE-52) column, prepared in a Pasteur pipette and equilibrated with a buffer containing 8 *M* urea, 50 m*M* Tris·HCl (pH 7.4), 0.1 *M* NaCl, and 5 m*M* $MgSO_4$. After washing with several column volumes of buffer, about 90% of the counts can be eluted in 1–2 ml of the above buffer made 0.5 *M* in NaCl. The RNA is then ethanol-precipitated, phenol-extracted, and reprecipitated twice as in method 1 above. Such RNA can be yet further purified from contaminating tRNA and rRNA fragments by preparative electrophoresis on polyacrylamide gels (see Steitz and Steege[35]).

## Sequence Analysis of Ribosome Binding Sites

$^{32}P$-labeled ribosome-protected regions are analyzed using the elegant RNA sequencing methods originally developed by Sanger and his colleagues,[36] incorporating later refinements in technology.[37] These can be

[35] J. A. Steitz and D. A. Steege, *J. Mol. Biol.* **114,** 545 (1977).

[36] B. G. Barrell, *in* "Procedures in Nucleic Acid Research" (G. L. Cantoni and D. R. Davies, eds.), Vol. 2, p. 751. Harper, New York, 1971.

[37] C. Squires, F. Lee, K. Bertrand, C. L. Squires, M. J. Bronson, and C. Yanofsky, *J. Mol. Biol.* **103,** 351 (1976).

applied directly to material isolated from the trimmed 70 S initiation complexes. Alternatively, fractionation of different mRNA regions from one another or of different size variants of the same region can be undertaken before sequence analysis. Methods for fractionation of sites by polyacrylamide gel electrophoresis or by homochromatography are described elsewhere (see Model and Robertson[22]). Obviously, two-dimensional systems give maximum resolution.

If a mixture of different ribosome binding sites is analyzed directly, then the existence of several "linkage groups" will become apparent during the analysis of partial products. Clues that more than one protected region have been isolated will come from the complexity of the first T1 RNase fingerprint. So far, no more than 3 different ribosome binding sites (a total of about 100 nucleotides) have been analyzed without prior fractionation[2]; this is probably the limit.

Likewise, it is well to remember that upon direct analysis of the total mRNA fragments derived from 70 S ribosomes, different yields of oligonucleotides arising from various portions of a single protected region will almost always be observed in T1 or pancreatic RNase fingerprints. That is, molar yields of oligonucleotides containing the initiator triplet and the region complementary to 16 S rRNA[19] can be up to 5-fold higher than distal oligonucleotides. Both low yield and unexpected 3′ termini (arising from the original nuclease treatment on the ribosomes) readily identify oligonucleotides derived from the periphery of the protected mRNA segment. (In one case a 3′ terminus not known to be the product of either $T_1$ or pancreatic RNase was observed.[2])

## Concluding Remarks

Finally, although it may be possible to isolate ribosome-protected mRNA regions in a form suitable for analysis by the recent "rapid RNA sequencing" techniques,[38-40] this is not a trivial task. Prerequisites here are (1) that the mRNA fragment has a defined 3′ or 5′ terminus (to be labeled by polynucleotide kinase[41,42] or RNA ligase,[43] respectively) and (2) that the

[38] H. Dones-Keller, A. M. Maxam, and W. Gilbert, *Nucl. Acids Res.* **4,** 2527 (1977).

[39] A. Simonsits, G. G. Brownlee, B. S. Brown, J. R. Rubin, and H. Guilley, *Nature (London)* **269,** 833 (1977).

[40] R. E. Lockard, B. Alzner-deWeerd, J. E. Heckman, J. MacGee, M. W. Tabor, and U. L. RajBhandary, *Nucl. Acids Res.* **5,** 37 (1978).

[41] J. H. Van de Sande, K. Kleppe, and H. G. Khorana, *Biochemistry* **12,** 5050 (1973).

[42] J. R. Lillehaug and K. Kleppe, *Biochemistry* **14,** 1225 (1975).

[43] T. E. England, R. I. Gumport, and O. C. Uhlenbeck, *Proc. Natl. Acad. Sci. U.S.A.* **74,** 4839 (1977).

fragment be *chemically* pure. $^{32}P$-labeled binding sites prepared as described above are sufficiently clean by *radiochemical* standards to permit analysis using Sanger methods; their contamination by unlabeled rRNA and tRNA fragments, which arise during the nuclease trimming of the 70 S initiation complexes, presents no problems when applying this technique. On the other hand, rigorous methods to eliminate such reaction by-products from the protected mRNA fragment must be developed before the new rapid sequencing methodology can be used. Meanwhile, the procedures described above and in volume LX [28][22] remain the most feasible for current analysis of ribosome binding sites from prokaryotic mRNAs.

Acknowledgments

The author was supported by Grants AI 10243 and CA 16038 from the National Institutes of Health. R. Young and M. Lerner made helpful comments on the manuscript.

# [32] Characterization of Translational Initiation Regions from Eukaryotic Messenger RNAs

*By* MARILYN KOZAK and AARON J. SHATKIN

Definition of the translational initiation region in a eukaryotic messenger RNA is somewhat more complicated than the corresponding definition for prokaryotic messages. The ribosome-protection assay, which was first successfully employed by Steitz in identifying coliphage R17 ribosome binding sites,[1] enables one to recognize the initiation region as the nuclease-resistant portion of an mRNA molecule, obtained when the messenger RNA is incorporated into an initiation complex with ribosomes and the exposed regions of the message are subsequently hydrolyzed with nuclease. An interesting complication encountered when this assay was employed with a number of messages in eukaryotic systems, however, is that the small (40 S) subunit of eukaryotic ribosomes protected a significantly larger fragment of the message than that protected by 80 S initiation complexes.[2-5] The 80 S-protected sequences were a subset of the 40 S-protected sequences. This is in contrast with results obtained with pro-

[1] J. A. Steitz, *Nature (London)* **224,** 957 (1969).

[2] S. Legon, *J. Mol. Biol.* **106,** 37 (1976).

[3] M. Kozak and A. J. Shatkin, *J. Biol. Chem.* **251,** 4259 (1976).

[4] M. Kozak and A. J. Shatkin, *J. Mol. Biol.* **112,** 75 (1977).

[5] M. Kozak and A. J. Shatkin, *J. Biol. Chem.* **252,** 6895 (1977).





ISBN 0-12-504180-2

karyotic translational components, where the protected mRNA sequences recovered from small (30 S) subunits and from 70 S initiation complexes were identical.[6] In the eukaryotic system, since interaction with 40 S subunits precedes formation of 80 S complexes, it seems reasonable to consider the "extra" sequences, protected by 40 S but not 80 S ribosomes, as part of the initiation region. In support of this point of view, we have shown that the slightly larger capped fragments protected by 40 S subunits (40 S-protected fragments) rebind to ribosomes with much greater efficiency than the smaller 80 S-protected fragments.[7] In addition to the disparity between 40 S- and 80 S-protected regions of the message, a further caveat that must be taken seriously in interpreting ribosome-protection data is that analysis of nuclease-resistant sequences reveals only where the (40 S or 80 S) ribosome is positioned at a particular stage in initiation when *stable* complexes accumulate—leaving open the possibility that the initial contact with the ribosomal subunit may have occurred elsewhere on the message. An alternative approach, obviating some of the problems involved in interpreting the protection assay, is to incubate fragmented messenger RNA with ribosomes, and then identify the (smallest) mRNA fragment that is able to associate with ribosomes. This ribosome-selection assay, like the ribosome-protection assay, provides a functional definition of a translational initiation region.

A less rigorous definition of an eukaryotic translational initiation region is that it consists of (some or all of) the sequences in between the 5′ terminus and the AUG initiation triplet (which can be recognized if the N-terminal amino acid sequence is known, for the nascent polypeptide). The justification for such a broad definition is that all well-characterized eukaryotic messages are functionally monocistronic, with initiation apparently confined to a region toward the 5′ end of the message. Furthermore, under certain conditions described below for reovirus mRNA, the 5′ terminus (blocked by m7G) was included within the portion of the message protected by 40 S ribosomes. A substantive body of evidence further supports the view that, with capped messages, the 5′-terminal m7G moiety promotes translational initiation.[8] The value of accepting this broader definition of an eukaryotic translational initiation region is that it enables one to exploit recent elegant techniques for rapid sequencing of nucleic acids; that is, one can obtain information about "initiation regions" by directly sequencing the 5′ ends of messages, without employing a ribosome binding assay. Thus, direct 5′-terminal sequencing of mRNA is one of the

[6] S. Legon, P. Model, and H. D. Robertson, *Proc. Natl. Acad. Sci. U.S.A.* **74,** 2692 (1977).
[7] M. Kozak and A. J. Shatkin, *Cell* **13,** 201 (1978).
[8] A. J. Shatkin, *Cell* **9,** 645 (1976).

methods considered below for structural analysis of eukaryotic initiation sites.

Deduction of mRNA structure from sequence studies carried out with DNA is more appropriately considered in the chapter on DNA sequencing. Here we might note that this powerful approach is straightforward when the cDNA to be sequenced has been transcribed using purified mRNA as template.[9-13] However, when sequencing is carried out directly on the DNA genome, the interpretation is immensely complicated[14] because a single messenger RNA molecule may contain sequences derived from multiple noncontiguous portions of the genome. Thus, it may be difficult to reconstruct the exact structure of an eukaryotic mRNA even after the DNA has been fully sequenced.

In the following discussion of eukaryotic initiation regions, we make no distinction between cellular messages and the mRNAs of viruses that infect eukaryotic cells. It seems likely that observations made with viral RNAs, which are often easier to obtain than cellular messages, accurately reflect the basic features of the eukaryotic translational machinery.

## Preparation of Appropriately Radiolabeled Messenger RNA

*In vivo* $^{32}$P-labeled viral mRNAs, suitable for sequence analysis of ribosome binding regions,[15,16] have been obtained by growing infected cells in the presence of inorganic $^{32}PO_4$. Owing to differences in intracellular nucleotide pools, the ideal of a uniform distribution of $^{32}$P at all internal positions in the mRNA chain is often not achieved. Thus, apparent yields may have to be corrected[16] for differences in the intracellular specific activity of ATP, GTP, etc. Appropriate precautions are also needed to minimize degradation of the mRNA during extraction and to ensure the purity of the final product.

With many animal viruses, $^{32}$P-labeled mRNA can be synthesized *in vitro,* exploiting the virion-associated transcriptase and related enzymes. RNA labeled to a high specific activity ($> 10^7$ cpm/$\mu$g) can be readily made *in vitro,* requiring far smaller amounts of radioactive precursors than would be needed for comparable *in vivo* labeling. Furthermore, *in vitro* one can

[9] F. E. Baralle, *Cell* **10,** 549 (1977).
[10] F. E. Baralle, *Nature* (*London*) **267,** 279 (1977).
[11] J. C. Chang, G. F. Temple, R. Poon, K. H. Neumann, and Y. Wai Kan, *Proc. Natl. Acad. Sci. U.S.A.* **74,** 5145 (1977).
[12] J. W. Szostak, J. I. Stiles, C. P. Bahl, and R. Wu, *Nature* (*London*) **265,** 61 (1977).
[13] G. G. Brownlee and E. M. Cartwright, *J. Mol. Biol.* **114,** 93 (1977).
[14] M. L. Celma, R. Dhar, J. Pan, and S. M. Weissman, *Nucl. Acids Res.* **4,** 2549 (1977).
[15] R. Dasgupta, D. S. Shih, C. Saris, and P. Kaesberg, *Nature* (*London*) **256,** 624 (1975).
[16] J. K. Rose, *Proc. Natl. Acad. Sci. U.S.A.* **74,** 3672 (1977).

synthesize mRNA labeled separately with each of the four [$\alpha$-$^{32}$P]-ribonucleoside triphosphates. Use of single-labeled RNA preparations greatly facilitates the subsequent sequence analysis by permitting identification of the nearest-neighbor residue.[17] While it is difficult to generalize from one system to another with respect to the amount of RNA required for sequence analysis, determination of the nucleotide sequences of six reovirus initiation regions[4,5,18] required a total of ten *in vitro* viral RNA preparations: two labeled with all four [$\alpha$-$^{32}$P]nucleoside triphosphates, and eight single-labeled preparations (two preparations made with [$\alpha$-$^{32}$P]GTP as the sole labeled precursor, two with [$\alpha$-$^{32}$P]ATP, etc.). Each RNA preparation (10–50 $\mu$g of RNA, specific activity 5 to 12 $\times$ $10^6$ cpm/$\mu$g) was a mixture of ten different messenger species, with four messages from the small-size class and three from the medium-size class as predominant components. Although recovery of ribosome-protected fragments (see below) from certain of the reovirus messages was about 5-fold less efficient than from other messages, we were able to sequence initiation regions from six of the reovirus messages by beginning with the indicated amount of RNA.

It is difficult to obtain large amounts of individual labeled cellular messenger RNAs. Therefore, procedures have been sought for introducing radioactivity *in vitro* in ways that would permit subsequent sequence analysis of cellular messages. Iodination with $^{125}$I has been employed in a number of cases. While this approach allowed preliminary analysis of interesting mRNA species[2,19,20] that were, at the time, difficult to label by other techniques, the studies that can be done with $^{125}$I-labeled RNA are limited. Since only cytidine residues are labeled under the usual conditions and iodine-substituted oligonucleotides have altered electrophoretic mobility, sequence analyses are difficult. Nevertheless, this approach generated extremely useful information about the ribosome binding site in rabbit $\beta$-globin mRNA.[2]

With the advent of new sequencing techniques, extensive sequence information can be obtained using RNA that carries only a 5′-terminal label. In most eukaryotic messages[8] the 5′ terminus is blocked by the structure m7G, in an inverted linkage with the penultimate nucleoside forming the cap structure, m7G(5′)ppp(5′)X . . . . If the naturally occurring terminal m7G moiety is removed, however, either by $\beta$-elimination[21]

[17] G. Pieczenik, P. Model, and H. D. Robertson, *J. Mol. Biol.* **90,** 191 (1974).

[18] M. Kozak, *Nature* (*London*) **269,** 390 (1977).

[19] J. H. Chen and A. Spector, *Biochemistry* **16,** 499 (1977).

[20] D. W. Leung, C. W. Gilbert, R. E. Smith, N. L. Sasavage, and J. M. Clark, Jr., *Biochemistry* **15,** 4943 (1976).

[21] R. E. Lockard and U. L. RajBhandary, *Cell* **9,** 747 (1976).

or by digestion with specific nucleases,[22,23] followed by removal of the 5′-phosphate residues with alkaline phosphatase, the exposed 5′ terminus can then be labeled *in vitro* by phosphate transfer from [$\gamma$-$^{32}$P]ATP, in a reaction catalyzed by polynucleotide kinase.[21] Care must be taken to use intact messenger RNA as the substrate, since the kinase-catalyzed reaction will introduce label at the site of any internal nicks. Furthermore, for reactions in which a 5′-terminal label is introduced *in vitro*, the messenger RNA substrate must be of highest purity, since contaminating RNA species would also become labeled. These potential problems can be obviated by using the following alternative labeling scheme. After removal of the m7G moiety by $\beta$-elimination, the (5′)pppX-ended molecules can be reblocked and labeled using [$\alpha$-$^{32}$P]GTP, in a reaction catalyzed by enzyme(s) purified from vaccinia virions.[24] With messages such as those of tobacco mosaic virus and brome mosaic virus, in which the 5′ penultimate residue (X in m7GpppX) is normally not methylated, the $\beta$-elimination step can be omitted. That is, by incubating untreated message with [$^3$H]methyl-*S*-adenosylmethionine and vaccinia enzyme(s),[25] [$^3$H]methyl radioactivity can be introduced to the 2′ position of X.

Messenger RNA labeled specifically at the 5′ terminus is usually sequenced directly, without using ribosomes to select the initiation region. With internally labeled mRNA, however, the limited portion of the message that interacts with ribosomes during translational initiation must first be separated from the rest of the RNA chain, as discussed in the following section.

## Purification of Ribosome Binding Fragments of mRNA

### *Conditions for Initiation Complex Formation in Cell-Free Extracts*

The large scale required for purification of mRNA ribosome binding sites usually necessitates use of crude cell-free extracts. Consequently, the precise requirements (initiation factors, etc.) for selection or protection of ribosome binding regions have not been fully defined, although a requirement for both ATP and GTP cofactors has been established. In both crude wheat germ[26] and reticulocyte lysates,[27] 40 S complexes containing Met-

[22] M. Zan-Kowalczewska, M. Bretner, H. Sierakowska, E. Szczesna, W. Filipowicz, and A. J. Shatkin, *Nucl. Acids Res.* **4,** 3065 (1977).

[23] H. Shinshi, M. Miwa, T. Sugimura, K. Shimotohno, and K. Miura, *FEBS Lett.* **65,** 254 (1976).

[24] S. Muthukrishnan, B. Moss, J. A. Cooper, and E. S. Maxwell, *J. Biol. Chem.* **253,** 1710 (1978).

[25] B. Moss, *Biochem. Biophys. Res. Commun.* **74,** 374 (1977).

[26] A. R. Hunter, R. J. Jackson, and T. Hunt, *Eur. J. Biochem.* **75,** 159 (1977).

[27] S. Legon, H. D. Robertson, and W. Prensky, *J. Mol. Biol.* **106,** 23 (1976).

tRNA and mRNA have been shown to be intermediates in 80 S initiation complex formation. Although addition of polyamines has been shown to enhance elongation in many eukaryotic cell-free extracts,[28,29] an effect on initiation is not usually noted, and therefore polyamines generally are not added during isolation of mRNA ribosome binding sites.

In reactions employing crude cell-free extracts, accumulation of 80 S initiation complexes is usually accomplished by adding an inhibitor of elongation. Three highly efficient inhibitors that have been used are diphtheria toxin (4000 mouse letal doses/ml, plus 0.5 m*M* NAD),[27] anisomycin (313 $\mu$g/ml)[16] and sparsomycin (200 $\mu M$).[3] The inhibitory effect of sparsomycin in mammalian extracts seems to be less complete than in wheat germ extracts. This is convenient from one point of view; namely, the limited peptide bond formation that occurs in mammalian extracts in the presence of high concentrations of sparsomycin allows accumulation and identification of N-terminal di- and tripeptides.[27,30] On the other hand, if the focus is on identification of ribosome-protected regions of mRNA, inhibition of elongation should be absolute. The more complete inhibition by sparsomycin in wheat germ extracts may account for the observation that there is less fraying of the protected mRNA regions recovered from wheat germ as opposed to reticulocyte ribosomes. For example, we observed that from each reovirus message, wheat germ ribosomes protected a single discrete fragment[4,5,18]—in contrast with the reticulocyte system, in which a single message yielded a series of overlapping protected fragments, due to partial nuclease sensitivity of sequences near the edges of the 80 S ribosome binding site.[31] Extensive fraying also occurred with the protected fragments of rabbit $\beta$-globin mRNA recovered from reticulocyte 40 S complexes.[2] Occurrence of fraying complicates the subsequent fractionation of the protected RNA fragments and reduces the yield of each species. In this regard, therefore, wheat germ extracts proved to be more efficient than mammalian cell-free extracts, at least for purification of reovirus ribosome binding fragments.

Since 40 S initiation complex formation precedes 80 S complex formation, and since 40 S complexes often protect a larger portion of the mRNA than do 80 S ribosomes, knowledge of the 40 S-protected sequences might permit a more complete understanding of the basis for ribosome recognition and binding. In eukaryotic systems, GTP hydrolysis is necessary for joining of 60 S ribosomal subunits.[32] Consequently, with many messages 40

[28] A. R. Hunter, P. J. Farrell, R. J. Jackson, and T. Hunt, *Eur. J. Biochem.* **75,** 149 (1977).

[29] J. F. Atkins, J. B. Lewis, C. W. Anderson, and R. F. Gesteland, *J. Biol. Chem.* **250,** 5688 (1975).

[30] A. E. Smith and D. T. Wigle, *Eur. J. Biochem.* **35,** 566 (1973).

[31] S. G. Lazarowitz and H. D. Robertson, *J. Biol. Chem.* **252,** 7842 (1977).

[32] D. H. Levin, D. Kyner, and G. Acs, *J. Biol. Chem.* **248,** 6416 (1973).

S complexes can be accumulated by adding the nonhydrolyzable analog GMPPCP to crude cell-free extracts, from which the endogenous GTP has been removed (or substantially reduced) by passage over a Sephadex G-25 column. Interestingly, the effectiveness of the GMPPCP-block varies somewhat from one message to another. We found, for example, that two reovirus messages from the small size class (designated Y and Z by Kozak and Shatkin[5]) form primarily 80 S complexes under conditions (0.48 m*M* GMPPCP) that block 40–60% of other reovirus messages in 40 S complexes.

In the case of reticulocyte extracts, which have a natural excess of 40 S over 60 S subunits, addition of saturating amounts of mRNA results in accumulation of some 40 S complexes without the necessity of employing an inhibitor.[27] Use of the peptide antibiotic edeine is a third approach that has been used to accumulate 40 S complexes.[27] We have found, however, that edeine mediates formation of abnormal complexes in which random regions of reovirus mRNA are protected by 40 S ribosomes.[33] Thus, edeine-induced complexes are inappropriate for characterization of initiation regions in eukaryotic messenger RNAs. A further note of caution is that the 40 S complexes formed with a given message and inhibitor appear to be slightly different depending on the source of the cell lysate. Specifically, with wheat germ 40 S ribosomes (at both 23° and 30°) in the presence of GMPPCP there was quantitative protection against RNase of the 5′-terminal cap on all reovirus messages,[3] while ascites or reticulocyte 40 S ribosomes, also blocked with GMPPCP and incubated at 30°, protected the cap on some reovirus messages but allowed pancreatic RNase cleavage within 3- to 6-nucleotides of the cap on other messages.[34] It may be that wheat germ 40 S complexes (i.e., the 40 S ribosome plus associated factors) are slightly bigger than the corresponding complexes from mammalian cells. From a practical standpoint, inclusion of the cap on ribosome-protected fragments is immensely helpful in that it enables one to know the position of the ribosome-protected region within the RNA chain. In this regard, the wheat germ system was advantageous for studies with reovirus messages.

While one eukaryotic cell-free system may be deemed preferable to another, based on minor differences in behavior, the major features involved in recognition of mRNA initiation regions appear to be the same for wheat germ, ascites, and reticulocyte ribosomes. With several reovirus messages, ribosome protection assays were carried out in all three of these systems. In each case, the protected mRNA fragments were found to derive from approximately the same 5′-proximal region of the mRNA.[4,31,34]

[33] M. Kozak and A. J. Shatkin, *J. Biol. Chem.* **253,** 6568 (1978).

[34] M. Kozak and A. J. Shatkin, unpublished observations.

*General Considerations for Purification of mRNA Initiation Regions*

One approach for separating the translational initiation region from the remainder of the mRNA sequences is to incubate fragmented mRNA with ribosomes, under initiation conditions, then separate the bound from the unbound fragments by sucrose gradient centrifugation. With the messenger RNA for brome mosaic virus coat protein (RNA 4) this *ribosome-selection assay* yielded overlapping fragments, all of which contained the 5′ terminus of the mRNA.[15] (The observed inability of internal regions of the fragmented message to associate with ribosomes is an impressive demonstration of the specificity of messenger/ribosome interaction.) The sequence of the smallest ribosome-selected fragment is shown in Table I. A ribosome-protection assay carried out with brome mosaic virus RNA 4 yielded a series of fragments containing the same initiation sequence.[15] Protection of multiple overlapping fragments in that study was probably due to use of a low concentration of anisomycin, resulting in incomplete inhibition of elongation.[35]

With a number of other eukaryotic mRNAs that have been studied, the *ribosome-protection assay* gave efficient recovery of discrete initiation regions, suitable for nucleotide sequence analysis. In order for the protection assay to be meaningful, the nuclease digestion conditions should not be chosen arbitrarily. If the nuclease concentration is too low or if the mRNA sequence lacks moieties sensitive to a particular RNase (e.g., if a G-poor RNA is digested with T1 RNase), the apparent extent of protection will be exaggerated. On the other hand, use of unnecessarily high nuclease concentrations may cause fraying and underestimation of protection. With reovirus messages, appropriate nuclease digestion conditions were selected by examining the size of both 40 S- and 80 S-ribosome-protected fragments following treatment with T1 RNase, over a range of concentrations, and also following digestion with pancreatic RNase.[3] For several reovirus messages, the T1 RNase-resistant fragments were slightly smaller than the pancreatic RNase-resistant fragments, and the size of the T1 RNase-resistant fragments remained constant when the nuclease concentration was varied from 120 to 460 units/ml. Thus, digestion of initiation complexes with T1 RNase at 300 units/ml (for 15 min at 23°) was chosen as the routine condition for protection assays with reovirus mRNAs. It is important to recognize that the large size of 40 S ribosome-protected fragments is probably a meaningful characteristic of eukaryotic systems, and is not due merely to underdigestion with nuclease. This was established for reovirus messages by varying the nuclease concentration, by testing more than one RNase, and by showing that identical nuclease digestion

[35] P. Kaesberg and D. S. Shih, personal communication, 1977.

conditions yielded 80 S ribosome-protected fragments approximately 25 to 35 nucleotides in length, in contrast with 40 S-protected fragments that were 45 to 55 nucleotides long,[3] depending on which message was used.

In cases where separation of individual messenger RNA species is difficult, it is possible to begin by binding a mixture of messages to ribosomes, trimming the initiation complexes with nuclease, and subsequently fractionating the protected initiation regions into individual species, suitable for sequencing. With reovirus messages, fractionation of initiation regions was accomplished by electrophoresis through 20% polyacrylamide gels followed in some cases by two-dimensional homochromatography[5] using unhydrolyzed yeast RNA (homo mixture b).[36] With a mixture of five vesicular stomatitis virus messenger RNAs, adequate resolution of all five ribosome binding sites was obtained by two-dimensional polyacrylamide gel electrophoresis.[16] Since fragments larger than 35 nucleotides are difficult to resolve by two-dimensional homochromatography, even using homomixture b, two-dimensional gels are probably more efficient, especially for fractionation of the large mRNA fragments that may be recovered from 40 S complexes.

*Preparation of the Reovirus Ribosome Binding Sites*

Both 40 S and 80 S ribosome-protected fragments from six reovirus messages have been sequenced, as shown in Table I. $^{32}$P-Labeled reovirus messages, synthesized *in vitro* using the virion-associated transcriptase,[3,37] were separated by centrifugation into small (*s*RNA, 4 messages), medium (*m*RNA, 3 messages), and large size classes. A given set of messages (either *s*- or *m*RNA) was incubated at 5–20 μg/ml in a cell-free extract prepared from commercial (Niblack) raw wheat germ. The wheat germ S23 extract (i.e., 23,000 *g* supernatant) was prepared and preincubated as described by Roberts and Paterson.[38] Small aliquots were stored in plastic vials in a liquid nitrogen freezer, with no detectable loss of activity over a 12-month period. For binding reactions, the extract was used at a final concentration of 35–45 $A_{260}$ units/ml, and was supplemented with the following components, at the indicated final concentration:

HEPES buffer, 30 m*M*, pH 7.4
KCl, 72 m*M*
Magnesium acetate, 2.8 m*M*
Creatine phosphate, 8 m*M* (dipotassium salt, Calbiochem)
Creatine phosphokinase, 120 μg/ml (Calbiochem, A grade)

[36] G. G. Brownlee, "Determination of Sequences in RNA" (T. S. Work and E. Work, eds.), North-Holland Publ., Amsterdam, 1972.

[37] G. W. Both, S. Lavi and A. J. Shatkin, *Cell* **4**, 173 (1975).

[38] B. E. Roberts and B. M. Paterson, *Proc. Natl. Acad. Sci. U.S.A.* **70**, 2330 (1973).

Dithiothreitol, 2 m*M* (Calbiochem)
ATP, 1 m*M*, pH 7
GMPPCP, 0.48 m*M* (Miles)
Sparsomycin, 200 $\mu M$ (from the Drug Research and Development Division of the National Cancer Institute)

The reaction conditions outlined above permitted accumulation of both 40 S and 80 S initiation complexes (Fig. 1), since GMPPCP at 0.48 m*M* only partially prevents 60 S subunit joining. After a 10-min incubation at 23° to allow initiation complex formation, the extract was treated with T1 RNase (300 units/ml) for 15 min at 23°, then chilled and layered onto 12 ml, 10% to 30% glycerol gradients containing 20 m*M* Tris(hydroxymethyl)-aminomethane (pH 7.4), 75 m*M* KCl, 3 m*M* magnesium acetate, and 0.2 m*M* ethylenediaminetetraacetate. Reaction volumes up to 0.5 ml could be applied to each gradient. The gradients were centrifuged for 4 hr in a Beckman SW 41 rotor at 39,000 rpm at 4°, and were then collected directly into tubes containing water-saturated phenol. The 40 S- and 80 S-protected mRNA fragments (comprising about 2% of the input radioactivity, as shown in Fig. 1) were located by counting an aliquot of each gradient fraction. The ribosome-protected material was then extracted twice with phenol at 4°. (For extraction of the 80 S-protected material, a 4-fold dilution was required to reduce the glycerol concentration.) The protected fragments were concentrated by ethanol precipitation, then dissolved in 100–200 $\mu$l of 8 *M* urea in electrophoresis buffer, heated for 3 min at 55° to dissociate aggregates, and applied to a cylindrical (6 × 100 mm) 20% polyacrylamide gel containing 8 *M* urea, as described previously.[3] Electrophoresis was carried out at 130 V for 24 hr at 23° with a buffer containing 90 m*M* boric acid, 90 m*M* tris(hydroxymethyl)aminomethane (Tris), and 2.5 m*M* disodium ethylenediaminetetraacetate. The gels were then sliced manually into 1.3-mm sections, which were monitored for Cerenkov radioactivity. The $^{32}$P-labeled material was eluted by soaking the appropriate gel slices overnight at room temperature in 0.5 ml of 0.2% sodium dodecyl sulfate and 20 m*M* Tris (pH 7.5), with 10–20 $\mu$g of transfer RNA as carrier. It should be noted that the polyacrylamide gel electrophoresis step served both to remove the large amount of ribosomal RNA, which would interfere with subsequent fingerprinting, and also to achieve a partial (or, in some cases, complete) fractionation of the protected mRNA fragments. An alternative procedure, avoiding the preparative polyacrylamide gel electrophoresis step, has also been used.[31]

## Sequence Determination of mRNA Initiation Regions

With ribosome-protected or -selected mRNA fragments that are internally labeled with $^{32}$P, sequence analysis usually involves two-dimensional

## Nucleotide Sequences of Initiation Regions from Eukaryotic Messenger RNAs

| Initiator codon ↓ | Messenger RNA | Basis for identification as an initiation region[a] |
|---|---|---|
| m7GpppGUAUUAAUA**AUG**UCGACUUCAG... | Brome Mosaic Virus coat protein[b] | 1,4 |
| m7GpppGUUUUUAUUUUUAAUUUUCUUUCAAAUACUUCCAUC**AUG**AGUUCUUCACAAAAGAAAGCUGGUGGGAAAGCUGG... | Alfalfa Mosaic Virus coat protein[c] | 4,5 |
| m7GpppG$^{m}$CUAUUUUGCCUCUUCCCAGACGUUGUCGCA**AUG**GAGGUGUGCUUGCCCAACG...<br>m7GpppG$^{m}$CUAAAGUCACGCCUGUCGUCGUCACU**AUG**GCUUCCUCACUCAG...<br>m7GpppG$^{m}$CUAUUCGCUGGUCAGUU**AUG**GCUCGCUGCGCGUUCCUAUUCAAG... | Reovirus sRNA[d,e] | 2 |
| m7GpppG$^{m}$CUAAUCUGCUGACCGUUACUCUGCAAAG**AUG**GGAACGCUCUUCCUAUCG...<br>m7GpppG$^{m}$CUAAAGUGACCGUGGUC**AUG**GCUUCAUUCAAGGGAUUCUCCG...<br>m7GpppG$^{m}$CUAUUCGCGGUC**AUG**GCUUACAUCGCAG... | Reovirus mRNA[e,f] | 2 |
| <br>m7Gppp(m$^{6}$)A$^{m}$ACAGUAAUCAA**AUG**UCUGUUACAGUCAAG...<br>m7Gppp(m$^{6}$)A$^{m}$ACAGAUAUC**AUG**GAUAAUCUCACAAAG...<br>...UUUCCUUGACACU**AUG**AAGUGCCUUUUGUACUUAG... | Vesicular Stomatitis Virus[g]<br>--N protein<br>--NS protein<br>--G protein | 3 |
| m7GpppACUCUUCUGGUCCCCACAGACUCAGAGAGAACCCACC**AUG**GUGC...<br>m7Gppp(m$^{6}$)A$^{m}$CACUUCUGGUCCAGUCCGACUGAGAAGGAACCACC**AUG**GUGC...<br>m7GpppACAUUUGCUUCUGACACAACUGUGUUCACUAGCAACCUCAAACAGACACC**AUG**GUGCACCUGACUCCU...<br>m7Gppp(m$^{6}$)A$^{m}$CACUUGCUUUUGACACAACUGUGUUUACUUGCAAUCCCCCAAAACAGACAGA**AUG**GUGCAUCUGUCCAGU... | α globin, human[h,i]<br>α globin, rabbit[j,k]<br>β globin, human[h,i]<br>β globin, rabbit[k-n] | 4,5<br>4,5<br>4,5<br>3,4,5 |
| ...CCUGUACGGAAGUGUUACUUCUGCUCUAAAAGCUUAUGAAG**AUG**GCCCCAACAAAA... | Simian Virus 40 VP1[o,p] | 4 |

[a] The following code has been used: (1) the sequence represents an mRNA fragment selected by binding to 80 S ribosomes; (2) the sequence was protected against nuclease by 40 S ribosomes; (3) the sequence was protected against nuclease by 80 S ribosomes under conditions of polypeptide chain initiation; (4) the nucleotide sequence following the AUG initiator codon corresponds to the N-terminal amino acid sequence; (5) the sequence was obtained by direct analysis of the 5′ portion of the message or of cDNA complementary to the 5′ portion of the message.

[b] R. Dasgupta, D. S. Shih, C. Saris, and P. Kaesberg, *Nature (London)*, **256**, 624 (1975).

[c] E. C. Koper-Zwarthoff, R. E. Lockard, B. Alzner-deWeerd, U. L. RajBhandary, and J. F. Bol, *Proc. Natl. Acad. Sci. U.S.A.* **74**, 5504 (1977).

[d] M. Kozak and A. J. Shatkin, *J. Biol. Chem.* **252**, 6895 (1977). Regions of ambiguity in these and other sequences in the table are underscored with a dashed line.

[e] M. Kozak, *Nature* (*London*) **269**, 390 (1977).

[f] M. Kozak and A. J. Shatkin, *J. Mol. Biol.* **112,** 75 (1977).
[g] J. K. Rose, *Proc. Natl. Acad. Sci. U.S.A.* **74,** 3672 (1977).
[h] J. C. Chang, G. F. Temple, R. Poon, K. H. Neumann, and Y. Wai Kan, *Proc. Natl. Acad. Sci. U.S.A.* **74,** 5145 (1977).
[i] F. E. Baralle, *Cell* **12,** 1085 (1977).
[j] F. E. Baralle, *Nature* (*London*) **267,** 279 (1977).
[k] R.E. Lockard and U. L. RajBhandary, *Cell* **9,** 747 (1976).
[l] F. E. Baralle, *Cell* **10,** 549 (1977).
[m] A. Efstratiadis, F. C. Kafatos, and T. Maniatis, *Cell* **10,** 571 (1977).
[n] S. Legon, *J. Mol. Biol,* **106,** 37 (1976).
[o] A. Van de Voorde, R. Contreras, R. Rogiers, and W. Fiers, *Cell* **9,** 117 (1976).
[p] J. Pan, V. B. Reddy, B. Thimmappaya, and S. M. Weissman, *Nucl. Acids Res.* **4,** 2539 (1977).

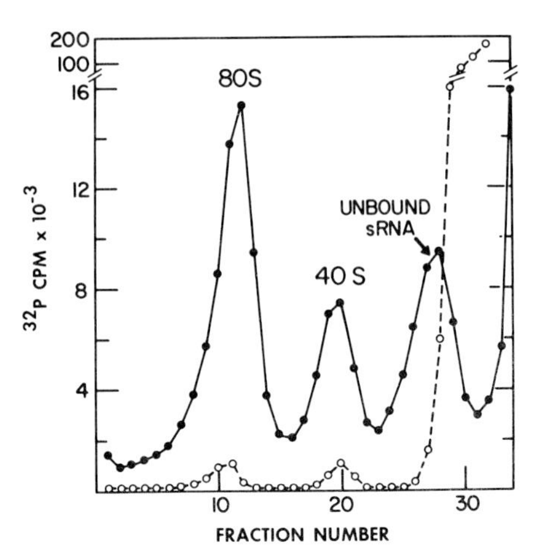


FIG. 1. Binding and protection of [$\alpha$-$^{32}$P]GTP-labeled reovirus sRNA in a wheat germ S23 extract, supplemented with GMPPCP, sparsomycin, and other components as listed in the text. Incubation was for 10 min at 23°. Equal aliquots of the reaction mixture were centrifuged before (●——●) and after (○---○) incubation with pancreatic RNase (3 $\mu$g/ml, 15 min, 23°). The glycerol gradients were centrifuged (right to left) for 4 hr at 39,000 rpm, 4°, in the SW 41 rotor.

fingerprinting of T1 and pancreatic RNase digests.[36,39] Standard techniques[36] are used to determine the sequence of each T1 (or pancreatic) RNase limit product, and the oligonucleotides are then ordered to reconstruct the complete sequence of the ribosome binding site. The only special consideration, when these classical radioactive RNA sequencing methods are applied to eukaryotic ribosome binding sites, concerns characterization of the m7GpppX cap. Techniques for analyzing the structure of the cap have been described in detail.[40] The only cautionary notes that might be emphasized here are (1) that the m7G moiety is alkali-labile,[41] undergoing conversion to the ring-opened structure 2-amino-4-hydroxy-5(*N*-methyl)-carboxamide-6-ribosylaminopyrimidine which has an altered electrophoretic mobility; (2) that cap-containing oligonucleotides have abnormally low mobility during homochromatography on PEI-cellulose[5]; and (3) that mRNA's synthesized *in vivo* are usually heterogeneous with respect to methylation of the cap-adjacent residues. That is, a given mRNA species usually yields both type 1 (m7GpppX$^{m}$) and type 2 (m7GpppX$^{m}$Y$^{m}$) caps.[21] Thus, the 5′-terminal oligonucleotide migrates as a doublet in two-dimensional fingerprints of ribosome binding sites.[16]

[39] G. Volckaert, W. MinJou, and W. Fiers, *Anal. Biochem.* **72,** 433 (1976).

[40] Y. Furuichi, M. Morgan, S. Muthukrishnan, and A. J. Shatkin, *Proc. Natl. Acad. Sci. U.S.A.* **72,** 362 (1975).

[41] S. Hendler, E. Furer, and P. R. Srinivasan, *Biochemistry* **9,** 4141 (1970).

New sequencing methods have recently been introduced for RNAs labeled exclusively in the 5′ terminus. The wandering-spot technique, involving two-dimensional fractionation after partial digestion with a nonspecific nuclease, has yielded sequence information for portions of the rabbit $\alpha$- and $\beta$-globin mRNA initiation regions[21] and for the complete initiation region from RNA 4 of alfalfa mosaic virus,[42] as indicated in the table. A one-dimensional sequencing technique for terminally labeled RNA, using base-specific nucleases, has also been described recently.[43-45] The simplicity of these new techniques promises that sequence information on many more eukaryotic messages will soon be forthcoming.

## Confirmation of the Function of Putative Initiation Regions

Within each ribosome-protected[2,16,18] or ribosome-selected[15] mRNA fragment, the presence of a single AUG triplet strongly points to that codon as the functional initiator. Nevertheless, at some point the nucleotide sequence following the presumptive initiator codon should be compared with the N-terminal amino acid sequence of the (nascent or unprocessed) polypeptide in order to confirm the identification. This criterion has been satisfied with initiation regions from brome mosaic virus RNA 4,[15] rabbit $\alpha$- and $\beta$-globin messages,[10] human $\alpha$ and $\beta$ globin mRNAs,[11,46] and RNA 4 of alfalfa mosaic virus.[42]

A more difficult task is to identify exactly which sequences, surrounding the initiating AUG, should properly be considered part of the initiation region. In the introduction we argued that determination of the entire 5′ sequence up to the AUG codon is a valid approach to analyzing eukaryotic translational initiation regions, since this sequence can be expected to include the ribosome binding site. At some point, however, structural analyses must be supplemented by functional studies demonstrating that the putative initiation region actually can associate with ribosomes. Ideally, progressively smaller fragments derived from the region in question should be tested for binding, in an attempt to pinpoint the functionally significant sequences.

A similar functional test—ability to reassociate with ribosomes—is required with initiation regions isolated as ribosome-protected fragments. The critical question is whether the limited region of the message protected

[42] E. C. Koper-Zwarthoff, R. E. Lockard, B. Alzner-deWeerd, U. L. RajBhandary, and J. F. Bol, *Proc. Natl. Acad. Sci. U.S.A.* **74,** 5504 (1977).

[43] A. Simoncsits, G. G. Brownlee, R. S. Brown, J. R. Rubin, and H. Guilley, *Nature (London)* **269,** 833 (1977).

[44] H. Donis-Keller, A. M. Maxam, and W. Gilbert, *Nucl. Acids Res.* **4,** 2527 (1977).

[45] R. C. Gupta and K. Randerath, *Nucl. Acids Res.* **4,** 3441 (1977).

[46] F.-E. Baralle, *Cell* **12,** 1085 (1977).

by the bound ribosome (blocked at a *late* stage in initiation complex formation) includes *all* the information required for ribosome recognition and attachment. The criterion of rebinding has been met with a 5′-terminal 23 nucleotide fragment from brome mosaic virus RNA 4,[15] with protected fragments from calf lens mRNA(s),[19] and with 40 S ribosome-protected capped fragments from several reovirus messages.[4,7] In contrast, the uncapped reovirus RNA fragments protected by 80 S ribosomes showed only minimal ability to rebind to ribosomes, as shown in Fig. 2 and, more extensively, by Kozak and Shatkin.[7] Thus, while the sequences flanking the AUG codon, and therefore protected by 80 S ribosomes, must by definition be considered as a "ribosome binding site," they may in fact represent only a portion of the sequences required for efficient interaction with the ribosome. This would be in contrast with prokaryotic systems, where the available evidence suggests that all sequences that contribute

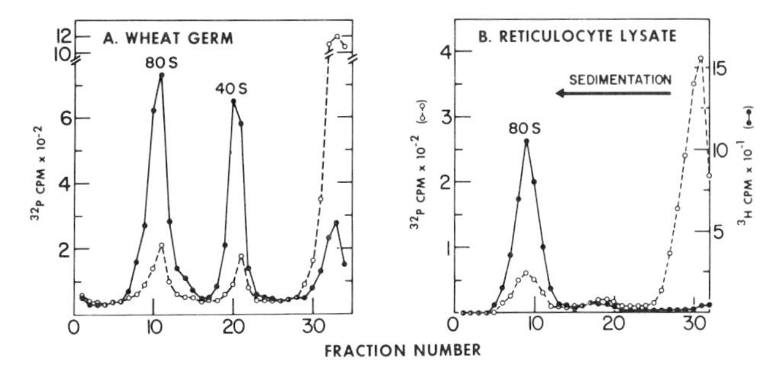


FIG. 2. Rebinding of capped (40 S-protected) and uncapped (80 S-protected) initiation fragments from reovirus messenger RNA's. (A) Binding reaction mixtures (50 $\mu$l) containing wheat germ S23 extract, GMPPCP, sparsomycin, and other components as listed in the text, were incubated at 23° for 10 min, then chilled and centrifuged (from right to left) through glycerol gradients. Parallel gradients were run, one containing 5600 $^{32}$P cpm of the capped and methylated 40 S-protected fragment (●——●), and the other containing the same amount of the uncapped, 80 S-protected fragment (○---○) from one of the small-size class messages. In other experiments with wheat germ in which GMPPCP was replaced with GTP (not shown), the 40 S peak of bound RNA fragments was absent; only 80 S complexes accumulated. (B) Six hundred counts per minute of the purified 40 S-protected 52-nucleotide fragment (labeled with [$^3$H]methyl *S*-adenosylmethionine, ●——●) and 2100 cpm of the uncapped 35-nucleotide 80 S-protected fragment (labeled with [$\alpha$-$^{32}$P]GTP, ○---○) from one of the medium-sized reovirus messages were incubated together under conditions permitting initiation complex formation. The 100-$\mu$l binding reaction mixture, containing 60 $\mu$l of hemin-treated rabbit reticulocyte lysate, was supplemented to obtain final concentrations of 75 m*M* K acetate, 2.1 m*M* Mg acetate, 12 m*M* HEPES (pH 7.4), 2 m*M* dithiothreitol, 0.2 m*M* GTP, 1 m*M* ATP, 20 $\mu$*M* methionine, 200 $\mu$*M* sparsomycin, 8 m*M* creatine phosphate, and 40 $\mu$g of creatine phosphokinase per milliliter. Incubation was for 6 min at 32° followed by chilling and centrifugation through a glycerol gradient. The extent of binding did not increase when the incubation time was prolonged to 10 min.

(positively) to initiation are included within the 70 S ribosome-protected portion of the message. One precaution when evaluating the ability of purified mRNA initiation fragments to rebind to ribosomes is to show that random RNA fragments (for example, the 80 S-protected material obtained when $^{32}P$-labeled mRNA is bound in the absence of sparsomycin or other inhibitors of elongation) do not stick to ribosomes under the conditions used to demonstrate binding of the authentic initiation sites. If artifactual sticking of random RNA fragments is observed, it can be abolished by including excess nonradioactive yeast RNA (50–100 $\mu$g/ml) during the binding. The carrier RNA does not reduce binding of authentic 5′-terminal reovirus initiation fragments.

Nucleotide sequence determinations coupled with functional studies on additional messages should help to clarify the principles underlying recognition of initiation regions by eukaryotic ribosomes.

# [33] Determination of the Number of Ribosomal Binding Sites on the RNAs of Eukaryotic Viruses

*By* LYDA NEELEMAN and LOUS VAN VLOTEN-DOTING

*Principle.* The assay measures the number of ribosomes that can be bound to the RNA in the absence of protein chain elongation. For these experiments one needs an active cell-free protein-synthesizing system. In principle, active systems derived from any source can be used. We will describe the procedure for the wheat germ (or embryo) system and for the nuclease-treated reticulocyte system, since these two eukaryotic cell-free systems are both devoid of endogenous messengers.[1-4] The protein chain elongation can be inhibited by a number of antibiotics.[5] We found sparsomycin to be very reliable. However, this antibiotic is not yet commercially available (sometimes it can be obtained from the National Institutes of Health, Bethesda, Maryland 20014). For a number of experiments we have used fusidic acid. However, the results obtained with this inhibitor are sometimes less clear-cut.

*Procedure.* The following experimental steps are required: (1) preparation of cell-free system; (2) determination of the appropriate concentration

[1] A. Marcus, B. Luginbill, and J. Feeley, *Proc. Natl. Acad. Sci. U.S.A.* **59,** 1243 (1968).
[2] B. E. Roberts and B. M. Paterson, *Proc. Natl. Acad. Sci. U.S.A.* **70,** 2330 (1973).
[3] J. W. Davies and P. Kaesberg, *J. Virol,* **12,** 1434 (1973).
[4] H. R. B. Pelham and R. J. Jackson, *Eur. J. Biochem.* **67,** 247 (1976).
[5] D. Vásquez. *FEBS Lett.* **40,** S63 (1974).



ISBN 0-12-504180-2

of inhibitor; (3) binding of labeled RNA to ribosomes and analysis of polyribosome profile.

## Preparation of Cell-Free Systems

### *Wheat Germ*

*Reagents*

Wheat germ (General Mills Inc., Vallejo, California, or from other sources; not all preparations are equally active)
Cyclohexane–carbon tetrachloride mixture, ratio 1:2.5
Grinding buffer: 3 m*M* Mg acetate, 90 m*M* KCl
Tris-acetate, 1 *M*, pH 7.6
Mg acetate, 0.1 *M*

*Procedure*. The wheat germ cell-free system can be prepared either as described by Marcus *et al.*[6] or as described below.

Wheat germ is purified by flotation on a cyclohexane–carbon tetrachloride mixture. As soon as the particles have found their own density, the upper layer is poured into a Büchner funnel and sucked dry. The germ is left to dry in the air. Purified germ can be stored at 4° in airtight plastic containers. Dry germ is ground in a precooled mortar and pestle with an equal amount of sand and twice the volume of grinding buffer. After a while more grinding buffer is added up to five times the amount of germ. The slurry is centrifuged for 10 min at 23,000 *g*. The supernatant, with the exception of the upper lipid layer, is removed with a Pasteur pipette. To this supernatant 0.01 volume each of Tris and Mg acetate is added. The suspension is recentrifuged for 10 min at 23,000 *g*. Pellicle material, the cloudy portion of the supernatant, and the upper lipid layer are discarded. The resulting supernatant (S23) can be stored for months at −80°.

### *Rabbit Reticulocytes*[4]

*Reagents*

Rabbits, about 2.5 kg
1-Acetyl-2-phenylhydrazine in sterile water, 1.25% (w/v). Each time, a fresh solution should be made. To dissolve the acetylphenylhydrazine the water should be heated to about 50°.
Heparin solution: 5000 units per milliliter of thromboliquine, pH 7.5
Saline: 0.13 *M* NaCl, 5 m*M* KCl, and 7.5 m*M* $MgCl_2$

[6] A. Marcus, D. Efron, and D. P. Weeks, this series, Vol. 30, p. 749.

*Procedure.* To obtain a reticulocyte cell-free system rabbits should first be made anemic by injection with acetylphenylhydrazine (injection on days 1, 2, 3, and 4 with 0.9 ml/kg of body weight. On day 10 the rabbits are bled. The blood is collected in a precooled beaker that has been rinsed with heparin (about 0.6 ml of this solution should remain in the beaker for each rabbit). Everything should be kept in an ice bath. Centrifuge for 10 min at 1000 *g* in the cold. Cells are washed three times with saline and lysed with 1.5 volumes of ice-cold water per volume of packed cells. Centrifuge 15 min at 30,000 *g*. The supernatant (S30) can be stored for months at −80°.

## Amino Acid Incorporation[6]

### *Wheat Germ*

*Reagents*

- Elution buffer: 1 m*M* Tris-acetate, pH 7.6, 50 m*M* KCl, 2 m*M* Mg acetate, and 4 m*M* 2-mercaptoethanol
- HEPES–KOH, 1 *M*, pH 7.6; HEPES = *N*-2′-hydroxyethylpiperazine-2-ethanesulfonic acid
- ATP, 0.1 *M*, pH 7.6
- GTP, 10 m*M*
- Creatine phosphate, 0.5 *M*
- Creatine phosphatase, 4 mg/ml
- Dithioerythritol, 0.3 *M*
- Spermine, 2.5 m*M*, pH 7.1
- KCl, 1 *M*
- K acetate, 0.5 *M*
- Amino acids minus leucine (or any other that will be given labeled), 0.5 m*M* each
- [$^{14}$C]Leucine (or another labeled amino acid), 50 $\mu$Ci/ml
- mRNA, 1 mg/ml
- S23
- Trichloroacetic acid (TCA), 7%

*Procedure.* Prior to use, 0.6 ml of the S23 is passed (in the cold) through a Sephadex G-25 (coarse) column (18 × 0.8 cm), equilibrated, and eluted with elution buffer. The first three cloudy five-drop fractions are used. To tubes in an ice bath is added (volume given in $\mu$l) 2, HEPES; 1, ATP; 0.2, GTP; 1.6, creatine phosphate; 0.5, creatine phosphatase; 0.8, dithioerythritol; 3.2, spermine; 0.75, KCl; 16, K acetate; 1.1, Mg acetate; 6, amino acids; 25, purified S23; 39, $H_2O$; 1, leucine; and 2, RNA (total volume 100 $\mu$l). The optimum concentration of $K^+$, $Mg^{2+}$, spermine, and mRNA has to be determined for each RNA.

Reaction mixtures are incubated for 1 hr at 30°; the reaction is stopped by the addition of 3 ml of 7% TCA, and the mixture is heated at 90° for 15 min. After cooling, the mixture is passed through a Whatman GF/C or GF/A glass filter; the filter is washed with 7% TCA and dried, and the radioactivity is determined.

*Reticulocytes*[4]

*Reagents*

Hemin solution: 1 m*M* hemin, 50 m*M* Tris·HCl, pH 8.0, 90% ethylene glycol. The hemin should first be dissolved in a small volume of 0.5 *M* KOH with the amount of Tris·HCl needed to obtain the correct final concentration. Add about three-fourths of the final volume of ethylene glycol. Stir, and adjust the pH to 8.0 with 1 *M* HCl (be careful: below pH 7.5 the hemin precipitates). Adjust the volume with ethylene glycol.
Creatine phosphatase, 5 mg/ml
KCl, 4 *M*
$MgCl_2$, 20 m*M*
Creatine phosphate, 0.2 *M*
$CaCl_2$, 100 m*M*
Micrococcal nuclease (EC 3.1.4.7.) (Boehringer), 6000 units/ml
S30
Amino acids minus leucine (or any other that will be given labeled), 0.5 m*M* each
[$^{14}$C]Leucine (or another labeled amino acid), 50 $\mu$Ci/ml
EGTA [ethylene glycol bis(2-aminoethyl ether)-*N*, *N*′-tetraacetic acid], 100 m*M*
Wheat germ tRNA, 1 mg/ml (for preparation see below)
mRNA, 1 mg/ml
NaOH, 0.3 *M*; $H_2O_2$, 0.15 *M*
TCA, 25%
TCA, 7%

*Procedure*. Prior to use 0.8 ml of the S30 is thawed in the presence of 12 $\mu$l of hemin solution and 8 $\mu$l of creatine phosphatase. Subsequently the following solutions (volumes are given in microliters) are added: 25, KCl; 25, $MgCl_2$; 50, creatine phosphate; 10, $CaCl_2$; 10, nuclease; 60, amino acids. The optimum concentration of $K^+$, $Mg^{2+}$, spermine, and mRNA has to be determined for each RNA and *for each lysate*. The mixture is incubated for 10 min at 20°. Add 20 $\mu$l of EGTA and place in ice.

To tubes in an ice bath is added (in $\mu$l): 17.8, $H_2O$; 1.75, tRNA; 48, treated S30; 1, leucine; and 1.4, mRNA. Reaction mixtures are incubated for 2 hr at 30°. The reaction is stopped by the addition of 1.5 ml of $NaOH \cdot H_2O_2$. Heat the mixture for about 5 min at 90° (the solution becomes yellowish). Add 1 ml of 25% TCA, and place the tubes in ice. After cooling, the mixture is passed through a Whatman GF/C or GF/A glass filter; the filter is washed with 7% TCA, and the radioactivity is determined.

## Preparation of Wheat Germ tRNA[7]

*Reagents*

Sodium dodecyl sulfate (SDS), 10%
Phenol, water saturated
tRNA buffer: 20 m*M* Tris · HCl, 375 m*M* NaCl; 8 m*M* $MgCl_2$, pH 7.6

*Procedure.* Prepare S23, centrifuge for 2 hr at 135,000 *g*. Collect with a Pasteur pipette the upper two-thirds of the supernatant (volume = *X*). Add 0.1 volume of SDS and an equal volume of phenol, shake vigorously, and centrifuge for 15 min at 23.000 *g*. Add to the water layer NaCl to a final concentration of 0.1 *M* and an equal volume of phenol; shake, and centrifuge. Add to the water layer twice the volume of cold ethanol and place the solution for 2 hr at −20°. Collect the RNA by centrifugation for 30 min at 12,000 *g*. Resuspend in 0.7 *X* ml of tRNA buffer, precipitate the RNA again with ethanol. Resuspend the RNA in 0.07 *X* ml of tRNA buffer. The tRNA is purified on a Sephadex G-50 (coarse) column (25 × 0.8 cm) equilibrated and eluted with tRNA buffer. The peak fractions are combined and dialyzed against water in the presence of bentonite. Afterward the bentonite is removed by centrifugation, and the tRNA is stored at −20°.

## Determination of the Appropriate Concentration of the Antibiotic

*Reagents.* See reagents for amino acid incorporation.

0.1 *M* fusidic acid } solutions can be stored at −20°
4 m*M* sparsomycin }
Or another inhibitor of protein chain elongation

*Procedure.* Before the binding assay can be performed, it is necessary to determine the concentration of inhibitor that will completely suppress the amino acid incorporation. This concentration may differ for different cell-free systems, and even for different reticulocyte *lysates* (Fig. 1). For procedure, see amino acid incorporation. Incubation time for wheat germ is 10 min at 30°; for reticulocytes, 60 min at 30°.

[7] K. L. Tao and T. C. Hall, *Biochem. J.* **125,** 975 (1971).

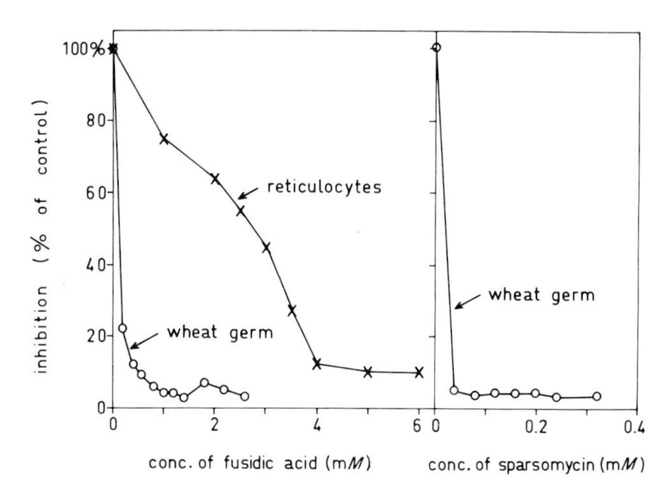


FIG. 1. Inhibition of translation of RNA 4 from alfalfa mosaic virus in wheat germ (O——O) or in reticulocytes (X——X) by fusidic acid or sparsomycin. Incorporation in the absence of inhibitor was about 8 × $10^3$ cpm.

## Binding of RNA to Ribosomes and Analysis of Polyribosome Profile

*Reagents.* See reagents for amino acid incorporation.

mRNA containing approximately 3 × $10^4$ cpm/$\mu$g

Inhibitor of protein chain elongation (see above)

Gradient buffer: 40 m*M* Tris·acetate, 20 m*M* KCl, 10 m*M* Mg acetate, pH 8.5

Sucrose density gradients from 12.5% to 50% sucrose in gradient buffer

*Procedure.* The incubation mixture for the binding reaction is identical to that of the amino acid incorporation except that inhibitor is present and that now the mRNA is labeled instead of one of the amino acids. Incubation is for 10 min at 30°. The reaction is stopped by the addition of twice the volume of cold gradient buffer. The samples are layered on a sucrose density gradient. Centrifugation is for 3 hr at 36,000 rpm in an SW 41 rotor. The $A_{260}$ of the gradients is monitored with a recording spectrophotometer, and the radioactivity of 0.3-ml fractions is determined after the addition of scintillation liquid (Fig. 2).

## Possibilities and Limitations of the Method

The ribosome binding method can also be used to isolate and analyze the ribosome binding sites.[8] It is also possible to use this method with

[8] M. Kozak and A. J. Shatkin, this volume [32].

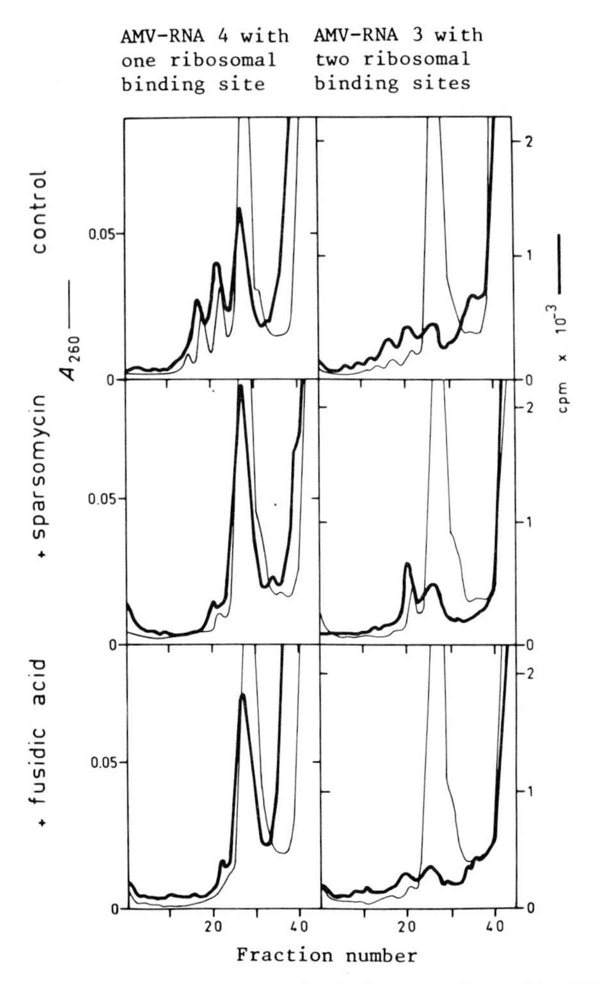


FIG. 2. Sucrose density gradient profiles of polyribosomes directed by [$^3$H]RNA 4 and [$^3$H]RNA 3 from alfalfa mosaic virus in wheat germ. Similar patterns were obtained with the reticulocyte cell-free system.

unlabeled RNA and [$^{35}$S]*N*-formylmethionine so that the initiating dipeptides can be analyzed.[9]

The determination of the number of ribosomal binding sites is based on the possibility of inhibiting completely the protein chain elongation. The concentration of inhibitor should not be taken higher than necessary, since at high concentrations other processes than elongation might be inhibited too. We found that in the wheat germ cell-free system at concentrations of fusidic acid above 1 m*M* the amount of RNA (even that of the monocis-

[9] A. E. Smith, *Eur. J. Biochem.* **33,** 301 (1973).

tronic control RNA) bound to ribosomes diminishes. Such an effect was not seen with concentrations of sparsomycin up to 0.2 m*M*. A low percentage of elongation might occasionally permit a second ribosome to bind to the same initiation site. For this reason a control experiment with a proven monocistronic mRNA should be performed. In the absence of added mRNA only monosomes are found with the wheat germ cell-free system. Although the mRNA of the reticulocyte cell-free system has been degraded, there are still (probably inactive) polysomes present. This makes the interpretation of the optical density patterns obtained with wheat germ ribosomes easier than those obtained with reticulocyte ribosomes.

When only one ribosome can be bound to an RNA this does not necessarily prove that all ribosomes start at the same position. There could be two overlapping initiation sites, e.g., Met-$X_1$-$X_2$ . . . $X_{12}$-Met-$X_{15}$-$X_{16}$ . . . $X_n$. An extraordinary case of two overlapping initiation sites is found upon translation of RNA 4 of alfalfa mosaic virus in *Escherichia coli*. On this RNA the protein chain can either start with *N*-acetylphenylalanine or with *N*-formylmethionine.[10] The UUU initiation triplet is located 36 bases ahead of the AUG triplet.[11] Initiation on the UUU triplet does not seem to take place in the wheat germ cell-free system.[12]

On the other hand, the finding of two ribosomal binding sites is not sufficient to prove that two different cistrons are read from one messenger. It has been reported that upon incubation *in vitro* of the RNA from bacteriophage Qβ[13] or of the mRNA of the tryptophan operon of *E. coli*[14] with ribosomes from *B. stearothermophilus* or *E. coli*, respectively, besides the initiator regions a noninitiator region is also protected from nuclease digestion. Up to now such a situation has not been reported for eukaryotic mRNAs. Nevertheless the finding that more than one ribosome can be bound to an RNA should be correlated with the appearance of more than one protein upon translation of this RNA.[15] Furthermore, different initiating (di)peptides or extensive variation in the ratio of formation of the products will demonstrate that the different products are not derived from a common precursor.

[10] A. Castel, B. Kraal, P. Kerklaan, J. Klok, and L. Bosch, *Proc. Natl. Acad. Sci. U.S.A.* **75,** 5509 (1978).

[11] E. C. Koper-Zwarthoff, R. E. Lockhard, B. Deweerd, U. L. RajBhandary, and J. F. Bol, *Proc. Natl. Acad. Sci. U.S.A.* **74,** 5504 (1977).

[12] L. van Vloten-Doting, J. Bol, L. Neeleman, T. Rutgers, D. van Dalen, A. Castel, L. Bosch, G. Marbaix, G. Huez, E. Hubert, and Y. Cleuter, *in* "Nucleic Acids and Protein Synthesis in Plants" (L. Bogorad and J. H. Weil, eds.), NATO Advanced Study Institute Series 12A, p. 387. Plenum, New York, 1977.

[13] J. A. Steitz, *J. Mol. Biol.* **73,** 1 (1973).

[14] T. Platt, C. Squires, and C. Yanofsky, *J. Mol. Biol.* **103,** 411 (1976).

[15] L. Neeleman, T. Rutgers, and L. van Vloten-Doting, *in* "Translation of Natural and Synthetic Polynucleotides" (A. B. Legocki, ed.), p. 292. Publ. House, Univ. of Agriculture, Poznan, Poland, 1977.

# [34] Methods for Analyzing Messenger Discrimination in Eukaryotic Initiation Factors

*By* CLAIRE H. BIRGE, FRED GOLINI, and ROBERT E. THACH

In this chapter we discuss methods for detecting messenger specificity in initiation factors. We use the terms "specific," "selective," and "discriminatory" interchangeably; however, we prefer the latter since it does not imply absolute specificity. While conflicting claims have been made about the existence of factors that show absolute specificity for one type or class of mRNAs (see Lodish[1] for a review of this question), we do not deal with the distinction between slight preferences among mRNAs versus absolute specificity. The reason for avoiding this issue is that at present it is difficult or impossible to prove absolute specificity in an initiation factor even if it exists, simply because one can never rule out the possibility that the factor required in large amounts for the translation of one mRNA might be also required in minute quantities for translation of all other mRNAs, and might be present as a contaminant in the cell-free system.

In operational terms, let us define an mRNA discriminatory factor as one that will enhance the initiation rate of one mRNA relative to another. The most obvious way of detecting messenger preference in a factor is to add the factor to a cell-free system in which two mRNAs ($mRNA_A$ and $mRNA_B$) are translated simultaneously, and in which initiation is rate limiting. If the addition of the factor being tested causes an increase in the relative rate of initiation on $mRNA_B$, for example, then the most straightforward interpretation of this result is that the factor preferentially recognizes $mRNA_B$ in some way and enhances its rate of initiation relative to that of $mRNA_A$.

While this simple experimental model seems straightforward, Lodish[1] has pointed out that it is susceptible to artifacts which may invalidate the above interpretation of the result. In particular, he has shown that under certain circumstances any factor which nonspecifically enhances the rate of initiation on *all* mRNAs may increase the relative rate of translation of $mRNA_B$; conversely, under the same conditions any factor that nonspecifically enhances the elongation rate on *all* mRNAs may produce the same result for $mRNA_A$. The chief condition that must be met for these artifactual results to occur is that $mRNA_A$ and $mRNA_B$ must possess intrinsically "fast" and "slow" initiation rates, respectively. Thus before any firm conclusions can be drawn as to whether a factor is truly messenger discriminatory, it is essential to determine whether the "elongation limitation

[1] H. F. Lodish, *Annu. Rev. Biochem.* **45,** 39 (1976)..



ISBN 0-12-504180-2

artifact," as this phenomenon will be called, is responsible. As it turns out this can be done quite easily for virtually any cell-free translation system, pure or crude, "active" or "inactive." The elongation limitation artifact itself is a significant problem only under a very special set of circumstances which, while known to occur in at least one case *in vivo*,[2] probably occurs only in a very few types of protein-synthesizing systems *in vitro*. (Chief among these are the concentrated reticulocyte lysate,[2] and in some instances fractionated systems of the Schreier–Staehelin type.)[3] These circumstances include an initiation rate for the "fast" mRNA so high (or an elongation rate so low) that the elongation rate partially limits the overall rate of protein synthesis in the sense that ribosomes travel away from the initiation region at too slow a rate to allow all potential initiation events from occurring. To see how this set of circumstances can lead to erroneous conclusions about the mRNA specificity of a factor, consider what effect the addition of a nonspecific initiation factor may have on the relative translation rates of $mRNA_A$ and $mRNA_B$. The initiation rate on $mRNA_A$ may be so fast, relative to the elongation rate, that the latter partially limits the former, that is, that the initiation signal on $mRNA_A$ is blocked a significant proportion of the time by a ribosome that has initiated but has not had sufficient time to move away from the region of this signal. The initiation rate on $mRNA_B$, on the other hand, is so much slower than the elongation rate that this type of elongation limitation does not occur. When these conditions pertain, the addition of any factor that increases nonspecifically the rate of initiation on all mRNAs (that is, increases $R^*$ in Lodish's terminology) will enhance the translation rate of $mRNA_B$ much more than that of $mRNA_A$. This is true because initiation on $mRNA_A$ is already partially limited by the elongation rate, and further enhancement of the initiation rate by an increase in $R^*$ will be largely prevented by this elongation limitation (that is, by ribosomes blocking the initiation signal on $mRNA_A$). In contrast, initiation on $mRNA_B$ will not be so limited, and hence will increase in direct proportion to $R^*$. Thus, the overall translation of $mRNA_B$ will be increased relative to that of $mRNA_A$, and this differential effect might be erroneously ascribed to some messenger specific property of the added factor. A similar enhancement of the translation rate of $mRNA_A$ relative to that of $mRNA_B$ would be produced by the addition of any factor which nonspecifically increases the elongation rate of the system.

Most cell-free systems have initiation rates that are much slower than the subsequent elongation steps, hence elongation does not significantly limit the initiation rate. (This can be deduced from the fact that polysomes

[2] H. F. Lodish, *Nature (London)* **251,** 385 (1974).

[3] M. B. Schreier and T. Staehelin, *Nature (London), New Biol.* **242,** 35 (1975).

produced in most *in vitro* systems are much smaller than their counterpart *in vivo,* even in the absence of "early quitting.") Thus the elongation limitation artifact is seldom a serious problem. Fortunately, Lodish has derived an equation that allows one to determine quantitatively the extent to which elongation limits initiation in any given system, and thereby to assess the likelihood of incurring artifacts by the addition of nonspecific factors to the system.[2] For this purpose the most useful form of this equation is $Q = mR^*K_1(1-nL)$, where $Q$ is the overall rate of protein synthesis, $m$ is the mRNA concentration, $R^*$ is the concentration of the rate-limiting nonspecific component required for initiation, $K_1$ is the overall rate constant for the interaction of mRNA with the small ribosome subunit and the formation of the 70 S or 80 S complex, $L$ is the number of codons on the mRNA blocked by an attached ribosome, and $n$, for our present purpose, is equal to the number of ribosomes attached to an mRNA in the steady state divided by the number of amino acids encoded by the mRNA (assuming monocistronic mRNAs). The extent to which protein synthesis on a given type of mRNA is limited by elongation is given by the term $(1-nL)$. When this term approaches unity, elongation is so fast that it does not seriously limit initiation, which then becomes rate limiting for overall protein synthesis. This would be the case for $mRNA_B$. For $mRNA_A$, however, $(1-nL)$ might be significantly less than unity, and would in fact be numerically equal to the probability that the initiation signal was free from blockage by a ribosome. The values of the term $(1-nL)$ with and without the factor being analyzed can be measured for each mRNA at various times in the course of the protein synthetic reaction by sucrose gradient sedimentation analysis of polysome size.[4] [The value of $(1-nL)$ is equal to $1-$(average number of ribosomes per mRNA $\div$ number of codons per mRNA) $\times$ 12. The average number of ribosomes per mRNA can be determined by following the distribution of labeled mRNA, labeled nascent polypeptide, or UV absorption of ribosomes.] From these data it can be immediately established which of the two mRNAs is the faster initiator [by determining which has smaller values for $(1-nL)$], and whether the factor stimulates initiation (thereby increasing polysome sizes) or elongation (reducing polysome sizes). In addition, analysis of the size distribution of polypeptides synthesized by polyacrylamide gel electrophoresis at each time point enables one to determine whether a quasi-steady-state balance between initiation and elongation (or termination) rates exists for the entire period of time during the reaction being analyzed.[4] This is important, because only under these conditions is the ribosome spacing along the

[4] F. Golini, S. S. Thach, C. H. Birge, B. Safer, W. C. Merrick, and R. E. Thach, *Proc. Natl. Acad. Sci. U.S.A.* **73,** 3040 (1976).

message more or less random. This condition is prerequisite for the validity of the equation.

Having thus determined the values of $(1-nL)$ for the two mRNAs in the presence of the putative mRNA discriminatory initiation factor, a simple arithmetical calculation will reveal whether its effect on the $mRNA_A/mRNA_B$ translation ratio is artifactual, or truly due to a preferential stimulation of initiation on $mRNA_B$. Let us assume for the moment that the value of $(1-nL)$ for $mRNA_B$ in the presence or absence of factor is close to unity, which is likely to be true for all but the most actively initiating cell-free systems. If $(1-nL)$ for $mRNA_A$ in the presence of factor is, say, 0.70, in its absence the value of $(1-nL)$ can at most be 1.0. Therefore, if the factor in question is nonspecific it must stimulate translation of $mRNA_A$ by at least 0.70 times the amount of stimulation seen with $mRNA_B$, which is equal to the increase in $R^*$ purportedly caused by the factor if it is in fact nonspecific. For example, if addition of the factor stimulates translation of $mRNA_B$ 2-fold (increases $R^*$ by a factor of 2) it must stimulate that of $mRNA_A$ by at least $0.70 \times 2 = 1.4$-fold. If translation of $mRNA_A$ is stimulated less than this amount, then it may be concluded that the factor discriminates between the two mRNAs, and preferentially stimulates $mRNA_B$ relative to $mRNA_A$. However, if $mRNA_A$ is stimulated by 1.4-fold or more, then the apparent preference for $mRNA_B$ (2-fold stimulation relative to 1.4-fold for $mRNA_A$) may be entirely artifactual.

In cases where $(1-nL)$ for $mRNA_B$ is significantly less than unity, one must first solve the two equations (one for values with added factor, the other for values without) for the actual increase in $R^*$. This increase which may be expressed as $R^{*+}/R^*$, is given by $Q^+_B(1-nL)_B \div Q_B(1-nL)^+_B$, where $K_1$ is assumed unchanged and the plus sign indicates values obtained in the presence of added factor. This value of $R^{*+}/R^*$ is then used in the similar two equations for $mRNA_A$, to solve for the theoretical value of $Q^+_A$, which is equal to $(R^{*+}/R^*)\ Q_A\ (1-nL)^+_A \div (1-nL)_A$. If this theoretical value is significantly greater than the experimentally measured value of $Q^+_A$, then it may be concluded that the factor actually shows preference for stimulating initiation on $mRNA_B$. If not, then the factor must be assumed to be nonspecific.

While the above approach relies on techniques that are readily available in any laboratory that studies protein synthesis, there is in fact a much simpler experimental approach to this question, which may be employed provided that the factor being tested is known to stimulate only initiation. This relies on the use of paper electrophoretic analysis of dipeptides produced in the presence of excess amounts of elongation inhibitors, such as cycloheximide.[5] If the initial dipeptides encoded by each message are

[5] F. Golini, S. Thach, C. Lawrence, and R. E. Thach, *ICN-UCLA Symposium on Animal Virology*, p. 717 (1976).

clearly resolvable, then this approach offers the advantage that, since ribosomes never leave the initiation signal on an mRNA, the question of elongation limitation and its attendant problems never arises. Thus, if addition of an initiation factor enhances the synthesis of dipeptide on $mRNA_B$ more than on $mRNA_A$, then it may be immediately concluded that the factor has messenger specificity.

Similar analyses and arguments may be used in the case of factors that stimulate elongation. In this case, however, the dipeptide assay may or may not show a differential effect. An unequivocal answer may be obtained from measurements of $(1-nL)$ values, as described above. In brief, the result of this analysis must reveal that the factor increases the translation ratio of $mRNA_A$ to $mRNA_B$ by more than the value of $1 \div (1-nL)_A{}^+$ [assuming that $(1-nL)_B$ is unity].

It should be noted that elongation limitation is not the only type of artifact that may be encountered in such systems. Aside from the trivial questions of whether the stimulating factor may have nuclease or protease activities, or whether it is contaminated with essential cations (such as $Mg^{2+}$ or polyamines), the matter of the experimental conditions employed for the assay is very important. It is well known, for example, that different mRNA's are translated optimally at different $K^+$ and $Mg^{2+}$ concentrations. Thus ideally the above type of analysis should be conducted twice, under optimum salt conditions for each mRNA.

A final important point involves the case where the two mRNAs actually compete with each other for a binding site on the initiation factor being studied, or on some other limiting component within the cell-free system. Experimentally this can be easily demonstrated, for the presence of one mRNA will reduce the translation rate of the other. While it is commonly assumed that the original equations of Lodish[2] deal with the case of competition, this is in fact not the case. Therefore, at present there is no complete mathematical expression that allows one to calculate the effects of messenger discriminatory factors on translation rates of competing mRNAs. The best that can be done at present is to make limited conclusions from the existing equations, such as was done in the case of viral versus host mRNA competition.[4] In this case, it was shown that the relief of viral mRNA suppression of host mRNA translation by the addition of an initiation factor (eIF-4B) could not be explained in terms of an elongation limitation artifact. The fact that the addition of viral mRNA actually suppressed host mRNA translation supported the concept that the initiation factor was not only messenger discriminatory (preferring viral over host mRNA), but also was competed for by the two mRNAs.

The details of the experimental techniques employed in studies of this kind have all been published elsewhere. Inasmuch as none of them were specifically developed for the purposes described here, it seems inappropriate to list particulars.

# [35] Isolation of Protein Kinases from Reticulocytes and Phosphorylation of Initiation Factors

*By* GARY M. HATHAWAY, TINA S. LUNDAK, STANLEY M. TAHARA, and JOLINDA A. TRAUGH

The postribosomal supernatant from rabbit reticulocytes contains multiple enzymic activities that modify the protein-synthesizing complex.[1] Among these are protein kinases, phosphoprotein phosphatases, and acetyltransferases. The protein kinases have been categorized historically on the basis of the stimulatory response (or lack of it) displayed toward 3′, 5′-cyclic AMP (cAMP).[2] Cyclic AMP-regulated protein kinases have been extensively studied in several tissues (for review, see Rubin and Rosen[3]). Cyclic AMP-independent protein kinases which modify casein have been observed in a variety of tissues and have been purified recently from rabbit erythrocytes, mouse plasmocytoma, rat liver, and Novikoff ascites tumor cells.[4-8] Still other protein kinases are found that are not stimulated by cAMP and do not phosphorylate casein.

The physiological role of phosphorylation in the regulation of protein synthesis has remained speculative, although many protein kinase activities modify proteins directly involved in translation. Both 40 S ribosomal subunits[9-12] and a polypeptide of initiation factor 3 (eIF-3)[13] are phosphorylated by the cAMP-regulated enzymes. A cAMP-independent protein kinase which phosphorylates the small subunit of initiation factor 2 (eIF-2) has been postulated to be the hemin controlled repressor (HCR).[14-16] In

[1] J. A. Traugh and S. B. Sharp, *J. Biol. Chem.* **252,** 3738 (1977).
[2] J. A. Traugh, C. D. Ashby, and D. A. Walsh, this series Vol. 38, p. 290.
[3] C. S. Rubin and O. M. Rosen, *Annu. Rev. Biochem.* **44,** 831 (1975).
[4] R. Kumar and M. Tao, *Biochim. Biophys. Acta* **410,** 87 (1975).
[5] M. E. Maragoudakis and H. Hankin, *Biochim. Biophys. Acta* **480,** 122 (1977).
[6] W. Thornburg and T. J. Lindell, *J. Biol. Chem.* **252,** 6660 (1977).
[7] M. Schmitt, J. Kempf, and C. Quirin-Stricker, *Biochim. Biophys. Acta* **481,** 438 (1977).
[8] M. E. Dahmus and J. Natzle, *Biochemistry* **16,** 1901 (1977).
[9] M. L. Cawthon, L. F. Bitte, A. Krystosek, and D. Kabat, *J. Biol. Chem.* **249,** 275 (1974).
[10] A. M. Gressner and I. G. Wool, *J. Biol. Chem.* **251,** 1500 (1976).
[11] J. A. Traugh and G. G. Porter, *Biochemistry* **15,** 610 (1976).
[12] U. K. Schubart, S. Shapiro, N. Fleischer, and O. M. Rosen, *J. Biol. Chem.* **252,** 92 (1977).
[13] J. A. Traugh and T. S. Lundak, *Biochem. Biophys. Res. Commun.* **83,** 379 (1978).
[14] G. Kramer, J. M. Cimadevilla, and B. Hardesty, *Proc. Natl. Acad. Sci. U.S.A.* **73,** 3078 (1976).
[15] P. J. Farrell, K. Balkow, T. Hunt, R. J. Jackson, and H. Trachsel, *Cell* **11,** 187 (1977).
[16] D. H. Levin, R. S. Ranu, V. Ernst, and I. M. London, *Proc. Natl. Acad. Sci. U.S.A.* **73,** 3112 (1976).



ISBN 0-12-504180-2

addition, other cAMP-independent protein kinases have been observed to phosphorylate the factors already mentioned as well as initiation factors 4B (eIF-4B) and 5 (eIF-5).[17-19] This article describes the identification and resolution of multiple forms of cAMP-independent protein kinase activities from rabbit reticulocytes that modify initiation factors, and the extensive purification of two of these enzymes using casein as substrate.

## Preparation of γ-Labeled Nucleotide Triphosphates[20]

*Reagents*

- Tris · HCl, 0.5 *M*, pH 8.0 (30°) (Sigma)
- $MgCl_2$, 1.0 *M*
- $KH_2PO_4/K_2HPO_4$, pH 7.4, 10 m*M*
- Cysteine, free base (Sigma)
- ATP, disodium salt (P-L Biochemicals)
- GTP, disodium salt (P-L Biochemicals)
- 3-Phosphoglycerate, sodium salt (Sigma)
- β-Nicotinamide adenine dinucleotide (NAD, Sigma)
- 3-Phosphoglycerate kinase/glyceraldehyde-3-phosphate dehydrogenase in $(NH_4)_2SO_4$ (Boehringer Mannheim)
- [$^{32}$P]Orthophosphate in HCl solution (ICN); neutralized with NaOH
- $NH_4OH$, 0.15 *M* in 50% ethanol
- HCl, 1 *N*
- Norite (acid-washed)

*Procedure.* To a 15 × 150 mm test tube containing 5.0 ml of doubly distilled water the following are added in order: 1 ml of Tris · HCl; 0.06 ml of $MgCl_2$; 0.2 ml of $KH_2PO_4/K_2HPO_4$; 2.4 mg of cysteine; 25 mg of disodium ATP; 5.4 mg of 3-phosphoglycerate; 0.1 mg of NAD; 0.2 ml (1.2 mg) of glyceraldehyde-3-phosphate dehydrogenase phosphoglycerate kinase mixture; 25 mCi of neutralized [$^{32}$P]orthophosphate; and sufficient distilled water to give a final volume of 10 ml. The mixture is swirled gently and incubated for 1 hr at 30° in a beaker of water. The mixture is acidified by the addition of 1 ml of HCl, and 250 mg of Norite are added. The solution is mixed well and allowed to stand for 10 min at room temperature. The

[17] J. A. Traugh, S. M. Tahara, S. B. Sharp, B. Safer, and W. C. Merrick, *Nature (London)* **263,** 163 (1976).

[18] O.-G. Issinger, R. Benne, J. W. B. Hershey, and R. R. Traut, *J. Biol. Chem.* **251,** 6471 (1976).

[19] S. M. Tahara, J. A. Traugh, S. B. Sharp, T. S. Lundak, B. Safer, and W. C. Merrick, *Proc. Natl. Acad. Sci. U.S.A.* **75,** 789 (1978).

[20] Prepared by the method of I. M. Glynn and J. B. Chappell, *Biochem. J.* **90,** 147 (1964) as modified by D. A. Walsh, J. P. Perkins, C. O. Brostrom, E. S. Ho, and E. G. Krebs, *J. Biol. Chem.* **246,** 1968 (1971).

suspension is filtered by suction through two 0.45 $\mu$m Millipore filters, and the charcoal is washed with two 10-ml portions of distilled water. The [$\gamma$-$^{32}$P]ATP is then eluted with 30 ml of ammoniated ethanol. The eluate is taken to dryness with a rotary evaporator, 50 ml of distilled water are added, and the solvent is evaporated a second time. The residue is taken up in 5 ml of distilled water and stored at $-20°$.

The preparation of [$\gamma$-$^{32}$P]GTP is performed in a manner analogous to the preparation of ATP. The following are added in order in a 15 × 150 ml test tube: 1.0 ml of water; 0.2 ml of Tris · HCl; 0.02 ml of $MgCl_2$; 0.03 ml of $KH_2PO_4/K_2HPO_4$; 1.2 mg of cysteine; 9 mg of GTP; 1.5 mg of 3-phosphoglycerate, 0.1 mg of NAD; 0.2 ml of glyceraldehyde-3-phosphate dehydrogenase/phosphoglycerate kinase mixture; 25 mCi of [$^{32}$P]-orthophosphate (neutralized), and distilled water to yield 2.0 ml. The mixture is incubated 3.0 hr at 30°, after which the reaction is stopped by adding 0.2 ml of HCl and 150 mg of Norite. The remaining steps are identical to the procedure outlined for the preparation of [$\gamma$-$^{32}$P]ATP.

Nucleotide triphosphate concentrations are determined in triplicate by spectral analysis after diluting the stock solution 1:100.[21] In order to determine specific activity, stock solutions are diluted 1:1000, and aliquots of 0.01 and 0.02 ml are dried on 1 × 2 cm strips of filter paper (Whatman ET 31) and counted in triplicate. Purity of the preparations is checked by thin-layer chromatography as described elsewhere.[22] Our preparations usually contain less than 1% inorganic phosphate and no detectable ADP. Yields of 50–80% of the radioactive phosphate are routinely obtained with specific activities averaging 700–1000 mCi/mmol.

## Purification of cAMP-Independent Protein Kinases Using Casein

*Reagents and Buffers*

[$\gamma$-$^{32}$P]ATP 45–135 mCi/mmol
[$\gamma$-$^{32}$P]GTP 45–135 mCi/mmol
DEAE-cellulose (Whatman DE-52)
Phosphocellulose (Whatman or Sigma)
Cellulose (Sigma-cell 50)
Hydroxyapatite (Bio-Rad HT)
Sulfopropyl-Sephadex (Pharmacia C-50-120)
Buffer A: 20 m$M$ tris(hydroxymethyl)aminomethane (Tris) · HCl, pH 7.4 (4°); 1 m$M$ EDTA; 10 m$M$ 2-mercaptoethanol; 0.02% $NaN_3$

[21] *In* "Ultraviolet Absorption Spectra of 5′-Ribonucleotides," pp. 2–14. P-L Biochemicals, Inc. Circular OR-10, Milwaukee, Wisconsin, 1956.

[22] G. A. Floyd and J. A. Traugh, Vol. LX [46].

Buffer B: 25 m$M$ $KH_2PO_4/K_2HPO_4$, pH 7.0; 10 m$M$ EDTA; 10 m$M$ 2-mercaptoethanol; 0.02% $NaN_3$

Buffer C: Same as buffer B, only 400 m$M$ $K_2HPO_4/KH_2PO_4$, pH 6.8

Toluene scintillation fluid: 5 g of 2,5-diphenyloxazole (PPO); 0.3 g of 1,4-bis[2(4-methyl-5-phenyloxazolyl)]benzene (dimethyl POPOP); 1 liter of toluene

Toluene–Triton scintillation fluid: 12 g of PPO; 0.6 g of dimethyl POPOP; 2 liters of toluene; 1 liter of Triton X-100 (Rohm and Haas)

Casein premixture: 87.5 m$M$ Tris · HCl, pH 7.0 (30°); 196 m$M$ KCl; 17.5 m$M$ $MgCl_2$; 7.51 mg/ml dephosphorylated casein

Histone premixture: 87.5 m$M$ Tris · HCl, pH 7.0 (30°); 17.5 m$M$ $MgCl_2$; 2.45 $\mu M$ cAMP; 7.51 mg/ml histone (Sigma type IIA)

Bovine serum albumin (Sigma), 1.0 mg/ml

*Assay Procedure for the Protein Kinases*

A typical assay for protein kinase activity would contain 0.04 ml of the appropriate assay "premix," 0.02 ml of the enzyme (generally at 1:20 dilution of column fractions with BSA), 0.14 m$M$ [$\gamma$-$^{32}$P]ATP, and distilled water to yield a final volume of 0.07 ml. It is convenient to start the assay by the addition of a solution containing radioactive ATP and "premix" to a tube that already contains enzyme (0°). The assay samples are initiated at fixed intervals, mixed with a vortex mixer, and transferred to a 30° water bath. After 15 min, the tubes are placed once again in an ice-bath, and 0.05-ml aliquots are removed at intervals equal to those used to initiate the reaction. The aliquots are spotted on 1 × 2 cm strips of Whatman ET-31 filter papers, dropped into 10% TCA (10 ml per assay), and gently agitated with a magnetic stirring bar for 10 min. The washing is repeated twice with 5% TCA and finally with 95% ethanol (5 ml per assay). The papers are dried with a heat lamp and then placed in polyester bags (3 ml, Nalge Corp.). After adding 0.2 ml of scintillation cocktail, the bags are sealed and counted in a liquid scintillation counter. Counts from a control assay which lacks kinase are subtracted. Our experiments have shown that this assay follows first-order kinetics with respect to the enzyme when less than 500 pmol of [$^{32}$P]inorganic phosphate are incorporated into casein per 15-min period. One unit of activity (EU) is defined as that amount of enzyme which incorporates 1 pmol of phosphate into protein per minute at 30°.

*Purification Procedure*

All steps are carried out at 0°–4°.

*Step 1. DEAE-Cellulose "Batch" Chromatography.* Two liters of the postribosomal supernate from a reticulocyte lysate are prepared as described elsewhere,[23] and mixed with 228 ml of 10-fold concentrated buffer A and 1 liter (settled volume) of DEAE-cellulose. The mixture is stirred with a glass rod and allowed to stand for 5 min. The stirring and settling are repeated 3 times, and the slurry is then filtered with suction through Whatman No. 1 filter paper. Care must be taken to ensure that the resin is not allowed to be sucked dry. The mixture is washed eight times with 250 ml portions of cold buffer A. The washed gel is then transferred to a 60 × 5 cm column and elution begun with buffer A. After about 200 ml have been collected, a 3.5 liter, linear gradient ranging from 0 to 0.5 *M* KCl in buffer A is applied, and the proteins are eluted at a flow rate of 150 ml/hr.

Fractions (22 ml each) are diluted 1:20 and assayed for protein kinase activity with casein and histone as substrate. Two peaks of cAMP-independent protein kinase activity are observed with casein and identified in order of elution (Fig. 1). Fractions of casein kinase I (CK I) ranging from 75 m*M* to 130 m*M* KCl are pooled (approximately 350 ml), dialyzed against 6 liters of buffer B, and applied directly to the next column. Fractions of

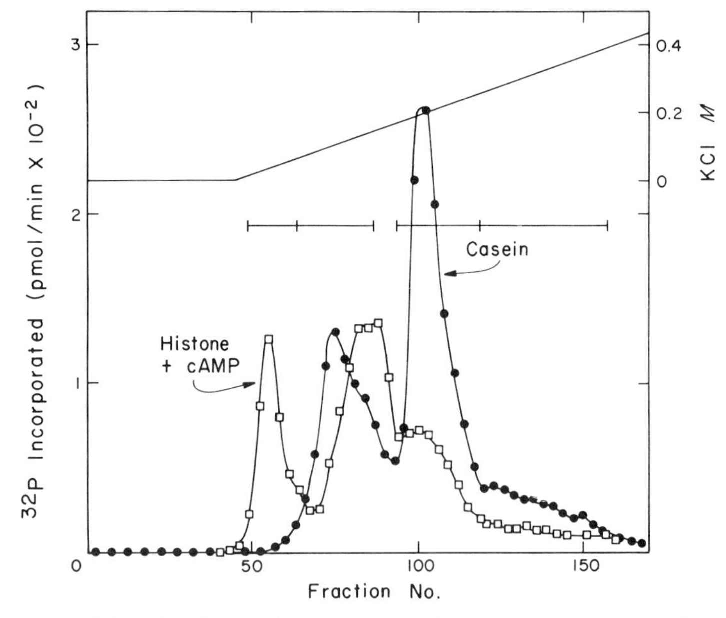


FIG. 1. Batch DEAE-cellulose chromatography of postribosomal supernate from rabbit reticulocytes. Fractions were assayed for kinase activity with casein (●——●) and with histone plus cAMP (□——□) as described in the text.

[23] R. G. Crystal, N. A. Elson, and W. F. Anderson, this series, Vol. 30, p. 101.

casein kinase II (CK II) ranging from 175 m$M$ to 220 m$M$ KCl are pooled, brought to 20 m$M$ 2-mercaptoethanol and 11 m$M$ EDTA, and solid $(NH_4)_2SO_4$ is added to 80% saturation. The pH is controlled by the addition, dropwise, of 2 $N$ $NH_4OH$. The precipitate is collected by centrifugation at 10,400 $g$ for 30 min, redissolved in buffer B, and dialyzed against buffer B.

Two peaks of cAMP-regulated protein kinase are observed using histone as substrate. The type I enzyme is eluted by 30 m$M$ to 60 m$M$ KCl. The type II enzyme appears in the 100 m$M$ to 160 m$M$ KCl fractions. The two activities are pooled separately and precipitated with ammonium sulfate as described for CK II.

*Step 2. Phosphocellulose Chromatography.* The dialyzed pool of CK I is applied to phosphocellulose (10 × 2.5 cm) and washed with one column volume of buffer B. The column is then developed with a 400-ml linear gradient of 0 to 1.25 $M$ NaCl in buffer B. CK I is eluted between 500 m$M$ and 750 m$M$ NaCl.

The concentrated CK II fractions are applied to the first phosphocellulose column under conditions identical to those for CK I. Under these conditions (low ion strength) the bulk of the activity (75–90%) appears in the void volume, and although only a small purification is obtained, a contaminant remains throughout the following steps if this first phosphocellulose column is eliminated. The activity from the first column is then pooled and dialyzed against buffer B which contains 0.25 $M$ NaCl and reapplied to a second phosphocellulose column equilibrated at the higher salt concentration. This column is then developed with a 400-ml gradient ranging from 0.25 $M$ to 1.25 $M$ NaCl in buffer B. CK II is found in fractions that range from 700 m$M$ to 850 m$M$ NaCl.

The type I and type II cAMP-regulated enzymes do not bind to phosphocellulose under the conditions of which the casein kinases are purified. Therefore these protein kinases are readily removed from both CK I and CK II at this step in the procedure. The partially purified cAMP-regulated kinases are present in the flow-through volumes of the phosphocellulose columns and are stored at 4° as ammonium sulfate suspensions.

*Step 3. Sulfopropyl-Sephadex Chromatography.* The CK I activity from the phosphocellulose step is pooled and dialyzed against buffer B. The enzyme is concentrated by batch elution from a 3–5 ml of column of phosphocellulose with 1.25 $M$ NaCl in buffer B, dialyzed against buffer B at pH 6.8, and applied to a sulfopropyl-Sephadex column (10 × 2.5 cm) equilibrated with buffer B. Casein kinase I is eluted between 300 m$M$ and 500 m$M$ NaCl by a linear gradient of 0 to 1.25 $M$ NaCl in buffer B (400 ml). It is important that the pH be no greater than 6.8 for the enzyme to bind to the resin. In addition, the CK I fractions must be assayed and concentrated quickly (batched off 3–5 ml of phosphocellulose) since this enzyme loses activity rapidly in a dilute state after sulfopropyl-Sephadex chromatography.

*Step 4. Hydroxyapatite Chromatography.* We mix 25 ml of hydroxyapatite with 200 ml of a 1% slurry of cellulose in 5 m$M$ potassium phosphate buffer, pH 7.0, to improve flow rates. After decanting fines twice, columns are poured and are then developed at a flow rate of 150 ml/hr. Both CK I and CK II bind to hydroxyapatite columns at pH 7.0. The individually pooled fractions from the previous step are dialyzed and

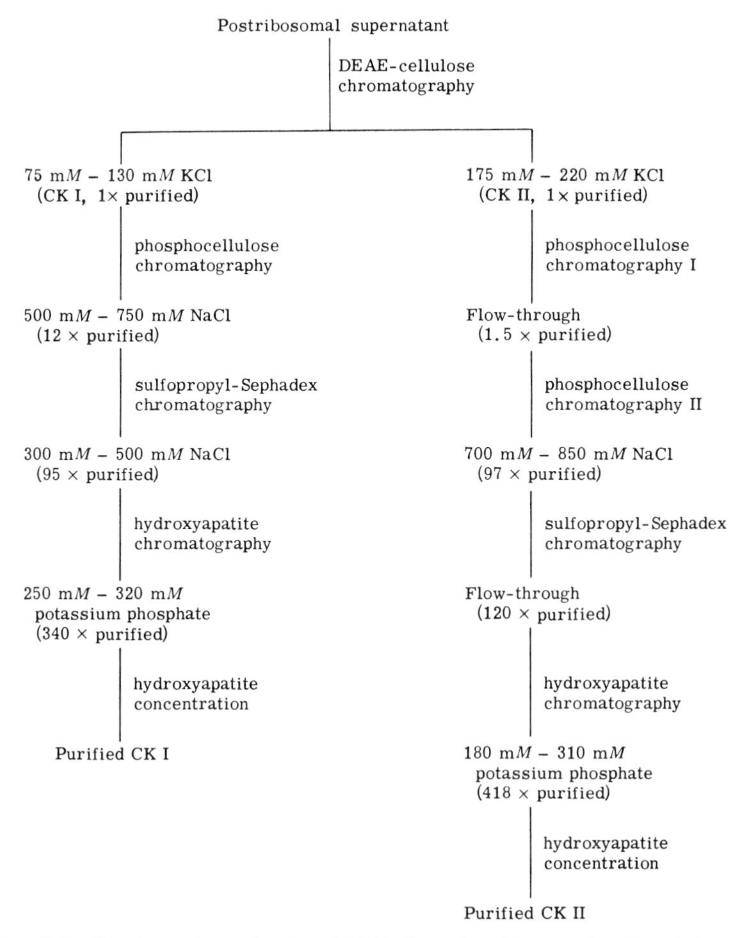


FIG. 2. Purification scheme for the cAMP-independent kinases phosphorylating casein. Purification factors were obtained with the pooled DEAE-cellulose fractions taken as unit purification. This was necessary because the individual activities in the postribosomal supernatant were not readily obtainable.

pumped onto 5 × 1.6 cm columns of hydroxyapatite and developed with 300-ml linear gradients of 25 m*M* to 0.5 *M* potassium phosphate, pH 6.8, which contain 1 m*M* EDTA, 10 m*M* 2-mercaptoethanol, and 0.02% sodium azide. Fractions are diluted 1:20 prior to assay in order to decrease the phosphate concentration which inhibits significantly at levels greater than 20 m*M*. Casein kinase I elutes from hydroxyapatite between 250 and 320 m*M* potassium phosphate. Casein kinase II elutes between 180 and 310 m*M*. Active fractions are pooled, dialyzed briefly against buffer B, and applied to 1.0 ml columns of hydroxyapatite. The enzymes are then removed in 2.0 ml of buffer C and stored at 4°. A typical purification scheme is shown in Fig. 2.

*Criteria of Purity: Polyacrylamide Gel Electrophoresis and Sedimentation Analysis*

Both CK I and CK II were analyzed by gel electrophoresis as described elsewhere in this article. Preparations of CK I show a major protein band of molecular weight 37,000 when stained with Coomassie Blue R-250. Preincubation of CK I with [$\gamma$-$^{32}$P]ATP and 10 m*M* $MgCl_2$ prior to analysis by gel electrophoresis resulted in the phosphorylation of this protein when visualized by autoradiography.

When CK II was electrophoresed under the same denaturing conditions, three major bands appeared with molecular weights of 42,500, 38,000, and 24,000. Incubation of the enzyme with 0.14 m*M* [$\gamma$-$^{32}$P]ATP and 10 m*M* $MgCl_2$ followed by electrophoresis and autoradiography showed that the 24,000 molecular weight band was being phosphorylated.

Sedimentation velocity experiments were performed in the optical centrifuge. The centrifugation was performed by layering 7.5 $\mu$g of CK II in 0.01 ml over a solution of 25 m*M* potassium phosphate buffer, pH 7.1, which also contained 0.50 *M* NaCl. The enzyme solution contained the same buffer with the concentration of NaCl reduced to 0.25 *M* in order to stabilize the boundary. A double-sector centerpiece was employed that contained a well similar to the type described by Vinograd.[24] The Spinco Model E analytical ultracentrifuge was equipped with scanner, UV monochromator, mirror optics, and a cylindrical lens that allowed routine detection at a wavelength of 236 nm. A single, symmetrical peak was detected for CK II. Data were obtained from the motion of the center of gravity of the distribution. Correction of the data to 20° and water yielded a value of 7.50 ± 0.05 S for the sedimentation coefficient of CK II.

[24] J. R. Vinograd, R. Bruner, R. Kent, and J. Weigle, *Proc. Natl. Acad. Sci. U.S.A.* **49,** 902 (1963).

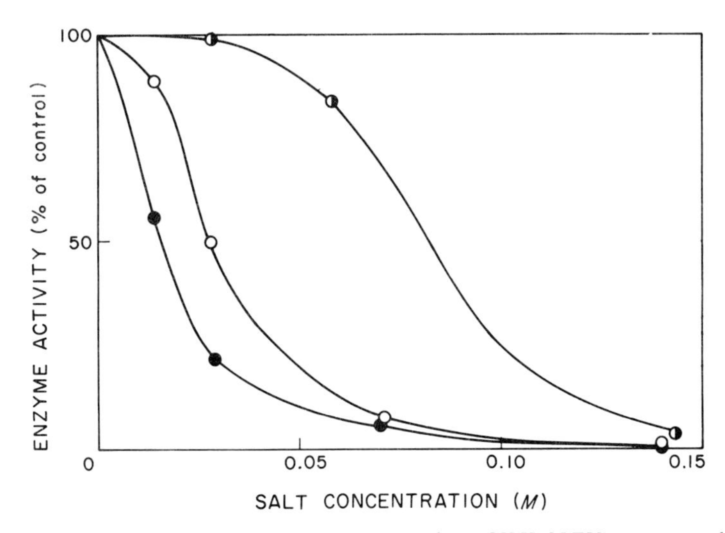


FIG. 3. Inhibition of CK II by sulfate and phosphate ions. CK II (25 EU) was assayed in the standard assay which also contained ammonium sulfate (●—●), ammonium phosphate (○—○), or ammonium chloride (◑—◑).

## Catalytic Properties

Values of $K_m$ for both ATP and GTP were determined from Lineweaver–Burk plots of data obtained under assay conditions zero order with respect to casein. For CK I we obtained $K_m$ values of 13 $\mu M$ and 900 $\mu M$ for ATP and GTP, respectively. Corresponding values for CK II were 10 $\mu M$ for ATP and 40 $\mu M$ for GTP. Thus CK I is readily distinguished from CK II on the basis of the ability to utilize GTP in the assay.

Both CK I and CK II show substrate inhibition with respect to nucleotide triphosphate at levels above about 150 $\mu M$. Both enzymes show a requirement for monovalent cations with an optimum between 100 and 250 m$M$ and for divalent cations with an optimum of approximately 5 m$M$. Casein kinase II is inhibited 50% by phosphate at a concentration of 30 m$M$ (pH 7.0) and by sulfate at 15 m$M$ (Fig. 3). Therefore, in order to obtain accurate estimates of activity, it is necessary to keep phosphate concentration low in the assay and to be certain all sulfate is removed after precipitation with ammonium sulfate.

## Purification of cAMP-Independent Protein Kinases Using Initiation Factors

### *Reagents and Buffers*

Buffer D: 20 m$M$ Tris · HCl, pH 7.4 (4°); 5 m$M$ 2-mercaptoethanol

Initiation factors: supplied by Dr. W. C. Merrick and purified as described in this volume[25]

Solution A: Acrylamide-bisacrylamide: 22.2% acrylamide; 0.6% *N*, *N*′-methylenebisacrylamide. The bisacrylamide was recrystallized in acetone prior to use

Solution B: Resolving gel buffer: 0.4% SDS; 1.5 *M* Tris · HCl, pH 8.8

Solution C: Stacking gel buffer: 0.4% SDS; 0.5 *M* Tris · HCl, pH 6.8

Solution D: Running (tray) buffer: 2.8% glycine; 0.6% Trizma base; 0.1% SDS; final pH of the buffer is 8.5

Solution E: Ammonium persulfate: 0.75% ammonium persulfate. Make the solution just prior to use

Solution F: TEMED (*N*,*N*,*N*′,*N*′-tetramethylethylenediamine)

Solution G: Sample buffer: 15% 2-mercaptoethanol; 6% SDS; 0.5 *M* Tris · HCl, pH 6.8; 30% glycerol; 0.2% bromophenol blue

Solution H: Destaining solution: 50% methanol (v/v); 7.5% acetic acid (v/v)

Solution I: Staining solution: 0.1% Coomassie Brilliant Blue in Solution H

*Phosphorylation of Initiation Factors.* Initiation factors are phosphorylated in a 0.045-ml reaction volume. Each reaction contains the following final concentrations: 50 m*M* Tris · HCl, pH 7.4 (30°) 10 m*M* $MgCl_2$; 0.14 m*M* [$\gamma$-$^{32}$P]ATP or GTP; initiation factor; and protein kinase. The pooled fractions of protein kinase are dialyzed against buffer D prior to use. The mixtures are kept on ice while the additions are made. The reaction is initiated with the radioactive nucleotide, vortexed, and incubated for 30 min at 30°. The reaction is terminated by the addition of 0.02 ml of solution G. The samples are then analyzed by polyacrylamide gel electrophoresis immediately or stored overnight at −20°.

The concentration of initiation factor for the assay is selected as the minimal amount of material that can be easily visualized on polyacrylamide gels. Thus the following microgram amounts of the individual initiation factors are used in the assay: eIF-2, 1.5–2.5; eIF-2A, 1–2; eIF-3, 10–20; eIF-4A, 1–2; eIF-4B, 1–2; eIF-4C, 1.5; eIF-4D, 1.5; eIF-5, 1–2.

*Identification of Phosphorylated Sites by Gel Electrophoresis.* Phosphorylation of the individual subunits of the initiation factors is monitored by polyacrylamide gel electrophoresis followed by autoradiography. A modification of the discontinuous gel system of Laemmli[26] is utilized in a slab apparatus similar to that described by Studier.[27] The glass plates used for the gel mold are 160 × 170 mm. One glass plate contains a 25 mm-deep

[25] W. C. Merrick, Vol. LX [8].

[26] U. K. Laemmli, *Nature* (*London*) **227**, 680 (1970).

[27] F. W. Studier, *J. Mol. Biol.* **79**, 234 (1973).

notch, in a U configuration. Teflon spacers, 0.75 mm thick, are secured on sides and bottom of plates with binder clips.

The following solutions are combined to construct the resolving gel: 4.4 ml of solution A; 2.5 ml of solution B; 0.5 ml of solution E; 2.4 ml of distilled water. The above mixture is deaerated for 5 min with a water aspirator. Solution F (0.01 ml) is added, mixed by gentle swirling, and immediately poured into the gel mold. The gel is overlayered with double-distilled water and allowed to polymerize 10–20 min. The dimensions of the resolving gel after polymerization are 8 cm to 8.5 cm long, 11 cm wide, and 0.75 mm thick.

The stacking gel is prepared by combining the following reagents: 1.0 ml of solution A; 1.13 ml of solution C; 0.33 ml of solution E; 2.25 ml of double-distilled water, and deaerated for 5 min. Solution F (0.005 ml) is added, and mixed by gentle swirling. The water used for overlayering the resolving gel is completely removed and the stacking gel is immediately poured over resolving gel. Ten lanes are made in the stacking gel by inserting a Teflon comb having teeth 0.9 cm wide, and 2.8 cm long, and the gel is allowed to polymerize. Following polymerization, the comb is carefully removed and the lanes are rinsed several times with double-distilled water to remove residual, unpolymerized solution. The stacking gel is 2.5–3 cm in length.

To prepare a sample for loading, 0.02 ml of solution G is added to yield a maximum volume of 0.065 ml. The sample is then heated at 65° for 15 min or 100° for 1 min, and loaded into the gel lane with a syringe. Solution D is placed in the top and bottom wells of the apparatus. Electrophoresis at constant voltage is performed at 100 V until the dye front enters the resolving gel (approximately 1 hr). The voltage is then raised to 150 V until the dye front is 1 cm from the bottom of the resolving gel (approximately 3 hr).

Postelectrophoretic procedures include staining the gel in solution I for 30 min, followed by destaining with several changes of solution H for a total time of 1 hr. If the destaining time is excessive, proteins in the gel leach out. The gel is mounted on Whatman ET-31 filter paper and dried under vacuum over boiling water as described by Maizel.[28]

The dried gel is covered with Saran Wrap and placed against Kodak No-Screen Medical X-ray film. The time of exposure is dependent on the amount of radioactivity incorporated and the specific activity of the phosphate, and ranges from 1 hr to several days.

The amount of phosphate incorporated into the individual subunits of the initiation factors is monitored by two different procedures. With the first method, the desired gel bands are excised from the dried gel using

[28] J. V. Maizel, *Methods Virol.* **5**, 179 (1971).

scissors. Each band is placed in a scintillation vial and allowed to stand overnight in 0.5 ml of 1% SDS. Toluene-Triton scintillation cocktail (7.5 ml) is added prior to counting. The second method of quantitation is accomplished by scanning densitometry of the autoradiogram. A time series of autoradiograms are used to ensure that the degree of exposure of the film by the radioactivity is linearly related to the amount of radioactive phosphate. The representative peaks from the scans are traced, and the areas excised and weighed.

*Identification of Protein Kinase Activities Modifying Initiation Factors*

Enzymes that phosphorylate initiation factors are obtained from the various purification steps used in the preparation and purification of the protein kinases that phosphorylate exogenous substrates. The latter enzymes, which can be easily monitored, are utilized as "markers." This eliminates the necessity of using large amounts of highly purified factor to identify the chromatographic positions of the enzymes. The initial step in the purification is to identify the fractions containing the various protein kinase activities following "batch" chromatography of the postribosomal supernatant fraction on DEAE-cellulose. After the assays with casein and histone are conducted, the eluent is divided and pooled into four fractions as shown in Fig. 1. The pooled fractions are assayed for protein kinase activity using the individual initiation factors as substrate. Since a considerable amount of endogenous phosphorylation occurs at this stage of purification, it is essential to analyze phosphate incorporation into endogenous proteins as well as monitor incorporation into the initiation factors. Thus, pooled fractions are incubated in the presence and the absence of initiation factor and in the presence and the absence of cAMP, and the reaction is analyzed by gel electrophoresis followed by autoradiography.

The fractions shown to contain protein kinase activities that phosphorylate any of the four initiation factors are purified further by phosphocellulose chromatography. The eluents are assayed using casein and histone as substrate, and then analyzed using the purified initiation factors. CK I and CK II are purified further by chromatography of these enzyme fractions on sulfopropyl-Sephadex and hydroxyapatite, and the protein kinase activities copurifying with these enzymes are identified by assaying with initiation factors as described above. The fractions containing the initiation factor kinases are stored in 80% ammonium sulfate at 4°.

*Nomenclature*

It is apparent that a number of chromatographically different forms of phosphotransferase activity are present in rabbit reticulocytes. Some of these cochromatograph with the protein kinase activities that modify the exogenous substrates casein and histone, whereas others do not. In addition many of these protein kinase activities modify different subunits of the same initiation factor or differentially utilize ATP and GTP in the phosphotransferase reaction. Thus it becomes necessary to devise a designation that adequately identifies the individual protein kinase activities while not being too restrictive. The latter is important, since an impure fraction could be composed of one or more activities that will be resolved by further purification.

Therefore we have utilized three pieces of information to identify the individual protein kinase activities that modify initiation factors. First is the name of the initiation factor that is modified [e.g., eIF-2]. If the factor is multimeric, the molecular weight of the subunit ($\times 10^{-3}$) is given in parenthesis [e.g., eIF-2(53)]. Last, the first letter of the nucleotide triphosphate that is preferentially utilized as the phosphate donor in the reaction is designated [e.g., eIF-2(53)A]. If the enzyme effectively utilizes both ATP and GTP as phosphate donor molecules, then both A and G are enumerated. It is important to note that phosphorylation with GTP is due to direct utilization of that nucleotide in the reaction, and not to exchange with the terminal phosphate of ATP and the subsequent utilization of the latter. When the products of the phosphotransferase reactions are analyzed by thin-layer chromatography, the remaining nucleotide has not been altered, and radioactive phosphate is associated with the nucleotide initially added to the reaction mixture.

Thus the three chromatographically distinct phosphotransferase activities that modify eIF-2 are designated eIF-2(53)A, eIF-2(53)AG, and eIF-2(38)AG. The protein kinase activities modifying eIF-3 have been designated eIF-3(all)AG, eIF-3(35)AG, and eIF-3(69, 130)G. The (all) refers to the four subunits in eIF-3 with molecular weights of 35,000, 69,000, 110,000, and 130,000 that are modified. Preliminary evidence indicates that more than one enzyme may be present in the preparation modifying all four subunits. The single polypeptide chain of eIF-4B is modified by eIF-4B A, eIF-4B AG, and eIF-4B G, and eIF-5 is phosphorylated by eIF-5 $A_1$, eIF-5 $A_2$, eIF-5 AG, and eIF-5 G. The subscripts indicate two chromatographically distinct forms that modify the initiation factor and use ATP.

## Results and Discussion

Analysis of the pooled DEAE-cellulose fractions and the eluents obtained by further chromatography of these fractions on additional ion-exchange resins have shown that four of the eight initiation factors are phosphorylated. These are initiation factors 2, 3, 4B, and 5. The chromatographic profile of the protein kinase activities using the individual initiation factors as substrate are summarized in Fig. 4. It can be observed from this figure that all four of the initiation factors are modified by more than one chromatographically distinct protein kinase activity, which in many instances can be further distinguished by substrate specificity with respect to the polypeptide chain comprising the factor, or the nucleotide involved in the reaction. The position of elution for each of the identified protein kinase activities is indicated for the DEAE-cellulose and phosphocellulose col-

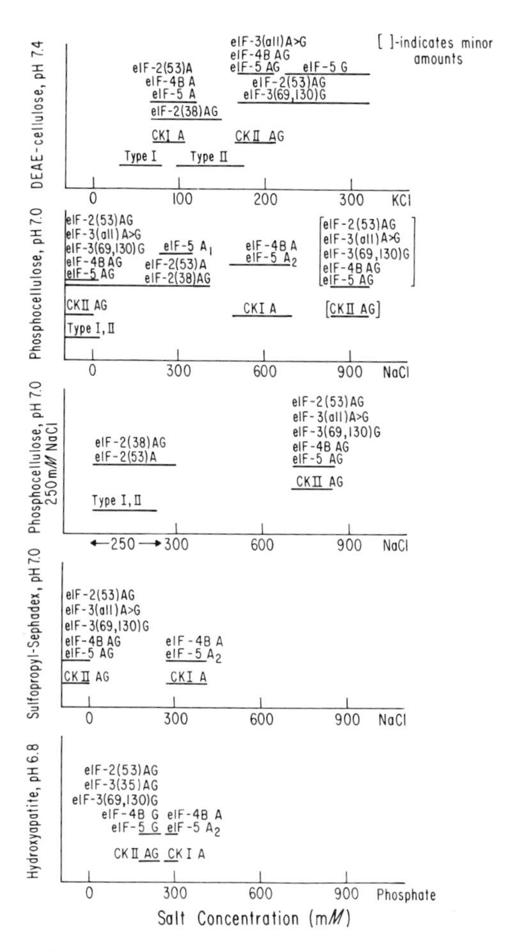


FIG. 4. Diagram of the elution profiles of the protein kinase activities which phosphorylate initiation factors. Aliquots (0.005–0.01 ml) of every second or third fraction of the various column eluents were assayed, using the individual initiation factors as substrate, and analyzed by gel electrophoresis followed by autoradiography.

umns. The protein kinase activities copurifying with CK I and CK II at this stage were purified further on sulfopropyl-Sephadex and hydroxyapatite.

All the activities phosphorylate at least two different initiation factors. It is not clear whether one or more enzymes are responsible for the multiple activities. Homogeneous preparations of the enzymes will be required before this can be determined, since even small amounts of protein kinase activity are detectable. Thus our highly purified preparations of CK I (greater than 75% pure) modify both eIF-4B and eIF-5 with ATP. This could be a single enzymic activity. CK II with a purity of greater than 85% phosphorylates all four factors (Fig. 5). In the phosphorylation of the 53,000-dalton subunit of eIF-2 and the 35,000-dalton subunit of eIF-3, both ATP and GTP are utilized as phosphate donors. In the phosphorylation of eIF-4B, eIF-5 and the 69,000- and 130,000-dalton subunits of eIF-3, GTP is used preferentially. This suggests that either two protein kinase activities are present in the preparation, or the nucleotide donor for the reaction is substrate dependent.

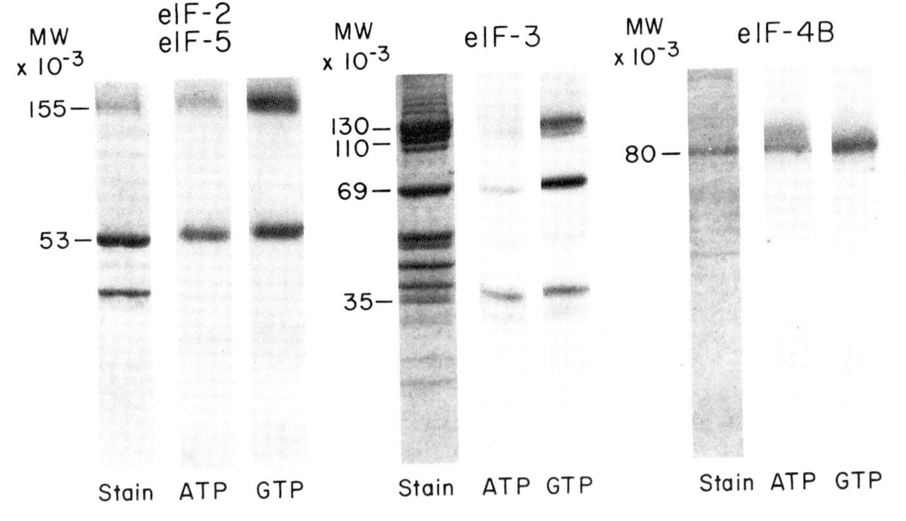


FIG. 5. Identification of initiation factors phosphorylated with purified CK II by gel electrophoresis and autoradiography. The reactions were carried out as described in the text with 110 EU of CK II purified through the hydroxyapatite step, and either 2.0 $\mu$g of eIF-2 and 1.0 $\mu$g of eIF-5, 11 $\mu$g of eIF-3, or 2.0 $\mu$g of eIF-4B. The specific activity of the ATP and GTP was 139 cpm/pmol. The phosphorylated subunits of the initiation factors are indicated by molecular weight ($\times 10^{-3}$). Lanes 1–9 are shown from left to right. Lane 1, stained protein pattern of eIF-2 and eIF-5; 2, autoradiogram with [$\gamma$-$^{32}$P]ATP; 3, autoradiogram with [$\gamma$-$^{32}$P]GTP. Lane 4, stained protein pattern of eIF-3; 5, autoradiogram with ATP; 6, autoradiogram with GTP. Lane 7, stained protein pattern of eIF-4B; 8, autoradiogram with ATP; 9, autoradiogram with GTP.

Further studies are needed to identify whether the different protein kinase activities phosphorylate the same site or distinct sites on the same molecule. Purification of the enzymes to homogeneity will be required to determine whether multiple enzymes are present, each specific for a single initiation factor, or if one protein kinase can phosphorylate two or more factors.

# Section IV

# Polypeptide Chain Elongation

# [36] Isolation of the Protein Synthesis Elongation Factors EF-Tu, EF-Ts, and EF-G from *Escherichia coli*

*By* PETER WURMBACH and KNUD H. NIERHAUS

Since Lipmann and co-workers separated protein factors for complementary stimulation of *in vitro* protein synthesis,[1,2] these factors have been characterized and various preparation procedures have been described for their isolation.[3-11]

Three soluble protein factors participate in a sequence of highly ordered events in the amino acid polymerization elongation cycle (Fig. 1).[12]

1. EF-Tu and EF-Ts: EF-Tu helps to bind AA-tRNA (AA = aminoacyl) via a "ternary complex" of AA-tRNA·EF-Tu·GTP to the ribosomal A-site. After EF-Tu-dependent GTP hydrolysis, the EF-Tu·GDP complex dissociates from the ribosome and is recycled to EF-Tu·GTP, probably via an EF-Tu·EF-Ts intermediate.

The peptidyltransferase, an integral part of the large ribosomal subunit, transfers the peptidyl residue from the peptidyl-tRNA to the aminoacyl-tRNA in the A-site, forming a peptide bond.

2. EF-G: EF-G catalyzes the movement of the ribosome and the codon–anticodon linked mRNA·peptidyl-tRNA complex relative to each other. The result is the removal of deacylated tRNA and the location of peptidyl-tRNA in the P-site. The new codon in the A-site directs the binding of the next ternary complex.

The multifunctionality of EF-Tu is notable: The factor provides at least three binding sites for GTP/GDP, aminoacyl-tRNA, EF-Ts, and the ribosome (the binding sites for GTP/GDP and EF-Ts possibly overlap). Furthermore, extraribosomal activities have been reported: EF-Tu is possibly

[1] J. E. Allende, R. Monro, and F. Lipmann, *Proc. Natl. Acad. Sci. U.S.A.* **51,** 1211 (1964).
[2] J. Lucas-Lenard and F. Lipmann, *Proc. Natl. Acad. Sci. U.S.A.* **55,** 1562 (1966).
[3] Y. Nishizuka, F. Lipmann, and J. Lucas-Lenard, this series, Vol. 12, p. 708.
[4] J. Gordon, J. Lucas-Lenard, and F. Lipmann, this series, Vol. 20, p. 281.
[5] A. Parmeggiani, C. Singer, and E. M. Gottschalk, this series, Vol. 20, p. 291.
[6] P. Leder, this series, Vol. 20, p. 302.
[7] K. Arai, M. Kawakita, and Y. Kaziro, *J. Biol. Chem.* **247,** 7029 (1972).
[8] Y. Kaziro, N. Inoue-Yokosawa, and M. Kawakita, *J. Biochem.* **72,** 853 (1972).
[9] M. S. Rohrbach, M. E. Dempsey, and J. W. Bodley, *J. Biol. Chem.* **249,** 5094 (1974).
[10] A. V. Furano, *Proc. Natl. Acad. Sci. U.S.A.* **72,** 4780 (1975).
[11] G. R. Jacobson and J. P. Rosenbusch, *FEBS Lett.* **79,** 8 (1977).
[12] K. H. Nierhaus and J. Weber, *Umschau* **72,** 346 (1972).



ISBN 0-12-504180-2

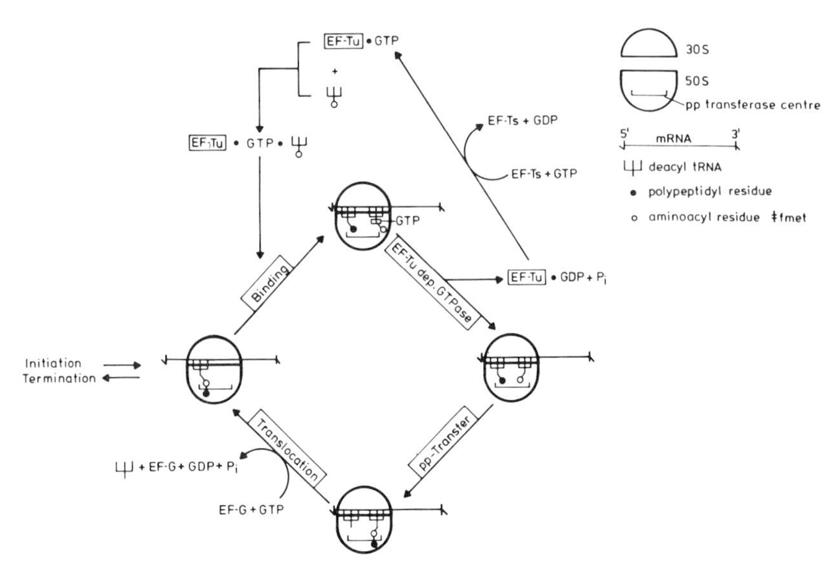


FIG. 1. Participation of bacterial elongation factors in the elongation cycle modified from K. H. Nierhaus and J. Weber, *Umschau* **72,** 346 (1972).

involved in the regulation of RNA synthesis,[13] EF-Tu (like EF-Ts) is one of the four subunits of Q$\beta$ replicase,[14] and EF-Tu is present in the cell membrane.[15] Two loci on the *E. coli* chromosome were found for this factor.[16,17] In this paper we propose a simple procedure for the isolation of both highly purified EF-Tu and EF-G. The method described avoids some difficulties we have had with the already published procedures.

## Elongation Factor Assays

### *Test for EF-Tu: EF-Tu-Dependent GDP Binding*

*Principle.* EF-Tu binds GDP with a binding constant of 4.6 m*M* (for comparison, the corresponding constants for the EF-Tu · EF-Ts and EF-Tu · GTP complexes are 10 n*M* and 0.36 $\mu M$, respectively).[18] The EF-Tu · [$^3$H]GDP complex can be trapped on nitrocellulose filters, in contrast to noncomplexed [$^3$H]GDP.[19]

[13] A. A. Travers, *Nature (London)* **244,** 15 (1973).

[14] T. Blumenthal, T. A. Landers, and K. Weber, *Proc. Natl. Acad. Sci. U.S.A.* **69,** 1313 (1972).

[15] G. R. Jacobson and J. P. Rosenbusch, *Nature (London)* **261,** 23 (1976).

[16] S. R. Jaskunas, L. Lindahl, M. Nomura, and R. R. Burgess, *Nature (London)* **257,** 458 (1975).

[17] S. Pedersen, S. V. Reek, J. Parker, R. J. Watson, J. D. Friesen, and N. P. Fiil, *Mol. Gen. Genet.* **144,** 339 (1976).

[18] K.-I. Arai, M. Kawakita, and Y. Kaziro, *J. Biochem.* **76,** 293 (1974).

[19] J. E. Allende and H. Weissbach, *Biochem. Biophys. Res. Commun.* **28,** 32 (1967).

*Materials*

[$^3$H]GDP solution (Amersham, England, Cat. No. TRK.335; specific activity: 12.7 Ci/mmol; in 50% aqueous ethanol). Before use dilute 20-fold with $H_2O$.

Tu-GDP-Mix: 200 m*M* Tris · HCl, pH 7.5 (0°); 40 m*M* Mg acetate; 600 m*M* $NH_4Cl$, 4 m*M* dithiothreitol (DTT)

Buffer A: 10 m*M* Tris · HCl, pH 7.8 (0°); 10 m*M* Mg acetate; 100 m*M* KCl; 5 m*M* β-mercaptoethanol; 0.58 m*M* ≙ 50 mg/l) phenylmethanesulfonylfluoride (PMSF); 20% (v/v) glycerol. The protease inhibitor PMSF has a low solubility; therefore, we add 50 mg per liter and stir the buffer overnight at 0°.

Dilution buffer: 10 m*M* Tris · HCl, pH 7.5; 10 m*M* Mg acetate

Nitrocellulose filters (Sartorius, Göttingen, Germany; Cat. No. 111–113)

Infrared lamp

Instagel (Packard)

*Procedure*. Tu-GDP-Mix, 10 μl, is added to 5 μl of [$^3$H]GDP solution and to 25 μl of buffer A containing the sample (usually we test the sample in 5 μl per column fraction from the factor isolation; see below). After 20 min at 0°, the reaction mixture is diluted with 1 ml of cold dilution buffer and then passed through a nitrocellulose filter. The filters are dried under an infrared lamp and counted in the presence of 5 ml of Instagel.

## *Test of EF-G: EF-G-Dependent GTPase Activity*

*Principle*. The EF-G factor catalyzes the cleavage of the γ-phosphate from GTP in the presence of ribosomes, but in the absence of other components of protein synthesis.[20] Two test systems are available. In the test system A, GTP-γ-$^{32}$P is used, and the released $^{32}P_i$ is chelated with molybdate in an acidic milieu, extracted with isopropyl acetate, and counted.[21] In the test system B (which we prefer for large numbers of assays) GTP-α-$^{32}$P is used. After GTP hydrolysis the resulting GDP-α-$^{32}$P is separated from GTP by cellulose thin-layer chromatography. This system has the advantage that both the nonhydrolyzed GTP and the hydrolysis product GDP can be assayed in the same sample. For a qualitative judgment it is sufficient to autoradiograph the thin-layer plates.

*Materials*

GTPase mix: 480 m*M* Tris · HCl, pH 7.8 (0°); 440 m*M* $NH_4Cl$; 65 m*M* $MgCl_2$; 140 m*M* β-mercaptoethanol

GTP solution (0.8 m*M*): Test system A: GTP-γ-$^{32}$P (Amersham,

[20] Y. Nishizuka and F. Lipmann, *Arch. Biochem. Biophys.* **116,** 344 (1966).

[21] B. E. Wahler and A. Wollenberger, *Biochem. Zh.* **329,** 508 (1958).

England; specific activity: about 6 Ci/mmol; Cat. No. P.B. 141) was diluted with a cold GTP solution (3 m*M*) to a specific activity of 30 mCi/mmol (e.g., 120 $\mu$l of cold GTP (3 m*M*) mix with 20 $\mu$l of GTP-$\gamma$-$^{32}$P (6 Ci/mmol) and 460 $\mu$l of $H_2O$). Test system B: GTP-$\alpha$-$^{32}$P (Amersham, England; specific activity: about 3 Ci/mmol, Cat. No. P.B. 146) was diluted with a cold GTP solution (3 m*M*) in a similar way to a specific activity of 30 mCi/mmol.

70 S ribosomes, prepared as described elsewhere.[22] Before use the ribosomes are washed with a buffer containing 10 m*M* Tris·HCl, pH 7.5; 10 m*M* Mg acetate; 1 *M* $NH_4Cl$; 6 m*M* $\beta$-mercaptoethanol. When (30 S + 50 S) subunits were used, the 50 S subunits should be isolated from sucrose solution (e.g., zonal gradient fractions) by high-speed centrifugation (16 hr at 100,000 *g*), but not by polyethylene glycol [PEG-6000] precipitation,[22] because PEG-treated ribosomes have lost their EF-G-dependent GTPase activity as tested in this system (our observation).

*Procedure*. In test system A, 45 $\mu$l of $H_2O$ are mixed with 15 $\mu$l of GTPase mix, 15 $\mu$l of GTP solution (about 3–5 × $10^5$ cpm), 3 $A_{260\text{ nm}}$ units of 70 S ribosomes in about 10 $\mu$l, and 5 $\mu$l containing EF-G or the sample to be tested. After incubation (10 min at 37°) the assays are made as follows: 5 $\mu$l of 50 m*M* $KH_2PO_4$ and 75 $\mu$l of 1 *M* $HClO_4$ are added to the reaction mixture. The aggregates are pelleted (5 min at 2000 rpm), and 100 $\mu$l of the supernatant are added to 400 $\mu$l of isopropyl acetate and 300 $\mu$l of 20 m*M* $Na_2MoO_4$. After shaking for 1 min the phases are separated by low-speed centrifugation; 150 $\mu$l of the upper organic phase are removed and counted.

In test system B, the reaction is performed in 50 $\mu$l under identical ionic conditions: 20 $\mu$l of $H_2O$ are mixed with 8 $\mu$l of GTPase mix, 8 $\mu$l of GTP mix, 3 $A_{260\text{ nm}}$ units of 70 S ribosomes in 10 $\mu$l, and 5 $\mu$l with EF-G or the sample. After incubation (10 min at 37°) the reaction is stopped by the addition of 2 $\mu$l of a mixture containing 50% formic acid and 10% TCA (1:1). After 10 min at 0° the aggregates are pelleted (2 min at 3000 rpm) and 10 $\mu$l of the supernatant is spotted on a 10 × 20 cm sheet of PEI-cellulose (Polygram CEL 300; Machery and Nagel, Düren, Germany) with a 10-$\mu$l micropipette (Brand, Wertheim, Germany; Cat. No. 708709). The spots are dried with a hair dryer (no heating), fixed by rinsing with methanol and dried again. The samples are developed by ascending chromatography in 1.5 *M* $KH_2PO_4$ (30 min), dried with a hair dryer (with heating) and exposed to an X-ray film (CURIX, RP1, AGFA) for 2 hr in the dark. After development the result can be judged qualitatively by inspection of the film. If the precise counts are needed, the GTP and GDP spots are marked on the

[22] K. H. Nierhaus and F. Dohme, this series, Vol. 59, [37].

cellulose sheet, cut and counted. GTP appears with an $R_f$ value of about 0.5, GDP of about 0.7.

### *Test for EF-Tu, EF-G, and EF-Ts: (A) Poly(U)-Dependent Poly(Phe) Synthesis with Phe-tRNA*

*Principle.* In the presence of Phe-tRNA and a $Mg^{2+}$ concentration of 12–25 m*M*, efficient poly(Phe) synthesis at times up to 30 min depends on poly(U), ribosomes, GTP, and elongation factors. Providing a 3- to 10-fold excess of EF-Tu over ribosomes is present (optimal amount of EF-Tu), full activity can be obtained by addition of EF-G and E-Tu only (test for EF-G and EF-Tu). If the amount of EF-Tu is reduced to about 5% of the optimal amount, then only nearly background activity is found in the presence of EF-G. Under these conditions the poly(Phe) synthesis depends on the presence of EF-Ts (test for EF-Ts). Thus, EF-Ts can be tested under these conditions, but only when EF-G and EF-Tu are available and free of EF-Ts. (Alternatively, the EF-Ts dependent GDP exchange in a preformed EF-Tu·[$^3$H]GDP complex can be measured.[23] For routine assays we prefer the poly(U) assay demonstrated here, which is more sensitive.)

*Materials.* S150 enzymes were prepared as described elsewhere in this volume.[22] Before use, the S150 enzymes were freed of endogenous tRNA by passage through a Sephadex-DEAE A-50 column. The column was equilibrated with a buffer containing 10 m*M* Tris·HCl, pH 7.5, 10 m*M* Mg acetate, 150 m*M* KCl. After applying the S150 fraction, the proteins including tRNA ligases were eluted with 200 m*M* KCl and tRNA, and other nucleic acids with 500 m*M* KCl in the same buffer.

[$^{14}$C]Phe-tRNA was prepared as follows: 50 $A_{260\ nm}$ units of $tRNA^{Phe}$ or 40 mg of $tRNA^{E.coli}$ (both from Boehringer, Mannheim, Germany) in 7.5-ml reaction volume (50 m*M* Tris·HCl, pH 7.3, 10 m*M* Mg acetate, 100 m*M* KCl, 3.2 m*M* ATP-$Na_2$, and 6 m*M* $\beta$-mercaptoethanol) are mixed with 0.25 mCi [$^{14}$C]phenylalanine (Amersham, England; Cat. No. CFB. 70; specific activity 495 mCi/mmol). After checking that the pH is 7.5, S150 enzymes (1 ml) are added and the mixture is incubated for 15 min at 37°. The reaction is stopped by chilling in an ice bath, and 0.375 ml of 20% (w/v) sodium acetate, pH 5.4, is added. After checking the input radioactivity, the proteins are extracted by vigorously shaking with 7.5 ml of 75% phenol/water (freshly distilled phenol is stored in 10-ml portions at −20°; before use 3.3 ml $H_2O$ is added). The phases are separated by centrifugation (5 min at 5000 rpm). After adding 5 ml of $H_2O$ to the phenol phase, the extraction is repeated. The two aqueous phases are combined, and the nucleic acid

[23] D. L. Miller and H. Weissbach, this series, Vol. 30, p. 219.

fraction is precipitated with 2 volumes of ethanol, keeping it 3 hr at −20°. After low-speed centrifugation (10 min at 10,000 rpm) the pellet is washed with ethanol, dried in a desiccator, and dissolved in 1 ml of $H_2O$. Absorption at 260 nm and radioactivity are measured, and 200-μl aliquots are stored at −80°. Other materials for the poly(U) system are as described elsewhere in this series.[22]

*Procedure.* The poly(U) system is performed as described elsewhere in this series,[22] except that tRNA and [$^{14}$C]phenylalanine are replaced by [$^{14}$C]Phe-tRNA (about 100,000–200,000 cpm per assay) and the S150 enzymes by appropriate amounts (e.g., 5 μl for routine tests of column fractions) of the purified factors EF-Tu, EF-Ts, and EF-G.

*Test for EF-Tu, EF-G, and EF-Ts: (B) R17-Dependent Protein Synthesis with AA-tRNA*

*Principle.* The principles for analyzing the various factors are identical to those explained above for the poly(U) system.

*Materials.*

Premix: 150 m*M* Tris·HCl, pH 7.8, 180 m*M* $NH_4Cl$, 30 m*M* Mg acetate, 3 m*M* DTT

70 S ribosomes were isolated as described elsewhere in this series.[22] The ribosomes are in 10 m*M* Tris·HCl, pH 7.5; 10 m*M* Mg acetate; 60 m*M* $NH_4Cl$, and 5 m*M* β-mercaptoethanol

Aminoacyl-tRNA: 40 mg of $tRNA^{E.coli}$ were treated precisely as described in the poly(U) system with two exceptions: (1) Instead of [$^{14}$C]phenylalanine, 200 μl of a cold amino acid mixture were added containing all 20 amino acids, each at 2.5 m*M*. When radioactive aminoacyl tRNA was desired, 5 ml of a [$^3$H]amino acid mixture (Amersham, England; about 5 mCi; TRK. 440) was added in addition to the cold amino acids. (2) To the total volume of 7.5 ml the formyl donor (10-formyltetrahydrofolate; 1 ml with 15 μmol) was added. The formyl donor was prepared following the procedure of Dubnoff and Maitra.[24] The final volume of 8.5 ml was treated as described for the poly(U) system. The tritiated aminoacyl-tRNA preparation contained about 1000 $A_{260\ nm}$ units/ml with $0.7 \times 10^6$ cpm per $A_{260\ nm}$ unit.

Main mix: 150 m*M* Tris·HCl, pH 7.8, 16 m*M* Mg acetate, 50 m*M* $NH_4Cl$, 6 m*M* DTT, 0.6 m*M* GTP, 5 m*M* ATP, 18 m*M* phosphoenol pyruvate, 2μg pyruvate kinase, 10% (w/v) polyethylene glycol-6000 (Merck, Darmstadt, Germany)

[24] J. S. Dubnoff and U. Maitra, this series, Vol. 20, p. 248.

Crude initiation factors (0.1–5 mg/ml) in 20 m$M$ Tris·HCl, pH 7.5 (0°); 1 m$M$ EDTA; 100 m$M$ $NH_4Cl$; 0.5 m$M$ DTT; 10% glycerol. The crude initiation factors were obtained in the following way: Ribosomes pelleted from the crude extract (S-30) as described[22] were resuspended overnight in 10 m$M$ Tris·HCl, pH 7.5, 1 $M$ $NH_4Cl$, and 6 m$M$ $\beta$-mercaptoethanol at a concentration of 0.5 g of ribosomes per milliliter. The ribosomes were pelleted (5 hr at 150,000 $g$). The supernatant was dialyzed against two changes of the glycerol buffer indicated above, stored at −80° and used as crude initiation factor preparation.

R17 RNA (from the RNA phage R17; Boehringer, Mannheim, Germany; Cat. No. 166065) in 10 m$M$ Tris·HCl, pH 7.5, 6 m$M$ $\beta$-mercaptoethanol

*Procedure*. A preincubation is performed to get rid of the endogenous mRNA. The preincubation mixture consists of 25 $\mu$l of premix. 1.5 $A_{260\ nm}$ units of 70 S ribosomes in 20 $\mu$l, 0.2 $A_{260\ nm}$ units of cold aminoacyl tRNA in 10 $\mu$l, 10 $\mu$l of EF-G, and 10 $\mu$l of EF-Tu (0.1–5 $\mu$g per factor). After incubation at 37° for 30 min, 25 $\mu$l was removed and mixed with 40 $\mu$l of main mix, 15 $\mu$l of crude initiation factors (2–10 $\mu$g), 1 $A_{260\ nm}$ unit R17 RNA in 15 $\mu$l and 5 $A_{260\ nm}$ units of [$^3$H]aminoacyl-tRNA in 5 $\mu$l. After incubation for 30 min at 37° the samples were treated in the same way as described for the poly(U) system.[22]

## Isolation of Elongation Factors

### *Isolation of EF-Tu and EF-G*

*Principle.* Gel filtration on Sephadex G-100 columns separates the factors EF-Tu and EF-G to a large extent in the presence of GDP. A complete separation of these two factors without any cross contamination is achieved by two successive G-100 gel filtration steps. Other proteins are removed by DEAE-Sephadex A-50 chromatography, in the absence of GDP. The pure factors are concentrated on a small DEAE-Sephadex A-50 column using a stepwise elution with 1 $M$ KCl. Ammonium sulfate precipitation is not used for the fractionation of the factors (this method does not work satisfactorily in our hands), but only for concentrating them (at 80% saturation, see Fig. 2).

*Materials*

Buffer A: 10 m$M$ Tris·HCl, pH 7.8 (0°), 10 m$M$ Mg acetate, 100 m$M$ KCl, 5 m$M$ $\beta$-mercaptoethanol, 0.58 m$M$ phenylmethanesulfonyl fluoride (PMSF; 50 mg/liter; added as last component, stir overnight), 20% (v/v) glycerol

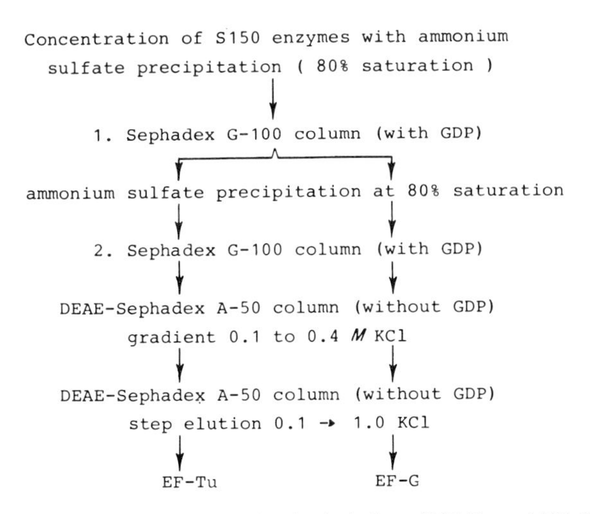


FIG. 2. Preparation scheme for the isolation of EF-Tu and EF-G.

Two Sephadex G-100 (fine) columns (3 × 200 cm) equilibrated with buffer A in the presence of 20 $\mu M$ GDP
Ammonium sulfate
Two DEAE-Sephadex A-50 columns (2 × 60 cm) equilibrated with buffer A
Two DEAE-Sephadex A-50 columns (1 × 30 cm) equilibrated with buffer A
Buffer B: 50 m$M$ Tris · HCl, pH 7.8, 2 m$M$ $\beta$-mercaptoethanol, 50% (v/v) glycerol

*Procedure.* The preparation is performed at 0–4°. S150 enzymes are prepared from 100–150 g of *E. coli* K 12 strain A 19 as described elsewhere in this series.[22] To 100–150 ml of S150 enzymes 1 $M$ Tris · HCl, pH 7.8 (2–3 ml) is added to raise the buffer capacity. The solution is brought to 80% saturation with solid $(NH_4)_2SO_4$ (42.5 g in 100 ml), stirred 30 min and centrifuged for 30 min at 30,000 g. The pellet is resuspended in a minimum (5–7 ml) of buffer A containing 20 $\mu M$ GDP (buffer A/GDP) and dialyzed twice until the solution is clear (about 2 hr) against 1 liter of the same buffer A/GDP.

The sample is mixed with 1/10 volume of 60% sucrose and applied to a Sephadex G-100 column (3 × 200 cm): The sample is cautiously layered on the gel beneath the overlay buffer by means of a L-shaped glass capillary connected to a 5-ml pipette with a 25-cm silicon tube (diameter 2 mm). Fractions of about 7 ml are collected (100 drops per fraction; 20 sec per drop), and the absorbance at 230 and 280 nm, the EF-Tu-dependent GDP binding and the EF-G-dependent GTPase activity are determined from every third fraction (Fig. 3, top row). EF-G appears in the elution volume from 450 to 600 ml, EF-Tu from 550 to 750 ml, both well separated from the bulk of material absorbing at 280 nm.

EF-G- and EF-Tu-containing fractions are pooled, avoiding severe cross contamination, as indicated, concentrated with $(NH_4)_2SO_4$ (80% saturation), resuspended in 4 ml of buffer A/GDP, and dialyzed against the

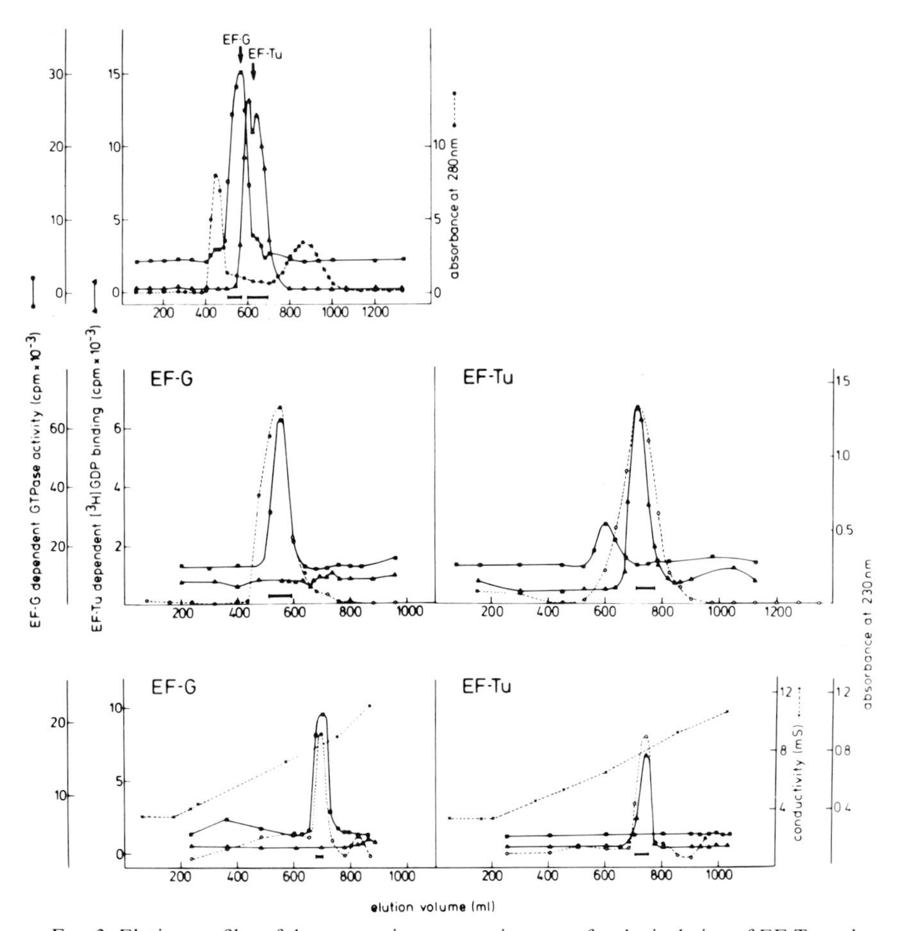


FIG. 3. Elution profiles of the successive preparation steps for the isolation of EF-Tu and EF-G. Top row: elution profile of S150 enzymes on Sephadex G-100; middle row, the second gel filtration step; bottom row, the factors are further purified on DEAE-Sephadex A-50. ●---●, Absorbance at 280 nm; ○---○, absorbance at 230 nm; ▲——▲, EF-Tu-dependent [³H]GDP binding; ■—■, EF-G-dependent GTPase activity; x---x, conductivity; ├──┤, pooled fractions.

same buffer. After mixing with 1/10 volume of 60% sucrose, the samples are subjected to the second gel filtration step (column: 3 × 200 cm). After gel filtration the EF-G- and EF-Tu-containing fractions can be pooled conveniently without any cross contamination (Fig. 3, middle row).

The pooled fractions are applied directly (without any concentrating and dialysis steps) to DEAE-Sephadex A-50 columns (2 × 60 cm) equilibrated in buffer A (without GDP). The proteins are eluted with a linear 1000-ml gradient from 0.1 to 0.4 *M* KCl in buffer A. It is advantageous to collect the fractions (100 drops per fraction) with respect to volume as opposed to time because the flow rate increases drastically. The fractions containing EF-G and EF-Tu, respectively, are collected (Fig. 3, lower row), dialyzed against buffer A and again applied to a DEAE A-50 column (1 × 30 cm). The factors are concentrated by a step elution using buffer A containing 1 *M* KCl. The fraction containing the factor is dialyzed three times for at least 60 min against 1 liter of buffer B, the absorption at 230 nm is determined (1 $A_{230nm}$ unit $\hat{=}$ 250 $\mu$g of protein), and the purified factors are stored at −20°.

### *Isolation of EF-Ts (and EF-Tu)*

*Principle.* This technique follows essentially that of Arai *et al.*[7] S150 enzymes are chromatographed on a DEAE-Sephadex A-50 column in the absence of GDP. Two Tu-containing peaks are eluted (detected by the GDP binding assay); the peak eluted with lower ionic strength consists of the EF-Tu · EF-Ts complex, the other is free EF-Tu. The EF-Tu · EF-Ts complex is pooled, dissociated with GDP, and subjected to a second chromatography step on DEAE-Sephadex A-50 in the presence of GDP. EF-Ts is eluted at lower ionic strength and EF-Tu at higher ionic strength than the nonfactor proteins, thus leading to purified EF-Ts and EF-Tu factors (see Fig. 4).

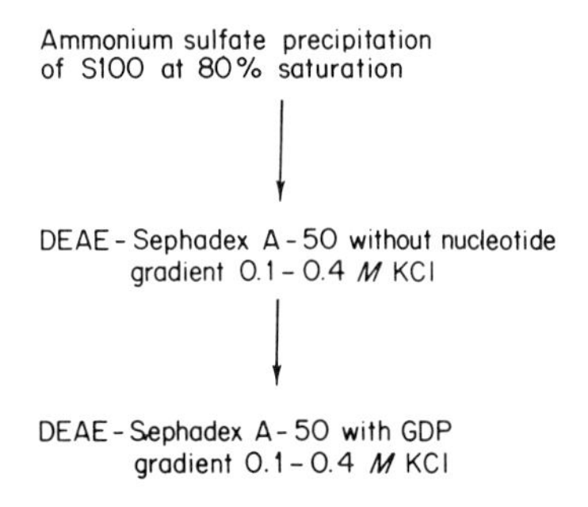


FIG. 4. Preparation scheme for the isolation of EF-Ts (and EF-Tu).

*Materials*

Ammonium sulfate

Two DEAE-Sephadex A-50 columns (2 × 60 cm) equilibrated with buffer A. The buffer of one column contains in addition 20 $\mu M$ GDP.

*Procedure.* S150 enzymes (100–150 ml) are prepared from 100–150 g of cells (*E. coli* K12, strain A19) as described elsewhere.[22] Tris buffer is added, and the proteins are precipitated with $(NH_4)_2SO_4$ following the method given for EF-G and EF-Tu above. The pellet is resuspended in 10–20 ml of buffer A without GDP, dialyzed against the same buffer and applied to a DEAE-Sephadex A-50 column equilibrated without GDP. The proteins are eluted with a 1000-ml linear gradient from 0.1 to 0.4 $M$ KCl in buffer A. Fractions containing 100 drops are collected and EF-Tu is traced by the GDP binding assay. The first peak (Fig. 5A) eluted at lower ionic

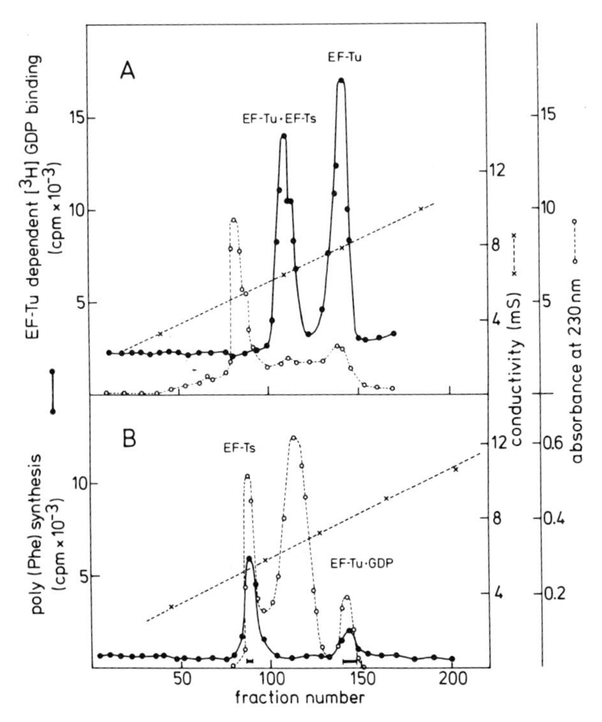


FIG. 5. Elution profiles of the preparation steps for the isolation of EF-Ts (and EF-Tu). (A) Chromatography (DEAE-Sephadex A-50) of S150 enzymes in the absence of GDP; (B) second chromatography on DEAE-Sephadex A-50 in the presence of GDP. ○---○, Absorbance at 230 nm; ●——●, EF-Tu-dependent GDP binding (A) and poly(Phe) synthesis in the presence of EF-G and under nonsaturating EF-Tu amounts (B), respectively; x---x, conductivity; ⊢—⊣, pooled fractions.

strength represents the EF-Tu·EF-Ts complex. This peak is pooled, dialyzed against buffer A containing 20 $\mu M$ GDP and applied to a second DEAE-Sephadex A-50 column equilibrated in the presence of GDP. The proteins are eluted with the same gradient as above but with GDP. Three peaks appear (Fig. 5B); the peak eluted at the lowest ionic strength (5.5 mS) is pure EF-Ts, the next peaks are the nonfactor proteins, and the last one (8 mS) is pure EF-Tu·GDP. Both factors are dialyzed against buffer B, the absorption at 230 nm is determined (1 $A_{230nm}$ unit $\hat{=}$ 250 $\mu$g protein), and the factors are stored at $-20°$.

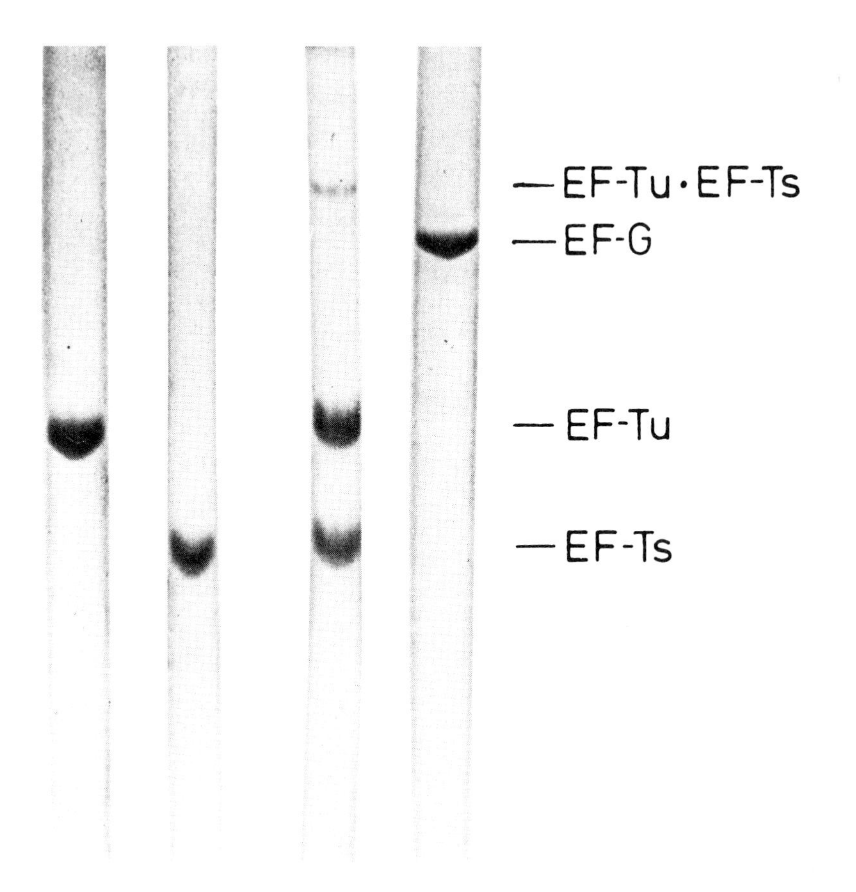


FIG. 6. Sodium dodecyl sulfate SDS–gel electrophoresis [10% polyacrylamide, 0.27% bisacrylamide, 6 *M* urea; buffer 0.2 *M* Na phosphate, pH 7; 0.2% (w/v) SDS, 90 V for 3.5 hr] of purified factors. Twenty micrograms of factors in buffer B were mixed with 50 $\mu$l of gel buffer, heated for 5 min at 90°, and applied to the gels. The bands were stained with Coomassie Brilliant Blue.

## Results

Figure 6 demonstrates the purity of the isolated elongation factors after polyacrylamide gel electrophoresis. All the isolated factors show one band only; when EF-Tu and EF-Ts are mixed, a faint third band appears, indicating the EF-Tu · EF-Ts complex.

The activity of the purified EF-G and EF-Tu factors is demonstrated in Table I. EF-G alone shows no activity, and EF-Tu 2.5% activity, in the poly(U) system. The low activity seen with EF-Tu is not due to traces of EF-G in the EF-Tu preparation but rather to small amounts of EF-G still remaining on the 70 S ribosomes, which are washed only once (data not shown). Using saturating amounts of both factors (EF-Tu was added in a 10-fold molar excess over ribosomes), an activity can be obtained that is comparable to that found with S150 enzymes. In the R17 system a high background is seen due to significant amounts of EF-Tu and EF-G present in the crude initiation factor preparation. Nevertheless more than 80% of the activity depends on the presence of added factors (190,000 compared to 23,000 cpm). S150 enzymes show slightly higher activity than (EF-G + EF-Tu), possibly owing to the presence of tRNA ligases which may recharge deacylated tRNA.

When EF-Ts is tested the amount of EF-Tu must be reduced drastically. Under these conditions, only small amounts of the ternary complex EF-Tu · AA-tRNA · GTP are available for a particular codon in the A-site. This situation reflects the *in vivo* condition: Although EF-Tu is present in a 10-fold molar excess over ribosomes in the cell,[10] only a small number of ternary complexes is available for any specific codon in the A site (statistically less than 0.5 complex per codon). Under this EF-Tu condition the

TABLE I
ACTIVITIES OF EF-Tu AND EF-G IN PROTEIN SYNTHESIS[a]

| Elongation factors | | Poly(U) | |
|---|---|---|---|
| EF-Tu | EF-G | system (cpm) | R17 system (cpm) |
| – | – | 309 | 23,396 |
| + | – | 1,251 | 62,423 |
| – | + | 259 | 48,032 |
| + | + | 50,334 | 190,034 |
| Controls | | | |
| + | + | — | Minus IF 2,143 |
| + | + | — | Minus R17 RNA 7,048 |
| S150 enzymes | | 51,092 | 261,172 |

[a] EF-Tu was added in a 10-fold molar excess over the ribosomes. [$^{14}$C]Phe-tRNA was used in the poly(U) and [$^{3}$H]aminoacyl-tRNA in the R17 system. IF means crude initiation factor.

TABLE II
ACTIVITY OF EF-Ts IN THE POLY(U) SYSTEM[a]

| Elongation factors | | | |
|---|---|---|---|
| EF-Tu (undersaturated) | EF-Ts | EF-G | Poly(U) system (cpm) |
| − | − | − | 370 |
| + | − | − | 3,391 |
| − | + | − | 337 |
| − | − | + | 439 |
| + | + | − | 4,724 |
| − | + | + | 495 |
| + | − | + | 2,797 |
| + | + | + | 36,491 |
| S150 enzymes | | | 75,232 |

[a] In this experiment an undersaturating amount of EF-Tu was added (5% of that amount added in the experiment of Table I).

poly(Phe) synthesis activity is almost completely dependent on the presence of EF-Ts. This is in agreement with older findings[2] (see also Table II: 797 and 36,491 cpm, respectively), underlining the importance of EF-Ts for protein synthesis.

# [37] Preparation of Homogeneous *Escherichia coli* Translocation Factor G

*By* PHILIP LEDER

G factor is one of two soluble *E. coli* factors required for the elongation of polypeptide chains.[1] Specifically, it is required for the translocation reaction in which peptidyl-tRNA is thought to be shifted from a recognition to a holding site on the ribosome. During this reaction, mRNA is also thought to be shifted a single codon frame allowing recognition of the next codon in sequence.[2] The purification scheme for G factor which is described here relies in its early steps upon that described by Lucas-Lenard and Lipmann.[3] Subsequent steps have been described by Leder, Skogerson, and Nau.[4] The product obtained by this purification, though only 30- to 50-fold purified with respect to crude

[1] Y. Nishizuka and F. Lipmann, *Arch. Biochem. Biophys.* **116**, 344 (1966).
[2] R. W. Erbe and P. Leder, *Biochem. Biophys. Res. Commun.* **31**, 798 (1968).
[3] J. Lucas-Lenard and F. Lipmann, *Proc. Nat. Acad. Sci. U. S.* **55**, 1562 (1966).
[4] P. Leder, L. E. Skogerson, and M. M. Nau, *Proc. Nat. Acad. Sci. U. S.* **62**, 454 (1969).



ISBN 0-12-504180-2

extracts, is homogeneous by several criteria. This reflects the very high concentration of G factor in rapidly growing *E. coli*.[5]

## Assay Methods

*Principle.* Although G factor can be assayed through its ribosomal-dependent GTPase activity or, more conveniently, using immunochemical techniques, the specificity of its role in complementing the function of T factor in polyphenylalanine synthesis makes this assay the most generally applicable. The assay is based on the requirement for G factor for the synthesis of [$^{14}$C]polyphenylalanine in the presence of polyuridylic acid, ribosomes, [$^{14}$C]phenylalanyl-tRNA, GTP, and partially purified T factor.

*Reagents*

- Five-times standard buffer (5 × SB) containing
  - Tris·acetate, 0.25 *M*, pH 7.2
  - $NH_4Cl$, 0.25 *M*
  - Magnesium acetate, 50 m*M*
- Polyuridylic acid, 20 mg/ml
- Ribosomes,[6] 500 $A_{260}$/ml
- [$^{14}$C]Phe-tRNA,[7] 1 nmole/ml
- GTP, 10 m*M*
- T factor,[6] 2 mg/ml
- Trichloroacetic acid, 10%
- Trichloroacetic acid, 5%

*Procedure.* Reactions are carried out in a volume of 0.05 ml and assembled at 0°C. Each reaction mixture contains 5 × SB, 0.01 ml; polyuridylic acid, 0.01 ml; ribosomes, 0.002 ml; T factor, 0.005 ml; and the G factor aliquot. Reactions are started by the simultaneous addition of [$^{14}$C]Phe-tRNA, 0.01 ml, and GTP, 0.005 ml. Incubation is at 23° for 10 minutes, so that initial rates may be measured. Reactions are stopped by the addition of 1 ml of 10% trichloroacetic acid, heated to 90° for 20 minutes, cooled, and passed through nitrocellulose filters.[8] These are washed three times with 3-ml portions of ice cold 5% trichloroacetic acid. If a limiting amount of G factor is used, specific activity may be expressed

[5] P. Leder, L. E. Skogerson, and D. Roufa, *Proc. Nat. Acad. Sci. U. S.* **62**, 928 (1969).

[6] *E. coli* ribosomes and T factor are prepared according to R. W. Erbe, M. M. Nau, and P. Leder, *J. Mol. Biol.* **39**, 441 (1969).

[7] *E. coli* B tRNA (General Biochemicals) is prepared according to G. von Ehrenstein, this series, Vol. XII, Part A, p. 588 (1967).

[8] M. W. Nirenberg, this series, Vol. VI, p. 17 (1963).

as $\mu\mu$moles [$^{14}$C]Phe incorporated in 10 minutes per microgram of protein.

## Purification Procedure

*Reagents*

Standard buffer: Tris-acetate, 50 m$M$, pH 7.2; $NH_4Cl$, 50 m$M$; magnesium acetate, 10 m$M$; dithiothreitol, 1 m$M$
DNase (Worthington), 1 mg/ml
$(NH_4)_2SO_4$ (enzyme grade)
Elution buffer A: Tris-Cl, 10 m$M$; pH 8.0; KCl, 0.15 $M$; dithiothreitol, $10^{-4}$ $M$
Elution buffer B: Tris·HCl, 10 m$M$, pH 7.0; KCl, 0.65 $M$; dithiothreitol, $10^{-4}$ $M$
Elution buffer C: Potassium cacodylate, 50 m$M$, pH 6.3; KCl, 50 m$M$; dithiothreitol, $10^{-4}$ $M$
Elution buffer D: potassium cacodylate, 50 m$M$, pH 6.3; KCl, 0.4 $M$; dithiothreitol, $10^{-4}$ $M$
DEAE-Sephadex A50 (Pharmacia)
DEAE-cellulose DE52 (Whatman)
*E. coli* MRE-600, early log phase

All steps are carried out at 4°. Dithiothreitol is added to solutions just before use.

*Step 1. Crude Extract.* Fifty grams of frozen, early log phase *E. coli* MRE-600 are thawed in 150 ml of standard buffer and passed once through a French pressure cell at 12,000 psi; 0.15 ml of DNase solution is added, and the extract is allowed to stand at 4° for 30 minutes and centrifuged at 25,000 $g$ (maximum) in a Sorvall Model RC2-B centrifuge for 20 minutes. The supernatant is again centrifuged at 106,000 $g$ in a Spinco Model L2-65B centrifuge for 120 minutes, poured off, and saved.

*Step 2. Ammonium Sulfate Fractionation.* The 106,000 $g$ supernatant is brought to 40% saturation with solid $(NH_4)_2SO_4$ (22.6 g/100 ml) while the pH is maintained at 7.2 with 0.1 $N$ KOH. After 20 minutes the solution is centrifuged at 30,000 $g$ for 20 minutes and the precipitate discarded. Additional ammonium sulfate is added to the supernatant to bring it to 65% saturation (17.2 g/100 ml). After 20 minutes the solution is centrifuged at 30,000 $g$ for 20 minutes, the supernatant discarded and the precipitate dissolved in about 10 ml elution buffer A.

*Step 3. DEAE-Sephadex.* After overnight dialysis against 2 liters of elution buffer A, the sample is applied to a 3 × 30 cm DEAE-Sephadex column previously equilibrated with elution buffer A and eluted with a linear gradient between 1 liter of elution buffer A and 1 liter of elution buffer B; 10-ml fractions are collected every 10 minutes. Under these conditions, T factor (assayed by [$^{14}$C]GTP binding[9]) is resolved into two peaks, both separated from G factor. The first T factor peak is frozen in 0.1-ml aliquots for use in further complementation assays for G factor. The fractions are assayed for G factor, and the peak fractions are pooled.

*Step 4. Acid DEAE-Cellulose.* The pooled fractions are dialyzed against 2 liters of elution buffer C overnight, applied to a 3 × 30 cm DEAE-cellulose column previously equilibrated with elution buffer C and eluted by a linear gradient between 1 liter of elution buffer C and 1 liter of elution buffer D; 10 ml fractions were collected every 10 minutes. The column is assayed for G factor by complementation and the most active fractions pooled. The factor is frequently quite pure at this point, often requiring no further purification as judged by polyacrylamide gel electrophoresis. Occasionally, a minor band will persist and can be removed by the crystallization step noted below, or by repetition of step 4.

*Step 5. Crystallization.* The procedure followed is that outlined by Jakoby.[10] The pooled fractions are precipitated by the addition of 43.6 g/100 ml of solid $(NH_4)_2SO_4$ to 70% saturation, allowed to stand 20 minutes and centrifuged at 30,000 *g* for 20 minutes. The precipitate is successively back-eluted for 30 minutes at 0° in 50, 45, 40, and 35% saturated $(NH_4)_2SO_4$ and the undissolved portion removed by centrifugation. The most active eluates, generally 45 and 40%, may be used directly as a source of homogeneous G factor. Frequently, upon standing in the cold, microcrystals form in the eluates and these may be removed by centrifugation. The process of crystallization is accelerated by allowing the eluates to stand undisturbed at *room temperature.* The crystals thus formed can be redissolved in elution buffer C and retain activity when stored at 4° at a concentration of 2 mg/ml, or at lower concentrations when frozen in liquid nitrogen.

*Remarks.* In contrast to T factor, G factor is quite stable during preparation. The process may be stopped at any point after step 2 and the preparation frozen without significant loss of activity. Very pure preparations in dilute solution (<2 mg/ml) are not stable and are best frozen in small aliquots which can be used as needed. An average preparation at the precrystalline eluate stage will show a 30–50-fold increase in specific activity over crude extract, will be homogeneous by gel electrophoresis and immunoelectrophoresis and will retain about 30% of the original activity.

[9]See J. M. Ravel and R. L. Shorey this volume [39].
[10]W. B. Jakoby, *Anal. Biochem.* **26**, 295 (1968).

# [38] The Binding of *Escherichia coli* Elongation Factor G to the Ribosome

*By* JAMES W. BODLEY, HERBERT WEISSBACH, and NATHAN BROT

*Principle.* Elongation factor G (EF G) is one of the soluble factors involved in the process of peptide chain elongation. It is required for translocation to occur, i.e., the movement of the ribosome relative to the mRNA. GTP is required for EF G to function in the overall reaction, and GTP is hydrolyzed to GDP and $P_i$ (for a recent review see Lucas-Lenard and Lipmann[1]). In addition, EF G in the presence of ribosomes, but in the absence of other components of protein synthesis, catalyzes the hydrolysis of GTP.[2] Both the EF G-dependent translocation and the "uncoupled" GTP hydrolysis are inhibited by the steroid antibiotic fusidic acid.[3] Recent studies have shown that EF G binds to the ribosome in a reaction dependent on either GTP or GDP to form a labile ribosome·EF G·GDP complex. The inhibition by fusidic acid of uncoupled GTP hydrolysis appears to result from the stabilization of the ribosome·EF G·GDP complex[4-6] through the binding of the antibiotic.[7] The resulting ribosome·EF G·GDP·fusidic acid complex dissociates relatively slowly and hence inhibits the catalytic action of EF G in the hydrolytic reaction. As mentioned above, the binding of EF G to ribosomes requires either GDP or GTP, and in the latter case, complex formation is associated with immediate and irreversible cleavage of GTP and release of $P_i$:

$$\text{Ribosome} + \text{EF G} + \text{GDP (GTP)} + \text{fusidic acid} \rightleftharpoons \text{Ribosome·EF G·GDP·fusidic acid} \ (+P_i)$$

The ribosome complex is sufficiently stable so that the EF G-dependent binding to the ribosome of either radiolabeled GDP or fusidic acid (or 24,25-dihydrofusidic acid) can be detected by a variety of means. The most convenient of these is by filtration through nitrocellulose filters (Millipore Corp.), a technique widely used in the quantitation of ribosome

[1] J. Lucas-Lenard and F. Lipmann, *Annu. Rev. Biochem.* **40**, 409 (1971).
[2] Y. Nishizuka and F. Lipmann, *Arch. Biochem. Biophys.* **116**, 344 (1966).
[3] N. Tanaka, T. Kinoshita, and H. Masukawa, *Biochem. Biophys. Res. Commun.* **30**, 278 (1968).
[4] J. W. Bodley, F. J. Zieve, L. Lin, and S. T. Zieve, *J. Biol. Chem.* **245**, 5656 (1970).
[5] J. W. Bodley, F. J. Zieve, and L. Lin, *J. Biol. Chem.* **245**, 5662 (1970).
[6] N. Brot, C. Spears, and H. Weissbach, *Arch. Biochem. Biophys.* **143**, 286 (1971).
[7] A. Ikura, T. Kinoshita, and N. Tanaka, *Biochem. Biophys. Res. Commun.* **41**, 1545 (1970).



ISBN 0-12-504180-2

complexes. The method described here is sufficiently rapid, specific, and sensitive so that it may be used as a routine assay for EF G.

*Materials*

Reaction Buffer: 0.1 *M* Tris·HCl (pH 7.4), 0.1 *M* $NH_4Cl$, 0.1 *M* magnesium acetate, 10 m*M* dithiothreitol

Wash Buffer: 10 m*M* Tris·HCl (pH 7.4), 10 m*M* $NH_4Cl$, 10 m*M* magnesium acetate, 10 m*M* fusidic acid (sodium salt)

EF G, 100–2000 units/ml.[8] See references cited in footnotes 6 and 8 for purification procedures.

Ribosomes, 750 $A_{260}$/ml, either 70 S or 50 S subunits washed at least three times with buffer containing either 0.5 or 1.0 *M* $NH_4Cl$. See Brot *et al.*[6] for isolation of ribosomes.

Nitrocellulose filters, Type HAWP 0.45 μm, 25 mm (Millipore Corp.)

For nucleotide binding:

$^3$H-GDP or $^3$H-GTP, 0.01 m*M* (1–5 Ci/mmole) (New England Nuclear Corp.)

Fusidic acid, 30 m*M* (sodium salt)

For [$^3$H]24,25-dihydrofusidic acid binding

[$^3$H]24,25-Dihydrofusidic acid (sodium salt), prepared from fusidic acid by reductive tritiation by the method of Godtfredsen and Vangedal,[9] 1 μ*M*, at least 10 Ci/mmole

GDP, 1 m*M*

*Procedure.* The incubation, final volume 50 μl, contains the following components which are added at 0°: 25 μl water and 5 μl each of reaction buffer, EF G, ribosomes, and either 1 m*M* GDP:GTP (for $^3$H-24,25-dihydrofusidic acid binding[10]) or 30 m*M* fusidic acid (for $^3$H-GDP binding). The reaction is initiated by the addition of 5 μl of the appropriate radiolabeled ligand. Generally the binding reaction is conducted at 0° for 5 minutes, but the ribosome·EF G·GDP·fusidic acid complex forms rapidly and is quite stable so that, when large numbers of routine assays are involved, reaction times may be varied up to 1 hour or more.

[8] One unit of EF G is defined as that amount of protein which, in the presence of excess ribosomes, causes the retention of 1 pmole of GDP to Millipore filters: J. H. Highland, L. Lin, and J. W. Bodley, *Biochemistry* **10**, 4404 (1971).

[9] W. O. Godtfredsen and S. Vangedal, *Tetrahedron* **18**, 1029 (1962).

[10] The details of the binding of [$^3$H]24,25-dihydrofusidic acid binding to the ribosome·EF G·GDP complex will be reported elsewhere: N. Richman, G. R. Willie, and J. W. Bodley, *Fed. Proc., Fed. Amer. Soc. Exp. Biol.* **31**, 898 (1972); G. R. Willie and J. W. Bodley (in preparation).

The reaction is terminated by the addition of approximately 3 ml of ice-cold wash buffer, and the diluted reaction is immediately filtered through a Millipore filter which has been previously rinsed with wash buffer. The reaction tube is rinsed once with wash buffer, and then two more 3-ml aliquots are passed through the filter. Because the complex dissociates during dilution and filtration, it is important to perform these operations rapidly and uniformly. Best results are obtained by dispensing the wash buffer from a 500-ml wash bottle. Under most circumstances repetitive filtrations can be performed at intervals of 30 to 60 seconds. Following the last wash the filter is immediately removed from the filter housing, dried, and assayed for radioactivity in a toluene-based fluor or the wet filter can be dissolved directly in the scintillation fluid described by Bray.[11]

*Comments.* In order to apply the present assay technique to the study of the interaction between the ribosome and EF G, the ribosome must first be washed free of bound EF G by repeated exposure to "high salt." When adequately treated in this way, 250 $\mu$g (3.75 $A_{260}$) of ribosomes in the absence of exogenous EF G bind negligible amounts of either [$^3$H]GDP or [$^3$H]24,25-dihydrofusidic acid.

The binding of GDP to the ribosomes is linear over a relatively wide range of EF G concentrations. While the maximum amount of GDP binding varies with different ribosome preparations, binding over the linear range is quite constant. The binding of [$^3$H]24,25-dihydrofusidic acid to the ribosome is hyperbolically related to EF G concentration, presumably because of the higher dissociation constant.

It is now apparent that translocation factors from eukaryote sources can bind GTP or GDP in the absence of ribosomes.[12-15] *E. coli* EF G, however, does not cause the retention of either GDP, GTP or 24,25-dihydrofusidic acid in the absence of ribosomes. However, in order to obtain this result EF G must be free of EF Tu, which does bind GDP and GTP; therefore, [$^3$H]GDP binding to ribosomes does not provide a good estimate of EF G in crude extracts which contain EF Tu. Similarly, although [$^3$H]24,25-dihydrofusidic acid appears to bind only to the ribosome·EF G·GDP complex, for reasons that are not entirely clear this assay too requires at least partial purification of EF G.

[11] G. A. Bray, *Anal. Biochem.* **1**, 279 (1960).
[12] S. Raeburn, R. S. Goor, J. A. Schneider, and L. S. Maxwell, *Proc. Nat. Acad. Sci. U.S.* **61**, 1428 (1968).
[13] J. W. Bodley, L. Lin, M. L. Salas, and M. Tao, *FEBS (Fed. Eur. Biochem. Soc.) Lett.* **11**, 153 (1970).
[14] L. Montanaro, S. Sperti, and A. Mattioli, *Biochim. Biophys. Acta* **238**, 493 (1971).
[15] E. Bermek and H. Matthaei, *Biochemistry* **10**, 4906 (1971).

# [39] GTP-Dependent Binding of Aminoacyl-tRNA to *Escherichia coli* Ribosomes[1-5]

*By* JOANNE M. RAVEL and ROSEANN L. SHOREY

*Principle.* Two transfer factors, $TI_s$ and $TI_u$,[6] and GTP are required for the binding of aminoacyl-tRNA at the acceptor site of a ribosome-mRNA complex. Binding of [$^{14}C$]- or [$^{3}H$]aminoacyl-tRNA to the ribosomes is conveniently measured by adsorption of the ribosomal complex on a nitrocellulose (Millipore) filter.[7] The adsorptive properties of Millipore filters can also be utilized to measure the intermediate steps in the binding of aminoacyl-tRNA to ribosomes. A mixture of $TI_s$ and $TI_u$ interacts with [$^{3}H$]GTP to form a $^{3}H$-labeled complex (complex I) that is adsorbed on a Millipore filter[8]; under these conditions both $TI_s$ and $TI_u$ are retained by the filter. In the presence of aminoacyl-tRNA, $TI_s$ and $TI_u$ interact with GTP to form an aminoacyl-tRNA·$TI_u$·GTP complex (complex II) that is not retained by a Millipore filter; $TI_s$ is still retained by the filter under these conditions. Complex II is separated from $TI_s$ by passage of the reaction mixture through a Millipore filter and is separated from excess GTP by chromatography of the Millipore filtrate on Sephadex G-25. The binding of the phenylalanyl-tRNA moiety of complex II (prepared with [$^{14}C$]phenylalanyl-tRNA and a mixture of [$^{3}H$]GTP and [$\gamma$-$^{32}P$]GTP) to a ribosome·poly (U) complex occurs at low concentrations of $Mg^{2+}$ and at low temperatures. The cleavage of the GTP moiety of complex II occurs as a result of the binding reaction and is measured by the liberation of $^{32}P_i$ and by

[1] J. M. Ravel, *Proc. Nat. Acad. Sci. U. S.* **57**, 1811 (1967).
[2] J. M. Ravel, R. L. Shorey, and W. Shive, *Biochem. Biophys. Res. Commun.* **29**, 68 (1967).
[3] J. M. Ravel, R. L. Shorey, S. Froehner, and W. Shive, *Arch. Biochem. Biophys.* **125**, 514 (1968).
[4] J. M. Ravel, R. L. Shorey, and W. Shive, *Biochem. Biophys. Res. Commun.* **32**, 9 (1968).
[5] R. L. Shorey, J. M. Ravel, C. W. Garner, and W. Shive, *J. Biol. Chem.* **244**, 4555 (1969).
[6] In an effort to establish a uniform nomenclature for the transfer factors, the following abbreviations are used: $TI_s$, $TI_u$, and TII. These factors are the same as those previously designated $FI_s$, $FI_u$, and FII (see footnotes 1 and 5) and have enzymatic activities comparable to the factors designated $T_s$, $T_u$, and G by J. Lucas-Lenard and F. Lipmann, [*Proc. Nat. Acad. Sci. U. S.* **55**, 1562 (1966)] and to the factors designated $S_1$, $S_3$, and $S_2$ by A. Skoultchi, Y. Ono, H. M. Moon, and P. Lengyel [*Proc. Nat. Acad. Sci. U. S.* **60**, 675 (1968)].
[7] M. W. Nirenberg and P. Leder, *Science* **145**, 1399 (1964).
[8] R. Ertel, N. Brot, B. Redfield, J. E. Allende, and H. Weissbach, *Proc. Nat. Acad. Sci. U. S.* **59**, 861 (1968).



ISBN 0-12-504180-2

the formation of a $[^3H]GDP \cdot TI_u$ complex (complex III). Complex III is adsorbed by a Millipore filter with the ribosomal complex; however, it can be separated from the ribosomal complex by centrifugation or by chromatography on Sephadex G-100. When the phenylalanyl-tRNA moiety of complex II is bound at the acceptor site of a ribosome·poly (U) complex carrying *N*-acetylphenylalanyl-tRNA at the donor site, peptide bond formation occurs.

The steps in the GTP-dependent binding of phenylalanyl-tRNA (Phe-tRNA) to a ribosome·poly (U) complex carrying *N*-acetylphenylalanyl-tRNA (acPhe-tRNA) at the donor site are summarized in the following equations:

$$TI_s + TI_u + GTP \xrightleftharpoons{Mg^{2+}} \text{Complex I} \quad (1A)$$

$$TI_s + TI_u + GTP + \text{Phe-tRNA} \xrightarrow{Mg^{2+},\ NH_4^+} \underset{\text{Complex II}}{[\text{Phe-tRNA} \cdot TI_u \cdot GTP]} + TI_s \quad (1B)$$

$$\text{Complex II} + \text{AcPhe-tRNA} \cdot \text{poly (U)} \cdot \text{ribosome} \xrightarrow{Mg^{2+},\ K^+,\ NH_4^+}$$

$$\begin{matrix} \text{tRNA} \cdot \\ \text{AcPhe-Phe-tRNA} \cdot \end{matrix} \text{poly (U)} \cdot \text{ribosome} + \underset{\text{Complex III}}{[\text{GDP-}TI_u]} + P_i \quad (2)$$

## General Procedures

*Growth of E. coli W.* The growth medium contains the following (in grams per liter): $MgSO_4 \cdot 7\ H_2O$, 0.1; $(NH_4)_2SO_4$, 1.0; $K_2HPO_4$, 7.0; $KH_2PO_4$, 3.0; sodium citrate·$2H_2O$, 0.5; Difco yeast extract, 1.0; Sheffield N-Z case, 1.0; and glucose, 2.0. A typical preparation consists of two 20-liter carboys each containing 10 liters of water. The carboys are sterilized in an autoclave and cooled to 37° in a water bath. The solid components of the medium are dissolved in the sterile water, and 1 liter of an overnight culture of *E. coli* W is added to each carboy. The medium is vigorously aerated for 2.5–3 hours at 37°. The carboys are then placed in an ice bath, and the aeration is continued for about 1 hour. The cells are harvested by means of a Sharples centrifuge, washed once with about 300 ml of Buffer C (10 m*M* Tris·HCl, pH 7.5; 1 m*M* dithiothreitol; and 10 m*M* $MgCl_2$) and stored as a paste at −90°. Approximately 30 g (wet weight) of cells are obtained per 10 liters of medium.

*Preparation of Ribosomes.* About 200 g of cells obtained as described above are suspended in 400 ml of Buffer C and are disrupted at 3000 psi in a Gaulin Laboratory Homogenizer. DNase (Worthington), 1 mg, is added, and whole cells and cell debris are removed by centrifugation at 30,000 *g* for 30 minutes. This and all subsequent operations are carried out at 4°. Ribosomes are sedimented by centrifugation at 78,000 *g* for 5 hours in the No. 30 rotor of a Spinco Model L centrifuge. The supernatant solution (Supernatant Solution 1) is saved and is the source of the transfer factors isolated by the procedure described below. The ribosomal pellet is resuspended in 100 ml of Buffer C, and the suspension is clarified by centrifugation at 15,000 *g* for 20 minutes. The ribosomes are sedimented by centrifugation at 150,000 *g* for 2.5 hours in the No. 50 rotor of the Spinco Model L, and the pellets are stored overnight at 10°. The ribosomes are resuspended in 100 ml of buffer (10 m*M* Tris·HCl, pH 7.5, containing 0.5 *M* $NH_4Cl$, 0.1 m*M* $MgCl_2$, and 1 m*M* dithiothreitol) and harvested by centrifugation at 150,000 *g* for 3 hours in the No. 50 rotor of the Spinco Model L. The supernatant solution (Supernatant 2) is retained as a source of the initiation factors. The ribosomes are washed three more times with 0.01 *M* Tris·HCl buffer, pH 7.5, containing the following supplements: first, 0.5 *M* $NH_4Cl$ and 1 m*M* $MgCl_2$; second, 0.5 *M* $NH_4Cl$ and 10 m*M* $MgCl_2$; and third, 10 m*M* $MgCl_2$. After each wash, the ribosomes are collected by centrifugation at 150,000 *g* for 2.5 hours. After the last wash the ribosomes are suspended in approximately 25 ml of 10 m*M* Tris·HCl, pH 7.5, containing 10 m*M* $MgCl_2$ and are clarified by centrifugation at 30,000 *g* for 30 minutes. Assuming that 14.4 optical density units at 260 mμ is equivalent to 1 mg of ribosomes per milliliter, approximately 3 g of ribosomes are obtained by this procedure. The salt-washed ribosomes are stored at −90° for as long as 6 months without appreciable loss in activity.

A 5% solution of potassium deoxycholate is prepared by dissolving 5 g of deoxycholic acid in 10 ml of ethanol, diluting with water to about 50 ml, adjusting the pH to 7.4–7.6 by the addition of 1 *N* KOH, and finally diluting to 100 ml with water. To an aliquot of the salt-washed ribosomes (1 ml containing about 125 mg of ribosomes) is added 2.6 ml of 10 m*M* Tris·HCl, pH 7.5, containing 10 m*M* $MgCl_2$, and 0.4 ml of 5% potassium deoxycholate. The mixture is stirred gently with a magnetic stirring bar for 20 minutes at 30°; the ribosomes are diluted to 40 ml with 0.01 *M* Tris·HCl, pH 7.5, containing 10 m*M* $MgCl_2$, and harvested by centrifugation at 150,000 *g* for 2 hours. The ribosomes are suspended in 4 ml of 10 m*M* Tris·HCl containing 10 m*M* $MgCl_2$ and clarified by centrifugation at 30,000 *g* for 20 minutes. The DOC-treated ribosomes are stored at −90° at a concentration of about 30 mg/ml for as long as 3 weeks without appreciable loss in activity. Ribosomes prepared in this manner appear to be free of initiation factors 1 and 2 and transfer factors $TI_s$ and $TI_u$; they exhibit only a small amount of GTPase activity (10 pmoles of GTP hydrolyzed per milligram of ribosomes in 5 minutes at 30° in the presence or absence of poly (U) and tRNA).

*Preparation of [$^{14}$C]- or [$^3$H]Phenylalanyl-tRNA.* Soluble RNA (tRNA) from *E. coli* B is obtained from General Biochemicals. A sonic extract of *E. coli* W prepared as previously described[9] is stored at −20° and used as a source of phenylalanyl-tRNA synthetase. An aliquot (5 ml) of the sonic extract is dialyzed overnight at 4° against 1 liter of 40 m*M* Tris·HCl, pH 7.7, containing 6 m*M* $\beta$-mercaptoethanol. It is imperative that the sonic extract be dialyzed just prior to use; otherwise the yield of labeled phenylalanyl-tRNA is greatly reduced, probably due to the presence of unlabeled phenylalanine in the sonic extract. The reaction mixture contains in a total volume of 10 ml: Tris·HCl, pH 7.7, 0.1 *M*; $MgCl_2$, 10 m*M*; potassium ATP, 2 m*M*, from a freshly prepared solution; $\beta$-mercaptoethanol, 20 m*M*; tRNA, 20 mg, assuming that 24.4 optical density units at 260 m$\mu$ is equivalent to 1 mg of RNA; 1 ml of freshly dialyzed sonic extract containing approximately 10 mg of protein; and either [$^{14}$C]phenylalanine, 10 $\mu$*M*, specific activity 100 Ci/mole, or [$^3$H]phenylalanine, 10 $\mu$*M*, specific activity, 500 Ci/mole. After 15 minutes of incubation at 37°, 1 ml of 2 *M* potassium acetate, pH 6.0, and 11 ml of 90% phenol are added and the two phases are thoroughly mixed by the use of a Vortex mixer. The phenol and aqueous phases are thoroughly mixed intermittently over a 15-minute period and the phases are clarified by centrifugation at 30,000 *g* for 15 minutes. The upper aqueous phase is removed and the phenol layer is extracted a second time with 2 ml of water. The aqueous layers are combined and made 0.2 *M* in potassium acetate, and the tRNA is precipitated by the addition of 2 volumes of cold absolute ethanol. After 2 hours at −20°, the precipitate is collected by centrifugation at 30,000 *g* for 15 minutes. The precipitate is dissolved in 5 ml of 0.2 *M* potassium acetate and the tRNA is precipitated a second time by the addition of 10 ml of ethanol. The phenylalanyl-tRNA is stored as a pellet at −90° for as long as 3 months. To remove small molecular weight material, 40 mg of the tRNA charged with phenylalanine is dissolved in 1.5 ml of 10 m*M* potassium succinate, pH 6, and is applied to a column (1 × 33 cm) of Sephadex G-25 equilibrated in 10 m*M* potassium succinate, pH 6. The column is washed with the same buffer; the fractions containing the tRNA are pooled, made 0.2 *M* in potassium succinate, and divided into 1–2-ml aliquots. To each aliquot is added 2 volumes of ethanol, and the mixture is allowed to stand for 20 minutes at −20°; the tRNA is collected by centrifugation and stored as a pellet at −90°. Just prior to use a pellet is dissolved in 10 m*M* potassium succinate, pH 6, and the amount of phenylalanyl-tRNA is determined by measuring the radioactivity in a

[9] J. M. Ravel, S. Wang, C. Heinemeyer, and W. Shive, *J. Biol. Chem.* **240**, 432 (1965).

small aliquot or in a cold 5% trichloroacetic acid precipitate. This procedure routinely yields tRNA charged with 550–650 pmoles of phenylalanine per milligram of tRNA. Solutions of phenylalanyl-tRNA are stored at −90° for a few days; however, after a week significant amounts of deacylation are observed.

*Preparation of N-Acetyl-[$^{14}$C]phenylalanyl-tRNA.* [$^{14}$C]Phenylalanyl-tRNA, 40 mg, prepared as described above (prior to passage through Sephadex G-25) is treated with acetic anhydride by the method of Haenni and Chapeville.[10] The acetylated preparation is passed through Sephadex G-25, precipitated, and stored as described above.

*Assays for $TI_s$, $TI_u$, and TII.* The activities of $TI_s$ and $TI_u$ are most easily measured by their interaction with GTP to form a complex that is adsorbed on a Millipore filter. The reaction mixture contains in a total volume of 0.2 ml: Buffer A-10 (Tris·HCl, pH 7.6, 50 m*M*; dithiothreitol, 5 m*M*; KCl, 80 m*M*; $NH_4Cl$, 80 m*M*; and $MgCl_2$, 10 m*M*); [$^3$H]GTP, 2 $\mu M$ (specific activity 150, approximately 100 cpm/pmole); and enzyme fractions. To assay for $TI_s$, the reaction mixture is supplemented with a preparation of $TI_u$ obtained as described below (5 $\mu$g protein containing approximately 30 units of $TI_u$). To assay for $TI_u$, the reaction mixture is supplemented with a preparation of $TI_s$ obtained as described below (4 $\mu$g protein). After 10 minutes of incubation at 0°, the reaction mixture is diluted with about 5 ml of cold Buffer A-10 (without dithiothreitol) and passed through a Millipore filter (25 mm in diameter, 0.45-$\mu$ pore size) previously soaked in Buffer A-10. The filter is washed with two additional 5-ml portions of cold Buffer A-10 and is dried by suction. The damp filter is placed in a vial containing 0.5 ml of 1 *N* NaOH and 15 ml of scintillation fluid. The scintillation fluid is prepared by mixing in the following order: 800 ml of dioxane, 200 ml of toluene, 5 g of 2,5-diphenyloxazole, 100 g of naphthalene, and 36 g of Cab-O-Sil. The contents of the vial are thoroughly mixed by the use of a Vortex mixer, and the amount of radioactivity is determined in a liquid scintillation counter. A unit of $TI_u$ is defined as the amount that binds 1 pmole of GTP in the presence of an excess of $TI_s$ under the conditions described above, and specific activity is defined as units per milligram of protein.

TII is measured by its ability to catalyze polyphenylalanine synthesis in the presence of $TI_s$ and $TI_u$. The reaction mixture contains in a total volume of 0.5 ml, Buffer A-10; ribosomes, 0.1 mg; poly (U), 10 $\mu$g; [$^{14}$C]phenylalanyl-tRNA, 0.1 mg charged with 65 pmoles of phenylalanine; GTP, 1 m*M*; $TI_s$, 2 $\mu$g of protein; $TI_u$, 10 $\mu$g of protein; and an

[10] A.-L. Haenni and F. Chapeville, *Biochim. Biophys. Acta* 114, 135 (1966).

aliquot of the enzyme fraction. After 5 minutes of incubation at 37°, the reaction is terminated by the addition of 5 ml of 5% trichloroacetic acid. The mixture is heated at 90° for 10 minutes, and the polyphenylalanine is adsorbed on a Millipore filter. The filter is washed with two 5-ml portions of 1% trichloroacetic acid and finally with 5 ml of water. The filter is dried by suction, and the amount of radioactivity on the filter is measured as described above.

*Separation of* $TI_s$, $TI_u$, *and TII by DEAE-Sephadex Chromatography*. To 400 ml of Supernatant Solution 1 obtained as described above, 77g of ammonium sulfate is added slowly with stirring. The pH is maintained at approximately 7.5 by the addition of 1 *N* KOH. This and all subsequent operations are carried out at 4°. After 20 minutes, the precipitate is collected by centrifugation at 30,000 *g* for 30 minutes and is discarded. To the supernatant fluid an additional 60 g of ammonium sulfate is added, and the precipitate is collected as before. The precipitate is resuspended in a minimum amount of 10 m*M* Tris·HCl, pH 7.5, containing 1 m*M* DTT, and is dialyzed overnight against two 1-liter portions of the same buffer. This 35–60% ammonium sulfate fraction (approximately 50 ml containing 50 mg of protein per milliliter) is divided into aliquots containing 800 mg of protein and is stored at −90° for as long as several months.

An aliquot of the dialyzed 35–60% ammonium sulfate fraction (16 ml containing 50 mg of protein per milliliter) is made 0.15 *M* in KCl and is applied to a column of DEAE-Sephadex A-50 (1.5 × 55 cm) equilibrated with 10 m*M* Tris·HCl (pH 7.5) containing 1 m*M* dithiothreitol and 0.15 *M* KCl. The column is washed with about 20 ml of the same buffer, and a linear KCl gradient is applied to the column. The mixing chamber and reservoir contain 500 ml of 10 m*M* Tris·HCl (pH 7.5) and 1 m*M* dithiothreitol, supplemented with 0.15 *M* KCl and 0.35 *M* KCl, respectively. The flow rate is adjusted to 0.25 ml per minute and fractions of 20 ml are collected. To stabilize the $TI_u$ that is eluted in fractions 35–45, 0.1 ml of a solution of 0.2 m*M* GTP and 0.4 *M* $MgCl_2$ is added to the tubes before the fractions are collected. Aliquots of each fraction (0.01 ml) are assayed for $TI_s$, $TI_u$, and TII activity as described above. In a typical elution, $TI_s$ is recovered primarily in fractions 26–32 (0.25–0.27 *M* KCl); $TI_u$ in fractions 36–39 (0.29–0.31 *M* KCl); and TII in fractions 41 to 49 (0.32–0.33 *M* KCl). Selected fractions are pooled and concentrated in a Diaflo model 50 ultrafiltration cell equipped with a UM-1 membrane. Only those fractions that do not contain detectable amounts of TII are pooled and used as a source of $TI_u$. A typical preparation of $TI_u$ (approximately 6 ml containing 10 mg of protein per milliliter) has a specific activity of 6000, is free of TII, but still contains a small amount of $TI_s$. The interaction of the $TI_u$ preparation with GTP under the conditions described above is stimulated 10- to 20-fold by the addition of $TI_s$. There is no detectable loss in $TI_u$ activity when this preparation is stored at $-90°$ in the presence of 2–10 m$M$ $Mg^{2+}$ and 1 $\mu M$ GTP for as long as 3 months. The preparation of $TI_s$ (4 ml containing 9 mg of protein per milliliter) contains a small amount of $TI_u$; the preparation of TII (2 ml containing 15 mg of protein per milliliter) appears to be free of both $TI_s$ and $TI_u$. $TI_s$ and TII can be stored at $-90°$ for 6 months or longer without appreciable loss in activity.

*Preparation of Complex II.* The reaction mixture contains in a final volume of 5 ml: Buffer B-5 (Tris·HCl, pH 7.5, 0.05 $M$; $NH_4Cl$, 0.16 $M$; $MgCl_2$, 5 m$M$; and dithiothreitol, 1 m$M$); $TI_s$, 0.3 mg of protein; $TI_u$, 1.5 mg of protein containing 9000 units; GTP, 50 nmoles (either [$^3H$]GTP, specific activity 300 Ci/mole, or [$\gamma$-$^{32}P$]GTP, specific activity 50–100 Ci/mole, or a mixture of the two); and tRNA, 1.5 mg, charged with 9–10 nmoles of [$^{14}C$]- or [$^3H$]phenylalanine. The [$\gamma$-$^{32}P$]GTP is purchased from International Chemical and Nuclear Corporation. After 5 minutes of incubation at 25°, the reaction mixture is passed through a stack of ten Millipore filters by the use of gentle suction, and the filters are washed with 1 ml of cold Buffer B-5. All subsequent operations are conducted at 4°. The filtrate and wash are applied to a column of Sephadex G-25 (1.5 × 60 cm) equilibrated with Buffer B-5, and the column is developed with Buffer B-5. The complex is not retarded by Sephadex G-25 and is eluted from the column at the void volume ($V_0 = 43$ ml). After the first 40 ml of effluent, fractions of 2 ml are collected. An aliquot (0.05 ml) of each fraction is removed and the amount of radioactivity present in each fraction is determined. The fractions containing the complex (8–10 ml total) are pooled and divided into 1- to 2-ml aliquots. To each aliquot are added two volumes of cold, absolute ethanol, and the mixture is allowed to stand at $-20°$ for 30 minutes. The precipitated complex is collected by centrifugation at 30,000 $g$ for 20 minutes at $-20°$, and the resulting pellet is stored at $-90°$ for as long as one month without appreciable loss in activity. Just prior to use, the pellet is dissolved in 1–2 ml of Buffer A-5; and from the radioactivity present in a small aliquot, the amounts of GTP and phenylalanyl-tRNA in the preparation are calculated. Some deacylation of the phenylalanyl-tRNA occurs during the preparation of the complex and only 6–7 nmoles (60–70%) of the phenylalanyl-tRNA are recovered. The amount of GTP in the preparation is between 8 and 9 nmoles and the ratio of GTP to phenylalanyl-tRNA is generally about 1.5 to 1.

### Binding of Phenylalanyl-tRNA to Ribosome·Poly(U) Complexes

*Preparation of Ribosome·Poly(U) Complexes.* Ribosome·poly(U) complexes are prepared by incubating for 5 minutes at 37°, 2 mg of ribosomes and 0.2 mg of poly(U) in 1 ml of Buffer A-10 (10 m$M$ $Mg^{2+}$) or Buffer A-5 (5 m$M$ $Mg^{2+}$). To prepare ribosome·poly(U) complexes carrying nonenzymatically bound phenylalanyl-tRNA, the incubation mixture is supplemented with 0.6 mg of tRNA charged with 0.4 nmoles of phenylalanine. The ribosomal complexes either are used directly or are harvested by centrifugation and stored at −90°. To prepare ribosome·poly(U) complexes carrying *N*-acetylphenylalanyl-tRNA, the reaction mixture contains in a total volume of 2.5 ml, Buffer A-5; ribosomes, 5 mg; poly(U), 0.5 mg; initiation factor 1, 150 $\mu$g protein; initiation factor 2, 225 $\mu$g of protein (both obtained by chromatography on DEAE-cellulose as described by Erbe, Nau, and Leder[11]); GTP, 10 $\mu M$; and *N*-acetyl-[$^{14}$C]phenylalanyl-tRNA, 1.5 nmoles. The reaction mixture is incubated for 15 minutes at 25° and layered on 7 ml of 10% sucrose in Buffer 5. The ribosomes are collected by centrifugation at 150,000 g for 3 hours, stored as a pellet at −90°, and resuspended in Buffer A-5 just prior to use. Approximately 8–10 pmoles of *N*-acetyl phenylalanyl-tRNA are bound per 0.1 mg ribosomes by this procedure.

*Binding of Phenylalanyl-tRNA to Ribosomes Catalyzed by $TI_s$ and $TI_u$ in the Presence of GTP.* The reaction mixture contains in 0.25 ml, Buffer A-5 or A-10; $TI_s$, 2 $\mu$g of protein; $TI_u$, 10 $\mu$g of protein containing 60 units; GTP, 2 $\mu M$; tRNA, 0.1 mg, charged with 60 pmoles of [$^{14}$C]- or [$^{3}$H]phenylalanine; and ribosome·poly(U) complexes, 0.1 mg, prepared as described above. The reaction mixture is incubated at 0°, 25°, or 37° for 5 minutes, diluted with about 5 ml of cold Buffer A-5 or A-10 (without dithiothreitol), and passed through a Millipore filter. The filter is washed with 2 5-ml portions of cold buffer, dried by suction, and counted as described above. The amount of phenylalanyl-tRNA bound to the ribosomes in the absence of $TI_s$ and $TI_u$ and GTP is designated nonenzymatic binding; the increase in the amount of phenylalanyl-tRNA bound in the presence of $TI_s$, $TI_u$, and GTP is designated enzymatic binding.

*Binding of Phenylalanyl-tRNA from Complex II to Ribosomes.* The reaction mixture contains in 0.25 ml, Buffer A-5 or Buffer A-10; Complex II containing between 10 and 60 pmoles of phenylalanyl-tRNA and between 15 and 90 pmoles of GTP; and 0.1 mg of the ribosome·poly(U) complexes. The reaction mixture is incubated at 0° or 25° for 5 minutes and the amount of phenylalanyl-tRNA bound to the ribosomes is determined as described above.

[11] R. W. Erbe, M. M. Nau, and P. Leder, *J. Mol. Biol.* **38**, 441 (1969).

*Hydrolysis of the GTP Moiety of Complex II.* The hydrolysis of the GTP moiety of Complex II that occurs as a result of the binding of the phenylalanyl-tRNA moiety of the complex to the ribosomes is measured by the formation of $^{32}P_i$ and by the formation of a [$^3$H]GDP·$TI_u$ complex (Complex III) that is retained by a Millipore filter. Duplicate reaction mixtures are prepared containing Buffer A-5 or A-10, 0.1 mg of ribosome·poly(U) complexes, and Complex II (containing [$^3$H]- and [$\gamma$-$^{32}$P]-GTP and [$^{14}$C]phenylalanyl-tRNA) in a final volume of 0.25 ml. After incubation at 0° or 25°, the amount of $^{32}P_i$ liberated is determined with one of the reaction mixtures by the procedure described by Conway and Lipmann.[12] The other reaction mixture is diluted with cold buffer and the amounts of [$^{14}$C]phenylalanyl-tRNA bound to the ribosomes and [$^3$H]GDP·$TI_u$ complex retained by a Millipore filter are measured. In each assay the amounts of $P_i$ and Complex III formed during the incubation in the absence of the ribosome·poly(U) complexes are determined. At low concentrations of Complex II (containing 10–40 pmoles of GTP) the amounts of $P_i$ and Complex III formed in the absence of ribosomes·poly(U) are low (approximately 0.5–2 pmoles), and the increases in the amounts of $P_i$ and Complex III formed in the presence of the ribosome·poly(U) complexes are equal to the amount of phenylalanyl-tRNA bound to the ribosomes. When the ribosomes are collected by centrifugation rather than by adsorption on a Millipore filter, Complex III is in the supernatant solution; it is adsorbed by a Millipore filter in the absence of the ribosomal complex. Complex III also can be separated from the ribosomal complex by chromatography on Sephadex G-100 in Buffer A-5 or A-10.

*Identification of Products Bound to Ribosomes.* The reaction mixture contains in a total volume of 2 ml, Buffer A-5, 2.0 mg ribosome·poly(U) complexes, and either Complex II (containing 0.6–1.2 nmoles of phenylalanyl-tRNA) or a mixture of $TI_s$, 40 $\mu$g protein; $TI_u$, 200 $\mu$g protein (1200 units); GTP, 2 $\mu M$; and phenylalanyl-tRNA, 0.6–1.2 nmoles. After incubation, the ribosomes are collected by centrifugation at 150,000 $g$ for 2.5 hours. The ribosomal pellet is suspended in 0.5 ml Buffer A-5, and aliquots are removed and the amounts of phenylalanyl-tRNA present in the pellet and bound to the ribosome are determined by a direct count and by the Millipore filter technique.

The products bound to the ribosomes (0.2 ml containing 0.6–0.8 mg of ribosomes) are released by the addition of 0.02 ml 4 $N$ KOH. A mixture of phenylalanine, diphenylalanine, and triphenylalanine (20 $\mu$g of each) is added as carrier. After heating at 50° for 15 minutes, 0.02 ml of 5 $N$ acetic acid is added and the hydrolyzate is applied to a column of

[12] See Vol. XII, Part B [154], p. 713.

Sephadex G-15 ($1.1 \times 15$ cm) equilibrated in 0.5 *N* acetic acid. The column is developed by 0.5 *N* acetic acid as described by Bretthauer and Golichowski.[13] After the first 10 ml of effluent, fractions of 1 ml are collected and the amounts of phenylalanine, *N*-acetylphenylalanine, and their di- and tripeptides are calculated from the radioactivity present in each of the peaks.

*Comments on the Formation and Properties of Complex II.* Complex II is formed with most aminoacyl-tRNA's but is not formed with deacylated tRNA, *N*-acetylphenylalanyl-tRNA, *N*-formylmethionyl-$tRNA_f$, or methionyl-$tRNA_f$.[2,14,15] The formation of Complex II is dependent upon $Mg^{2+}$ (a minimum of 1 m*M*), is stimulated by $NH_4^+$ (40 m*M*), and occurs over a pH range (6.5–8.0). $TI_s$ and $TI_u$ interact with GDP to form a GDP·protein complex, but no subsequent interaction with aminoacyl-tRNA is observed. The formation of an isolable complex with $TI_s$, $TI_u$, phenylalanyl-tRNA, and the GTP analog, 5′-guanylylmethylenediphosphonate, has not been observed.

In the presence of $Mg^{2+}$, the GTP moiety of Complex II does not dissociate from $TI_u$ during filtration on Sephadex G-100 and is not exchangeable with free GTP, GDP, or $P_i$ at 37°. The aminoacyl-tRNA moiety of Complex II dissociates from the GTP·$TI_u$ complex during filtration on Sephadex G-100, and exchanges with free aminoacyl-tRNA slowly at 0° and rapidly at 25°. In the absence of $Mg^{2+}$, both the GTP and the aminoacyl-tRNA dissociate from the $TI_u$.

*Comments on the Binding of Phenylalanyl-tRNA to Ribosome·Poly(U) Complexes.* Nonenzymatic binding of phenylalanyl-tRNA to ribosome·poly(U) complexes, i.e., binding which occurs in the absence of $TI_s$, $TI_u$, and GTP, increases linearly with increasing concentrations of $Mg^{2+}$. In contrast, maximal binding of phenylalanyl-tRNA from Complex II or a mixture of $TI_s$, $TI_u$, and GTP occurs at 5 m*M* $Mg^{2+}$. The binding of the phenylalanyl-tRNA moiety of Complex II to the ribosome·poly(U) complexes occurs rapidly (in less than 5 minutes) at 0° as well as at elevated temperatures (25°–37°). Enzymatic binding of phenylalanyl-tRNA at the acceptor site on the ribosomes is not inhibited by tRNA but is inhibited by chlortetracycline. GDP inhibits the binding of phenylalanyl-tRNA to ribosomes catalyzed by $TI_s$ and $TI_u$ in the presence of GTP, but has no effect on the binding of the phenylalanyl-tRNA moiety of Complex II to the ribosomes. With ribosome·poly(U) or ribosome·poly(U) complexes carrying nonenzymatically bound phenyl-

[13] R. K. Bretthauer and A. M. Golichowski, *Biochim. Biophys. Acta* **155**, 1462 (1964).
[14] J. Gordon, *Proc. Nat. Acad. Sci. U. S.* **59**, 179 (1968).
[15] Y. Ono, A. Skoultchi, A. Klein, and P. Lengyel, *Nature (London)* **220**, 1304 (1968).

alanyl-tRNA, about 4–6 pmoles of phenylalanyl-tRNA are bound enzymatically at the acceptor site per 0.1 mg of ribosomes. Neither the nonenzymatically nor enzymatically bound phenylalanyl-tRNA is exchangeable with free phenylalanyl-tRNA. With ribosome·poly(U) complexes carrying *N*-acetylphenylalanyl-tRNA bound at the donor site with initiation factors and GTP, about 8–10 pmoles of phenylalanyl-tRNA are enzymatically bound per 0.1 mg of ribosomes and about 60–70% of this phenylalanyl-tRNA interacts with the bound *N*-acetylphenylalanyl-tRNA to form dipeptide at the acceptor site. No detectable tripeptide is formed. Binding of phenylalanyl-tRNA at the acceptor site is also obtained with a mixture of $TI_s$, $TI_u$, and the GTP analog 5′-guanylylmethylenediphosphonate; however, $Mg^{2+}$ concentrations of 8–10 m*M* are required for maximal binding and no dipeptide is formed.

## [40] Polypeptide Chain Elongation Factors from Pig Liver

*By* KENTARO IWASAKI and YOSHITO KAZIRO

The elongation of ribosome-bound polypeptides in eukaryotic cells requires two complementary soluble factors,[1,2] EF-1 and EF-2.[3] EF-1 catalyzes the binding of aminoacyl-tRNA to the aminoacyl site on the ribosomes, while EF-2 translocates the peptidyl-tRNA bound to the aminoacyl site to the peptidyl site. Purification of these factors has been reported with a number of eukaryotic tissues including pig liver,[4-9] rat

[1] R. Haselkorn and L. B. Rothman-Denes, *Annu. Rev. Biochem.* **42,** 397 (1973).
[2] H. Weissbach and S. Ochoa, *Annu. Rev. Biochem.* **45,** 191 (1976).
[3] T. Caskey, P. Leder, K. Moldave, and D. Schlessinger, *Science* **176,** 195 (1972).
[4] K. Iwasaki, K. Mizumoto, M. Tanaka, and Y. Kaziro, *J. Biochem. (Tokyo)* **74,** 849 (1973).
[5] K. Iwasaki, S. Nagata, K. Mizumoto, and Y. Kaziro, *J. Biol. Chem.* **249,** 5008 (1974).
[6] S. Nagata, K. Iwasaki, and Y. Kaziro, *J. Biochem. (Tokyo)* **82,** 1633 (1977).
[7] K. Iwasaki, K. Motoyoshi, S. Nagata, and Y. Kaziro, *J. Biol. Chem.* **251,** 1843 (1974).
[8] K. Motoyoshi, K. Iwasaki, and Y. Kaziro, *J. Biochem. (Tokyo)* **82,** 145 (1977).
[9] K. Mizumoto, K. Iwasaki, M. Tanaka, and Y. Kaziro, *J. Biochem. (Tokyo)* **75,** 1047 (1974).



ISBN 0-12-504180-2

liver,[10-12] reticulocytes,[13-15] calf brain,[16] calf liver,[17] Krebs ascites cells,[18] *Artemia salina* cysts,[19,20] silk gland,[21,22] wheat embryos,[23-26] and yeast.[27,28] Purified EF-2 from these tissues has similar properties; it is a single polypeptide of a molecular weight of about 100,000, is inactivated by the incubation with thiol reagents or with the diphtheria toxin in the presence of NAD, and requires GTP for its activity. On the contrary, physicochemical as well as enzymic properties of EF-1 are more complicated. At least two forms of EF-1 differing in their molecular weights are identified in a variety of tissues[29]; the one designated as the high-molecular-weight form or EF-$1_H$ has a molecular weight of over 100,000 and the other designated as the low-molecular-weight form or EF-$1_L$ has a molecular weight of about 50,000.[16,17] Further complication was brought by the observations that EF-1 from pig liver,[4] silk gland,[21] *Artemia salina* cysts,[20] and wheat embryos[25] consists of more than one component, whereas that from reticulocytes,[13] calf brain,[16] and Krebs ascites cells[18] is composed of one component.

In the course of purification of EF-1 from pig liver, we found that it is resolved into two complementary factors, EF-1$\alpha$ and EF-1$\beta\gamma$.[4] EF-1$\alpha$ has a molecular weight of 53,000 and presumably corresponds to EF-$1_L$ from other tissues,[6] whereas EF-1$\beta\gamma$ has a molecular weight of about 90,000 and consists of two different subunits of the molecular weights of about 30,000

[10] M. Schneir and K. Moldave, *Biochim. Biophys. Acta* **166,** 58 (1968).
[11] J. F. Collins, H.-M. Moon, and E. S. Maxwell, *Biochemistry* **11,** 4187 (1972).
[12] S. Raeburn, J. F. Collins, H.-M. Moon, and E. S. Maxwell, *J. Biol. Chem.* **246,** 1041 (1971).
[13] W. L. McKeehan and B. Hardesty, *J. Biol. Chem.* **244,** 4330 (1969).
[14] W. M. Kemper and W. C. Merrick, *Arch. Biochem. Biophys.* **174,** 603 (1976).
[15] W. C. Merrick, W. M. Kemper, J. A. Kantor, and W. F. Anderson, *J. Biol. Chem.* **250,** 2620 (1975).
[16] H.-M. Moon, B. Redfield, S. Millard, F. Vane, and H. Weissbach, *Proc. Natl. Acad. Sci. U.S.A.* **70,** 3282 (1973).
[17] C. K. Liu, A. B. Legocki, and H. Weissbach, *in* "Lipmann Symposium: Energy, Regulation and Biosynthesis in Molecular Biology" (D. Richter, ed.), p. 384. de Gruyter, Berlin, 1974.
[18] J. Drews, K. Bednarik, and H. Grasmuk, *Eur. J. Biochem.* **41,** 217 (1974).
[19] C. Nombela, B. Redfield, S. Ochoa, and H. Weissbach, *Eur. J. Biochem.* **65,** 395 (1976).
[20] L.I. Slobin and W. Möller, *Eur. J. Biochem.* **69,** 351 (1976).
[21] S. Ejiri, K. Murakami, and T. Katsumata, *FEBS Lett.* **82,** 111 (1977).
[22] H. Taira, S. Ejiri, and K. Shimura, *J. Biochem. (Tokyo)* **72,** 1527 (1972).
[23] B. Golińska and A. B. Legocki, *Biochim. Biophys. Acta* **324,** 156 (1973).
[24] G. A. Lanzani, R. Bollini, and A. N. Soffientini, *Biochim. Biophys. Acta* **335,** 275 (1974).
[25] R. Bollini, A. N. Soffientini, A. Bertani, and G. A. Lanzani, *Biochemistry* **13,** 5421 (1974).
[26] T. Twardowski and A. B. Legocki, *Biochim. Biophys. Acta* **324,** 171 (1973).
[27] D. Richter and F. Lipmann, *Biochemistry* **9,** 5065 (1970).
[28] L. L. Spremulli and J. M. Ravel, *Arch. Biochem. Biophys.* **172,** 261 (1976).
[29] S. Nagata, K. Iwasaki, and Y. Kaziro, *J. Biochem. (Tokyo)* **80,** 73 (1976).

and 55,000, which were named EF-1$\beta$ and EF-1$\gamma$, respectively.[7,8] EF-1$\alpha$ catalytically promotes the binding of an aminoacyl-tRNA to ribosomes in the presence of GTP, while stoichiometrically in the presence of guanyl-5′-yl imidodiphosphate [GMP-P(NH)P], a nonhydrolyzable analog of GTP.[5,30] EF-1$\beta\gamma$ as well as EF-1$\beta$ stimulates the exchange of the guanine nucleotide bound to EF-1$\alpha$ with the exogenous one.[7,8,31,32] Thus, EF-1$\alpha$ functionally corresponds to bacterial EF-Tu and EF-1$\beta\gamma$ or EF-1$\beta$ to bacterial EF-Ts. Recently, we purified the high-molecular-weight form of EF-1 from pig liver and found that it is a complex of EF-1$\alpha$ and EF-1$\beta\gamma$, or EF-1$\alpha\beta\gamma$, which is functionally similar to the bacterial EF-Tu · EF-Ts complex. The methods presented below describe the purification procedures of these factors from pig liver. Briefly, the postmitochondrial supernatant from pig liver is fractioned by an aqueous two-phase separation method[6,33] employing dextran and polyethylene glycol, followed by ammonium sulfate precipitation. The crude fraction thus obtained is fractionated with ammonium sulfate into two fractions, i.e., 30–50% and 50–80% saturated ammonium sulfate fractions. The former fraction contains the high-molecular-weight form of EF-1, or EF-1$\alpha\beta\gamma$, and is used as the starting material for the purification of EF-1$\alpha\beta\gamma$. EF-1$\beta\gamma$ is also purified from this fraction through a separate procedure,[8] and its subunits, EF-1$\beta$ and EF-1$\gamma$, are obtained by treating purified EF-1$\beta\gamma$ with guanidine hydrochloride.[32] The latter fraction, on the other hand, contains both EF-1$\alpha$ and EF-2 which are separated by CM-Sephadex column chromatography,[6] and purified separately thereafter through a few steps.[6,8,9]

*Reagents*. All reagents used are the reagent grade.

- Trichloroacetic acid, 5%
- NaClO, 1%
- Sodium cholate, 20%; Sigma Chemical Co., St. Louis, Missouri
- Poly(U); Yamasa Shoyu Co. Ltd., Chiba, Japan
- tRNA from *Escherichia coli* B; Schwarz/Mann, Orangeburg, New York
- GTP; Yamasa Shoyu Co. Ltd., Chiba, Japan
- Guanyl-5′-yl imidodiphosphate [GMP-P(NH)P]; C. F. Boehringer & Sohne GmbH, Mannheim, West Germany
- [$^{14}$C]Phenylalanine (382 Ci/mol); Daiichi Pure Chemicals, Tokyo, Japan

[30] S. Nagata, K. Iwasaki, and Y. Kaziro, *Arch. Biochem. Biophys.* **172,** 168 (1976).

[31] S. Nagata, K. Motoyoshi, and K. Iwasaki, *Biochem. Biophys. Res. Commun.* **71,** 933 (1976).

[32] K. Motoyoshi and K. Iwasaki, *J. Biochem. (Tokyo)* **82,** 703 (1977).

[33] P.-Å. Albertson, "Partition of Cell Particles and Macromolecules." Wiley (Interscience), New York, 1971.

[$^3$H]GDP (12,700 Ci/mole), Radiochemical Centre, Amersham, England

*Solutions*

Solution A: 35 m$M$ Tris · HCl (pH 7.5), 9 m$M$ $Mg(CH_3COO)_2$, 70 m$M$ KCl, 0.1 m$M$ EDTA, and 10 m$M$ 2-mercaptoethanol

Solution B: 20 m$M$ Tris · HCl (pH 7.5), 3 m$M$ $Mg(CH_3COO)_2$, and 0.3 $M$ KCl

Solution C: 50 m$M$ Tris · HCl (pH 8.0), 4 m$M$ $Mg(CH_3COO)_2$, 5 m$M$ 2-mercaptoethanol, 25 m$M$ KCl, and 0.35 $M$ sucrose

Solution D: 20 m$M$ Tris · HCl (pH 7.5), 0.1 m$M$ EDTA, and 10 m$M$ 2-mercaptoethanol

Solution E: same as solution D except that the pH of Tris · HCl is 9.0

Solution F: same as solution D except that the pH of Tris · HCl is 7.2

Solution G: 20 m$M$ Tris · HCl (pH 7.5), 0.1 m$M$ EDTA, 10 m$M$ 2-mercaptoethanol, 0.25 $M$ sucrose, and 5% (v/v) glycerol

Solution H: same as solution G except that the pH of Tris · HCl is 8.0

Solution I: 10 m$M$ potassium phosphate buffer (pH 7.5), 5 m$M$ 2-mercaptoethanol, and 5% (v/v) glycerol

Solution J: 40 m$M$ Tris · HCl (pH 7.5), 40 m$M$ 2-mercaptoethanol, and 5 $M$ urea

Solution K: 10 m$M$ potassium phosphate buffer (pH 7.5), 10 m$M$ 2-mercaptoethanol, 2.5 m$M$ EDTA, and 25% (v/v) glycerol

## Preparation of Biological Materials

### *Washed Ribosomes from* Artemia salina *Cysts*

Washed ribosomes from *A. salina* cysts are prepared essentially according to Zasloff and Ochoa,[34] with a few modifications.[9] All procedures are carried out at 0°–4°. About 30 g of *A. salina* cysts (Aquarium Stock Co. Inc., 31 Warren Street, New York, New York) are suspended in 100 ml of ice-cold 1% NaClO in a 2-liter beaker, stirred for 5 min and then diluted with 1 liter of ice-cold water. Upon standing for a few minutes to sediment the cysts from the suspension, the supernatant together with floating damaged cysts is aspirated off. The sediment is then suspended in about 400 ml of ice-cold water. After settling the suspension for a few minutes, the supernatant is again aspirated off. Washing with water is repeated 10 times, and then the sediment is suspended in about 200 ml of solution A containing 0.25 $M$ sucrose. The suspension is transferred into a mortar and the supernatant obtained upon standing is completely removed by aspiration. The sediment is then ground for about 20 min with the occasional microscopic

[34] M. Zasloff and S. Ochoa, *Proc. Natl. Acad. Sci. U.S.A.* **68,** 3059 (1971).

inspection. When the cysts are completely broken, they are suspended in 100 ml of solution A containing 0.25 *M* sucrose. The suspension is filtered through 4 layers of cheesecloth, and the filtrate is centrifuged at 15,000 *g* for 15 min. The supernatant is again filtered through 4 layers of cheesecloth. The filtrate is then centrifuged for 15 min at 30,000 rpm in a type 30 Spinco rotor (Spinco Division, Beckman Instruments, Inc., Palo Alto, California) to remove glycogen. After removal of the floating reddish orange materials with a spatula, the supernatant is combined and finally centrifuged for 90 min at 50,000 rpm in a type 60 Ti Spinco rotor. The surface of the pellet is carefully rinsed with solution A and then suspended in 6 ml of the same solution in a hand homogenizer to give a crude ribosomal preparation. To this suspension is added one-third of its volume of 2 *M* KCl to make a final concentration of 0.5 *M*. After mixing, 3 ml of the suspension are layered over a discontinuous sucrose gradient consisting of 4 ml of 1 *M* and 3 ml of 0.4 *M* sucrose in solution B, and centrifuged for 16 hr at 50,000 rpm in a type 65 Spinco rotor. The surface of the pellet is carefully rinsed, and the ribosomal pellet is thoroughly suspended with a hand homogenizer in solution A containing 0.25 *M* sucrose using an about equal volume to the pellet. Insoluble materials are then removed by centrifugation at 15,000 *g* for 15 min and the opalescent supernatant is used as washed ribosomes. The preparation can be stored at −80° for several months without any appreciable loss of the activity.

### *Ribosome-Poly(U)-tRNA Complex*[4]

The reaction mixture contains, in a final volume of 1.5 ml, 50 m*M* Tris·HCl (pH 7.5), 12 m*M* $Mg(CH_3COO)_2$, 0.1 *M* $NH_4Cl$, 30 m*M* KCl, 0.2 m*M* dithiothreitol, 3 mg of poly(U), 1.5 mg of deacylated *E. coli* tRNA and 150 $A_{260}$ units of washed *A. salina* ribosomes. After incubation at 37° for 15 min, the mixture is divided into small aliquots and stored at −80°.

### *Comments*

It is desirable to use "runoff" 80 S ribosomes for the study of protein biosynthesis dependent on the exogenous mRNA, and a number of procedures have been reported for their preparation from various eukaryotic tissues. However, after the report by Zasloff and Ochoa,[34] which indicated that ribosomes in *A. salina* cysts exist mainly as free 80 S ribosomes and are easily purified to the state free from any of the polypeptide chain elongation factors, we have used them in the most of the experiments because of their high activity and of the ease of preparation. For instance, ribosomes prepared from *A. salina* cysts as above are almost 100% active both in the enzymic binding of phenylalanyl-tRNA dependent on added poly(U) under optimum conditions[5] and in the binding reaction of EF-2 to ribosomes in the presence of guanine nucleotides.[35] As described above, the procedure of preparing washed ribosomes from *Artemia salina* cysts is as simple as that from *E. coli*, and highly purified preparation is obtained in a large scale and in a short time, from the cysts or "brine shrimp eggs" which are commercially available.

As reported by Hardesty *et al.*,[36] the first step of polyphenylalanine synthesis from phenylalanyl-tRNA is the binding of $tRNA^{Phe}$ to the ribosomal peptidyl site dependent on added poly(U). Thus, when ribosomes, poly(U) and phenylalanyl-tRNA, including uncharged $tRNA^{Phe}$, are separately added to the reaction mixture, a lag time is observed in the binding of phenylalanyl-tRNA to ribosomes as well as in polyphenylalanine synthesis from phenylalanyl-tRNA, making the accurate assay of the elongation factors rather difficult. However, when these reactions are carried out with the use of the ribosome-poly(U)-tRNA complex, which has been prepared by preincubating ribosomes, poly(U) and tRNA in the presence of a high concentration of $Mg^{2+}$, the reactions start immediately without any lag time. Accordingly, under these conditions the accurate assays of the factors become possible.

We have used tRNA from *E. coli* B rather than tRNAs from eukaryotic tissues, such as liver and yeast. In the beginning of this investigation, we examined the activities of tRNAs from rat liver and *E. coli* and found them to be equally active in polyphenylalanine synthesis, when they are charged with phenylalanine by aminoacyl-tRNA synthetases from respective sources. As far as we have examined, lysyl-tRNA and leucyl-tRNA from *E. coli*, in addition to phenylalanyl-tRNA, can be used in our system, with equal efficiency to those from rat liver.

## Assays

Three types of the assay are used for measuring the activities of EF-1$\alpha$, EF-1$\beta\gamma$, and EF-2. The EF-1$\alpha$ activity is determined by the extent of enzymic binding of [$^{14}$C]phenylalanyl-tRNA to ribosomes in the presence of GMP-P(NH)P.[6,30] The assay is stoichiometric to the amount of EF-1$\alpha$ added, since no cyclic reuse of EF-1$\alpha$ is possible in the absence of hydrolysis of the guanine nucleotide (stoichiometric binding assay). The EF-1$\beta\gamma$ or EF-1$\beta$ activity is determined by the rate of exchange of [$^3$H]GDP bound to EF-1$\alpha$ with exogenous GTP in the presence of EF-1$\beta\gamma$ or EF-1$\beta$

[35] K. Mizumoto, K. Iwasaki, and Y. Kaziro, *J. Biochem. (Tokyo)* **76,** 1296 (1974).

[36] B. Hardesty, W. Culp, and W. McKeehan, *Cold Spring Harbor Symp. Quant. Biol.* **34,** 331 (1969).

to the reaction mixture (nucleotide exchange assay).[8] Finally, the EF-2 activity is measured by the EF-2 dependent synthesis of polyphenylalanine from [$^{14}$C]phenylalanyl-tRNA in the presence of a saturating amount of EF-1$\alpha$ free of contaminating EF-2 (polymerization assay).[9] The standard conditions of these assays are as follows.

*Standard Stoichiometric Binding Assay*

The reaction mixture contains, in a final volume of 0.1 ml, 50 m*M* Tris·HCl (pH 7.5), 7.5 m*M* $Mg(CH_3COO)_2$, 75 m*M* KCl, 0.2 m*M* dithiothreitol, 0.1 m*M* GMP-P(NH)P, 10 $\mu$g of bovine serum albumin, 16 pmol of [$^{14}$C]phenylalanyl-tRNA (382 Ci/mol), 1.0 $A_{260}$ unit of ribosome-poly(U)-tRNA complex, and 0–10 pmol of EF-1$\alpha$. Incubation is carried out for 5 min at 37°. The reaction is stopped by the addition of 3 ml of chilled washing buffer containing 20 m*M* Tris·HCl (pH 7.5), 5 m*M* $Mg(CH_3COO)_2$, and 100 m*M* $NH_4Cl$, and the diluted sample is poured onto a nitrocellulose membrane filter (pore size 0.45 $\mu$m, Sartorius-Membranfilter GmbH, 34 Göttingen, West Germany). The filter is washed twice with 3 ml of the washing buffer, dried, and counted in a liquid scintillation spectrometer. Values for binding without added EF-1$\alpha$ are routinely determined (about 0.6 pmol) and subtracted from the total amount of phenylalanyl-tRNA bound to ribosomes. Under the conditions described, the amount of [$^{14}$C]phenylalanyl-tRNA enzymically bound to ribosomes is proportional to that of EF-1$\alpha$ present in the reaction mixture for net values up to 10 pmol of phenylalanyl-tRNA.

*Standard Nucleotide Exchange Assay*

The reaction mixture contains, in a final volume of 90 $\mu$l, 20 m*M* Tris·HCl (pH 7.5), 0.2 m*M* dithiothreitol, 4.5 m*M* $Mg(CH_3COO)_2$, 66.7 m*M* $NH_4Cl$, 8.5% (v/v) glycerol, 167 $\mu$*M* GTP, 24 $\mu$g of *E. coli* tRNA charged with 20 kinds of amino acids, 10 $\mu$g of bovine serum albumin, and an appropriate amount of EF-1$\beta\gamma$ or EF-1$\beta$. The mixture is preincubated at 37° for 5 min to convert EF-1$\alpha$ present in the EF-1$\beta\gamma$ or EF-1$\beta$ sample into the ternary complex containing EF-1$\alpha$, GTP, and aminoacyl-tRNA. To the preincubated mixture is added 10 $\mu$l of EF-1$\alpha$·[$^3$H]GDP complex, which has been prepared by incubating a mixture containing 20 m*M* Tris·HCl (pH 7.5), 0.2 m*M* dithiothreitol, 10 m*M* $Mg(CH_3COO)_2$, 100 m*M* $NH_4Cl$, 7 $\mu$*M* [$^3$H]GDP (500 Ci/mol), 1 mg/ml of bovine serum albumin, 25% (v/v) glycerol, and 0.33 mg/ml of purified EF-1$\alpha$. The solution is rapidly mixed and incubated at 0° for 45 sec. After incubation the mixture is diluted with 0.5 ml of the ice-cold dilution buffer containing 20 m*M* Tris·HCl (pH 7.5),

10 m*M* $Mg(CH_3COO)_2$, 100 m*M* $NH_4Cl$, 10 m*M* 2-mercaptoethanol, 100 μg of bovine serum albumin per milliliter, and 25% (v/v) glycerol to stop the reaction. To determine the amount of the EF-1α·[$^3$H]GDP complex remaining, the diluted sample is poured onto a nitrocellulose membrane filter, which is then carefully washed twice with 0.5 ml of the dilution buffer without glycerol and bovine serum albumin. After drying the filter, the radioactivity retained is measured with a liquid scintillation spectrometer. Values for the amount of EF-1α·[$^3$H]GDP remaining in the absence of EF-1βγ or EF-1β are routinely determined and the net values of the decrease in the amount of EF-1α·[$^3$H]GDP is defined as the amount of [$^3$H]GDP exchanged with GTP. Under the conditions described, it is proportional to the amount of EF-1βγ or EF-1β present in the reaction mixture for net values below 15 pmol of [$^3$H]GDP exchanged.

*Standard Polymerization Assay*

The reaction mixture contains, in a final volume of 0.1 ml, 50 m*M* Tris·HCl (pH 7.5), 5 m*M* $Mg(CH_3COO)_2$, 50 m*M* KCl, 50 m*M* $NH_4Cl$, 0.2 m*M* dithiothreitol, 0.1 m*M* GTP, 0.5 $A_{260}$ unit of ribosome-poly(U)-tRNA complex, 40 pmol of [$^{14}$C]phenylalanyl-tRNA (100 Ci/mol), 2 μg of purified EF-1α, and an appropriate amount of EF-2. After incubation for 20 min at 37°, the reaction is stopped with 3 ml of 5% trichloroacetic acid, followed by 0.1 ml of 1 mg/ml bovine serum albumin solution. The samples are then heated at 90° for 15 min and cooled in ice; the precipitates are collected on a glass-fiber filter (Whatman GF/C, W. & R. Balston Ltd., Springfield Mill, Maidstone, Kent, England). The filter is washed three times with 3 ml of 5% trichloroacetic acid and once with 1 ml of acetone, dried, and counted in a liquid scintillation spectrometer. Values for incorporation without added EF-2 are almost nil when salt-washed ribosomes from *A. salina* cysts and homogeneously purified EF-1α from pig liver are used. The amount of polyphenylalanine synthesized is proportional to the amount of EF-2 present in the reaction mixture for the values below 20 pmol of phenylalanine polymerized.

*Comments*

Polyphenylalanine synthesis from phenylalanyl-tRNA is dependent on three factors: EF-1α, EF-1βγ, and EF-2.[4] However, when the saturating amount of EF-1α is present in the reaction mixture, the reaction is no longer dependent on EF-1βγ but only on EF-2, since the function of EF-1βγ is to stimulate the cyclic reuse of EF-1α.[8] Accordingly, the dependence of the reaction on EF-1βγ can be shown only in the presence of a limited amount of EF-1α. Thus, under the conditions for the standard polymerization assay, i.e., in the presence of 2 μg of purified EF-1α, the reaction is dependent only on EF-2, and is not influenced by the presence or the absence of EF-1βγ.

It is possible to determine the EF-1α activity by the polymerization assay, when the sample is free of EF-1βγ.[5] For this purpose, the reaction mixture should contain a saturating amount of EF-2, namely 2 μg of purified EF-2 from pig liver, and an appropriate amount of EF-1α. It is also possible to determine the EF-1α activity by the binding of [$^{14}$C]phenylalanyl-tRNA to ribosomes in the presence of 0.1 m*M* GTP instead of GMP-P(NH)P, provided that EF-1α is free of both EF-2 and EF-1βγ. Since EF-1α functions catalytically in the presence of GTP, this assay is about 5 times as sensitive when compared to the standard stoichiometric binding assay with GMP-P(NH)P. However, it is not recommended to use this assay for the determination of the EF-1α activity in the crude fraction that has been treated with *N*-ethylmaleimide to inactivate the contaminating EF-2 as reported by McKeehan and Hardesty,[13] because EF-1α is also inactivated with this reagent.

The EF-1βγ activity can be determined by the stimulation of polyphenylalanine synthesis, when the reaction mixture contains a saturating amount of EF-2 (2 μg of purified EF-2) and a limited amount of EF-1α (40 ng of purified EF-1α).[8] Under these conditions the reaction is almost completely dependent on added EF-1βγ and a good proportionality is obtained between the amount of EF-1βγ present in the reaction mixture and that of polyphenylalanine synthesized. However, when the sample to be assayed is contaminated with EF-1α, the determination of the EF-1βγ activity by this method tends to give an overestimation, because polyphenylalanine synthesis is stimulated not only by EF-1βγ but also by EF-1α, since only a limited amount of EF-1α is present in the reaction mixture.

## Purification of Polypeptide Chain Elongation Factors

### *Preparation of Crude Factors*[6]

All procedures are carried out at 0°–4°.

*Crude Extract.* Fresh pig liver (1.3 kg) obtained from a slaughterhouse is cut into small pieces and homogenized with 2 liters of solution C for 3 min in a Waring blender at the top speed. The homogenate is centrifuged at 15,000 *g* for 15 min and the precipitate is discarded. The supernatant obtained (the postmitochondrial supernatant, 2350 ml) contains 72 mg of protein per milliliter.

*Aqueous Two-Phase Separation.* To 2 liters of the supernatant from the previous step are added 562.5 ml of 30% (w/w) polyethylene glycol No. 6000 (Nakarai Chemicals, Kyoto, Japan) and 187.5 ml of 20% (w/w) dextran T500 (Pharmacia Fine Chemicals, Uppsala, Sweden). The mixture is stirred for about 10 min in a Waring blender, and the phases are separated by centrifugation for 15 min at 15,000 *g*. The lower dextran phase is saved because it contains all the polypeptide chain elongation factors. To the lower phase are added 412.5 g of solid $NH_4Cl$ and the polyethylene glycol phase that has been prepared in a similar manner as above except that the supernatant is replaced by 2 liters of solution C. After dissolving $NH_4Cl$ thoroughly, the mixture is stirred for 10 min, and then the two phases are again separated by centrifugation. Since all the polypeptide chain elongation factors are transferred into the polyethylene glycol phase by this procedure, the upper polyethylene glycol phase is collected. Polyethylene glycol in this phase is removed by the use of an aqueous two-phase separation system consisting of polyethylene glycol and ammonium sulfate. To 2500 ml of the upper phase obtained as above are added 440 g of solid $(NH_4)_2SO_4$. After dissolving $(NH_4)_2SO_4$, the mixture is stirred for 10 min and centrifuged for 15 min at 8000 *g* using a horizontal rotor to separate the three phases: the top (polyethylene glycol), the middle (precipitate), and the bottom (ammonium sulfate) phases. The bottom phase containing all the polypeptide chain elongation factors is carefully saved by aspiration and the proteins are precipitated by the addition of 416 g (20 g/100 ml) of $(NH_4)_2SO_4$. After stirring overnight, the precipitate is collected by centrifugation for 30 min at 15,000 *g*, and used as the crude factors.

### *Separation of EF-1αβγ from EF-1α and EF-2 by Ammonium Sulfate Fractionation*[6]

The precipitated proteins from the previous step (crude factors) are extracted three times with 200 ml, 150 ml, and 150 ml of 50% saturated ammonium sulfate solution in solution C [29.1 g of $(NH_4)_2SO_4$ per 100 ml of solution C]. The undissolved materials after the extraction are dissolved in solution D containing 25% (v/v) glycerol, and this fraction is designated as the 30–50% saturated ammonium sulfate fraction. The proteins extracted with 50% saturated ammonium sulfate solution are precipitated by adding $(NH_4)_2SO_4$ (19 g/100 ml) to give 80% saturation, and the precipitate is collected by centrifugation for 30 min at 15,000 *g*. The pellet is dissolved in solution D containing 25% (v/v) glycerol and designated as the 50–80% saturated ammonium sulfate fraction. It is found that the 50–80% saturated ammonium sulfate fraction contains the low-molecular-weight form of EF-1 (EF-1α) together with EF-2, while the 30–50% saturated ammonium sulfate fraction contains the high-molecular-weight form of EF-1 (EF-1$\alpha\beta\gamma$). Therefore, the former is used as the crude fraction of EF-1$\alpha$ and EF-2, while the latter as that of EF-1$\alpha\beta\gamma$ as well as EF-1$\beta\gamma$. Routinely, the procedure up to this step is repeated and the materials from two preparations are combined and used for further purification.

### *Separation of EF-1α and EF-2 by CM-Sephadex Column Chromatography*[6]

The 50–80% saturated ammonium sulfate fraction obtained as above (100 mg of protein per milliliter, 150 ml) is extensively dialyzed against solution D containing 50 m*M* $NH_4Cl$ and 25% (v/v) glycerol, and applied to a column (3 × 47 cm) of CM-Sephadex C-50 (Pharmacia Fine Chemicals, Uppsala, Sweden) previously equilibrated with solution D containing 40 m*M* $NH_4Cl$ and 25% (v/v) glycerol. The column is then developed successively with each 1 liter of solution D containing 25% (v/v) glycerol and increasing concentrations of $NH_4Cl$ (58, 165, and 350 m*M*) at a flow rate of 20 ml/hr and 10-ml fractions are collected. As shown in Fig. 1, most of the EF-2 activity is found in the fractions eluted with solution D containing 165 m*M* $NH_4Cl$ and 25% (v/v) glycerol, while the EF-1$\alpha$ activity is located in the fractions eluted with solution D containing 350 m*M* $NH_4Cl$ and 25% (v/v) glycerol. The active fractions are separately combined and used as the crude EF-2 fraction (330 ml, 2.8 mg of protein per milliliter) and the crude EF-1$\alpha$ fraction (210 ml, 2.4 mg of protein per milliliter), respectively, for further purification.

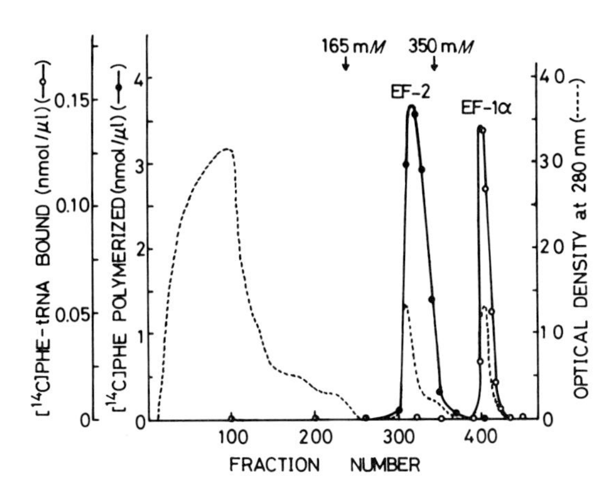


FIG. 1. Separation of EF-2 and EF-1$\alpha$ by CM-Sephadex column chromatography. The EF-2 and EF-1$\alpha$ activities are determined by standard polymerization assay with 0.02 μl of the fraction and by standard stoichiometric binding assay with 0.07 μl of the fraction, respectively. ●——●, EF-2 activity; ○——○, EF-1$\alpha$ activity; ----, optical density at 280 nm.

*Purification of EF-2*[8,9]

The crude EF-2 fraction obtained as above is dialyzed overnight against 5 volumes of solution D containing 10% (v/v) glycerol and 603 g of $(NH_4)_2SO_4$ per liter. The precipitated materials are collected by centrifugation and the precipitate is dissolved in a small volume of solution H containing 85 m*M* KCl. The solution is dialyzed overnight against the same buffer-salt solution. After dialysis, 11.5 ml of the solution containing 66 mg of protein per milliliter are obtained.

*DEAE-Sephadex Column Chromatography*. The dialyzed material is applied to a column (1.6 × 55 cm) of DEAE-Sephadex A-50 (Pharmacia Fine Chemicals, Uppsala, Sweden) which has been equilibrated with solution H containing 80 m*M* KCl, and the column is washed with 100 ml of solution H containing 90 m*M* KCl. Elution is then carried out with a linear gradient of KCl from 90 to 200 m*M* in solution H with a total volume of 1.5 liters at a flow rate of 15 ml/hr and 5-ml fractions are collected. EF-2 is eluted from the column at about 0.15 *M* KCl. The active fractions are combined (194 ml), and the proteins are precipitated by adding 100 g of solid ammonium sulfate. The precipitate collected by centrifugation is dissolved in a small volume of solution G containing 45 m*M* KCl and the solution is dialyzed overnight against 1 liter of the same buffer–salt solution with one change of the buffer. After dialysis, 3.5 ml of the solution containing 10 mg/ml of protein are obtained.

*CM-Sephadex Column Chromatography*. The dialyzed material obtained as above is applied to a column (1.6 × 50 cm) of CM-Sephadex C-50, which has been equilibrated with solution G containing 40 m*M* KCl. After washing the column with 100 ml of solution G containing 50 m*M* KCl, it is developed with a linear gradient of KCl from 50 to 160 m*M* in solution G with a total volume of 1 liter. The flow rate is 12 ml/hr and 4-ml fractions are collected. EF-2 is eluted from the column as a single and almost symmetrical peak at about 0.1 *M* KCl. The fractions containing EF-2 activity are combined (102 ml) and dialyzed overnight against 2 liters of solution H containing 559 g of $(NH_4)_2SO_4$ per liter. After dialysis, the precipitate is collected by centrifugation and dissolved in a small volume of solution H and dialyzed overnight against 1 liter of the same solution. The dialyzed material (2.1 ml) contains 16 mg of protein per milliliter and is used as purified EF-2. The overall purification is about 800-fold starting from the postmitochondrial supernatant with the yield of 30%.

Purified EF-2 obtained as above is homogeneous when examined by polyacrylamide gel electrophoresis in the presence of sodium dodecyl sulfate (SDS) and can be stored at −80° without appreciable loss of the activity for at least several months. Analyses of purified EF-2 by several methods indicate that the $s^{\circ}_{20,w}$ is approximately 5.32 S and that the molecular weight is 100,000.[37] EF-2 catalyzes the translocation reaction of peptidyl-tRNA from the ribosomal aminoacyl to the peptidyl site.[38]

*Purification of EF-1α*[6]

The crude EF-1α fraction obtained as described in the previous section is dialyzed overnight against 5 liters of solution D containing 10% (v/v) glycerol and 80% saturated ammonium sulfate [65 g $(NH_4)_2SO_4$ per 100 ml]. The precipitate formed after dialysis is collected by centrifugation and dissolved in a small volume of solution E containing 5 m*M* $NH_4Cl$ and 25% (v/v) glycerol. The solution is then dialyzed overnight against the same buffer.

*DEAE-Sephadex Column Chromatography*. The dialyzed material containing 504 mg of protein is applied to a column (1.2 × 16 cm) of DEAE-Sephadex A-25 previously equilibrated with solution E containing 25% (v/v) glycerol, and the column is washed with solution E containing 5 m*M* $NH_4Cl$ and 25% (v/v) glycerol at a flow rate of 10 ml/hr, and 3.5-ml fractions are collected. The flow-through fractions containing the EF-1α activity are pooled, and dialyzed overnight against solution D containing 180 m*M* $NH_4Cl$ and 25% (v/v) glycerol. After dialysis, 27 ml of the fraction are obtained, which contains 12.3 mg of protein per milliliter.

*CM-Sephadex Column Chromatography*. The dialyzed solution from the previous step is applied to a column (1.6 × 75 cm) of CM-Sephadex C-50 equilibrated with solution D containing 170 m*M* $NH_4Cl$ and 25% (v/v) glycerol. After washing the column with solution D containing 180 m*M* $NH_4Cl$ and 25% (v/v) glycerol, elution is carried out with a linear gradient of $NH_4Cl$ from 185 to 315 m*M* in solution D containing 25% (v/v) glycerol with a total volume of 1.5 liters. The flow rate is 5 ml/hr, and 2.5-ml fractions are collected. As shown in Fig. 2, EF-1α is eluted as four peaks at 225, 245, 270, and 290 m*M* $NH_4Cl$, which are designated as peaks I, II, III, and IV, respectively. Each peak is separately combined, and the pooled fractions are dialyzed against 5 liters of solution D containing 10% (v/v) glycerol and 80% saturated ammonium sulfate prepared as above. The precipitate formed is collected by centrifugation and dissolved in a small volume of solution D containing 25% (v/v) glycerol, and the solution is dialyzed overnight against solution D containing 0.3 *M* $NH_4Cl$ and 25% (v/v) glycerol. After removing the insoluble materials by centrifugation at

[37] K. Mizumoto, K. Iwasaki, Y. Kaziro, C. Nojiri, and Y. Yamada, *J. Biochem. (Tokyo)* **75,** 1057 (1974).

[38] M. Tanaka, K. Iwasaki, and Y. Kaziro, *J. Biochem. (Tokyo)* **82,** 1035 (1977).

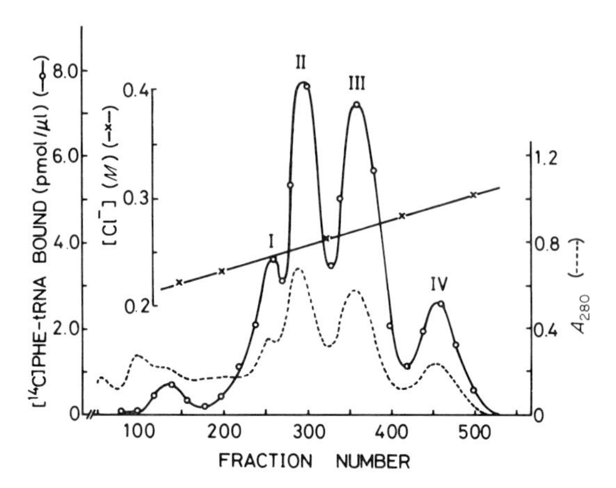


FIG. 2. CM-Sephadex C-50 column chromatography of EF-1$\alpha$ [S. Nagata, K. Iwasaki, and Y. Kaziro, *J. Biochem.* (*Tokyo*) **82,** 1633 (1977)]. The active fractions obtained by chromatography on a DEAE-Sephadex A-25 column are dialyzed and applied to a column of CM-Sephadex C-50 as described in the text. ○——○, EF-1$\alpha$ activity; ×——×, concentration of $NH_4Cl$; and ----, optical density at 280 nm.

15,000 *g* for 20 min, the clarified supernatant is obtained as the purified EF-1$\alpha$ from each peak. Thus, 2.6 ml of peak I (12.3 mg of protein per milliliter), 8.3 ml of peak II (10 mg of protein per milliliter), 4.6 ml of peak III (17.5 mg of protein per milliliter), and 4.2 ml of peak IV (5.8 mg of protein per milliliter) are obtained. The overall purification is about 340-fold starting from the postmitochondrial supernatant with a total yield of 24%. The purified EF-1$\alpha$ can be stored at $-80°$ without any appreciable loss of the activity for at least one year, provided that the protein concentration is higher than 1 mg/ml.

When the four peaks are examined for their purity by polyacrylamide gel electrophoresis in the presence of SDS, it is found that peaks II, III, and IV are homogeneous, but peak I is contaminated slightly with other proteins. The molecular weight of EF-1$\alpha$ in each peak is identical, 53,000, and no difference is detected in their enzymic properties. The only difference observed between them is their isoelectric points, i.e., 9.3, 9.5, 9.7, and 9.9 for peaks I, II, III, and IV, respectively, but its nature is unknown.[6]

## *Purification of EF-1$\beta\gamma$*[8]

The 30–50% saturated ammonium sulfate fraction described in the previous section is dissolved in a small volume of solution I and used as the crude fraction of EF-1$\beta\gamma$. From 3 kg of pig liver, 100 ml of this crude fraction are obtained, which contain 129 mg of protein per milliliter.

*Cholate Treatment.* To 100 ml of the crude fraction is added 10 ml of 20% (w/v) sodium cholate solution. The mixture is stirred for 15 min, then 72.5 ml of solution I saturated with ammonium sulfate at 4° are added. After stirring for another 15 min, the mixture is centrifuged at 10,800 $g$ for 20 min. Slightly yellowish materials floating on the surface are carefully removed with a spatula, and the yellow supernatant is discarded. The wall of tubes is wiped carefully, then the precipitate is dissolved in solution I to a final volume of 100 ml. This treatment with sodium cholate is repeated twice, and the last pellet is dissolved in a small volume of solution I. The solution is dialyzed against 2 liters of solution I containing 0.1 $M$ KCl with one change of the buffer. The insoluble material appearing after dialysis is removed by centrifugation and the supernatant (100 ml) containing 74 mg/ml protein is saved.

*First DEAE-Sephadex Column Chromatography.* The solution obtained as above is diluted 1.5-fold with solution I containing 0.1 $M$ KCl to make a protein concentration to 50 mg/ml, and the diluted solution is applied onto a column (2.5 × 75 cm) of DEAE-Sephadex A-50, which has been equilibrated with solution I containing 0.1 $M$ KCl. The column is washed with 450 ml of solution I containing 0.11 $M$ KCl. The flow rate is 25 ml/hr, and 12-ml fractions are collected. After washing the column, proteins are eluted in two steps: first with 700 ml of solution I containing 0.3 $M$ KCl, then with 700 ml of solution I containing 0.6 $M$ KCl. The EF-1$\beta\gamma$ activity is found only in the fractions eluted with 0.6 $M$ KCl. The active fractions are combined (370 ml) and the combined fractions are dialyzed overnight against 5 liters of solution I saturated with ammonium sulfate at 4°. The precipitates appearing after dialysis are collected by centrifugation, dissolved in a small volume of solution I containing 0.3 $M$ KCl, and dialyzed overnight against the same buffer-salt solution. After dialysis, 8 ml of solution containing 33 mg/ml of protein is obtained.

*Second DEAE-Sephadex Column Chromatography.* The material obtained by the first DEAE-Sephadex column chromatography is applied to a second column (1.6 × 70 cm) of DEAE-Sephadex A-50, which has been equilibrated with solution I containing 0.3 $M$ KCl. The flow rate is adjusted to 10 ml/hr, and 5-ml fractions are collected. After washing the column with 200 ml of solution I containing 0.3 $M$ KCl, elution is carried out with a linear gradient of KCl from 0.3 to 0.66 $M$ in solution I with a total volume of 1.4 liters. EF-1$\beta\gamma$ is eluted from the column as a single and almost symmetrical peak at about 0.45 $M$ KCl. The fractions containing EF-1$\beta\gamma$ activity are combined (155 ml) and dialyzed overnight against 5 liters of solution I containing 90% saturated ammonium sulfate. The precipitate appearing after dialysis is collected by centrifugation, dissolved in a small volume of a solution containing 10 m$M$ potassium phosphate buffer (pH 7.5), 5 m$M$

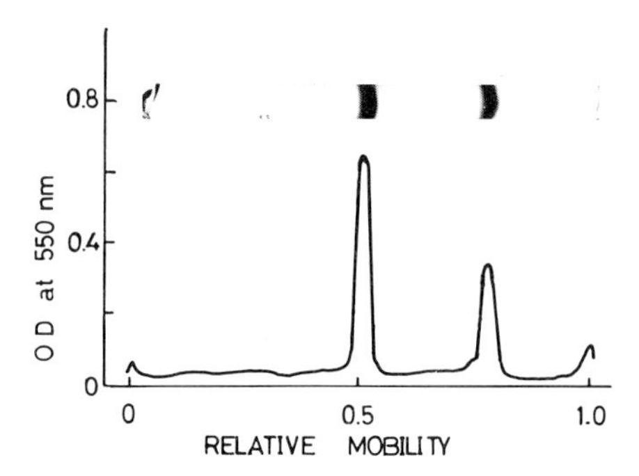


FIG. 3. Polyacrylamide gel electrophoresis of EF-1$\beta\gamma$ in the presence of sodium dodecyl sulfate [K. Motoyoshi, K. Iwasaki, and Y. Kaziro, *J. Biochem. (Tokyo)* **82,** 145 (1977)]. The direction of electrophoresis is from left to right. After electrophoresis, the gel is stained with Coomassie Brilliant Blue and the stained gel is scanned with a Shimadzu CS 900 densitometer. The picture of the stained gel (upper) and the densitometric tracing are shown.

2-mercaptoethanol, 0.3 *M* KCl, and 20% (v/v) glycerol, and dialyzed overnight against 1 liter of the same solution with one change of the solution. After dialysis, 3.5 ml of the material containing 11 mg/ml of protein is obtained, which is used as purified EF-1$\beta\gamma$. Electrophoretically, purified EF-1$\beta\gamma$ is nearly homogeneous and can be stored at $-80°$ for several months without any loss of the activity. The overall purification is about 30-fold starting from the ammonium sulfate fraction with a recovery of 10%.

The molecular weight of purified EF-1$\beta\gamma$ is estimated to be about 90,000 by gel filtration with Sephadex G-200. When it is analyzed by polyacrylamide gel electrophoresis in the presence of SDS, two clear bands are observed as shown in Fig. 3, the molecular weights of which are estimated to be 30,000 and 55,000. Thus, the former is designated as EF-1$\beta$ and the latter as EF-1$\gamma$. The densitometric tracing of the stained gel shown in Fig. 3 reveals the ratio of the amount of protein in EF-1$\beta$ to that in EF-1$\gamma$ is 0.5. This result indicates that the molar ratio of these two proteins is one to one, assuming their molecular weights as above. From these observations it is concluded that EF-1$\beta\gamma$ has a molecular weight of about 90,000, and consists of two unequal subunits having molecular weights of 30,000 (EF-1$\beta$) and 55,000 (EF-1$\gamma$) in the ratio of one to one.[8]

### *Resolution of EF-1$\beta$ and EF-1$\gamma$*[32]

Homogeneously purified EF-1$\beta\gamma$ (10 mg) prepared as above is concentrated to 0.35 ml under reduced pressure, then 0.7 ml of a solution containing 8 *M* guanidine hydrochloride and 50 m*M* 2-mercaptoethanol is added.

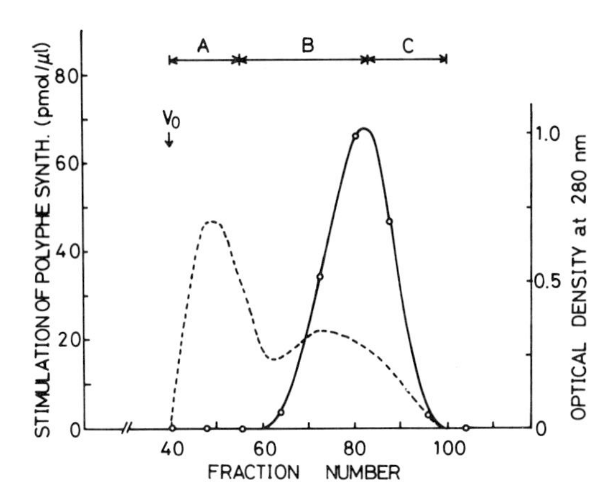


FIG. 4. Sephadex G-200 column chromatography of guanidine hydrochloride-treated EF-1$\beta\gamma$ in the presence of urea [K. Motoyoshi and K. Iwasaki, *J. Biochem. (Tokyo)* **82**, 703 (1977)]. After diluting an aliquot, its activity to stimulate polyphenylalanine synthesis is determined. ○——○, Stimulation of polyphenylalanine synthesis; ----, optical density at 280 nm. Fractions corresponded to A, B, and C as indicated in the figure are combined separately. $V_0$ indicates the position of the void volume.

The sample is kept at room temperature for 1 hr, then dialyzed overnight at 4° against 100 ml of solution J with one change of the solution. The dialyzed material (0.35 ml) is applied to a column (0.9 × 70 cm) of Sephadex G-200 (Pharmacia Fine Chemicals, Uppsala, Sweden) previously equilibrated with solution J. The column is developed with solution J, and fractions of 0.4 ml are collected at a flow rate of 4 ml/hr. The elution profile of proteins determined from the optical density at 280 nm shows two peaks (Fig. 4), suggesting that EF-1$\beta\gamma$ has been resolved into two subunits. When an aliquot from every third fraction is examined by polyacrylamide gel electrophoresis in the presence of SDS, the first peak is found to contain only EF-1$\gamma$, while the earlier half of the second peak contains both EF-1$\beta$ and EF-1$\gamma$, and the later half contains only EF-1$\beta$. Furthermore, when the fractions are diluted 30-fold with a solution containing 10 m$M$ potassium phosphate buffer (pH 7.5), 5 m$M$ 2-mercaptoethanol, 0.3 $M$ KCl, 1 mg/ml bovine serum albumin, and 20% (v/v) glycerol, then assayed for stimulation of polyphenylalanine synthesis in the presence of both a limited amount of EF-1$\alpha$ and a saturating amount of EF-2 (see the comments described under *Assays*), the activity is observed only with the fractions corresponding to the second peak, although the peak of the activity is eluted slightly behind that of the absorption at 280 nm. Thus, fractions Nos. 40–55, 56–83, and 84–100 are separately combined, and the combined fractions are designated as fractions A, B, and C, respectively. The combined fractions are then dialyzed against a buffer solution containing 40 m*M* Tris·HCl (pH 7.5), 40 m*M* 2-mercaptoethanol, and 2 *M* urea for 18 hr at 4°, and for another 18 hr at 4° against a buffer solution consisting of 10 m*M* potassium phosphate buffer (pH 7.5), 5 m*M* 2-mercaptoethanol, and 20% (v/v) glycerol. After dialysis, the purity of each fraction is examined by polyacrylamide gel electrophoresis in the presence of SDS to reveal that fractions A and C contain EF-1$\gamma$ and EF-1$\beta$, respectively, while fraction B is a mixture of the two subunits.

Thus, fractions A and C are used as separated EF-1$\gamma$ and EF-1$\beta$, respectively. As far as we have examined, the functions of EF-1$\beta$ are similar to those of EF-1$\beta\gamma$. Furthermore, they are equally effective in stimulating three reactions, i.e., the nucleotide exchange reaction, the EF-1$\alpha$-dependent binding of phenylalanyl-tRNA to ribosomes in the presence of GTP, and polyphenylalanine synthesis in the presence of a limited amount of EF-1$\alpha$, provided that their molar concentrations are the same.[31] It is also shown that EF-1$\beta\gamma$ as well as EF-1$\beta$ forms a complex with EF-1$\alpha$ in a molar ratio of one to one, namely, EF-1$\alpha\beta\gamma$ and EF-1$\alpha\beta$ complexes.[39] However, little is known about the function of EF-1$\gamma$.

## *Purification of the High-Molecular-Weight Form of EF-1 (EF-1$\alpha\beta\gamma$)*

Purification of the high-molecular-weight form of EF-1, i.e., EF-1$\alpha\beta\gamma$, is still under investigation. The following is a procedure of its purification preliminarily examined, which may require a few modifications to increase its recovery.

*Gel Filtration with Sephadex G-200.* A portion of the 30–50% saturated ammonium sulfate fraction described in the previous section is dissolved in a small volume of solution C (39 ml, 105 mg of protein per milliliter) and applied to a column (6 × 47 cm) of Sephadex G-200 previously equilibrated with solution D containing 100 m*M* $NH_4Cl$ and 25% (v/v) glycerol. The column is eluted with the same solution at a flow rate of 40 ml/hr and 10-ml fractions are collected. The EF-1$\alpha$ activity determined by the standard stoichiometric binding assay appears in two peaks; one in the void volume and the other in the elution volume coinciding with free EF-1$\alpha$. In addition to the EF-1$\alpha$ activity, the former contains the EF-1$\beta\gamma$ activity when each fraction is assayed by the standard nucleotide exchange assay, and thus, the active fractions eluted in the void volume are combined. To the combined fraction (312 ml) are added 2 volumes of solution D saturated with $(NH_4)_2SO_4$ at 4°. The precipitate formed is collected by centrifugation, dissolved in a small volume of solution F containing 20 m*M* $NH_4Cl$ and 25%

[39] S. Nagata, K. Motoyoshi, and K. Iwasaki, *J. Biochem. (Tokyo)* **83,** 423 (1978).

(v/v) glycerol, and dialyzed overnight against 2 liters of the same solution with two changes of the solution. After dialysis, 38 ml of the material containing 35.5 mg/ml of proteins are obtained.

*CM-Sephadex C-50 Column Chromatography.* The material obtained as above is applied to a column (2.8 × 27 cm) of CM-Sephadex C-50, previously equilibrated with solution F containing 20 m*M* $NH_4Cl$ and 25% (v/v) glycerol. After washing the column with solution F containing 30 m*M* $NH_4Cl$ and 25% (v/v) glycerol, proteins are successively eluted with solution F containing 175 m*M* $NH_4Cl$ and 25% (v/v) glycerol and then with a solution containing 350 m*M* $NH_4Cl$ and 25% (v/v) glycerol at a flow rate of 10 ml/hr, and 5-ml fractions are collected. Since both EF-1$\alpha$ and EF-1$\beta\gamma$ activities are found in the fractions eluted with solution F containing 175 m*M* $NH_4Cl$ and 25% (v/v) glycerol, the active fractions are combined and proteins are precipitated with solution D saturated with ammonium sulfate as described above. The precipitate collected by centrifugation is dissolved in a small volume of solution K containing 140 m*M* $(NH_4)_2SO_4$, and the dissolved material is dialyzed against the same solution. After dialysis, 15.9 ml of the solution are obtained, which contain 18.3 mg/ml of protein.

*Hydroxyapatite Column Chromatography.* The material obtained in the previous step is applied to a column (1.6 × 60 cm) of hydroxyapatite (Hypatite C, Clarkson Chemical Co. Inc., Williamsport, Pennsylvania) previously equilibrated with solution K containing 140 m*M* $(NH_4)_2SO_4$. After washing the column with 300 ml of the same solution at a flow rate of 25 ml/hr, the adsorbed proteins are eluted with solution K containing 330 m*M* $(NH_4)_2SO_4$ at a flow rate of 15 ml/hr. The EF-1$\alpha$ together with EF-1$\beta\gamma$ activities are found in this eluate. The combined active fraction is dialyzed overnight against solution D containing 5% (v/v) glycerol which has been saturated with $(NH_4)_2SO_4$ to 80%. The precipitate formed is collected by centrifugation, dissolved in a small volume of solution D containing 100 m*M* $NH_4Cl$ and 5% (v/v) glycerol, and dialyzed overnight against the same solution. Thus, 6.5 ml of the solution containing 9.23 mg/ml protein are obtained.

*Ammonium Sulfate Fractionation.* To the solution obtained as above (6.5 ml) are added 5.5 ml of solution D saturated with $(NH_4)_2SO_4$ to make a final saturation of $(NH_4)_2SO_4$ to 46%. The precipitate appearing upon standing for 1 hr is removed by centrifugation, and 2.4 ml of the saturated $(NH_4)_2SO_4$ solution as above are added to the supernatant to make a final saturation to 55%. After standing the mixture for 1 hr, the precipitated materials are collected by centrifugation, dissolved in a small volume of solution D containing 100 m*M* $NH_4Cl$, and dialyzed overnight against the same solution. After dialysis, 1.18 ml of the solution containing 23.5 mg/ml of protein are obtained.

*Glycerol Density Gradient.* A 50-$\mu$l aliquot of the solution obtained from the previous step is layered on the top of the 10–25% (v/v) linear glycerol gradient in solution D containing 100 m$M$ $NH_4Cl$ and the gradient is centrifuged at 40,000 rpm for 27 hr in an SW 41 Spinco rotor at 2.2°. The gradient is then fractionated from the top into 20 fractions (0.6 ml per fraction). The EF-1$\alpha$ as well as EF-1$\beta\gamma$ activity is located in the middle of the fractions (Nos. 10 and 11), and these fractions are used as the purified high-molecular-weight form of EF-1. The preparation thus obtained appears to be over 85% pure. The overall purification from the ammonium sulfate fraction determined by the EF-1$\beta\gamma$ activity is about 20-fold with a recovery of about 10%.

Judging from the results obtained by polyacrylamide gel electrophoresis in the presence of SDS, isoelectric focusing in the presence of urea in a polyacrylamide gel, and the estimation of the contents of both EF-1$\alpha$ and EF-1$\beta\gamma$ from their enzymic activities, the high-molecular-weight form of EF-1 consists of EF-1$\alpha$, EF-1$\beta$, and EF-1$\gamma$ in a molar ratio of one to one to one. The values of $s^{\circ}_{20,w}$ and the Stokes radius of the high-molecular-weight form of EF-1 are 6.25 S and 49 Å, respectively. From these values its molecular weight is roughly estimated to be about 130,000, which is close to the sum of the molecular weights of EF-1$\alpha$ (53,000), EF-1$\beta$ (30,000), and EF-1$\gamma$ (55,000), or 138,000. Thus, the identity of the high-molecular-weight form of EF-1 as an EF-1$\alpha\beta\gamma$ complex is established.[40]

[40] S. Hattori, K. Iwasaki, and Y. Kaziro, unpublished results, 1978.

# [41] Separation of Cytoplasmic and Mitochondrial Elongation Factors from Yeast

*By* DIETMAR RICHTER and FRITZ LIPMANN

Only relatively recently has it been firmly established that the eukaryotic organism contains two sets of protein-synthesizing apparatus: one, located in the cytoplasm, which represents the typical eukaryotic system, and another in mitochondria, which closely resembles the prokaryotic system.[1-3] Functionally and in chromatographic behavior, the

[1] A. W. Linnane, J. M. Haslam, H. B. Lukins, and P. Nagley, *Annu. Rev. Microbiol.* **26**, 163 (1972).
[2] P. Borst, *Annu. Rev. Biochem.* **41**, 333 (1972).
[3] H. Küntzel, *Curr. Top. Microbiol. Immunol.* **54**, 94 (1971).



ISBN 0-12-504180-2

elongation factors from the cytoplasm and mitochondria are quite different.[4] They can be isolated in two ways: (A) mitochondria and cytoplasm are separated by discontinuous centrifugation, and then the mitochondria are lysed mechanically or by osmotic shock to yield mitochondrial factors; (B) cells are disrupted by high shear forces which cause the breakage not only of cell walls but also of the organelles. Thus, these homogenates contain cytoplasmic as well as mitochondrial elongation factors which can be separated by chromatography.[4] Method B is rapid and convenient since it avoids isolation of the organelles; method A is essential, however, for identification.

This article describes both methods of isolation of mitochondrial and cytoplasmic elongation factors from yeast cells. A brief method for the isolation of mitochondrial ribosomes is also included. Since the isolation of the two cytoplasmic elongation factors from yeast has been published elsewhere,[5] we have put more emphasis on the separation of the two sets of elongation factors as well as on the isolation and purification of the mitochondrial elongation factors. Because of parallels between bacterial and mitochondrial elongation factors, the nomenclature for prokaryotic elongation factors is used for the latter: EF T is the aminoacyl-tRNA binding factor, and EF G is the peptidyl translocase. For the cytoplasmic elongation factors, EF 1 and EF 2 are used, as generally applied in eukaryotic protein synthesis.

It appears that mitochondrial ribosomes and elongation factors are interchangeable with their bacterial, but not with their cytoplasmic, counterparts in the eukaryotic cell.[4,6-8] The only case where a mitochondrial (or bacterial) elongation factor is compatible with eukaryotic ribosomes is EF T; however, its cytoplasmic counterpart, EF 1, does not react with mitochondrial or prokaryotic ribosomes.[4,9,10] For analysis during isolation of mitochondrial factors it is preferable to complement them with the more easily available ribosomes and elongation factors from *Escherichia coli*.

*Reagents and Materials*

Buffer 1: 20 m*M* Tris·HCl, pH 7.4, 1 m*M* dithiothreitol

[4] D. Richter and F. Lipmann, *Biochemistry* **9**, 5065 (1970).
[5] D. Richter and F. Klink, this series, Vol. 20, p. 349.
[6] M. Grandi and H. Küntzel, *FEBS* (*Fed. Eur. Biochem. Soc.*) *Lett.* **10**, 25 (1970).
[7] A. Perani, O. Tiboni, and O. Ciferri, *J. Mol. Biol.* **55**, 107 (1971).
[8] A. H. Scragg, *FEBS* (*Fed. Eur. Biochem. Soc.*) *Lett.* **17**, 111 (1971).
[9] I. Krisko, J. Gordon, and F. Lipmann, *J. Biol. Chem.* **244**, 6117 (1969).
[10] D. Richter, *Biochem. Biophys. Res. Commun.* **38**, 864 (1970).

Buffer 2: 5 m*M* K phosphate, pH 7.2, 1 m*M* dithiothreitol. All buffers are adjusted at 4°.
Hydroxylapatite (Clarkson Chemical Co., Hypatite C)
DEAE-cellulose (Bio-Rad)
[$^{14}$C]Phenylalanine, specific activity 460 mCi/mmole
Dithiothreitol (RSA Corporation)
Yeast tRNA (Miles Laboratories)
Triton X-100 (Rohm and Haas)

Unless otherwise stated, all operations are carried out at 4°.

*Strains*

*Saccharomyces fragilis,* ATCC No. 10022.
*Saccharomyces cerevisiae,* strain 18A; the mitochondrial DNA-depleted "petite" mutants II-1-40 and III-1-7 can be obtained by treatment of the cells with ethidium bromide.[11,12]

## General Procedures

*Assay Method for Polyphenylalanine Synthesis*

The function of the two elongation factors from yeast cytoplasm can be studied with 80 S cytoplasmic ribosomes from yeast or mammals; mitochondrial elongation factors can be combined with either mitochondrial or bacterial ribosomes.

Polyphenylalanine synthesis is measured by determination of $^{14}$C radioactivity incorporated into hot trichloroacetic acid-insoluble protein. A typical assay (125 $\mu$l) contains: 50 m*M* Tris·HCl, pH 7.4, 100 m*M* $NH_4Cl$, 60 m*M* KCl, 10 m*M* $Mg(CH_3COO)_2$, 1 m*M* dithiothreitol, 2 m*M* GTP, 100–150 $\mu$g of ribosomes, 30 $\mu$g of poly(U), 10–20 pmoles of [$^{14}$C]Phe-tRNA (350 pmoles of [$^{14}$C]Phe per milligram of yeast tRNA), and 50–70 $\mu$g of each of the enzymes. Incubation is carried out at 37° for 10 minutes; the radioactive protein is analyzed as described.[5] Protein concentration can be estimated by the method of Lowry *et al.*[13] or of Warburg and Christian.[14]

[11] E. S. Goldring, L. J. Grossman, D. Krupnick, D. R. Cryer, and J. Marmur, *J. Mol. Biol.* **52**, 323 (1970).
[12] P. O. Slonimski, G. Perrodin, and J. H. Croft, *Biochem. Biophys. Res. Commun.* **30**, 232 (1968).
[13] O. H. Lowry, W. J. Rosebrough, A. L. Farr, and R. J. Randall, *J. Biol. Chem.* **193**, 265 (1951).
[14] O. Warburg and W. Christian, *Biochem. Z.* **310**, 384 (1941).

*Growth Conditions*

The following yeast strains can be used: *S. cerevisiae* (wild-type 18A and two mutants[11] lacking mitochondrial DNA, II-140 and III-1-7), and *S. fragilis*. The yeast cells are grown in a medium containing, per liter: 5 g of yeast extract, 10 g of peptone, 6 g of $(NH_4)_2HPO_4$, 2 g of $MgSO_4$, 9 g of KCl, and 33 ml of a 60% lactate syrup. The pH is adjusted to 4.5 with concentrated HCl, and the medium is sterilized at 15 psi for 20 minutes. A 10-liter carboy is inoculated with 300 ml of an overnight culture containing 7.0 $A_{450}$ (absorbance at 450 nm) units/ml. Cells are grown at 30° in a New Brunswick fermentor under aeration (2 liters of air per minute) with stirring (800 rpm). The culture is grown to a turbidity of 8.0 as measured from the 450-nm absorbance, and is quickly cooled by passing it through a copper cooling coil. Cells are harvested in a Sharples continuous flow centrifuge. The "petite" mutants are grown under similar conditions except that lactate is replaced by 1.5% glucose. The yield for all strains varies between 7 and 12 g of cells (wet weight) per liter of medium. The well-packed cells are kept overnight at 4° without loss of activity of the factors.

## Isolation of the Elongation Factors

*Method A*

*Elongation Factors from Isolated Mitochondria.* For the isolation of mitochondria, spheroplasts from 1 kg of yeast cells (wet weight) are prepared[15] and gently homogenized for 15 seconds in a Waring Blendor; the mitochondrial particles are isolated and washed according to the method of Mattoon and Balcavage.[16] Washed mitochondria are suspended in buffer 1 with 10 m*M* $Mg(CH_3COO)_2$, and yield 4–5 mg of mitochondrial protein per milliliter of solution. The mitochondrial suspension is passed through a French press at 6000 psi. The homogenate is centrifuged at 105,000 *g* for 2 hours, the resulting S100 fraction is treated with 43 g of $(NH_4)_2SO_4$/100 ml, pH 6.9, stirred for 15 minutes, and centrifuged at 20,000 *g* for 15 minutes. The pellet is dissolved in 5 ml of buffer 2 and dialyzed against 2 liters of the same buffer. About 100 mg of protein are applied to a hydroxyapatite column (0.8 × 12 cm) previously equilibrated with buffer 2, and the column is washed with 100 ml of the same buffer. Mitochondrial EF G elutes at 30 m*M* and EF T at 70 m*M* phosphate buffer; all buffers contain 1 m*M* dithiothreitol, and the pH is adjusted to 7.2. Fractions containing either EF T or EF G are combined and concen-

[15] E. A. Duell, S. Inoue, and M. F. Utter, *J. Bacteriol.* **88**, 1762 (1964).
[16] J. R. Mattoon and W. X. Balcavage, this series, Vol. 10, p. 135.

trated using a Diaflo Model 50 ultrafiltration cell with a PM-10 membrane. Both factors can be stored for several months in liquid nitrogen. For further purification steps of the mitochondrial elongation factors see method B.

*Preparation of Mitochondrial Ribosomes.* Washed mitochondria from 1000 g of yeast cells (wet weight) are diluted with an equal volume of 40 m*M* Tris·HCl buffer, pH 7.4, and 20 m*M* $Mg(CH_3COO)_2$ and lysed by adding 1/20 of the volume of a 20% Triton X-100 solution. The ribosomes are pelleted at 105,000 *g* for 2 hours and redissolved in 20 m*M* Tris·HCl (pH 7.4) and 10 m*M* $Mg(CH_3COO)_2$. The $A_{260}:A_{280}$ of this preparation is 1.93; the yield is 10–15 mg of ribosomal protein.

*Method B*

*Separation of Cytoplasmic and Mitochondrial Elongation Factors from Yeast Homogenates.* Yeast cells (200 g wet weight) are resuspended in 400 ml of buffer 1 and disrupted in a Manton-Gaulin mill.[17] The homogenate is centrifuged at 5000 *g* for 10 minutes, and the pellet is reextracted twice with 300 ml of buffer 1. The supernatant fractions are combined and further clarified by centrifugation at 18,000 *g* for 20 minutes. The pH of the supernatant fluid is readjusted with 1 *M* Tris·HCl to 7.4. After centrifugation at 78,000 *g* for 2 hours, two-thirds of the supernatant fractions are collected; the combined fractions are referred to as S100. The yield is 1040 ml or 9310 mg of protein.

*Ammonium Sulfate Fractionation.* The pH of the S100 fraction is adjusted to 6.8 with 10% acetic acid, and 70 g of ammonium sulfate/100 ml of fluid are added. The slurry is stirred for 1 hour, then centrifuged at 18,000 *g* for 1 hour. The supernatant fluid is decanted and the protein pellet is reextracted three times with 150 ml of 25% (step 1), three times with 22% (step 2), and finally three times with 18% (step 3) ammonium sulfate solutions (w/w). All solutions contain 1 m*M* dithiothreitol, and the pH is adjusted to 6.8 with $NH_4OH$. The supernatant fractions of each step are combined and reprecipitated with 20 g of ammonium sulfate/100 ml of fluid. The precipitates are collected by centrifugation at 18,000 *g* for 15 minutes, dissolved in buffer 1, and dialyzed against 3 liters of the same buffer. The yield for step 1 is 20 ml, or 935 mg of protein; for step 2, 47 ml, or 2700 mg of protein; and for step 3, 56 ml, or 1860 mg of protein. Step 1 contains no mitochondrial elongation factors, but cytoplasmic EF 1 and some cytoplasmic EF 2. The second step contains most of the mitochondrial EF T but is almost free of its complementary factor, EF G; this step also contains the bulk of cytoplasmic EF 1 and EF 2. Mitochon-

[17] J. Gordon, J. Lucas-Lenard, and F. Lipmann, this series, Vol. 20, p. 281.

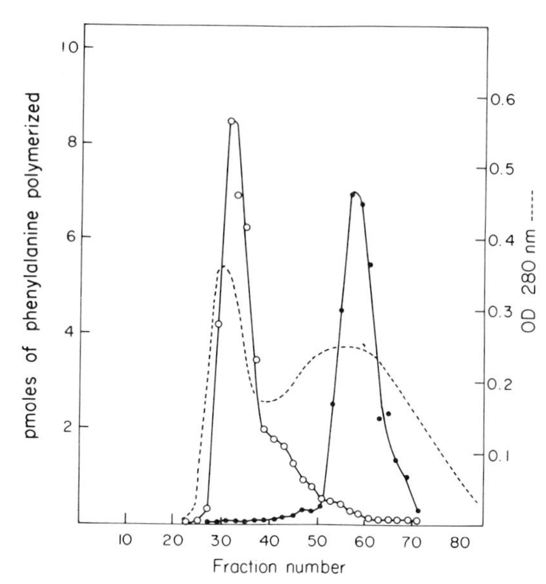


FIG. 1. Separation of mitochondrial EF T (●—●) and cytoplasmic EF 1 (○—○).

drial EF G is found in step 3 together with some EF T. The latter step also contains some cytoplasmic EF 2. The ammonium sulfate fractions can be stored for several months in liquid nitrogen without loss of activity.

*Separation of Mitochondrial EF T and Cytoplasmic EF 1.* The cytoplasmic EF 1 can be isolated from either step 1 or step 2 of the ammonium sulfate fractionation by gel filtration on Sephadex G-200 columns. The molecular weight of this EF 1 is high (220,000), and hence it can be separated rather easily from the rest of the elongation factors which are of lesser size (between 60,000 and 90,000). Figure 1 shows typical elution and activity profiles of cytoplasmic EF 1 compared with those of its mitochondrial counterpart, EF T. From the ammonium sulfate step 2, 80 mg of protein are layered on top of a Sephadex G-200 column (4.5 × 80 cm) equilibrated with buffer 1. Elution is carried out with the same buffer; 4.5 ml per fraction are collected. Aliquots (10 $\mu$l) of the eluted fractions are assayed for polyphenylalanine synthesis with an excess of ribosomes and complementary factors. The cytoplasmic EF 1 is free of cytoplasmic EF 2 and mitochondrial elongation factors. The mitochondrial EF T is contaminated by small amounts of cytoplasmic EF 2 and mitochondrial EF G. For further purification of the mitochondrial EF T, see below.

*Separation of Mitochondrial EF G and Cytoplasmic EF 2.* The ammonium sulfate step 3 contains most of the mitochondrial EF G as well as

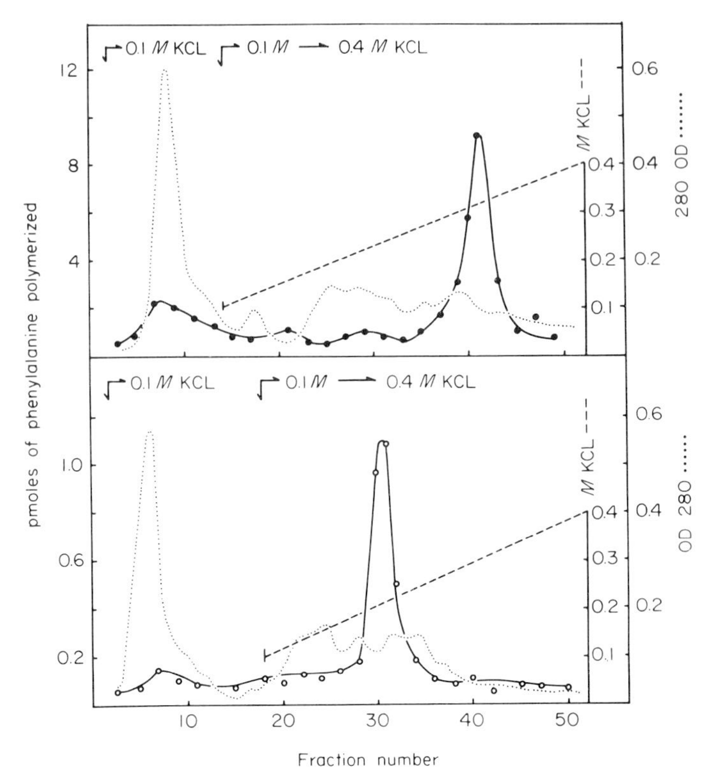


FIG. 2. Chromatography of mitochondrial EF G (○—○) and cytoplasmic EF 2 (●—●) on DEAE-cellulose columns.

some cytoplasmic EF 2. The molecular weights of both factors are very close and do not separate on a Sephadex column. However, they both have different affinities for DEAE-cellulose. Figure 2 shows that mitochondrial EF G is recovered from the column at about 0.2 *M* KCl, whereas the cytoplasmic EF 2 comes off at 0.3 *M* KCl. DEAE-cellulose columns (1.2 × 15 cm) are equilibrated with buffer 1 containing 0.1 *M* KCl. From the ammonium sulfate steps 2 or 3, 50–80 mg of protein are dialyzed against the same buffer and applied to the columns, which are then washed with 100 ml of the same KCl concentration. Linear gradients from 0.1–0.4 *M* KCl in buffer 1 (50 × 50 ml) are used; 3.5 ml per fraction are collected. Aliquots (10 $\mu$l) of the eluted fractions are analyzed for polyphenylalanine activity in the presence of ribosomes and complementary factors as described above. Both cytoplasmic EF 2 and mitochondrial EF G are free of contamination by other elongation factors.

*Purification of Mitochondrial EF T.* The ammonium sulfate step 2, which has most of the mitochondrial EF T, is used for its further purification. This includes protamine sulfate treatment, stepwise chromatography on hydroxylapatite columns, anion-exchange chromatography on DEAE-cellulose, and gel filtration on Sephadex G-200. The steps are summarized in Table I.

*Protamine Sulfate Step.* Neutralized protamine sulfate solution, 0.8 ml, is added per 10 ml of protein solution (step 2 of the ammonium sulfate fractionation). The solution is stirred for 20 minutes and the resulting precipitate is removed by centrifugation at 15,000 *g* for 15 minutes. The supernatant fluid is dialyzed against 2 liters of buffer 2. The yield is 45 ml, or 2480 mg of protein (Table I).

*Hydroxylapatite Step.* A hydroxylapatite column (4.5 × 9 cm) is packed and equilibrated with buffer 2. The protein fraction of the protamine sulfate step containing mitochondrial EF T is then applied to this column; the latter is washed with 500 ml each of 10 m*M* and 30 m*M* phosphate buffers, pH 7.2, both of which contain 1 m*M* dithiothreitol. Mitochondrial EF T is eluted with 70 m*M* phosphate buffer with 1 m*M* dithiothreitol. Fractions with mitochondrial EF T activity are combined and 43 g of $(NH_4)_2SO_4$/100 ml of solution are added. The slurry is stirred for 20 minutes and then centrifuged at 18,000 *g* for 20 minutes. The

TABLE I
PURIFICATION STEPS OF THE MITOCHONDRIAL EF T AND EF G FROM *Saccharomyces cerevisiae*, STRAIN 18A[a]

| | Total protein (mg) | | Specific activity (nmoles/10 min/mg) | | Total enzyme units (nmoles/10 min) | |
|---|---|---|---|---|---|---|
| | T | G | T | G | T | G |
| 1. S100 | 9310 | | 0.100 | | 931 | |
| 2. Ammonium sulfate | 2720 | 1860 | 0.187 | 0.250 | 509 | 465 |
| 3. Protamine sulfate | 2480 | 1815 | 0.190 | 0.252 | 471 | 457 |
| 4. Hydroxylapatite | 424 | 211 | 0.420 | 0.430 | 178 | 90.7 |
| 5. DEAE-cellulose | 31.1 | 8.4 | 5.5 | 7.0 | 171 | 58.8 |
| 6. Sephadex G-150 | 12.1 | 3.2 | 9.2 | 12.7 | 111 | 40.6 |

[a] Activities for EF G and EF T are determined by polyphenylalanine synthesis in the presence of an excess of the complementary *Escherichia coli* factor and *E. coli* ribosomes [D. Richter and F. Lipmann, *Biochemistry* **9,** 5065 (1970)]. The activities are derived from experiments with a linear dependence on either EF G or EF T concentration.

precipitate is dissolved in buffer 1 and dialyzed against 2 liters of the same buffer. The yield is 12 ml, or 424 mg of protein (Table I).

*DEAE-Cellulose Step.* A DEAE-cellulose column (1.3 × 23 cm) is equilibrated with buffer 1 containing 0.1 *M* KCl. The mitochondrial EF T from the hydroxylapatite step is applied to the column, which is then washed with 200 ml of buffer 1 containing 0.1 *M* KCl; the mitochondrial EF T is eluted with a linear gradient from 0.1 to 0.5 *M* KCl in buffer 1 (200 × 200 ml), and 4 ml/tube are collected. Fractions containing mitochondrial EF T are combined and concentrated in a Diaflo Model 50 ultrafiltration cell with a UM-10 membrane. The yield is 1.5 ml, or 31 mg of protein.

*Sephadex G-150 Step.* Mitochondrial EF T is further purified by gel filtration through a Sephadex G-150 column (1.3 × 75 cm). Buffer 1 is used to equilibrate the column and to elute mitochondrial EF T; 2.5 ml/tube are collected. The tubes with mitochondrial EF T are combined and concentrated as described for the DEAE-cellulose step. The yield is 1.5 ml, or 12 mg of protein.

*Further Purification of Mitochondrial EF G.* Step 3 of the ammonium sulfate fractionation is used to purify mitochondrial EF G, using a similar method to that described for mitochondrial EF T, but with the following exceptions: mitochondrial EF G is eluted from the hydroxyapatite column (4.5 × 6.0 cm) with 30 m*M* phosphate buffer, and DEAE-chromatography (column size 1.2 × 24 cm) is carried out with a linear gradient from 0.1 to 0.4 *M* KCl in buffer 1 (150 × 150 ml). The results of the various purification steps of mitochondrial EF T and EF G are summarized in Table I.

TABLE II
PURIFICATION OF MITOCHONDRIAL EF T AND EF G FROM *Saccharomyces cerevisiae* STRAIN 18A AND THE "PETITE" MUTANTS II-1-40 AND III-1-7

| | Enzyme units (nmoles/10 min)/g of total S100 protein | | | | | |
|---|---|---|---|---|---|---|
| | 18A | | II-1-40 | | III-1-7 | |
| Purification steps | T | G | T | G | T | G |
| 1. S100 | 100 | 100 | 76 | 94 | 65 | 82 |
| 2. Ammonium sulfate | 55 | 50 | 45 | 47 | 37 | 43 |
| 3. Protamine sulfate | 51 | 49 | 39 | 45 | 32 | 31 |
| 4. Hydroxylapatite | 19 | 10 | 12 | 8 | — | — |
| 5. DEAE-cellulose | 18 | 6 | 10 | 5 | — | — |
| 6. Sephadex G-150 | 12 | 4 | 6 | 3 | — | — |

*Mitochondrial Elongation Factors from "Petite" Mutants*

Mitochondrial elongation factors can be obtained from various yeast strains including mutants lacking mitochondrial DNA. This shows that mitochondrial EF T and EF G are coded by nuclear, not by mitochondrial, DNA.[18] Table II compares mitochondrial EF T and EF G from the wild-type strain 18A and from the two "petite" mutants II-1-40 and III-1-7. Mitochondrial elongation factors from all these strains are identical in their functional, protein chemical, and immunological properties.[18]

[18] D. Richter, *Biochemistry* **10**, 4422 (1971).

# [42] Soluble Protein Factors and Ribosomal Subunits from Yeast. Interactions with Aminoacyl-tRNA

*By* A. Toraño, A. Sandoval, and C. F. Heredia

The elongation of the peptide chain on the ribosomes requires the participation of protein factors which are present in the soluble fraction of both prokaryotic and eukaryotic cells. Yeast has two different protein synthesizing systems: a mammalian-type one in the cytoplasm[1] and other in the mitochondria with properties similar to those from bacteria.[2,3] Two elongation factors, EF-1 and EF-2, have been found in both the cytoplasm[4,5] and the mitochondria[2,3] of yeast cells. EF-1 promotes a GTP-dependent binding of aminoacyl-tRNA to the ribosomes. EF-2 is involved in the translocation of the growing peptide chain from the acceptor to the donor site of the ribosome.

*In vitro* assays for these factors are usually based on their complementarity for the polymerization of amino acids from [$^{14}$C]aminoacyl-tRNA by purified ribosomes. EF-1 is also assayed by its capacity to catalyze the binding of [$^{14}$C]phenylalanyl-tRNA to the ribosomes·poly(U) complex in a GTP-dependent reaction. This factor can discriminate against

[1] R. K. Bretthauer, L. Marcus, J. Chaloupka, H. O. Halvorson, and R. M. Bock, *Biochemistry* **2**, 1079 (1963).
[2] D. Richter and F. Lipmann, *Biochemistry* **9**, 5065 (1970).
[3] A. H. Scragg, H. Morimoto, V. Villa, J. Nekkorocheff, and H. O. Halvorson, *Science* **171**, 908 (1971).
[4] F. Klink and D. Richter, *Biochim. Biophys. Acta* **114**, 431 (1966).
[5] M. S. Ayuso and C. F. Heredia, *Biochim. Biophys. Acta* **145**, 199 (1967).



ISBN 0-12-504180-2

*N*-acetylphenylalanyl-tRNA for this enzymatic binding reaction.[6] Recent experiments have shown the existence in the soluble fraction of yeast extracts of an activity, different from the two elongation factors, which promotes a GTP-dependent binding of *N*-acetylphenylalanyl-tRNA to the ribosome·poly(U) complex.[7]

This article describes the preparation of a supernatant fraction from the yeast *Saccharomyces fragilis* × *Saccharomyces dobzanskii* and the interactions of this fraction with aminoacyl and *N*-acetylaminoacyl-tRNA, the isolation from this supernatant fraction of partially purified elongation factors, and the preparation of ribosomes and active ribosomal subunits and their interactions with aminoacyl-tRNA.

*Reagents*

Buffer A: 10 m*M* Tris·HCl (pH 7.5), 5 m*M* $Mg(CH_3—COO)_2$, 10 m*M* mercaptoethanol, 10 m*M* KCl

Buffer B: 0.5 *M* potassium phosphate (pH 6.5), 10 m*M* mercaptoethanol

Buffer C: 30 m*M* Tris·HCl (pH 7.2), 0.5 *M* $NH_4Cl$, 0.1 *M* $Mg(CH_3—COO)_2$, 0.25 *M* sucrose and 5 m*M* mercaptoethanol

Buffer D: 50 m*M* Tris·HCl (pH 7.7), 12 m*M* $Mg(CH_3—COO)_2$, 20 m*M* mercaptoethanol, 0.8 *M* KCl

Buffer E: 35 m*M* Tris·HCl (pH 7.7), 9 m*M* $Mg(CH_3—COO)_2$, 0.25 *M* sucrose, 70 m*M* KCl

## Assay Methods

*Binding of Aminoacyl-tRNA to the Ribosomes.* The complete reaction mixture for the binding of aminoacyl-tRNA to the ribosomes contains in a final volume of 0.1 ml the following components: 50 m*M* Tris (adjusted to pH 6.5 with acetic acid), 100 m*M* $NH_4(CH_3—COO)$, 50 $\mu$g of poly(U) (Boehringer), 1 m*M* GTP, 10 m*M* $Mg(CH_3—COO)_2$, 15–30 pmoles of [$^{14}$C]phenylalanyl or *N*-acetyl-[$^{14}$C]phenylalanyl-tRNA (200–300 pmoles of [$^{14}$C]amino acid per milligram of tRNA), 50–100 $\mu$g of soluble proteins and 0.5–2 $A_{260}$ units of ribosomes or ribosomal subunits as indicated. The reaction is started by addition of the aminoacyl-tRNA. After incubation at 30°, usually for 20 minutes, the mixture is diluted with 3 ml of cold Tris·acetate buffer containing 100 m*M* $NH_4(CH_3—COO)$ and 10 m*M* $Mg(CH_3\text{-}COO)_2$ and filtered through Millipore filters[8] (0.45 $\mu$m, 2.5 cm).

[6] M. S. Ayuso and C. F. Heredia, *Eur. J. Biochem.* **7**, 111 (1968).

[7] A. Toraño, A. Sandoval, C. SanJosé, and C. F. Heredia, *FEBS* (*Fed. Eur. Biochem. Soc.*) *Lett.* **22**, 11 (1972).

[8] M. Nirenberg and P. Leder, *Science* **145**, 1399 (1964).

The filters are washed three times with 3 ml each time of the same buffer, dried, and counted in a liquid scintillation spectrometer. Preparation of [$^{14}$C]phenylalanyl-tRNA is carried out by charging deacylated crude yeast tRNA (Sigma) with [$^{14}$C]phenylalanine (200 $\mu$Ci/$\mu$mole) followed by phenol extraction and ethanol precipitation.[9] Acetylation of phenylalanyl-tRNA was performed as described elsewhere.[10]

*Polymerization of Amino Acids.* The polymerization of amino acids is followed by determination of the radioactivity incorporated into hot trichloroacetic acid-insoluble material. The reaction mixture contains, in a final volume of 0.1 ml, the following components: 50 m*M* Tris (adjusted to pH 6.5 with acetic acid), 100 m*M* $NH_4(CH_3—COO)$, 50 $\mu$g of poly(U), 1 m*M* GTP, 10 m*M* $Mg(CH_3—COO)_2$, 15–30 pmoles of [$^{14}$C] phenylalanyl-tRNA, 50–100 $\mu$g of "soluble protein fraction" and 0.5–2 $A_{260}$ units of ribosomes or ribosomal subunits as indicated. Incubation is at 30° usually for 10 minutes. After this time, 0.05 ml of a solution of bovine albumin (4 mg/ml) is added as carrier followed by 3 ml of cold 5% trichloroacetic acid. The mixture is heated at 90° for 15 minutes, cooled, and filtered through glass fiber disks (Whatman GF/C, 2.5 cm). The precipitate is washed three times with 3 ml each time of cold 5% trichloroacetic acid, dried, and counted in a liquid scintillation spectrometer.

## Preparative Procedures

*Growth of the Yeast.* The yeast hybrid *Saccharomyces fragilis* × *Saccharomyces dobzanskii* has been used for the preparation of the soluble protein factors and the ribosomes. The yeast cells are grown at 30° in a medium containing, per liter, the following components: 2 g of $KH_2PO_4$, 8 g of $(NH_4)_2SO_4$, 0.016 g of $MgCl_2$, 0.3 g of yeast extract (Difco), 20 g of glucose and 0.5 ml of a solution containing per liter: 16.5 g of $MgCl_2$, 2 g of $MgSO_4 \cdot 7H_2O$, 1 g of NaCl, 0.5 g of $FeSO_4 \cdot 7H_2O$, 0.5 g of $ZnSO_4 \cdot 7H_2O$, 0.5 g of $MnSO_4 \cdot 7H_2O$, 0.05 g of $CuSO_4 \cdot 5H_2O$, and 10 ml of 0.1 *N* $H_2SO_4$. The solutions of salts and glucose are sterilized separately by autoclaving at 120° and 2.5 atm for 20 minutes and mixed at the time of inoculation.

Yeast cells from an agar slant are inoculated in 25 ml of a medium containing 2% glucose and 0.3% yeast extract and incubated with vigorous agitation at 30° for 24 hours. Appropriate amounts of this inoculum are transferred to 10-liter flasks containing 7 liters of the culture medium described above and incubated at 30° with aeration. When the cultures have

[9] C. F. Heredia and H. O. Halvorson, *Biochemistry* **5**, 946 (1966).
[10] A. L. Haenni and F. Chapeville, *Biochim. Biophys. Acta* **114**, 135 (1966).

reached an optical density of 0.9 at 660 nm they are cooled by addition of ice. The cells are harvested by centrifugation in a continuous-flow rotor at 4° and washed three times by centrifugation with cold buffer A. The pelleted cells are placed on a porous disk for 30–60 minutes at 4°, and they are kept frozen at −20° until used.

*Preparation of the Crude Extracts.* Frozen yeast cells are suspended (25% w/v) in cold buffer A, and they are broken in a Ribi cell fractionator (Sorvall Model RF-1) at 20,000 psi and 5°. Alternatively the yeast cells can be broken by grinding in a cold mortar with 2–3 times its weight of sand or glass beads and extracted with 2.5 volumes of buffer A. Alumina must be avoided for breaking the cells since most of the amino acid polymerization factors are adsorbed on it. The homogenates are centrifuged in the cold, first at 8000 *g* for 10 minutes to eliminate unbroken cells, debris, and sand or glass beads if the second procedure is used, and then twice at 20,000 *g* for 20 minutes. The 20,000 *g* supernatant is further centrifuged at 105,000 *g* for 90 minutes. The supernatant after this centrifugation is referred to as S105 fraction. The pellet contains the crude ribosomes.

*Purification of the Soluble Protein Factors.* One volume of the S105 supernatant is adjusted to a protein concentration of 6.5 mg/ml with cold buffer A and then treated with 0.1 volume of a suspension of alumina $C_\gamma$ (Sigma) containing 10 mg/ml, dry weight. After stirring for 10 minutes at 4°, the gel is separated by centrifugation in the cold at 5000 rpm for 10 minutes, and the supernatant is discarded. The gel is washed with cold 0.1 *M* potassium phosphate buffer pH 6.5 containing 10 m*M* mercaptoethanol. The proteins that remain adsorbed in the gel are eluted by suspending the gel in 0.5 *M* potassium phosphate buffer (buffer B) (about one-tenth of the original S105 volume). After stirring for 5 minutes at 4°, the suspension is centrifuged in the cold at 5000 rpm for 10 minutes, and the pellet was discarded. The supernatant (0.5 *M* potassium phosphate eluate) is dialyzed for 3–4 hours against cold buffer A. The resulting preparation is referred to as soluble protein fraction. The specific activity of this fraction is 15–20 times greater than that of the original S105, as determined by the poly(U)-directed polymerization of [$^{14}$C]phenylalanine from [$^{14}$C]phenylalanyl-tRNA using purified yeast ribosomes.[5]

A simple method to obtain preparations of EF-1 with low EF-2 activity from the 0.5 *M* potassium phosphate eluate is as follows: 1–2 ml of the 0.5 *M* potassium phosphate eluate (adjusted to 1–2 mg of protein per milliliter with buffer B) are placed in a test tube, and the tube is immersed in a water bath (60°) with agitation. When the temperature inside the tube has reached 55°, the tube is maintained at this temperature for 3–4 minutes and then rapidly cooled in an ice bath. The preparation is then dialyzed for 3–4 hours against cold buffer A. By this treatment usually about 90% of the EF-2 activity originally present in the preparation is inactivated, as measured by the loss of phenylalanine polymerization from [$^{14}$C]phenylalanyl-tRNA while factor EF-1 remains active. With preparations of different protein content it is advisable to make a time course heat inactivation curve at 55° to establish the optimal conditions.

Factor EF-2 is obtained by filtration of the 0.5 *M* potassium phosphate eluate through Sephadex G-200. For this purpose the protein concentration is adjusted with buffer B to about 2 mg/ml and 3–5 ml of this preparation is passed through a Sephadex G-200 column (2 cm × 45 cm) equilibrated with buffer A. The proteins are eluted with buffer A. Fractions of 4 ml are collected at a flow rate of 0.2 ml/min, and EF-2 activity is determined in 0.1 ml aliquots of each fraction by its complementation with EF-1 for the poly(U)-directed polymerization of phenylalanine from [$^{14}$C]phenylalanyl-tRNA. EF-2 is quite unstable and has to be used soon after preparation. A method for the isolation of elongation factors from *Saccharomyces cerevisiae* has been published.[11]

A summary of the interactions of the soluble protein factors with phenylalanyl-tRNA and *N*-acetylphenylalanyl-tRNA is shown in Table I.

TABLE I
BINDING OF AMINOACYL-tRNA TO PURIFIED YEAST RIBOSOMES PROMOTED BY SUPERNATANT FACTORS. EFFECT OF *N*-ETHYLMALEIMIDE (NEM)[a]

| Expt. No. | Additions | Aminoacyl-tRNA bound: *N*-(Ac)-Phe-tRNA (pmoles) | Aminoacyl-tRNA bound: Phe-tRNA (pmoles) |
|---|---|---|---|
| 1 | None | 0.3 | 0.4 |
| | Soluble protein fraction | 2.3 | — |
| | NEM-treated soluble protein fraction | 0.3 | 4.3 |
| 2 | Elongation factor 1 | — | 3.1 |
| | NEM-treated factor 1 | — | 2.7 |
| 3 | None | 0.5 | 0.4 |
| | Soluble protein fraction | 2.5 | — |
| | Elongation factor 1 | 0.4 | 2.5 |
| | Elongation factor 2 | 0.4 | — |

[a] Conditions as described in the text using 2 $A_{260}$ units of ribosomes and approximately 50 μg of each of the protein fractions. When indicated protein fractions were preincubated with *N*-ethylmaleimide (20 m*M*) at 30° for 5 minutes. From A. Toraño, A. Sandoval, C. SanJosé, and C. F. Heredia, *FEBS* (*Fed. Eur. Biochem. Soc.*) *Lett.* **22,** 11 (1972).

[11] D. Richter and F. Klink, this series, Vol. 20, p. 349.

EF-1 interacts with phenylalanyl-tRNA and promotes binding of this compound to the ribosomes. *N*-acetylphenylalanyl-tRNA cannot be substituted for phenylalanyl-tRNA as substrate for this binding reaction. There is, however, another activity in the "soluble protein fraction" which promotes binding of *N*-acetylphenylalanyl-tRNA to the ribosomes. This activity, in contrast with EF-1, is sensitive to treatment with *N*-ethylmaleimide. These two factor-dependent binding reactions require poly(U), GTP, ammonium and magnesium ions.[6,7]

*Purification of the Ribosomes.* For the purification of the ribosomes two different procedures can be used. One is a modification of the method developed by Bruenning and Bock.[12] Crude ribosomes are suspended in cold buffer C to an optical density of about 300 $A_{260}$ units/ml. Nine milliliters of this ribosomal suspension are layered over 3 ml of buffer C in which the concentration of sucrose has been increased to 2 *M*. After centrifugation at 105,000 *g* in a Spinco 40 rotor for 4 hours, three layers can be observed. An upper layer, an interphase, and a lower layer which contains the ribosomes. The two first layers are carefully removed with a syringe and discarded. The bottom layer is diluted twice with cold buffer A and passed through a Sephadex G-25 column equilibrated with the same buffer. The ribosomes are eluted with this buffer. Fractions are collected, and the optical density at 260 nm is recorded. The fractions containing the ribosomes are pooled, diluted with 2–3 volumes of buffer A, and centrifuged at 105,000 *g* for 2 hours, the ribosomal pellet is suspended in the appropriate amount of buffer A. The ribosomal suspension is centrifuged at 10,000 *g* for 10 minutes to remove denatured material and then adjusted with buffer A to the desired concentration. The ribosomes are distributed in small aliquots and kept either under liquid nitrogen or at $-70°$. Under these conditions they remain fully active for the polymerization of amino acids for at least 1 month. The ribosomes obtained in this way are completely dependent on the supernatant factors for the polymerization of amino acids from aminoacyl-tRNA.[6]

Purified ribosomes can also be obtained by an alternative procedure[6] in which crude ribosomes are suspended in cold buffer C to an optical density of about 300 $A_{260}$ units/ml. The suspension is stirred overnight at 4°, and the ribosomes are diluted 1:1 with buffer A, sedimented by centrifugation at 105,000 *g* for 2 hours, resuspended in buffer A, and sedimented again by centrifugation. Finally, they are suspended in buffer A, centrifuged (10,000 *g* for 10 minutes) to remove denatured material, adjusted to the desired concentration, and kept in small aliquots as indicated above.

[12] G. Bruenning and R. M. Bock, *Biochim. Biophys. Acta* **149**, 377 (1967).

*Preparation of Ribosomal Subunits.* A procedure to obtain active ribosomal subunits from other species of *Saccharomyces* has been previously reported.[13] The method described here is a slight modification of that developed for rat liver[14] and *Artemia salina*[15] ribosomes. Crude ribosomes are suspended in a dissociation buffer (buffer D). The ribosomal suspension is adjusted to an optical density of 400 $A_{260}$ units/ml. About 2 ml of this ribosomal suspension is layered on the top of a 50-ml linear sucrose gradient (15 to 30%) made in the same buffer and centrifuged at 2–5° in the SW 25.2 rotor of the Spinco ultracentrifuge for 12–14 hours at 22,500 rpm. After centrifugation fractions of the gradient (2 ml) are collected and their optical density at 260 nm is measured. Fractions corresponding to the 40 S and 60 S regions are pooled separately and diluted with one volume of 20 m*M* β-mercaptoethanol, 30 m*M* $Mg(CH_3—COO)_2$. The subunits are sedimented by centrifugation at 40,000 rpm and 2–5° for 14 hours in the Spinco 50 rotor. At this stage the 40 S subunits are quite pure, but the 60 S subunits are still contaminated with both 40 S subunits and 80 S monomers. For a more extensive purification, the pellets corresponding to each subunit are separately resuspended in about 1 ml of the buffer D and rerun on separate sucrose gradients as indicated above. After recording the optical densities of the gradients at 260 nm, the fractions corresponding to each peak are pooled, and the subunits are pelleted by centrifugation at 40,000 rpm for 14 hours at 2–5° in the Spinco 50 rotor. The pellets are suspended in buffer E and stored in 50% glycerol at −20°. Under these conditions the subunits retain full activity at least for several weeks. Cross contamination is less than 5% in the case of 40 S and about 10% in the case of the 60 S as estimated by the percentage of phenylalanine polymerized by each fraction as compared with the control (Table II).

TABLE II
RECONSTITUTION OF 80 S RIBOSOMES FROM THEIR SUBUNITS[a]

| Particles | Phenylalanine polymerized (pmoles) |
|---|---|
| 80 S ribosomes | 12.0 |
| 40 S subunits | 0.5 |
| 60 S subunits | 1.6 |
| 40 + 60 S subunits | 10.3 |

[a] Conditions as described in the text using 2, 1.2, and 0.7 $A_{260}$ units of 80 S, 60 S, and 40 S particles, respectively.

[13] E. Battaner and D. Vazquez, this series, Vol. 20, p. 446.
[14] T. E. Martin and I. G. Wool, *Proc. Nat. Acad. Sci. U.S.* **60**, 569 (1968).
[15] M. Zasloff and S. Ochoa, *Proc. Nat. Acad. Sci. U.S.* **68**, 3059 (1971).

TABLE III
BINDING OF PHE-tRNA AND N-(Ac)-PHE-tRNA TO RIBOSOMES AND RIBOSOMAL SUBUNITS

| Expt. No.[a] | Ribosomal particle | Additions | Aminoacyl-tRNA bound: N-(Ac)-Phe-tRNA (pmoles) | Aminoacyl-tRNA bound: Phe-tRNA[b] (pmoles) |
|---|---|---|---|---|
| 1 | 80 S | — | 2.8 | 4.2 |
| | 40 S | — | 3.1 | 2.5 |
| | 60 S | — | <0.1 | <0.1 |
| 2 | 40 S | — | 0.2 | 0.2 |
| | 40 S | Soluble protein fraction | 0.4 | 0.5 |
| | 60 S | — | <0.1 | <0.1 |
| | 60 S | Soluble protein fraction | 0.2 | 0.3 |
| | 40 S + 60 S | — | 0.7 | 0.8 |
| | 40 S + 60 S | Soluble protein fraction | 2.3 | 3.6 |

[a] Expt. 1 was carried out at 20 m*M* $Mg^{2+}$ and 1 m*M* spermidine, without added GTP and soluble factors, using 1.4, 1, and 0.5 $A_{260}$ units of 80 S, 60 S, and 40 S particles, respectively. Expt. 2 was carried out at 10 m*M* $Mg^{2+}$ and 1 m*M* spermidine using 2.3 and 1.3 $A_{260}$ units of 60 S and 40 S particles, respectively.

[b] The soluble protein fraction used for the binding of Phe-tRNA was preincubated with *N*-ethylmaleimide as indicated in Table I, to inactivate EF-2 and the *N*-(Ac)-Phe-tRNA binding activity.

The results in Table III summarize the interactions of the ribosomal particles with aminoacyl-tRNA. At high magnesium concentration (Expt. 1) both the 40 S subunits and the 80 S ribosomes bind phenylalanyl-tRNA or *N*-acetylphenyl-tRNA[16] in a reaction which requires poly(U) but not GTP or soluble protein factors. This binding reaction does not occur with the 60 S subunits alone. At lower magnesium concentrations (Expt. 2) both phenylalanyl-tRNA or *N*-acetylphenylalanyl-tRNA bind preferentially to the 80 S ribosomes (Table I) or to the combination of 40 S plus 60 S subunits (Table III) in response to the soluble factors described above.

## Acknowledgments

We are indebted to Miss C. Moratilla for expert technical assistance and to Miss C. Estévez for the typing of the manuscript.

[16] D. Vazquez, E. Battaner, R. Neth, G. Heller, and R. E. Monro, *Cold Spring Harbor Symp. Quant. Biol.* **34**, 369 (1969).

# [43] Separation and Characterization of Yeast Elongation Factors

*By* LAWRENCE SKOGERSON

*Saccharomyces cerevisiae* provides an opportunity unique among eukaryotic organisms to apply methods of genetics and biochemistry to the solution of complex metabolic problems. Conditional lethal mutants defective in translation of mRNA[1] are potentially useful tools in studies on the mechanism and regulation of ribosomal reactions. Analysis of the mutants requires the availability of well defined *in vitro* assays for the many different protein synthetic elements. Because of the availability of well characterized assays, the elongation factors are the most straightforward elements to identify. Consequently factors from yeast required for the poly(U)-directed formation of polyphenylalanine were isolated. Three factors from the postribosomal fraction were required rather than the two observed in other eukaryotic systems.[2] Two of the factors corresponded to EF-1 and EF-2 on the basis of activity in Phe-tRNA binding and diphtheria toxin-dependent ADP-ribosylation, respectively. Furthermore, the EF-1 and EF-2 fractions from yeast could be interchanged with those from rat liver.[3] The third component, temporarily designated EF-3, was more active in promoting the ribosome-dependent hydrolysis of GTP than EF-2,[2] and stimulated the binding of Phe-tRNA to ribosomes with either yeast or liver EF-1.[3] The requirement for EF-3 was observed only with yeast ribosomes, not with those from rat liver, HeLa cells, or *Artemia salina*.[3,4]

The initial method used to isolate the three factors employed an aqueous two-phase extraction with dextran and polyethylene glycol.[2] Although this procedure was rapid, residual polyethylene glycol frequently interfered with subsequent chromatographic steps. Another problem was that substantial factor activity remained in the dextran phase and could not readily be recovered. Because of these problems the purification procedure was redesigned to begin directly with the postribosomal supernatant. Three chromatographic steps were used to separate and partially purify the three factors. Essentially homogeneous EF-1 was obtained with greater than 50% yield of activity. The other two factors, EF-2 and EF-3, were each

[1] L. H. Hartwell and C. S. McLaughlin, *J. Bacteriol.* **96,** 1664 (1968).

[2] L. Skogerson and E. Wakatama, *Proc. Natl. Acad. Sci. U.S.A.* **73,** 73 (1976).

[3] L. Skogerson and D. Engelhardt, *J. Biol. Chem.* **252,** 1471 (1977).

[4] L. Skogerson, unpublished results, 1977.




ISBN 0-12-504180-2

completely free of the other and were each about 30% pure. Alternative methods for separation of yeast elongation factors have been described.[5,6]

*Reagents*

[$^{14}$C]Phenylalanine, 250 mCi/mmol; New England Nuclear
[*adenine*-$^{14}$C]NAD, 174 mCi/mmol; Amersham
Polyuridylic acid, GTP, and ATP from Miles Laboratories
Diphtheria toxin 2000 LF units/ml, Connaught Laboratories

## Assays

*Polyphenylalanine Synthesis.* Reaction mixtures of 50 $\mu$l contained 60 m$M$ Tris-acetate, pH 7.0, 50 m$M$ $NH_4Cl$, 5 m$M$ $MgCl_2$, 2 m$M$ spermidine, 20 m$M$ dithiothreitol, 50 $\mu$g bovine serum albumin, 0.04 $A_{260}$ unit of ribosomes,[2] 10 pmol of [$^{14}$C]Phe-tRNA(5500 dpm),[7] 87 $\mu M$ GTP, 0.3 m$M$ ATP, 8 $\mu$g of poly(U), 0.2 $\mu$g of EF-1, 0.04 $\mu$g of EF-2, and 0.2 $\mu$g of EF-3. After an appropriate incubation period at 30°, reactions were stopped by addition of 0.5 ml of 10% (w/w) $CCl_3COOH$ and the mixtures were heated at 90° for 10 min. Precipitable radioactivity was collected on glass fiber filters (Whatman GF/C) and washed with 5% $CCl_3COOH$. The filters were analyzed in a liquid scintillation spectrometer at 80% efficiency with Liquifluor (New England Nuclear) counting fluid.

*Diphtheria Toxin-Catalyzed ADP-Ribosylation.* The amount of EF-2 was determined in reactions containing 40 m$M$ Tris · HCl, pH 7.5, 40 m$M$ $NH_4Cl$, 40 $\mu$g of bovine serum albumin, 20 m$M$ DTT, 1.6 $\mu M$ [*adenine*-$^{14}$C]NAD, and 2 LF units of diphtheria toxin in a volume of 50 $\mu$l. After 10 min at 30°, proteins were precipitated with 10% $CCl_3COOH$, the precipitate was collected and radioactivity was measured as described for polyphenylalanine synthesis.

## Purification and Separation

### *Growth of Cells and Preparation of Extracts*

*Saccharomyces cerevisiae* strains D-587-4B or 2180-1A were grown to early log phase (5 g wet weight per liter) in media containing 5 g of yeast extract, 10 g of Bacto peptone (Difco), and 20 g of glucose per liter. Cells were stored frozen as the wet paste at −100°.

[5] D. Richter and F. Klink, this series, Vol. 20, p. 349.
[6] L. L. Spremulli and J. M. Ravel, *Arch. Biochem. Biophys.* **172,** 261 (1976).
[7] L. Skogerson, C. McLaughlin, and E. Wakatama, *J. Bacteriol.* **116,** 818 (1973).

For preparation of extracts cells were thawed in an equal volume of 60 m*M* Tris-acetate, pH 7.0, 50 m*M* $NH_4Cl$, 10% glycerol, and 5 m*M* β-mercaptoethanol. Acid-washed glass beads (0.2 mm) equivalent to five times the weight of cells were added, and the mixture was stirred at top speed in an Osterizer blender for 1 min. The supernatant was aspirated and the beads were washed with the original amount of fresh buffer. Cell debris was removed by centrifugation at 30,000 *g* for 20 min. The resulting supernatant was adjusted to 0.5 *M* KCl by the addition of a 3 *M* solution and then centrifuged for a sufficient period to pellet 40 S subunits. The usual time with a Beckman 50.2 Ti rotor was 2 hr at 50,000 rpm.

### *DEAE-Sephadex Chromatography*

Proteins were concentrated from the 100,000 *g* supernatant (S100) by addition of 472 g of $(NH_4)_2SO_4$ per liter (70% saturation at 4°). The precipitate was collected by centrifugation at 30,000 *g* for 15 min and dissolved in 50 m*M* Tris·HCl, pH 7.5, 50 m*M* KCl, 10 m*M* β-mercaptoethanol, and 25% glycerol. The solution was desalted either using a Sephadex G-25 column of appropriate size or by overnight dialysis against the same buffer. The usual yield from 50 g cells was about 8 ml with 20 mg/ml of protein.

DEAE-Sephadex A-50 was resuspended in 50 m*M* Tris·HCl, pH 7.5, 100 m*M* KCl and heated in a boiling water bath for 2 hr. Fine particles were removed by decantation 2–3 times and the adsorbent was equilibrated with the starting buffer which contained 50 m*M* Tris · HCl, pH 7.5, 100 m*M* KCl, 10 m*M* β-mercaptoethanol, and 25% glycerol. A large-diameter column (5 cm) was used in order to obtain reasonable flow rates. The final column volume was 1–1.5 ml of applied S100 protein per milligram. The sample was diluted with starting buffer to a concentration of 5–6 mg/ml and applied at a rate of about one-third of a column volume per hour. The sample was washed in with an equal volume of starting buffer, and a gradient with a volume of 10 times the column volume (100 m*M* to 600 m*M* KCl) was applied.

A representative DEAE-Sephadex profile is shown in Fig. 1. Under the conditions used here, EF-1 did not adsorb to DEAE-Sephadex and eluted with the initial flow-through protein. A rather significant trailing of activity was always observed and was reduced by starting the gradient early. As previously observed, EF-2 and EF-3 eluted close together between 220 and 260 m*M* KCl.[2]

If adequate flow rates are maintained, all three factors are completely eluted during an overnight run. The early elution of EF-1 permits its almost immediate recovery. Proper resolution should be obtained if the column volume is not decreased below 1 ml/mg applied protein. Therefore, the

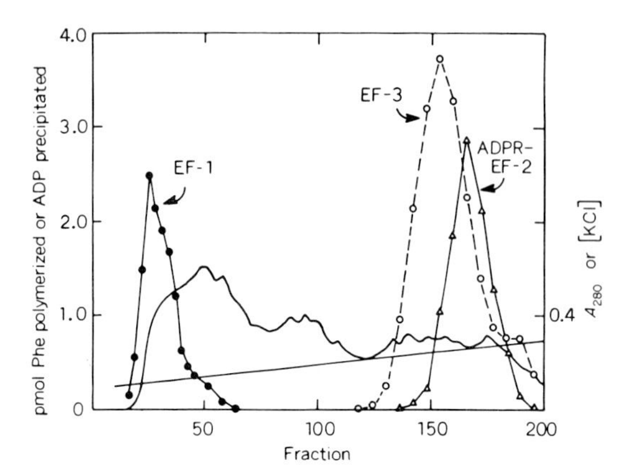


FIG. 1. DEAE-Sephadex chromatography of yeast S100. Solid curve represents protein as determined by absorbance at 280 nm; straight line represents applied gradient. ●, EF-1 activity; ○, EF-3 activity as determined by polyphenylalanine synthesis; △, EF-2 as the substrate of diphtheria toxin-catalyzed ADP-ribosylation.

main limitation on the amount of material that can be chromatographed is the size of available columns. Satisfactory results have been obtained using S100 that was stored frozen at $-100°$.

### *CM-Sepharose Chromatography of EF-1*

Purification of EF-1 to near homogeneity was achieved by a single additional step. CM-Sepharose was equilibrated with 50 m*M* Tris · HCl, pH 7.5, 100 m*M* $NH_4Cl$, 0.1 m*M* EDTA, 10 m*M* $\beta$-mercaptoethanol, and 25% glycerol. The column used had a diameter of 2.5 cm and a bed volume 25% that of the DEAE-Sephadex column. Pooled peak fractions of EF-1 activity from the DEAE-Sephadex column were applied directly at a flow rate of 1 column volume per hour. Protein was eluted with a gradient (total volume 20 times column volume) of $NH_4Cl$ from 100 to 400 m*M*.

A typical CM-Sepharose profile is shown in Fig. 2. The bulk of protein did not adsorb or eluted early in the gradient; the bulk of EF-1 activity eluted at about 0.2 *M* $NH_4Cl$. As seen in the figure a small shoulder of activity preceded the main peak. The appearance of this shoulder in different preparations was quite variable, ranging from 0% to 30% of the total activity. No explanation can be given for this heterogeneity at the present time.

Fractions containing the peak of EF-1 activity were pooled and concentrated approximately 10-fold using an Amicon PM-10 membrane. The concentrated sample was then dialyzed against 50 m*M* Tris · HCl, pH 7.5, 50

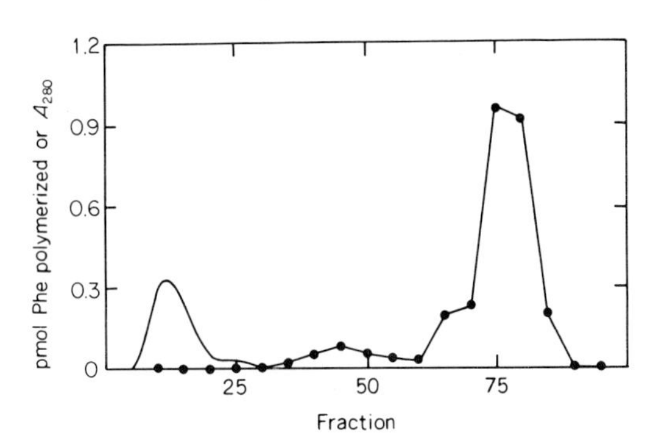


FIG. 2. Carboxymethyl-Sepharose chromatography of yeast EF-1. Solid curve represents protein as determined by absorbance at 280 nm; circles represent EF-1 activity in polyphenylalanine synthesis.

m$M$ $NH_4Cl$, 0.1 m$M$ EDTA, 2 m$M$ dithiothreitol, and 30% glycerol. This fraction was quite stable when frozen at −100°; no loss in activity was observed over a 6-month period.

Analysis of the final sample by sodium dodecyl sulfate (SDS)–gel electrophoresis[8] revealed a major band with a molecular weight of about 50,000, which comprised at least 90% of the stained material. Several minor bands were also seen.

A nonlinear response of polyphenylalanine synthesis to S100 rendered quantitation of factors in that fraction unreliable. As shown in Table I, the specific activity of the CM-Sepharose fraction was about 5-fold higher than the DEAE-Sephadex material, with about a 65% recovery. The rather large amount of EF-1 is expected in view of the high concentration of ribosomes in yeast.

By this procedure, nearly homogeneous yeast EF-1 can be prepared in about 3 days using two chromatographic steps. With little increase in time,

TABLE I
PURIFICATION OF YEAST EF-1

| Fraction | Volume (ml) | Protein (mg) | Specific activity[a] | Total activity[b] |
|---|---|---|---|---|
| S-100 | 8 | 168 | ND | — |
| DEAE-Sephadex | 48 | 19.7 | 2.2 | 43,300 |
| CM-Sepharose | 10 | 2.4 | 11.8 | 28,300 |

[a] Defined as picomoles of Phe polymerized per minute per microgram of protein.
[b] Obtained by multiplying specific activity by total protein.

[8] K. Weber and M. Osborn, *J. Biol. Chem.* **244,** 4406 (1969).

a much larger starting sample could be processed. The protein profile of Fig. 2 suggests that a similar column could accommodate a much larger sample with similar results.

*Separation of EF-2 and EF-3 on CM-Sepharose*

Separation of EF-2 and EF-3 is based on the observation that EF-3 is rather strongly adsorbed to cation exchange columns whereas EF-2 is not. Very little separation of the two components is achieved on DEAE-Sephadex, seen in Fig. 1 and reported previously,[2] or on DEAE-cellulose.

Fractions containing the bulk of EF-2 and EF-3 from the DEAE-Sephadex column were pooled and concentrated 10-fold using an Amicon PM-30 membrane. The concentrated sample was dialyzed against 50 m*M* Tris·HCl, pH 7.5, 30 m*M* $NH_4Cl$, 0.1 m*M* EDTA, 10 m*M* β-mercaptoethanol, and 25% glycerol. A column of CM-Sepharose was prepared identical to that used for EF-1 and equilibrated with the same buffer as used for dialysis. The same column can be used for both steps if washed in between with 1 *M* $NH_4Cl$. The sample was applied, and the column was run at a flow rate of 0.5 to 1 column volume per hour. Proteins were eluted with a gradient (20 times column volume) of 30 m*M* to 380 m*M* $NH_4Cl$. As seen in Fig. 3, EF-2 eluted just behind the main protein peak and EF-3 eluted as a broad peak of activity between 110 m*M* and 140 m*M* $NH_4Cl$. Peak fractions of activity were pooled, concentrated by Amicon filtration, and dialyzed against 50 m*M* Tris·HCl, pH 7.5, 50 m*M* $NH_4Cl$, 0.1 m*M* EDTA, 2 m*M* dithiothreitol, and 30% glycerol.

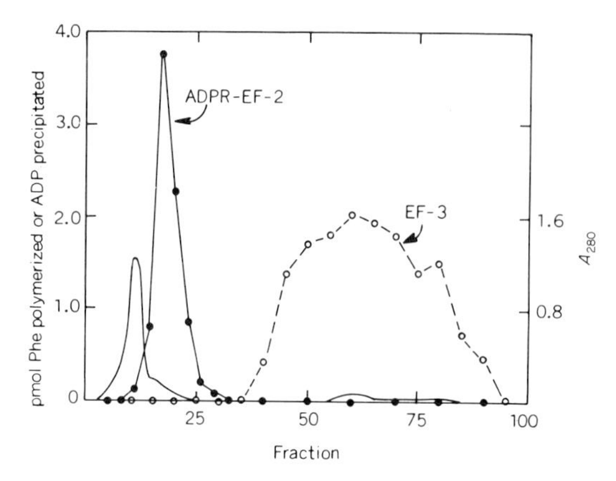


FIG. 3. Carboxymethyl-Sepharose separation of EF-2 and EF-3. Solid curve represents protein as determined by absorbance at 280 nm; ●, EF-2 as the substrate of diphtheria toxin-catalyzed ADP-ribosylation; ○, EF-3 activity in polyphenylalanine synthesis.

Analysis of the two fractions by SDS–gel electrophoresis[8] showed a major band in each and a number of minor bands. In each case the major band corresponded to 30–50% of the total staining material. The mobilities of the major bands suggested molecular weights of about 100,000 for EF-2 and 125,000 for EF-3. Similar molecular weights were obtained by gel filtration chromatography of EF-2 and EF-3 mixtures on Sephadex G-100 (data not shown); presumably each factor exists as a single polypeptide chain.

Quantitation of factor activities in the S100 was not possible owing to the nonlinear response of polyphenylalanine synthesis to that fraction. For this reason recoveries of EF-2 were determined using diphtheria toxin-catalyzed ADP-ribosylation. These data are shown in Table II. The two chromatographic steps resulted in about a 25-fold purification with about 30% recovery. The specific activity of the final fraction, 2.85 pmol/$\mu$g, is equivalent to one ADP-ribose per 350,000 daltons. This value is consistent with a purity of about 30%, assuming a molecular weight of 100,000 for EF-2.

Each of the two fractions obtained by this procedure was demonstrably free of the other component, either by activity or gel electrophoresis. A procedure for purification of EF-2 using chromatography on hydroxyapatite, DEAE-Sephadex, DEAE-cellulose, and a second hydroxyapatite step was previously described.[9] The yield from this procedure was very low, on the order of 1%. Combination of the two methods might produce more favorable yields.

### *Possible Requirement for ATP in Polyphenylalanine Synthesis*

Specific steps of protein synthesis require specific nucleotide triphosphates. Formation of aminoacyl-tRNA requires ATP, and several ribosomal reactions are dependent on GTP. An unexpected effect of ATP

TABLE II
SUMMARY OF EF-2 RECOVERIES

| Fraction | Volume (ml) | Protein (mg) | Specific activity[a] | Total activity[b] |
|---|---|---|---|---|
| S100 | 8 | 168 | 0.11 | 18,500 |
| DEAE-Sephadex | 15 | 34 | 0.39 | 13,300 |
| CM-Sepharose | 13 | 2.3 | 2.85 | 6,600 |

[a] Defined as picomoles of ADP-ribose precipitated per microgram of protein.
[b] Obtained by multiplying specific activity by total protein.

[9] U. Somasundaran and L. Skogerson, *Biochemistry* **15**, 4760 (1976).

on polyphenylalanine synthesis was observed with the yeast system, suggesting that ATP as well as GTP may be required for a ribosomal reaction. The initial observation was that a very high concentration of GTP was required for polyphenylalanine synthesis relative to that previously reported. These data are presented in Fig. 4, which shows that at least 1 m*M* GTP was required for saturation. In attempting to explain this phenomenon we observed that ATP exerted a sparing effect on the requirement for GTP. In the presence of 0.3 m*M* ATP, less than 0.1 m*M* GTP was fully saturating. Less total nucleotide was required in the presence of ATP than in the absence.

A possible explanation of the ATP effect was that the GTP was contaminated with inhibitory amounts of GDP or GMP. In fact, PEI-cellulose thin-layer chromatography[10] revealed trace amounts of GMP and about 10% GDP in the GTP sample. To rule out this possibility the GTP was purified by chromatography on DEAE-Sephadex.[11] The purified material had no detectable GDP or GMP as determined by PEI-cellulose thin-layer analysis. The effect of ATP on polyphenylalanine synthesis was the same whether or not the GTP had been purified.

A second possible explanation was that GTP was being hydrolyzed and ATP was necessary for its regeneration. To test this idea polyphenylalanine synthesis was carried out with [$^3$H]GTP in the presence or the absence of ATP. Analysis of the nucleotide products by PEI-thin-layer chromatog-

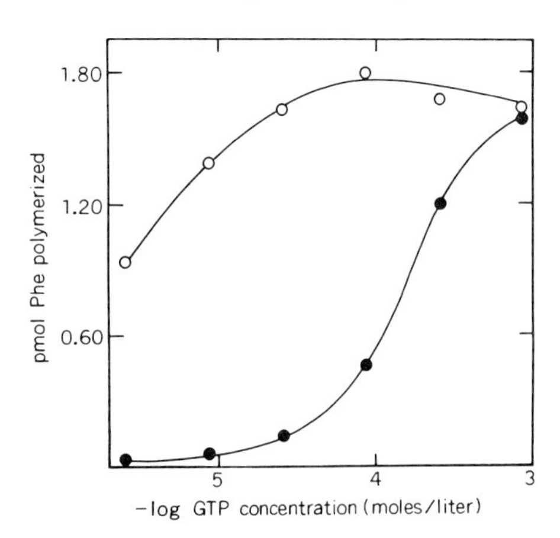


FIG. 4. Polyphenylalanine synthesis as a function of GTP concentration in the presence or the absence of ATP. Standard reactions were carried out for 5 min as described in the text except that the GTP concentration was varied as indicated. ●, No ATP was present; ○, 0.3 m*M* ATP was added.

[10] M. Cashel and B. Kalbacher, *J. Biol. Chem.* **245,** 2309 (1970).
[11] P. F. Schendel and R. D. Well, *J. Biol. Chem.* **248,** 8319 (1973).

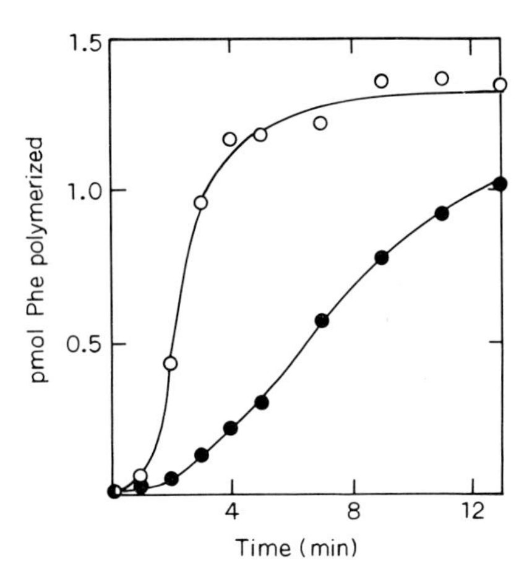


FIG. 5. Polyphenylalanine synthesis as a function of time in the presence or the absence of ATP. Standard reaction conditions were used except that the GTP concentration was 10 $\mu M$ and the reaction volumes were 0.6 ml. At indicated times 50-$\mu$l samples were removed and analyzed for [$^{14}$C]polyphenylalanine. ●, No ATP; ○, 0.3 m$M$ ATP.

raphy showed that about 10% of the GTP was converted to GDP in either case (data not given). Presumably rapid hydrolysis of GTP was not the reason for the stimulation by ATP at lower GTP concentrations.

These observations suggested an effect of ATP on this system at the ribosomal level that was independent of the GTP requirement. The question remains what phase of the reaction was primarily affected by ATP. In order to determine whether the effect might be during an initiating step or during elongation itself, the effect of ATP on the rate of the reaction was measured at a low GTP concentration. The data shown in Fig. 5 indicated that the effect of ATP was greater on the initial rate of the reaction than on the extent. These results indicate that the ATP effect was either on an early step of the reaction or that it increased the rate of one of the elongation reactions. We cannot say whether the ATP requirement is absolute or not. Small amounts of ATP may be present in the system or the ATP requirement may be fulfilled by high concentrations of GTP. Whether the ATP requirement observed here is related to that seen in formation of the 80 S initiation complex requires further experiments.[12]

## Summary

Methods are described for separation of three factors from yeast that are required for polyphenylalanine synthesis. Two chromatographic steps

[12] T. Staehelin, H. Trachsel, B. Erni, A. Boschetti, and M. H. Schreier, *FEBS Proc. Meet. 10th* **39,** 309 (1975).

result in a nearly homogeneous preparation of EF-1 in relatively high yield. An additional chromatographic step yields EF-2 and EF-3 fractions, each of which is free of the other activity and about 30% pure. The entire procedure can be carried out in 3 days and could readily be scaled up to prepare large amounts of the factors.

An unusual property of the yeast system is a marked stimulation of the initial rate of polyphenylalanine synthesis by ATP.

The methods described here have been used to isolate factors from both D587-4B and 2180-1A with identical results.

# Section V

# Polypeptide Chain Termination

# [44] Release Factors: *in Vitro* Assay and Purification

*By* C. T. CASKEY, E. SCOLNICK, R. TOMPKINS,
G. MILMAN, and J. GOLDSTEIN

Peptide chain termination can be investigated *in vitro* by the sequential translation of initiator and terminator trinucleotide codons to bind f[$^3$H]Met-tRNA to ribosomes and subsequently release formyl[$^3$H]-methionine.[1] The release of formyl[$^3$H]methionine from fMet-tRNA·AUG·ribosome intermediates is analogous to the release of peptides upon peptide chain termination and requires both terminator codon and protein release factor, R.[1]

$$\text{f[}^3\text{H]Met-tRNA} + \text{AUG} + \text{ribosome} \rightleftharpoons \text{f[}^3\text{H]Met-tRNA}\cdot\text{AUG}\cdot\text{ribosome}$$

$$\text{f[}^3\text{H]Met-tRNA}\cdot\text{AUG}\cdot\text{ribosome} + \text{R} + \text{terminator codon} \rightarrow \text{formyl[}^3\text{H]methionine}$$

The release of formylmethionine is directed by one of three mRNA codons (UAA, UAG, or UGA) in reactions containing *Escherichia coli* B cellular components. The expression of each codon requires one of two protein release factors[2] (R1 or R2). These protein molecules apparently recognize trinucleotide codons for they participate in formylmethionine release[2] and bind to ribosomes[3] with codon specificity. Release factor R1 recognizes UAA or UAG; R2 recognizes UAA or UGA. The method of quantitating R activity and the purification of R1 and R2 are related in this report.

## Preparation of fMet-tRNA·AUG·Ribosome Intermediate

Formylmethionine-tRNA is the species of tRNA participating in initiation of *Escherichia coli* protein synthesis. This species of tRNA was chosen as an analog of peptidyl-tRNA because of its high affinity for the ribosomal P site in the presence of its trinucleotide codon, AUG.[4] Aminoacyl-tRNA or fMet-tRNA bound to the ribosomal A site[5] are unreactive ribosomal intermediates for *in vitro* release. The fMet-tRNA·AUG·ribosome intermediate is routinely formed at 10 m*M* $Mg^{2+}$

[1] C. T. Caskey, R. Tompkins, E. Scolnick, T. Caryk, and M. Nirenberg, *Science* **162**, 135 (1968).

[2] E. Scolnick, R. Tompkins, T. Caskey, and M. Nirenberg, *Proc. Nat. Acad. Sci. U. S.* **61**, 768 (1968).

[3] E. Scolnick and T. Caskey, *Proc. Nat. Acad. Sci. U. S.*, **64**, 1235 (1969).

[4] M. S. Bretscher and K. A. Marcker, *Nature* (*London*) **211**, 380 (1966).

[5] R. Tompkins, E. Scolnick, and T. Caskey, *Proc. Nat. Acad. Sci. U. S.*, **65**, 702 (1970).



ISBN 0-12-504180-2

and thus obviates the need for GTP and initiation protein factors.[6] Since crude initiation factors contain R activity, they should be partially purified if required for special peptide release studies.

The precise level of ribosomes and fMet-tRNA used for fMet-tRNA·AUG·ribosome intermediate formation varies with the ribosome activity (25–60%) for fMet-tRNA binding. Therefore preliminary fMet-tRNA binding studies are necessary to optimize conditions for each ribosome preparation. On the basis of these studies, levels of ribosomes are employed that result in at least 80% of available fMet-tRNA being ribosomal bound, fMet-tRNA·AUG·ribosome. A typical reaction is incubated 15 minutes at 30°C and contains, in 0.10 ml: 120–200 pmoles f[$^3$H]Met-tRNA; 10–20 $A_{260}$ units *E. coli* B ribosomes; 2.1 nmoles AUG (0.1 $A_{260}$ unit); 50 m*M* Tris·acetate, pH 7.2; 10 m*M* magnesium acetate; and 50 m*M* ammonium or potassium acetate. The reaction is placed on ice; 0.40 ml of cold buffer is added, which contains: 50 m*M* Tris·acetate, pH 7.2; 50 m*M* ammonium or potassium acetate; and 35 m*M* magnesium acetate. Aliquots of 0.020 ml are examined for quantity of f[$^3$H]Met-tRNA·AUG·ribosome intermediate by the technique of Nirenberg and Leder[7]; f[$^3$H]Met-tRNA by trichloroacetic acid precipitation; and formyl[$^3$H]methionine as described below. Typical values are: 4–6 pmoles f[$^3$H]Met-tRNA·AUG·ribosome intermediate; 5–7 pmoles f[$^3$H]Met-tRNA; and 0.15–0.40 pmoles formyl[$^3$H]-methionine. These values do not change with storage for months at −170° or with several freeze–thaw cycles. This quantity of f[$^3$H]Met-tRNA·AUG·ribosome intermediate is adequate for 30–40 release determinations.

*Materials. Escherichia coli* ribosomes are prepared by 4–7 successive 0.5 *M* ammonium chloride washes as described by Lucas-Lenard and Lipmann[8] and are devoid of initiation, elongation, and release factors. If initiation factors are desired, $F_1$, $F_3$, and portions of $F_2$ are separated from R1 and R2 by DEAE-Sephadex[9] or cellulose[10] column chromatography. The f[$^3$H]Met-tRNA is prepared by a partially purified preparation containing fMet-tRNA synthetase and transformylase.[11] Each reaction for the formation of f[$^3$H]Met-tRNA is incubated 30° for 30 minutes and contains, in 1 ml: 0.24 mg leucovorin (Lederle Co.);

[6] M. Salas, M. J. Miller, A. J. Wahba, and S. Ochoa, *Proc. Nat. Acad. Sci. U. S.* **57**, 1865 (1967).
[7] M. Nirenberg and P. Leder, *Science* **145**, 1399 (1964).
[8] J. Lucas-Lenard and F. Lipmann, *Proc. Nat. Acad. Sci. U. S.* **55**, 1562 (1966).
[9] Z. Vogel, A. Zamir, and D. Elson, personal communication.
[10] R. W. Erbe, M. M. Nau, and P. Leder, *J. Mol. Biol.* **39**, 441 (1969).
[11] G. Milman, J. Goldstein, E. Scolnick, and T. Caskey, *Proc. Nat. Acad. Sci. U. S.* **63**, 183 (1969).

5.4 or 25 $A_{260}$ unit of tRNA$^{fMet}$ or unfractionated tRNA, respectively; 0.5 mg enzyme protein; 0.1 *M* potassium cacodylate, pH 6.9; 1 m*M* ATP; 10 m*M* magnesium chloride; 4 m*M* reduced glutathione; and 0.1 m*M* [$^3$H]methionine and 19 other nonradioactive amino acids. The f[$^3$H]Met-tRNA is deproteinized, isolated, and stored[11] as previously described. Preparations of the tRNA$^{fMet}$ are obtained by benzoylated DEAE[12] or reverse phase column chromatography[13]; these are devoid of tRNA$^{Met}$ and have accepted between 330 and 1300 pmoles of formylmethionine per $A_{260}$ unit. Although unfractionated f[$^3$H]Met-tRNA is acceptable for use, it binds less efficiently to ribosomes and contains [$^3$H]Met-tRNA$^{Met}$. Purified f[$^3$H]Met-tRNA$^{fMet}$ is routinely used. All trinucleotide codons were prepared and isolated, and their base composition was determined as previously described.[14,15]

## Formyl[$^3$H]methionine Release from f[$^3$H]Met-tRNA·AUG·Ribosome

The release of formyl[$^3$H]methionine from fMet-tRNA·AUG·ribosome intermediates, dependent upon terminator codon (Fig. 1),

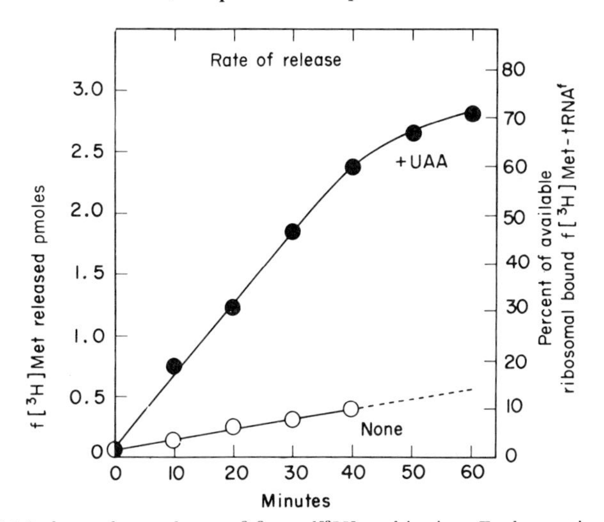


FIG. 1. UAA-dependent release of formyl[$^3$H]methionine. Each reaction is incubated 30°C for the indicated time interval and contains in 0.05 ml: 3.8 pmoles f[$^3$H]Met-tRNA·AUG·ribosome intermediate; 0.8 $\mu$g R1 (Fraction IV); as indicated, 7.5 nmoles UAA; and additional components as indicated in the text. The amount of formyl[$^3$H]methionine present at zero time (0.30 pmole) was subtracted from each value.

[12] I. Gillam, S. Millward, D. Blew, M. von Tigerstrom, E. Wimmer, and G. M. Tener, *Biochemistry*, **6**, 3043 (1967).

[13] J. F. Weiss, R. L. Pearson, and A. D. Kelmers, *Biochemistry* **7**, 3479 (1968).

[14] P. Leder, M. F. Singer, and R. L. C. Brimacombe, *Biochemistry* **4**, 1561 (1965).

[15] R. E. Thach and P. Doty, *Science* **148**, 632 (1965).

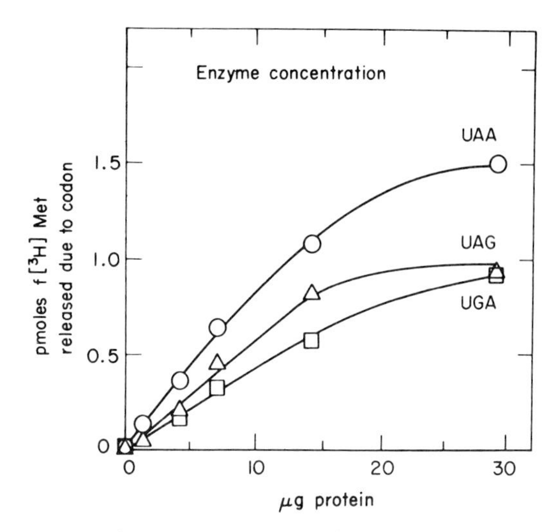


FIG. 2. Concentration of R and rate of formyl[$^3$H]methionine release. Each reaction is incubated at 24° for 10 minutes and contains, in 0.05 ml: 4.8 pmoles f[$^3$H]Met-tRNA·AUG·ribosome intermediate; indicated quantity of fraction II; 2.5 nmoles terminator trinucleotides; and other components as indicated in the text. The amount of formyl-[$^3$H]methionine formed without trinucleotides was subtracted from each value.

corresponds to *in vitro* peptide chain termination. Formylmethionine release occurring in the absence of codon is due to chemical and enzymatic deacylation unrelated to R activity. This latter release is high when reactions contain crude R but equals chemical deacylation when purified R preparations are used (Fractions IV and V).

A typical release reaction is incubated for 15 minutes at 24° and contains in 50 μl: 4–6 pmoles fMet-tRNA·AUG·ribosome intermediate; 2.5 nmoles UAA, UAG, or UGA (0.1 $A_{260}$ unit); 50 m*M* Tris·acetate, pH 7.2; 75 m*M* ammonium or potassium acetate; 30 m*M* magnesium acetate; and initiated by the addition of 0.01–15.0 μg R1 or R2 protein. Additions are made at 4° in the order listed prior to transfer to a 24° bath. Although codon-directed release occurs under a variety of conditions,[10] optimal rates of formylmethionine release occur at 75 m*M* ammonium acetate (3-fold greater than potassium acetate), 30 m*M* magnesium acetate (the ribosomal intermediate dissociates below 15 m*M* $Mg^{2+}$), and 24° (4-fold greater than 0° or 40°). The 50 m*M* trinucleotide ($K_m$ = 50–80 μ*M*[16]) and 4–6 pmoles fMet-tRNA·AUG·ribosome levels are arbitrarily fixed. The initial rate of formyl[$^3$H]methionine release is proportional to R (Fig. 2). A microunit of R corresponds to the amount of R protein releasing 1 pmole of formylmethionine per minute under the above optimal conditions.

Each reaction is terminated by the addition of 0.250 ml of 0.1 *N* HCl and 1.5 ml of ethyl acetate, and agitated 15 seconds by a Vortex mixer. One milliliter of ethyl acetate (upper phase) is transferred to 10 ml of Bray's[16] or BioSolv (3 liters toluene, 100 ml Beckman BioSolv, and 300 ml Beckman Fluoralloy) counting fluid, and radioactivity is determined by scintillation counting in a Beckman Model LS-250 at 25 and 45% counting efficiency, respectively. Since formyl[$^3$H]methionine is extracted into ethyl acetate at 70% efficiency, total formylmethionine release is 2.15 × (cpm-bkgd). We report total formyl-[$^3$H]methionine release.

## Purification of Release Factors

Both R1 and R2 are found in low levels[11] ($10^2$ per cell) and are identified in ribosomal (20%) and supernatant (80%) fractions of *E. coli* extracts. Each release factor has been extensively purified, R1 homogeneous and R2 80% pure by acrylamide gel analysis. Each of these highly purified release factors is active with UAA for formylmethionine release; only R1 is active with UAG; and only R2 is active with UGA.

*Escherichia coli* B are grown to early log at 32° with aeration in 1% dextrose and 0.8% Difco nutrient broth. All purification steps are performed at 4°. Details of purification and recovery at each step in purification from 1 lb of *E. coli* B are given in the table.

PURIFICATION OF RELEASE FACTORS[a]

| R factor | Fraction | Protein (mg) | | Specific activity | Recovery | Purification |
|---|---|---|---|---|---|---|
| | | Applied | Recovered | | | |
| R1 | II. Ammonium sulfate | – | 8000 | 0.7 | 1.00 | 2 |
| | III. Sephadex G-100 | 8000 | 2000 | 2.5 | 0.91 | 7 |
| | IV. DEAE-Sephadex | 2000 | 139 | 46.1 | 1.18 | 136 |
| | VI. CM-Sephadex | 82 | 5 | 480 | 0.64 | 1410 |
| R2 | II. Ammonium sulfate | – | 8000 | 1.2 | 1.00 | 2 |
| | III. Sephadex G-100 | 8000 | 2000 | 4.6 | 0.97 | 8 |
| | V. DEAE-Sephadex | 2000 | 45.5 | 160 | 0.80 | 274 |
| | VII. Hydroxylapatite | 16.3 | 1.6 | 670 | 0.42 | 1150 |

[a]The details of R1 and R2 purification are given in the text. Fractions were assayed for R1 with 2.5 nmoles UAG and for R2 with 2.5 nmoles UGA at 24°. Specific activity is the number of microunits (see text) per milligram of protein. Recovery is the ratio of input to output activity for each purification step. Purification is the ratio of the specific activity at the end of each purification step to that of fraction I.

[16]G. Bray, *Anal. Biochem.* 1, 279 (1960).

One pound of *E. coli* cells is suspended in 681 ml of buffer A (0.1 *M* Tris·acetate, pH 8.0; 50 m*M* potassium chloride; 14 m*M* magnesium chloride; and 1 m*M* dithiothreitol. Worthington DPFF DNase is added (3.0 μg/ml), and cells are lysed with a French pressure cell at 18,000 psi. After 20 minutes of 30,000 *g* centrifugation of the extract, the upper two-thirds of the supernatant fraction is subjected to 5 hours 137,000 *g* centrifugation. The upper two-thirds of supernatant from this centrifugation corresponds to fraction I. A reliable R specificity activity is difficult to obtain with this fraction.

Both R1 and R2 are precipitated from fraction I by the addition of ammonium sulfate (32.5 g/100 ml). The precipitated R (fraction II, 8.0 g) is collected by centrifugation (15 minutes, 30,000 *g*); dissolved in 75 ml of buffer B (50 m*M* Tris·HCl, pH 8.0; 0.2 *M* potassium chloride; 1 m*M* EDTA; and 1 m*M* dithiothreitol), and applied to a 5 × 100 cm column packed with Sephadex G-100 (bead) equilibrated with buffer B. Each 12-ml fraction is eluted at a flow rate of 70 ml/hour with R activity eluting in tubes 45–60. Since R1 and R2 do not separate by this procedure, all tubes are examined for release activity in the presence of UAA. The active tubes are pooled (fraction III, 2 g) and applied to a 5 × 67 cm column packed with *de-fined* DEAE-Sephadex G-50 equilibrated in buffer B. The protein is eluted with 4200 ml of buffer B that contains a linear potassium chloride gradient (0.2–0.7 *M*). R1 elutes at ~0.3 *M* (fraction IV) potassium chloride and R2 (fraction V) at ~0.4 *M* as shown in Fig. 3. Since T factor consistently elutes slightly before R1, T is identified by [$^{14}$C]GTP binding[17] before the tubes are examined for release activity. A 5–10 μl addition of column fractions to a release reaction with the corresponding trinucleotide is adequate for identification of R1 and R2. These fractions of R1 (IV) and R2 (V) are not cross-contaminated; they are stable in storage at −170° but do contain tRNA and elongation factors T and G. The following procedures remove these components and further purify both R1 and R2.

Release factor R1 (fraction IV) is equilibrated in buffer C (50 m*M* imidazole, pH 6.0; 0.1 *M* potassium chloride; 3 m*M* β-mercaptoethanol; and 0.1 m*M* EDTA) by G-25 column chromatography and applied to a 40 × 2.5 cm column packed with CM-Sephadex equilibrated in buffer C. From 0.1 to 1.5 g R1 protein has been applied to a column of this size. The column is washed with 200 ml of buffer C, and R1 is eluted by 600 ml of buffer C in a linear potassium chloride gradient (0.1–0.8 *M*) at a flow rate of 45 ml/hour. Release activity of each 7.5-ml fraction

[17] R. Ertel, N. Brot, B. Redfield, J. E. Allende, and H. Weissbach, *Proc. Nat. Acad. Sci. U. S.* **59**, 862 (1968).

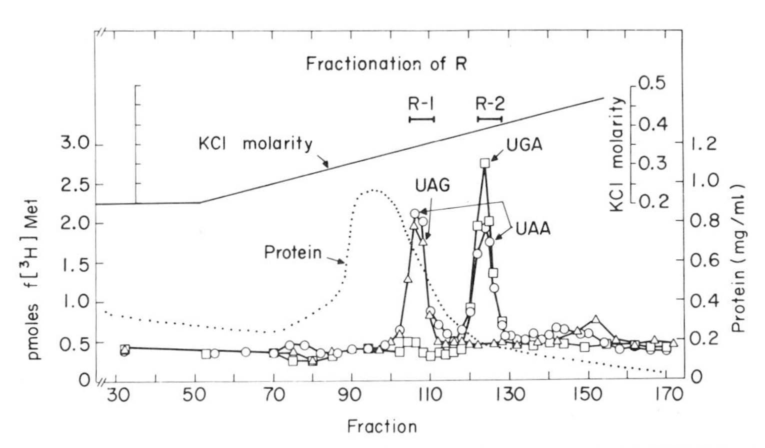


FIG. 3. Separation of R1 and R2. A 36 × 2.6 cm column packed with DEAE-Sephadex was used to fractionate 450 mg of Fraction II with a 1300-ml gradient (0.2–0.6 *M* potassium chloride). Each 11.0-ml tube was examined for release activity with each of the terminator trinucleotides. Each reaction was incubated for 25 minutes at 30° and contained in 50 $\mu$l: 4.4 pmoles f[$^3$H]Met-tRNA·AUG·ribosome intermediate; 10 $\mu$l of each column fraction; 7.5 nmoles of indicated terminator trinucleotide; and additional components as described in the text. The amount of formyl[$^3$H]methionine present at zero time (0.30 pmole) was subtracted from each value.

(Fig. 4) is assessed in the presence of UAA or UAG, and the pooled active tubes correspond to fraction VI.

Release factor R2 is equilibrated in buffer D (10 m*M* potassium phosphate, pH 7.2; 3 m*M* $\beta$-mercaptoethanol) by G-25 column chromatography, and applied to a 9 × 2 cm column packed with hydroxylapatite (Clarkson Chemical Co.) equilibrated in buffer D. From 16 to 500 mg of R2 protein has been applied to a column of this size. The column is washed successively with 60 ml of 1 m*M*, 70 ml of 35 m*M*, and 70 ml of 80 m*M* potassium phosphate in buffer D at a flow rate of 120 ml/hour. Release activity of each 7-ml fraction is assessed in the presence of UAA or UGA (Fig. 4). Since phosphate buffer is inhibitory to the release assay, fractions should be dialyzed prior to assay, or, alternately, column additions to release reactions be kept below 2 $\mu$l. Fractions containing R2(80 m*M* potassium phosphate) are pooled and correspond to fraction VII.

Release factor preparations (IV through VII) are concentrated by pressure filtration (Amico Co., Lexington, Massachusetts), dialyzed against buffer E (50 m*M* Tris·HCl, pH 8.0; 50 m*M* potassium chloride; and 3 m*M* dithiothreitol) and stored at −170°. Fractions IV through VII are stable for 9 months or more when stored at −170° and tolerate several freeze–thaw cycles well. Fractions VI (R1) and VII (R2) are >1000-fold purified, devoid of tRNA, elongation factor T, S protein,

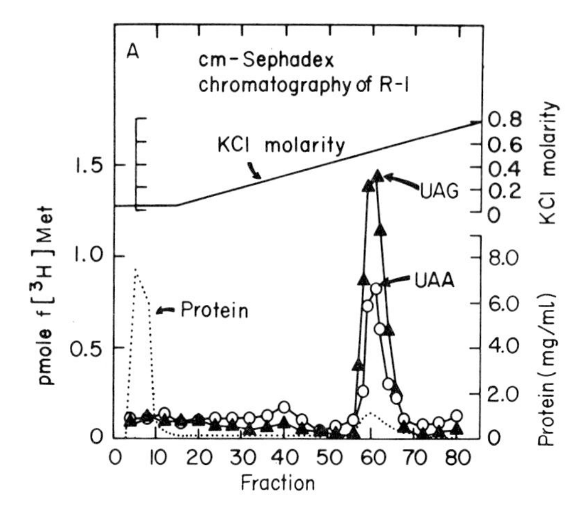


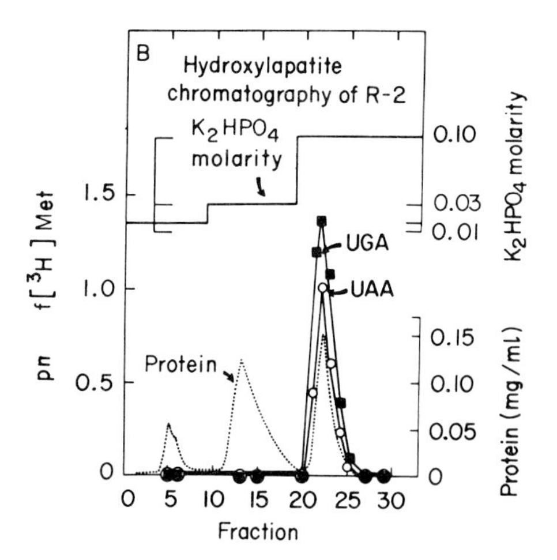


FIG. 4. Purification of R Factors. (A) Each release assay was incubated 25 minutes at 30° and contained, in 50 μl: 5.7 pmoles f[$^3$H]Met-tRNA·AUG·ribosome intermediate; 5 μl of indicated column fraction; as indicated, 2.5 nmoles UAA or UAG; and additional components as indicated in the text. The amount of formyl[$^3$H]methionine present at zero time (0.30 pmole) was subtracted from each value. (B) Each release assay was incubated 5 minutes at 24°C and contained, in 50 μl: 5.3 pmoles of f[$^3$H]Met-tRNA·AUG·ribosome intermediate; 2 μl of indicated fraction; 2.5 nmoles of UAA or UGA as indicated; and additional components as described in the text. The amount of formyl[$^3$H]methionine at zero time (0.25 pmole) was subtracted from each value.

and contain 0.3% elongation factor G.[11] Both R1 and R2 can be further purified by preparative disc gel electrophoresis with good recovery (>50%) and will be described in detail in another report. Acrylamide gel analysis of R1, fraction VI and R2, fraction VII, indicate, however, that they contain 25% and 50% R factor, respectively.

# [45] Mammalian Release Factor; *in Vitro* Assay and Purification[1]

*By* C. T. CASKEY, A. L. BEAUDET, and W. P. TATE

Peptide chain termination and several of its intermediate steps can be studied in mammalian extracts. The codon directed release of peptides from ribosomal bound peptidyl-tRNA is investigated in mammalian extracts by modification of the formylmethionine (fMet) release assay previously described for bacterial extracts.[2] As shown in reaction (1), f[$^3$H]Met-

$$\text{f[}^3\text{H]Met-tRNA} + \text{reticulocyte ribosomes} \rightarrow \text{f[}^3\text{H]Met-tRNA·ribosome}$$
$$\text{f[}^3\text{H]Met-tRNA·ribosome} + \text{RF} + \text{terminator codon} + \text{GTP} \rightarrow \text{f[}^3\text{H]Met} \quad (1)$$

tRNA can be bound to reticulocyte ribosomes at a magnesium ion concentration of 55 m*M* without protein factors, GTP, polynucleotides, or oligonucleotides.[3] The release of fMet from these ribosomal complexes is stimulated by one of three oligonucleotides of defined sequence (UAAA, UAGA, and UGAA) or randomly ordered polynucleotide templates containing the codons UAA, UAG, or UGA.[3,4] Thus far the triplets UAA, UAG, or UGA have not stimulated fMet release with mammalian extracts. The codon directed release of fMet is stimulated by GTP and requires a mammalian protein release factor (RF). The *in vitro* release of fMet from fMet-tRNA · ribosome intermediates is believed to be analogous to the release of peptides upon peptide chain termination in mammalian cells.

Two additional methods for investigation of mammalian peptide chain termination are also available. The first studies fMet-tRNA hydrolysis as

[1] This work was supported by Grant GM-18682-02.

[2] C. T. Caskey, R. Tompkins, E. Scolnick, T. Caryk, and M. Nirenberg, *Science* **162**, 135 (1968).

[3] J. L. Goldstein, A. L. Beaudet, and C. T. Caskey, *Proc. Nat. Acad. Sci. U.S.* **67**, 99 (1970).

[4] A. L. Beaudet and C. T. Caskey, *Proc. Nat. Acad. Sci. U.S.* **68**, 619 (1971).



ISBN 0-12-504180-2

a partial reaction of the total process.[4] The hydrolysis of ribosomal bound f[$^3$H]Met-tRNA which occurs in reactions containing RF and 20% ethanol does not require terminator codon recognition and is not stimulated by GTP. This partial reaction of peptide chain termination, outlined below (reaction 2), differs, therefore, from reaction (1) in its specificity.

$$\text{RF} + \text{f[}^3\text{H]Met-tRNA}\cdot\text{ribosome} + 20\%\ \text{ethanol} \rightarrow \text{f[}^3\text{H]Met} \quad (2)$$

The RF-mediated hydrolysis of f[$^3$H]Met-tRNA on ribosomes in this reaction has been separated from the intermediate steps of reaction (1) which require the participation of GTP and oligonucleotide codons.

The binding of RF to ribosomes can be determined indirectly with radioactive oligonucleotides. While UA[$^3$H]AA is used exclusively in these studies, it appears likely that UAG[$^3$H]A and UG[$^3$H]AA would also be effective. The ribosomal binding of RF can be examined by the quantitation of the amount of radioactive RF·UA[$^3$H]AA·ribosome formed.[5]

$$\text{RF} + \text{ribosomes} + \text{UA[}^3\text{H]AA} + \text{GDPCP} \xrightarrow{\text{ETOH}} \text{RF}\cdot\text{UA[}^3\text{H]AA}\cdot\text{ribosomes} \quad (3)$$

The RF·UA[$^3$H]AA·ribosome complex is stabilized for isolation by addition of 20% ethanol to all reactions. Following this stabilization the radioactively labeled ribosomal complex can be isolated and quantitated by centrifugation, Sephadex column chromatography, or by Millipore filtration. The recovery of the complex on Millipore filters is enhanced 2-fold by GTP and 5-fold by equivalent amounts of GDPCP. It is not known whether GTP or GDPCP is a component of the ribosomal complex.

By use of the three *in vitro* methods outlined above, total and partial reactions of peptide chain termination can be investigated in mammalian extracts. This report outlines the methodology of the *in vitro* assays and the purification and characterization of rabbit reticulocyte RF.

*Preparation of fMet-tRNA·Ribosome Intermediates*

Formyl methionine-tRNA, the initiator tRNA species of *Escherichia coli* protein biosynthesis is used as a peptidyl-tRNA analog.[6] The fMet-tRNA has high affinity for the ribosomal site and the quantitation of f[$^3$H]Met release is simple. Both mammalian and *E. coli* f[$^3$H]Met-tRNA (formylated with *E. coli* transformylase) bind equally well to reticuloctye ribosomes at a magnesium ion concentration of 55 m*M*. The f[$^3$H]Met-tRNA·ribosome complexes are reactive with puromycin yielding 95–100% of the radioactivity as f[$^3$H]Met-puromycin in less than 5 minutes.[3] The ribosomal binding and puromycin reactivity of mammalian and *E. coli*

[5] W. P. Tate, A. L. Beaudet, and C. T. Caskey, in preparation.

[6] M. S. Bretscher and K. A. Marcker, *Nature* (*London*) **211**, 380 (1966).

fMet-tRNA were indistinguishable. Highly purified *E. coli* $tRNA_f^{Met}$ (1400 pmoles of methionine acceptance per $A_{260}$ unit) (available through NIGMS) is routinely used. The [$^3$H]Met-$tRNA_f$ binds to reticulocyte ribosomes as well as f[$^3$H]Met-$tRNA_f$, but the [$^3$H]Met-tRNA·ribosome complexes do not react with puromycin to form [$^3$H]Met-puromycin or RF to form [$^3$H]Met without other factors.

The precise level of ribosomes and f[$^3$H]Met-tRNA needed for f[$^3$H]Met-tRNA·ribosome intermediate formation varies with f[$^3$H]Met-tRNA binding activity of the ribosomal preparation. Therefore, preliminary studies are necessary to determine the quantity of ribosomes needed to bind 85% of the f[$^3$H]Met-tRNA. A typical binding reaction is incubated 15 minutes at 30° and contains in 5.0 ml: 50 m*M* Tris·HCl, pH 7.0; 55 m*M* $MgCl_2$; 80 m*M* $NH_4Cl$; 300–500 $A_{260}$ units of reticulocyte ribosomes and 2 nmoles *E. coli* f[$^3$H]Met-tRNA. Aliquots of 0.01 ml are examined for quantity of f[$^3$H]Met-tRNA·ribosome intermediate formed by the technique of Nirenberg and Leder[7]; f[$^3$H]Met-tRNA precipitable by trichloroacetic acid; and free f[$^3$H]Met by ethyl acetate extraction.[8] Typical values are: 3–5 pmoles f[$^3$H]Met-tRNA·ribosome; 3–5 pmoles f[$^3$H]Met-tRNA; and 0.3–0.5 pmole f[$^3$H]Met. These values do not change with storage at −170° over several months. This 5.0 ml quantity of f[$^3$H]Met-tRNA·ribosome intermediate is adequate for 500 determinations. It is stored at −170° in aliquots of 0.3 ml.

*Synthesis and Purification of Oligonucleotides*

The oligonucleotide sets (UAA and UAAA; UGA and UGAA) are synthesized enzymatically by polynucleotide phosphorylase from the doublets UA or UG and ADP.[9] The UAGA is synthesized in two steps. The UAG is synthesized and isolated as previously described.[10] The UAGA is synthesized enzymatically by polynucleotide phosphorylase, UAG and ADP. The oligonucleotides are isolated from the reaction mixture by DEAE-cellulose chromatography as described by Petersen and Reeves[11] except that urea is omitted. The tri- and tetranucleotides isolated by this method have less than 2% contaminating ultraviolet absorbing material as analyzed by T2 ribonuclease digestion and base ratio analysis.[9] Synthesis and isolation of UA[$^3$H]AA is performed as described above except that reactions contained [$^3$H]ADP (20.3 Ci/mmole) and nonradioactive UA.

[7] M. W. Nirenberg and P. Leder, *Science* **145**, 1399 (1964).

[8] P. Leder and H. Bursztyn, *Biochem. Biophys. Res. Commun.* **25**, 233 (1966).

[9] P. Leder, M. F. Singer, and R. L. C. Brimacombe, *Biochemistry* **4**, 1561 (1965).

[10] R. Thach, *in* "Procedures in Nucleic Acid Research" (G. L. Cantoni and D. R. Davies, eds.), p. 520. Harper, New York, 1966.

[11] G. B. Petersen and J. M. Reeves, *Biochim. Biophys. Acta* **129**, 438 (1966).

*Codon-Directed f[$^3$H]Met Release*

Mammalian RF from rabbit reticulocytes, guinea pig, and Chinese hamster liver participate in fMet release from fMet-tRNA·ribosome substrates with oligonucleotides UAGA, UAGG, UAAA, or UGAA but not with AAAA, UUUU, UAA, UAG, or UGA.[4] Furthermore, randomly ordered poly(U,G,A), ($U,A_3$), and ($U,A_3,G_{0.5}$) stimulate release of f[$^3$H]Met. A typical release reaction is incubated 15 minutes at 24° and contains in 50 μl; 3.0 pmoles of f[$^3$H]Met-tRNA·ribosome intermediate; 20 m*M* Tris·HCl pH 7.4; 60 m*M* KCl; 11 m*M* $MgCl_2$; 0.1 m*M* GTP; 0.1 $A_{260}$ unit UAAA, UAGA, or UGAA; or alternatively 0.2 $A_{260}$ units of poly(U,G,A), ($U,A_3$), ($U,A_3,G_{0.5}$), and partially purified mammalian RF.

Reactions are terminated by addition of 0.25 ml of 0.1 *N* HCl and 1.5 ml of ethyl acetate. The aqueous and ethyl acetate phases after 10 seconds of vortex mixing are separated by centrifuging in a desk-top clinical centrifuge for 5 minutes. One milliliter of ethyl acetate (upper phase) is transferred to 10 ml of Biosolv (3 liter toluene, 100 ml Beckman Biosolv, and 300 ml Beckman Fluoralloy) counting fluid, and the radioactivity is quantitated in a Beckman LS-233. Total f[$^3$H]Met release is 2.15 × (cpm-bgd) since only a 1.0-ml aliquot is counted and 70% of the released f[$^3$H]Met is extracted into ethyl acetate. We routinely report total f[$^3$H]Met released.

The rate of f[$^3$H]Met release directed by tetranucleotides is markedly stimulated (3–4-fold) by GTP, inhibited by GDPCP, and unaffected by GDP. The rate of this release directed by tetranucleotides is markedly inhibited in reactions with a potassium ion concentration above 60 m*M*. The rate of f[$^3$H]Met directed by polynucleotides is also stimulated by GTP (1.5–2.0-fold), but is not inhibited by potassium ions up to 150 m*M*. Assay of RF in column fractions (which contain variable potassium ion concentration) is most commonly determined with polynucleotides.

*Ethanol-Directed f[$^3$H]Met Release*

The hydrolysis of ribosomal bound fMet-tRNA with RF can occur in the absence of terminator codon recognition if reactions contain 20% ethanol, methanol, or acetone. The addition of ethanol stimulates ribosomal binding of RF and thus eliminates requirement for codon. Release of f[$^3$H]Met under these conditions is not stimulated by GTP and is partially inhibited by potassium ion concentrations above 120 m*M*. The f[$^3$H]Met release requires both ethanol and partially purified RF (fraction IV). Little release is observed with crude RF.

A typical reaction is incubated 20 minutes at 24° and contains in 0.05 ml: 20 m*M* Tris·HCl; 60 m*M* potassium chloride, 11 m*M* magnesium

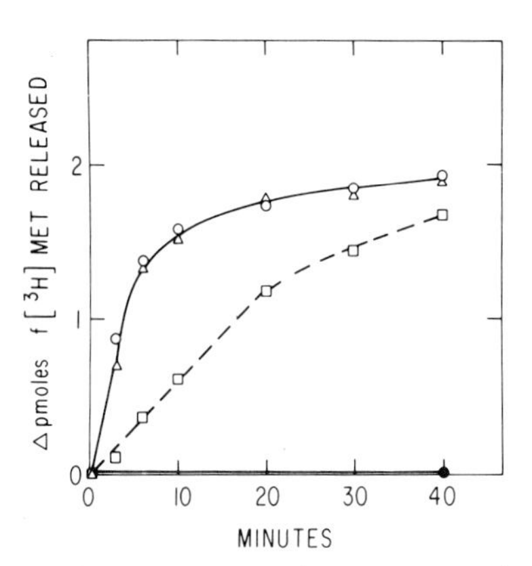


FIG. 1. Reticulocyte release factor activity. At the indicated time intervals, f[$^3$H]Met released from a 0.03-ml portion of a 0.25-ml reaction was determined. Each tube was incubated at 24° and contained: 15.25 pmoles of f[$^3$H]Met-tRNA$_f$· ribosome intermediates; 65 μg of reticulocyte RF (fraction V); 0.1 m*M* GTP; 1.0 $A_{260}$ unit poly($A_3$,U) (△), or 0.5 $A_{260}$ unit UAAA (○), or 20% ethanol (v/v) (□), or no template (●); and other components as described. The zero time value of 0.18 pmole was subtracted from all values.

chloride; 3.0 pmoles of f[$^3$H]Met-tRNA·ribosome; variable quantities of partially purified RF; and 20% ethanol. The quantity of f[$^3$H]Met released is determined as described above.

The relative rates of RF dependent f[$^3$H]Met release as directed by UAAA, poly(U,$A_3$), and 20% ethanol are shown in Fig. 1.

*Reticulocyte RF Binding to Ribosomes*

The ribosomal binding of *E. coli* RF1 (UAA or UAG) and RF2 (UAA or UGA) to *E. coli* ribosomes has been shown to occur with the indicated codon specificity.[12] This ribosomal binding is rapidly quantitated using radioactive terminator codons. The complex RF·UA[$^3$H]AA·ribosome (reaction 3) is quantitated by retention of the radioactively labeled UAAA on Millipore filters.[12] Two modifications of the procedure permitted its application to mammalian extracts. Mammalian RF binds to reticulocyte ribosomes with UA[$^3$H]AA but not with UA[$^3$H]A, and this RF binding is stabilized by GDPCP. The requirements for retention of the radioactive complex on Millipore filters is given in Table I.

A typical reaction is incubated at 4° for 15 minutes and contains in 0.05 ml; 1.8 $A_{260}$ units of reticulocyte ribosomes; 20.0 pmoles of UA[$^3$H]AA (10.0 Ci/mmole); 0.1 m*M* GDPCP; 50 m*M* Tris·acetate pH 7.4; 20 m*M* magnesium acetate; 0.1 *M* ammonium acetate; 20% ethanol;

[12] E. M. Scolnick and C. T. Caskey, *Proc. Nat. Acad. Sci. U.S.* **64**, 1235 (1969).

TABLE I
REQUIREMENTS FOR UA[$^3$H]AA BINDING WITH RETICULOCYTE RELEASE FACTOR (RF)[a]

| Condition | UA[$^3$H]AA bound (pmoles) |
|---|---|
| Complete | 2.06 |
| −RF | 0.07 |
| −Ribosomes | 0.02 |
| −Ethanol | 0.07 |
| −GDPCP | 0.40 |
| −GDPCP, +GTP | 0.85 |
| −GDPCP, +GDP | 0.43 |

[a] Requirements for UA[$^3$H]AA binding with reticulocyte RF; reactions contain as indicated 1.8 $A_{260}$ unit ribosomes, 15 μg of reticulocyte RF, 20% v/v ethanol, 0.1 m*M* guanine nucleotide, and 11 pmoles of UA[$^3$H]AA.

and purified RF (fraction IV). The reaction is terminated by addition of 0.5 ml of cold buffer containing 50 m*M* Tris·acetate pH 7.4; 20 m*M* magnesium acetate; 0.1 *M* ammonium acetate; and 10% ethanol. The radioactive complex is collected on a Millipore filter, washed with 10 ml of the above cold buffer, dried, and counted in 10 ml of POPOP counting fluid.

The methods employed to study the partial reactions of peptide chain termination described above have been useful in probing the intermediate events involved and the possible role of peptidyl transferase in this process.[3-5,13]

*Purification of Rabbit Reticulocyte Release Factor, and Ribosomes*

*Lysate.* New Zealand white rabbits (about 4–5 pounds) are injected intramuscularly with 0.1 mg of vitamin $B_{12}$ and 1 mg of folic acid in 0.9% NaCl pH 7.0. For six successive days they are injected subcutaneously with 2.5% phenylhydrazine (0.1 ml/lb), pH 7.0.[14] The rabbit's lateral ear vein is lacerated, and 50–70 ml of blood is obtained under a vacuum of 5–7 psi. The cells from the heparinized blood are collected by centrifuging at 10,000 *g* for 10 minutes at 4°, washed twice by suspending in 2 volumes of buffer (0.14 *M* NaCl, 50 m*M* KCl, and 5 m*M* $MgCl_2$) and collected by repeated centrifuging. The packed cells are lysed by adding 4 volumes of hypotonic buffer (2 m*M* $MgCl_2$, 1 m*M* dithiothreitol, 0.1 m*M* EDTA, pH 7.0). The suspension is stirred for several minutes at 4° and centrifuged at 15,000 *g* for 30 minutes at 4°. This supernatant fraction

[13] C. T. Caskey, A. L. Beaudet, E. M. Scolnick, and M. Rosman, *Proc. Nat. Acad. Sci. U.S.* **68**, 3163 (1971).

[14] J. M. Gilbert and W. F. Anderson, this series, Vol. 20, p. 542.

can be stored at $-170°$ without loss of RF activity. Typically a preparation of lysate from 20 rabbits is needed for RF isolation and purification.

*Ribosomes.* The rabbit reticulocyte lysate is centrifuged at 105,000 *g* for 3 hours. The ribosomal pellet is suspended in 0.1 of the original lysate volume in 50 m*M* Tris·HCl pH 7.4, 0.5 m*M* KCl, 30 m*M* $\beta$-mercaptoethanol, 2 m*M* $MgCl_2$, and 15% (v/v) glycerol by stirring 16 hours at 4°. The ribosomes are collected by centrifuging at 105,000 *g* for 3 hours at 4°. The ribosomal pellet is resuspended in 3 volumes of buffer (50 m*M* Tris·HCl pH 7.4, 0.25 *M* sucrose, 2 m*M* $MgCl_2$, 3 m*M* $\beta$-mercaptoethanol), and after a low speed centrifugation (30,000 *g* for 15 minutes) is stored at $-170°C$. This ribosomal preparation is used in the preparation of f[$^3$H]Met-tRNA·ribosome intermediates and is free of RF activity.

*RF Purification.* A preparation of rabbit reticulocyte lysate from 20 animals is used for the following purification scheme. The lysate is subjected to 105,000 *g* centrifugation for 3 hours at 4° to pellet the ribosomes. The RF activity in the supernatant of the lysate (fraction I) is partially purified by ammonium sulfate precipitation (42.4 g/100 ml) at 4° maintaining the pH at 7.5. The precipitate containing RF (fraction II) is collected by centrifuging (30,000 *g*, 15 minutes, at 4°) and dissolved in 50–100 ml of buffer A (0.1 *M* KCl, 20 m*M* Tris·HCl pH 7.8, 1 m*M* dithiothreitol, and 0.1 m*M* EDTA). Fraction II is dialyzed against 50 volumes of buffer A for 18 hours. Since fractions I and II hydrolyze fMet-tRNA independent of terminator codon, the RF activity determinations on these fractions are difficult and unreliable. Fraction II is applied to a column ($2.5 \times 50$ cm) packed with DEAE-Sephadex equilibrated with buffer A. A column of this size could accommodate 40 g of protein in fraction II. After sample application, the column is washed with buffer A (2.5 column volumes) followed by 1600–2000 ml of a solution containing all the components of buffer A and a linearly increasing concentration gradient of potassium chloride (0.1 *M* to 0.7 *M*) (Fig. 2A). Fractions of 10 ml are collected at a flow rate of 30–40 ml/hr. The RF activity of column fractions can be determined using UAAA, UAGA, UGAA, or poly(U,G,A) (Fig. 2A). Recently the reticulocyte RF has fractionated with apparent heterogeneity on DEAE-Sephadex using 2000 ml of solution A containing a linearly increasing concentration gradient of potassium chloride (0.1 to 0.5 *M*). The trailing shoulder of RF (Fig. 2A) can be more widely spread by such a gradient but has not been completely separated from the major RF peak at this time. The peak RF activities elute at 0.31 and 0.34 KCl under these conditions. Since the characterization of the later eluting RF activity is not complete at this time, the remaining purification will deal with the major RF fraction eluting at 0.31 *M* KCl. The fractions (fraction III) containing RF activity are pooled, concentrated by pressure filtration, and dialyzed against buffer B (50 m$M$ Tris·HCl pH 7.8, 0.1 $M$ KCl, 1 m$M$ DTT, and 0.1 m$M$ EDTA).

Fraction III can be further purified by use of phosphocellulose column chromatography. Fraction III is applied to a 30 × 1.5 cm (or 30 × 2.5 cm) column packed with phosphocellulose equilibrated in buffer B. After sample application, the column is washed with 2–3 column volumes

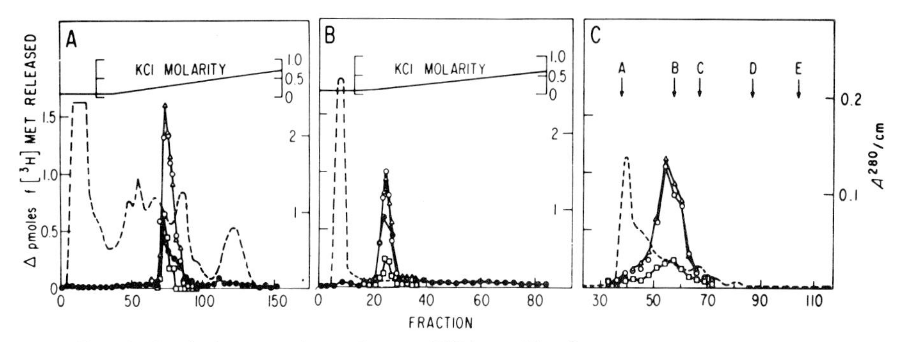


FIG. 2. Reticulocyte release factor (RF) purification.

(A) Fraction II (3.36 g of protein in 82 ml) was applied to a 43 × 2.5-cm DEAE-Sephadex column; washed with 230 ml of buffer A; and eluted with a KCl gradient (2000 ml). Fractions were 15.5 ml. All release reactions were 10 minutes, 24°, and contained: 0.2 $A_{260}$ unit poly(A,G,U) (●); 3.05 pmoles of f[$^3$H]Met-tRNA$_f$·ribosome intermediates; 0.1 m$M$ GTP; 0.015 ml of each column fraction, and other components described. Activity was determined with 0.1 $A_{260}$ unit UAAA (○), UGAA (△), or UAGA (□) under the identical conditions except that all fractions were dialyzed against buffer C prior to assay with 0.025 ml of each fraction. Values for f[$^3$H]Met release are stimulation by oligonucleotide. The peak $A_{260}$ value (fraction 12) is 26.5 $A_{260}$/cm. Note that the left-hand scale for f[$^3$H]Met release is constant in panels A, B, and C, while the right-hand scale for $A_{260}$/cm is 10-fold less in panel C than in A and B. - - -, $A_{280}$.

(B) Fraction III (300 mg of protein in 59 ml) was applied to a 33 × 2.5-cm phosphocellulose column, washed with 100 ml of buffer B, and eluted with a linear (1200 ml) KCl gradient. Fractions, 15.5 ml, were assayed as described in (A) except that 0.015 ml, rather than 0.025 ml, of each dialyzed fraction was assayed with tetranucleotide.

(C) Fraction IV (5.84 mg of protein in 4.0 ml) was applied to a 92 × 2.5-cm Sephadex G-200 column and eluted. Fractions were 3.8 ml. Release assays were incubated 10 minutes, 24° and contained: 2.92 pmoles of f[$^3$H]Met-tRNA$_f$·ribosome intermediates; 0.1 m$M$ GTP; where indicated 0.1 $A_{260}$ unit UAAA (○), UGAA (△), or UAGA (□); 0.025 ml of each fraction; and other components as indicated. The zero time value (0.24 pmole) is subtracted. Each purified protein marker was applied separately in 4.0 ml, and its position of elution was determined by absorbance at 280 nm or 416 nm (cytochrome $c$). Arrow A indicates dextran blue (void volume, 142.5 ml); arrow B, pyruvate kinase (217 ml, MW 237,000); arrow C, aldolase (252 ml, MW 158,000); arrow D, ovalbumin (327 ml, MW 43,000); and arrow E, cytochrome $c$ (392 ml, MW 12,523).

of buffer B followed by 400 ml (or 1200 ml if the larger column is used) of buffer B containing a linear concentration gradient of KCl (0.1–0.4 *M*) (Fig. 2B). The 5-ml fractions are eluted at a rate of 30 ml per hour. The RF activity is determined as described above and elutes at 0.21 *M* KCl (Fig. 2B). The active fractions are pooled and concentrated by pressure filtration (fraction IV).

Fraction IV can be further purified by several alternative methods. Sephadex G-200 column chromatography has proved most convenient and reliable. Fraction IV (4.0 ml) is applied to a 92 × 2.5 cm Sephadex G-200 column equilibrated with buffer C (20 m*M* Tris·HCl pH 7.8, 50 m*M* KCl, 1 m*M* DTT, and 0.1 *M* EDTA). The elution position of RF with respect to marker proteins is given in Fig. 2C. This elution position for RF would give an estimated molecular weight of 255,000. In separate studies where the G-200 chromatography was done as upward flow in 0.3 *M* KCl, the molecular weight estimate was 200,000. The fractions containing RF are pooled and concentrated (fraction V) by pressure filtration. Fraction V has no detectable EF1 or EF2 activity.[4] A summary of the results of the above purification procedures is given in Table II.

Two purification procedures, hydroxylapatite and sucrose gradient sedimentation, have been found successful with reticulocyte RF. These procedures are recommended when RF of high purity is needed. Use of the above procedures and judicious pooling of only the most active fractions can achieve purification of a homogeneous fraction, as judged by acrylamide gel analysis.

TABLE II
RETICULOCYTE RELEASE FACTOR (RF) PURIFICATION[a]

| Fraction | Protein (mg) | Specific activity[b] | Yield (%) |
|---|---|---|---|
| I S100 | 14,796 | — | — |
| II Ammonium sulfate | 1,822 | — | — |
| III DEAE-Sephadex | 294 | 13.9 | 100 |
| IV Phosphocellulose | 7.2 | 198 | 35 |
| V Sephadex G-200 | 1.2 | 572 | 17 |

[a] For calculation of specific activity, the rate of f[$^3$H]Met release was determined at 15 minutes, 24°, in reactions containing limiting levels of RF, 2.92 pmoles of f[$^3$H]Met-tRNA$_f$·ribosome intermediates, 0.2 $A_{260}$ unit of poly(A,G.U), 0.1 m*M* GTP, and other components as described. The percent yield is the cumulative recovery based on the activity of fraction III. Activity was calculated as the increase in f[$^3$H]Met release due to the addition of oligonucleotide. Activity determinations on fractions I and II are difficult because of the presence of a codon-independent fMet-tRNA hydrolase activity in these fractions.

[b] Picomoles of f[$^3$H]Met per milligram· in 15 minutes.

A purification of RF by hydroxylapatite column chromatography has been found similar to Sephadex G-200. The RF fraction is dialyzed against buffer D (10 m$M$ $KH_2PO_4$ pH 7.4, 1 m$M$ DTT, and 0.1 m$M$ EDTA) prior to application to a 10 × 2 cm column of hydroxylapatite equilibrated in buffer D. The column is washed successively with 4–5 column volumes

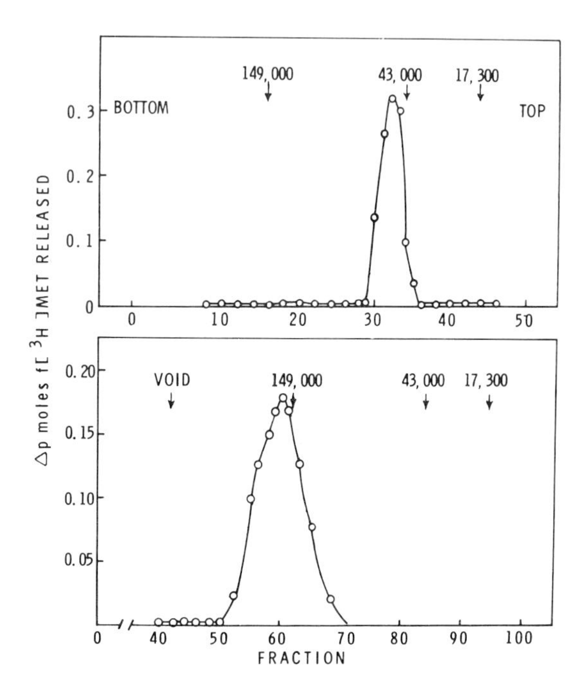


FIG. 3. (A) The major fraction of partially purified reticulocyte release factor (RF) (○—○) was sedimented through a 5 to 20% sucrose gradient (13 ml) in an SW 41 rotor of Beckman L265B ultracentrifuge at 39,000 rpm for 25 hours at 2°. Proteins, aldolase (149,000), ovalbumin (43,000), and myoglobin (17,800) were used as molecular weight markers (arrows). The sedimentation profiles have been superimposed in this diagram. Fractions were 0.25 ml. All release reactions were 15 minutes at 24° and contained: 0.1 $A_{260}$ units of UAAA, 3.00 pmoles of f[³H]met-$tRNA_f$·ribosome intermediates, 0.1 m$M$ GTP, 0.010 ml of each fraction, and buffer components as described.

(B) The major fraction of partially purified reticulocyte RF (4-ml sample) was chromatographed on a Sephadex G-200 column (100 × 2.5 cm) and eluted with 50 m$M$ Tris·HCl, 0.3 $M$ KCl, 1 m$M$ EDTA, and 0.1 m$M$ DTT. Fractions were 4 ml. Release assays were incubated 15 minutes at 24° and contained 0.1 $A_{260}$ units of UAAA, 3.0 pmoles of f[³H]Met-$tRNA_f$·ribosome intermediate; 0.1 m$M$ GTP, 0.010 ml of each fraction, and buffer components as described. Each purified protein marker was applied separately in 4 ml, and its position of elution was determined by absorbance at 280 nm or 410 nm (myoglobin). The void volume as determined from the elution volume of dextran blue is also shown.

of buffer D, buffer D containing 0.1 *M* $KH_2PO_4$, pH 7.4, and buffer D containing 0.2 *M* $KH_2PO_4$, pH 7.4. The RF activity elutes in the final buffer.

Reticulocyte RF can be purified by sucrose gradient sedimentation. A 0.5 ml fraction of RF is sedimented through a 5–20% sucrose gradient (13 ml) in an SW 41 rotor of Beckman L265B ultracentrifuge at 39,000 rpm for 25 hours at 4°. The RF activity of 0.25-ml fractions are given in Fig. 3 together with the sedimentation positions of proteins of known molecular weight. The estimated molecular weight of RF based on these studies is 54,000. This is a useful final purification step because of the strikingly different migration of RF (on the basis of molecular weight) on Sephadex G-200 and when sedimented in a sucrose gradient (Fig. 3A, 3B).

*Polyacrylamide Gel Analyses of Reticulocyte RF*

The composition of the highly purified major RF fraction (fraction V, sedimented through a 5–20% sucrose gradient as described—legend Fig. 3A) is examined by 0.1% sodium dodecyl sulfate-10% polyacrylamide gel electrophoresis as described by Laemmli.[15] After electrophoresis of the bromophenol dye to the end of an 11 cm gel (3 hours at 2 mA/gel), the gel was removed from the glass tube and stained with 0.2% Coomassie Brilliant Blue in 50% methanol, 7% acetic acid for 2 hours at 37°. The gels were destained with 7% acetic acid, 5% methanol over 30 minutes at 1 A in a quick gel destainer apparatus (Canalco). These studies suggest the major fraction of reticulocyte RF contains a single component of estimated molecular weight 54,000.

[15] U. K. Laemmli, *Nature* (*London*) **227**, 680 (1970).

# Section VI

# Ribosomes

# [46] Sedimentation Velocity Analysis in Accelerating Gradients

*By* RAYMOND KAEMPFER and MATTHEW MESELSON

Zone sedimentation is extensively employed for the separation and analysis of macromolecules and macromolecular complexes. In order to obtain stably sedimenting bands, preformed density gradients are usually used as supporting media. Sucrose and glycerol are the solutes most commonly chosen for the construction of such gradients. However, a shortcoming of these substances is that the viscosity of their solutions increases considerably with concentration. For example, in linear sucrose gradients as often employed, the concentration increase is accompanied by a viscosity rise so pronounced that the velocity of sedimenting particles actually decreases as they move down the gradient, even though the centrifugal acceleration to which they are subject increases. This deceleration, which clearly diminishes the separation between two sedimenting zones, can be avoided by the use of appropriately constructed exponential sucrose gradients in which sedimentation velocity is constant.[1]

Even better separation can be achieved with gradients described here, in which the viscosity *decreases* with increasing concentration. In these accelerating gradients, the combination of increasing centrifugal force and decreasing viscous drag allows a substantial increase in sedimentation velocity as a function of distance sedimented. The faster sedimenting particle is accelerated over the slower sedimenting one, leading to a separation greater than proportional to the difference in their sedimentation coefficients.

## Principle

*Viscosity of CsCl Solutions.* The relative viscosity of solutions of CsCl in $D_2O$ and $H_2O$ at 0.0° is shown in Fig. 1. In both solvents, the viscosity decreases markedly with increasing concentration up to a density of 1.5–1.6 $g/cm^{-3}$. An initial viscosity decrease also occurs at higher temperatures, but it is less pronounced.[2,3]

This unusual viscosity behavior of CsCl solutions allows the con-

[1] H. Noll, *Nature (London)* **215**, 360 (1967).
[2] P. A. Lyons and J. F. Riley, *J. Amer. Chem. Soc.* **76**, 5216 (1954).
[3] R. Bruner and J. Vinograd, *Biochim. Biophys. Acta* **108**, 18 (1965).



ISBN 0-12-504180-2

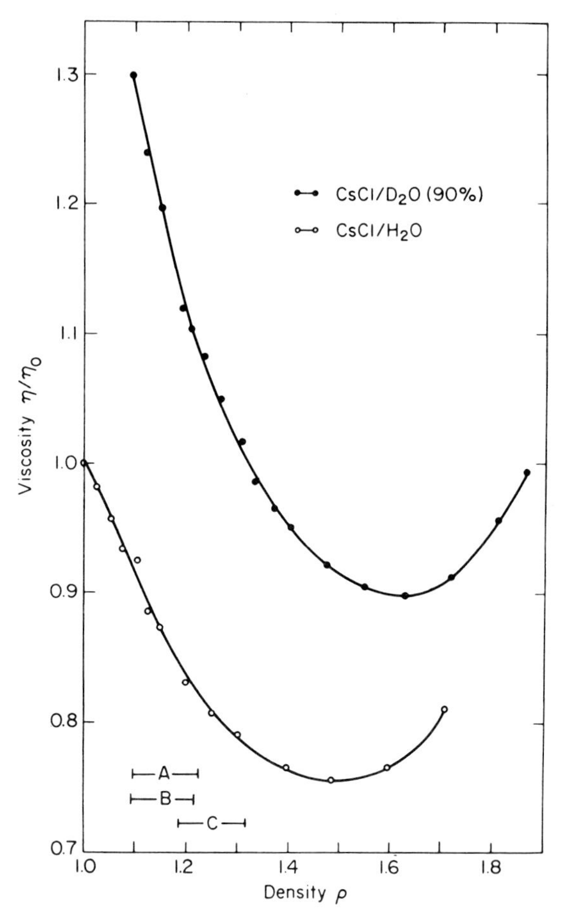


FIG. 1. Relative viscosities of cesium chloride solutions at 0°. The density of the solutions in $D_2O$ and $H_2O$ were determined by pycnometry. Viscosities relative to $H_2O$ ($\eta/\eta_0$) were determined at $0.0 \pm 0.1°$ in an Ostwald-Cannon-Fenske viscometer. The effect of kinetic energy loss is negligible.

struction of gradients in which particles experience less viscous drag the farther they sediment.

*Sedimentation Velocity.* The velocity $V$ of a sedimenting particle depends upon its distance $r$ from the axis of rotation according to the relation

$$V \sim \frac{(1 - \bar{v}\rho)\, r}{\eta} \tag{1}$$

where the density $\rho$ and the viscosity $\eta$ are themselves functions of $r$. The partial specific volume $\bar{v}$ may usually be taken to be independent of $r$.

In order to illustrate the characteristics of accelerating CsCl gradients, we have calculated values for the above expression for various exponential gradients at 0°, using experimentally determined values of $\rho$ and taking $\eta$ from the curves in Fig. 1. The partial specific volume used for these illustrative calculations is 0.591 $g^{-1}cm^3$, a value determined for *Escherichia coli* ribosomes and subunits in a dilute potassium chloride solution.[4] The results of the calculations are shown in Fig. 2. If the velocity of a particle at the top of the gradient ($r = 6$ cm) is taken as unity, the curves give the relative velocity at any point in the gradient. It can be seen that the calculated sedimentation velocity increases almost 2.5-fold across these gradients, both in $H_2O$ and $D_2O$ solutions. The composition of each gradient is shown in the figure and the density ranges are indicated in Fig. 1.

*Zone Separation.* The curves of Fig. 2 show that the relative sedimentation velocity is very nearly a linear function of sedimentation distance, allowing us to express the sedimentation velocity as

$$V = V_0 (1 + kL) \tag{2}$$

where $L$ is the distance sedimented expressed as a fraction of the total length of the gradient, $k$ is a constant, and $V_0$ is the initial velocity at $L = 0$. At time $t$, the distance sedimented, $L$, will be

$$L = \frac{e^{V_0 kt} - 1}{k} \tag{3}$$

In a constant velocity gradient, two particles with sedimentation coefficients $S_1$ and $S_2$ will travel distances $L_1$ and $L_2$ such that

$$\frac{L_1}{L_2} = \frac{S_1}{S_2} \tag{4}$$

In an accelerating gradient as considered above, however,

$$\frac{L_1}{L_2} = \frac{e^{S_1/S_2 \ln (1 + kL_2)} - 1}{kL_2} \tag{5}$$

This expression allows calculation of $L_1$ for any value of $L_2$. When the faster sedimenting particle (sedimentation coefficient $S_2$) reaches the bottom of the gradient, $L_2 = 1$. For that case, Eq. (5) becomes

$$\frac{L_1}{L_2} = L_1 = \frac{e^{S_1/S_2 \ln (1 + k)} - 1}{k} \tag{6}$$

[4] W. E. Hill, G. P. Rossetti, and K. E. van Holde, *J. Mol. Biol.* 44, 263 (1969).

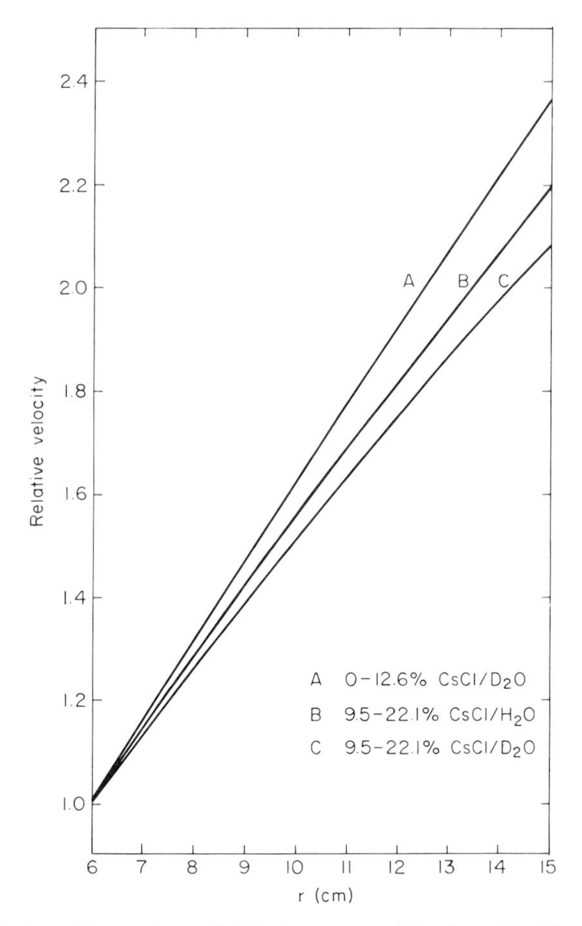


FIG. 2. Relative sedimentation velocities in exponential cesium chloride gradients at 0°. See text. Velocities were calculated using the dimensions of the International SB 283 rotor.

The values of $L_1/L_2$ versus $S_1/S_2$ for a 0–12.6% exponential $CsCl/D_2O$ gradient have been calculated using Eq. (6) and data from curve *A* in Fig. 2. The value of $k$ is taken from the curve as 1.36 at $r = 15$ cm. The results are shown in Fig. 3.

The ordinate in Fig. 3 represents the position $L_1$ of the slower sedimenting particle at the moment that the faster sedimenting particle has reached the bottom of the gradient ($L_2 = 1$). It may be seen that for the entire range of $S_1/S_2$, the separation between the two species is greater in the accelerating gradient than in the isokinetic gradient (broken line). It should be noted, however, that the separation increase

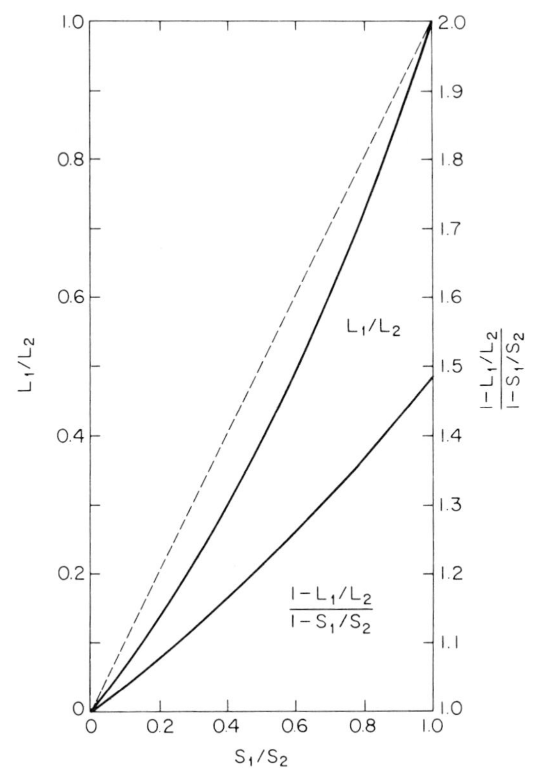


FIG. 3. Particle separation in exponential 0–12.6% $CsCl/D_2O$ gradients at 0°. The separation between particles with $\bar{v} = 0.591\ g^{-1}\ cm^3$, calculated as described in the text, is indicated by the line $L_1/L_2$ for the case that the faster species has reached the bottom of the gradient ($L_2 = 1$). For comparison, the separation in an isovelocity gradient is also shown (broken line). The value of $(1 - L_1/L_2)/(1 - S_1/S_2)$ represents the calculated increase in particle separation in a CsCl gradient relative to that in an isokinetic gradient.

relative to the isokinetic gradient, expressed as $(1 - L_1/L_2)/(1 - S_1/S_2)$, is greater, the closer $S_1$ approximates $S_2$. Thus, the accelerating gradient is most useful in separating particles possessing closely similar sedimentation coefficients.

## Procedure

*Preparation of the Gradients.* Exponential gradients are prepared in a simple gradient maker.[1] A burette containing the heavier starting solution (CsCl concentration $C_h$) is connected through a rubber stopper to a mixing vessel containing a volume $V_m$ of the lighter starting solution (concentration, $C_l$). A needle passing through the rubber stopper into

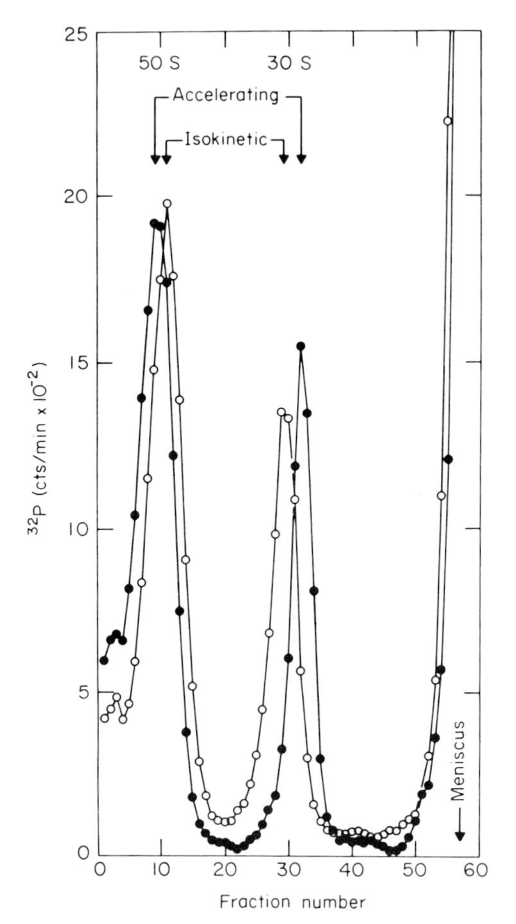


FIG. 4. Sedimentation distribution of 30 S and 50 S ribosomal subunits in accelerating and isokinetic gradients. A mixture of $^{32}P$-labeled *Escherichia coli* 30 S and 50 S ribosomal subunits was sedimented through a 12.5 ml 0–12.6% $CsCl/D_2O$ gradient and, separately, through a 12.5–19.6% exponential sucrose/$H_2O$ gradient. The gradients were identical in size. Both gradients contained Tris, $Mg^{2+}$, and gelatin as described in the text. The gradients were sedimented for 3.9 hours in an International SB 283 rotor at 41,000 rpm, at a temperature close to 0°. At this time, the $CsCl/D_2O$ gradient was fractionated, while centrifugation of the sucrose gradient was continued to a total of 6.2 hours. Fractions of constant volume were collected from the bottom of the tubes. The meniscus was delineated by a marker dye, added after centrifugation. Fractions were collected on filter paper squares which were dried and counted in a liquid scintillation spectrometer. The two gradients are plotted in the same figure to facilitate comparison. (●——●) $CsCl/D_2O$; (○——○) sucrose/$H_2O$.

this solution is connected on the outside to a length of thin tubing reaching to the bottom of a 14 ml centrifuge tube for the International SB283 or Spinco SW 40 Ti rotor. As the more concentrated solution is dripped into the mixing vessel, the gradient is formed upward from the bottom of the tube. During this process, the solution in the mixing vessel is stirred magnetically.

After $V$ ml has dripped from the burette, the concentration $C_v$ in the mixing vessel is

$$C_V = C_h - (C_h - C_l)\, e^{-V/V_m} \tag{7}$$

The 12.5 ml gradients A, B, and C are all produced using $V_m = 12.5$ ml.

The concentration $C_l$ and $C_h$ are 0% (w/w) and 20% for gradient A and 9.5% and 29.5% for gradients B and C. The gradient solutions for analysis of ribosomes contained 10 m$M$ Tris·HCl, pH 7.4, 10 m$M$ magnesium acetate, and 0.1 mg/ml gelatin. $D_2O$ remaining in the mixing vessel may be recovered by distillation.

The gradients are allowed to cool for 1–2 hours and centrifuged at 0° in an International SB 283 rotor at 41,000 rpm. A Spinco SW 40 Ti rotor can also be used.

EXAMPLE. The properties of one type of accelerating gradient are illustrated in Fig. 4. A mixture of 30 S and 50 S ribosomal subunits from *Escherichia coli* was sedimented in a 0–12.6% exponential CsCl/$D_2O$ gradient. In parallel, the same mixture was sedimented through a 12.5–19.6% exponential sucrose/$H_2O$ gradient. The latter gradient is isokinetic. Both gradients had identical dimensions and were fractionated into aliquots of identical volume. It can be seen that the separation between the 30 S and 50 S peaks in the CsCl/$D_2O$ gradient is 1.24–1.27 times greater than that in the isokinetic gradient. The 50 S particles have moved an approximate distance $L_2 = 0.89$ of the total gradient length. Using this value to calculate $L_1$ according to Eq. (5), the separation increase, $(1 - L_1/L_2)/(1 - S_1/S_2)$, for 30 S and 50 S particles ($S_1/S_2 = 0.60$) is 1.22. Thus, the value observed is in good agreement with the calculations given above. A larger separation increase is observed in cases where the ratio $S_1/S_2$ is higher than 0.60.

Other applications of accelerating gradients may be found in reference 5.

*Resolution in Accelerating Gradients.* It may be noted from Fig. 4 that the bandwidths of 30 S and 50 S ribosomal subunit peaks in a CsCl/$D_2O$ gradient are essentially identical to those in an isokinetic sucrose gradient. Therefore, the increase in separation represents an actual increase in resolution.

[5] R. Kaempfer, *Proc. Nat. Acad. Sci. U. S.* 61, 106 (1968).

# [47] Preparation of Ribosomal Subunits by Large-Scale Zonal Centrifugation

*By* Paul S. Sypherd and John W. Wireman

The increase in studies on ribosome structure and function has placed a large demand on purified ribosomal subunits. The introduction of large-scale zonal rotors was an important step in solving this problem, and the more recent application of hyperbolic sucrose gradients has increased the capacity up to 2 g of applied ribosomes while maintaining good resolution. We describe in this section a modification of the procedure of Eikenberry *et al.*[1] which permits easy loading and unloading of the zonal rotor with relatively simple equipment. The procedure is described for a Beckman Ti 15 rotor and ribosomes from *Escherichia coli.*

### *Reagents*

2000 ml TM-4 buffer ($10^{-2}$ *M* Tris·HCl, pH 7.4; $10^{-4}$ *M* $MgCl_2$)
800 ml 50% (w/v) sucrose in TM-4 buffer
700 ml 7.4% (w/v) sucrose in TM-4 buffer
200 ml 45% (w/v) sucrose in TM-4 buffer
2000 ml 60% (w/v) sucrose in TM-4 buffer
Ribosomes in TM-4 buffer (7000–30,000 $OD_{260}$ units, representing 0.5–2 g of ribosomes)

### *Equipment*

Beckman Ti 15 rotor
2-Chamber mixing device with 100-ml capacity
2-Chamber mixing device with 2000-ml capacity (see Fig. 1)
Peristaltic pump to deliver 1 liter/hour

### *Procedure*

1. All solutions and the rotor should be cooled to 5°. Completely fill the Ti 15 rotor with TM-4 buffer and secure the top. The rotor is brought to 3000–4000 rpm with the upper bearing and seal assembly in place.

2. The sample is applied in an inverse linear gradient. Fill the mixing

[1] E. F. Eikenberry, T. A. Bickle, R. R. Traut, and C. A. Price, *Eur. J. Biochem.* **12**, 113–116 (1970).



ISBN 0-12-504180-2

side of a simple gradient-mixing device with the 50 ml of the ribosome suspension and the other side with 50 ml of 5% sucrose in TM-4.

3. The sample is pumped into the rotor at the rotor edge with a peristaltic pump. The sample will layer beneath the buffer and displace buffer through the rotor center. Stop the pump just as the mixing chamber is emptied and with sample still filling the line. Avoid air in the input lines.

4. If a gradient-producing pump is not available, a simple 2-chamber gradient mixer can be used (Fig. 1). The mixing side should be filled with 700 ml of 7.4% sucrose in TM-4 buffer and sealed, and the second chamber filled with 800 ml of 50% sucrose in TM-4 buffer.

5. While mixing, connect the large mixing chamber to the tube, which now contains some sample solution, open the port connecting the 2 chambers and start the peristaltic pump. The gradient will be pumped beneath the sample and push the sample toward the center of the rotor. Buffer will be displaced from the rotor through the center vein. The volume of liquid in the mixing side of the chamber should remain constant. Pump 700 ml of gradient into the rotor, i.e., until 100 ml of 50% of sucrose remain in the outer chamber. The loading is then completed by pumping 200 ml of 45% sucrose through the outer vein as a cushion. The gradient pitch and position in the rotor is shown in Fig. 2. The gradient may conveniently be run for 5.5 hours at 35,000 rpm, 8 hours at 30,000 rpm, or 15 hours at 22,000 rpm for bacterial ribosomes. Shorter times will be necessary for the ribosomes of eukaryotes.

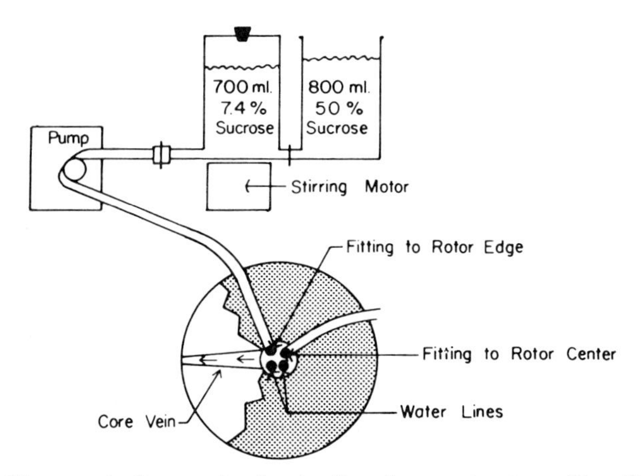


FIG. 1. Placement of apparatus for loading the zonal rotor with a 700-ml 7.4% to 34% hyperbolic gradient. The gradient mixer has a total capacity of 2 liters, and has the left side sealed after filling with sucrose. The gradient is pumped into the rotor with a peristaltic pump which delivers 1000–1500 ml per hour. Both the sample and the gradient are pumped into the rotor at the rotor edge.

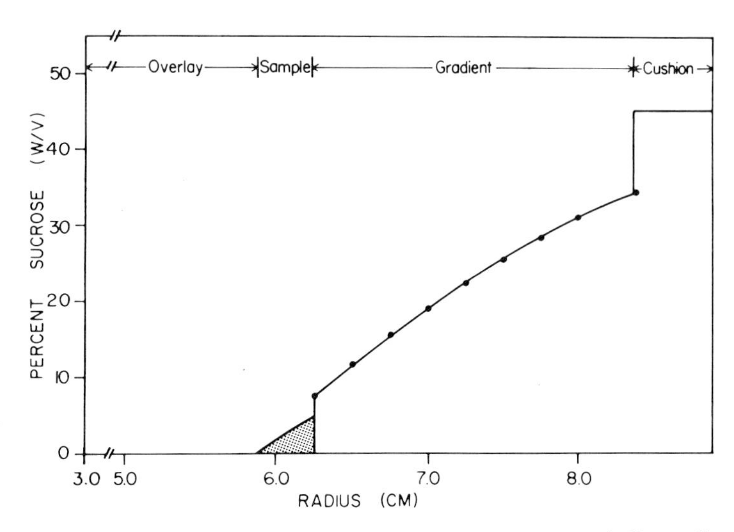


FIG. 2. The placement and shape of the 7.4% to 34% hyperbolic gradient. The sample is loaded as an inverse gradient with respect to ribosome concentration, in 0–5% sucrose concentration. The sample is denoted by the shaded area.

6. To recover the separated subunits, the rotor is slowed to 3000–4000 rpm and the upper bearing and seal assembly repositioned. Pump 60% sucrose in buffer into the tube leading to the rotor edge. The rotor contents will be displaced through the rotor center. The first 800 ml from the rotor may be discarded, and then 20-ml fractions collected until 60% sucrose emerges from the rotor (about 50 fractions). Although $OD_{260}$ is the most accurate wavelength to locate the subunit peaks, in practice $OD_{305}$ may be used ($OD_{260}/OD_{305}$ = 45 for ribosomes). A typical $OD_{305}$ profile is shown in Fig. 3.

7. The ribosomal subunits may be concentrated from the fractions in any of several ways. We have routinely recovered particles after first reducing the sucrose concentration by dialysis against buffer, followed by ammonium sulfate precipitation. After dialysis, add 1 *M* $MgSO_4$ to make the final concentration $10^{-2}$ *M*. Add solid $(NH_4)_2SO_4$ in the ratio of 60 g/100 ml of ribosome suspension. Allow the precipitate to form in the cold, then centrifuge for 20 minutes at 12,000 *g*. Resuspend the precipitate in TM-2 buffer ($10^{-2}$ *M* Tris·HCl, pH 7.4, $10^{-2}$ *M* $Mg^{2+}$), and dialyze into TM-2. The magnesium concentration must be maintained at 10 m*M* until the $(NH_4)_2SO_4$ is removed. Concentration of pooled fractions can also be done by low pH precipitation or by pressure filtration.

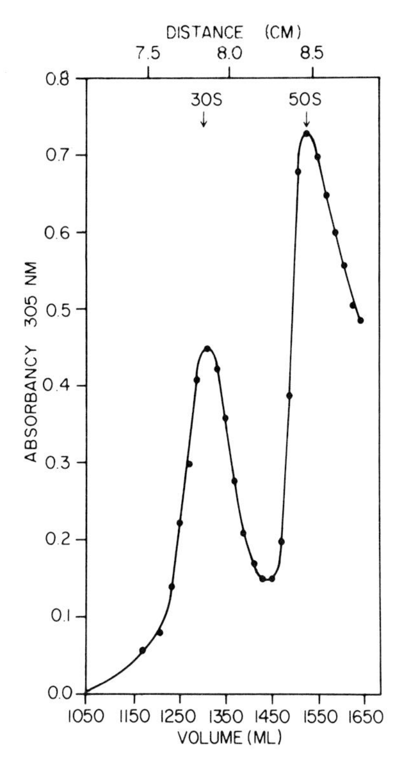


FIG. 3. Separation of *Escherichia coli* ribosomes. A total of 15,000 $A_{260}$ units (1 g) was run for 5.5 hours at 34,000 rpm in the Beckman Ti 15 rotor.

## *Comments*

In much of the earlier work in separating ribosomes on zonal rotors, a linear gradient was used. A gradient which is formed linearly will actually be concave, or exponential, as a function of sucrose concentration versus radius. The hyperbolic gradient is far superior in terms of resolution and capacity.[2] Eikenberry *et al.*[1] provided a useful discussion of some theoretical aspects of the hyperbolic zonal gradient, including the effects of increasing the length of the gradient and sample size. As they pointed out, the system they described is a compromise of several variables which produces the optimum separation. There are several advantages in placing the sample at the 6 cm position rather than at the rotor center. The sample at the 6 cm position occupies a narrow symmetrical band,

[2] A. S. Berman, *Nat. Cancer Inst. Monogr.* **21**, 41 (1966).

rather than a wide asymmetrical band adjacent to the rotor core. In addition, the running time, and therefore diffusion, is reduced by the higher centrifugal force at the 6 cm position. We have investigated several different configurations of the hyperbolic gradient, with no improvement in resolution over that shown in Fig. 3 for the 7.4%–34% gradient.

The procedure described here employs a simple and inexpensive device for forming the hyperbolic gradient. The gradient can be formed by a programmed gradient pump (e.g., Beckman Model 141) by following the manufacturer's recommendations (see, also Eikenberry *et al.*[1]). In earlier experiments we formed the gradient first, moved it to the center of the rotor with 45% sucrose, and then layered the ribosome gradient through the inside of the rotor, followed by the 800 ml buffer overlay. This necessitated pumping solutions through the outer, then inner channels of the rotor core. We have found that the rotating rulon seal must be fitted precisely to allow pumping into the center of the core against a rotor full of sucrose solution. The method outlined here permits complete loading and unloading by pumping only through the outer veins of the rotor core, and virtually eliminates sample loss. Loss of sample is a common experience when it is injected through the core center.

The ribosome separation shown in Fig. 3 is typical when the running time has been chosen correctly. Significantly longer times will result in driving the smaller subunits into the area occupied by the larger ones and reduce the purity of both. This is easily diagnosed by the loss of good separation of the peaks. It is not uncommon to find on the heavy side of the 50 S particles a sharp peak, which consists of 50 S ribosomes that have piled up at the interface between the end of the gradient and the 45% sucrose cushion.

Separations like that shown in Fig. 3 yield 30 S ribosomes of high purity (i.e., greater than 95%), as determined by sedimentation analysis, polyuridylic acid-directed polyphenylalanine synthesis, and disc gel analysis of proteins. The method of ribosome preparation frequently affects the purity of the 50 S particles. We have found that in some cases ribosomes prepared by ammonium sulfate precipitation and separated in TM-4 buffer yield 50 S preparations which appear homogeneous by sedimentation analysis, but which have appreciable 30 S backgrounds in protein-synthesizing systems. This apparently is due to the presence of 30 S aggregates, probably dimers, in the 50 S peak. The aggregate can easily be seen by analysis in acrylamide-agarose gels. This problem may be eliminated by using TMK buffer ($10^{-2}$ *M* Tris, $10^{-1}$ *M* KCl, $10^{-3}$ *M* $Mg^{2+}$) in the centrifugation procedures instead of TM-4. When separations are performed in sucrose gradients which contain salt, shorter running times are required. This is probably due to the ribosomes assuming a more compact configuration, and consequently sedimenting at a faster rate.

# [48] A Gel Electrophoretic Separation of Bacterial Ribosomal Subunits with and without Protein S1

*By* ALBERT E. DAHLBERG

Gel electrophoresis is a simple and rapid method for the characterization of bacterial polyribosomes and ribosomal subunits.[1,2] The fine resolution of this method has permitted the separation of 30 S subunits differing by a single ribosomal protein, S1,[3,4] which is thought to have a significant role in the initiation of protein synthesis. This separation may be of special use to investigators in this area and is therefore described here in some detail.

The content of S1 in preparations of ribosomal proteins can be determined by two-dimensional protein gel electrophoresis,[5] but quantitation is not simple; 30 S subunits, as often prepared, are deficient in S1, since S1 is easily removed from ribosomes. The S1 content of a ribosome preparation may be determined quite readily with a small amount of sample by the gel electrophoretic method described below, in which S1-deficient subunits (F-30 S) migrate significantly faster in the gel than S1-containing subunits (S-30 S) (see slot 1 of Fig. 1).

## *Gel Electrophoresis Conditions*

Several different gel conditions can be used to separate S-30 S and F-30 S subunits.[3,4] In our laboratory we routinely use a slab gel system (E. C. Apparatus Corp., St. Petersburg, Florida) in which 8 or 16 samples may be analyzed simultaneously, requiring about 0.05 to 0.1 OD unit of 30 S subunits per sample. The subunits are stained with Stains-all.[1] Maximum separation is obtained in a composite gel containing 3% acrylamide and 0.5% agarose, with a 90 m*M* Tris·HCl, 2.5 m*M* EDTA, 90

[1] A. E. Dahlberg, C. W. Dingman, and A. C. Peacock, *J. Mol. Biol.* **41,** 139 (1969).
[2] A. Talens, O. P. Van Diggelen, M. Brongers, L. M. Popa, and L. Bosch, *Eur. J. Biochem.* **37,** 121 (1973).
[3] A. E. Dahlberg, *J. Biol. Chem.* **249,** 7673 (1974).
[4] W. Szer and S. Leffler, *Proc. Natl. Acad. Sci. U.S.A.* **71,** 3611 (1974).
[5] E. Kaltschmidt and H. G. Wittmann, Ann. Biochem. **36,** 401 (1970).



ISBN 0-12-504180-2

m$M$ boric acid, pH 8.3, buffer,[6] and electrophoresis at 0°, 200 V for 6–10 hr as in Fig. 1.[7] Although the EDTA in the buffer unfolds the subunits,[8] S1 remains associated with the 30 S subunits causing them to migrate as S-30 S. S1 also remains associated with 30 S subunits in a gel containing 0.1 m$M$ $MgCl_2$ where the subunit is less unfolded (as shown in Fig. 2, slots 5–8). In both buffer systems all ribosomes (polyribosomes and 70 S ribosomes) are dissociated into subunits and migrate together with free subunits. Polyribosomes and 70 S ribosomes are electrophoresed intact

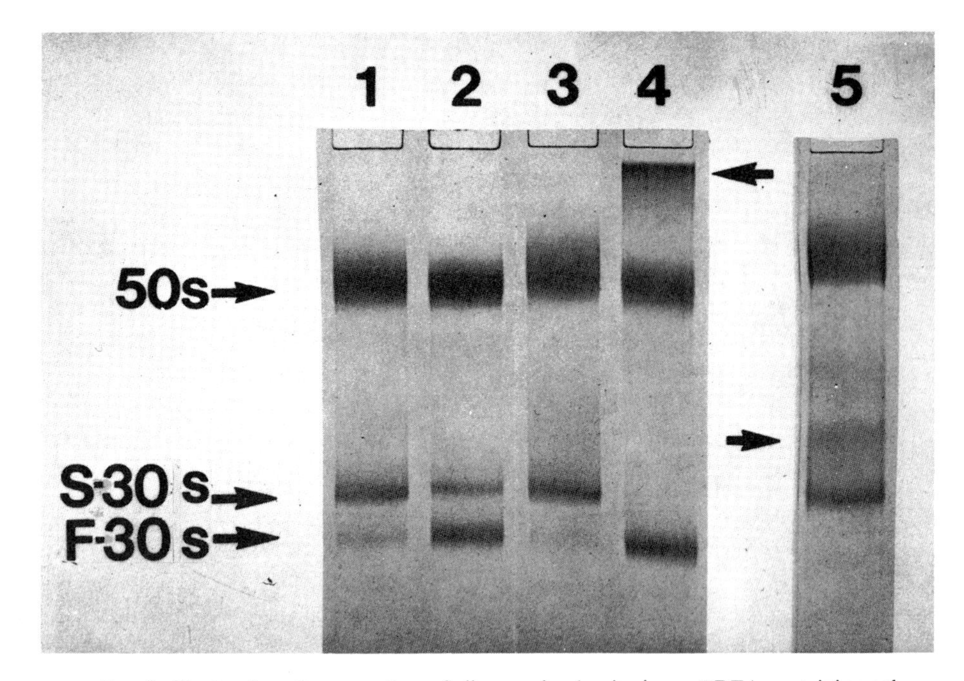


FIG. 1. Electrophoretic separation of ribosomal subunits in an EDTA-containing gel. Ribosomes were prepared from slow-cooled cells (slot 1), subsequently washed in high salt (slot 2), incubated with a 2-fold excess of ribosomal protein S1 for 5 min at 0° (slot 3), and subsequently incubated with poly(U) for 1 min at 0° (slot 4) as described in the text. The ribosomes in slot 5 received a 4-fold excess of ribosomal protein S1. The samples were subjected to electrophoresis in a composite gel containing 3% acrylamide, 0.5% agarose, and a Tris-EDTA-borate pH 8.3 buffer for 6 hr at 200 V and 0°.

[6] A. C. Peacock and C. W. Dingman, *Biochemistry* **7**, 668 (1968).

[7] An excellent, detailed description of the preparation and electrophoresis of composite gels is available from Miss Sylvia L. Bunting and Dr. A. C. Peacock, N.C.I., N.I.H., Bethesda, Maryland 20014.

[8] A. E. Dahlberg, F. Horodyski, and P. B. Keller, *Antimicrob. Agents Chemother.* **13**, 331 (1978)

and still achieve separation of free S-30 S and F-30 S subunits using a buffer containing 2 m$M$ $MgCl_2$, 6 m$M$ KCl, and 25 m$M$ Tris·CHl, pH 7.5.[3] Composite gels with a lower acrylamide concentration (2.5%) give less separation of S-30 S and F-30 S subunits but require a shorter time of electrophoresis (see Fig. 2, slots 1–4). At a higher (4%) acrylamide concentration agarose may be omitted and good separation achieved, but the gel is less stable and more difficult to handle.

*Analysis of Ribosome Samples*

Gel electrophoretic analysis of ribosome samples containing different proportions of the S-30 S and F-30 S subunits is shown in Fig. 1. Slot 1 contains subunits from 0.2 OD unit of washed 70 S ribosomes prepared from a sonicated lysate of slowly cooled *Escherichia coli* cells by pelleting (60,000 rpm for 1 hr at 3° in the 65 rotor in an L2-65B Beckman ultracentrifuge) and resuspension in 25 m$M$ Tris·HCl, 60 m$M$ KCl, 10 m$M$ $MgCl_2$ pH 7.5 buffer. Most of the 30 S subunits contain S1 and migrate as S-30 S in slot 1. A "high salt" wash of the ribosomes (in 1 $M$ $NH_4Cl$, 20 m$M$ $MgCl_2$, 25 m$M$ Tris·HCl pH 7.5 buffer) prior to electrophoresis in slot 2 removed most of the S1 from the 30 S subunits giving mainly F-

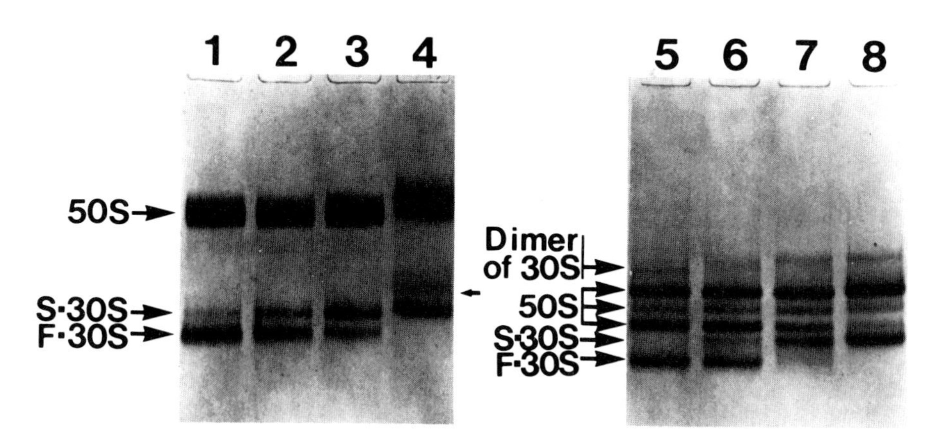


FIG. 2. Effect on S1 on ribosomal subunits electrophoresed in two different buffer systems. Salt-washed 70 S ribosomes, as in Fig. 1, slot 2, were incubated with protein S1 at 0° for 10 min at the following concentrations: no S1 (slots 1 and 5), 0.2 molar equivalent S1 (slots 2 and 6); 0.6 molar equivalent S1 (slots 3 and 7); 2 molar equivalents (slots 4 and 8). Both gels contained 2.5% acrylamide, 0.5% agarose. Gel on left (slots 1–4) contained Tris-EDTA-borate pH 8.3 buffer and was electrophoresed for 4 hr at 200 V and 0°. Gel on right (slots 5–8) contained 0.1 m$M$ $MgCl_2$, 25 m$M$ Tris·HCl pH 8.0 buffer and was electrophoresed for 3 hr at 200 V and 0° with constant recirculation of buffer. Buffer was changed after 1.5 hr of electrophoresis.

30 S in slot 2. The addition of protein S1 (approximately 2-fold excess) to the washed ribosomes for 5 min at 0° prior to electrophoresis in slot 3 converted F-30 S to S-30 S. The subsequent addition of poly(U) (0.5 $\mu$g) to the S1-containing, washed ribosomes of slot 3 for 1 min at 0° yielded the F-30 S form once again (slot 4). It has been proposed[3] that the S1, initially bound only to the 30 S subunit, forms a stronger association with the poly(U), and then dissociates from the 30 S subunit and remains associated preferentially with the poly(U) when the subunit and poly(U) are dissociated by gel electrophoresis in EDTA [see arrow for poly(U), slot 4]. The presence of mRNA in any ribosome sample consistently yields F-30 S in an EDTA-containing gel, regardless of whether the mRNA was added *in vitro* [as poly(U)] or was present *in vivo* (on polyribosomes). Caution is thus warranted when using the gel method to determine the S1 content of a ribosome preparation if the possibility exists of mRNA contamination.

### *S1 Binding Sites on the 30 S Subunit*

There are two binding sites for S1 on 30 S subunits[9,10] although most probably there is only one copy of S1 per 30 S subunit *in vivo*.[11] Evidence for two sites was first seen by the appearance of an even slower migrating band of 30 S subunits than S-30 S upon addition of a molar excess S1 to ribosomes (see arrows in Fig. 1, slot 5 and Fig. 2, slot 4). As with S-30 S, the addition of poly(U) converts this slower band to F-30 S (data not shown). The significance of the two different sites is not yet determined. It is known that one site involves the binding of S1 to the 3′ end of the 16 S rRNA. It is possible to remove this S1 binding site from the subunit by cleavage of a 49-nucleotide fragment from the 3′ terminus by colicin E3. The S1 molecule bound to the RNA fragment can then be separated from the ribosomal subunit and isolated.[12] The colicin E3-treated subunit is now able to bind only one S1 molecule; the double S1 form seen in Fig. 1, slot 5 and Fig. 2, slot 4 is not observed with these subunits.[9]

### *S1 Binding to the 50 S Subunit*

Evidence that S1 also binds the 50 S ribosomal subunits is seen in Fig. 2. The mobility of the 50 S subunit is retarded when S1 is added in 2-fold excess to 70 S ribosomes (slot 4). This effect is not apparent at

[9] M. Laughrea and P. B. Moore, *J. Mol. Biol.* **112**, 399 (1977).

[10] A. E. Dahlberg, unpublished results.

[11] J. Van Duin and P. H. Van Knippenberg, *J. Mol. Biol.* **84,** 185 (1974).

[12] A. E. Dahlberg and J. E. Dahlberg, *Proc. Natl. Acad. Sci. U.S.A.* **72,** 2940 (1975).

lower concentrations of S1 (slots 2 and 3), where S1 binding to F-30 S is favored. The binding of S1 to 50 S subunits is demonstrated even more dramatically in the gel containing 0.1 m$M$ $MgCl_2$ (slots 5-8), where the 50 S subunits are separated into three distinct bands. This gel shows that addition of excess S1 converts the fastest migrating 50 S form to the slowest migrating 50 S form in a way analogous to the conversion of F-30 S to S-30 S. The precise stoichiometry of this reaction and the way in which binding of S1 to both 30 S and 50 S subunits brings about a conformational rearrangement in the 70 S ribosomal structure is under investigation.[13]

[13] A. E. Dahlberg and A. Wahba, unpublished results.

# [49] Reconstitution of Ribosomes from Subribosomal Components

By P. Traub, S. Mizushima, C. V. Lowry, and M. Nomura

Several different approaches can be taken to elucidate the function of the molecular components of ribosomes. One is through chemical modification of ribosomes and attempts to correlate the resulting functional alterations with the chemical alteration of component molecules. Another approach is genetic. Identification of the altered components of mutant ribosomes may give information on the role of the normal components. A third approach makes use of reconstitution techniques.[1-5] It is useful in analyzing the functional requirements for each ribosomal component in a systematic way. It is especially powerful in combination with the two former approaches. For example, it provides an excellent means of identifying mutationally or physiologically altered ribosomal components.[6-10]

[1] K. Hosokawa, R. Fujimura, and M. Nomura, *Proc. Nat. Acad. Sci. U. S.* **55**, 198 (1966).
[2] T. Staehelin and M. Meselson, *J. Mol. Biol.* **16**, 245 (1966).
[3] P. Traub and M. Nomura, *Proc. Nat. Acad. Sci. U. S.* **59**, 777 (1968).
[4] M. Nomura, P. Traub, and H. Bechmann, *Nature* (London) **219**, 793 (1968).
[5] M. Nomura, S. Mizushima, M. Ozaki, P. Traub, and C. Lowry, *Cold Spring Harbor Symp. Quant. Biol.* **34**, 49 (1969).
[6] P. Traub and M. Nomura, *Science* **160**, 198 (1968).
[7] M. Ozaki, S. Mizushima, and M. Nomura, *Nature (London)* **222**, 333 (1969).
[8] A. Bollen, J. Davies, M. Ozaki, and S. Mizushima, *Science* **165**, 85 (1969).
[9] E. A. Birge and C. G. Kurland, *Science* **166**, 1282 (1969).
[10] J. Konisky and M. Nomura, *J. Mol. Biol.* **26**, 181 (1967).



ISBN 0-12-504180-2

Besides several techniques for the isolation of *Escherichia coli* ribosomes, this chapter describes methods for the dissociation of ribosomal subunits into their macromolecular constituents and for the reconstitution of ribosomal particles from these constituents.

## Isolation of 70 S Ribosomes and Their Subunits

### Growth of Cells

*Escherichia coli* strain Q13,[11] which lacks RNase I and has a defect in polynucleotide phosphorylase, is grown with forced aeration at 37° in Tryptone broth (1.3% Tryptone, 0.7% NaCl). The cells are harvested in the exponential phase of growth, at a density of 5 to $9 \times 10^8$ cells/ml. In order to arrest further cellular metabolism, the cell suspension is poured over crushed ice and immediately centrifuged in a Sharples continuous flow centrifuge. The cell paste is frozen without further washing and stored at −70°C.

### Preparation of Cell Extracts

*Solution* I: 10 m*M* Tris·HCl, pH 7.5 (20°), 30 m*M* $NH_4Cl$, 10 m*M* $MgCl_2$, 6 m*M* β-mercaptoethanol

*Grinding with Alumina*

If no automatically operating device for cell breakage is available, or if only a small amount of ribosomes is needed, the cells are disrupted by grinding them with a mortar and pestle for several minutes with 2.5 parts (w/w) of alumina powder. The process is considered complete when the paste gets noticeably thinner. The broken cells are extracted with 4 volumes of Solution I. In order to reduce the viscosity of the extract, RNase-free deoxyribonuclease is added to a final concentration of 2 μg/ml. Alumina and coarse cell debris are removed by centrifugation for 10 minutes at 18,000 rpm (39,000 *g*) in the SS34 rotor of the Servall refrigerated centrifuge. Fine cell debris are sedimented by spinning the supernatant at 30,000 rpm for 35 minutes in a Spinco fixed-angle rotor 30. (Temperature of centrifugation is always about 4° unless otherwise stated.) The pellets are discarded.

*Shaking with Glass Beads*[12]

One hundred grams of frozen cell paste is thawed, suspended in

[11] R. F. Gesteland, *J. Mol. Biol.* 16, 67 (1966).
[12] W. Doerfler, W. Zillig, E. Fuchs, and M. Albers, *Hoppe-Seyler's Z. Physiol. Chem.* **330**, 96 (1962).

200 ml of Solution I, mixed with 600 ml of dry glass beads (0.1 mm diameter, Schueler Company, New York) and shaken in a water-cooled cell mill (Edmund Bühler Company, Tübingen, Germany) for 5–10 minutes at high frequency (3700 rpm). Glass beads and cell homogenate are separated by suction on a coarse sintered-glass funnel, and the dry glass beads are washed 3 times with 100 ml of Solution I. Homogenate and wash solutions are combined and centrifuged at 30,000 rpm for 35 minutes in a Spinco fixed-angle rotor 30. The sediment is discarded. No DNase is used in this procedure in the authors' laboratory.

## Isolation and Purification of 70 S Ribosomes

*Solution*: 30% (w/v) sucrose in Solution I

In order to separate ribosomes from other cellular constituents, the crude extract, free of cell debris, is centrifuged at 30,000 rpm for 6 hours in a Spinco fixed-angle rotor 30. The ribosome free supernatant is decanted and saved for preparation of a partially purified enzyme fraction (see below). When the glass beads shaking method is used to prepare extracts, a grayish jelly like material containing DNA is present on top of the ribosome pellets. This is carefully removed with a spatula. The ribosomal pellets from 100 g wet weight of cells are dissolved in approximately 100 ml of Solution I. The resuspension is facilitated by use of a Teflon homogenizer to disperse the pellet in the centrifuge tube. Insoluble material is sedimented by centrifugation at 18,000 rpm for 10 minutes in the SS34 rotor of the Servall refrigerated centrifuge. Ten-milliliter portions of the ribosome solution are layered on top of 30% sucrose in Solution I and centrifuged for 18 hours at 30,000 rpm in a Spinco fixed-angle rotor 30. The pellets are resuspended and the purification step is repeated. The pelleted material is again dissolved in Solution I, the resulting solution freed from slight turbidity by low-speed centrifugation, and the volume is adjusted to approximately 80 ml by the addition of Solution I.

## Separation of 30 S and 50 S Ribosomal Subunits

*Solution* II: 10 m*M* Tris·HCl, pH 7.5 (20°C), 30 m*M* $NH_4Cl$, 0.3 m*M* $MgCl_2$, 6 m*M* β-mercaptoethanol

The dissociation of 70 S ribosomes into 30 S and 50 S subunits is accomplished by dialysis against Solution II (containing 5% sucrose when differential centrifugation is intended). The subunits can be separated by either differential or sucrose gradient centrifugation.

[13] C. G. Kurland, *J. Mol. Biol.* **18**, 90 (1966).

For large-scale separation, zonal centrifugation using a special rotor, such as Spinco zonal rotor BIV, may be useful.[14] Without such special equipment, the following methods are useful for large-scale separation.

*Differential Centrifugation*

A mixture of ribosomal subunits (approximately 30 mg/ml) in a modified Solution II containing 5% sucrose and 0.6 m*M* $Mg^{2+}$ is centrifuged at 15,000 rpm for 10 minutes to remove insoluble materials. The supernatant is then centrifuged at 40,000 rpm for 5.5 hours at 3° in a Spinco SW 40 Ti rotor. That partial separation of the ribosomal particles actually has taken place is easily detected in a dark cold room with a light beam entering the centrifuge tube at an angle of 90° relative to the observer. Because of the Tyndall scattering of the ribosomal particles, 4 areas can be observed from the top to the bottom of the centrifuge tube: a dark field free of ribosomes; an area of 30 S subunits; an optically denser field containing 30 S and 50 S ribosomes; and a clear pellet also consisting of both ribosomal subunits. The boundaries between the different fields are marked by an abrupt change in the refractive index. After removal of the ribosome-free top solution by means of a syringe equipped with a long, fine, blunt hypodermic needle, the layer containing pure 30 S ribosomes is carefully taken without disturbing the boundary of the lower layer. Then the centrifuge tubes are filled with the modified Solution II and the pellets are redissolved with the aid of a homogenizer. Under optimal conditions, each one of the following centrifugation steps will yield pure 30 S particles in a yield of 30–35% of those present in the centrifuge tube. Theoretically, a little less than 40% of the 30 S are separated from 50 S if the 50 S are all just sedimented at the end of each run. In order to approach the optimum time of centrifugation, the temperature must be taken into account as well as the changing concentration of ribosomes after each successive run. As a rule the first two runs are 5.5 hours and the second two are 5 hours. After the four centrifugations, 60–80% of the 30 S ribosomes are recovered practically free of 50 S. The enriched 50 S ribosomes are further purified by sucrose gradient centrifugation; employing a Spinco SW 25.1 or SW 27 rotor (see below). The $Mg^{2+}$ ion cencentration of the 30 S ribosome solution is adjusted to 10 m*M* by the addition of 1 *M* $MgCl_2$. After centrifugation for 4.5 hours at 50,000 rpm in a Spinco fixed-angle rotor 50.1, the sedimented 30 S ribosomes are redissolved in Solution I. In general pelleted 30 S ribosomes are difficult to resuspend in Solution I, presumably because of

[14] N. G. Anderson and C. L. Burger, *Science* 136, 646 (1962).

their tendency to aggregate in high $Mg^{2+}$. For this reason it is convenient first to dissolve the pellet in Solution II and then raise the $Mg^{2+}$ concentration to that of Solution I. The ribosome solution is clarified by low speed centrifugation, adjusted to a concentration of 500 $OD_{260}$ units/ml, divided into aliquots, frozen in dry ice–acetone, and stored at −70°.

*Sucrose Gradient Centrifugation*

*Solutions*

10% (w/v) sucrose in Solution II
30% (w/v) sucrose in Solution II

The sucrose solutions are freed from contaminating ribonucleases by treatment with bentonite. All ingredients, except $MgCl_2$, are dissolved in water, stirred for 1 hour with 2% (w/v) bentonite and centrifuged at 10,000 rpm (16,000 *g*) for 1 hour in a GSA rotor of the Servall RC2 centrifuge. The volume of the supernatant is adjusted by the addition of water and the sucrose solution is supplemented with $MgCl_2$. It is stored at 4°.

For large-scale preparations, fixed angle rotors such as an angle rotor 30 are preferable to conventional swinging-bucket rotors like the SW 25.1. Four milliliters of a ribosome mixture in Solution II, with a concentration of about 600 $OD_{260}$ units/ml, are carefully layered, through the center hole of the Spinco cap, on top of each 32 ml of 10 to 30% linear sucrose gradient in tubes for an angle rotor 30 and centrifuged for 12 hours at 30,000 rpm. Because of the Tyndall effect of the ribosomal particles, two bands consisting mainly of 30 S and 50 S ribosomes, respectively, are easily recognizable in a dark cold room with laterally incident light. A long, blunt hypodermic needle attached to a 10-ml syringe is inserted into the centrifuge tube through the center hole of the cap. The empty top solution is taken and discarded. Next, the zone containing 30 S material is removed and saved. The centrifuge tubes are then filled with Solution II and centrifuged for 12 hours at 30,000 rpm. The pellets containing 50 S ribosomes and some 30 S ribosomes are redissolved in Solution II (4 ml per tube) and rerun on 10 to 30% sucrose gradients. The fractionation of the sucrose gradients and the concentration of the enriched 50 S is carried out as before. The 30 S ribosome solution, which is virtually free of 50 S particles, is adjusted to 10 m*M* $Mg^{2+}$ by the addition of 1 *M* $MgCl_2$ and concentrated by a 7-hour centrifugation at 50,000 rpm in a Spinco fixed-angle rotor 50.1. The sedimented ribosomes are dissolved in Solution I and their concentration, after removal of insoluble material, is adjusted to

500 $OD_{260}$ units/ml. Aliquots are frozen in dry ice–acetone and stored at −70°. For most purposes, especially for the preparation of 30 S ribosomal protein, 30 S ribosomes obtained in this way are sufficiently pure.

Because of the characteristic distribution of components in sucrose gradients, the 50 S ribosome fraction obtained as above is usually heavily contaminated with 30 S particles. For this reason the 50 S ribosome preparation obtained as described above has to be purified by two additional cycles of sucrose gradient centrifugation. Two-milliliter portions of the 50 S ribosome preparation, containing 300 $OD_{260}$ units/ml Solution II, are layered on top of 26-ml 10 to 30% linear sucrose gradients and centrifuged for 11 hours at 25,000 rpm (63,600 *g*) in a Spinco SW 25.1 rotor. The fractionation of the gradients is carried out either as described above by visually locating the 50 S band or by immersing a needle from the top to the bottom of the gradient and passing the content of the centrifuge tube under constant pressure slowly through a 1 mm LKB flow-through cell. The separation is followed and recorded by a combination of a Beckman spectrophotometer and a Gilford multiple-sample absorbance recorder at 290 mμ. Fractions containing 50 S particles are pooled, and the particles (in Solution II) are concentrated at 50,000 rpm (190,000 *g*) for 6 hours in a Spinco fixed-angle rotor 50.1. After a second cycle of sucrose gradient centrifugation, the pelleted 50 S particles are dissolved in TMA I. Insoluble material is removed by low speed centrifugation. The ribosome concentration is adjusted to 500 $OD_{260}$ units/ml and the solution divided into aliquots, frozen in dry ice–acetone, and stored at −70°.

For several purposes, especially when the ribosomal particles are to be analyzed for their functional properties, the 30 S ribosomes must be purified by an additional sucrose gradient centrifugation to remove small amounts of contaminating 50 S particles. The final 30 S ribosome pellet of the crude separation outlined above is dissolved in Solution II and the concentration adjusted to approximately 300 $OD_{260}$ units/ml. Two milliliters of this solution are layered on each 26-ml 10 to 30% sucrose gradient and centrifuged for 12 hours at 25,000 rpm in a Spinco SW 25.1 rotor. The gradients are fractionated and analyzed utilizing the continuous flow technique with automatic recording as described above. Fractions containing 30 S ribosomes are pooled, and the particles are collected by centrifugation at 50,000 rpm (190,000 *g*) for 9 hours in a Spinco fixed-angle rotor 50.1. The sediment is resuspended in Solution I, and the solution is clarified by low speed centrifugation. Aliquots containing 500 $OD_{260}$ units/ml are frozen in dry ice–acetone and stored at −70°.

## Partial Reconstitution of 30 S and 50 S Ribosomal Subunits

The technique of partial reconstitution is based upon the discovery that isolated ribosomal subunits subjected to CsCl density-gradient equilibrium centrifugation lose a distinct fraction of their original protein (approximately 30 to 40%)[15]. The 50 S ribosomes are converted into 40 S core particles and split proteins (SP50); 30 S ribosomes into 23 S core particles and split proteins (SP30).[1,2] As revealed by polyacrylamide gel disc electrophoretic analysis of the protein-deficient particles and their corresponding split proteins,[16-18] the dissociation is a nonrandom process. Under defined conditions, only a certain number of the multiple protein species are detached from the ribosomes. Both 40 S and 23 S core particles or split proteins are completely inactive by themselves in several known ribosomal functions.[1,2,18-20] The various activities, however, can be almost fully restored when the components are mixed under defined conditions.[1,2,18,20] The split proteins can be separated into individual proteins and their functional roles can be examined.[19]

### Preparation of Core Particles and Split Protein[1,2,15]

*Solutions*

CsCl solution: 64% (w/v) CsCl in 20 m*M* Tris·HCl, pH 7.6 (4°) (density $\rho = 1.890$ g/cm$^3$, refractive index $n = 1.4167$)

Dilution buffer: 10 m*M* Tris·HCl, pH 7.5 (20°), 30 m*M* $NH_4Cl$, 10 m*M* $MgCl_2$, 0.005% gelatin

LiCl-buffer solution: 1 *M* LiCl in Solution I

Since ammonium sulfate treatment of 70 S ribosomes or washing with 1 *M* $NH_4Cl$ results in a substantial loss of 23 S core particles during CsCl density-gradient centrifugation, 30 S ribosomes without such treatment are used. 50 S particles are not affected by salt treatment in this respect. The dissociation of 30 S and 50 S ribosomal subunits into core particles and split protein is performed employing a Spinco SW 50 rotor. 3.4 ml of CsCl solution are mixed with 0.2 ml of 1 *M*

[15] M. Meselson, M. Nomura, S. Brenner, C. Davern, and D. Schlessinger, *J. Mol. Biol.* **9**, 696 (1964).

[16] P. Traub, M. Nomura, and L. Tu, *J. Mol. Biol.* **19**, 215 (1966).

[17] R. F. Gesteland and T. Staehelin, *J. Mol. Biol.* **24**, 149 (1967).

[18] P. Traub and M. Nomura, *J. Mol. Biol.* **34**, 575 (1968).

[19] P. Traub, K. Hosokawa, G. R. Craven, and M. Nomura, *Proc. Nat. Acad. Sci. U. S.* **58**, 2430 (1967).

[20] H. J. Raskas and T. Staehelin, *J. Mol. Biol.* **23**, 89 (1967).

$MgCl_2$ and 0.02 ml of 1.2 *M* β-mercaptoethanol. Then 1.4 ml of a ribosome solution containing up to 500 $OD_{260}$ units per milliliter of Solution I is rapidly mixed in. The mixture (5 ml) is centrifuged at 40,000 rpm for 36 hours in a Spinco SW 50 rotor. The gradients are fractionated from the bottom of the centrifuge tubes; about 15 fractions are collected per gradient. A small aliquot (5 to 20 μl) is taken from each fraction, mixed with 1 ml of dilution buffer and analyzed for its optical density at 260 mμ. The buoyant density of the peak fraction should be 1.65 to 1.66 $g/cm^3$. Fractions containing core particles are pooled and kept on ice. While the CsCl solution is running out from the centrifuge tube, the split protein meniscus usually becomes attached to the tube wall. To recover the split protein, the hole in the bottom of the empty tube is sealed with a piece of tape, the protein disk dissolved in 1 ml of LiCl-buffer solution and the tube rinsed twice with 1 ml of the same solution. The solutions are combined along with the 2 to 3 top fractions of the CsCl gradient and kept on ice. The protein concentration of the split protein solution is expressed in terms of $OD_{260}$ equivalents/ml. One $OD_{260}$ equivalent is defined as the amount of split protein which is obtained from 1 $OD_{260}$ unit of the original 30 S or 50 S particles.

## Reconstitution of Ribosomal Subunits from Core Particles and Split Proteins[1,2,18]

In earlier studies,[1,2,10] core particles (either 23S, or 40S core particles) dissolved in CsCl solution (about 5 *M*) were mixed with corresponding split proteins (SP30 proteins or SP50 proteins) dissolved in Solution I containing 1 *M* LiCl in a ratio of 1 equivalent to about 1.2 equivalents. The mixture was dialyzed first against Solution I containing 0.2 *M* LiCl and then against Solution I. The resultant particles (reconstituted 50S or 30S particles) are concentrated by centrifugation. The pelleted ribosomes are suspended in Solution I and clarified by low speed centrifugation and their concentrations are suitably adjusted. With this earlier method, the activity of the reconstituted 30S particles was generally 50 to 85% of that of native 30S particles, and the reconstituted 50S particles showed about 30 to 60% of the activity of native 50S particles, when they were assayed in the presence of native 50S and 30S particles, respectively, for their poly(U)-dependent polyphenylalanine synthesizing activity.

Better activity recovery can be obtained[21] using the same condition

[21] A. Atsmon, P. Spitnik-Elson, and D. Elson, *J. Mol. Biol.* 45, 125 (1969).

as that developed for the reconstitution of 30S particles from 16S RNA and total 30S proteins.[3,22] Core particles are dialyzed against Solution I. Split proteins are dialyzed against Solution V (see a later section). They are then combined, and the final ionic strength of the mixture is adjusted to 0.37. The mixture is incubated at 40° for 15–20 minutes and then cooled to 0°. The reconstituted particles are recovered by centrifugation.

## Total Reconstitution of 30S Ribosomal Subunits[3–5,22]

Reconstitution of 30S ribosomal subunits from purified 16S rRNA and a mixture of 30S ribosomal proteins is now a routine procedure.[3–5] Using this system the functional role of 16S rRNA as well as that of each of the separated 30S protein components can be analyzed.[3–5]

### Preparation of 16S RNA

*Solutions*

Solution III: 5 m*M* $H_3PO_4$ adjusted to pH 7.5 with KOH at 4°, 20 m*M* $MgCl_2$, 0.5 *M* KCl, 6 m*M* β-mercaptoethanol. To make this solution, all the components but $MgCl_2$ are mixed first, and the pH is adjusted to about 7.7 in close to the final volume. The required amount of 1 *M* $MgCl_2$ solution is then added dropwise under vigorous shaking.

Solution IV: the same as Solution III, but the concentration of KCl is 0.25 *M*

Saline-citrate EDTA: 0.15 *M* NaCl, 15 m*M* Na citrate, 10 m*M* EDTA, pH 7.0

5% (w/v) sucrose solution: 5% (w/v) sucrose in saline-citrate-EDTA.

20% (w/v) sucrose solution: 20% (w/v) sucrose in saline-citrate-EDTA. (The sucrose solutions are freed from ribonuclease as described above.)

Phenol: purified by distillation, and equilibrated with suitable buffers just before use

*16S RNA from 23S Core Particles*

Because of their high purity with respect to contaminating nonribosomal proteins, especially ribonucleases, 23S core particles are a good starting material for the preparation of 16S RNA. The original 23S core particle solution is first dialyzed against Solution I to remove

[22] P. Traub and M. Nomura, *J. Mol. Biol.* 40, 391 (1969).

CsCl, and then against Solution III. Equal volumes of phenol saturated with Solution III and core particle solution, containing up to 200 $OD_{260}$ units/ml, are vigorously shaken for 5 minutes on a Vortex mixer in a cold room. The mixture is separated into two layers by centrifugation at 10,000 rpm (12,000 *g*) for 5 minutes in an SS34 rotor of the Servall RC2 centrifuge. The aqueous layer is carefully taken without disturbing the interphase and the phenolization is repeated 6 times. To recover as much RNA as possible, the 7 phenol layers are washed once with one volume of Solution III, in the following way: Phenol layer and wash solution are shaken on a Vortex mixer and centrifuged as described above. The aqueous phase is transferred to the next tube and the washing of this and the 5 following phenol layers is performed in exactly the same way, with the washing proceeding in the same order as the main extraction. The clear main and wash solution are combined and dialyzed against Solution IV until no absorbing material at 280 m$\mu$ can be detected in the outer solution. The optical density of the RNA solution is measured at 260 m$\mu$. The protein content, as determined by the method of Lowry,[23] should be less than 1% of that of the original core particles. RNA can be concentrated by alcohol precipitation as described below. This and also the following RNA preparations can be frozen in dry ice-acetone and thawed several times without losing their reconstitution activity. They are stored at −70°.

*16S RNA from 30S Ribosomal Subunits*

To one volume of a solution of 30S ribosomes ($OD_{260}$ 200–600/ml) in Solution I is added an equal volume of saline-citrate-EDTA buffer (0.15 *M* NaCl, 15 m*M* Na Citrate, 10 m*M* EDTA, pH 7.0), 1/5 volume of 2% bentonite, and 1/25 volume of 10% sodium dodecyl sulfate. Bentonite is included to prevent breakdown of the RNA by traces of ribonucleases in the ribosome preparation. The solution is shaken with an equal volume of phenol previously equilibrated with saline-citrate-EDTA in a cold room. Phenol and aqueous layer are separated by centrifugation at 10,000 rpm for 5 minutes with the SS34 rotor of the Servall RC2 centrifuge. The phenol treatment is repeated twice, always including 0.2% (w/v) bentonite. To recover all RNA, the phenol layers may be washed once with 5 m*M* EDTA, pH 7. Finally the RNA solution is dialyzed against Solution IV.

*16 S RNA from 70 S Ribosomes*

Ribosome pellets, 70 S, obtained by high speed centrifugation of the

[23] O. H. Lowry, N. J. Rosebrough, A. L. Farr, and R. J. Randall, *J. Biol. Chem.* **198**, 265, (1951).

crude cell-extract (see above) are dissolved in Solution II at a concentration of 200–400 $OD_{260}$ units/ml. To this is added an equal volume of saline-citrate-EDTA solution, 1/10 volume of 10% sodium dodecyl sulfate solution (final concentration 0.43%), and 1/5 volume of 2% bentonite suspension (final concentration 0.17%). The mixture is kept for about 5 minutes on ice with occasional shaking. The mixture is then shaken with an equal volume of phenol previously equilibrated with saline-citrate-EDTA solution.

The separation of the mixture into two layers follows the same procedure described above. Bentonite is added to the water phase, and the phenolization is repeated two more times. The RNA is then precipitated with 2 volumes of ethanol (precooled to −20°). After 1 hour or longer at −20°, the precipitate is collected by centrifugation at 10,000 rpm for 5 minutes in an SS34 rotor of the Servall RC2 centrifuge. It is dissolved in saline-citrate-EDTA, reprecipitated with 2 volumes of cold ethanol, dissolved in saline-citrate-EDTA, and subjected to sucrose gradient centrifugation. Two milliliters of RNA solution, containing 200 $OD_{260}$ units/ml, are layered on each 26-ml 5 to 20% sucrose gradient and centrifuged for 19 hours at 27,000 rpm in a Spinco SW 27 rotor. The gradients are fractionated, then analyzed by the continuous flow technique with automatic recording. Fractions containing 16S RNA are pooled and mixed with 2 volumes of cold ethanol. The precipitated RNA is collected by centrifugation, dissolved in saline-citrate, and if necessary, purified by a second cycle of sucrose gradient centrifugation. The final solution of 16S RNA in saline citrate is dialyzed against Solution IV and checked for its optical density at 260 mμ.

## Preparation of Total 30 S Ribosomal Protein

*Solutions*

Urea-LiCl solution: 8 *M* urea, 4 *M* LiCl

Solution V: the same as solution III and IV, but the KCl concentration is 1 *M*

A solution of 30 S ribosomes containing 300–500 $OD_{260}$ units per milliliter of Solution I or II, is mixed with an equal volume of urea-LiCl solution and kept on ice for at least 15 hours.[24] The precipitated RNA is sedimented by centrifugation at 18,000 rpm (34,000 *g*) for 10 minutes in the SS34 rotor of the Servall RC2 centrifuge. To remove urea and LiCl, the supernatant is extensively dialyzed against Solution V. Its protein concentration is expressed in terms of "30 S equivalents"/ml;

[24] P. Spitnik-Elson, *Biochim. Biophys. Acta* **80**, 594 (1964).

one "30 S equivalent" is defined as the protein content of 1 $OD_{260}$ unit of 30 S ribosomes. The protein concentration in these units is thus calculated from the amount of native ribosomes used and the volume of the protein solution. Proteins from core particles can be prepared in the same way.

## Reconstitution of 30 S Ribosomal Subunits

Usually 30 $OD_{260}$ units of 16 S RNA dissolved in 10 ml of Solution IV are mixed with 36 equivalents of total 30 S ribosomal protein dissolved in 0.6 ml of Solution V. This volume ratio should be adhered to since it results in an ionic strength of 0.37, which is optimal for reconstitution. The RNA solution is preheated to 40° in a 40° water bath. The protein solution is then added while the mixture is constantly shaken. The resulting mixture is incubated at 40° for an additional 15–20 minutes. The reconstitution mixture is cooled to 4° and concentrated by centrifugation at 50,000 rpm (160,000 *g*) for 3.5 hours in a Spinco fixed-angle rotor 50. The pellet is dissolved in Solution II, and the solution is centrifuged for 5 minutes at 18,000 rpm (39,000 *g*) in an SS34 rotor of the Servall RC2 centrifuge. The $Mg^{2+}$ concentration of the clarified ribosome solution is adjusted to 10 m*M*, and the final ribosome concentration to 50 $OD_{260}$ units/ml. The reconstituted particles are frozen in dry ice–acetone and stored at −70°. They retain their biological activity for several months.

For the reconstitution of 30 S ribosomal subunits, the presence of phosphate is not essential. In the above procedures, Solutions IV and V may be substituted by Solutions VI and VII, respectively.[25] Composition of the latter buffers is as follows:

Solution VI: 30 m*M* Tris·HCl, pH 7.4, at 30°, 20 m*M* $MgCl_2$, 1 *M* KCl, 6 m*M* β-mercaptoethanol

Solution VII: the same as Solution VI, but the KCl concentration is 0.25 *M*

## Reconstitution of Protein-Deficient Subribosomal Particles from 16 S rRNA and a Mixture of Separated Proteins

The functional role of individual proteins can be studied by performing reconstitution using 16 S rRNA and a mixture of separated proteins with omission of the protein being studied.[5,7] The properties of the resulting particles may then be examined.

[25] V. Erdmann and M. Nomura, unpublished experiments.

Proteins from 30 S subunits are purified[5,7] by a method similar to that described by Kurland and his co-workers (see this volume [40]).

The following proteins, using our nomenclature,[5] are soluble in Solution V; P1, P2, P3, P3a, P3b, P3c, P4, P4a, P4b, P10, P10a, P10b, P11, P13, P14, P15, and can be stored in this buffer at −70°. Other proteins (P5, P6, P7, P8, P9, and P12) are not so soluble in Solution I, and therefore a suitable amount of 8 *M* urea solution is added to these proteins to solubilize them. Such solutions are also kept at −70°. The method of reconstitution of protein deficient particles is as follows:

All except one of the proteins are mixed in equivalent amounts and dialyzed briefly against 6 *M* urea buffer (6 *M* urea, 10 m*M* $H_3PO_4$ neutralized to pH 8.0 with methylamine at 4°C, 3 m*M* $\beta$-mercaptoethanol) to dissolve the small amount of precipitate which is often present, and then dialyzed against Solution V overnight at 4°.

Usually 30 $OD_{260}$ units of 16 S rRNA dissolved in Solution IV are mixed with the protein mixture obtained as above, which contains 45 equivalents of each of the proteins. The ionic strength is adjusted to 0.37, and the mixture is incubated at 40° for 20 minutes as described above for the reconstitution of 30 S particles from 16 S rRNA and a total 30 S protein mixture. Reconstituted particles are purified by centrifugation through a sucrose gradient (5% to 20% sucrose in the following solution: 0.35 *M* KCl, 20 m*M* $MgCl_2$, 10 m*M* potassium phosphate buffer, pH 7.5, at 45,000 rpm for 150 minutes in a Spinco SW 50 rotor). Fractions are collected from the bottom, and aliquots are analyzed for $OD_{260}$. Peak fractions that contain RNA are pooled, and particles are recovered by centrifugation in a Spinco fixed-angle rotor 50 at 50,000 rpm for 10 hours.

## Physical Characterization of Ribosomes, Ribosomal Components, and Reconstituted Ribosomal Particles

### Analysis of Ribosomal Proteins by Polyacrylamide Gel Disc Electrophoresis

*Solutions*

A: 0.48 *N* NaOH, 17.2% (v/v) acetic acid, 4% (v/v) *N,N,N′,N′*-tetramethylethylenediamine, 48% (w/v) urea

B: 0.48 *N* NaOH, 2.88% (v/v) acetic acid, 0.46% (v/v) *N,N,N′,N′*-tetramethylethylenediamine, 48% (w/v) urea

C: 13.3% (w/v) acrylamide, 0.2% (w/v) *N,N′*-methylenebisacrylamide, 48% (w/v) urea

D: 5% (w/v) acrylamide, 1.25% (w/v) *N,N'*-methylenebisacrylamide, 48% (w/v) urea
E: 1 mg riboflavin in 50 ml 8 *M* urea
F: Ammonium persulfate: 1.12% in 8 *M* urea

About 3 OD units of 30 S or 4 OD units of 50 S particles in 30 μl of Solution II are mixed with 5 μl each of 0.2 *M* Na-EDTA (pH 7.0), pancreatic ribonuclease (50μg/ml, Worthington Biochem. Corp.) and T2 ribonuclease (50μg/ml, Miles Laboratories). The mixture is incubated at 37° for 20 minutes. The resultant solution is applied directly on polyacrylamide gels.

The procedure used for polyacrylamide gel electrophoresis is a modification[26] of the method of Reisfeld, *et al.*[27] The technique for polymerization of the gel columns is comprehensively described in an instruction manual for the Canalco Model 6 (from Canalco, 4935 Cordell Avenue, Bethesda 14, Maryland). The different gel solutions are mixed from the stock solutions in the following way: Separating gel: 1 part A, 6 parts C, 1 part ammonium persulfate; stacking gel (for spacer and sample gel): 1 part B, 4 parts D, 1 part E, 2 parts 8 *M* urea. The polymer content of the separating gel is 10%, and that of the stacking gel, 2.5%.

The electrophoretic runs are carried out in glass tubes, 85 mm long with an inner diameter of 6 mm. The volume of the mixture for the separating gel is 1.7 ml, for the stacking gel 0.25 ml, and for the sample gel 0.25 ml. The separating gel is first polymerized in the tubes. Ammonium persulfate in the mixture is the catalyst for polymerization. Before polymerization takes place, water is layered on top of the gelation solution. This makes a flat boundary between the separating gel and the spacer gel. After completion of the polymerization (about 20 minutes), the water on top is removed and the spacer gel solution is applied. Polymerization is induced by irradiation of the gel with a fluorescent light.

Finally, the protein from 3 $OD_{260}$ units 30 S particles or 4 $OD_{260}$ units 50 S particles in a volume of 5–50 μl is mixed with the sample gel mixture, which is then applied on the top of the spacer gel. Buffer is carefully layered over the mixture, and polymerization is again induced by irradiation.

Although high concentration of β-mercaptoethanol prevents the polymerization of the sample gel, about 0.5% of β-mercaptoethanol is usually added to keep proteins in reduced form.

[26] P. S. Leboy, E. C. Cox, and J. G. Flaks, *Proc. Nat. Acad. Sci. U. S.* **52**, 1374 (1964).
[27] R. A. Reisfeld, U. J. Lewis, and D. E. Williams, *Nature (London)* **195**, 281 (1962).

The electrophoresis is performed at 4°, using 0.07 *M* $\beta$-alanine acetic acid buffer, pH 4.5, in the upper and lower buffer tray. With the anode at the top, the current is kept constant at 4 mA/tube. The progress of the electrophoresis is indicated by a trace of pyronine red. When the tracking dye has reached the bottom of the gels, the current is turned off. The gels are removed from the glass tubes by injecting water between the glass wall and gel column. Gels are stained for at least 1 hour by immersion in a 1% solution of amido black in 7.5% (v/v) acetic acid. The amido black-loaded gels are washed with 7.5% acetic acid for 2–3 hours, then destaining is carried out in 10 mm (i.d.) tubes at room temperature at a current of 10 mA/tube with the anode at the bottom and with 7.5% acetic acid in the buffer trays. Under these conditions, the separation of ribosomal proteins takes about 3 hours, the electrophoretic destaining about 1.5 hour. The destained polyacrylamide gels can be stored in 7.5% acetic acid at 0°.

## Functional Characterization of Subribosomal Particles and Reconstituted Ribosomes

### Polypeptide Synthesis Directed by Poly (U)

The biological activity of reconstituted ribosomal particles and of native ribosomes is generally assayed in a poly (U)-dependent phenylalanine incorporation system.[28] The enzyme fraction necessary for RNA-directed protein synthesis is prepared in the following way[29]: The ribosome-free supernatant obtained by high-speed centrifugation of the crude extract from 20 g of *E. coli* Q13 cells (see above) is diluted twofold with Solution I and stirred with 12 g of dry DEAE-cellulose (0.8 meq/g). The DEAE-cellulose has been equilibriated with Solution I and washed with distilled water before drying. The suspension is poured on a coarse sintered-glass funnel and washed with approximately 1 liter of Solution I. The cake of DEAE-cellulose is suspended in the same buffer, packed in a column, and freed from excess buffer under weak pressure. Absorbed proteins are eluted with 0.25 *M* $NH_4Cl$ in Solution I in a sharp band. The elution is followed with an LKB Uvicord absorption meter. The protein solution (concentration: 6–8 mg protein/ml) is divided into 1-ml aliquots, frozen in dry ice–acetone and stored at −70°. Employing this procedure, material of low molecular weight, basic proteins, and most of the nucleic acid are removed from the ribosome-free supernatant.

[28] M. Nirenberg and J. H. Matthaei, *Proc. Nat. Acad. Sci. U. S.* 47, 1589 (1961).
[29] P. Traub and W. Zillig, *Hoppe-Seyler's Z. Physiol. Chem.* 343, 246 (1966).

The incorporation of [$^{14}$C]phenylalanine into trichloroacetic acid-precipitable polypeptides is carried out in a final volume of 0.15 ml. Each incubation mixture contains: 75 μl of Solution I containing 1 $OD_{260}$ unit of particles derived from 30 S ribosome and 2 $OD_{260}$ units of particles derived from 50 S ribosomes; 3 μl of pyruvate kinase (1 mg/ml); 12 μl of enzyme fraction (70–100 μg of protein;[30] 5 μl of [$^{14}$C]phenylalanine (10 Ci/mole, 1.5 m*M*); 10 μl of poly(U) (4 mg/ml); and 45 μl of Mix I. Three hundred microliters of Mix I contains 5 μl of Tris·HCl (1 *M*, pH 7.4, at 30°), 5 μl of 1 *M* $Mg(OAc)_2$, 17 μl of 1.2 *M* $NH_4Cl$, 5 μl of a mixture of all the amino acids except tyrosine at 10 m*M*, 10 μl of 5 m*M* tyrosine, 25 μl of 40 m*M* ATP, 10 μl of 3 m*M* GTP, 3 μl of 1.2 *M* β-mercaptoethanol, 5 μl of 0.2 *M* dithiothreitol, 50 μl of 0.1 *M* phosphoenolpyruvic acid tricyclohexylammonium salt, 40 μl of tRNA mixture (General Biochemicals, 25 mg/ml) and 125 μl of $H_2O$.

The mixtures are incubated at 37° for 10 minutes in the absence of poly(U). The phenylalanine incorporation is then started by the addition of the template. After an additional 30-minute incubation period at 37°, the reaction is stopped by the addition of 3 ml of 5% trichloroacetic acid. The acid precipitates are heated in a boiling water bath for 15 minutes, brought to room temperature, and filtered into Reeve Angel Ultra filters (984H). The filters are dried, and the radioactivity is measured in a low-background gas-flow counter.

## Protein Synthesis Directed by Natural mRNA

The assay is performed using the same incorporation system as that described above, except that poly(U) is replaced by the RNA of the *E. coli* phage f2 and [$^{14}$C]phenylalanine by [$^{14}$C]valine. Usually 20 μl of f2 RNA (150 $OD_{260}$ units/ml and 10 μl of [$^{14}$C]valine solution (50 Ci/mole, 0.75 m*M*) are used per 0.15 ml incubation mixture. Since the native 30 S ribosomes and their derived reconstituted particles have lost the major part of the polypeptide chain initiation factors during their purification and reconstitution, the incubation mixture must be

[30]The assay system should be tested in order to be certain that the amount of phenylalanine incorporation is directly proportional to the amount of ribosomes added. The most likely source of nonlinearity would be an insufficient amount of enzyme, and therefore the amount of enzyme should be optimized. The system for measuring 30 S ribosome activity contains standard 50 S ribosomes, and vice versa. It is necessary to include a blank without added 30 S ribosomes since the 50 S preparation may contain significant amounts of 30 S ribosomes as contaminant. The corresponding blank should be included when one is measuring 50 S ribosome activity.

supplemented with a suitable amount of initiation factors (see the next section).

A significant stimulation of [$^{14}$C]valine incorporation has been observed when the incubation mixture also contains leucovorin (5-formyltetrahydrofolic acid). Usually 5 $\mu$l of Ca-leucovorin solution (300 $\mu$g/ml, Lederle Laboratories) is included per 0.15 ml incubation mixture.

## Preparation of Crude Initiation Factors

Crude initiation factors are prepared as follows: Twenty grams of frozen cells are broken in Solution I with glass beads by the same method as that described before (see section "Shaking with Glass Beads). Again, no DNase is added. The extract is centrifuged for 20 minutes at 40,000 rpm in a Spinco fixed-angle rotor 40, and the pellets are discarded. The supernatant fluid is centrifuged for 3–3.5 hours at 60,000 rpm in a Spinco fixed-angle rotor 65. The whole pellet, including ribosomes and "DNA layer" are well suspended in 20 ml of a solution containing 1 *M* $NH_4Cl$, 30 m*M* $MgCl_2$, 20 m*M* Tris·HCl (pH 7.8 at 4°), and 6 m*M* of $\beta$-mercaptoethanol, and are kept at 4° overnight. The solution is then centrifuged for 3–3.5 hours at 60,000 rpm in a Spinco fixed-angle rotor 65. The upper four-fifths of the supernatant is collected. To 3 volumes of supernatant fluid is added 7 volumes of saturated ammonium sulfate solution (pH 7.0), and the solution is kept on ice for 1 hour. The resultant precipitate is recovered by centrifugation at 15,000 rpm for 10 minutes, dissolved in 6 ml of Solution I, and dialyzed overnight against Solution I containing an additional 0.25 mole of $NH_4Cl$ per liter. The solution is passed through a small DEAE-cellulose column (0.8 cm × 8 cm) and unabsorbed proteins are recovered in the same volume as that of the original solution. Aliquots of this preparation, frozen in dry ice–acetone and stored at −70°, are used as crude factor. The optimal amount needed for the incorporation assay should be determined for each preparation.

## Phe-tRNA Binding to Ribosomal Particles

*Solutions*

5X Binding buffer: 0.5 *M* Tris·HCl (pH 7.4 at 30°), 0.1 *M* $MgCl_2$, and 0.25 *M* LiCl.

Binding buffer: the 5× binding buffer is diluted 5-fold with water.

The binding of $^{14}$C-labeled Phe-tRNA directed by poly(U) is assayed employing the Millipore filter technique of Nirenberg and Leder.[31] Sixty microliters of each reaction mixture contains: 30 μl of Solution I containing 0.2 $OD_{260}$ units of 30 S (and, if desired, 0.4 $OD_{260}$ units of 50 S), 10 μl of 5 × binding buffer, 10 μl of poly(U) (2 mg/ml) and 10 μl [$^{14}$C]Phe-tRNA (0.5 μCi/ml, about 2.5 mg/ml). [$^{14}$C]Phe-tRNA can be prepared by charging the tRNA mixture with [$^{14}$C]phenylalanine using the supernatant enzyme preparation (see Section "Polypeptide Synthesis Directed by Poly(U)". A commercial preparation (e.g., New England Nuclear Corp. product) can also be used. The mixtures are incubated at 30° for 20 minutes. The reaction is stopped by the addition of 3 ml of ice-cold binding buffer. The reaction mixture is filtered through a Millipore filter (HA 0.45 μm, Millipore Filter Corp.), and washed 4 times with 3 ml of ice cold binding buffer. The filters are dried, and the radioactivity is measured in a low-background gas-flow counter or in a liquid scintillation counter.

[31] M. Nirenberg and P. Leder, *Science* **145**, 1399 (1964).

# [50] Total Reconstitution of 50 S Subunits from *Escherichia coli* Ribosomes

*By* KNUD H. NIERHAUS and FERDINAND DOHME

The 30 S subunit from *E. coli* ribosomes consists of 21 proteins and 16 S RNA, and the 50 S subunit consists of 34 proteins and two RNA molecules.[1] In spite of the large number of ribosomal components, which are all present in one copy per ribosome[2] (except L7/L12: 3 or 4 copies), the subunits can be self-assembled *in vitro* by a technique known as total reconstitution.

The total reconstitution of the 30 S subunit was achieved by a single incubation of the dissociated components,[3] the rate-limiting step being a conformational change in a reconstituted intermediate formed during the course of assembly.[4] In contrast, the total reconstitution of the large subunit from *E. coli* ribosomes requires a two-step incubation of the

[1] E. Kaltschmidt and H. G. Wittmann, *Proc. Natl. Acad. Sci. U.S.A.* **67**, 1276 (1970).
[2] J. S. Hardy, *Mol. Gen. Genet.* **140**, 253 (1975).
[3] P. Traub and M. Nomura, *Proc. Natl. Acad. Sci. U.S.A.* **59**, 777 (1968).
[4] P. Traub and M. Nomura, *J. Mol. Biol.* **40**, 391 (1969).



ISBN 0-12-504180-2

components (44°, 4 m*M* $Mg^{2+}$ →50°, 20 m*M* $Mg^{2+}$).[5,6] In both incubations a conformational change in reconstituted intermediates is the rate-limiting step, and these conformational changes have different ionic and incubation optima.[6]

In addition to its importance for assembly studies, the technique of reconstitution has been used successfully for the elucidation of the functional role of a number of ribosomal components.[7-10]

## Preparation of Ribosomal Subunits

*Materials*

L-$H_2O$ medium[11]: One liter contains 10 g of Bacto-tryptone (Difco), 5 g of yeast extract (Difco), 5 g of NaCl, 1 ml of 1 *N* NaOH

Glucose, 20% (w/v)

Alumina A-305 (Alcoa)

Dissociation buffer: 10 m*M* K phosphate, pH 7.5 (0°), 1 m*M* $MgCl_2$, 6 m*M* β-mercaptoethanol

DNase (RNase free; Worthington, Freehold, New Jersey Cat. No. 6330).

TMNSH buffer: 10 m*M* Tris·HCl, pH 7.5 (0°), 10 m*M* $MgCl_2$, 60 m*M* $NH_4Cl$, 6 m*M* β-mercaptoethanol

40% (w/v) sucrose (Merck, Darmstadt, Germany; Cat. No. 2654) in dissociation buffer (2 liters per zonal centrifugation in a Beckman Ti 15 rotor)

50% (w/v) sucrose in dissociation buffer (2 liters per zonal centrifugation)

Polyethylene glycol-6000 (Merck-Schuchardt, Hohenbrunn, Germany; Cat. No. 807491)

Mg acetate, 1 *M*

*Procedure*. Sterilized 20% glucose, 25 ml, is added to 2.5 liters of sterilized L-$H_2O$ medium. The medium is inoculated with *E. coli*, strain A19[12] (RNase $I^-$, $Met^-$, rel $A^-$) and shaken overnight at 37°. The over-

[5] K. H. Nierhaus and F. Dohme, *Proc. Natl. Acad. Sci. U.S.A.* **71**, 4713 (1974).

[6] F. Dohme and K. H. Nierhaus, *J. Mol. Biol.* **107**, 585 (1976).

[7] M. Nomura and W. A. Held, *in* "Ribosomes" (M. Nomura, A. Tissières, and P. Lengyel, eds.) p. 193. Cold Spring Harbor Laboratory, Cold Spring Harbor, New York, 1974.

[8] R. Werner, A. Kollak, D. Nierhaus, G. Schreiner, and K. H. Nierhaus, *in* "Topics in Infectious Diseases" (J. Drews and F. E. Hahn, eds.), p. 219. Springer-Verlag, Berlin and New York, 1974.

[9] K. H. Nierhaus and V. Montejo, *Proc. Natl. Acad. Sci. U.S.A.* **70**, 1931 (1973).

[10] S. Spillmann, F. Dohme, and K. H. Nierhaus, *J. Mol. Biol.* **115**, 513 (1977).

[11] E. S. Lennox, *Virology* **1**, 190 (1955).

[12] R. F. Gesteland, *J. Mol. Biol.* **16**, 67 (1966).

night culture is poured into 100 liters of L-$H_2O$ medium supplemented with 1 liter of 20% glucose, and vigorously stirred at 37° until an optical density of $A_{650\ nm}$ = 0.5 is reached (after about 2.5 hr; generation time ~33 min). Growth is stopped by adding 25 liters of ice-cold water; the cells are centrifuged, collected in plastic sacks, and stored at −80°. All subsequent procedures are performed at 0°–4°.

For preparation of 70 S ribosomes, the cells are washed with dissociation buffer (2 ml ger pram of cells) containing 1 μg of DNase per milliliter. The pelleted cells (15 min at 30,000 *g*) were mixed with Alcoa (2 g per gram of cells) and ground in the Retsch-mill KMI (Retsch, Haan, Germany; Cat. Nos. 231, 232, and 233) for 30 min. About 100 g of cells are a convenient quantity to use. Dissociation buffer is added (2 ml per gram of cells), and the paste is homogenized for 10 min in the mill. Alumina and cell debris are removed by two subsequent low speed centrifugations (each 20 min at 30,000 *g*), and the supernatant (S30) is centrifuged at high speed (5 hr at 150,000 *g*). The pellet contains 70 S ribosomes and is resuspended in dissociation buffer. After low speed centrifugation (in order to remove aggregates; 5 min at 10,000 *g*) the optical density is measured ($A_{260\ nm}$). The ribosomes are divided into portions containing about 7000 $A_{260}$ units and stored at −80°. Usually, 600 $A_{260}$ units of 70 S ribosomes are obtained per gram of cells.

The supernatant from the high speed run is dialyzed overnight against two changes of a buffer containing 10 m*M* Tris·HCl, 10 m*M* Mg acetate, and 6 m*M* β-mercaptoethanol, and again centrifuged for 5 hr at 150,000 *g*. The upper three-quarters of this supernatant is referred to as the S150 enzymes fraction, and this is stored at −80° in 500-μl portions.

To separate 30 S + 50 S subunits, about 7000 $A_{260}$ units of ribosomes are applied to a B15 Ti zonal rotor containing a hyperbolic gradient[13] (800 ml from 6% to 38% sucrose) made from dissociation buffer and 40% sucrose in dissociation buffer. After centrifugation (17 hr at 25,000 rpm) the gradient is pumped out with 50% sucrose in dissociation buffer. Fractions (about 18 ml) containing either 30 S or 50 S subunits are collected, the $Mg^{2+}$ concentration is raised to 10 m*M*, and 10% (w/v) of ground polyethylene glycol-6000 is added.[14] After stirring for about 20 min at 0° the subunits are pelleted (30 min at 30,000 *g*) and dissolved to concentrations of about 400 $A_{260}$ units/ml in the TMNSH buffer. After low speed centrifugation (5 min at 10,000 *g*) the absorption at 260 nm is determined; the subunit suspensions are stored at −80°.

[13] E. F. Eickenberry, T. A. Bickle, R. R. Traut, and C. A. Price, *Eur. J. Biochem.* **12**, 113 (1970).

[14] A. Expert-Bezançon, M. F. Guérin, D. H. Hayes, L. Legault, and J. Thibault, *Biochimie* **56**, 77 (1974).

## Preparation of RNA from the Large Subunits

*Materials*

Phenol 70%: freshly distilled phenol is stored at −20° in 10 ml portions; before use 4.3 ml of glass-distilled water are added per portion.

Bentonite-SF (Serva, Heidelberg, Germany; Cat. No. 14515)

Sodium dodecyl sulfate (SDS) (Serva, Heidelberg, Germany; Cat. No. 20760)

TM-4 buffer: 10 m*M* Tris·HCl, pH 7.6 (0°), 4 m*M* Mg acetate

RNA buffer: 10 m*M* Tris·HCl, pH 7.6 (0°), 50 m*M* KCl, 1% MeOH

*Procedure*. Activity of the 50 S subunits and their contamination with 30 S particles are tested in the poly(U) system (see below), and the intactness of the 23 S RNA is checked by RNA gel electrophoresis.[15] The RNA is isolated by phenol extraction from 50 S subunits containing intact 23 S RNA; sterilized tubes and pipettes are used in all steps.

The preparation, a simplification of the procedure previously described,[5,6] is as follows. To 50 S subunits, at a concentration of less than 400 $A_{260}$ units/ml, 1/10 volume of 10% SDS, 1/20 volume of 2% bentonite, and 1.2 volume of 70% (v/v) phenol are added; the mixture is shaken vigorously for 8 min and centrifuged for 10 min at 10,000 *g*. The aqueous phase is mixed with 1.2 volume of 70% phenol, shaken for 5 min, and centrifuged; the extraction is repeated a third time. The RNA is precipitated from the aqueous phase at −20° overnight by addition of 2 volumes of cooled (−20°) ethanol. After centrifugation (45 min at 10,000 *g*) the pellet is washed with ethanol (about 1 ml per 400 $A_{260}$ units of 50 S particles) and again centrifuged (5 min at 10,000 *g*). The ethanol washing procedure is repeated until no phenol odor is perceived. If the unfractionated RNA is to be used in reconstitution experiments, the RNA is resuspended in TM-4 buffer at a concentration of about 200 $A_{260}$ units/ml and stored in small portions at −80°.

If the RNA is to be separated into 23 S and 5 S RNA, the RNA pellet is resuspended in RNA buffer (600–800 $A_{260}$ units/ml) after the ethanol washing procedure. About 3000 $A_{260}$ units are applied to a Sephadex G-100 column (dimension 2 × 200 cm; equilibrated with the same buffer). Fractions containing 23 S RNA or 5 S RNA are collected; the RNA is precipitated with ethanol as described above, dissolved in TM-4 buffer to concentrations of about 400 $A_{260}$ units/ml (23 S RNA), or 40 $A_{260}$ units/ml (5 S RNA), and stored at −80°C.

[15] M. Ceri and P. Y. Maeba, *Biochim. Biophys. Acta* **312**, 337 (1973).

## Preparation of the Total Proteins (TP50) from the Large Subunit

*Materials*

Mg acetate, 1 *M*
Glacial acetic acid
Bentonite-SF (Serva, Heidelberg, Germany; Cat. No. 14515)
Tris, 1 *M*, unbuffered; wash with bentonite-SF before use; add 1 g of bentonite per liter, stir at 4° for 1 hr, and filter through two layers of Selecta filters [Schleicher and Schüll, Dassel, Germany; 595 1/2, diameter 240 mm] without vacuum pump.
TM-4 buffer: 10 m*M* Tris·HCl, pH 7.6 (0°), 4 m*M* Mg acetate
Neutralit (indicator strips, pH 5–10; Merck, Darmstadt, Germany; Cat. No. 9533)

*Procedure.* To the 50 S suspension (made 0.1 *M* in magnesium by addition of 1/10 volume of Mg acetate) 2 volumes of glacial acetic acid are added. After stirring for 45 min at 0° and centrifugation at 30,000 *g* for 30 min,[16] 5 volumes of acetone are added to the protein supernatant. The precipitated proteins are separated by centrifugation at 30,000 *g* for 30 min, dried in a desiccator for 30–60 min, and dissolved in TM-4 buffer containing 6 *M* urea to a concentration of about 150 equivalent units per milliliter (1 equivalent unit of protein is the amount of protein extracted from 1 $A_{260}$ unit of 50 S subunits). Unbuffered Tris (1 *M*) is added to a final concentration of 40–60 m*M* until the pH of the protein solution reaches about 6 (checked with indicator paper). The solution is dialyzed overnight against TM-4 buffer containing 6 *M* urea, once for 45 min against a 500-fold volume of TM-4 buffer, and three times for 45 min against a 500-fold volume of TM-4 buffer containing 0.01% acetic acid. (The final pH of the solution is 6, which is raised to about 7.2 by the Tris present in the reconstitution mixture.) The solution is centrifuged at 6000 *g* for 5 min, the absorption at 230 nm is determined, and the solution is stored in small portions at −80°. We use the following relationships: 1 $A_{230}$ unit of a protein solution is equivalent to 250 μg. 1 $A_{230}$ unit of a TP50 solution is equivalent to 10 equivalent units of TP50.

## Reconstitution Procedure

*Materials*

Reconstitution mix: 110 m*M* Tris·HCl, pH 7.7 (0°), 4 *M* $NH_4Cl$, 4 m*M* Mg acetate, 0.2 m*M* EDTA, 20 m*M* β-mercaptoethanol
TM-4 buffer: 10 m*M* Tris·HCl, pH 7.6 (0°), 4 m*M* Mg acetate

[16] S. J. S. Hardy, C. G. Kurland, P. Voynow, and G. Mora, *Biochemistry* **8**, 2897 (1969).

*Procedure.* TM-4 buffer, 90 μl, containing 2.5 $A_{260}$ units of 23 S RNA, 0.1 $A_{260}$ unit of 5 S RNA, and 3 to 4 equivalent units of TP50, is treated with 10 μl of the reconstitution mix. The order of addition is RNA, 10 μl of reconstitution mix, TM-4, and TP50 solution. The final concentrations are 20 m*M* Tris·HCl (actual pH 7.2–7.4), 4 m*M* $Mg^{2+}$, 400 m*M* $NH_4^+$, 0.02 m*M* EDTA, and 2 m*M* β-mercaptoethanol. After incubation for 20 min at 44°, the $Mg^{2+}$ concentration is raised to 20 m*M* by addition of 4 μl of 400 m*M* Mg acetate; the second incubation (90 min at 50°) follows. Aliquots (40 μl) of the resulting solution can be tested for structural[6] or functional properties, e.g., for peptidyltransferase activity (fragment assay) or poly(Phe) synthesis activity (see below).

## Poly(U)-Dependent Poly(Phe) Synthesis

*Materials*

Energy mix: 20 m*M* Tris·HCl, pH 7.8 (0°), 65 m*M* $NH_4Cl$, 20 m*M* Mg acetate, 4 m*M* ATP, 0.13 m*M* GTP, 20 m*M* phosphoenolpyruvate (adjusted to pH 7.6 (0°) with 1 *N* KOH; usually 50–80 m*N* in the energy mix is sufficient); divided into 0.9-ml portions (sufficient for 40 assays) and stored at −20°

Pyruvate kinase (Boehringer-Mannheim, Germany, Cat. No. 15744), 1 mg per milliliter of water

$tRNA^{E.\ coli}$, 20 mg/ml

Poly(U)-[$^{14}$C]Phe mix: 1 volume of poly(U) solution (4 mg/ml) is mixed with 1 volume of 1.5 m*M* phenylalanine and 1 volume of [$^{14}$C]phenylalanine (Amersham, 440 Ci/mol; Cat. No. CFB 70, diluted with water to about 200,000 cpm/10 μl); divided into 0.6-ml portions (sufficient for 40 assays) and stored at −20°C

S150 enzymes (see section "Preparation of Ribosomal Subunits, Procedure")

Reconstitution buffer: 20 m*M* Tris·HCl, pH 7.6, 20 m*M* Mg acetate, 400 m*M* $NH_4Cl$, 6 m*M* β-mercaptoethanol

Trichloroacetic acid (TCA), 5% (w/v)

Whatman 3 MM filter (diameter 2.5 cm)

Ether–ethanol, 1:1

Ether

Infrared lamp

Insta-Gel (Packard Instrument, Cat. No. 6003059)

*Procedure.* Poly(U) mix sufficient for at least 20 assays is freshly made for each experiment in the following way (here for 40 assays):

| | | |
|---|---|---|
| Poly(U)-[$^{14}$C]Phe mix | 600 μl | |
| Energy mix | 900 μl | |
| S150 enzymes | ca. 400 μl | (optimized for each preparation) |
| Pyruvate kinase (1 mg/ml) | 60 μl | |
| tRNA (20 mg/ml) | 40 μl | |
| Poly(U) mix | 2000 μl | |

Poly(U) mix, 50 μl, is added to 40 μl of reconstitution buffer containing 1 $A_{260}$ unit of 50 S subunits and 0.5 $A_{260}$ unit of 30 S subunits. After incubation at 30° for 45 min, the samples are cooled in an ice bath and 70 μl from each sample is applied to a Whatman 3 MM filter (the dry filter is numbered with a pencil near the edge according to the tube number and carries a neddle that is pierced through the filter to prevent filter stacking). The filters are thrown into a beaker containing 5% TCA (at least 2 ml per filter). After 10 min the TCA is removed (vacuum pump); the same amount of 5% TCA is added, and the beaker is heated for 15 min in a 90° water bath. The filters are washed 2–3 times with 5% TCA, then ether–ethanol is added (same amount as that for TCA). After 5 min the filters are washed with ether, dried under an infrared lamp, and counted in the presence of 4 ml of Insta-Gel.

The batchwise technique described[17] is advantageous for large numbers of samples.

## Concluding Remark

The two-step reconstitution leads to particles that are structurally indistinguishable from native 50 S subunits[5,6] and show 50–100% of the activity of native subunits in various functional tests.[18]

[17] R. J. Maus and G. D. Novelli, *Arch. Biochem. Biophys.* **94,** 48 (1961).

[18] F. Dohme and K. H. Nierhaus, *Proc. Natl. Acad. Sci. U.S.A.* **73,** 221 (1976).

# [51] Purification of Ribosomal Proteins from *Escherichia coli*[1]

*By* C. G. KURLAND, S. J. S. HARDY, and G. MORA

Electrophoretic fractionations on starch gel provided the first evidence that there are many proteins in the *E. coli* ribosome.[2] This heterogeneity was also reflected in the sedimentation equilibrium data obtained with unfractionated ribosomal proteins; these have apparent molecular weights that are distributed between 10,000 and 40,000.[3] Such observations suggested that a purification scheme for the ribosomal proteins could exploit differences in the molecular weight as well as the charge of individual proteins. Thus, chromatographic fractionation on carboxymethyl cellulose[2,4] and electrophoretic fractionation in polyacrylamide gels that also function as molecular sieves[5,6] have been used to obtain purifications of some of the ribosomal proteins. We have used a combination of ion exchange chromatography on phosphocellulose and molecular sieve chromatography in Sephadex for the purifications of the ribosomal proteins.[7] It would be useful to consider some of the limitations of the present methodology before describing the procedures in detail.

The purification of the individual ribosomal proteins is not an end in itself. Rather, the purified proteins are required for detailed structural and functional analyses. Therefore, one highly desirable feature of a purification scheme for these proteins would be that they may be obtained easily in large amounts. Unfortunately, only a relatively small number of the ribosomal proteins can be obtained in a pure state by a single chromatographic or electrophoretic fractionation step.[4-7] The rest are obtained in groups that require additional fractionation before they can be resolved into individual proteins. Therefore, the procedures currently in use for routine preparative scale purification of ribosomal

[1]Supported by funds from the U. S. Public Health Service (GM12411, GM13832).
[2]J. P. Waller, *J. Mol. Biol.* **10**, 319 (1964).
[3]W. Moller and A. Chrambach, *J. Mol. Biol.* **23**, 317 (1969).
[4]P. B. Moore, R. R. Traut, H. Noller, P. Pearson, and H. Delius, *J. Mol. Biol.* **31**, 441 (1968).
[5]E. Kaltschmidt, M. Dzionara, D. Donner and H. G. Wittmann, *Mol. Gen. Genet.* **100**, 364 (1967).
[6]S. Fogel and P. S. Sypherd, *Proc. Nat. Acad. Sci. U. S.* **59**, 1329 (1968).
[7]S. J. S. Hardy, C. G. Kurland, P. Voynow, and G. Mora, *Biochemistry* **8**, 2897 (1969).



ISBN 0-12-504180-2

proteins should be viewed as quite crude ones that must be improved upon if not replaced as soon as possible.

The fractionation techniques that have been developed for use in the *E. coli* system will undoubtedly be used to study the ribosomal proteins of other systems. Therefore, it will be necessary in each new situation to employ some rigorous criterion to establish the purity of the isolated proteins. The electrophoretic homogeneity of samples has been most often employed as a criterion of purity for the ribosomal proteins. However, this technique is a dangerous one to rely on in the absence of other evidence of purity. For example, it is well known that more than half of the 30 S ribosomal proteins electrophorese in the standard polyacrylamide gels[8] as single bands which contain two or more proteins (see Fig. 1). Furthermore, we have identified mixtures of proteins that are pure according to the criteria that they appear as a single electrophoretic component in polyacrylamide gels and that sedimentation equilibrium analysis reveals only one molecular species. Nevertheless, these samples contained two chemically unique proteins.[9] For these reasons, it is necessary to rely on chemical criteria to establish the purity of the protein samples. There are a number of ways to do this. We have compared the physical molecular weight, determined by equilibrium sedimentation analysis, with chemical molecular weights, calculated from the amino acid composition and number of tryptic peptides.[9] When such calculations agree, it is possible to conclude that the protein sample is at least 80% pure. However, if such a chemical molecular weight estimate is approximately twice as large as the physical molecular weight estimate, the sample must contain at least two different proteins.

## Protein Extraction

We have used three different procedures to obtain ribosomal proteins that are free of RNA and able to reconstitute ribosomes when incubated with RNA. One of these is the procedure of Spahr[10] for autodigestion of ribosomes by the contaminating endonuclease (latent ribonuclease) in buffered solutions of urea. Here, the degraded RNA fragments are dialyzed away and the proteins, which are soluble in urea, remain in the dialysis bag. A second procedure is that of Leboy *et al.* which employs LiCl to precipitate the RNA; the proteins remain in solution with urea.[8] Finally, we have most often relied on a modification of the acetic acid extraction that was described by Waller and Harris.[11]

[8] P. S. Leboy, E. C. Cox, and J. G. Flaks, *Proc. Nat. Acad. Sci. U. S.* **52**, 1367 (1964).
[9] G. R. Craven, P. Voynow, S. J. S. Hardy, and C. G. Kurland, *Biochemistry* **8**, 2906 (1969).
[10] P. F. Spahr, *J. Mol. Biol.* **4**, 395 (1962).
[11] J. P. Waller and J. I. Harris, *Proc. Nat. Acad. Sci. U. S.* **47**, 18 (1961).

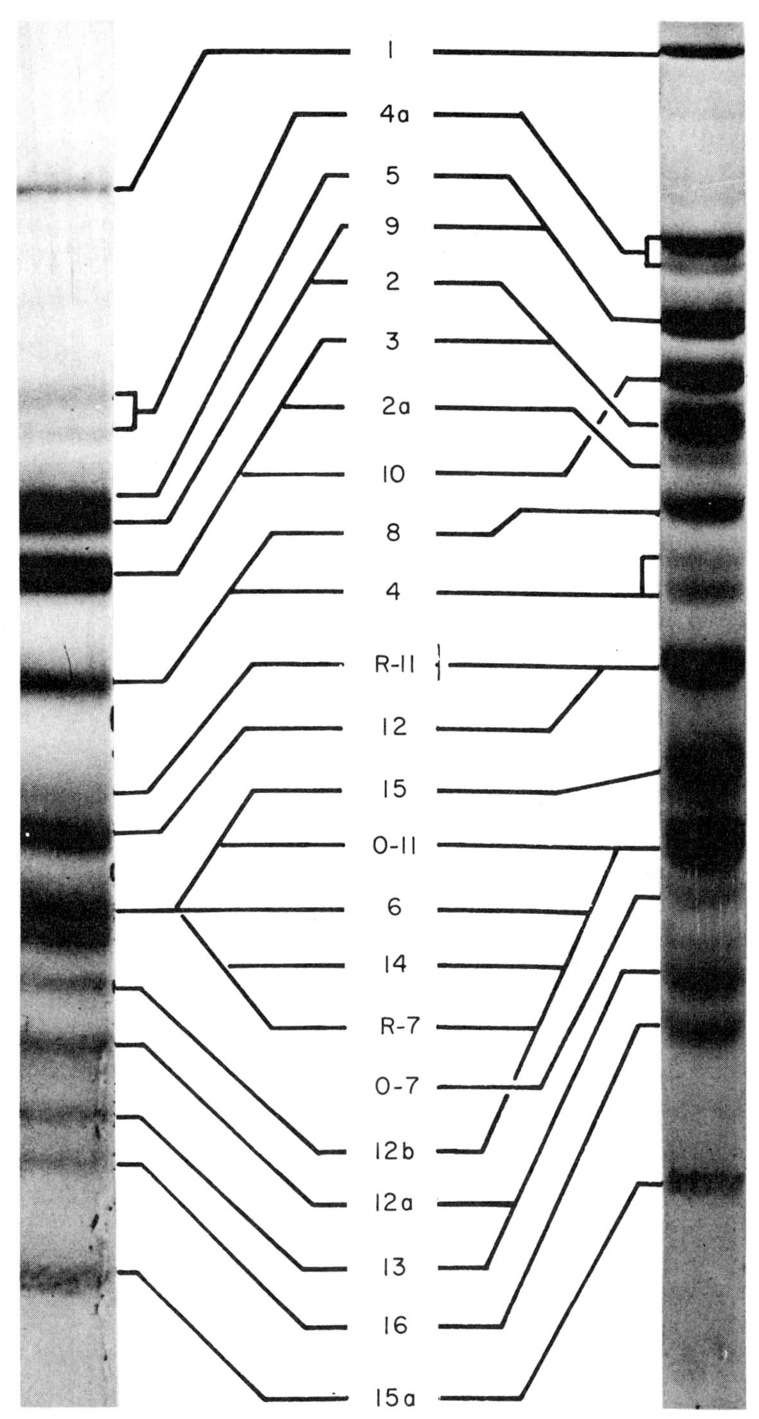


FIG. 1. The disc electrophoresis patterns obtained with the 30 S proteins fractionated on "soft" polyacrylamide gel (left), which contain 10% acrylamide, 0.15% methylenebisacrylamide, and on "hard" polyacrylamide gel (right), which contain 10% acrylamide, 0.75% methylenebisacrylamide.

Our version of the acetic acid extraction employs high concentrations of $MgCl_2$ to reduce the contamination of the proteins by RNA that would otherwise not precipitate in the absence of excess divalent cations.[7] One-tenth volume of 1 *M* $MgCl_2$ is added to one volume of ribosomes (200 $A_{260}$–1500 $A_{260}$) followed in rapid succession by two volumes of glacial acetic acid; all this is done at 0–2° with vigorous stirring. The suspension is left to stir in the cold for 45 minutes. The RNA precipitate is removed by centrifugation at 20,000 *g* for 10 minutes. The RNA pellet is washed with 67% acetic acid containing 32 m*M* $MgCl_2$, and the two supernatants are pooled for dialysis. More than 90% of the protein is recovered with less than 0.5% contamination by RNA.

## Phosphocellulose Chromatography

The protein samples must be exhaustively dialyzed before they are applied to the chromatographic columns. The acetic acid extracted proteins are dialyzed for 2 or 3 days against successive 2-liter volumes of buffer (6 *M* urea, 50 m*M* $NaH_2PO_4$, 12 m*M* methylamine at pH 6.5 or pH 5.8 and containing 50 $\mu$l $\beta$-mercaptoethanol per liter) until the pH of the fresh dialysis buffer is unchanged after 12 hours of dialysis. The dialysis tubing is pretreated by heating it in a dry state at 70° for 3 or more days and then boiling it in EDTA before washing it in buffer.

We usually monitor the column eluates by measuring the extinction of chromatographic samples at 230 m$\mu$. Therefore, it is necessary to be sure that the extinction of the buffered urea is less than 0.1 $A_{230}$ compared to water. When the $A_{230}$ of buffer is too high, the stock urea solutions (7.5 *M*) can be treated with 50 g/liter of activated charcoal; this reduces the $A_{230}$ of the buffered urea to an acceptable level.[7]

The phosphocellulose that we have employed for fractionation of ribosomal proteins has been obtained from a variety of sources, but Mannex P (0.9 meq/g) has been used most often. The phosphocellulose is yellow before it is washed; we have found that it is necessary to completely decolorize the phosphocellulose in order to obtain good fractionation with this material. The phosphocellulose is fined before washing. This is done by suspending 50 g in 2 liters of water and decanting the suspension four or five times until roughly half the material is discarded. The remaining material is suspended in 0.1 *N* NaOH for a period that varies with different lots of phosphocellulose. Some lots take 15 minutes, others 30 minutes, to remove the yellow color. The NaOH is removed by washing the phosphocellulose on a filter with four successive 1-liter volumes of water. Then the phosphocellulose is washed in 0.1 *N* HCl followed by 4 liters of water. Finally the phosphocellulose is suspended in 1.5 liters of buffer to equilibrate overnight before the column is poured. A column of 90 × 1.8 cm can be poured with the fined phosphocellulose obtained from an initial 50 g batch.

The columns are packed under pressure (5 psi) at a flow rate dependent on the size of the column; the rule of thumb that we have used is that the column can be packed with a flow rate roughly ten times that used to develop the column after the sample is adsorbed (see below). After the column is packed, a disk of filter paper is placed on the top surface and the column is washed with standard buffer until the pH of the effluent is the same as that of the buffer.

The sample, containing roughly 400 mg of ribosomal proteins is run on to the column (2.8 × 60 cm) and this is followed by one void volume of standard buffer (pH 5.8 or pH 6.5). Then the column is developed with 6 liters of a 0–0.6 *M* NaCl gradient in standard buffer and eluted at a flow rate of 45 ml/hour. The elution profile for 30 S proteins chromatographed at pH 5.8 is shown in Fig. 2. At least seven of the 30 S proteins can be obtained in a reasonably pure state after a single passage on phosphocellulose (see the table). The remaining proteins must be rechromatographed either on phosphocellulose or Sephadex. The initial chromatography can be done at pH 5.8, which is preferable for the least basic proteins, or at pH 6.5, which is preferable for the most basic proteins.

## Purification of the Proteins

When a chromatographic fraction contains two or more components that are well separated by polyacrylamide gel electrophoresis, it is

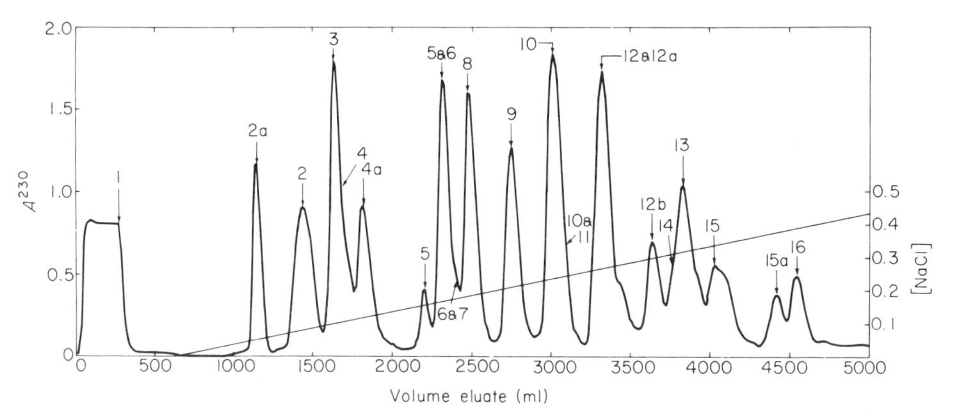


FIG. 2. The elution profile of 30 S ribosomal protein that has been chromatographed on phosphocellulose as described in the text.

usually found that the proteins have different molecular weights but similar charge at the pH of chromatography. Therefore such protein mixtures can be resolved by molecular sieve chromatography in Sephadex (G-100). In most cases the protein samples obtained from phosphocellulose are too dilute for chromatography on Sephadex, so they are concentrated by ultrafiltration before being rerun. We have obtained good separations on a 85 × 2.8 cm G-100 column in standard buffer. The sample is placed on the column in a volume of 10 ml and the flow rate of elution is 15 ml/hr.

When the chromatographic fractions contain components that migrate on polyacrylamide gels in a tight cluster, the proteins usually have similar molecular weights. Therefore, these proteins cannot be resolved in Sephadex, but differences in their pK's can be exploited. This is done by rerunning the mixtures on phosphocellulose at the same or at a different pH with a NaCl gradient, which is shallower than that employed in the initial separation.

The rechromatography on phosphocellulose is usually performed on columns somewhat smaller than those used for the initial fractionation. The chromatographic fraction containing 4–30 mg of protein is applied to a 60 × 1.6 cm phosphocellulose column. Then approximately one-half void volume of standard buffer is passed through the column followed by a 1-liter NaCl gradient with a range of 0.2 *M* and an initial concentration of NaCl that is 50 m*M* lower than the salt concentration at which the sample was originally eluted from the column. The flow rate for development of such columns is 15 ml/hour.

All of the 30 S ribosomal proteins except two can be obtained by a single rerun either in Sephadex or on phosphocellulose. The procedures that we have used to purify 21 30 S proteins are described in the table. Nineteen of these have unique amino acid compositions, tryptic peptides, and are pure according to the criteria that physical molecular weights (measured by sedimentation equilibrium) agree reasonably well with chemical molecular weights (estimated from the amino acid composition and number of tryptic peptides).[9] Two proteins (30 S-5 and 30 S-9) have very similar amino acid compositions, tryptic peptides, cyanogen bromide peptides, and molecular weights. We believe that these are altered forms of a single protein, so that 30 S-5 and 30 S-9 are counted as a single protein. Finally, we have occasionally obtained a contaminant of 30 S-15 that may represent another unique ribosomal protein which we have failed to purify routinely. Therefore, there are at least twenty and possibly 21 proteins that can be obtained from the 30 S ribosomal subunit of *E. coli*.

Purification Procedures for the Proteins of the 30 S Subunit

| Protein | Major contaminants after initial chromatography at | | First rechromatography conditions | Second rechromatography conditions |
|---|---|---|---|---|
| | pH 5.8 | pH 6.5 | | |
| 1 | None | None | – | – |
| 2 | None | None | – | – |
| 2a | None | None | – | – |
| 3 | 4a | 4a | Sephadex G-100 | – |
| 4 | 3 and 4a | 3 and 4a | pH 7.5, 0.04–0.24 *M* NaCl | – |
| 4a | 4 | 4 | pH 6.5, 0.05–0.25 *M* NaCl | – |
| 5 | None or 6 and 7 | None or 6 and 7 | Sephadex G-100 | – |
| 6 | 5 and 7 | 5 and 7 | Sephadex G-100 to remove 5 | pH 5.5, 0.1–0.3 *M* NaCl[a] |
| 7 | 5 and 6 | 5 and 6 | Sephadex G-100 to remove 5 | pH 5.5, 0.1–0.3 *M* NaCl[a] |
| 8 | None | None | – | |
| 9 | None | None | – | |
| 10 | None | None | – | |
| 11 | 10 or 12 | 10 | pH 6.5, 0.2–0.4 *M* NaCl for 12, Sephadex G-100 for 10 | – |
| 12 | 12a | None | pH 6.5, 0.2–0.4 *M* NaCl | – |
| 12a | 12 | None | pH 6.5, 0.2–0.4 *M* NaCl | – |
| 12b | None or 12a | None | pH 6.5, 0.2–0.4 *M* NaCl | – |
| 13 | 14 | 14 | pH 6.5, 0.26–0.46 *M* NaCl[a] | – |
| 14 | 13 | 13 | pH 6.5, 0.26–0.46 *M* NaCl[a] | – |
| 15 | 13 | 13 | pH 6.5, 0.3–0.5 *M* NaCl[a] | – |
| 15a | 15 | 15 | Sephadex G-100 | – |
| 16 | None | None | – | – |

[a] These proteins are initially run in an unreduced state; then they are reduced with 2-mercaptoethanol before the indicated rechromatography step.

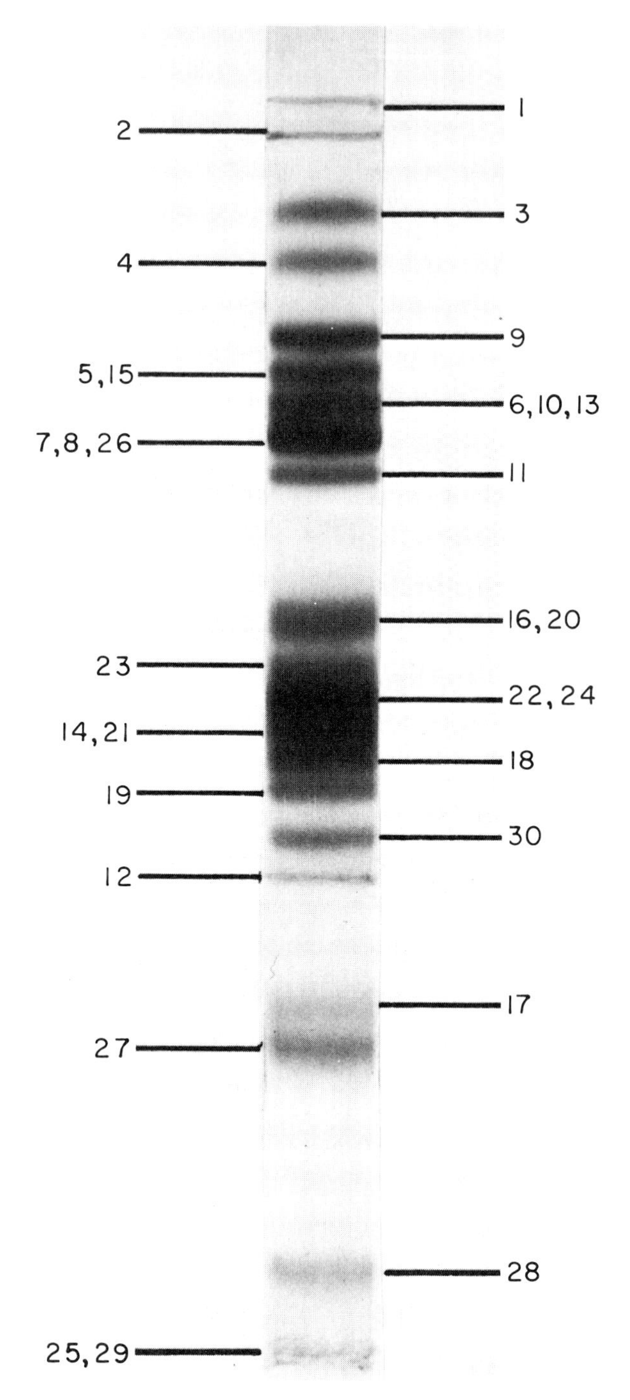


FIG. 3. The disc electrophoresis pattern obtained with the 50 S proteins fractionated on "soft" polyacrylamide gels, which contain 10% acrylamide, 0.15% methylenebis-acrylamide. The numbering of these proteins is arbitrary.

We have not yet completed the analysis of the 50 S ribosomal proteins. However, twenty 50 S proteins have been purified and these have unique amino acid compositions, tryptic peptides, and molecular weights. There are probably ten more 50 S proteins that remain to be characterized (see Fig. 3). We have used the same chromatographic techniques to purify the proteins of the 50 S subunit as were applied to the 30 S proteins. However, one major modification that we have employed with the 50 S proteins entails fractionating these proteins with ammonium sulfate[12] before chromatography on phosphocellulose. Such group fractionations provide a substantial simplification of the chromatographic profiles obtained during the first passage of the separate fractions on phosphocellulose.

The proteins are extracted from 50 S ribosomal subunits by the modified acetic acid procedure; then they are alkylated with iodoacetamide and finally they are dialyzed in standard buffer at pH 6.5. Then 176 g/liter of ammonium sulfate are added very slowly to a solution containing 1–2 g of protein per liter, and the suspension is stirred at 0°–2° for 30 minutes. The suspension is centrifuged for 30 minutes at 13,000 *g*, and the supernatant is recovered. (There rarely is any precipitate in this step.) After the addition of another 96 g of ammonium sulfate per liter, the suspension is stirred and centrifuged as above; the precipitate of this step is saved (fraction A). Then the precipitate is collected in the same manner after the addition of 99 g of ammonium sulfate per liter; this is fraction B. Finally the precipitate (fraction C) and the supernatant (fraction D) are obtained after the addition of 105 g of ammonium sulfate per liter. The pH of the solutions is adjusted to 6.5 before ammonium sulfate is added, and the precipitates are all redissolved in standard buffer. Once the fractions A, B, C, and D have been dialyzed against buffer, they are chromatographed as described above.

**Miscellaneous Precautions**

It is sometimes observed that the recovery of protein from phosphocellulose is quite low. This is a particularly serious problem when partially purified proteins are being rerun on phosphocellulose. Generally, low recoveries can be attributed to one of three problems. First, the phosphocellulose may not have been properly washed. Second, the amount of protein adsorbed to the column may have been too small. Finally, the column may not have been equilibrated properly with buffer; thus, at low pH's it is very difficult to elute basic proteins from phosphocellulose. When the phosphocellulose system is intended for the fractionation of ribosomal proteins from an organism other than

[12] P. Spitnik-Elson, *Biochim. Biophys. Acta* 74, 105 (1963).

*E. coli*, the column size and pH of chromatography will probably have to be adapted to the new system.

The sharpness of peaks eluted from phosphocellulose can be quite deceptive. For example, it is quite commonplace for an apparently well resolved, symmetrical chromatographic peak to contain substantial contamination by one or more proteins. Therefore, it is essential to scan each fraction in a chromatographic peak electrophoretically before the fractions are pooled.

It is possible to overestimate the number of different proteins extracted from the ribosomes if certain precautions are not taken. Thus, the proteins can be chemically modified by a number of agents, such as the decomposition products of urea, or by oxygen. The chemically modified protein will then appear to be chromatographically and electrophoretically unique.[9] For this reason we have included methylamine in the buffers to prevent the modification of proteins by cyanate, which is produced by the decomposition of urea. Similarly, the presence of reducing agents, such as dithiothreitol or mercaptoethanol, will prevent the formation of disulfide bonded aggregates. If the proteins are not intended for functional studies, it is preferable to alkylate them with iodoacetic acid or iodoacetamide in order to prevent such aggregation.

Finally, proteolytic enzymes that contaminate the ribosomes can produce fragments from native proteins, and these fragments will behave as though they are unique proteins.[9] Such artifacts can be detected by comparing the peptide maps of the purified proteins. When two proteins manifest a significant overlap of tryptic peptides, one or both may be degradation products such as 30 S-5 and 30 S-9.[9] By employing well washed ribosomes, by working rapidly in the cold, and by employing esterase inhibitors, the effects of proteolytic activity can be minimized.

# [52] Purification of Ribosomal Proteins from *Escherichia coli* under Nondenaturing Conditions

*By* JAN DIJK and JENNY LITTLECHILD

These methods have been developed to isolate proteins from the *Escherichia coli* ribosome that are in a more "native" state than those that have been previously isolated in the presence of acetic acid and



ISBN 0-12-504180-2

urea. The term "native" does not necessarily imply that the purified proteins are in the same conformational state as they are to be found *in situ* on the ribosome, since under these conditions protein–protein and protein–RNA interactions certainly play an important part. The conditions used for this purification procedure do not involve protein denaturants such as urea and extreme pH or lyophilization, but do employ a high-salt extraction with LiCl followed by fractionation in the presence of salt. High salt concentrations especially of LiCl are known to perturb the tertiary structure of proteins.[1] The procedure described here employs concentrations of salt below that known to cause protein denaturation at pH values of 5.6–9.0.

Since recent work[2] has provided evidence that these so-called "salt extracted" proteins do maintain more tertiary structural interactions and represent a more homogeneous population of protein molecules than the equivalent proteins isolated by the previously described methods (review[3]), they are of obvious importance for use in physical studies. By using these gentler methods of isolation potential protein–protein complexes can be isolated from the subunits which provide a useful insight into the little studied area of protein–protein interactions and the part they play in the ribosomal tertiary structure. The salt-extracted proteins are more soluble at high ionic strength and less soluble at low salt concentrations. This is the reverse of the solubility exhibited by previously prepared ribosomal proteins.

## Materials and Reagents

*Enzymes and Proteins*. Bovine serum albumin is obtained from Calbiochem; glyceraldehyde-3-phosphate dehydrogenase, myoglobin, lysozyme, chymotrypsinogen, carbonic anhydrase, cytochrome *c*, and pyruvate kinase from Boehringer; aldolase (rabbit) and ovalbumin from Serva; and DNase I (RNase-free) from Worthington.

*Chemicals*. Benzamidine hydrochloride, Bicine [*N*,*N*-bis(2-hydroxyethyl)glycine], dithioerythritol, dithiothreitol, *N*-2-hydroxyethylpiperazine-*N*′-2-ethanesulfonic acid (HEPES), phenylmethylsulfonyl fluoride, polyethylene glycol 20,000 (Aquacide III), and sodium deoxycholate are purchased from Calbiochem; CM-Sephadex C-25, DEAE-Sephadex A-25, Sephadex G-100 and G-150 are obtained from Pharmacia, and sucrose (RNase-free) from Schwarz/Mann.

[1] S. Maruyama, K. Kuwajima, K. Nitta, and S. Sugai, *Biochim. Biophys. Acta* **494,** 343 (1977).

[2] C. A. Morrison, E. M. Bradbury, J. Littlechild, and J. Dijk, *FEBS Lett.* **83,** 348 (1977).

[3] H. G. Wittmann, *in* "Ribosomes" (M. Nomura, A. Tissières, and P. Lengyel, eds.), p. 93. Cold Spring Harbor Laboratory, Cold Spring Harbor, New York, 1974.

Stock LiCl and $MgCl_2$ solutions must be treated with purified bentonite[4] and activated charcoal, and then filtered through standard Whatmann filters and finally through Millipore HAPW filters (0.45 $\mu$m pore size).

The use of normal commercial dialysis tubing (Visking, molecular weight cutoff of 15,000) leads to the loss of many ribosomal proteins. The use of especially treated dialysis tubing commercially available as Spectrapor tubing (Spectrum Medical Industries, Los Angeles) is recommended. The species designated Spectrapor 3 (molecular weight cutoff 3500) and Spectrapor 6 (molecular weight cutoff of 2000) are most suitable for use with ribosomal proteins.

*Bacteria. E. coli* A19 cells are grown in rich medium at 37° to late log phase (5 g/liter wet weight). They are harvested in a continuous-flow centrifuge.

Bacterial cells, ribosomes, subunits, and purified proteins are stored frozen at −80°. The entire fractionation is carried out at 0°–4°.

*Buffers for Ribosome and Subunit Preparation*

Buffer A: 10 m*M* Tris·HCl pH 7.5, 0.1 *M* KCl, 20 m*M* $MgCl_2$

Buffer B: 10 m*M* potassium phosphate pH 7.5, 1 m*M* $MgCl_2$

Buffer C: 10 m*M* Tris·HCl, pH 7.5, 70 m*M* KCl, 1 m*M* $MgCl_2$

2-Mercaptoethanol is added to the above buffers at a final concentration of 6 m*M*, just before use.

*Procedure*

A 500-g batch of cells is thawed at 4° and washed with 500 ml of buffer A. The cells are collected by centrifugation at 15,000 *g* for 30 min and resuspended in a further 500 ml of buffer A using a Waring Blendor, with the addition of 3 mg of DNase. The cells are then broken by passing them twice through a Manton–Gaulin press at 10,000 psi, with cooling between each cycle. The resultant suspension is centrifuged at 30,000 *g* for 10 min to remove unbroken cells, followed by a further centrifugation at 30,000 *g* for 30 min to remove the cell debris. The supernatant is then centrifuged at 100,000 *g* for 4 hr to pellet the ribosomes. After resuspension of this pellet into 200 ml of buffer A, the preparation is then clarified by a further centrifugation at 30,000 *g* for 30 min. The ribosomes can be pelleted from the resultant supernatant by centrifugation at 100,000 *g* for 4 hr. The pellets are then resuspended into 100 ml of buffer A and frozen until needed. Subunits are obtained by diluting the monosome suspension to 250 $A_{260}$ units/ml with buffer B and dialysis against two changes of this buffer. Samples equivalent to 10,000 $A_{260}$ units can then be applied to a

[4] H. Fraenkel-Conrat, B. Singer, and A. Tsugita, *Virology* **14,** 54 (1961).

15 to 38% sucrose gradient using a Ti 15 zonal rotor. The resultant subunit peaks are then collected and precipitated by a modification of the procedure of Expert-Bezançon *et al.*[5] Polyethylene glycol 6000 (Merck) is added to the subunits at a concentration of 11% (w/v), in the presence of 20 m*M* $MgCl_2$; the mixture is then stirred for 30 min, and the subunits are pelleted by centrifugation for 1 hr at 15,000 *g*. The 30 S subunits are finally resuspended in buffer C and the 50 S subunits in buffer A, both at a concentration of 200 $A_{260}$ units/ml. They are then frozen until needed.

## Proteins from the 30 S Ribosome Subunit

The following buffers are used for the 30 S ribosomal protein purification:

Buffer C: 10 m*M* Tris·HCl pH 7.5, 70 m*M* KCl, 1 m*M* $MgCl_2$
Buffer D: 50 m*M* sodium acetate pH 5.6
Buffer E: 50 m*M* sodium acetate, pH 5.6, 0.4 *M* LiCl

2-Mercaptoethanol is added to all three buffers at a final concentration of 6 m*M*. The protease inhibitors phenylmethylsulfonyl fluoride at a final concentration of 50 μ*M* and benzamidine at a final concentration of 0.1 m*M* are added to buffers D and E just before use. The former inhibitor was made 50 m*M* in absolute ethanol to be used as a stock solution, since it rapidly hydrolyzes on contact with water. A problem with proteolytic degradation is observed with several of the proteins during the purification procedure. This can be overcome in most instances by the presence of the above-mentioned protease inhibitors used throughout the protein isolation. When the level of these inhibitors is reduced or they are omitted, the protein most susceptible to cleavage is S 5.

### *Salt Extraction of 30 S Proteins*

An amount equivalent to 150,000 $A_{260}$ units of 30 S subunits can be conveniently processed at one time. The proteins are split into two main groups using LiCl.[6,7] An increase in the number of groups into which the proteins can be split is not advisable, since one protein is then present in several groups, making any further purification more tedious. For the same reason it is better not to increase the concentration of ribosomes

[5] A. Expert-Bezançon, M. F. Guérin, D. H. Hayes, L. Legault, and H. Thibault, *Biochimie* **56,** 77 (1974).
[6] I. Itoh, E. Otaka, and S. Osawa, *J. Mol. Biol.* **33,** 109 (1968).
[7] H. E. Homann and K. H. Nierhaus, *Eur. J. Biochem.* **20,** 249 (1971).

in the extraction above 50 $A_{260}$ units/ml, since this results in a similar problem of group overlap.

The frozen 30 S subunits are thawed at 4° and pelleted by centrifugation at 100,000 *g* for 10 hr to remove residual polyethylene glycol. The pellets are then resuspended into buffer C at a concentration of 100 $A_{260}$ units/ml. To this suspension an equal volume of 2 *M* LiCl in buffer C is added, and the mixture is made 1 m*M* with respect to EDTA. After stirring for 10 hr at 4° the core particles are pelleted by centrifugation at 100,000 *g* for a further 10 hr. Owing to a limitation of volume during the centrifugation step, the procedure must be repeated three times using 1 liter of extract on each occasion.

The supernatant is then diluted with an equal volume of buffer D and dialyzed against three changes of this buffer (volume of 10 liters each). The removal of LiCl results in some precipitation (10% of the protein), which is removed by centrifugation at 15,000 *g* for 30 min. This precipitate is rich in proteins S9, S10, and S6, Resolubilization with high salt (2 *M* LiCl) and high concentrations of dithiothreitol (up to 10 m*M*) is possible to a limited degree. The soluble extract is now ready for application to a CM-Sephadex C-25 column. The core particles are stored at −80° until a further extraction can be performed.

The proteins removed from the 30 S subunit during the first extraction are shown in Table I.

When convenient, the frozen core particles are thawed and reextracted in a similar manner, but on this occasion are directly resuspended into 2 *M* LiCl in buffer C at a concentration of 50 $A_{260}$ units/ml with the addition of EDTA, pH 7.0, to a final concentration of 10 m*M*. The extract so obtained, after centrifugation and removal of the core particles, is diluted with an equal volume of buffer and dialyzed against buffer D as described above. On removal of the LiCl, protein precipitation is hardly present for this extract so the clarification step is not necessary. The second extract is now ready for application onto the ion-exchange column. The resultant core particles are frozen at −80°. The proteins which should be removed from the 30 S subunit during the second extraction are shown in Table I.

A third extraction of these cores is performed by resuspending them into 4 *M* LiCl in buffer C with the addition of EDTA, pH 7.0, to a final concentration of 10 m*M*. Any remaining protein that may be present on the RNA core particles should be checked by an extraction with 67% acetic acid as described previously.[8] The third and final extract of the core particles contains only a small amount of protein and is usually not

[8] S. J. S. Hardy, C. G. Kurland, P. Voynow, and G. Mora, *Biochemistry* **8**, 2897 (1969).

TABLE I
DISTRIBUTION OF THE RIBOSOMAL PROTEINS FROM THE 30 S SUBUNIT OVER THE THREE EXTRACTS[a]

| Protein | 1 *M* LiCl, 1 m*M* EDTA, buffer C | 2 *M* LiCl, 10 m*M* EDTA, buffer C | 4 *M* LiCl, 1 m*M* EDTA, buffer C |
|---|---|---|---|
| S1 | ++ | (±) | |
| S2 | ++ | − | |
| S3 | ++ | − | |
| S4 | ++ | (±) | (±) |
| S5 | ++ | − | |
| S6 | + | − | |
| S7 | (±) | (±) | (±) |
| S8 | (±) | ++ | |
| S9 | + | − | |
| S10 | + | − | |
| S13 | − | ++ | |
| S14 | + | − | |
| S15 | (±) | ++ | |
| S16 | ++ | − | |
| S17 | (±) | + | |
| S19 | − | ++ | |
| S20 | + | ++ | |
| S21 | ++ | − | |

[a] The presence of large amounts of protein is indicated by ++, of smaller amounts by +, and negligible amounts by (±). S11, S12, and S18 were not detected in sufficient quantities (see text).

fractionated further (see Table I). After extraction of the RNA core with 67% acetic acid, a little S7 and a trace of S4 is removed.

## *Column Chromatography of 30 S Proteins*

The first extract is applied to the preequilibrated ion-exchange column (3 cm × 40 cm) at a flow rate of 100 ml/hr. This CM-Sephadex column is eluted with a linear gradient of LiCl (0.15 to 0.8 *M*); volume 7 liters. Fractions of 15 ml are collected and analyzed for protein quantity by measurement of the $A_{235}$ or by fluorescamine assays on 40-μl samples and for protein content by cylindrical polyacrylamide gel electrophoresis in 6 *M* urea at pH 4.5 using 200-μl aliquots or by SDS polyacrylamide slab gel electrophoresis using 100-μl aliquots as described in the section on electrophoretic analysis of protein samples. Proteins S9, S10, and S6 form the insoluble material that is removed from the extract by centrifugation as described above. When the precipitated proteins are not removed from the extracts before application to the ion-exchange column with the idea that they might solubilize with the increasing LiCl gradient, a large amount of protein is left on the top of the column, which, when redissolved, is representative of the total protein content of the extract, not just of the precipitated proteins. Also, small amounts of proteins S9 and S10 that do not redissolve are smeared across the column profile and do not elute as a single protein peak. This effect is also observed with protein S6, even when precipitated material is removed from the extract. The profile that should be obtained from this first ion-exchange column is shown in Fig. 1.

The second extract is later applied to a preequilibrated CM-Sephadex column as described above and is eluted with a convex gradient of LiCl (0.15 to 1.0 *M*); volume 7 liters, at a flow rate of 100 ml/hr. The convex gradient used in this second extraction increases the separation of proteins in the later part of the elution. The profile that should be obtained from the second ion-exchange column is shown in Fig. 2.

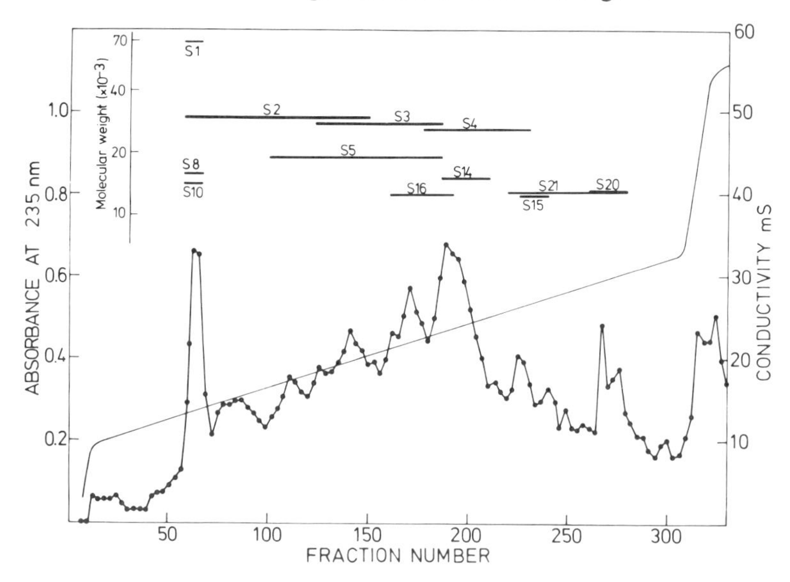


FIG. 1. Fractionation of proteins obtained from the 1 *M* LiCl extraction of 30 S subunits from *Escherichia coli* A19, by chromatography on CM-Sephadex C-25 in buffer D, pH 5.6. A salt gradient of LiCl is followed by a high-salt wash (2 *M* LiCl). This results in a mixture of aggregated protein (less than 10% of the total) being eluted from the column which is representative of all the proteins in the initial extract. Polyacrylamide gel electrophoresis demonstrates that the elution peaks contain the proteins indicated. The molecular weights used for these proteins are obtained from the sodium dodecyl sulfate–acrylamide gel system. [U. K. Laemmli and M. Favre, *J. Mol. Biol.* **80**, 575 (1973).]

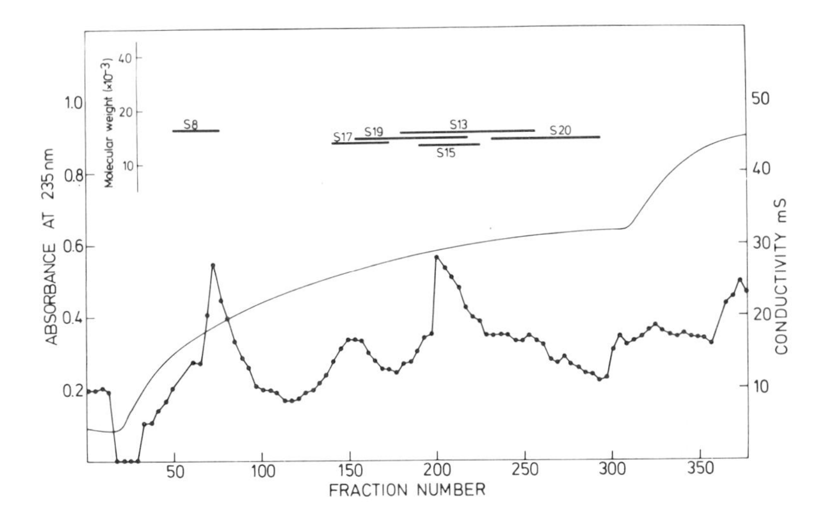


FIG. 2. Fractionation of proteins obtained from the 2 *M* LiCl extraction of 30 S subunits from *Escherichia coli* A19, by chromatography on CM-Sephadex C-25 in buffer D, pH 5.6. An exponential gradient of LiCl was followed by a further short gradient of LiCl, 1.0 to 2.0 *M*. Polyacrylamide gel electrophoresis demonstrates that the elution peaks contain the proteins indicated. The molecular weights used for these proteins were obtained as for Fig. 1.

It is found that, with a few exceptions, most fractions from the ion-exchange columns have to be subjected to a second purification step on Sephadex G-100 after protein concentration, which is considered separately in a later section. The protein mixtures, pooled to achieve maximum separation, are then applied to Sephadex G-100 columns (5 × 150 cm) that have been preequilibrated with buffer E.

The isolated proteins are listed in Table II together with their expected yield and the chromatographic steps used during their purification. Traces of protein S6 are found associated with proteins S4 and S5.

A small amount of protein S7 is often found associated with proteins S13 and S19 during the second fractionation procedure. This S7 cannot be separated from S13–S19 by Sephadex G-100 chromatography.

Several proteins appear to migrate together after the two chromatographic procedures employed. A list of these is given in Table III, and they are assumed to be potential protein–protein complexes. These proteins migrate together to a varying extent on the CM-Sephadex ion-exchange column and can be separated only to a limited degree during gel filtration on Sephadex G-100. Some examples of this effect are listed. (1) Protein S3, comigrates on the ion-exchange column with protein S5, but when this is applied to a Sephadex G-100 column most of the mixture

TABLE II
PURIFIED 30 S RIBOSOMAL PROTEINS[a]

| Protein | Yield (%) | Yield (mg) | Procedure |
|---|---|---|---|
| S1 | — | 26 | CM-S, G-100 |
| S2 | 31 | 75 | CM-S |
| S3 | 15 | 34 | CM-S, G-100 |
| S4 | 26 | 44 | CM-S, G-100 |
| S5 | 35 | 47 | CM-S, G-100 |
| S8 | 35 | 33 | CM-S |
| S13 | 18 | 19 | CM-S, G-100 |
| S14 | 9 | 10 | CM-S, G-100 |
| S15 | 21 | 16 | CM-S, G-100 |
| S16 | 13 | 12 | CM-S, G-100 |
| S17 | 10 | 9 | CM-S, G-100 |
| S19 | 19 | 19 | CM-S, G-100 |
| S20 | 16 | 12 | CM-S (G-100) |
| S21 | 15 | 9 | CM-S, G-100 |

[a] The yield of each protein is calculated for 100,000 $A_{260}$units of 30 S subunits using a protein concentration as determined by the nitrogen assay. The amounts given in milligrams of protein include that contained in the potential protein–protein complexes as shown in Table III. The purification steps are indicated by the following abbreviations: CM-S for CM-Sephadex C-25 chromatography; G-100 for gel filtration on Sephadex G-100.

TABLE III
POTENTIAL 30 S PROTEIN–PROTEIN COMPLEXES[a]

| Potential complex | Yield (mg) | Procedure |
|---|---|---|
| S2–S3 | 3 | CM-S, G-100 |
| S3–S5 | 37 | CM-S, G-100 |
| S3–S4-S5 | 40 | CM-S, G-100 |
| S13–S19 | 19 | CM-S, G-100 |
| S13–S20 | 12 | CM-S, G-100 |

[a] The yield of each potential complex is calculated for 100,000 $A_{260}$ units of 30 S subunits using a protein concentration as determined by the nitrogen assay.

does not separate but some protein S5 is eluted earlier, indicating that it is probably aggregated. (2) Approximately 30% of the proteins S3, S4, S5 comigrate on the ion-exchange column, although this amount varies from one extraction to another. Some S4 separates from the S3–S5 mixture, but no further separation of S4 occurs on Sephadex G-100. (3) Most of protein S13 migrates with protein S19 on the ion-exchange column along with protein S15, which is separated from S13–S19 on Sephadex G-100. The S13–S19 elutes as a single peak with an even distribution of the two proteins.[9]

## Proteins from the 50 S Subunit

The following buffers are used for the purification of the 50 S ribosomal proteins:

Buffer F: 10 m*M* HEPES (titrated to pH 7.0 with NaOH), 10 m*M* $MgCl_2$
Buffer G: 10 m*M* HEPES, pH 7.0, 10 m*M* EDTA
Buffer H: 5 m*M* HEPES, pH 7.0

To these buffers 2-mercaptoethanol, phenylmethylsulfonyl fluoride (dissolved in absolute ethanol), and benzamidine hydrochloride are added, just before use, to final concentrations of 6 m*M*, 50 μ*M*, and 0.1 m*M*, respectively.

### *Salt Extraction of 50 S Proteins*

The 34 proteins of the 50 S subunit are extracted into 4 groups by the addition of LiCl. The first two extractions, using 1 *M* and 2 *M* LiCl, respectively, are carried out in the presence of 10 m*M* $Mg^{2+}$. Thereafter, the subunits are unfolded by treatment with EDTA and two further extractions are performed using 1 *M* and 7 *M* LiCl, respectively. The cores after the fourth extraction contain hardly any protein, as can be shown by extraction with 67% acetic acid[8] and subsequent two-dimensional electrophoresis. The concentration of subunits during extraction has to be as low as possible, otherwise proteins will be present in more than one extract. In order to limit the number of subsequent ultracentrifugation runs a compromise value of 100 $A_{260}$units/ml is used. The extractions are performed in the following way:

An amount of approximately 300,000 $A_{260}$units of 50 S subunits is usually processed. To the subunit solution concentrated $NH_4Cl$ solution is added to a final concentration of 0.5 *M*, and the subunits are pelleted

[9] J. Dijk, J. Littlechild, and R. A. Garrett, *FEBS Lett.* **77**, 295 (1977).

by centrifugation for 12 hr at 100,000 *g*. This will remove most of the residual polyethylene glycol used during the previous precipitation step. The pellets are dissolved in buffer F and diluted to the appropriate concentration (see below). A concentrated LiCl stock solution is added under rapid stirring until the required concentration (1 *M*) is reached. The concentration of the subunits should be 100 $A_{260}$ units/ml at this stage. After standing for 12 hr, the cores and extracted proteins are separated by ultracentrifugation (12 hr at 100,000 *g*). The cores are dissolved in buffer F and stored at −80° until the next extraction (2 *M* LiCl) can be performed. For this extraction they are diluted with buffer F and LiCl is added to 2 *M*. Further treatment is the same as mentioned above. The cores, before the third extraction can be carried out, have to be dialyzed against buffer F in order to remove $Mg^{2+}$. They are then diluted with buffer G, and concentrated LiCl solution is added to 1 *M* final concentration. The cores are finally extracted with 7 *M* LiCl in buffer G. The extraction steps used are summarized in Table IV together with the extracted proteins.

*Column Chromatography of 50 S Proteins*

Each extract is applied to a CM-Sephadex C-25 column at low ionic strength. Usually a small amount of protein precipitates during dialysis against the low ionic strength buffer. This is removed by centrifugation; sometimes proteins can be recovered from this precipitate by dissolving it again in high ionic strength buffer (e.g., 2 *M* LiCl) and attempting another fractionation. The acidic proteins that do not bind to the CM-Sephadex column are absorbed onto a small DEAE-Sephadex A-25 column. Both columns are eluted with a salt gradient (LiCl); for the CM-Sephadex column a concave gradient gives better separation in the early part of the gradient. The first three extracts are treated in an identical manner. The fourth extract which mainly contains protein L4 is not processed since the recovery of this protein after chromatography is negligible. They are diluted with an equal volume of buffer H and dialyzed against a large volume of buffer H containing 70 m*M* LiCl. After the extract has reached a conductivity which is lower than that of the starting buffer for the CM-Sephadex column, the protein precipitate is removed by centrifugation (30 min, 10,000 *g*). The supernatant is applied to a CM-Sephadex column (3 × 45 cm, approximately 300 ml) equilibrated with 0.1 *M* LiCl in buffer H at a flow rate of 100 ml/hr.

After the sample has been applied the column is washed with 500–600 ml of starting buffer, after which the gradient is begun. The gradient is generated from 10 liters of starting buffer which is pumped onto the

TABLE IV
DISTRIBUTION OF THE RIBOSOMAL PROTEINS FROM THE 50 S SUBUNIT OVER THE FOUR EXTRACTS[a]

| Protein | 1 *M* LiCl/$Mg^{2+}$ | 2 *M* LiCl/$Mg^{2+}$ | 1 *M* LiCl/ EDTA | 7 *M* LiCl/EDTA |
|---|---|---|---|---|
| L1 | ++ | + | (±) | − |
| L2 | ++ | + | − | − |
| L3 | − | − | ++ | (±) |
| L4 | − | − | − | ++ |
| L5 | + | ++ | + | − |
| L6 | ++ | + | (±) | − |
| L7/12 | ++ | + | − | − |
| L9 | + | + | + | − |
| L10 | ++ | + | − | − |
| L11 | ++ | + | − | − |
| L13 | − | + | ++ | − |
| L14 | + | + | − | − |
| L15 | ++ | − | + | − |
| L16 | ++ | (±) | − | − |
| L17 | − | + | ++ | − |
| L18 | + | ++ | − | − |
| L19 | − | + | ++ | − |
| L21 | − | + | − | − |
| L22 | − | − | ++ | − |
| L23 | − | − | ++ | − |
| L24 | − | ++ | ++ | − |
| L25 | + | ++ | (±) | − |
| L27 | ++ | (±) | − | − |
| L28 | ++ | + | − | − |
| L29 | − | + | + | − |
| L30 | ++ | + | + | − |
| L32 | − | ++ | − | − |
| L33 | − | ++ | − | − |
| L34 | + | − | − | − |
| 5 S RNA | + | ++ | + | − |

[a] The presence of 50% or more of a protein in one extract is indicated by ++, of small amounts by +, and of negligible amounts by (±). In addition, the presence of 5 S RNA is indicated.

column at a rate of 100 ml/hr; a second pump delivers 1.2 *M* LiCl in buffer H into the starting buffer container at a rate of 30 ml/hr. Fractions of 20 ml are collected; the elution is recorded by measurements of the absorbance at 230 nm or by fluorescamine assays on 40-μl samples. The gradient is monitored by measurement of the conductivity. Proteins are located by analyzing 100-μl samples from every third fraction by SDS gel electrophoresis or 500-μl samples by gel electrophoresis in 6 *M* urea at pH 4.5.

The breakthrough volume of the CM-Sephadex column, containing unbound acidic proteins, is diluted with an equal volume of buffer H and applied to a DEAE-Sephadex A-25 column (2 × 15 cm, approximately 50 ml) equilibrated with 50 m*M* LiCl in buffer H at a flow rate of 60 ml/hr. After washing the column with 100 ml of starting buffer the proteins (mainly the L7/12–L10 complex) are eluted by a linear LiCl gradient from 50 m*M* to 0.5 *M*. The complex is eluted at 0.2 *M* LiCl. This step, besides providing some further purification of the L7/L12–L10 complex (5 S RNA present in the extracts is retained on the column) mainly serves to concentrate the protein complex, which is obtained in a large volume after the CM-Sephadex chromatography.

The distribution of ribosomal proteins over the fractions is judged from the electrophoretic patterns (Figs. 3–5) and fractions are pooled

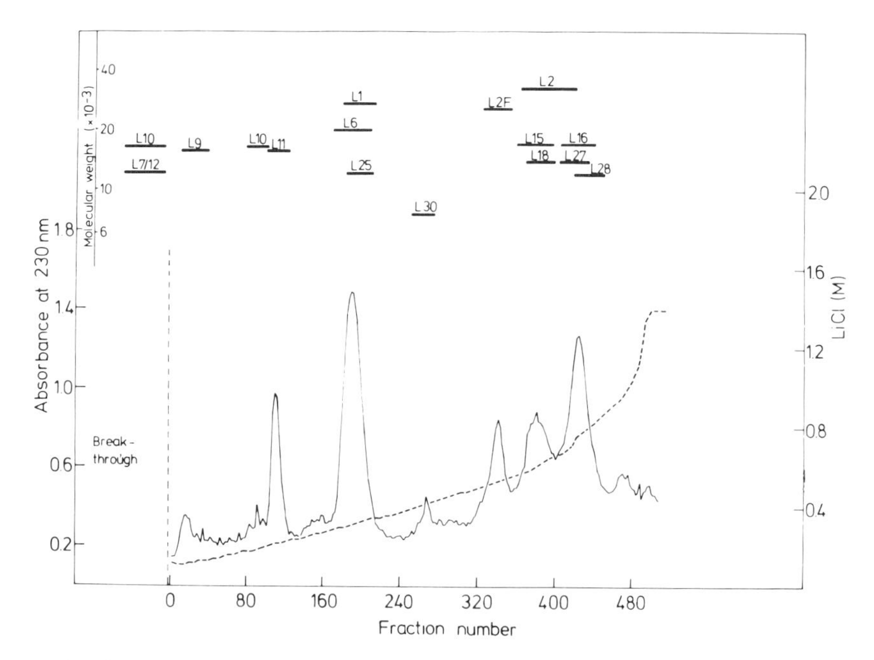


FIG. 3. Separation of the 50 S ribosomal proteins from extract 1 (1 *M* LiCl, 10 m*M* $MgCl_2$) on CM-Sephadex C-25 at pH 7.0. ——, Absorbance at 230 nm; - - -, LiCl gradient. The presence of ribosomal proteins, as determined by disc gel or sodium dodecyl sulfate (SDS) slab gel electrophoresis, is indicated by the solid bars. They are positioned according to their molecular weight as determined by SDS slab gel electrophoresis.

accordingly. They are either processed further by a second ion-exchange chromatographic step or are concentrated to approximately 50 ml volume (see section on concentration methods) for gel filtration on Sephadex G-100.

The conditions for a second ion-exchange step are generally taken from the results of the first CM-Sephadex column, but a less steep gradient is used. As a typical example, the separation of L16 and L27 is given in detail. The fractions 400–430 from the first extract fractionation (Fig. 3) are pooled, diluted with 2 volumes of buffer H and applied to a small CM-Sephadex C-25 column (2 × 15 cm) equilibrated with 0.5 *M* LiCl in buffer H. The proteins are eluted with a linear gradient of 0.5 *M* to 0.9 *M* LiCl in buffer H at a flow rate of 50 ml/hr. Protein L27 is eluted at 0.58 *M* LiCl, L16 at 0.65 *M* LiCl. As can be seen from this example, the proteins are eluted at lower salt concentrations than in the first ion-exchange chromatography. Proteins L1 and L6 (Figs. 3 and 4) can be separated only by a second ion-exchange chromatographic step at pH 9.0; L6 is eluted first at 0.17 *M* LiCl, L1 at 0.21 *M* LiCl. The L25 contamination can be removed from L6 by gel filtration on Sephadex G-100.

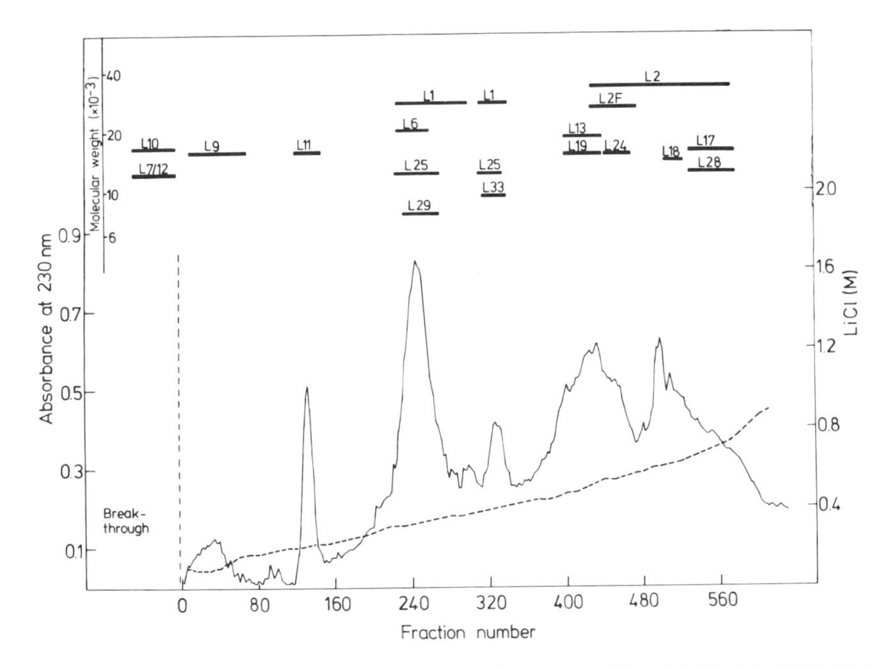


FIG. 4. Separation of 50 S ribosomal proteins from extract 2 (2 *M* LiCl, 10 m*M* $MgCl_2$) on CM-Sephadex C-25 at pH 7.0. Details as in Fig. 3.

Protein mixtures after concentration to approximately 50 ml are applied to a Sephadex G-100 column (5 × 150 cm, approx. 3000 ml), which is eluted with 0.5 *M* LiCl in buffer H at a flow rate of 60 ml/hr. Fractions of 15 ml are collected, and 100-$\mu$l samples are analyzed by SDS gel electrophoresis or 500-$\mu$l samples by gel electrophoresis in 6 *M* urea at pH 4.5. Fractions containing a single protein are pooled and concentrated by the Sephadex G-150 technique (see below). The 50 S proteins that can be isolated by these procedures are listed in Table V together with their yield and the chromatographic procedures used. Several proteins found in the extracts are not recovered after chromatography, e.g., L4, L5, L14, and L32. This is probably caused by aggregation and solubility problems with these particular proteins.

In addition to the ribosomal proteins listed, two other proteins are found reproducibly that do not correspond to any of the identified ribosomal proteins. They represent proteolytic degradation products of proteins L2 and L23, respectively, and are designated L2F and L23F (Figs. 3-5).

Several protein mixtures cannot be separated by the methods described. Since they migrate together with a constant ratio between the

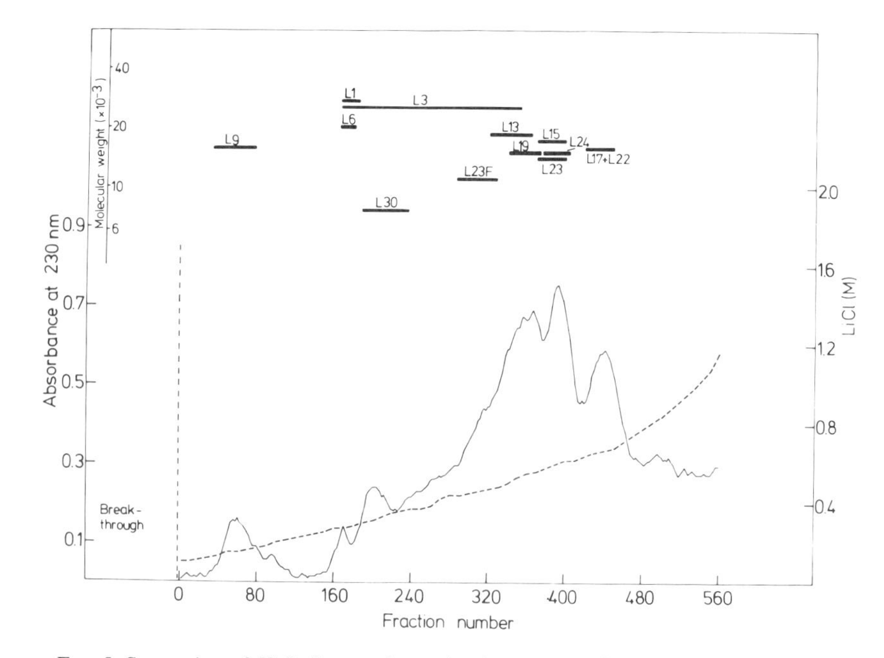


FIG. 5. Separation of 50 S ribosomal proteins from extract 3 (1 *M* LiCl, 10 m*M* EDTA) on CM-Sephadex C-25 at pH 7.0. Details as in Fig. 3.

TABLE V
PURIFIED 50 S RIBOSOMAL PROTEINS

| Protein | Yield (mg) | Purification steps |
|---|---|---|
| L1 | 67 | CM-S; CM-S (pH 9.0, 0.15–0.30 *M*) |
| L2 | 36 | CM-S; G-100 |
| L3 | 130 | CM-S; G-100 |
| L6 | 51 | CM-S; CM-S (pH 9.0, 0.15–0.30 *M*); G-100 |
| L9 | 65 | CM-S |
| L10 | 18 | CM-S |
| L11 | 167 | CM-S |
| L15 | 18 | CM-S; G-100 |
| L16 | 26 | CM-S; CM-S (pH 7.0, 0.4–0.9 *M*); G-100 |
| L17 | 10 | CM-S; G-100 |
| L18 | 10 | CM-S; G-100 |
| L22 | 26 | CM-S; G-100 |
| L23 | 22 | CM-S; G-100 |
| L24 | 74 | CM-S; G-100 |
| L25 | 24 | CM-S; G-100 |
| L27 | 21 | CM-S; CM-S (pH 7.0, 0.4–0.9 *M*); G-100 |
| L28 | 7 | CM-S; G-100 |
| L29 | 5 | CM-S; G-100 |
| L30 | 15 | CM-S; G-100 |
| L33 | 3 | CM-S; G-100 |
| L2F | 14 | CM-S; G-100 |

[a] The yield of each protein was calculated for the extraction of 210,000 $A_{260}$units (13 g) of 50 S subunits using a protein concentration determined by nitrogen assay. The abbreviations are as described in Table II, and in addition DEAE-S is used for DEAE-Sephadex A-25 chromatography. If conditions for ion-exchange chromatography were different from those described in sections on materials and methods or in Figs. 3–5, they are indicated in parentheses, mentioning the pH and the LiCl gradient used.

two protein bands, they are considered to be potential protein–protein complexes (Table VI). This list includes the very stable complex of L7/12–L10.[9,10]

## Electrophoretic Analysis of Protein Samples

Proteins can be analyzed for purity and identity by (1) a two-dimensional polyacrylamide gel electrophoresis system as described by

[10] I. Pettersson, S. J. S. Hardy, and A. Liljas, *FEBS Lett.* **64,** 135 (1976).

TABLE VI
POTENTIAL 50 S PROTEIN-PROTEIN COMPLEXES[a]

| Protein | Yield (mg) | Purification steps |
|---|---|---|
| L7/12–L10 | 225 | CM-S; DEAE-S (pH 7.0, 0.05–0.2 *M*); G-100 |
| L13–L19 | 70 | CM-S; CM-S (pH 7.0, 0.4–0.8 *M*); G-100 |
| L16–L27 | 15 | CM-S; CM-S (pH 7.0, 0.4–0.9 *M*); G-100 |
| L3–L23F | 21 | CM-S; G-100 |

[a] Yields were calculated and purification steps are described as in Table V.

Kaltschmidt and Wittmann[11] using the stain Amido Black; the gel size has been decreased to 15 × 15 cm; (2) a two-dimensional electrophoresis system using SDS in the second dimension[12,13]; (3) electrophoresis on cylindrical polyacrylamide gels in 6 *M* urea at pH 4.5[14]; (4) SDS slab gel electrophoresis on 15% polyacrylamide gels.[15] Both SDS gels are stained with Coomassie Brilliant Blue stain R250. Protein samples too dilute for direct application to the gel can be precipitated by the addition of an equal volume of 10% trichloroacetic acid in the presence of 10 μg of sodium deoxycholate per milliliter. After 1 hr at 0° the precipitate is collected by centrifugation in a Beckman Microfuge Model B for 1–2 min at 12,000 rpm. The pellet is either washed with diethyl ether to remove residual trichloroacetic acid or dissolved in sample solution for the electrophoresis system, which is at a higher pH to compensate for the acid in the sample pellet.

The purity and extent of any degradation occurring in the proteins is readily observed by the use of the discontinuous SDS acrylamide slab gel (Figs. 6–8). This method does not separate all the ribosomal proteins, but it is especially useful for the direct identification of several of the proteins of higher molecular weight. The stacking effect of the discontinuous system together with the sensitivity of the Coomassie Brilliant Blue stain reveals components that cannot be seen by the other gel methods employed.

[11] E. Kaltschmidt and H. G. Wittmann, *Anal. Biochem.* **36,** 401 (1970).
[12] L. J. Mets and L. Bogorad, *Anal. Biochem.* **57,** 200 (1974).
[13] A. Kyriakopoulos and A. R. Subramanian, *Biochim. Biophys. Acta* **474,** 308 (1977).
[14] I. Hindennach, G. Stöffler, and H. G. Wittmann, *Eur. J. Biochem.* **23,** 7 (1971).
[15] U. K. Laemmli and M. Favre, *J. Mol. Biol.* **80,** 575 (1973).

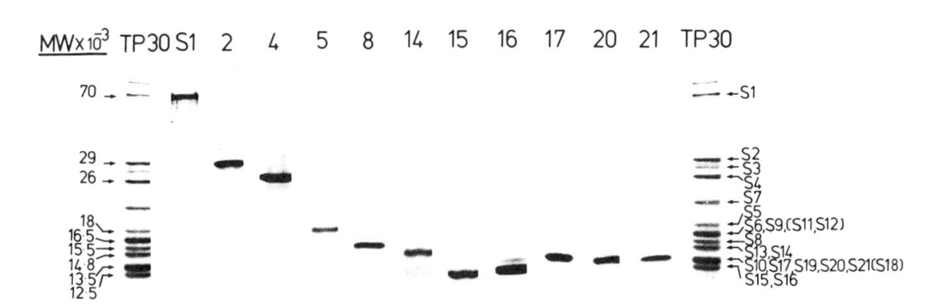


FIG. 6. Sodium dodecyl sulfate (SDS) polyacrylamide gel electrophoresis of the purified 30 S proteins obtained by the nondenaturing isolation procedure. Their position relative to that of the TP30, extracted with acetic acid [S. J. S. Hardy, C. G. Kurland, P. Voynow, and G. Mora, *Biochemistry* **8**, 2897 (1969)], is shown, together with the molecular weight obtained with reference to the standard proteins: bovine serum albumin (MW 68,000), pyruvate kinase (MW 57,000), ovalbumin (MW 43,500), aldolase (MW 40,000), glyceraldehyde phosphate dehydrogenase (MW 36,000), carbonic anhydrase (MW 29,200), chymotrypsinogen (MW 25,700), myoglobin (MW 16,957), lysozyme (MW 14,319), and cytochrome *c* (MW 11,748).

For most of the ribosomal proteins an overestimation of the molecular weight is observed using this method compared with that obtained from the primary sequence. The values obtained from the SDS gel electrophoresis are shown in Fig. 6 and 8.

### Storage of Purified Proteins

Purified 30 S proteins in buffer E and 50 S proteins in buffer H containing 0.5 *M* LiCl are concentrated to a protein concentration of 1–5 mg/ml. They are then dialyzed against their respective buffers, in which 2-mercaptoethanol has been substituted by 1 m*M* dithioerythritol and stored in small aliquots at −80°. The proteins are fairly soluble at these concentrations even after freezing and thawing. Prolonged storage at −20° leads to some precipitation of protein after thawing.

Protein S2 should be dialyzed against a buffer at pH 8.5 containing 50 m*M* Bicine, 0.4 *M* LiCl, 10 m*M* dithioerythritol, 50 μ*M* phenylmethylsulfonyl fluoride, 0.1 m*M* benzamidine before concentration, since the isoelectric point of this protein is at pH 6.7.[16] Precipitated protein is seen in S2 samples on standing at 4° for a week at pH 5.6. Protein S8, at concentrations above 1 mg/ml, is more soluble if the salt concentration in buffer E is raised to 0.6 *M* LiCl.

[16] E. Kaltschmidt, *Anal. Biochem.* **43**, 25 (1971).

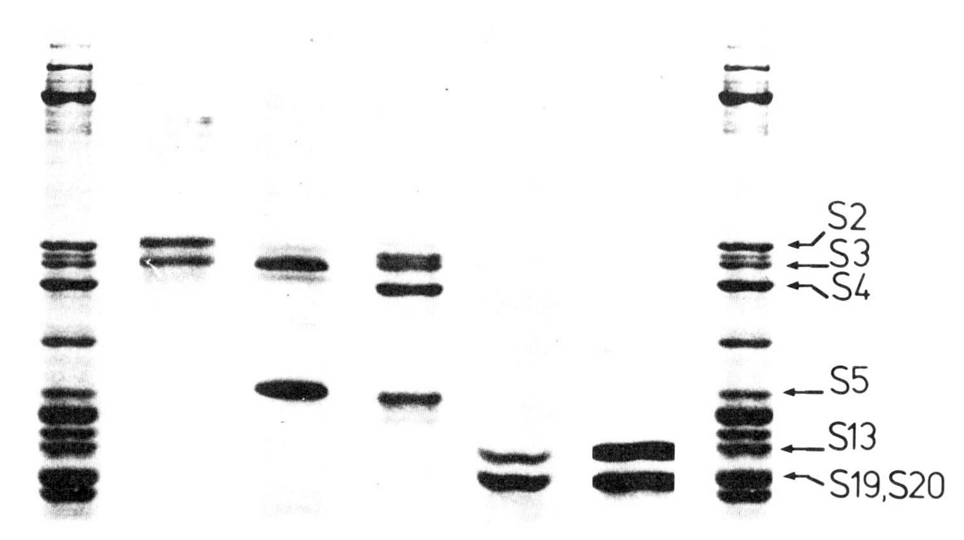


FIG. 7. Sodium dodecyl sulfate (SDS) polyacrylamide gel electrophoresis of potential protein–protein complexes obtained during the isolation procedure. These proteins could be separated to only a limited degree, as described in the text. Their positions relative to the TP30 are shown.

A few 50 S proteins have a limited solubility even at high ionic strength. Protein L10 is insoluble at protein concentrations above 1 mg/ml because of severe aggregation; L1 and L24 are soluble up to 1–2 mg/ml. In contrast, proteins L6, L7/12–L10, L11, and L30 are very soluble at concentrations of 10 mg/ml and higher.

## Concentration of Protein Solutions

Lyophilization should not be used since it causes protein denaturation. Salt-extracted proteins after lyophilization become very soluble in water but insoluble in salt-containing buffers.

Precipitation techniques with ammonium sulfate are not practical owing to (1) the low protein concentrations and large volumes involved; (2) a requirement by these proteins for a high salt concentration for precipitation.

The proteins can be concentrated by the following methods. (1) Dialysis against dry Sephadex G-150, which is slow and time consuming but is the gentlest method. (2) Dialysis against 15% polyethylene glycol

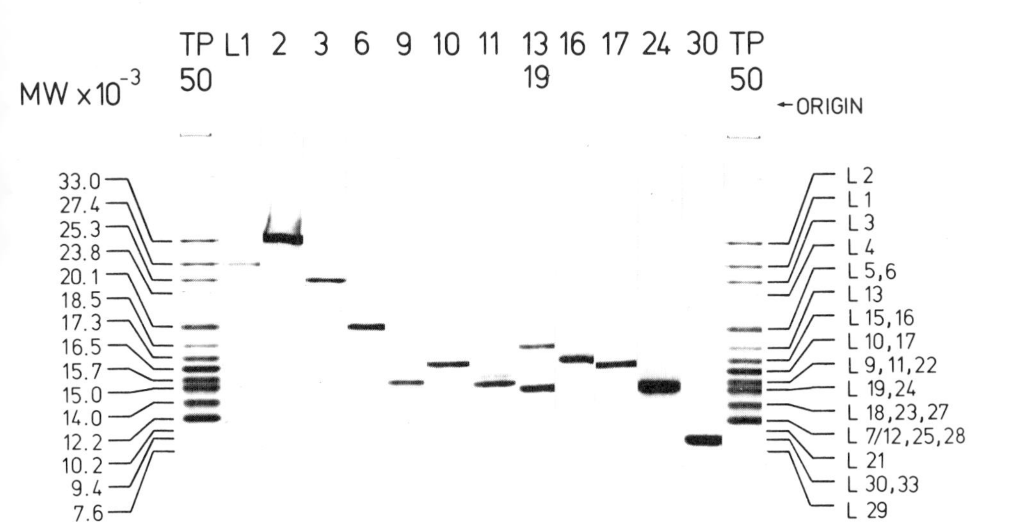


FIG. 8. Sodium dodecyl sulfate (SDS) polyacrylamide gel electrophoresis of some of the 50 S ribosomal proteins. For comparative purposes the separation of the mixture of 50 S subunit proteins (TP50) is also shown; the ribosomal proteins present in each band are identified, and their molecular weights are given. These were determined as described for Fig. 6.

20,000 in buffer E for 30 S proteins and 0.5 *M* LiCl in buffer H for 50 S proteins. This method allows a certain amount of polyethylene glycol to pass across the membrane that cannot be removed by subsequent dialysis against buffer solutions. (3) Pressure filtration using Millipore concentration cells (75 ml or 3 ml capacity) using Pellicon (PSAC 01310 and PSAC 04710) or Amicon (Diaflo UM-2) membrane filters both with a molecular weight cutoff of 1000. This method is good for the larger ribosomal proteins, although it has the tendency to cause some protein aggregation. Many of the ribosomal proteins pass totally or partially through these membranes despite the fact that the membrane should withhold proteins above MW 1000. (4) Reabsorption onto small CM-Sephadex columns followed by batch elution at high ionic strength. This method is gentler and results in little loss of protein if the high ionic strength buffer is allowed to remain in contact with the ion-exchanger for several hours. Methods (1) and (4) were found to be most favorable for proteins that are to be used for physical studies.

### Determination of Protein Concentration

Protein concentrations can be determined by three methods. (1) One is a modified fluorescamine assay[17] in which 40 μl of the protein solution are mixed with 260 μl of 50 m*M* sodium tetraborate buffer, pH 9.0, followed by the rapid addition of 50 μl of a fluorescamine solution in acetone (0.6 mg/ml). The fluorescence is measured using a 250-μl cuvette in an Aminco-Bowman fluorometer. Lysozyme is used to obtain a calibration curve in the range of 1–100 μg. This method although rapid to perform usually gives an overestimation of the protein concentration, probably owing to the relatively high content of lysine found in the ribosomal proteins. (2) A nitrogen assay,[18] where ammonium sulfate is used as a standard and the nitrogen content of each protein is calculated from its amino acid composition.[19,20] This method cannot be used in the presence of nitrogen-containing compounds such as HEPES buffer, which must be removed by dialysis. (3) The Lowry reaction.[21] This assay generally gives lower concentrations than that determined by method (2), and the presence of thiol reagents in the protein solutions creates errors in the assay even if the modification of Geiger and Bessman[22] is used. Therefore, method (2) is the best and most reliable way of determining the exact protein concentration. Method (1) is rapid and easy to perform, and a conversion factor can be calculated for each of the proteins to correct for the overestimated values obtained by this method. This factor can vary from 1.5 to 3.5.

## Yield of Ribosomal Proteins

Proteins that are obtained pure after one chromatographic step have increased yields. The reason why some proteins are absent from the lists is attributed to their limited solubility under the conditions used for fractionation. Once a protein precipitates from a mixture, other proteins tend to stick to this precipitate, and their yield is therefore reduced. The yields obtained by this new purification method are not very different from those obtained by the acetic acid/urea isolation procedure of Hindennach *et al.*[14,23]

[17] P. Böhlen, S. Stein, W. Dairman, and S. Udenfriend, *Arch. Biochem. Biophys.* **155,** 213 (1973).

[18] L. Jaenicke, *Anal. Biochem.* **61,** 623 (1974).

[19] E. Kaltschmidt, M. Dzionara, and H. G. Wittmann, *Mol. Gen. Genet.* **109,** 292 (1970).

[20] G. Stöffler and H. G. Wittmann, *in* "Molecular Mechanisms of Protein Biosynthesis" (H. Weissbach and S. Pestka, eds.), p. 117. Academic Press, New York, 1977.

[21] O. H. Lowry, N. J. Rosebrough, A. L. Farr, and R. J. Randall, *J. Biol. Chem.* **193,** 265 (1951).

[22] P. J. Geiger and S. P. Bessman, *Anal. Biochem.* **49,** 467 (1972).

[23] I. Hindennach, E. Kaltschmidt, and H. G. Wittmann, *Eur. J. Biochem.* **23,** 12 (1971).

## Variations in the Conditions of Protein Fractionation

The 30 S protein fractionation procedure is carried out in a sodium acetate buffer, pH 5.6, in the presence of LiCl since this system is found (1) to give good separation, (2) to keep the majority of the 30 S proteins in a soluble state, (3) to avoid problems with proteolysis. Several proteins not obtained during this procedure can be found if the fractionation is carried out in the presence of a phosphate-KCl buffer pH 7.0 containing the usual quantities of protease inhibitors and reducing agents. These proteins include S6 and S9, but under these conditions other proteins are more insoluble (for example protein S8) and the amount of proteolysis is increased so that several proteins cannot be obtained in an intact form. The extracts of proteins from the ribosome appear to be more soluble in the presence of a phosphate buffer at pH 7.0 than in other buffers, e.g., HEPES, at the same pH.

The 50 S proteins are purified using a LiCl-HEPES buffer at pH 7.0, which is found to give good separation for most of the proteins. Substitution by phosphate-KCl buffers at pH 7.0 leads to some improvement in the solubility and recovery of several proteins. As in the case of the 30 S proteins, protein mixtures are more soluble. In this system proteins L1, L6, L11, L16, L17, L21, L24, and L25 are obtained in higher yields whereas the yields of proteins L2, L3, L22, L27, and L28 are decreased. Some purified proteins, such as L1, L11,[24] and L24, are more soluble and less aggregated in the phosphate buffer whereas others (for example L3 and L30) are less soluble.

## Conclusion

Although these new methods of ribosomal protein isolation, avoiding the use of urea, acetic acid, and lyophilization, are very time consuming and involve the handling of large volumes of protein solutions, they do produce proteins that appear to be in a more "native" state.[25] For this reason they will be of importance for physical studies of ribosomal proteins.[24,26]

[24] L. Giri, J. Dijk, H. Labischinski, and H. Bradaczek, *Biochemistry*, **17**, 745 (1978).

[25] J. Littlechild and A. Malcolm, *Biochemistry*, **17**, 3363 (1978).

[26] R. Österberg, B. Sjöberg, and J. Littlechild, *FEBS Lett*. **93**, 115 (1978).

# [53] Two-Dimensional Polyacrylamide Gel Electrophoresis for Separation of Ribosomal Proteins

*By* H. G. WITTMANN

Ribosomes from all organisms studied to date are very complex. They consist of three RNA molecules and numerous proteins. A rapid, sensitive, and very reproducible method of determining the number of proteins in ribosomes is two-dimensional electrophoresis on slabs of polyacrylamide gel. This technique was first described for the separation of the 55 ribosomal proteins of *Escherichia coli*[1,2] and was later somewhat modified for other needs.[3,4] Because the original technique resulted in a very good separation not only of ribosomal proteins from prokaryotes,[2] but also of eukaryotes,[5-7] it will be described in this article in some detail. Drawings with the technical details of the various parts of the apparatus were given elsewhere[1] in order to make possible the construction of the apparatus in an institute workshop. Where this is not feasible it might be of interest that the apparatus is commercially available from Firma Desaga, Heidelberg, Germany, or from C. A. Brinckmann, Westbury, New York.

## Procedure

For separation in the first dimension, the protein sample is placed in the middle of a polyacrylamide rod. This is achieved in the following way: Close a glass tube (180 × 5 mm) at the lower end with a cap; fill cold acrylamide solution (buffer B of the table) up to the middle of the tube; overlayer with water; allow gel to polymerize; suck off water from the top of the polyacrylamide; add proteins dissolved in about 0.1–0.2 ml of buffer A (see table); overlayer with water; allow gel to polymerize with light; suck off water; fill the tube with cold acrylamide solution and allow gel to polymerize. After removal of the cap at the bottom, the filled tubes are placed into a rack which is constructed similarly to those

[1] E. Kaltschmidt and H. G. Wittmann, *Anal. Biochem.* **36**, 401 (1970).
[2] E. Kaltschmidt and H. G. Wittmann, *Proc. Nat. Acad. Sci. U.S.* **67**, 1276 (1970).
[3] H. Welfle, *Acta Biol. Med. Ger.* **27**, 547 (1971).
[4] O. H. W. Martini and H. J. Gould, *J. Mol. Biol.* **62**, 403 (1971).
[5] J. Delaunay and G. Schapira, *Biochim. Biophys. Acta* **259**, 243 (1972).
[6] C. C. Sherton and I. G. Wool, *J. Biol. Chem.* **247**, 4460 (1972).
[7] A. G. Lambertsson, *Mol. Gen. Genet.* **118**, 215 (1972).



ISBN 0-12-504180-2

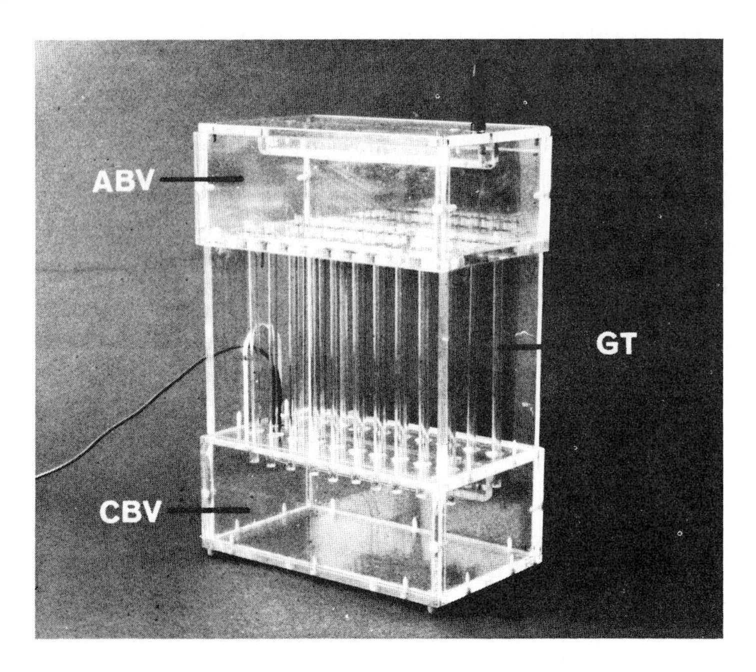


FIG. 1. Apparatus for protein separation in the first dimension. ABV: anode buffer vessel; CBV: cathode buffer vessel; GT: glass tube for polyacrylamide gel and protein sample. The apparatus can take 20 tubes.

used for one-dimensional disc electrophoresis but is high enough to take tubes of 18-cm length (Fig. 1).

The vessels for the anode and cathode buffers are located at the top and the bottom of the apparatus, respectively. They are filled with buffer C (see table). The electrophoretic run is at 90 V for 36 hours at room temperature. The buffer in the two vessels is renewed after approximately 15 hours. At the end of the run the gels are removed from the glass rods by injecting glycerin with a syringe between the gel and the glass rod and then applying pressure with another syringe filled with water. The buffer in the gel rods is replaced by placing the gels in a container (Fig. 2) filled with buffer D (see table). After 45 and 90 minutes the buffer is replaced by a new one, and after additional 45 minutes the gel rods are placed on top of the chamber (Figs. 3 and 4) used for the electrophoresis in the second dimension. Before this is done, the chamber is closed at the bottom with polyacrylamide as follows: The same acrylamide solution that is used for the electrophoretic run in the second dimension (buffer E, see table) is filled into a flat container and over-

## Composition of Buffers

Buffer A: for dissolving the protein sample

- 48.0 g urea
- 4.0 g acrylamide
- 0.2 g bisacrylamide
- 0.085 g $Na_2$-EDTA
- 0.32 g boric acid
- 0.06 ml TEMED
- $H_2O$ to make 99 ml

Add 1 ml of solution in which 0.5 mg of riboflavin and 5 mg of ammonium peroxodisulfate are dissolved. Dissolve protein (10–50 μg per protein species) in about 0.1–0.2 ml of buffer A and apply onto the polyacrylamide rod for separation in the first dimension

Buffer B: for separation in the first dimension (pH 8.6)

- 36.0 g urea
- 8.0 g acrylamide
- 0.3 g bisacrylamide
- 0.8 g $Na_2$-EDTA
- 3.2 g boric acid
- 4.85 g Tris
- 0.3 ml TEMED
- $H_2O$ to make 99.4 ml

Add 0.6 ml of a 7% ammonium peroxodisulfate solution for polymerization

Buffer C: electrode buffer (pH 8.6) for first dimension

- 360 g urea
- 2.4 g $Na_2$-EDTA
- 9.6 g boric acid
- 14.55 g Tris
- $H_2O$ to make 1 liter.

The same buffer is used as anode or cathode buffer

Buffer D: for dialyzing between the first and second dimension

- 480 g urea
- 0.74 ml glacial acetic acid
- 2.4 ml 5 *N* KOH
- $H_2O$ to make 1 liter

Buffer E: for separation in the second dimension (pH 4.6)

- 360 g urea
- 180 g acrylamide
- 5.0 g bisacrylamide
- 52.3 ml glacial acetic acid
- 9.6 ml 5 *N* KOH
- 5.8 ml TEMED
- $H_2O$ to make 967 ml

Add 33 ml of a 5% ammonium peroxodisulfate solution for polymerization

Buffer F: electrode buffer for second dimension

- 140 g glycine
- 15 ml glacial acetic acid
- $H_2O$ to make 10 liters

The same buffer is used as anode or cathode buffer

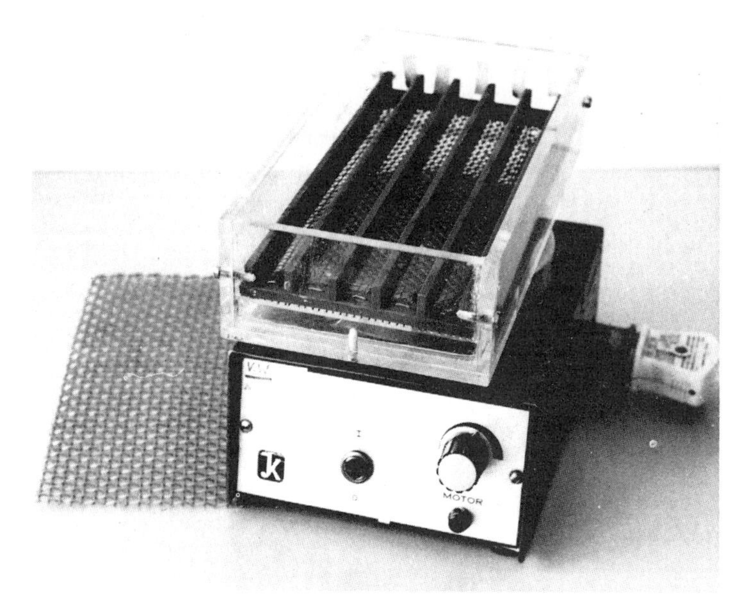


FIG. 2. Container for buffer D.

layered with water. Before the acrylamide solution polymerizes the chamber shown in Fig. 3 is carefully placed into it (Fig. 5).

After polymerization of the acrylamide solution in the flat container the water is removed from the top of the gel and new cold acrylamide solution (buffer E of Table 1) is filled into the five vertical slots. Then the gel rods which were dialyzed 3 × 45 minutes against buffer D of the table are placed on top of the vertical slots in such a way that they are almost embedded in the acrylamide solution. Care must be taken to remove air bubbles in the acrylamide solution. After polymerization the chamber is removed from the flat container, cleaned from residual polyacrylamide gel and placed into the vessel for the cathode buffer (Fig. 6). Then buffer F of the table is filled into the chamber, which serves as the anode buffer vessel and into the vessel for the cathode buffer. Finally a cover containing the platinum wires for the anode is put on top of the chamber. The electrophoretic run is done at room temperature for 24 hours at 100 V. Without voltage regulation the voltage increases during the run. Therefore the time for the run has to be shorter.

At the end of the run the chamber is taken out from the cathode container and dismantled. The gel slabs are removed, put into a rack (Fig. 7) and stained in solutions of Amido black or Coomassie blue. Destaining is done by removal of the stain with tap water for 1–2 hours

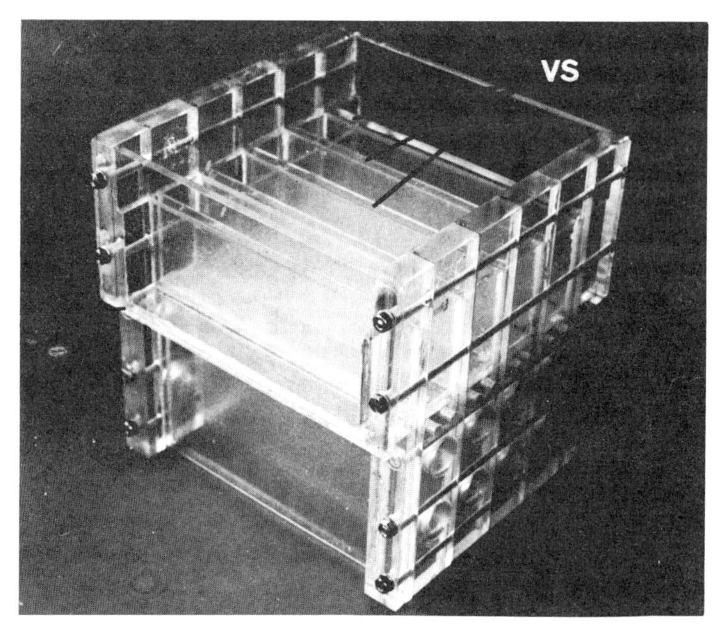


FIG. 3. Chamber for protein separation in the second dimension.

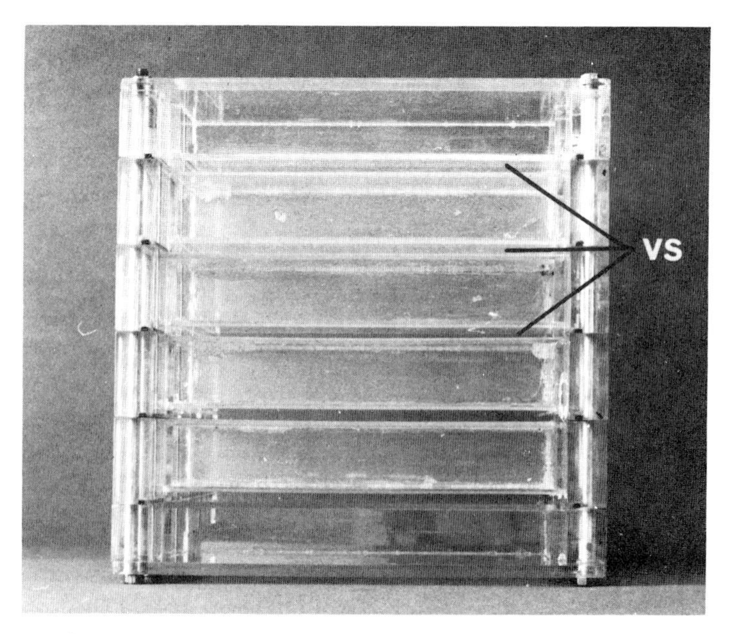


FIG. 4. Chamber (Fig. 3) seen from the top showing the vertical slots (VS) on which the polyacrylamide rods are placed.

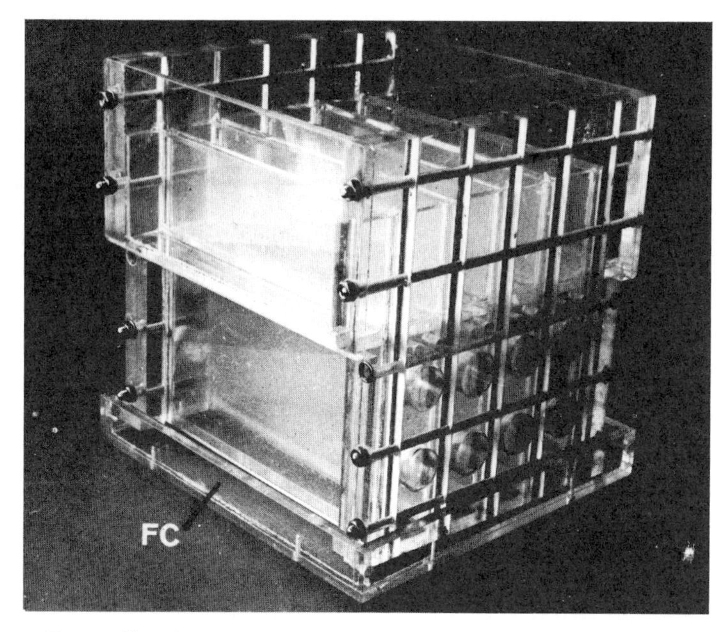


FIG. 5. Chamber (Fig. 3) with flat container (FC) at the bottom.

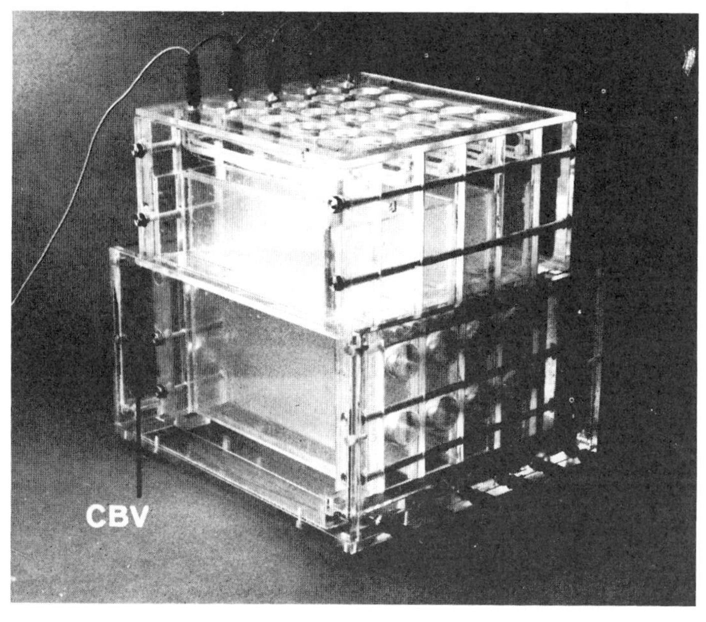


FIG. 6. Fully assembled apparatus during protein separation in the second dimension. CBV: cathode buffer vessel.

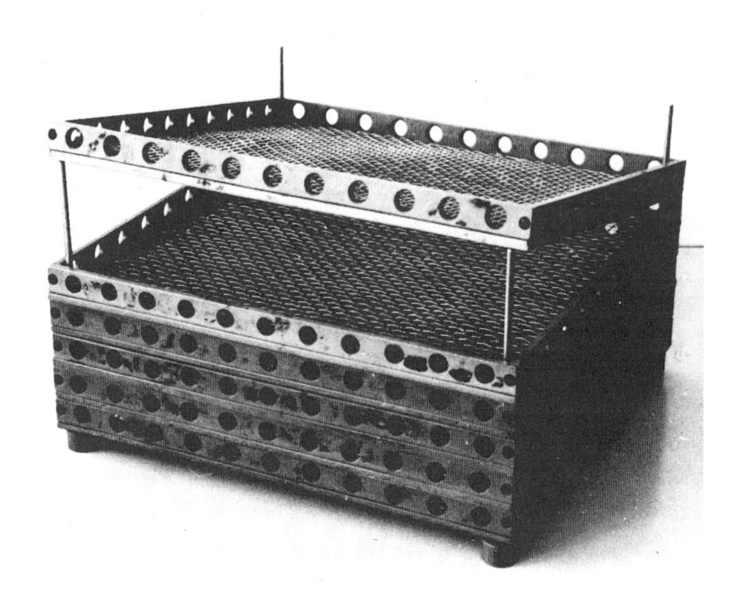

FIG. 7. Rack for staining and destaining of five polyacrylamide slabs.

and then with 1% acetic acid, which is either renewed every 12 hours or recycled through filters of charcoal. The destained and transparent polyacrylamide plates are photographed or stored wrapped in a very thin plastic sheet. The stained protein spots remain sharp for a period of several months if the plates are stored in the cold and dark.

## Applications

The two-dimensional polyacrylamide gel electrophoresis technique has been used for studying the following problems:

1. Number of ribosomal proteins in *E. coli* ribosomes (Fig. 8) and their subunits.[2]

2. Identification of isolated *E. coli* ribosomal proteins[8-10] and efficiency of various methods for extraction and fractionation of ribosomal proteins.[11,12]

[8] I. Hindennach, G. Stöffler, and H. G. Wittmann, *Eur. J. Biochem.* **23**, 7 (1971).
[9] I. Hindennach, E. Kaltschmidt, and H. G. Wittmann, *Eur. J. Biochem.* **23**, 12 (1971).
[10] G. Funatsu, K. Nierhaus, and B. Wittmann-Liebold, *J. Mol. Biol.* **64**, 201 (1972).
[11] E. Kaltschmidt and H. G. Wittmann, *Biochemie* **54**, 167 (1972).
[12] E. Schwabe, *Hoppe-Seyler's Z. Physiol. Chem.* **353**, 1899 (1972).

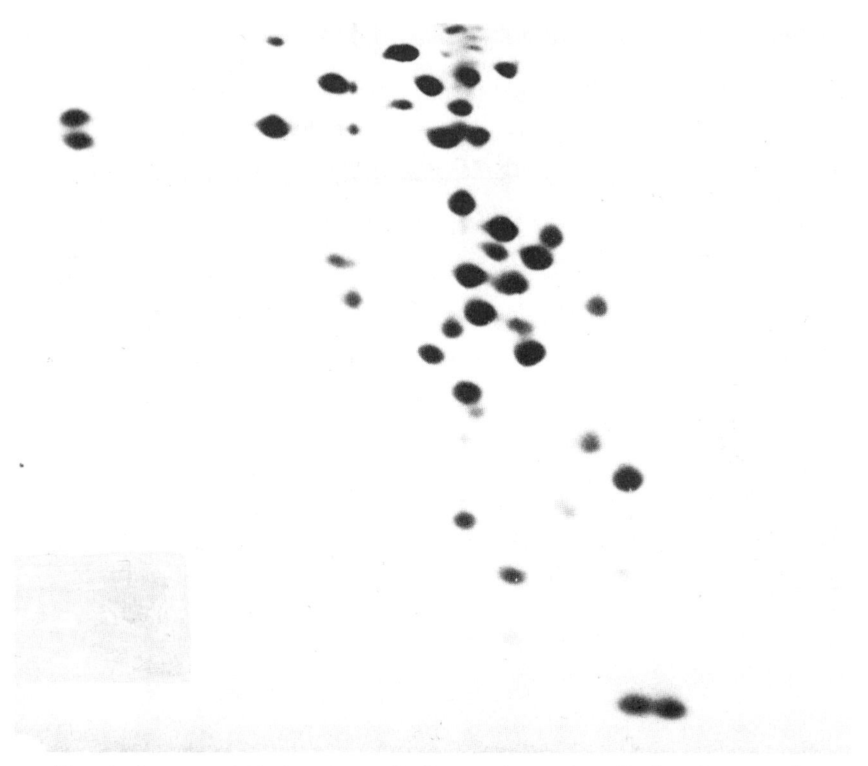

FIG. 8. Pattern of *Escherichia coli* ribosomal proteins (70 S) after two-dimensional polyacrylamide gel electrophoresis. For details see E. Kaltschmidt and H. G. Wittmann, *Proc. Nat. Acad. Sci. U.S.* **67**, 1276 (1970).

3. Isoelectric points of *E. coli* ribosomal proteins.[13]
4. Identification of the proteins which form a complex with 5 S RNA.[14,15]
5. Determination of which proteins are present in the various precursors of *E. coli* ribosomes and their core particles obtained by salt treatment.[16,17]
6. Stoichiometry of proteins in *E. coli* ribosomes.[18]

[13] E. Kaltschmidt, *Anal. Biochem.* **43**, 25 (1971).
[14] P. N. Gray, R. A. Garrett, G. Stöffler, and R. Monier, *Eur. J. Biochem.* **28**, 412 (1972).
[15] J. R. Horne and V. A. Erdmann, *Mol. Gen. Genet.* **119**, 337 (1972).
[16] H. E. Homann and K. H. Nierhaus, *Eur. J. Biochem.* **20**, 249 (1971).
[17] K. H. Nierhaus, K. Bordasch, and H. E. Homann, *J. Mol. Biol.* **74**, 587 (1973).
[18] H. J. Weber, *Mol. Gen. Genet.* **119**, 233 (1972).

7. Difference in ribosomal proteins from *E. coli* cells grown under different growth conditions.[19,20]

8. Topological studies on *E. coli* ribosomes.[21–24]

9. Determination which ribosomal proteins are different between *E. coli* wild type and mutants, e.g., revertants from streptomycin dependence to independence.[25–27]

10. Proteins different in ribosomes from various *E. coli* strains.[28]

11. Comparison of the ribosomal proteins isolated from bacteria belonging to the same or different families.[29]

12. Comparison of the proteins from cytoplasmic and chloroplast ribosomes of higher plants.[30]

13. Number of proteins from eukaryotic ribosomes and their subunits.[4–7,31]

[19] E. Deusser and H. G. Wittmann, *Nature (London)* **238**, 269 (1972).
[20] E. Deusser, *Mol. Gen. Genet.* **119**, 249 (1972).
[21] R. R. Crichton and H. G. Wittmann, *Mol. Gen. Genet.* **114**, 95 (1971).
[22] P. Spitnik-Elson and A. Breimann, *Biochim. Biophys. Acta* **254**, 457 (1971).
[23] L. Kahan and E. Kaltschmidt, *Biochemistry* **11**, 2691 (1972).
[24] R. R. Crichton and H. G. Wittmann, *Proc. Nat. Acad. Sci. U.S.* **70**, 665 (1973).
[25] E. Deusser, G. Stöffler, H. G. Wittmann, and D. Apirion, *Mol. Gen. Genet.* **109**, 298 (1970).
[26] G. Stöffler, E. Deusser, H. G. Wittmann, and D. Apirion, *Mol. Gen. Genet.* **111**, 334 (1971).
[27] R. Hasenbank, C. Guthrie, G. Stöffler, H. G. Wittmann, L. Rosen, and D. Apirion, *Mol. Gen. Genet.* in press.
[28] E. Kaltschmidt, G. Stöffler, M. Dzionara, and H. G. Wittmann, *Mol. Gen. Genet.* **109**, 303 (1970).
[29] M. Geisser, G. Stöffler, and H. G. Wittmann, manuscript submitted.
[30] H. G. Janda, C. Gualerzi, H. Passow, G. Stöffler, and H. G. Wittmann, manuscript submitted.
[31] H. Welfle, J. Stahl, and H. Bielka, *Biochim. Biophys. Acta* **243**, 416 (1971).

# [54] A Modified Two-Dimensional Gel System for the Separation and Radioautography of Microgram Amounts of Ribosomal Proteins[1]

*By* G. A. HOWARD and R. R. TRAUT

Polyacrylamide gel electrophoresis has been a highly useful technique for separating complex mixtures of proteins. Recently a two-dimensional

[1] Supported by research grants from the U.S. Public Health Service (GM 17924) and the Damon Runyon Memorial Fund for Medical Research (DRG-1140). R.R.T. is an established Investigator of the American Heart Association.



ISBN 0-12-504180-2

slab gel electrophoresis system was described by which all the ribosomal proteins of *Escherichia coli* were separated into distinct spots.[2,3] This system and others have been used to separate eukaryotic as well as prokaryotic ribosomal proteins.[4-8] The system developed by Kaltschmidt and Wittmann[3] gives excellent resolution of both prokaryotic and eukaryotic ribosomal proteins moreover, standard and highly useful nomenclature of *E. coli* ribosomal proteins has been defined in terms of this system.[9]

We have adopted the basic system of Kaltschmidt and Wittmann to use thinner and smaller slabs which require smaller amounts of sample, and which can be dried for radioautography of radioactive proteins. The advantage of the smaller two-dimensional slab described here are summarized below:

| | Miniature system | Kaltschmidt–Wittmann system[3] |
|---|---|---|
| Ribosomal protein required | 100–200 μg | 2–4 mg |
| Total electrophoresis time | 12 hr | 40–50 hr |
| Staining and destaining time | 3–5 hr | 30–40 hr |
| Radioautography | Very good—gel can be totally dried without distortion or shrinkage for accurate radioautograms | Must be done with wet slab |
| Cost of basic apparatus | Not yet commercially available (about $1150) | (Desaga-Brinkman) $1995.00 |

The resolution of *E. coli* ribosomal proteins with the miniature slab gel system described here is essentially identical to that reported by Kaltschmidt and Wittmann.[3] It has also been employed to separate eukaryotic ribosomal proteins and to identify by radioautography those phosphorylated by protein kinase. A two-dimensional gel system having many of the ad-

[2] H. G. Wittmann, this volume [53].

[3] E. Kaltschmidt and H. G. Wittmann, *Anal. Biochem.* **36**, 401 (1970).

[4] O. H. W. Martini and H. J. Gould, *J. Mol. Biol.* **62**, 403 (1971).

[5] T. Hultin and A. Sjöqvist, *Anal. Biochem.* **46**, 342 (1972).

[6] H. Van Tan, J. Delaunay, and G. Schapira, *FEBS (Fed. Eur. Biochem. Soc.) Lett.* **17**, 163 (1971).

[7] C. C. Sherton and I. G. Wool, *J. Biol. Chem.* **247**, 4460 (1972).

[8] H. Welfle, J. Stahl, and H. Bielka, *Biochim. Biophys. Acta* **243**, 416 (1971).

[9] H. G. Wittmann, G. Stöffler, I. Hindennach, C. G. Kurland, L. Randall-Hazelbauer, E. A. Birge, M. Nomura, E. Kaltschmidt, S. Mizushima, R. R. Traut, and T. A. Bickle, *Mol. Gen. Genet.* **111**, 327 (1971).

vantages of that described here has also been employed by Dr. David Elson (personal communication).

## Methods

### *First Dimension by Disc Gel Electrophoresis*

#### *Solutions*

Separating gel, pH 8.7 (modification of that reported by Kaltschmidt and Wittmann[3]):

Urea, 6.0 *M*, 360.0 g/l
Acrylamide, 4.0 wt. %, 40.0 g/l
Bisacrylamide, 0.13 wt. %, 1.33 g/l
EDTA-$Na_2$, 20 m*M*, 8.0 g/l
Boric acid, 0.52 *M*, 32.0 g/l
Tris, 0.4 *M*, 48.6 g/l
TEMED, 0.45 ml

The above solution is filtered and may be stored at 4° for several weeks. It is degassed just before polymerization, catalyzed with ammonium persulfate: 5 μl of a freshly prepared 10% (w/v) solution per 1.0 ml of gel solution.

Running buffer, pH 8.2
EDTA-$Na_2$, 10 m*M*, 2.4 g/l
Boric acid, 80 m*M*, 4.8 g/l
Tris, 60 m*M*, 7.25 g/l

Tracking dyes
Pyronine G, 0.5% in $H_2O$ (cationic)
Bromophenol blue, 0.1% in $H_2O$ (anionic)

Acrylamide (technical grade), bisacrylamide, and TEMED were obtained from Eastman Chemicals; EDTA-$Na_2$ and Tris from Sigma; and the remaining reagents from Mallinckrodt. Recrystallization of the reagents was unnecessary and results were the same with crude or purified reagents.

#### *Apparatus*

Electrophoresis in the first dimension was performed in either of two ways:

(a) Standard disc gel electrophoresis apparatus, as described by Davis,[10] in which gels are run in glass tubes either 0.45 cm ×

[10] B. J. Davis, *Ann. N.Y. Acad Sci.* **121**, 404 (1964).

9 cm, or 0.45 × 12.5 cm. Alternatively, tubes of smaller inner diameter (2–3 mm) can be employed in order to simplify insertion of the gel cylinder into the second dimension slab.

(b) Slab gel apparatus with preformed slots for individual samples as described by Reid and Bieleski,[11] and modified by Studier.[12,13] A slice of the slab containing the resolved sample in one of the slots is cut out and used for the second dimension. Further details of this method will not be presented here. The gel solution used is the same in (a) and (b).

*Sample Application and Electrophoresis Conditions.* Two methods of applying sample in the first dimension by method (a) above have been used in our laboratory. Both minimize the substantial loss of protein at the center-origin reported in other systems in which sample is polymerized in a sample gel containing acrylamide.[7,14]

METHOD 1. The sample (100–200 μg) is applied in agarose in the center of the first-dimensional gel; in this way proteins migrating both toward the anode and the cathode are resolved in the same run. For this method the longer disc gel tubes (0.45 cm × 12.5 cm) are used. The lower half of the separating gel is poured into the tube and overlayered with water according to standard methods. Agarose (1% in pH 8.2 running buffer) is liquefied in a boiling water bath, mixed in a 1:1 ratio with the protein sample and kept at 40° in a heating block until it is layered onto the flat surface of the polymerized lower gel. Water is again layered over the agarose-sample mixture until the agarose has hardened; it is then removed, and the tube is filled with more separating gel solution over the sample zone. Electrophoresis is carried out with the cathode above, with pyronine G as the tracking dye.

METHOD 2. Identical amounts of sample are applied at the top of each of two separate gels in the shorter tubes (0.45 cm × 9.0 cm). Electrophoresis of one of the gels is from the anode to the cathode with pyronine G as the tracking dye; electrophoresis of the other gel is from the cathode to the anode using bromphenol blue as the tracking dye.

By method 1 small losses of proteins occur as material immobilized in the agarose layer. These are much less, however, than those observed in other systems.[7,14] By method 2, essentially all the protein is recovered in the bands migrating from the origin. Thus, although ostensibly method 2 requires twice as much protein sample as method 1, in fact, because noth-

[11] M. S. Reid and R. L. Bieleski, *Anal. Biochem.* **22**, 374 (1968).

[12] F. W. Studier, *Science* **176**, 367 (1972).

[13] F. W. Studier, personal communication.

[14] E. Kaltschmidt and H. G. Wittmann, *Proc. Nat. Acad. Sci. U.S.* **67**, 1276 (1970).

ing is lost, the increased sample required is less than double that required when the sample is polymerized in the center of a single disc gel.

In both methods 1 and 2, electrophoresis is performed at 3 mA/gel tube for 30 minutes to facilitate the stacking of the protein bands, then the current is increased to 6 mA/gel and electrophoresis continued for 5–6 hours. The gels can be removed at once from the glass tubes in preparation for the second dimension, or stored in the tubes at 4° for 24–48 hours before use without significant diffusion of the protein bands or loss of resolution.

*Preparations of First-Dimensional Gel for the Second-Dimension*

*Solutions*

Dialysis buffer, pH 5.2 (Kaltschmidt and Wittmann,[3] "starting buffer")

Urea, 8.0 *M*, 480.0 g/l
Acetic acid, glacial, 40 m*M*, 0.74 ml g/l
KOH, 10 m*M*, 0.67 g/l

*Procedure.* The first-dimensional disc gels of larger diameter are sliced in half longitudinally in order to fit between the glass plates in the second dimension. This can be done by placing the gel in a halved piece of tygon tubing matching the diameter of the gel, and then slicing the gel with uniform downward pressure of a thin knife blade. The small diameter gels can be used directly.

The first-dimensional gels are then dialyzed against the pH 5.2 buffer for a total of 60 minutes with at least two changes of buffer.

*Second Dimension by Slab Gel Electrophoresis*

*Solutions*

Separating gel solution, pH 4.5 (modification of that reported by Kaltschmidt and Wittmann[3]):

Urea, 6.0 *M*, 360.0 g/l
Acrylamide, 18.0 wt. %, 180.0 g/l
Bisacrylamide, 0.25 wt. %, 2.5 g/l
Acetic acid, glacial, 0.92 *M*, 53.0 ml
KOH, 0.048 *N*, 2.7 g/l
TEMED, 5.8 ml

The above solution is filtered and may be stored at 4° for several weeks. It is degassed just before polymerization is catalyzed with ammonium persulfate: 30 μl of a 10% (w/v) solution/1.0 ml of gel solution.

Running buffer, pH 4.0 (modification of that reported by Kaltschmidt and Wittmann[3]):

Glycine, 0.18 *M*, 14.0 g/l
Acetic acid, glacial, 6 m*M*, 1.5 ml

Tracking dye:
Pyronine G, 0.5% in $H_2O$ containing 20% glycerol

Stains and destaining solution:
Coomassie brilliant blue, R-250, 0.1% in 7.5% glacial acetic acid:50% methanol:$H_2O$
Destaining solution, 50% methanol:7.5% glacial acetic acid: $H_2O$

*Apparatus.* The apparatus in the second dimension (Fig. 1) is an adaption of the thin-sheet gel apparatus referred to in first-dimensional method 1 above.[11-13] Two glass plates are sandwiched together with plexiglass spacers to allow the gel sheet of approximately 2 mm thickness to be formed. A thin coating of petroleum jelly on the spacers is used to prevent leakage of the gel solution.

*Procedure.* The first-dimensional slice is placed in position against the top spacer as shown (Fig. 1); any excess gel at the ends is trimmed off to allow the gel to fit in the defined width. When the first-dimension has been run as two separate short gels in opposite directions (Method 1), the two origins are placed adjacent to each other in the center as shown in Fig. 1. The two glass plates are clamped together at the sides and the top (where the first-dimensional slice is positioned) with foldback binder clips. The plates are inverted 180° and the cavity thus formed is filled with second-dimensional separating gel solution. After the gel has polymerized, the plates are again turned 180° so that the first-dimensional slice is on top. The spacer is removed, all excess petroleum jelly is wiped away, the binder clips are carefully removed, and the plates with the enclosed gel sheet are clamped to the apparatus previously described by Reid and Bieleski,[11] as modified by Studier.[12,13]

*Electrophoresis Conditions.* The tracking dye (0.1% pyronine G in 20% glycerol) is layered across the top of the first-dimensional slice, under the running buffer. Electrophoresis is carried out with the anode on top for 30–60 minutes at 40 V to allow stacking of the proteins in the first-dimensional gel slice, then the voltage is increased to 80–150 V and electrophoresis is continued for 6–12 hours. The exact voltage within the

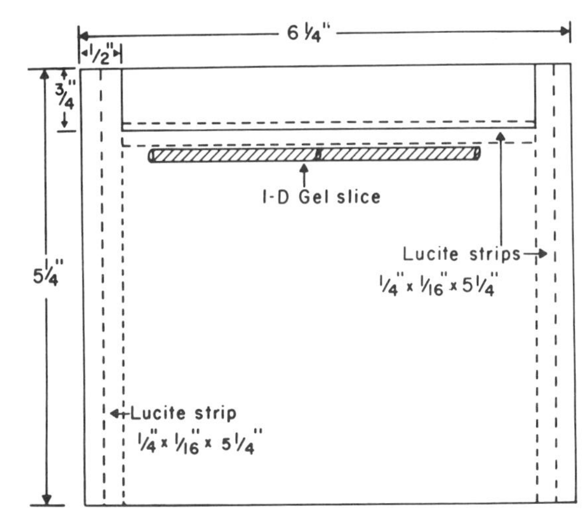


FIG. 1. Placement of the plexiglass spacers and the gel slice from the first dimension between the glass plates used for the second dimension. The two glass plates (one with a notch in the top as indicated) are made from ordinary double-strength window glass. The resulting "sandwich" is clamped in place in an apparatus like that previously described [M. S. Reid and R. L. Bieleski, *Anal. Biochem.* **22**, 374 (1968)], as modified by Studier [F. W. Studier, *Science* **176**, 367 (1972); F. W. Studier, personal communication].

limits tested makes no difference in the resulting separation of the proteins; thus convenience of running time is the major criterion for choosing a particular voltage. At 80 V the dye front moves 1 cm per hour.

Electrophoresis is stopped when the tracking dye is within 1 cm of the bottom of the gel. The top glass plate is loosened from the gel slab by carefully prying it upward with a wide spatula. The gel sheet is lifted off the lower plate and placed in a tray containing the stain solution, where it is left with occasional agitation for 1–4 hours. The gels are destained by slow shaking in a tray of the destaining solution on a mechanical shaking bath with several changes of the solution.

*Treatment of Gel after Destaining.* After destaining, the gel sheet can be photographed wet, or dried onto filter paper either for storage or for autoradiography. Drying is accomplished by using a vacuum and low heat as described by Maizel[15] (a modification of the procedure of Fairbanks *et al.*[16]). The thinness of the gels allows them to be totally dried in about 2 hours without distortion, shrinkage, or cracking.

[15] J. V. Maizel, Jr., *in* "Methods in Virology" (K. Maramorosch and H. Kaprowski, eds.), Vol. V, p. 179. Academic Press, New York, 1971.

[16] G. Fairbanks, C. Levinthal, and R. H. Reeder, *Biochem. Biophys. Res. Commun.* **20**, 393 (1965).

*Determination of Molecular Weights of Proteins Separated by Two-Dimensional Electrophoresis*

In addition to providing the molecular weight of the proteins contained in the individual spots resolved in the two-dimensional technique described, the analytical technique (SDS gel electrophoresis) which follows is also a criterion for the purity of the resolved spots.

*Solutions*

Acrylamide disc gels, 10%, containing sodium dodecyl sulfate (SDS) as described by Bickle and Traut[17]
Sample buffer, pH 7.2:

$NaH_2PO_4 \cdot H_2O$, 2.9 m*M*, 0.4 g/l
$Na_2HPO_4 \cdot 7H_2O$, 7.2 m*M*, 1.94 g/l
Glycerol, 20 vol %, 200 ml
$\beta$-Mercaptoethanol, 0.14 *M*, 10 ml

Tracking dye: bromophenol blue, 0.1% in $H_2O$
Stain and destaining solutions: same as for second dimension above

*Apparatus*

Standard polyacrylamide disc gel electrophoresis apparatus, as described by Davis,[10] with gels formed in 0.45 cm × 9.0 cm glass tubes

*Preparation of Sample.* The center regions of stained spots on the two-dimensional slab are cut out, rinsed with water and macerated in small culture tubes with a glass rod. From 0.2 to 0.3 ml of a solution containing 1% SDS and 6 *M* urea is added to each tube, which are then left at room temperature for 24 hours. Then, 50 μl of sample buffer solution, 5 μl of 0.1% bromophenol blue (tracking dye), plus sufficient 10 *N* NaOH to adjust the pH to about 7.0 (about 10 μl) is added to each tube. The contents of each tube are mixed thoroughly, heated for 10 minutes at 65°, then applied to the top of the SDS gels as previously described.[17] To determine accurately the molecular weight of the proteins removed from the second-dimensional gel, protein standards of known molecular weight can be added directly to the macerated gel before the heating step, or run on separate gels.

[17] T. A. Bickle and R. R. Traut, *J. Biol. Chem.* **246**, 6828 (1971).

*Electrophoresis Conditions*. Electrophoresis is run from the cathode to the anode at 4–5 mA/gel until all the tracking dye has entered the SDS gel; the current is then increased to 10 mA/gel and electrophoresis is continued until the tracking dye is about 1 cm from the bottom of the gel. The gels are removed from the tubes, soaked about 20 minutes in 7.5% acetic acid to remove the SDS, then stained in 0.1% Coomassie brilliant blue overnight. The excess stain in the gels is then removed by transverse electrophoresis as previously described.[17]

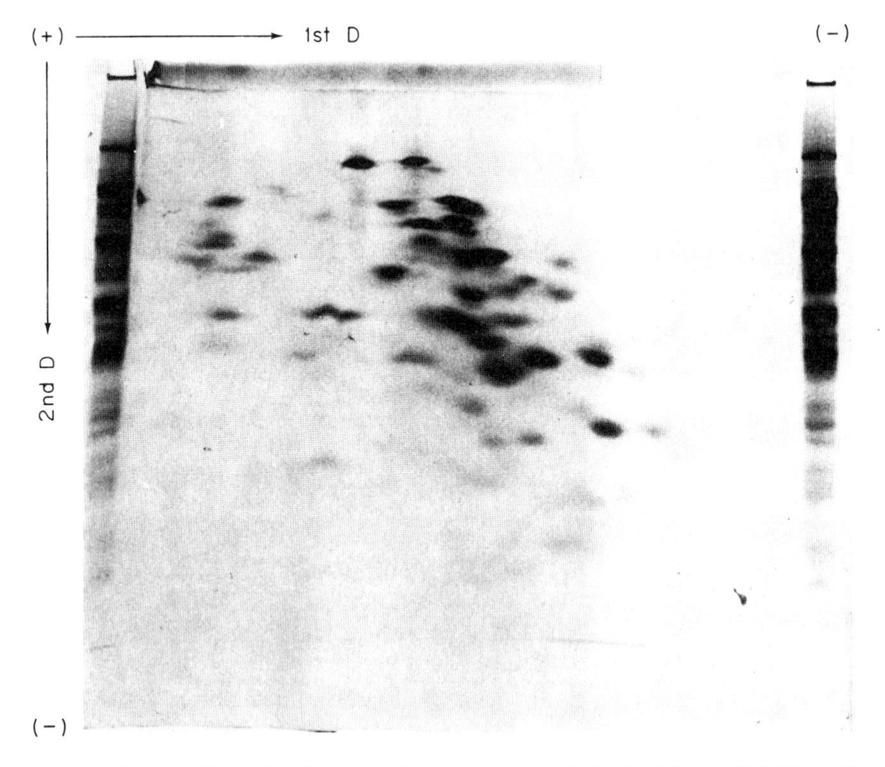


FIG. 2. Two-dimensional electrophoresis pattern of *Escherichia coli* MRE 600 50 S ribosomal subunit proteins extracted with 3 *M* LiCl–4 *M* urea [P. Spitnik-Elson, *Biochim. Biophys. Acta* **80**, 594 (1964); P. B. Moore, R. R. Traut, H. Noller, P. Pearson, and H. Delius, *J. Mol. Biol.* **31**, 441 (1968)]. The separated subunits were obtained by sucrose density gradient sedimentation of ribosomes washed once in 0.5 *M* $NH_4Cl$, 30 m*M* $MgCl_2$ [T. A. Bickle and R. R. Traut, *J. Biol. Chem.* **246**, 6828 (1971)]. Migration directions in each dimension are indicated by arrows. The O (top center) indicates the origin of the first-dimensional gel. In this case the sample was loaded in the center of the first-dimensional gel with agarose as described in the text. Gel and buffer conditions were as described in the text.

## Specific Applications of the Method

### *Separation of Prokaryotic Ribosomal Proteins*

As can be seen in Figs. 2 and 3, the methods described here give separation of the ribosomal proteins of *E. coli* 30 S and 50 S subunits comparable to that reported by Kaltschmidt and Wittmann.[2,3,14]

The two alternative methods described for applying the sample (methods 1 and 2) in the first dimension are illustrated with *E. coli* proteins. Both methods avoid the significant losses of protein at the center-origin as observed in the systems previously described.[7,14] In the first

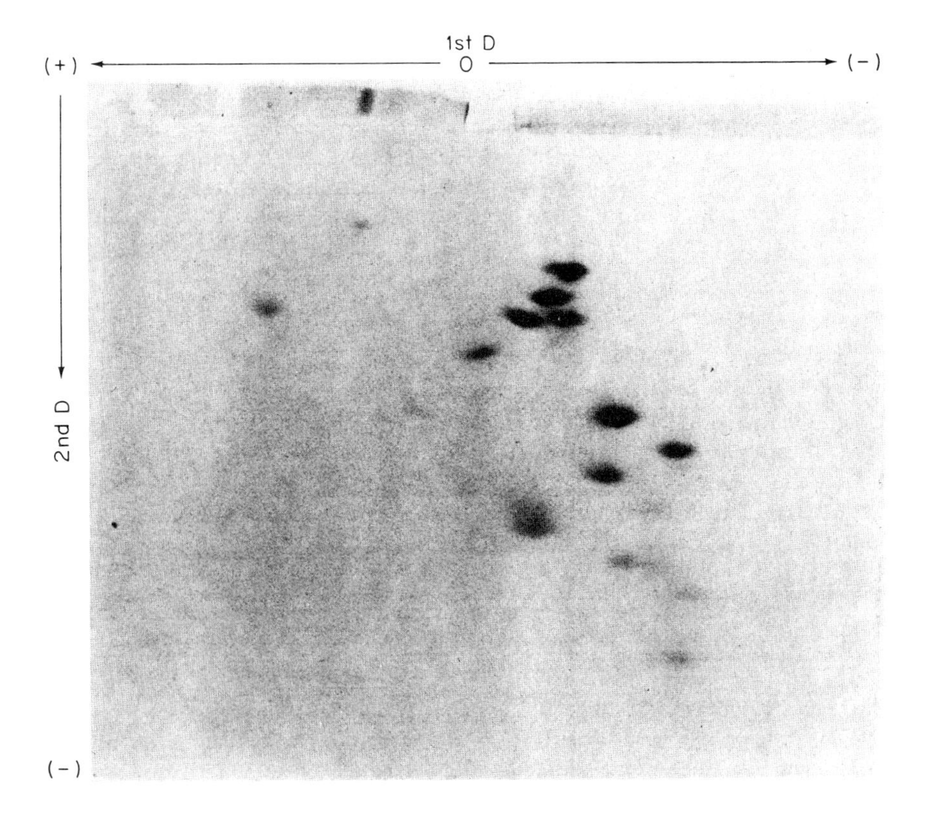


FIG. 3. Two-dimensional electrophoresis pattern of *Escherichia coli* MRE 600 30 S ribosomal subunit proteins extracted with 3 *M* LiCl–4 *M* urea. [P. Spitnik-Elson, *Biochim. Biophys. Acta* **80**, 594 (1964); P. B. Moore, R. R. Traut, H. Noller, P. Pearson, and H. Delius, *J. Mol. Biol.* **31**, 441 (1968)]. The separated subunits were obtained as in Fig. 2. In this case two short first-dimensional gels with identical top-loaded samples were run in opposite directions as described in the text. Gel and buffer conditions were as for Fig. 2.

method (Fig. 2) the sample was applied to the center of the first dimension in 0.5% agarose. Very little material remains stained at the origin. In the second method (Fig. 3) in which two separate, short first-dimensional gels were run in opposite directions no immobilized material appears at the origin. This method, although requiring twice as much material, is very useful in that it gives a pattern virtually identical to the single center-loaded gel and to that reported by Kaltschmidt and Wittmann,[3] while

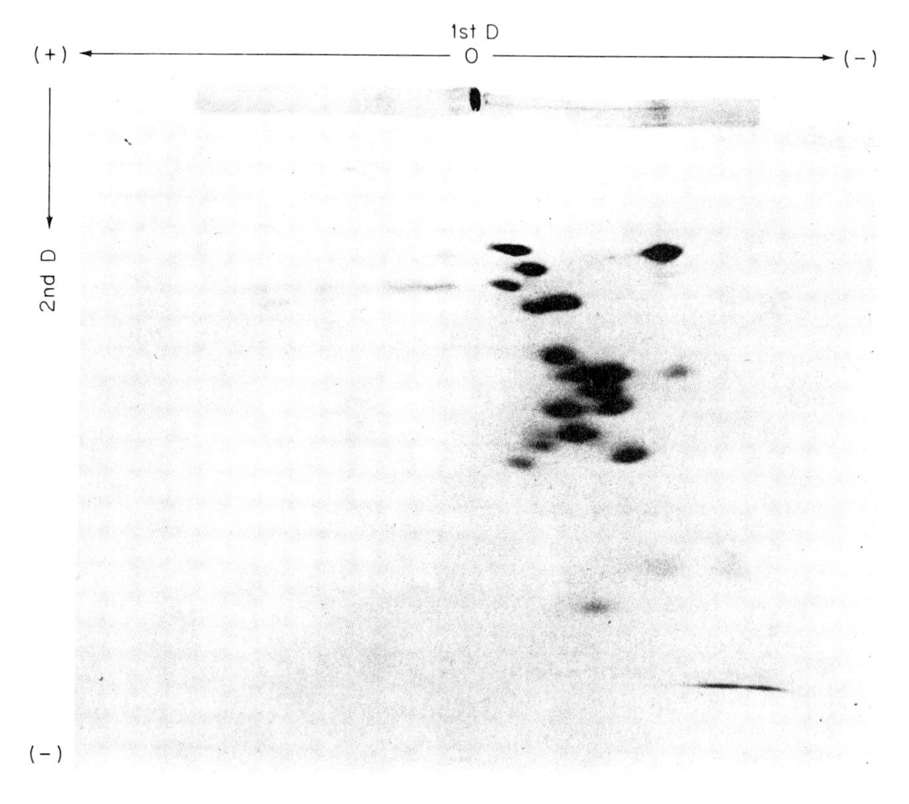


FIG. 4. Two-dimensional electrophoresis pattern of Novikoff hepatoma ascites cells 80 S ribosomal proteins extracted with 3 *M* LiCl–4 *M* urea [P. Spitnik-Elson, *Biochim. Biophys. Acta* **80**, 594 (1964); P. B. Moore, R. R. Traut, H. Noller, P. Pearson, and H. Delius, *J. Mol. Biol.* **31**, 411 (1968)]. The 80 S ribosomes were prepared as described by Busch [H. Busch, R. K. Busch, W. H. Spahn, J. Wikman, and Y. Daskal, *J. Exp. Biol. Med.* **137**, 1470 (1971)]. Migration direction in each dimension is indicated by an arrow. The first dimension was run in this case as a single top-loaded gel, thus the origin is at the upper left as indicated and the few acidic proteins in the mixture are not shown. The protein patterns at the sides of the second dimension were obtained by running samples of total protein loaded in wells at the top of the second dimension as described in the text. Gel and buffer conditions were as described in the text.

leaving essentially no protein remaining bound at the origin. Moreover, since our gel system requires only microgram amounts of ribosomal protein, the use of this "double" first dimension still uses only a fraction of the material required by the larger gel systems.

*Separation of Eukaryotic Ribosomal Proteins*

The method described here has been used in this laboratory to study various types of eukaryotic ribosomes. It has been especially useful in that it requires only small amounts of material. Figure 4 shows the separation of 69–70 ribosomal proteins from 80 S ribosomes of Novikoff hepatoma ascites cells. A further refinement to the gel system is also illustrated in Fig. 4. At the side of the gel slab samples of the total protein are run at the same time as the gel slice from the first dimension. This is done by using a plexiglass spacer at the top, when forming the second-dimensional gel, which has projections on it to make sample wells on each side of the first-dimensional slice. This has been very useful in doing comparative studies of ribosomal proteins from various sources, in that it makes it

FIG. 5. Radioautogram of Novikoff hepatoma ascites cells 80 S ribosomal proteins labeled with [$^{32}$P]ATP and protein kinase. The two-dimensional gel slab was dried as described in the text and the autoradiogram was made by placing Kodak No-Screen Medical X-ray film directly against the dried gel.

easier to determine where to look for differences in the two-dimensional gel patterns. This is particularly important in the case of eukaryotic ribosomes where one may be looking for only 1 or 2 protein differences out of a total of 65–70 proteins.

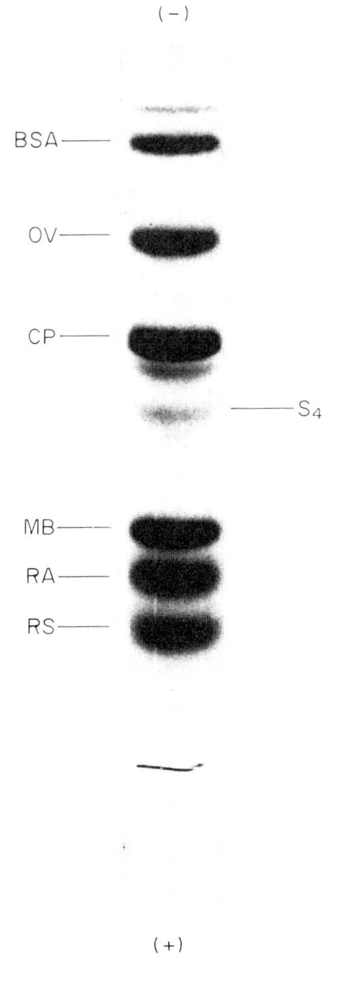


FIG. 6. Sodium dodecyl sulfate–10% acrylamide gel of *Escherichia coli* S4 protein. The stained protein spot taken from a two-dimensional gel slab like that in Fig. 3 was treated as described in the text before being applied onto the SDS gel. Standard proteins of known molecular weight were also applied onto the SDS gel for molecular weight calibration [K. Weber and M. Osborn, *J. Biol. Chem.* **244**, 404 (1964)]. BSA, bovine serum albumin; OV, ovalbumin; CP, carboxypeptidase A; MB, myoglobin; RA, ribonuclease A; RS, ribonuclease S.

*Autoradiography*

Figure 5 shows an autoradiogram of 80 S ribosomal proteins from Novikoff hepatoma ascites cells which were phosphorylated by protein kinase from rabbit skeletal muscle in a reaction in which the $^{32}P$ from [$\gamma$-$^{32}P$]ATP is transferred to the protein.[18] A comparison of Figs. 4 and 5 shows the specificity of the phosphorylation reaction. The totally dried gel was placed against the X-ray film. Although autoradiograms of $^{32}P$ can be made of wet gels,[19] those of $^{14}C$ or $^{35}S$ require drying of the gel.

*Molecular Weight Determination of Separated Proteins*

Figure 6 shows an example of the SDS gel analysis of one of the *E. coli* 30 S proteins separated on the two-dimensional gel. The S4 protein from *E. coli* 30 S subunits was cut from a sheet gel containing all 21 of the 30 S proteins, then treated as described above for the third dimension. Several proteins of known molecular weight were added as standards to give an accurate molecular weight value for the protein in question.[20] The molecular weight determined here of 27,000 for protein S4 compares quite favorably with that of 26,600 as reported by Traut *et al.*[21]

By means of this same method, molecular weights of other *E. coli* 30 S proteins have been determined which agree with previously reported values (Traut *et al.*[21]); see tabulation.

| Protein | Determined molecular weight | Previously reported molecular weight |
|---|---|---|
| S1 | 68,000 | 68,000 |
| S3 | 29,400 | 29,900 |
| S5 | 21,000 | 20,200 |
| S7 | 19,600 | 19,600 |
| S9 | 16,200 | 17,200 |

[18] E. G. Krebs, *Curr. Top. Cell. Regul.* **5**, 99 (1972).

[19] J. Stahl, H. Welfle, and H. Bielka, *FEBS* (*Fed. Eur. Biochem. Soc.*) *Lett.* **26**, 233 (1972).

[20] K. Weber and M. Osborn, *J. Biol. Chem.* **244**, 4406 (1969).

[21] R. R. Traut, H. Delius, C. Ahmad-Zadeh, T. A. Bickle, P. Pearson, and A. Tissières, *Cold Spring Harbor Symp. Quant. Biol.* **34**, 25 (1969).

# [55] Analytical Methods for Ribosomal Proteins of Rat Liver 40 S and 60 S Subunits by "Three-Dimensional" Acrylamide Gel Electrophoresis

*By* KIKUO OGATA and KAZUO TERAO

To study the metabolism and function of eukaryotic ribosomal proteins in detail, it is essential to identify the individual proteins of 60 S and 40 S subunits. Although the separation of ribosomal proteins especially by column chromatography was recently developed by us[1] and by Wool's group,[2-4] it is rather difficult to use it for the identification of small amounts of ribosomal proteins. For this purpose, two-dimensional acrylamide gel electrophoresis developed by Kaltschmidt and Wittmann,[5,6] may be most useful, owing to its high resolution capacity, and this technique has been used by many investigators for the identification of ribosomal proteins of animal cells.[7-19]

However, in the case of eukaryotic ribosomal subunits it is not settled whether each spot on two-dimensional gel represents one kind of protein. During analysis of the proteins of 60 S and 40 S subunits of rat liver

[1] K. Terao and K. Ogata, *Biochim. Biophys. Acta* **285,** 473 (1972).
[2] E. Collatz, A. Lin, G. Stöffler, K. Tsurugi, and I. G. Wool, *J. Biol. Chem.* **251,** 1808 (1976).
[3] K. Tsurugi, E. Collatz, I. G. Wool, and A. Lin, *J. Biol. Chem.* **251,** 7940 (1976).
[4] K. Tsurugi, E. Collatz, K. Todokoro, and I. G. Wool, *J. Biol. Chem.* **252,** 3961 (1977).
[5] E. Kaltschmidt and H. G. Wittmann, *Anal. Biochem.* **36,** 401 (1969).
[6] E. Kaltschmidt and H. G. Wittmann, *Proc. Natl. Acad. Sci. U.S.A.* **67,** 1276 (1970).
[7] O. H. W. Martini and H. J. Gould, *J. Mol. Biol.* **62,** 403 (1971).
[8] H. V. Tan, J. Delaunay, and G. Schapira, *FEBS Lett.* **17,** 163 (1971).
[9] C. C. Sherton and I. G. Wool, *J. Biol. Chem.* **249,** 2258 (1974).
[10] H. Welfle, J. Stahl, and H. Bielka, *FEBS Lett.* **26,** 228 (1972).
[11] K. Tsurugi, T. Morita, and K. Ogata, *Eur. J. Biochem.* **32,** 555 (1973).
[12] N. Hanna, G. Bellemare, and C. Godin, *Biochim. Biophys. Acta* **331,** 141 (1973).
[13] S. K. Chatterjee, M. Kazemie, and H. Matthaei, *Hoppe-Seyler's Z. Physiol. Chem.* **354,** 481 (1973).
[14] B. Peeters, L. Vanduffel, A. Depuydt, and W. Rombauts, *FEBS Lett.* **36,** 217 (1973).
[15] G. A. Howard and R. R. Traut, *FEBS Lett.* **29,** 177 (1973).
[16] S. M. Lastick and E. H. McConkey, *J. Biol. Chem.* **251,** 2867 (1976).
[17] A. M. Rebout, M. Buisson, M. J. Marion, and J. P. Rebout, *Biochim. Biophys. Acta* **432,** 176 (1976).
[18] R. Reyes, D. Vásquez, and J. P. G. Ballesta, *Biochim. Biophys. Acta* **435,** 317 (1976).
[19] K. Terao and K. Ogata, *Biochim. Biophys. Acta* **402,** 214 (1975).



ISBN 0-12-504180-2

ribosomes by two-dimensional acrylamide gel electrophoresis, it was found that the mobility of several proteins by SDS-acrylamide gel electrophoresis remained unchanged even after staining these proteins with Amido Black 10B. Therefore, it was thought to be possible to identify the individual ribosomal proteins and estimate their molecular weights by using two-dimensional acrylamide gel electrophoresis followed by SDS-acrylamide gel electrophoresis ("three-dimensional" electrophoresis). We have developed this method for analysis of individual proteins of rat liver subunits.[19] Similar studies was carried out by Lin and Wool.[20] We shall describe this technique in detail. Since ribosomes are known to absorb soluble proteins, it is necessary to use purified ribosomes or their subunits free from contamination with cell sap proteins. Therefore, we shall describe also our methods of preparing pure 40 S and 60 S ribosomal subunits almost free from contamination from cell sap proteins.[21]

## Isolation of Rat Liver Ribosomes

To minimize contamination with cell sap proteins, we prepared rat liver ribosomes from microsomes by the slightly modified methods of Rendi and Hultin,[22] which use high KCl medium during deoxycholate treatment, and discontinuous density gradient centrifugation after deoxy cholate treatment as follows.

### *Media*

Medium A: 0.25 *M* sucrose, 25 m*M* KCl, 5 m*M* $MgCl_2$, and 0.05 *M* Tris·HCl, pH 7.6

Medium H-l: 0.15 *M* sucrose, 25 m*M* KCl, 10 m*M* $MgCl_2$, 35 m*M* Tris·HCl, pH 7.8

Medium H-2: 0.3 *M* sucrose, 0.6 *M* KCl, 10 m*M* $MgCl_2$, 35 m*M* Tris·HCl, pH 7.8

Medium A′: 0.25 *M* sucrose, 50 m*M* KCl, 5 m*M* $MgCl_2$, 10 m*M* $KHCO_3$, 50 m*M* Tris·HCl, pH 7.8

Rats of the Wistar strain, weighing 200–250 g are starved for about 15 hr prior to sacrifice in order to remove glycogen, which interferes with the preparation of ribosomes. After decapitation of rats, the livers are immediately removed and placed in several volumes of ice-cold medium A, and all subsequent operations are performed in a cold room at 0°–3°.

[20] A. Lin and I. G. Wool, *Mol. Gen. Genet.* **134,** 1 (1974).

[21] K. Terao, K. Tsurugi, and K. Ogata, *J. Biochem.* **76,** 1113 (1974).

[22] R. Rendi and T. Hultin, *Exp. Cell Res.* **19,** 253 (1960).

The liver is blotted, weighed, and minced with scissors, then homogenized in 2 volumes (v/w) of ice-cold medium A with a loosely fitted Teflon-glass homogenizer (clearance 0.3-0.5 mm), which is kept in ice-cold water. Six strokes at 1000 rpm are applied. The homogenate is centrifuged at 10,000 *g* for 10 min. The supernatant is then taken carefully with a Pasteur pipette to avoid the turbid zone near the precipitate. The turbid zone and precipitate are dispersed into 1 volume (v/w) of medium A with 3 strokes of homogenization and recentrifuzed at 10,000 *g* for 10 min. Both supernatant fractions are then combined and centrifuged in a Beckman 60 Ti rotor at 176,000 *g* for 70 min.

The resulting pellet is homogenized in medium H-1 (1 volume of the original tissue weight). To 70 ml of microsomal suspension, 20 ml of 2.5 *M* KCl-10 m*M* $MgCl_2$ solution and 10 ml of 10% freshly prepared DOC are added.

An 18-ml aliquot of the suspension is layered over 20 ml of medium H-2 and centrifuged in a Beckman 60 Ti rotor at 176,000 *g* for 90 min. The supernatant is discarded, and the pellet is rinsed with a small volume of medium A′. The pellet is then suspended in medium A′ (2-3 mg of rRNA per 0.1 ml) and clarified by centrifugation at 20,000 *g* for 10 min. The yield is about 3 mg of ribosomal protein per gram of original rat liver. $A_{260}$:$A_{280}$ and $A_{260}$:$A_{235}$ of the ribosomes preparations are about 1.85 and 1.60, respectively. The sedimentation pattern with a Spinco analytical centrifuge shows that the main components of the ribosomal fraction are 110 S and 75 S subunits.[23] Our ribosomal preparation is almost free from the contamination with cell sap as judged from the fact that poly(U)-dependent polyphenylalanine synthesis is dependent on cell sap.[21]

## Preparation of Active 40 S and 60 S Subunits from Ribosomes

### *Media*

Medium II: 0.85 *M* KCl, 10 m*M* $MgCl_2$, 10 m*M* 2-mercaptoethanol, and 50 m*M* Tris·HCl, pH 7.6

Medium III: 50 m*M* KCl, 2 m*M* $MgCl_2$, 10 m*M* 2-mercaptoethanol, and 20 m*M* Tris·HCl, pH 7.6

Ribosomal subunits are prepared by treatment of ribosomes with puromycin, followed by sucrose density-gradient centrifugation in high KCl medium at relatively high temperature[24] as follows. Ribosomes in

[23] H. Sugano, I. Watanabe, and K. Ogata, *J. Biochem.* **61**, 778 (1967).

[24] T. E. Martin, F. S. Rolleston, R. B. Low, and I. G. Wool, *J. Mol. Biol.* **43**, 135 (1969).

medium A′ are incubated with 0.2 m$M$ puromycin at 37° for 10 min. Then 2.5 $M$ KCl, 0.1 $M$ 2-mercaptoethanol, and 1 $M$ $MgCl_2$ are added to the incubation mixture to make 1 $M$ KCl, 10 m$M$ $MgCl_2$, and 20 m$M$ 2-mercaptoethanol, respectively.

The suspension containing 10 mg–12.5 mg of RNA in 2 ml is layered onto a 15 to 30% linear sucrose density gradient containing medium II, and ribosomal subunits are separated by centrifugation at 95,000 $g$ for 5 hr in a Spinco SW 27 rotor at 26°. The 60 S subunits contain dimerized 40 S subunits (10–15% as RNA), and they are removed as follows.

The 60 S fraction is dialyzed against Medium III overnight at 0°. When necessary, the suspension is concentrated with a collodion bag (SM 13200, Sartorius-Membran filter GmbH). The suspension is layered onto a 15 to 30% sucrose gradient containing medium III and centrifuged at 27,000 rpm for 255 min at 26°. Since 40 S subunits in the 60 S fraction associate with 60 S subunits to form 80 S particles, 60 S subunits thus prepared are almost free from contamination by 40 S subunits.

## Extraction of Ribosomal Proteins

Ribosomal proteins are extracted with acetic acid by a modification of the procedure of Hardy *et al.*[25] One molar $MgCl_2$ is added to the subunit suspension to make the final concentration 20 m$M$, and an equal volume of cold 99% ethanol is then added.

The mixture is kept at 0° for 1 hr, and subunits are sedimented by centrifugation at 10,000 $g$ for 10 min. Precipitated subunits are suspended in 100 m$M$ $MgCl_2$ (about 10 mg of RNA per milliliter). After the addition of 2 volumes of glacial acetic acid, the mixture is stirred at 0° for 48 hr. Ribosomal proteins are obtained by centrifugation of this mixture at 59,000 $g$ for 30 min. The yield of extraction is more than 90%.

## Two-Dimensional Polyacrylamide Gel Electrophoresis

### *Media*

Separation gel (pH 8.6): urea, 54 g; boric acid, 4.8 g; acrylamide, 12 g; Tris, 7.3 g; bisacrylamide, 0.3 g; TEMED, 0.45 ml; EDTA-$Na_2$, 1.2 g; water to make 148.5 ml. For polymerizing, 1.5 ml of 7% ammonium peroxodisulfate solution is added.

Sample solution: sucrose, 10 g; urea, 48 g; boric acid, 0.32 g; EDTA-$Na_2$, 85 mg; TEMED, 0.06 ml; water to make 100 ml

Electrode buffer (pH 8.6): urea, 360 g; boric acid, 9.6 g; EDTA-$Na_2$, 2.4 g; Tris, 14.55 g; water to make 1 liter

[25] S. J. S. Hardy, C. G. Kurland, P. Voynow, and G. Mora, *Biochemistry* **8**, 2897 (1969).

Dialyzing buffer for the l-D gel: urea, 480 g; 5 *N* KOH, 2.4 ml; glacial acetic acid, 0.74 ml; water to make 1.0 liter

2-D separation gel, pH 4.6: urea, 360 g; glacial acetic acid, 52.3 ml; acrylamide, 150 g; 5 *N* KOH, 9.6 ml; bisacrylamide, 5 g; TEMED, 5.8 ml; water to make 967 ml. For polymerizing, 33 ml of a 10% ammonium peroxodisulfate solution is added.

Electrode buffer for 2-D: glycine, 140 g; glacial acetic acid, 15 ml; water to make 10 liters

*Procedure*

Two-dimensional acrylamide gel electrophoresis is carried out according to the method of Kaltschmidt and Wittman[5,6] except that 8% acrylamide gel is used for the first-dimensional electrophoresis and 15% acrylamide gel for the second electrophoresis.[19] Acetic acid-soluble proteins are precipitated with 10 volumes of acetone at $-20°$ for at least 3 hr, usually overnight, and the resulting precipitates (1- 1.5 mg of protein) are dissolved in the sample solution. The solution is reduced with 50 m*M* 2-mercaptoethanol to avoid aggregation due to disulfide bond formation. This treatment markedly reduces the amount of heavily staining material located near the origin. The ribosomal protein solution (1.5-2 mg of protein) is placed at a height of 150 mm from the bottom of separation gel. The gel is polymerized by the addition of ammonium peroxodisulfate solution in a glass tube (7 × 200 mm). Descending electrophoresis is performed for 24 hr at 4 mA per tube. After the first run the gel is removed by crushing the glass tube and is dialyzed against the dialyzing buffer for l-D gel. It is applied to 2-D separation gel, which is polymerized in a vertical chamber (200 × 200 × 4 mm) by the addition of ammonium peroxodisulfate as described above. A current of 50 mA per gel slab is applied at 4° for about 24 hr, using pyronine as a marker. For small amounts of samples (500 μg of 80 S proteins, 300 μg of 40 S or 60 S proteins), small-scale apparatus for acrylamide gel electrophoresis is used in which first-dimensional electrophoresis is carried out in 120 × 6 mm tube and 2.5 mA per tube is applied for 24 hr. For two-dimensional electrophoresis a vertical chamber 150 × 150 × 3 mm is used and a current of 25 mA per gel slab is applied. The gel is stained with 0.5% Amido Black 10B in 10% acetic acid at room temperature for 1 hr and destained in 7% acetic acid.

### Sodium Dodecyl Sulfate–Acrylamide Gel Electrophoresis

Sodium dodecyl sulfate (SDS) acrylamide gel electrophoresis is carried out according to the method of Weber and Osborn.[26]

[26] K. Weber and M. Osborn, *J. Biol. Chem.* **244,** 4406 (1969).

*Media*

Incubation medium 1 (I-1 medium): 1% SDS, 10 m*M* phosphate buffer, pH 7.0

Incubation medium 2 (I-2 medium): 5% sucrose, 0.1% SDS, 50 m*M* 2-mercaptoethanol, 10 m*M* phosphate buffer, pH 7.0, with bromophenol blue as a marker

Separation gel: gel buffer: 8.8 g of $NaH_2PO_4 \cdot 2H_2O$, 51.5 g of $Na_2HPO_4 \cdot 12H_2O$, 2 g of SDS made up to 1 liter with water; gel solution: 22.2 g of acrylamide, 0.6 g of methylenebisacrylamide, made up to 100 ml with water

Electrode buffer: gel solution diluted 1:1 with water

*Procedure*

For polymerizing, 20 ml of gel buffer, 18 ml of gel solution, 2 ml of 1.5% ammonium persulfate solution, and 0.06 ml of TEMED are mixed and poured into glass tubes (6 × 100 mm). A few drops of water are layered on the top of the gel solution. After the gel hardens, the water layer is sucked off; a few drops of electrode buffer are layered on the gel, and then the gel columns are allowed to stand overnight. Just before use, the buffer solution layered on the gel is removed.

Stained gel disks containing protein spots are immediately removed from the gel slab of two-dimensional gel electrophoresis with a stainless borer having an internal diameter of 6 mm. The following procedures should be carried out as soon as possible. The middle layers (2 mm thick) of the gel disks are incubated at 37° in 1 ml of I-1 medium for 30 min.[26a] These procedures are repeated three times by changing the same medium. Finally, they are incubated in 1 ml of I-2 medium at 37° for 30 min.

The incubation gel disks are then placed on the top of the SDS polyacrylamide gel columns (6 × 80 mm) containing 10% acrylamide and 0.1% SDS described above. Electrophoresis is carried out for 5-6 hr at 8 mA per tube, using bromophenol blue as a marker. After electrophoresis, the gel is placed in the solution containing 10% methanol and 7.5% acetic acid at room temperature for 30 min and then transferred to the staining solution containing 0.5 g of Coomassie Brilliant Blue in a mixture of 454 ml of 50% methanol and 46 ml of glacial acetic acid at room temperature for 2-3 hr. Destaining of the gel is performed in destaining solution (7.5% acetic acid and 5% methanol) at room temperature.

The molecular weights of stained materials are calculated by the method of Weber and Osborn.[26] As the internal standard, bovine serum albumin, ovalbumin, trypsin, soybean trypsin inhibitor, and horse heart cytochrome *c* are used.

[26a] When small-scale gel electrophoresis is carried out, whole gel disks are used (manuscript in preparation).

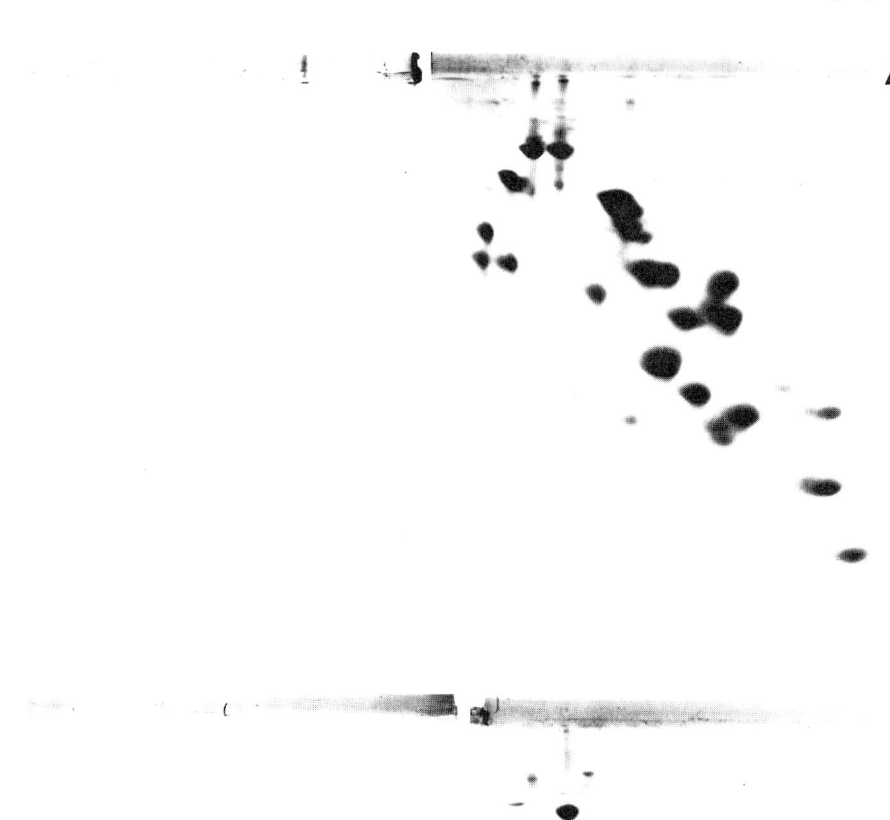
A
B

### Proteins of 60 S and 40 S Subunits

As shown in Fig. 1A and B, 35 and 29 protein spots are identified on two-dimensional acrylamide gel electrophoresis of 60 S and 40 S ribosomal protein, respectively. The patterns are represented schematically in Fig. 2A and B. The protein spots are numbered according to the system of Kaltschmidt and Wittmann.[6] Proteins of large and small subunits are marked with the letters L and S, respectively.

Disulfide reduction results in the disappearance of the spot L11 in addition to the decrease of aggregated materials near the origin. The faint L23 spot becomes distinct after this treatment, although it is uncertain whether the spot L11 is converted to the spot L23.

The patterns of SDS-acrylamide gel electrophoresis of 40 S proteins are shown in Fig. 3. Similar patterns are obtained with 60 S proteins. A number of ribosomal proteins show only one band, or in some cases two distinct bands, on SDS gel. Two bands are usually observed in the case of the partially overlapping spots on the two-dimensional gel. In Fig. 3, M is used to mark the main components and N to mark the components derived from the neighboring spot. It must be emphasized that the immediate application of SDS-acrylamide gel electrophoresis after two-dimensional gel electrophoresis prevents the formation of minor degraded or aggregated components in the case of high-molecular-weight proteins near the origin as described previously.[19]

The molecular weights of proteins of rat liver 60 S and 40 S subunits are summarized in Table I. The number average molecular weights for proteins of the 40 S and 60 S subunits are 23,000 and 23,900, respectively.

### Comments

Our three-dimensional polyacrylamide gel electrophoresis provides a convenient method for the identification of small amounts of ribosomal proteins. Especially, since stained spots on the two-dimensional gel are used for SDS-acrylamide gel electrophoresis as described above, the procedures may be easily carried out and very faint spots on the gel can be analyzed. The analysis of individual ribosomal proteins labeled *in vivo* and *in vitro* can be made by these procedures as described in this volume [42].

The purity of ribosomes or ribosomal subunits seems satisfactory, as discussed previously.[19] The two-dimensional electrophoretograms of 40

---

FIG. 1. Two-dimensional electrophoretograms of liver ribosomal proteins. (A) Proteins from 60 S subunits. (B) Proteins from 40 S subunits.

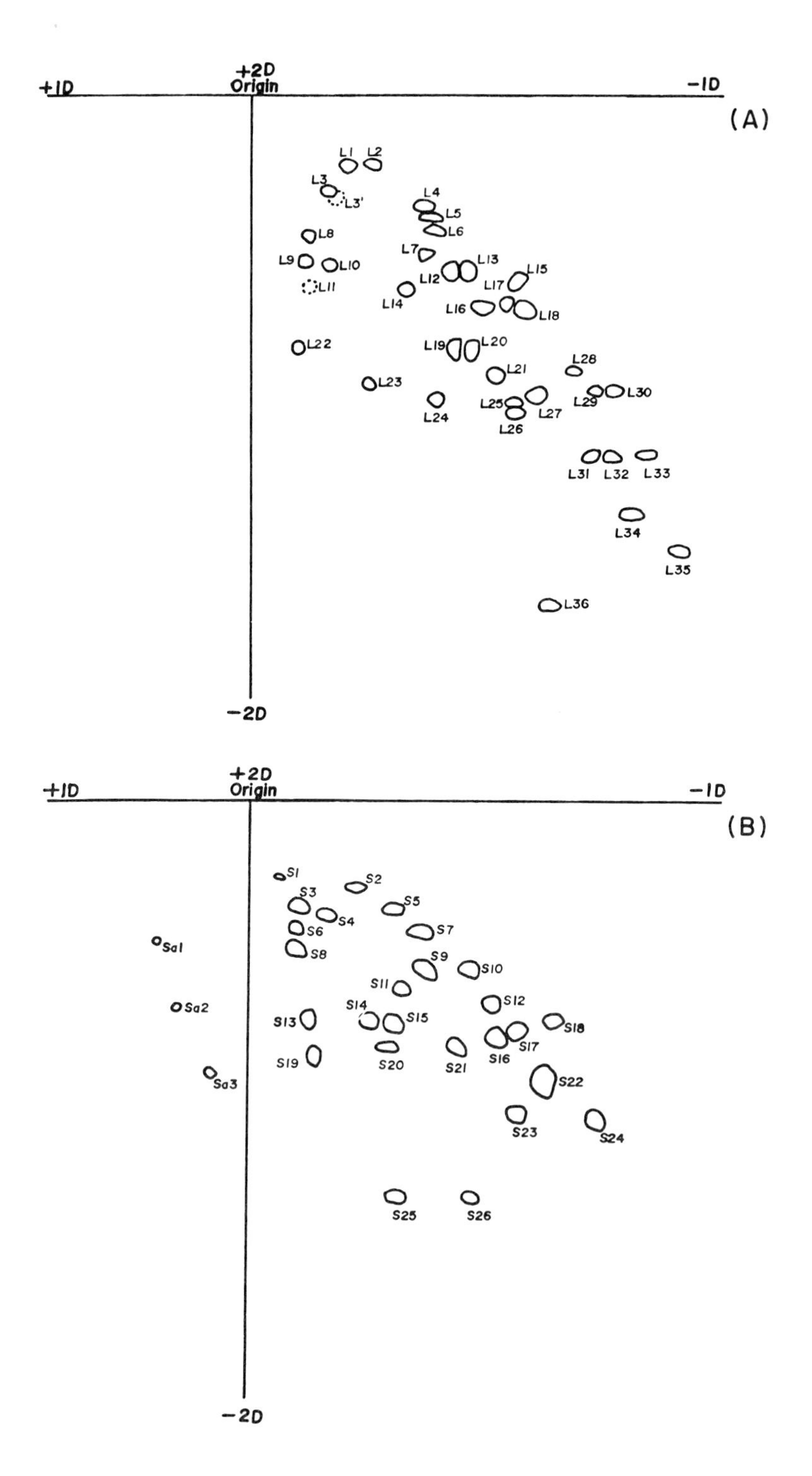


FIG. 2. Schema of the two-dimensional electrophoretograms of liver ribosomal proteins. (A) Proteins from 60 S subunits. (B) Proteins from 40 S subunits.

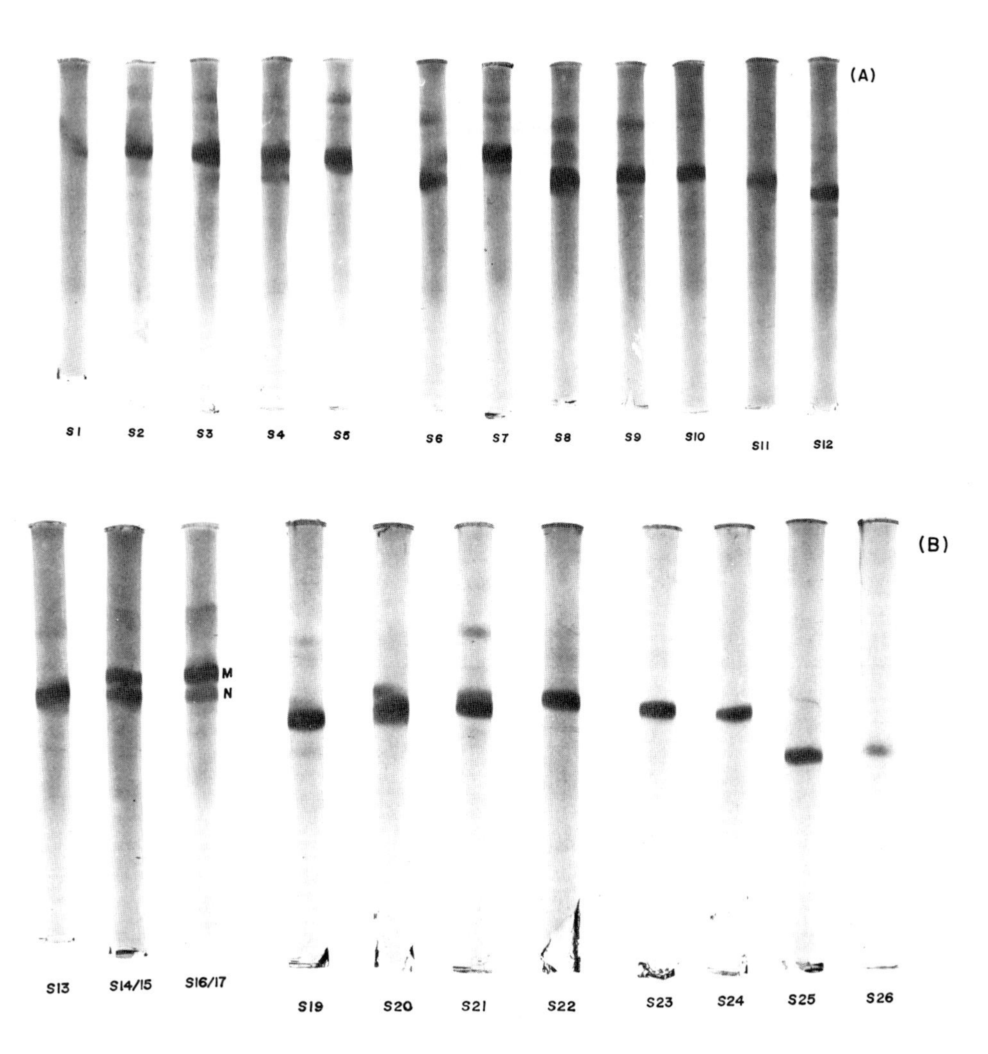


FIG. 3. Patterns of sodium dodecyl sulfate-gel electrophoresis of the stained 40 S proteins on two-dimensional gels.

TABLE I
MOLECULAR WEIGHTS OF RIBOSOMAL PROTEINS FROM 40 S AND 60 S SUBUNITS

| 40 S | | | | 60 S | | | |
|---|---|---|---|---|---|---|---|
| S1 | 34,000 | S19 | 17,000 | L1 | 54,000 | L19 | 22,000 |
| S2 | 38,000 | S20 | 17,000 | L2 | 60,000 | L20 | 22,000 |
| S3 | 35,000 | S21 | 20,000 | L3 | 39,000 | L21 | 19,000 |
| S4 | 35,000 | S22 | 21,000 | L4 | 40,000 | L22 | 15,000 |
| S5 | 32,000 | S23 | 21,000 | L5 | 32,000 | L23 | 14,000 |
| S6 | 26,000 | S24 | 21,000 | L6 | 30,000 | L24 | 15,000 |
| S7 | 37,000 | S25 | 11,000 | L7 | 25,000 | L25 | 20,000 |
| S8 | 23,000 | S26 | 14,000 | L8 | 24,000 | L26 | 17,000 |
| S9 | 25,000 | | | L9 | 19,000 | L27 | 16,000 |
| S10 | 28,000 | Sal | 22,000 | L10 | 22,000 | L28 | 19,000 |
| S11 | 24,000 | Sa2 | 14,000 | (L11 | 20,000) | L29 | 17,000 |
| S12 | 21,000 | Sa3 | 10,000 | L12 | 27,000 | L30 | 21,000 |
| S13 | 18,000 | | | L13 | 29,000 | L31 | 13,000 |
| S14 | 20,000 | | | L14 | 23,000 | L32 | 19,000 |
| S15 | 22,000 | | | L15 | 28,000 | L33 | 17,000 |
| S16 | 23,000 | | | L16 | 28,000 | L34 | 16,000 |
| S17 | 18,000 | | | L17 | 23,000 | L35 | 16,000 |
| S18 | 21,000 | | | L18 | 26,000 | L36 | 10,000 |
| $\Sigma M_i$ = | 668,000 | | | $\Sigma M_i$ = | 837,000 | | |
| $\overline{M_n}$ = | 23,000 | | | $\overline{M_n}$ = | 23,900 | | |

S, 60 S, and 80 S proteins are reproducible. Our electrophoretograms are similar to those reported by Sherton and Wool,[9] which were done on ribosomal protein from the same source. A greater number of their protein spots are comparable with ours. As discussed previously,[19] the differences between our and their electrophoretograms may generally be explained by the different conditions of electrophoresis. However, we must point out that acidic proteins are not observed in our preparation of 60 S proteins. Acidic proteins may be removed from the 60 S subunits during the purification procedures, especially the DOC treatment of microsomes in the presence of high KCl, although further experiments must be done to elucidate this point.

There are three methods generally used for the extraction of ribosomal proteins, as follows: (1) acetic acid extraction,[25] (2) LiCl-urea extraction,[27] and (3) RNase digestion[28] followed by urea treatment. It is notice-

[27] P. Spitnik-Elson, *Biochim. Biophys. Res. Commun.* **18**, 557 (1965).

[28] V. Mutolo, G. Giudice, V. Hopps, and G. Donatuti, *Biochim. Biophys. Acta* **138**, 214 (1967).

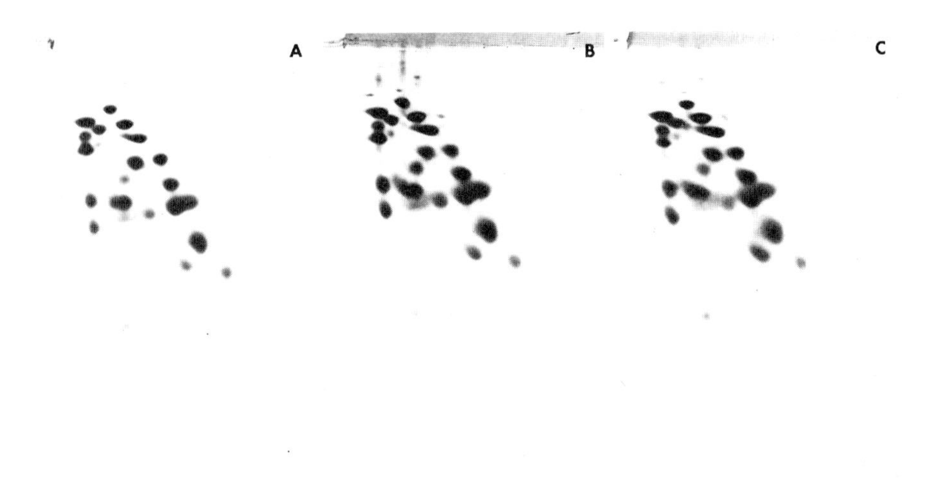


FIG. 4. Two-dimensional electrophoretograms of 40 S ribosomal proteins prepared from the same sample of 40 S subunits by three different kinds of extraction. (A) Extraction with acetic acid. Detailed methods are given in the text. (B) Extraction with LiCl-urea. Ribosomal subunits are suspended (2-3 mg/ml) in 50 m*M* Tris·HCl buffer, pH 7.6 and one volume of 4 *M* LiCl-8 *M* urea solution is added. The suspension is stirred for 48 hr at 0°. The RNA precipitate is collected by centrifugation at 59,000 *g* for 30 min. The supernatant is dialyzed against 0.2 *N* HCl for 1-2 hr, and 10 volumes of acetone are added as described in the text. The resulting precipitates are dissolved in the sample solution for electrophoresis. (C) RNase digestion. Ribosomal subunits are suspended (2-3 mg/ml) in 6 *M* urea, and pancreatic RNase (10 g/ml) and RNase Tl (100 units/ml) are added. The suspension is dialyzed against 6 *M* urea for 4-6 hr at room temperature (about 25°) and then overnight at 0° with two changes of 6 *M* urea. The suspension is further dialyzed against 0.2 *N* HCl for 1-2 hr, and 10 volumes of acetone are added. The resulting precipitates are dissolved in the sample solution for electrophoresis. Three kinds of ribosomal proteins thus prepared are subjected to two-dimensional gel electrophoresis simultaneously.

able that we get similar electrophoretograms of 40 S proteins among ribosomal proteins extracted by these three different methods (Fig. 4). Similar results are obtained with 60 S proteins. The results are somewhat different from those reported by Sherton and Wool.[29]

Concerning the molecular weights of 60 S and 40 S ribosomal proteins, our values are generally somewhat higher than those for the corresponding proteins described by Wool's group, which have been recently separated by chromatographic procedures.[2-4] It was reported that molecular weights of *E. coli* ribosomal proteins determined by SDS-

[29] C. C. Sherton and I. G. Wool, *Mol. Gen. Genet.* **135,** 97 (1974).

acrylamide gel electophoresis are somewhat greater than the actual values determined from the amino acid sequences.[30–37]

[30] R. Chen and B. Wittmann-Liebold, *FEBS Lett.* **52,** 139 (1975).
[31] K. G. Britarand and B. Wittmann-Liebold, *Hoppe-Seyler's Z. Physiol. Chem.* **356,** 1343 (1975).
[32] J. S. Vandekerckhove, W. Rombauts, B. Peeters, and B. Wittmann-Liebold, *Hoppe-Seyler's Z. Physiol. Chem.* **356,** 1955 (1975).
[33] B. Wittmann-Liebold, B. Grever, and R. Panvenbecker, *Hoppe-Seyler's Z. Physiol. Chem.* **356,** 1977 (1975).
[34] B. Wittmann-Liebold, E. Marzinzig, and A. Lehmann, *FEBS Lett.* **68,** 110 (1976).
[35] I. Heiland, D. Braner, and B. Wittmann-Liebold, *Hoppe-Seyler's Z. Physiol. Chem.* **357,** 1751 (1976).
[36] H. Lindemann and B. Wittmann-Liebold, *Hoppe-Seyler's Z. Physiol. Chem.* **358,** 843 (1977).
[37] J. S. Vandekerckhove, W. Rombauts, and B. Wittmann-Liebold, *Hoppe-Seyler's Z. Physiol. Chem.* **358,** 989 (1977).

# [56] Cross-Linking of Ribosomes Using 2-Iminothiolane (Methyl 4-Mercaptobutyrimidate) and Identification of Cross-Linked Proteins by Diagonal Polyacrylamide/Sodium Dodecyl Sulfate Gel Electrophoresis[1]

*By* JAMES W. KENNY, JOHN M. LAMBERT, and ROBERT R. TRAUT

Many biological structures contain assemblies of different proteins. It is frequently valuable to determine the spatial relationships among the different protein components of the multiprotein complex. Bifunctional reagents have been used effectively, to cross-link one protein component to others that occupy a suitably "neighboring" site in the structure or complex under investigation. A problem frequently encountered is that of identification of the monomeric components of cross-linked dimers or oligomers. The presence of a readily cleavable bond in the cross-linking reagent permits reversal of the cross-linking reaction and regeneration of monomeric components from isolated cross-linked complexes, thus facilitating their identification. Methods are described here that employ reversible cross-linking and analysis of a complex mixture of cross-linked products. They have been used successfully in the investigation of the protein topography of ribosomal subunits of *Escherichia coli*. They are

[1] Supported by a research grant from the U.S. Public Health Service (GM 17924).



ISBN 0-12-504180-2

of general applicability and are useful in the investigation of many other biological structures containing multiple protein components.

The reagent 2-iminothiolane, formerly called methyl 4-mercaptobutyrimidate[2,3] reacts with lysine amino groups in the intact ribosomal subunit to form amidine derivatives containing sulfhydryl groups. Disulfide bonds form when the modified subunit is subjected to oxidation. Some of the disulfide bonds are intramolecular, while others are intermolecular and represent "cross-links" that provide information on the relative spatial arrangement of the different ribosomal proteins. The term "cross-link," when used in the remainder of the article, will imply intermolecular disulfide-linked proteins. It is a prerequisite for any cross-linking procedure that it not alter the structure under study. Various physical properties of ribosomal subunits, treated as described here, are not detectably altered. The cross-linked subunits retain the capacity to reassociate to form 70 S ribosomes and retain up to 50% of their activity in polyphenylalanine synthesis.[4]

Methods for the separation and identification of cross-linked dimers are described. Of particular general applicability is the technique of diagonal polyacrylamide/sodium dodecyl sulfate (SDS) gel electrophoresis.[5] It is a two-dimensional electrophoretic separation, utilizing the size dependence of the mobility of proteins in SDS to distinguish cross-linked from monomeric proteins. The first electrophoresis is performed under nonreducing conditions, and the second under reducing conditions. This results in a pattern in which non-cross-linked proteins fall on a diagonal line and cross-linked proteins fall beneath the diagonal.

Schematic diagrams of the modification and cross-linking reactions (Fig. 1) and of diagonal gel electrophoresis (Fig. 2) are shown. The procedures will be described in detail as they have been applied to the 50 S ribosomal subunit of *Escherichia coli*. In addition to the two general methods already mentioned, techniques for the purification of the mixture of cross-linked protein from 50 S ribosomal subunits prior to diagonal gel electrophoresis will be described.

## Modification of 50 S Ribosomal Subunits with 2-Iminothiolane

*Solutions*

1. $NH_4Cl$, 100 m*M*; Tris·HCl, pH 7.2, 10 m*M*; $MgCl_2$, 10 m*M*; 2-mercaptoethanol, 14 m*M*

[2] R. R. Traut, A. Bollen, T. T. Sun, J. W. B. Hershey, J. Sundberg, and L. R. Pierce, *Biochemistry* **12,** 3266 (1973).

[3] R. Jue, J. H. Lambert, L. R. Pierce and R. R. Traut, *Biochemistry*, in press (1978).

[4] J. M. Lambert, R. Jue, and R. R. Traut, *Biochemistry*, in press (1978).

[5] A. Sommer and R. R. Traut, *Proc. Natl. Acad. Sci. U.S.A.* **71,** 3946 (1974).

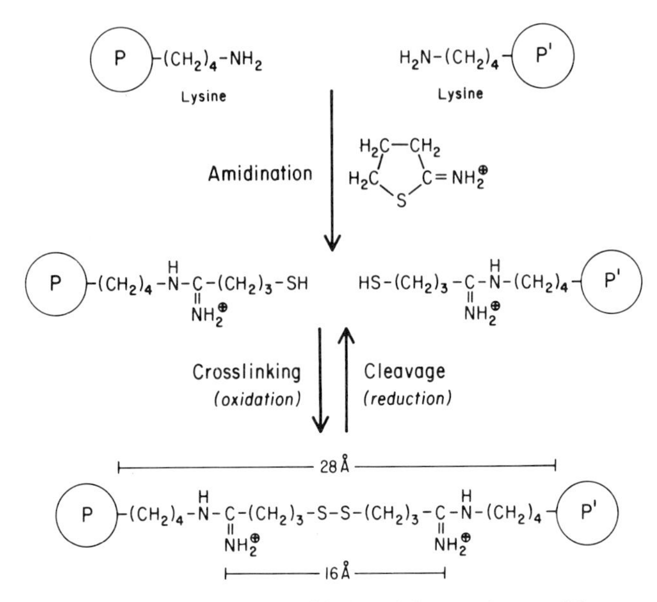


FIG. 1. Modification of proteins with 2-iminothiolane and reversible cross-linking by disulfide bond formation.

2. KCl, 50 m*M*; triethanolamine·HCl, pH 8.0, 50 m*M*; $MgCl_2$, 1 m*M*
3. Solution 2 with 5 m*M* dithiothreitol
4. 2-Iminothiolane, 500 m*M*; triethanolamine·HCl, pH 8.0, 500 m*M*; triethanolamine, free base, 500 m*M*

Tris was obtained from Sigma; 2-mercaptoethanol from BDH; dithiothreitol from Pierce; triethanolamine from Eastman and distilled under vacuum prior to use. 2-Iminothiolane was prepared as described[2] or was purchased from Pierce. All reagents were of reagent grade.

*Procedure*

Radioactive 50 S ribosomal subunits were isolated from *E. coli* MRE600 grown in the presence of [$^{35}$S]sulfate as described[6] and were more than 95% free of contaminating 30 S subunits as determined by analytical centrifugation. They were stored in solution 1 at −70°. The specific radioactivity of the 50 S ribosomal protein was 170 × $10^6$ cpm/mg.

The ribosomal subunits are reduced by incubation for 30 min at 30° in solution 1 to which 1% (v/v) 2-mercaptoethanol was added. The ribo-

[6] T. T. Sun, A. Bollen, L. Kahan, and R. R. Traut, *Biochemistry* **13,** 2334 (1974).

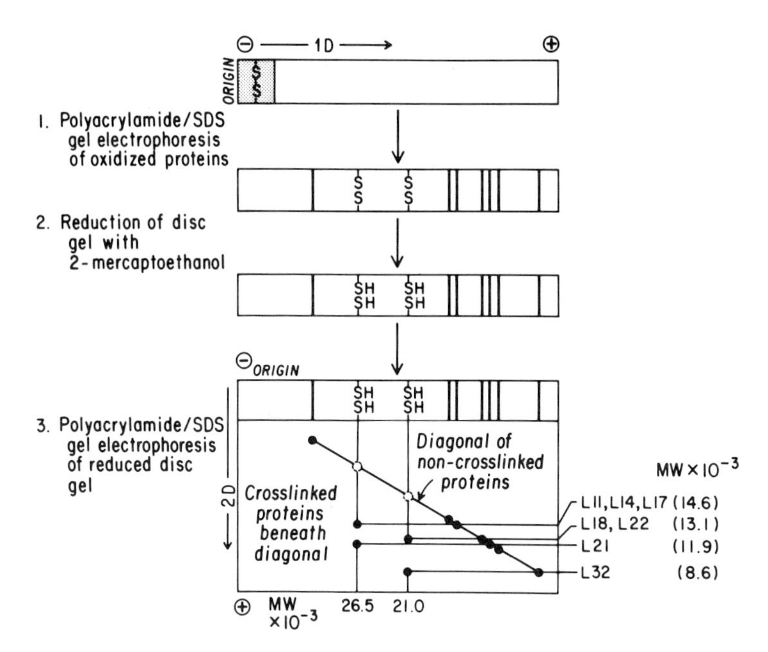


FIG. 2. Two-dimensional diagonal polyacrylamide/sodium dodecyl sulfate gel electrophoresis.

somal subunits are then passed through a BioGel P-2 column (15 cm × 0.7 cm, for a 1.0-ml sample equilibrated with solution 3) in order to remove free amines that might react with 2-iminothiolane. The concentration of ribosomal subunits is adjusted to an $A_{260}$ of 45 (1 mg of ribosomal protein per milliliter) with solution 3; 12 m$M$ 2-iminothiolane (24 $\mu$l solution 4 per milliliter of sample) is added, and the mixture is incubated for 2.5 hr at 0°. The pH of the modification reaction is 8.0. Under these conditions each ribosomal protein reacts with on average two molecules of 2-iminothiolane.[3] The modified subunits are incubated with 40 m$M$ hydrogen peroxide (4.5 $\mu$l of 30% hydrogen peroxide per milliliter of sample) at 0° for 30 min to promote cross-linking between adjacent sulfhydryl groups by disulfide bond formation. These reactions are represented in Fig. 1. Unreacted hydrogen peroxide is removed by the addition of catalase (15 $\mu$g of catalase per milliliter of sample) followed by incubation for 15 min at 0°. Unreacted 2-iminothiolane is removed either by passing the modified, oxidized sample through a BioGel P-2 column equilibrated with solution 2, or by dialysis against solution 2, in order to prevent reaction of the imidate with newly exposed amino groups in subsequent steps. Iodoacetamide is added to a concentration of 40 m$M$. The solution is incubated for 30 min at 30° to alkylate free sulfhydryl groups inaccessible to oxidation.

## Extraction of Protein from Cross-Linked Ribosomal Subunits

Cross-linked 50 S ribosomal subunits are mixed with an equal volume of a solution containing 8 $M$ urea (ultra pure), 6 $M$ LiCl and 40 m$M$ iodoacetamide (added immediately before use) and incubated at 0° for 24 hr. The precipitated RNA is removed by centrifugation at 10,000 rpm for 15 min. The supernatant protein fraction is dialyzed exhaustively against 6% acetic acid and lyophilized. Alternatively, the protein is precipitated by addition of 10 volumes of 10% (w/v) trichloroacetic acid and recovered by centrifugation at 10,000 rpm for 30 min at 4°. The precipitate is washed in ethanol/ether and dried under vacuum.

## Two-Dimensional Diagonal Polyacrylamide/SDS Gel Electrophoresis

The disulfide bonds formed by oxidation of modified ribosomal subunits are readily cleavable by reduction. Polyacrylamide/SDS diagonal gel electrophoresis uses this property of the cross-link to separate intermolecular cross-linked dimers from protein monomers containing only intramolecular disulfide bonds. First the sample is electrophoresed under nonreducing conditions to maintain disulfide bonds intact. The proteins are reduced in the gel to cleave the disulfide bonds and convert cross-linked complexes into monomeric proteins. Monomeric proteins that had migrated as disulfide-linked complexes in the first dimension migrate more rapidly in the second electrophoresis. Uncross-linked proteins have the same electrophoretic mobility in both electrophoretic separations. The resulting protein pattern is composed of a diagonal line of non-cross-linked proteins with a complex array of cross-linked proteins below the diagonal. Figure 2 shows a schematic diagram of diagonal gel electrophoresis.

The SDS gel system described here gives a linear relationship between apparent log (molecular weight) and mobility for both cross-linked protein dimers and monomers between 10,000 and 60,000 daltons as calibrated using individual monomeric 30 S ribosomal proteins or commercially available molecular weight standards.[7] Within this range, the sum of the apparent molecular weights of the monomer proteins below the diagonal arising from a putative dimer is within 7.5% of that of the cross-linked

[7] A. Sommer and R. R. Traut, *J. Mol. Biol.* **97**, 471 (1975).

complex. This additivity relationship together with the coincidence of the spots on the same vertical line provide the major criteria for identifying pairs of proteins originally cross-linked. The diagram in Fig. 2 shows these two criteria for assigning the monomeric proteins originating from a cross-linked dimer formed in the intact subunit.

*Solutions and Acrylamide Gel Composition*

5. SDS sample buffer, pH 6.8: SDS, 4% w/v; Tris·HCl, pH 6.8, 80 m*M*; iodoacetamide, 40 m*M*; glycerol, 10% v/v. The solution is filtered and stored at room temperature. Iodoacetamide is added immediately before use.
6. Upper gel, pH 7.8: acrylamide, 5% w/v; *N,N'*-Methylenebisacrylamide (MBA), 0.26% w/v; SDS, 0.25% w/v; Tris·HCl, pH 7.8, 125 m*M*; tetramethylethylenediamine (TEMED), 0.05% v/v. The components are mixed and filtered at room temperature. The solution is degassed and polymerization is initiated by addition of 5ml per liter of a freshly prepared solution of ammonium persulfate (10% w/v).
7. Separation gel, pH 8.7: acrylamide, 17.5% w/v; MBA, 0.35% w/v; SDS, 0.1% w/v; Tris·HCl, pH 8.7, 335 m*M*; TEMED, 0.033% v/v. The components are mixed, filtered, and degassed. Polymerization is initiated by addition of 6.6 ml per liter of a freshly prepared solution of ammonium persulfate (10% w/v). The final acrylamide:bisacrylamide ratio is 30:0.6.
8. Electrophoresis buffer, pH 8.7: Glycine, 2.8% w/v; SDS, 0.1% w/v; Tris base 0.58% w/v
9. Tracking dye: Bromphenol blue, in solution 5, 0.1% w/v

SDS was obtained from Serva and iodoacetamide from Sigma. Acrylamide (technical grade) and bisacrylamide were obtained from Eastman and used without recrystallization.

*First SDS Electrophoresis*

Approximately 1 mg of lyophilized or precipitated cross-linked protein is dissolved in 25–50 $\mu$l of solution 5 and incubated for 15 min at 65°. Tracking dye is added immediately before electrophoresis. The gels are poured in silicon-coated glass tubes 14 cm × 0.4 cm (i.d.). The separation gel (solution 7) and the upper gel (solution 6) are 10 cm and 1 cm, respectively. Electrophoresis toward the anode is at 2 mA/gel for 3.5 hr at room temperature using solution 8 as the electrolyte. In experiments for which it is desirable to have marker proteins, such as total 50 S protein, for the second electrophoresis (see Fig. 6), this is added to the origin of the gel 10 min prior to completion of electrophoresis. After electrophoresis, the gel is removed from the tube by smashing the glass and then soaked in 50 ml of solution 8 made 3% (v/v) in 2-mercaptoethanol, for 15 min at 65°. The gel is then incubated for 30 min at room temperature in another 50 ml of solution 8 in which the pH is adjusted to 6.8. The gel is now ready to be embedded as the origin of the second polyacrylamide gel, which is a slab.

*Second SDS Electrophoresis*

The apparatus is a modification of that described previously[8] and consists of two glass plates (24 cm × 12 cm) separated by Plexiglas spacers (0.4 cm) clamped to a Plexiglas unit with upper and lower reservoirs for electrolyte. The reduced gel from the first dimension is embedded at the origin of the gel slab by first squeezing it between the glass plates and pouring the separation gel on top of it. The composition of the gel is identical to that of the first electrophoresis. Tracking dye (solution 9) is applied just above the embedded gel cylinder prior to electrophoresis. Electrophoresis is carried out for 1 hr at 50 V followed by 30 hr at 90 V using solution 8 as the electrolyte. The gel slab is stained for 30 min in a solution containing methanol, glacial acetic acid, and water (5:1:5 by volume) with 0.55% (w/v) Amido black.

## Purification of Radioactive, Cross-Linked Ribosomal Proteins prior to Diagonal Gel Electrophoresis

Figure 3 shows stained diagonal gels for ribosomal proteins from both the 30 S and 50 S subunits. The patterns are complex. Many protein dimers can be identified by the criteria mentioned previously: the additivity of apparent molecular weights ($\text{monomer}_a + \text{monomer}_b = \text{cross-linked species}_c$), and the finding of a and b on the same vertical line descending from c (see Figs. 2 and 3). Many more protein dimers are present than can be readily identified. This is because many ribosomal proteins have the same or similar molecular weights. Accordingly, procedures were developed to simplify the samples analyzed by diagonal gel electrophoresis. The cross-linked subunits are first extracted with increasing concentrations of LiCl.[9] Then each extracted fraction is separated by electrophoresis in polyacrylamide/urea gels.[10] The gel is sliced

[8] G. A. Howard and R. R. Traut, this volume [54].
[9] A. Sommer and R. R. Traut, *J. Mol. Biol.* **106,** 995 (1976).
[10] U. C. Knopf, A. Sommer, J. Kenny, and R. R. Traut, *Mol. Biol. Rep.* **2,** 35 (1975).

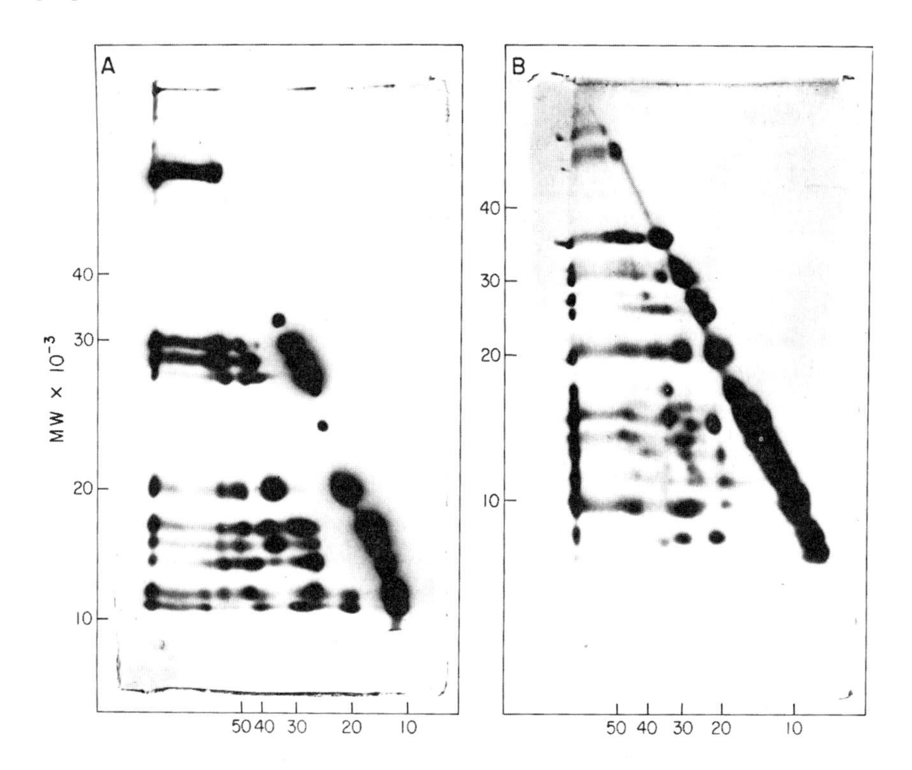


FIG. 3. Two-dimensional diagonal polyacrylamide/SDS gels of proteins extracted from ribosomal subunits modified with 2-iminothiolane and oxidized. (A) 30 S subunits. (B) 50 S subunits. Electrophoresis was as represented in Fig. 2 and described in the text, except that the acrylamide concentration for the 30 S subunits was 13.5%.

into 0.5-cm segments, each of which serves as one sample for diagonal gel electrophoresis.

*Extraction with LiCl*

*Solutions*

10. Solution 2 with 1.0 *M* LiCl
11. Solution 2 with 1.5 *M* LiCl
12. Solution 2 with 2.0 *M* LiCl

Cross-linked 50 S ribosomal subunits labeled *in vivo* with [$^{35}$S]sulfate[6] (180 $A_{260}$ units or 4 mg of protein) in 4.2 ml of solution 2 with 80 m*M* iodoaeetamide are adjusted to 0.2 m*M* EDTA and mixed with an equal volume of solution 10. The mixture is incubated for 5 hr at 4° and then centrifuged for 17 hr at 27,000 rpm in a Beckman SW 56 rotor. The radioactive supernatant fraction is mixed with 300 μg of nonradioactive, uncross-linked total 50 S ribosomal protein, dialyzed against 6% acetic acid, and lyophilized. The procedure is repeated on the pelleted protein-deficient ribosomal subunit fraction with solutions 11 and 12 successively to extract additional ribosomal protein fractions. At each step the protein-deficient "core" is first suspended in solution 2 containing 80 m*M* iodoacetamide and incubated for 30 min at 30° in order to alkylate any free sulfhydryl groups that might become exposed by removal of proteins. Free sulfhydryl groups are capable of undergoing either random intermolecular oxidation or disulfide interchange when the proteins are extracted from the intact ribosomal subunit or core particle. The final core particle is also alkylated and then treated with 66% acetic acid, 33 m*M* $MgCl_2$ to extract the remaining protein and precipitate the RNA. The protein is dialyzed against 6% acetic acid and lyophilized. All protein fractions are enriched for specific cross-links as well as monomeric proteins. The recovery of protein and radioactivity is shown in Table I.

*Electrophoretic Fractionation*

Lyophilized protein fractions (see Table I for amounts) are resuspended in 50 μl of a buffer containing 8 *M* urea and 40 m*M* iodoacetamide. Pyronine G is added to the solution as a tracking dye. Samples are fractionated by electrophoresis in 4% polyacrylamide gels (10 cm × 0.4 cm i.d.) containing 6 *M* urea and 38 m*M* bis-Tris·acetate, pH 5.5.[10] Electrophoresis is carried out for 5 hr at 1 mA per gel toward the cathode. A detailed description of this gel system follows in the next section. The gel is removed from the glass tube and sliced into 20 equal fractions of

TABLE I
EXTRACTION OF CROSS-LINKED $^{35}$S-LABELED 50 S SUBUNITS WITH LiCl[a]

| Protein fraction | Protein (mg) | Radioactivity (cpm × $10^{-6}$) | Percentage recovered |
|---|---|---|---|
| 0.00–0.50 *M* LiCl | 1.55 | 266 | 38.7 |
| 0.50–0.75 *M* LiCl | 0.76 | 130 | 18.9 |
| 0.75–1.00 *M* LiCl | 0.21 | 36 | 5.2 |
| 1.00 *M* LiCl core (extracted with acetic acid) | 0.23 | 40 | 5.8 |

[a] The total protein present prior to initial extraction was 4 mg (12 mg of 50 S subunits) containing 687 × $10^6$ cpm. Protein recovered was calculated from specific radioactivity as measured after dialysis against 6% acetic acid. The recovery was 68%.

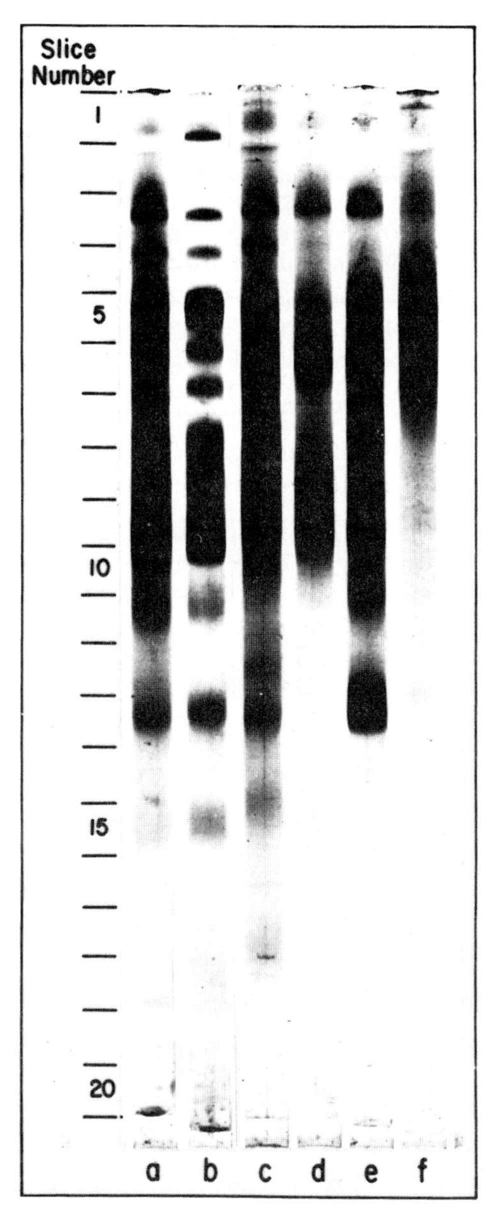


FIG. 4. Polyacrylamide/urea gel electrophoresis of protein fractions extracted from cross-linked 50 S ribosomal subunits at different concentrations of LiCl. The acrylamide concentration was 4% and the pH was 5.5. Details are given in the text. a, Cross-linked total 50 S protein; b, noncross-linked total 50 S protein; c–e, protein extracted from cross-linked 50 S subunits with LiCl: c, 0–0.5 *M*; d, 0.5–0.75 *M*; e, 0.75–1.0 *M*; f, proteins remaining after extraction with 1.0 *M* LiCl. The core particle was extracted with 66% acetic acid.

0.5 cm length. The gels are illustrated in Fig. 4, along with unfractionated cross-linked 50 S ribosomal protein and noncross-linked monomeric total 50 S protein.

Each 0.5-cm gel slice containing radioactive protein from cross-linked ribosomal subunits is inserted at one end of a 10.5 cm × 0.4 cm (ID) silicon-coated glass tube. The end containing the gel slice is covered with Parafilm. The tube is inverted and filled with a polyacrylamide/SDS gel solution containing 17.5% acrylamide (solution 7), and electrophoresed as previously described. The second electrophoresis is also carried out as previously described and is followed by staining the gel with Amido black. An example of a simplified diagonal gel pattern resulting from these fractionation steps is shown in Fig. 6A.

## Two-Dimensional Polyacrylamide/Urea Gel Electrophoresis for the Identification of Radioactive Proteins from Diagonal Gels

The position of a stained protein spot on a diagonal gel is insufficient, in many cases, for unambiguous identification of ribosomal proteins since many have similar molecular weights. It is for this reason that radioactive protein is used throughout the methods described here. Radioactive proteins beneath the diagonal are eluted, mixed with total noncross-linked, nonradioactive 50 S ribosomal protein, and analyzed as described below.

### *Solutions and Acrylamide Gel Composition*

*Protein Elution*

13. Tris·acetate, pH 7.8, 100 m*M*; SDS, 1% w/v; 2-mercaptoethanol, 1% v/v
14. Tris·acetate, pH 7.8, 50 m*M*; urea, 8.0 *M*; 2-mercaptoethanol, 1% v/v

*First electrophoresis*

15. Sample buffer: urea, 8.0 *M*; iodoacetamide, 40 m*M*
16. Upper gel, pH 4.7: acrylamide, 4% w/v; MBA, 0.066% w/v; urea, 6.0 *M*; Bis-Tris, 38 m*M*; TEMED, 0.02% v/v
17. Separation gel, as solution 16 except pH 5.5

The pH of solutions 16 and 17 is adjusted with glacial acetic acid after mixing the components. To catalyze polymerization of the gels, 5 ml per liter of ammonium persulfate (10% w/v) are added to the degassed gel solutions.

18. Electrophoresis buffers: (upper reservoir) Bis-Tris·acetate, pH 3.7, 20 m*M*; (lower reservoir) Bis-Tris·acetate, pH 7.0, 20 m*M*
19. Tracking dye: pyronine G, 0.5% w/v in solution 15

*Second Electrophoresis*

20. Separation gel, pH 4.6: acrylamide, 18% w/v; MBA, 0.25% w/v; urea, 6.0 *M*; glacial acetic acid, 920 m*M*; KOH, 48 m*M*; TEMED, 0.58% v/v

The second electrophoretic separation is a slight modification of the system of Kaltschmidt and Wittmann.[11] Polymerization is catalyzed by the addition of 30 ml per liter of ammonium persulfate (10% w/v).

21. Electrophoresis buffer, pH 4.0: glycine, 180 m*M*; acetic acid, 6 m*M*
22. Staining solution: trichloroacetic acid, 12.5% w/v; Coomassie Blue G-250, 0.0125% w/v

Urea "ultra pure" and Bis-Tris were purchased from Sigma, glycine from Eastman, trichloroacetic acid (analytical grade) from Mallinckrodt, and Coomassie Blue G-250 from Pierce.

*Elution of Proteins from Diagonal Gels*

The stained spots from diagonal gels are cut out. The gel is then macerated with a glass rod to make a fine slurry suspended in an adequate volume (200–500 μl depending on size of gel segment) of solution 13. The slurry is incubated for 15 min at 65° and then cooled to room temperature. Nonradioactive, noncross-linked total 50 S protein, 300 μg, is added. This carrier is added in order to decrease loss of radioactive protein and to act as marker on subsequent two-dimensional polyacrylamide/urea gel electrophoresis. The mixture of protein and gel particles is adjusted to approximately 8 *M* urea by adding solid crystals, approximately doubling the volume, and applied to a column (2.5 cm × 0.8 cm for a 1.0 ml sample) containing Bio-Rad Dowex 1-X8 (20–50 mesh, acetate form) equilibrated with solution 14 to remove SDS. The sample enters the column under gravity and is washed through with 0.5 ml of 66% acetic acid. SDS and stain are bound to the resin. The eluate is dialyzed against 6% acetic acid and lyophilized.

*First Electrophoresis*

Each sample is resuspended in 25–50 μl of solution 15, mixed with tracking dye (solution 19), and applied to silicon-coated glass tubes (12 cm × 0.3 cm i.d.) containing a separation gel (solution 17) and an upper gel (solution 16), which are 10 cm and 0.5 cm, respectively. The acrylamide concentration is low (4%), and the separation is predominantly due

[11] E. Kaltschmidt and H. G. Wittmann, *Anal. Biochem.* **36**, 401 (1970).

to differences in charge. Electrophoresis is carried out for 5 hr at 1 mA/gel at room temperature toward the cathode. The pH of the upper electrophoresis buffer is 3.7 and that of the lower, pH 7.0. The low pH of the upper buffer allows the entry of all ribosomal proteins including the acidic proteins L7 and L12 into the gel.

*Second Electrophoresis*

After completion of the first electrophoresis, the gel is removed from the tube and embedded at the origin of the second urea gel slab. The high acrylamide concentration (18%) results in a separation of the proteins based predominantly on size. The apparatus used is similar to that described earlier.[8] The dimensions of the gel are 10 cm × 12 cm × 0.3 cm. After polymerization of the gel (solution 20) and application of tracking dye (solution 19), electrophoresis is carried out toward the cathode for between 7 and 16 hr at 150 V to 65 V in a glycine buffer at pH 4.0 (solution 21). The gel slabs are stained for 30 min in 100 ml of solution 22 containing Coomassie Blue G-250.[12] The stained protein spots are intensified by soaking the gel in 100 ml of 6% acetic acid for 30–60 min. To clear the gel of trichloroacetic acid, which is required if the gel is to be dried in preparation for radioautography, the gel is transferred to another 100 ml of 6% acetic acid and slowly shaken for approximately 18 hr. The gel is then dried onto Whatman 3 MM paper under vacuum with heating using a commercially available gel slab dryer unit.

*Identification of Individual Ribosomal Proteins by Radioautography*

The gel is exposed to X-ray film (Kodak No-Screen medical X-ray film) by placing the gel directly against the film and clamping between two 1 cm-thick foam pads and two 15 cm × 20 cm × 0.7 cm plywood sheets. The exposure time depends on the radioactivity of the sample; for example, 10,000 cpm of $^{35}$S-labeled protein requires an exposure time of approximately 2 weeks. The X-ray film is developed using standard procedures.

## Results and Discussion

Figure 3 shows diagonal gels of both cross-linked 30 S and 50 S ribosomal subunits of *Escherichia coli*. The complexity of the patterns beneath the diagonal is apparent. An exhaustive analysis of such patterns is made difficult because of overlap of spots due to cross-linking among proteins, many of which have the same or similar molecular weights. The

[12] W. Diezel, G. Kopperschlager, and E. Hofmann, *Anal. Biochem.* **48,** 617 (1972).

specificity of the patterns is notable. The patterns, though difficult to analyze in detail, are characteristic "fingerprints" of the protein topography of each subunit. There are differences in position and intensities of spots: some proteins are frequently cross-linked and often to more than one neighboring protein; others are less frequently found in cross-links. Partial purification of the mixture of the proteins extracted from cross-linked 50 S ribosomal subunits was obtained by salt extraction and electrophoresis. The gels in Fig. 4 illustrate the fractionation achieved by extraction of the particle with increasing concentrations of LiCl. The horizontal lines indicate the 0.5-cm slices that were used for diagonal gel electrophoresis.

The polyacrylamide/urea gel electrophoresis system used for the final identification of ribosomal proteins eluted from diagonal gels is illustrated in Fig. 5. In this example total 70 S ribosomal protein was analyzed. However, it is clear that the system separates as discrete spots all the 50 S proteins. Thus the elution of a radioactive component from a diagonal gel, mixing with nonradioactive total 50 S protein, and electrophoresis in this system followed by staining and radioautography, leads to its unambiguous identification.

A diagonal gel of one of the purified fractions from cross-linked 50 S ribosomal subunits is shown in Fig. 6A. Comparison with Fig. 3B shows the degree of purification obtained. Two pairs of spots are indicated by arrows. $A_1$ and $A_2$ fall on the same vertical line whose intercept on the horizontal axis indicates a molecular weight for the cross-linked species of 31,000. The molecular weights of $A_1$ and $A_2$ given by the intercepts on the vertical axis are 20,800 and 10,100, respectively, giving a sum equal to that of the putative cross-linked dimer. Component A could be L5 and/or L6, judged from its mobility in SDS gels, and $A_2$ could be one of four different proteins. The radioactive components were eluted and analyzed by electrophoresis in the two-dimensional polyacrylamide/urea gel system. Figure 6B shows a radioautograph of the gel. The dark spots correspond to L5 and L25 as determined by superposition of the X-ray film on the stained gel. Similar analysis of $B_1$ and $B_2$ show them to have molecular weights consistent with their presence in a cross-linked dimer and their identity as L17 and L32. More than thirty protein dimers have been identified from the 50 S ribosomal subunit using the methods described here. While the purification procedures are time consuming, they simplify greatly the identification of components of dimers. Purification also facilitates identification of dimers formed in moderate to low yield. Important in this regard is the fact that all protein pairs identified using the purified fractions appear on diagonal gels like that shown in Fig. 3B of the total cross-linked protein.

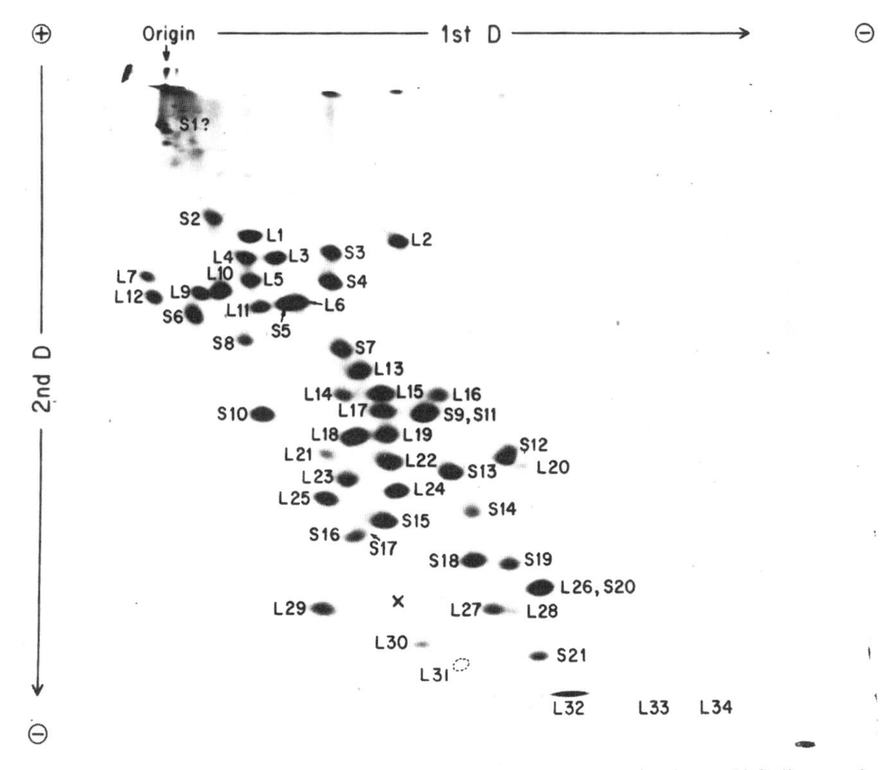


FIG. 5. Two-dimensional polyacrylamide/urea gel electrophoresis of total 70 S ribosomal protein of *Escherichia coli*. Protein was extracted from 70 S ribosomes with 66% acetic acid, dialyzed, and lyophilized. Electrophoresis was as described in Fig. 4 for the first electrophoresis and at pH 4.6 in 18% acrylamide for the second electrophoresis. See text for details. The numbering scheme conforms to that of E. Kaltschmidt and H. G. Wittmann, *Proc. Natl. Acad. Sci. U.S.A.* **67**, 1276 (1970). The system clearly resolves all proteins except S16/S17 and S9/S11. L26 and S20 are the same protein as that found on both subunits. The spot marked X is a 50 S protein not previously reported.

There are many possible explanations for the variability in the yield of cross-linked protein pairs: differences in reactivity of lysine residues and of the sulfhydryl groups introduced; the relative "isolation" of certain proteins, or lysines contained therein, from other proteins, possibly due to shielding by RNA; possible compositional and conformation heterogeneity in the population of ribosomal subunits; competition between dimer formation and the formation of higher cross-linked oligomers; and, perhaps most important, competition between intermolecular cross-linking and intramolecular disulfide bond formation.[3,4] It has been contended that disulfide cross-linking leads to artifacts: that cross-linked protein

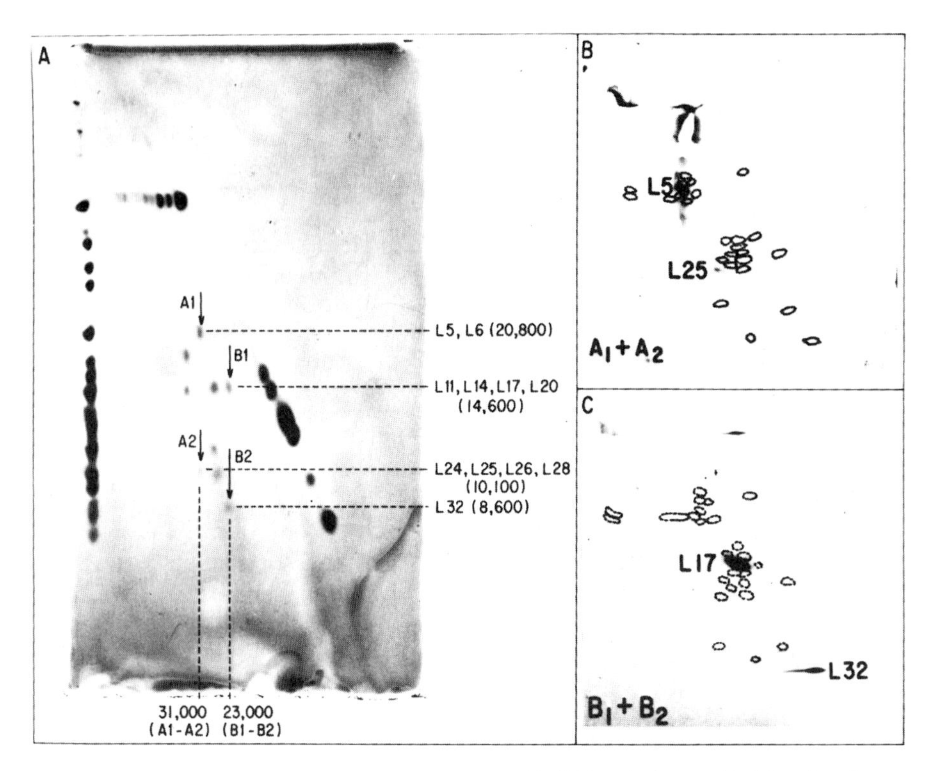


FIG. 6. Two-dimensional diagonal polyacrylamide/SDS gel of a purified fraction from cross-linked 50 S subunits, and radioautographs of two-dimensional polyacrylamide/urea gels illustrating identification of cross-linked monomeric proteins. Slice number 6 from panel f (Fig. 4) was analyzed by diagonal gel electrophoresis (A). The radioactive proteins A1, A2, B1, and B2 were eluted from the gel, mixed with nonradioactive total 50 S protein, and electrophoresed in the two-dimensional polyacrylamide/urea gel system shown in Fig. 5. (B) and (C) are radioautographs in which the dark spots represent the radioactive components and the dotted circles represent stained marker proteins whose positions were used to identify the proteins of interest.

dimers may be due to disulfide interchange and/or random oxidation and may not reflect protein neighborhoods existing in the intact ribosomal subunit.[13,14] The methods described here incorporate deliberate procedures to preclude such possible artifacts. Alkylation with iodoacetamide is included at each step of the procedure to minimize or exclude the presence of free sulfhydryl groups, necessary for either interchange or

[13] H. Peretz, H. Towbin, and D. Elson, *Eur. J. Biochem.* **63**, 83 (1976).

[14] C. G. Kurland, *in* "Molecular Mechanisms of Protein Biosynthesis" (H. Weissbach and S. Pestka, eds.), p. 81. Academic Press, New York, 1977.

nonspecific oxidation. The results, however, are nearly identical to those obtained earlier[5,7] when acidification alone was employed to prevent the possible occurrence of nonspecific events. The cross-links reflect reactions which take place in the intact ribosomal subunit. The procedures do not detectably alter the structure of the subunit.

Other cleavable cross-linking reagents have been employed in the study of ribosome protein topography and in other systems. They include dimethyl 3,3′-dithiobispropionimidate,[15,16] dithiobis(succinimidyl)propionimidate,[17] tartryl diazides,[18] and, *N,N′*-bis(2-carboxyimidoethyl) tartarimide.[19] Diagonal gel electrophoresis has also been employed by others, both with the disulfide reagents and the tartaric acid derivatives. A method for cleaving cross-links formed with dimethyl suberimidate has been reported.[20] A method quite similar to that described here has been used in an investigation of chromatin structure employing the reagent methyl 3-mercaptopropionimidate.[21]

The use of 2-iminothiolane to form interprotein cross-links has several distinct advantages: the disulfide cross-link is readily reversed by reduction; the compound is relatively stable in solution compared to other imidates, probably owing to its five-membered ring structure; it is available commercially; the cross-linking reaction is separated from the initial protein modification reaction with the consequence that any lysine that reacts with the reagent is a potential site for cross-linking. Other imidate cross-linking reagents are bifunctional imidates; e.g., dithiobispropionimidate, dimethylsuberimidate and may give a lower yield of cross-linking due to the likelihood that one functional group may react with protein while the second may hydrolyze prior to reaction with a second protein.

Protein cross-linking with 2-iminothiolane and oxidation to form disulfide linkages between neighboring proteins and analysis by diagonal gel electrophoresis are general techniques. Detailed methods have been described here as they have been applied to the investigation of the protein topography of ribosomes. However, with slight modifications the methods for cross-linking and for diagonal gel electrophoresis can be readily applied to the investigation of other protein complexes.

[15] K. Wang, and F. M. Richards, *J. Biol. Chem.* **249,** 8005 (1974).

[16] A. Ruoho, P. A. Bartlett, A. Dutton, and S. J. Singer, *Biochem. Biophys. Res. Commun.* **63,** 417 (1975).

[17] A. J. Lomant and G. Fairbanks, *J. Mol. Biol.* **104,** 243 (1976).

[18] L. C. Lutter, C. G. Kurland, and G. Stöffler, *FEBS Lett.* **54,** 144 (1975).

[19] J. R. Coggins, J. Lumsden, and A. D. B. Malcolm, *Biochemistry* **16,** 1111 (1977).

[20] A. Expert-Bezançon, D. Barritault, M. Milet, M.-F. Guérin, and D. H. Hayes, *J. Mol. Biol.* **112,** 603 (1977).

[21] J. O. Thomas and R. D. Kornberg, *FEBS Lett.* **58,** 353 (1975).

# [57] Photoaffinity Labeling of Ribosomes

*By* B. S. COOPERMAN, P. G. GRANT, R. A. GOLDMAN, M. A. LUDDY, A. MINNELLA, A. W. NICHOLSON, and W. A. STRYCHARZ

The *Escherichia coli* ribosome is a complex organelle composed of 54 different proteins and three RNA chains, two of which are quite large (1600 and 3300 bases long, respectively).[1] A principal goal of current research on the ribosome is the construction of a structure-function map in which different proteins and RNA regions would be located and assigned roles in the overall process of protein synthesis. The affinity-labeling approach, by which one forms a covalent bond between a ligand (usually radioactive) and its receptor site, permitting, on subsequent analysis, identification of the components of the ligand binding site, has yielded important information toward this end. The merits of this approach are that it can be applied to the intact ribosome, that the data obtained are easily interpretable, and that the sites of incorporation can in principle be determined to any desired level of detail, from individual proteins or extended RNA regions to individual amino acid or nucleotide residues.

Affinity-labeling studies on the ribosome have already provided evidence as to the locations of the peptidyltransferase center, the mRNA codon site, and the binding sites for several antibiotics and protein cofactors involved in protein synthesis. These results, as well as the problems that must be addressed in obtaining them, have been discussed elsewhere by ourselves and others in several recent review articles on affinity and photoaffinity labeling in general,[2,3] as well as on affinity labeling of ribosomes in particular.[4-6] We wish to present here the major conclusions that we have reached as to the performance of these experiments, referring in particular to past and ongoing work in our laboratory on the interaction of puromycin with the ribosome. We conclude with a

[1] H. G. Wittmann, *Eur. J. Biochem.* **61,** 1 (1976).

[2] B. S. Cooperman, "Aging, Carcinogenesis, and Radiation Biology: The Role of Nucleic Acid Addition Reactions" (K. C. Smith, ed.), p. 315. Plenum, New York, 1976.

[3] H. Bayley and J. R. Knowles, this series, Vol. 46 [8].

[4] B. S. Cooperman, *in* "Bioorganic Chemistry" (E. E. van Tamelen, ed.), Vol. 4, p. 81. Academic Press, New York, 1978.

[5] A. E. Johnson and C. R. Cantor, this series, Vol. 46 [15].

[6] A. Zamir, this series, Vol. 46 [73].



ISBN 0-12-504180-2

description of the synthesis of some new photolabile derivatizing reagents and their application in derivatizing several antibiotic molecules.

## Covalent Incorporation

Covalent bond formation in affinity-labeling reactions may proceed either via a photoinduced process involving, in general, bond insertion or radical coupling reactions, or via reaction of a nucleophile with an electrophile. The most common procedure is to prepare a photolabile or electrophilic derivative of the ligand whose binding site is being studied. If the derivative retains affinity for the binding site, then covalent bond formation can be achieved either upon irradiation of the ligand–receptor complex, or, in the case of electrophilic derivatives, spontaneously with nucleophilic groups at or accessible from the binding site. Alternatively, the native ligand–receptor complex may be subjected to a general cross-linking procedure, as, for instance, by exposure to ultraviolet irradiation (with or without a photosensitizer present) or to a bifunctional electrophilic reagent. For site localization on the ribosome, photoinduced affinity labeling is preferable to electrophilic labeling. The reactive intermediates generated in the former process generally have much lower chemical selectivity toward covalent bond formation than electrophiles, thus increasing the likelihood that covalent bond formation occurs at the site of noncovalent binding rather than at the most reactive nucleophilic position accessible from this site. As even photogenerated intermediates show some chemical specificity, it is desirable to use more than one type of photolabile derivative in affinity-labeling studies. Use of photolabile derivatives is preferable to a general photoinduced cross-linking procedure, since with an appropriate derivative it should be possible to perform the experiment in such a way that the major photochemical event is photolysis of the derivative, with little accompanying photochemical destruction of the receptor. This is a particularly important consideration for the ribosome, since ultraviolet irradiation is known to induce protein–protein and protein–RNA cross-links in both the 30 S and 50 S particles, to induce chain breaks in 16 S RNA, and to inhibit several ribosomal activities such as poly(U)-dependent polyphenylalanine synthesis, and mRNA-dependent tRNA binding.[7–9] On the other hand, direct photolysis of the native ligand–ribosome complex, when it leads to covalent incor-

[7] L. Gorelic, "Protein Crosslinking: Biochemical and Molecular Aspects" (M. Friedman, ed.), p. 611. Plenum, New York, 1977.

[8] H. Kagawa, H. Fukutome, and Y. Kawade, *J. Mol. Biol.* **26**, 249 (1967).

[9] B. S. Cooperman, J. Dondon, J. Finelli, M. Grunberg-Manago, and A. M. Michelsen, *FEBS Lett.* **76**, 59 (1977).

poration, offers two important practical advantages in that it does not require either the synthesis of a new derivative or the demonstration that the derivative faithfully mimics the native ligand. Furthermore, by examining the labeling results as a function of light fluence, one can determine whether incorporation occurs into the native ribosome or into one that has undergone a photoinduced change in structure.

We and others have found that incorporation levels useful for mapping studies can be achieved by direct or sensitized photolysis with the following ligands in radioactive form: puromycin,[10,11] blasticidin S,[12] tetracycline,[12] chloramphenicol,[13] and initiation factor 3 (IF-3).[9] This phenomenon is, however, not completely general, since only very low incorporation levels of [$^3$H]dihydrostreptomycin were achieved.[9] In such experiments, the amount of incorporation is typically proportional to light fluence (or nearly so), while ribosomal activity [e.g., poly(U)-directed polyphenylalanine synthesis, peptidyltransferase] usually declines with light fluence. Therefore an operational definition of incorporation levels useful for mapping studies is that a certain specific radioactivity of labeled ribosomes is achieved before the ribosome has lost too much (e.g., 50%) of its activity. As an example, we find it convenient to analyze 100 $\mu$g of total ribosomal protein (approximately 110 pmol) containing at least 1500 cpm of radioactivity by one-dimensional gel electrophoresis. In our experiments we typically use $^3$H-labeled antibiotics having specific radioactivities of 1–10 Ci/mmol. Assuming that half the incorporation is into protein, a counting efficiency of 20% for tritium, and a ligand specific radioactivity of 3 Ci/mmol, allows the conclusion that a one-dimensional analysis can be obtained if incorporation occurs into 2% of the ribosomes. Of course, even lower levels of labeled ribosomes can be analyzed if larger samples and/or ligands of higher specific radioactivity are used. The detailed photochemistry underlying the successful incorporation experiments mentioned above is not clear, although radical coupling reactions are likely candidates. Regardless of the underlying mechanisms, these results indicate that any photoaffinity-labeling study with ribosomes should begin with attempts to photoincorporate native ligand. The results from such studies could then be compared with those obtained using photolabile derivatives of the ligand.

[10] B. S. Cooperman, E. N. Jaynes, Jr., D. J. Brunswick, and M. A. Luddy, *Proc. Natl. Acad. Sci. U.S.A.* **72,** 2974 (1975).

[11] E. N. Jaynes, Jr., P. G. Grant, G. Giangrande, R. Wieder, and B. S. Cooperman, *Biochemistry* **17,** 561 (1978).

[12] R. A. Goldman and B. S. Cooperman, manuscript in preparation.

[13] N. Sonenberg, M. Wilchek, and A. Zamir, *Biochem. Biophys. Res. Commun.* **59,** 663 (1974).

Localization of Incorporation Sites

One of the disturbing aspects of the ribosomal affinity labeling literature is that different research groups working with derivatives of the same native ligand have reported different affinity-labeling results. This is in part due to the fact that different affinity-labeling reagents and procedures have been employed, but this cannot be the whole story since apparently contradictory results have been obtained in some cases even when similar or identical reagents have been used. There appear to be four reasons underlying these findings: first, the conformational lability of the ribosome, whose structure is known to be very sensitive to changes in pH, temperature, $Mg^{2+}$, and monovalent cation concentration; second, the lack of a standard preparation for homogeneous, active ribosomes—thus, the functional properties and protein compositions of ribosomes prepared from bacteria harvested at different stages of the growth phase, or using different cell-breaking methods, or different washing procedures, can differ significantly; third, in some cases, the presence of more than one binding site on the ribosome for a given ligand—thus, different labeling results can be obtained at different ligand concentrations; and fourth, the fact that different laboratories perform affinity-labeling experiments using different reaction conditions and different preparations of ribosomes.[4]

Because of these considerations it is obvious that, in order to get an accurate picture of a ligand binding site, it is necessary to examine the labeling pattern obtained as one varies the method of ribosome preparation, the reaction medium used for affinity labeling, the concentration of ligand, and the nature of the affinity-labeling procedure used. Thus, from a practical standpoint, a rapid and straightforward method of assessing the results of an affinity-labeling experiment is needed if definitive results are to be obtained. For ribosomal proteins, one-dimensional polyacrylamide gel electrophoresis allows partial resolution of ribosomal proteins which is often sufficient to detect variations in labeling pattern as a function of the variables discussed above. In our own laboratory we have been using slab gels which routinely permit analysis of 6–10 samples simultaneously. After electrophoresis the gels are sliced and the radioactivity in the slices is determined. We have been using a standard one-dimensional gel on which the migrations of virtually all the 54 ribosomal proteins are known. Thus from one-dimensional electrophoretic analysis we can localize peak(s) of radioactivity, see how the peaks respond to changes in the protocol of the experiment, and have an idea about which of the proteins (from a group of typically 4–6) are involved. In many cases this analysis can be performed on the total protein from the 70 S ribosome, although it is sometimes desirable to first separate the ribosome into 30 S and 50 S subunits by sucrose gradient centrifugation and analyze the proteins of each subunit separately. Definitive identification of labeled proteins can often be accomplished using the two-dimensional gel electrophoresis procedure of Kaltschmidt and Wittmann,[14] which essentially resolves all 54 proteins. Because this procedure is fairly time consuming, we generally employ it in analyzing only those affinity labeling experiments that have been shown to be the most interesting by one-dimensional analysis. We have been using the Howard and Traut modification of the procedure,[15] in which the first dimension is run as two separate gels, in order to eliminate the otherwise substantial loss of protein at the center-origin. Covalent incorporation of a ligand can lead to markedly altered gel electrophoretic mobilities of the modified ribosomal proteins. Moreover, in a given experiment any particular ribosomal protein may be modified to the extent of 1% or less. As a result, it is quite possible to have major radioactive peaks in a region of gel not staining for protein. To overcome this problem we typically run a set of identical two-dimensional gels. The first gel is cut into large pieces both surrounding each of the stained proteins and in regions of the gel even remotely close to proteins, and the radioactivity in each of the pieces is determined. In this way, regions of the gel containing important amounts of radioactivity are located. These regions are then cut more finely in the second gel, permitting an accurate placement of the peaks of radioactivity. Such placement is easy to compare from gel to gel, because the native protein staining pattern is highly reproducible and provides a large number of internal reference points.

Our work with puromycin[10,11] provides an illustration of the utility of this overall approach. The results of a typical one-dimensional gel electrophoretic analysis of total protein from a 50 S particle is shown in Fig. 1. The large peak seen in region IV is due primarily to incorporation into protein L23, as has been shown by two-dimensional gel electrophoretic analysis and specific immunoprecipitation (see below). In the experiment shown, the total radioactivity in each labeling mixture was kept constant as puromycin concentration was raised, so that the decrease in labeling seen in region IV is evidence that labeling is proceeding via a site-specific process. By contrast, labeling in region II is almost independent of puromycin concentration, indicating that here labeling is predominantly nonspecific. The data in Fig. 1 could be used to estimate a dissociation constant for specific puromycin binding of 0.7 m$M$, which agrees reasonably well with $K_m$ values obtained in the peptidylpuromycin and fragment

[14] E. Kaltschmidt and H. G. Wittmann, *Proc. Natl. Acad. Sci. U.S.A.* **67**, 1276 (1970).
[15] G. H. Howard, and R. R. Traut, *FEBS Lett.* **29**, 177 (1973).

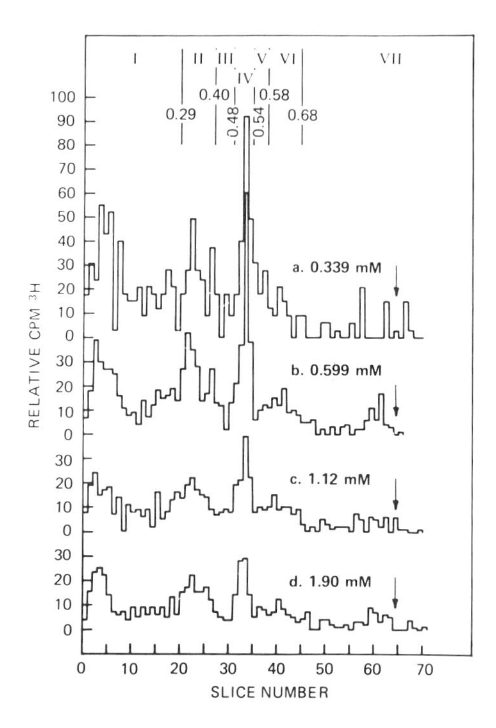


FIG. 1. One-dimensional polyacrylamide gel pattern of labeled proteins from 50 S particles as a function of puromycin concentration. Experimental conditions: 111 $A_{260}$ units per milliliter of ribosomes; photolysis was for 8 min at 254 nm. The specific activities of puromycin were: (a) 890 Ci/mol; (b) 504 Ci/mol; (c) 269 Ci/mol; (d) 159 Ci/mol. Reported counts per minute are for protein from 3.5 $A_{260}$ units of 50 S particles. Reprinted with permission from E. N. Jaynes, Jr., P. G. Grant, G. Giangrande, R. Wieder, and B. S. Cooperman, *Biochemistry* **17**, 561 (1978). Copyright by the American Chemical Society.

assays. These results were obtained using total protein from the 50 S particle which had been isolated from 70 S ribosomes, the target in the photoaffinity-labeling experiment. Qualitatively similar results were obtained when total protein from the 70 S particle was analyzed directly (Fig. 2).

One-dimensional patterns such as those shown in Fig. 1 and 2 could then be used to study changes in labeling as the experimental protocol was varied. Thus, labeling was clearly more specific for region IV at low rather than high light fluences, allowing the conclusion that L23 labeling occurs with native, rather than light-denatured, ribosomes. Similarly, high labeling of region IV was observed using ribosomes prepared in a number of ways and over a range of KCl concentrations, suggesting that L23 is a highly conserved part of the puromycin binding site. Region IV

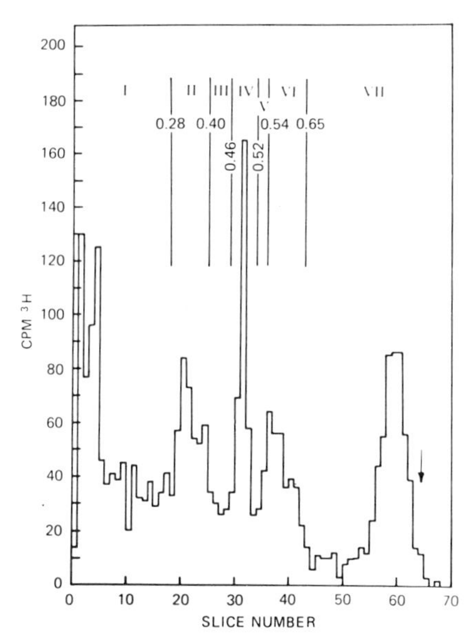


FIG. 2. One-dimensional polyacrylamide gel pattern of labeled proteins from 70 S particles. Experimental conditions: 111 $A_{260}$ units per milliliter of ribosomes; 0.074 m*M* puromycin (2013 Ci/mol); photolysis was for 8 min at 254 nm. Reported counts per minute are for protein from 5.6 $A_{260}$ units of 70 S particles. Reprinted with permission from E. N. Jaynes, Jr., P. G. Grant, G. Giangrande, R. Wieder, and B. S. Cooperman, *Biochemistry* **17**, 561 (1978). Copyright by the American Chemical Society.

labeling was also used to test whether other peptidyltransferase inhibitors bound competitively with puromycin.[11,16] From the results in Fig. 3 it is clear that two structural analogs of puromycin, *N*-phenylalanylpuromycin aminonucleoside (PhePANS) and cytidylyl-(3′-5′)-3′-*O*-phenylalanyladenosine (CAPhe), block region IV labeling whereas the other inhibitors, which are not structural analogs, do not. The most interesting results are seen with chloramphenicol, which has the effect of changing the labeling pattern so that a new protein, S14 (see below), which falls in region VI, now becomes the major site of labeling (Fig. 3, D5). The results with tetracycline are also interesting, because this antibiotic, which is not an inhibitor of peptidyltransferase, has the effect of vastly increasing the extent (note the radioactivity scale in D6) of region IV (L23 from two-dimensional gel analysis) labeling as well as its specificity.

We have explored the chloramphenicol effect in some detail. Puromycin blocking experiments, as described above, showed that the label-

[16] P. G. Grant, E. N. Jaynes, Jr., W. A. Strycharz, B. S. Cooperman, and M. Nomura, manuscript in preparation.

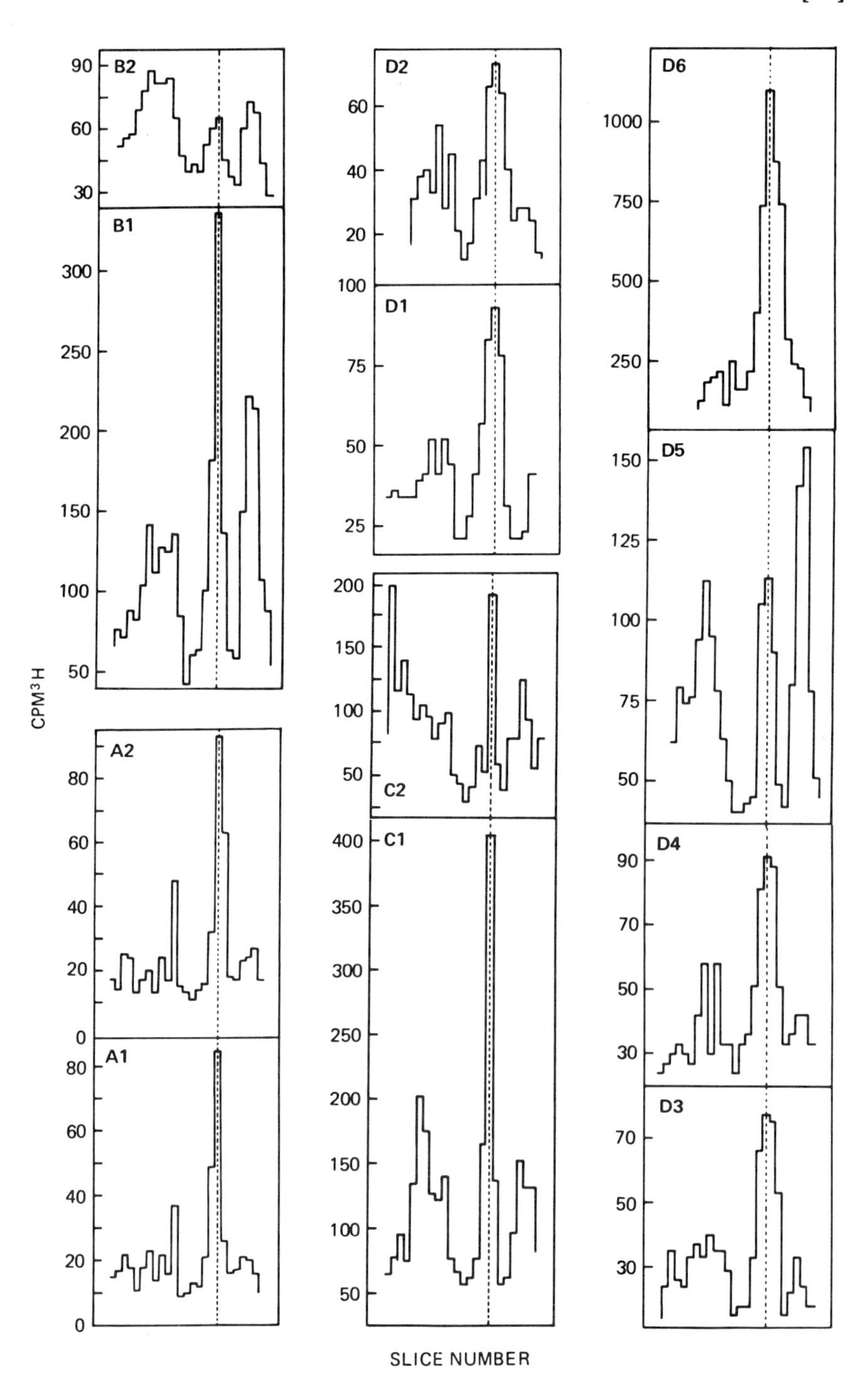
B2
B1
A2
A1
D2
D1
C2
C1
D6
D5
D4
D3
CPM ³H
SLICE NUMBER

ing in region VI arises from a site-specific process. That the change is dependent on specific chloramphenicol binding, rather than from a nonspecific photochemical effect arising from the nitrophenyl moiety, is shown by the experiments described in Fig. 4, from which it is clear that neither *p*-nitrophenyl acetate, nor L-*erythro*-2,2-dichloro-*N*-[β-hydroxy-α(hydroxymethyl)-*p*-nitrophenyl] acetamide (LECAM) is as effective as chloramphenicol (the D-*threo* isomer) in raising region VI labeling. Here it should be noted that LECAM is not an effective peptidyltransferase inhibitor. When the labeling patterns in the presence of chloramphenicol were examined as a function of light fluence, it was found that at low fluences the one-dimensional pattern obtained resembled that seen in the absence of chloramphenicol, whereas at higher fluences the pattern seen in Fig. 4d was obtained. Further detailed study by both one and two-dimensional gel electrophoresis showed that, in contrast to the results obtained with puromycin alone, the changes in pattern seen do not reflect labeling of the native ribosome, but rather one that has been altered by light in a specific, chloramphenicol-dependent reaction. These studies have provided strong evidence for the existence of two specific ribosomal sites for both puromycin and chloramphenicol, one each on both the 30 S and 50 S subunits.[16]

Identification of L23 and S14 as the major proteins labeled on irradiation of ribosomes and puromycin in the absence and in the presence

FIG. 3. Effect of the presence during photolysis of other peptidyltransferase inhibitors on puromycin incorporation into proteins from 70 S particles as measured by one-dimensional gel electrophoresis. The part of the electrophoretogram displayed corresponds to regions II–VI in Fig. 2. The dotted vertical line defines the position of the major peak seen in region IV of Fig. 2. (A1) puromycin alone; (A2) plus 1 m*M* lincomycin; (B1) puromycin alone; (B2) plus 2.1 m*M* PhePANS; (C1) puromycin alone; (C2) plus 0.25 m*M* CAPhe; (D1) puromycin alone; (D2) plus 50 μ*M* erythromycin; (D3) plus 50 μ*M* sparsomycin; (D4) plus 50 μ*M* blasticidin S; (D5) plus 100 μ*M* chloramphenicol; (D6) plus 50 μ*M* tetracycline. Experimental conditions are tabulated below.

| Series | Ribosome concentration ($A_{260}$/ml) | Puromycin concentration (m*M*) | Puromycin specific activity (Ci/mol) | Photolysis time (min) | Wave-length (m) |
|---|---|---|---|---|---|
| A | 99 | 0.044 | 3500 | 8 | 350 |
| B | 99 | 0.15 | 926 | 120 | 350 |
| C | 111 | 0.074 | 2013 | 8 | 254 |
| D | 100 | 0.10 | 750 | 20 | 350 |

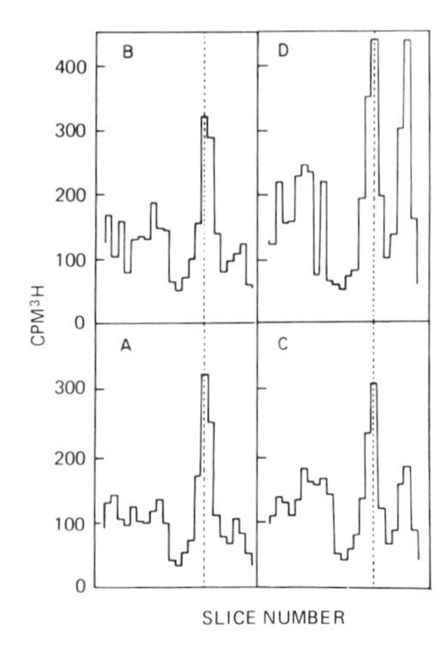


FIG. 4. Effect of chloramphenicol and other nitrophenyl-containing molecules on puromycin incorporation into proteins from 70 S particles. Data are plotted as in Fig. 3. Experimental conditions: 100 $A_{260}$ units per milliliter of ribosomes; 0.1 m*M* puromycin (1500 Ci/mol); photolysis was for 20 min at 350 nm. Reported counts per minute are per 6.45 $A_{260}$ units of 70 S protein applied to gel: (a) puromycin alone, (b) plus 50 $\mu M$ *p*-nitrophenyl acetate, (c) plus 50 $\mu M$ LECAM, (d) plus 50 $\mu M$ chloramphenicol.

of chloramphenicol, respectively, was first made on the basis of two-dimensional gel electrophoretic analysis (Fig. 5). Radioactivity coelectrophoresed with S14, which is a very well-resolved protein in the Kaltschmidt and Wittmann system. However, radioactivity is not coincident with L23 protein staining, so that an alternative method was needed to allow a definitive identification to be made. An immunoprecipitation method was employed, utilizing antibodies directed against individual ribosomal proteins. The results are shown in Table I and provide the necessary confirmation. The proteins listed include all those which migrate in a two-dimensional gel in the general vicinity of either L23 or S14. Other methods can also be used to identify labeled proteins. For example, one can first separate proteins from a single subunit into several different groups on the basis of how tightly they are bound to ribosomal RNA and then resolve the groups of proteins on SDS or sarkosyl-containing polyacrylamide gels, in which proteins are resolved on the basis of their molecular weights.[17] In addition, it should now be possible to identify a ribosomal protein by its primary structure. Full or partial amino acid sequences for almost all the ribosomal proteins are

[17] O. Pongs and E. Lanka, *Proc. Natl. Acad. Sci. U.S.A.* **72,** 1505 (1975).

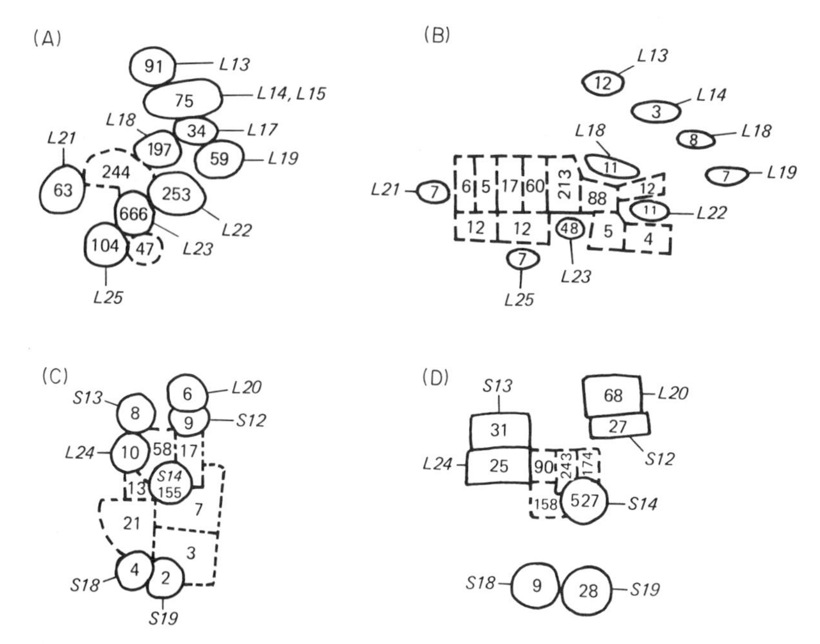


FIG. 5. Regions surrounding the major radioactive peaks on two-dimensional gel electrophoresis of proteins from 70 S proteins labeled in the absence (a, b) and in the presence (c, d) of chloramphenicol. Solid lines surround areas of protein staining. Block numbers indicate counts per minute. Numbers are not comparable from one gel to another; since different-size samples were used. In gels (a) and (c) large areas around each protein were cut out. In gels (b) and (d) areas were cut for each protein corresponding closely to the observed staining and only the basic proteins were analyzed, thus increasing the resolution in the region of interest. Experimental conditions: (a, b) 104 $A_{260}$ units per milliliter of ribosomes; 13 n*M* puromycin; photolysis was for 8 min at 254 nm. Essentially identical results were obtained on irradiation at 350 nm: (c, d) 100 $A_{260}$ units per milliliter of ribosomes; 0.1 m*M* puromycin; 0.1 m*M* chloramphenicol; photolysis was for 20 min at 350 nm.

now available,[18] and radiochemical methods now exist for characterizing microgram amounts of protein, for instance, by tryptic peptide analysis.[19]

Although several affinity-labeling experiments on ribosomes have been shown to proceed with dominant labeling of the RNA fraction, work on localizing the sites of labeling within limited regions (~50 nucleotides) of either the 16 S or 23 S RNA has been stymied by the lack of a rapid screening method corresponding to either the one- or two-dimensional gel electrophoretic analysis of ribosomal proteins. However, some recent

[18] G. Stöffler and H. G. Wittmann, "Molecular Mechanisms of Protein Biosynthesis" (H. Weissbach and S. Pestka, eds.), p. 117. Academic Press, New York, 1977.

[19] J. H. Elder, R. A. J. Hampton, and R. A. Lerner, *J. Biol. Chem.* **252,** 6510 (1977).

TABLE I
IMMUNOPRECIPITATION RESULTS

| % Labeled 50 S protein precipitated | | | % Labeled 30 S protein precipitated | | |
|---|---|---|---|---|---|
| Antibody to protein | Puromycin alone | Puromycin plus chloramphenicol | Antibody to protein | Puromycin alone | Puromycin plus chloramphenicol |
| L18 + L22 | 4 | 5 | S12 | 10 | 9 |
| L19 | 3 | 5 | S13 | 9 | 11 |
| L21 | 3 | 6 | S14 | 24 | 42 |
| L23 | 46 | 31 | S18 | 9 | 7 |
| L25 | 4 | 4 | S19 | 6 | 5 |

developments offer the hope that such methods can be developed in the near future. The endonuclease RNase III, when utilized at low ionic strength, has been shown to cleave long RNA chains, including ribosomal RNA, into a set of apparently well defined fragments, separable by gel electrophoresis.[20] Thus treatment of affinity-labeled ribosomal RNA with RNase III followed by electrophoretic analysis could provide the required screening method. Labeled fragments could then be identified within the primary sequence by standard fingerprinting techniques. In addition, ribosomal DNA is currently being mapped with restriction enzymes.[21] Localization of the site of affinity-labeled RNA could in principle be accomplished by determining the amount of incorporation into ribonuclease-resistant RNA hybridized to a series of DNA restriction fragments. Finally, several 30 S and 50 S proteins, when added to 16 S RNA or 23 S RNA, respectively, are known to protect specific RNA regions from endonuclease-catalyzed hydrolysis.[22,23] It should therefore be possible to at least partially localize sites of RNA affinity labeling by determining the amount of incorporation into RNA fragments made endonuclease resistant by the addition of specific RNA binding proteins.

## Photolabile Derivatives vs Direct Photolysis

To investigate the question of the extent to which the results detailed above with native puromycin depend on the specific photochemical reaction(s) leading to incorporation, we have begun a program to repeat the experiment with a series of different photolabile derivatives of puromycin. This work is still very much in progress, although interesting results have already been obtained with two puromycin derivatives: the arylazide analog, 6-dimethylamino-9-[3′-deoxy-3′-(*p*-azido-L-phenylalanylamino)-β-D-ribofuranosyl]purine(*p*-azidopuromycin), and *N*-ethyl-2-diazomalonyl puromycin (*N*-EDM puromycin). *p*-Azidopuromycin was prepared[24] by condensing *N*-tBOC-*p*-azidophenylalanine[25] with puromycin aminonucleoside, following a procedure used in synthesizing other puromycin analogs[26] and subsequently removing the BOC protecting group. It was synthesized in $^{3}$H form using radioactive puromycin aminonucleoside (Amersham). It is approximately as active as native puro-

[20] J. J. Dunn, *J. Biol. Chem.* **251,** 3807 (1976).
[21] M. Nomura, *Cell* **9,** 633 (1976).
[22] C. Branlant, J. Sriwidada, A. Krol, and J. P. Ebel, *Nucl. Acids Res.* **4,** 4323 (1977).
[23] A. Muto, C. Ehresmann, P. Fellner, and R. A. Zimmermann, *J. Mol. Biol.* **86,** 411 (1974).
[24] A. W. Nicholson and B. S. Cooperman, *FEBS Lett.* **90,** 203 (1978).
[25] R. Schwyzer and M. Caviezel, *Helv. Chim. Acta* **54,** 1395 (1971).
[26] R. J. Harris, J. F. B. Mercer, D. C. Skingle, and R. H. Symons, *Can. J. Biochem.* **50,** 918 (1972).

mycin in a peptidyl transferase assay, but as expected is more efficiently incorporated into ribosomes on irradiation at either 254 nm or 350 nm, presumably reflecting nitrene incorporation. One-dimensional polyacrylamide gel electrophoretic analyses of total protein from 50S and 30S particles isolated from labeled 70S ribosomes (Fig. 6), give patterns superficially quite similar to that obtained with puromycin [compare Fig. 6(a) and Fig. 1(a), or Fig. 6(a) plus Fig. 6(b) with Fig. 2]. Indeed, preliminary immunoprecipitation results show L23 to be the major labeled protein. However, in contrast to the labeling obtained with puromycin, *p*-azidopuromycin also labels one or both of the proteins L18 and L22, as well as the proteins S4 and S18, to fairly high levels. Protein S14 is also labeled appreciably. A preliminary report of our photoaffinity labeling results with *p*-azidopuromycin has already appeared.[24] Although our current results support most of our previous conclusions, two important differences should be noted. On the basis of radioactivity comigrating with native protein on two-dimensional gel electrophoretic analysis of labeled protein, we had tentatively concluded that L11 was the major labeled protein in the 50 S subunit, and that L23 was not a major labeled protein. The recent immunoprecipitation results, showing high L23 labeling, can be explained if it is assumed that modified L23 does not comigrate with native L23. Both the recent gel electrophoretic and immunoprecipitation analyses show little or no labeling of L11. A thorough reexamination of our methods has led to the following conclusions. Our initial analyses were conducted on proteins extracted from 70S ribosomes, and the washing procedures used were inadequate to fully remove

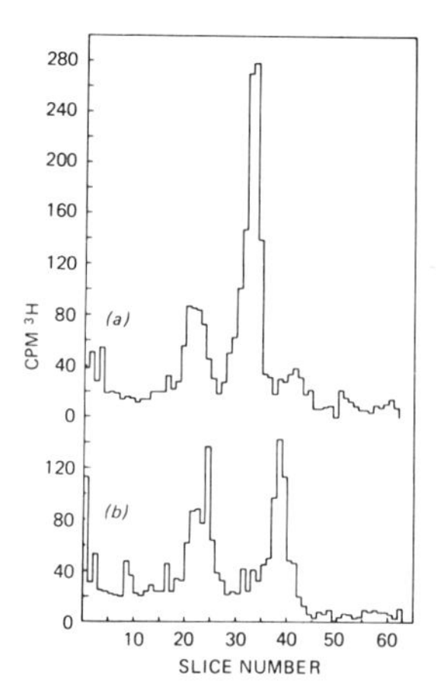


FIG. 6. *p*-Azidopuromycin incorporation into 30S and 50S proteins measured by one-dimensional gel electrophoresis. Experimental conditions: 100 $A_{260}$ units/ml ribosomes; 0.027 m$M$ *p*-azidopuromycin (1670 Ci/mole). Photolysis was for 4 min at 254 nm. (a) 50 S proteins, (b) 30 S proteins. Reported counts are for 315 pmole of protein applied to each gel.

nonincorporated photolyzed *p*-azidopuromycin. Furthermore, some of this nonincorporated material, by an unfortuante coincidence, migrates on gel electrophoresis with or close to protein L11. When 70 S ribosomes are first resolved into 30 S and 50 S subunits prior to protein extraction, nonincorporated photolyzed *p*-azidopuromycin is efficiently removed, and true L11 labeling is seen to be very low.

*N*-EDM puromycin was prepared by condensation of ethyl-2-diazomalonyl chloride with puromycin. It was synthesized in radioactive form from [$^3$H]puromycin (Amersham). This derivative, though not a substrate for peptidyl transferase, does have inhibitory activity. On photolysis in the presence of 70 S ribosomes, the two major labeled proteins found are S18 and S14, and no 50 S proteins are labeled to a significant extent. Thus, *p*-azidopuromycin mimics puromycin in labeling proteins at both the 50 S and 30 S binding sites, whereas *N*-EDM puromycin is directed more toward the 30 S site. That both puromycin and *p*-azidopuromycin photolabel L23 increases our confidence that this protein is a true component of the puromycin binding site on the 50 S subunit. For the 30S binding site, two proteins are strongly implicated, S14, which is labeled by puromycin and one or both of the derivatives, and S18, labeled by both derivatives. However, there is no independent evidence placing these proteins close to one another and labeling of S18 must always be considered with caution because of the high intrinsic nucleophilicity of this protein.[27] We plan to continue this study with other photolabile puromycin derivatives. Descriptions of the synthesis of several new $\alpha$-diazocarbonyl reagents useful in derivatizing puromycin and other antibiotic molecules are presented in the next section.

## Synthesis of $\alpha$-Diazocarbonyl Derivatives

Of the three reagent types most frequently used to date in photoaffinity labeling experiments, $\alpha$-diazocarbonyls, arylazides, and arylketones, the first, $\alpha$-diazocarbonyls, are in some ways the best.[2] Since we are currently studying the interaction of several antibiotics with the ribosome, we wanted to have available reagents permitting insertion of an $\alpha$-diazocarbonyl-containing group at a variety of positions. We therefore synthesized two alkylating reagents, ethyl-4-chloro-2-diazoacetoacetate (I) and ethyl-4-iodo-2-diazoacetoacetate (II), and one carbonyl reagent, 2-carbethoxy-2-diazoacethydrazide (III), to complement the acylating reagents that were previously available, ethyl-2-diazomalonyl chloride (IV)[28] and 2-diazo-3,3,3-trifluoropropionyl chloride (V).[29] Radioactive photola-

[27] P. B. Moore, *J. Mol. Biol.* **60,** 169 (1971).

[28] R. J. Vaughan and F. H. Westheimer, *J. Am. Chem. Soc.* **91,** 217 (1969).

[29] V. Chowdhry, R. Vaughan, and F. H. Westheimer, *Proc. Natl. Acad. Sci. U.S.A.* **73,** 1406 (1976).

bile derivatizing reagents are particularly useful, since they allow conversion of the native antibiotic into a radioactive photoaffinity label in a single step. What is needed is an efficient microscale synthesis, so that the reagents can be made at high specific radioactivity. We have developed such a synthesis for both reagents III and IV.

Below we describe the synthesis of reagents I–III, the microscale radioactive synthesis of reagents III and IV, and the application of reagents I–IV in derivatizing several antibiotics. Each of the new compounds described has been characterized by spectral and/or elemental analysis, the details of which will be published elsewhere. Their structures are given in Table II.

*Synthesis of Ethyl-4-chloro-2-diazoacetoacetate (I)*

Ethyl-2-diazoacetate (Aldrich) (0.35 mol) was combined with chloroacetyl chloride (Aldrich) (0.16 mol) in benzene (100 ml) and stirred overnight at room temperature. After removal of benzene by evaporation, pure compound (I) was obtained by vacuum distillation (78°/0.2 mm) in 50% yield.

*Synthesis of Ethyl-4-iodo-2-diazoacetoacetate (II)*

Compound (I) (0.02 mol) was combined with KI (0.03 mol) in 2-butanone (30 ml). After stirring overnight at 85° the reaction mixture was cooled and filtered. The filtrate was evaporated to dryness, and the residue was extracted with $CCl_4$. The $CCl_4$ solution was washed with 10% sodium thiosulfate, 5% $NaHCO_3$, and water, dried over $MgSO_4$, filtered and evaporated to dryness. The orange-yellow residue was taken up in hexane, and the hexane solution was decanted and evaporated to dryness, giving chromatographically pure (II) in 68% yield.

*Synthesis of 2-Carbethoxy-2-diazoacethydrazide (III)*

A solution of ethyl-2-diazomalonyl chloride (2 ml) in 60% ethanol (220 ml) was added dropwise to a solution made up of 95% hydrazine (6 ml), 95% ethanol (350 ml), and 1% aqueous HCl (170 ml) at 0°. After stirring for 20 min the solution was extracted with $CHCl_3$. The $CHCl_3$ layer was washed with water until the pH of the water layer was neutral, dried over $MgSO_4$, filtered, and evaporated to dryness. The residue was recrystallized from cyclohexane/benzene (5:2) yielding pale yellow needle crystals, m.p. 76°–78°. The yield was 80%.

*Microscale Synthesis of* [$^{14}$C]*Ethyl-2-diazomalonylchloride (IV)*

All procedures were performed under nitrogen in a glove bag (Instruments for Research and Industry) whose air outlet is connected to a 10%

NaOH trap followed by a Dry Ice trap. [$^{14}$C]Phosgene, obtained in a sealed tube (41 $\mu$mol, 250 $\mu$Ci, New England Nuclear), was placed with just the tip of tube in Dry Ice/acetone for 24 hr. This ensures that all the phosgene is in the tip. The tube was then snapped open, a 10% solution of ethyl diazoacetate (Aldrich) in methylene chloride (2- to 3-fold molar excess) was rapidly added, and the tube was stoppered with a cork. The reaction mixture was allowed to warm to room temperature, left standing for 24 hr, and applied to a 20 × 20 cm Brinkmann silica gel GF 250 TLC plate, which was developed with methylene chloride. The faster moving ultraviolet (UV) absorbing band ($R_f$ 0.55) was scraped off the plate. The silica gel was packed into a disposable Pasteur pipette fitted with a glass wool plug. Elution with methylene chloride gave (IV) in 90% yield. The methylene chloride solution of (IV) when stored at −80° in a well stoppered vessel is stable indefinitely.

*Microscale Synthesis of* [$^{14}$C]*-2-Carbethoxy-2-diazoacethydrazide (III)*

[$^{14}$C]Ethyl-2-diazomalonylchloride in methylene chloride (36 $\mu M$, 0.5 ml) was cooled to 0°, and 5 $\mu$l of hydrazine hydrate (65%) were added. The reaction mixture was allowed to sit for 10 min at 4°. Extraction of the organic layer with eight 200-$\mu$l portions of water, to remove the excess hydrazine, afforded pure $^{14}$C-labeled (III) in 85% yield.

*Synthesis of N-(3-Carbethoxy-3-diazo)acetonylgougerotin (VI)*

A 30-fold molar excess of (II) was added to a 10 m$M$ solution of gougerotin in 37% acetone/water buffered with 40 m$M$ sodium carbonate·HCl (apparent pH, 9.5) and the reaction mixture was maintained at 40° for 3 hr. The product, (VI), was purified by preparative TLC (silica gel, 95% ethanol, $R_f$ 0.45) and eluted off the plate with 30% ethanol/water in 70% yield.

*Synthesis of N-(3-Carbethoxy-3-diazo)acetonylblasticidin S (VII)*

A 10-fold molar excess of (II) was added to a 10 m$M$ solution of blasticidin S in 60% acetone/water buffered with 50 m$M$ sodium carbonate·HCl (apparent pH, 9.5), and the reaction mixture was maintained at 50° for 30 min. The product, (VII), was purified by preparative TLC (silica gel, ethanol:water, 7:3, $R_f$ 0.35) and eluted off the plate with 30% ethanol/water in 15% yield.

*Synthesis of S-(3-Carbethoxy-3-diazo)acetonyl-7-thiolincomycin (VIII)*

7-Thiolincomycin (0.076 mmol)[30] was incubated for 30 min at 25° in

[30] B. J. Magerlein and F. Kagan, *J. Med. Chem.* **12,** 974 (1969).

TABLE II
STRUCTURES OF α-DIAZOCARBONYL DERIVATIZING REAGENTS AND ANTIBIOTIC DERIVATIVES

*Derivatizing Reagents*

| | | |
|---|---|---|
| Ethyl-4-chloro-2-diazoacetoacetate | (I) | $ClCH_2C(=O)CN_2CO_2Et$ |
| Ethyl-4-iodo-2-diazoacetoacetate | (II) | $ICH_2C(=O)CN_2CO_2Et$ |
| 2-Carbethoxy-2-diazoacethydrazide | (III) | $NH_2NHC(=O)CN_2CO_2Et$ |
| Ethyl-2-diazomalonylchloride | (IV) | $ClC(=O)CN_2CO_2Et$ |
| 2-Diazo-3,3,3-Trifluoropropionyl chloride | (V) | $ClC(=O)CN_2CF_3$ |

*Antibiotics*

Structure: HO—$CH_2$ ribofuranose bearing $N^6$,$N^6$-dimethyl-adenine; 3′-NH—C(=O)—CH($CH_2$—C₆H₄—$R_1$)NH$R_2$; 2′-OH

| | $R_1$ | $R_2$ |
|---|---|---|
| Puromycin | $CH_3O$ | H |
| *p*-Azidopuromycin | $N_3$ | H |
| *N*-EDM puromycin | $CH_3O$ | $C(=O)CN_2CO_2Et$ |

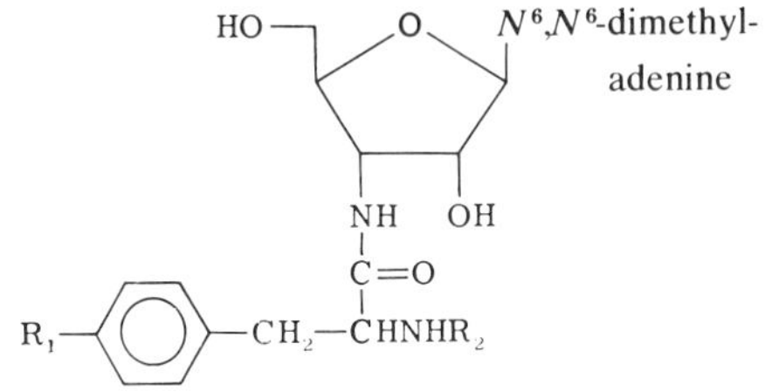

*N*-(3-Carbethoxy-3-diazo)acetonylgougerotin (VI)

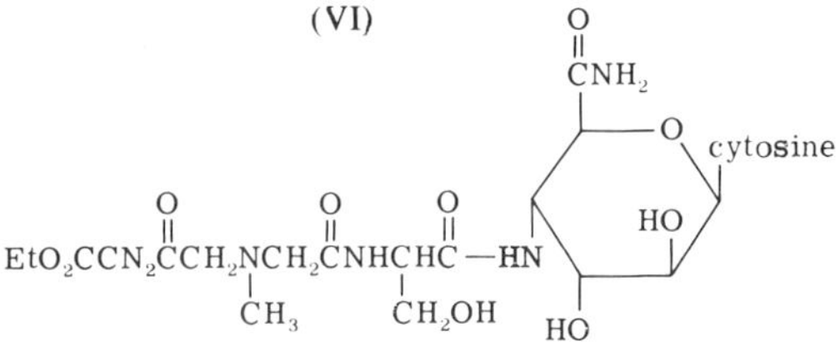

$$EtO_2CCN_2C(=O)CH_2N(CH_3)CH_2C(=O)NHCH(CH_2OH)C(=O)—HN—$$ (pyranose ring bearing $C(=O)NH_2$, HO, HO, and cytosine)

| | | R |
|---|---|---|
| *N*-(3-Carbethoxy-3-diazo)acetonylblasticidin S | (VII) | $CH_2C(=O)CN_2CO_2Et$ |
| *N*-(Ethyl-2-diazomalonyl)blasticidin S | (IX) | $C(=O)CN_2CO_2Et$ |

TABLE II (*continued*)

*S*-(3-Carbethoxy-3-diazo)acetonyl-7-thiolincomycin (VIII)

*N*-(Ethyl-2-diazomalonyl)streptomycyl hydrazone (X)

G = guanidinium

a solution consisting of 0.60 ml of ethanol, 0.13 ml of 2 *N* NaOH, and dithiothreitol (0.083 mmol). Reagent I (0.64 mmol) was then added, and the reaction mixture was stirred and incubated at 37° for 5 min. The product was purified by preparative TLC (silica gel, chloroform:methanol, 10:1, $R_f$ 0.67) and eluted off the plate with ethanol in 71% yield.

*Synthesis of N-(Ethyl-2-diazomalonyl)blasticidin S (IX)*

A 2-fold molar excess of (IV) was added to a 20 m*M* solution of blasticidin S in 50% acetone/water buffered with 0.125 *M* sodium bicarbonate·HCl (apparent pH, 8.5). The reaction proceeded to completion in 5 min at 0°. The product (IX) was purified by preparative TLC (silica gel, ethanol:water, 4:1, $R_f$ 0.30) and eluted off the plate with 30% ethanol/water in virtually quantitative yield. This procedure is similar to one published previously in the preparation of *N*-ethyl-2-diazomalonyl puromycin.[10] A procedure for O-acylation using reagent (IV) has previously been described in this series.[31]

*Synthesis of N-(Ethyl-2-diazomalonyl)streptomycyl hydrazone (X)*

A 10-fold molar excess of (III) was added to a 48 m*M* solution of streptomycin·3HCl in methanol and the reaction mixture was maintained at 37° for 3 hr. Addition of two volumes of tetrahydrofuran precipitated (X) in essentially quantitative yield.

*Microscale Synthesis of* $^{14}$*C-Labeled Compound (X)*

$^{14}$C-Labeled reagent (III) in methylene chloride (40 $\mu M$, 0.5 ml) was evaporated to dryness in portions in a rotary evaporator. To the residue was added 100 $\mu$l of a 0.1 *M* streptomycin sulfate solution in a 10 m*M* acetic acid/Na acetate buffer, pH 4.5. The reaction mixture was allowed to sit in an ice bath for 45 min, applied to a 20 × 20 Analtech silica gel GF 250 TLC glass plate, and developed with 10 m*M* acetic acid/Na acetate buffer, pH 4.5. The TLC was run at 4° to minimize Dimroth rearrangement to the corresponding triazole.[31] The slow-moving UV-absorbing spot ($R_f$ 0.05–0.10) corresponds to (X) and was eluted in 15% overall yield with the developing buffer. This low overall yield is undoubtedly a result of poor elution of (X) from the TLC plate, since analytical TLC of the reaction mixture indicated essentially quantitative conversion of (III) to (X) under the reaction conditions employed. In subsequent work with other streptomycin derivatives, we have been far more successful using 2 *N* KCl as an elutant, and this solvent should be employed in the future.

[31] B. S. Cooperman and D. J. Brunswick, this series, Vol. 38 [53].

# [58] Radioactive Chemical Labeling of Ribosomal Proteins and Translational Factors *in Vitro*

*By* CLAUDIO GUALERZI and CYNTHIA L. PON

Understanding the molecular mechanisms governing the process of protein synthesis and the functioning of the ribosomes and ribosomal factors has required in the past, and will require even more in the future, the availability of simple techniques that allow the obtaining of pure



ISBN 0-12-504180-2

components radioactively labeled to a high specific activity. In this chapter we outline some procedures routinely used in our laboratory to label *in vitro* ribosomal proteins and factors.

*Materials*

Buffer A: Tris·HCl, 10 m*M*, pH 7.7; $NH_4Cl$, 60 m*M*; Mg acetate, 10 m*M*; 2-mercaptoethanol, 6 m*M*
Buffer B: Na borate buffer, 0.1 *M*, pH 9 at 1°; KCl, 300 m*M*; 2-mercaptoethanol, 5 m*M*
Buffer C: Na borate buffer, 0.1 *M*, pH 8.5 at 1°; $MgCl_2$, 20 m*M* KCl, 100 m*M*; 2-mercaptoethanol, 5 m*M*
Buffer D: Tris·HCl, 20 m*M*, pH 7.7; $NH_4Cl$, 200 m*M*; EDTA, 1 m*M*; glycerol 10%
Formaldehyde, 35% aqueous solution (Merck)
[$^{14}$C]Formaldehyde (10–50 mCi/mmol, New England Nuclear)
Sodium borohydride (Merck)
Sodium boro[$^3$H]hydride (8 Ci/mmol, New England Nuclear)
Guanidinium HCl (puriss. C. Roth, Karlsruhe)
Ethanol, absolute (Merck)
*N*-ε-Monomethyllysine and *N*-ε-dimethyllysine (Bachem, Liestal)

## Reductive Alkylation

The addition of aliphatic or aromatic aldehydes to primary amino groups (i.e., ε-$NH_2$ of lysine) leads to the formation of Schiff bases that can be converted into stable *N*-methyl derivatives upon reduction with mild reducing agents, such as $NaBH_4$[1] and $NaBH_3CN$.[2] This method of modifying lysine ε-$NH_2$ groups is more specific than nucleophilic displacement or addition obtained with other reagents[2] and should therefore provide a useful tool in the study of the topography, conformational changes, and structure–function relationships in ribosomal proteins and translational factors.

The most commonly employed reductive alkylation of lysines, i.e., reductive methylation with formaldehyde and sodium borohydride, leads to the formation of both ε-monomethyllysine and ε-dimethyllysine.[1] This method was first introduced by Means and Feeney, who showed that it is possible to modify proteins *in vitro* with minimal changes in their gross physical properties.[1] Rice and Means[3] applied this technique to label

[1] G. E. Means and R. E. Feeney, *Biochemistry* **7**, 2192 (1968).
[2] M. Friedman, L. D. Williams, and M. S. Masri, *Int. J. Peptide Protein Res.* **6**, 183 (1974).
[3] R. R. Rice and G. E. Means, *J. Biol. Chem.* **246**, 831 (1971).

radioactively *in vitro* several proteins including chymotrypsin and pancreatic ribonuclease. The specific activity obtained was satisfactory (approximately 5000 cpm/$\mu$g), and, although some proteins became inactivated by this procedure, others (e.g., chymotrypsin) retained full catalytic activity. Subsequently, initiation factor IF-3 was labeled *in vitro* by this method without any loss of biological activity,[4,5] and since then, reductive methylation has been widely applied, in our laboratory as well as in several others, to label ribosomal proteins and other translational factors to high specific activity without apparent loss or modification of biological activity.

In this section we show the effect of a number of variables on the yield of the reductive methylation reaction. The examples refer primarily to the labeling of a specific protein, initiation factor IF-3, but can be extended directly to any other protein having comparable properties (e.g., ribosomal proteins).

The temperature (between 0° and 37°) and the time of incubation (between 30 sec and 30 min) have little or no effect on the yield of methylation (not shown). Since, however, the likelihood of inactivation of a protein increases when these two parameters are increased, it is advisable that, unless otherwise required, the reaction be run in an ice bath for periods between 30 sec and 3 min.

Reductive methylation shows a pH optimum between 9.8 and 10.2 (Fig. 1), which nearly coincides with the p$K$ (p$K_{a3}$) of the $\epsilon$-$NH_2$ group of lysine. Adequate labeling, however, can also be obtained at lower pH (e.g., pH 7.7). This is advantageous, as not all proteins can withstand high pH. The strong pH dependence of the reductive methylation reaction should be borne in mind, however, in experiments correlating structure and function, where, for instance, a protein (e.g., a ribosomal protein) is labeled in the presence and in the absence of nucleic acid (e.g., rRNA), because in this case local changes of the pH may be expected to play a role in determining the extent of methylation.

Concerning the buffer system used, Rice and Means[3] carried out reductive methylation in 0.2 *M* borate buffer, pH 9.0. In our laboratory, we obtain satisfactory results using a lower concentration (0.1 *M*) of this buffer. It has been our experience, however, that under these buffer conditions all three initiation factors tend to adhere more extensively to glass and plastic surfaces. Silicone coating of the glassware has been of some help in minimizing this problem, at least in the case of IF-3. If borate buffer is not suitable, it can be replaced by HEPES buffer, pH 9.0.[6]

[4] C. L. Pon, S. M. Friedman, and C. Gualerzi, *Mol. Gen. Genet.* **116,** 192 (1972).
[5] C. Gualerzi, M. R. Wabl, and C. L. Pon, *FEBS Lett.* **35,** 313 (1973).
[6] U. Kleinert and D. Richter, *FEBS Lett.* **55,** 188 (1975).

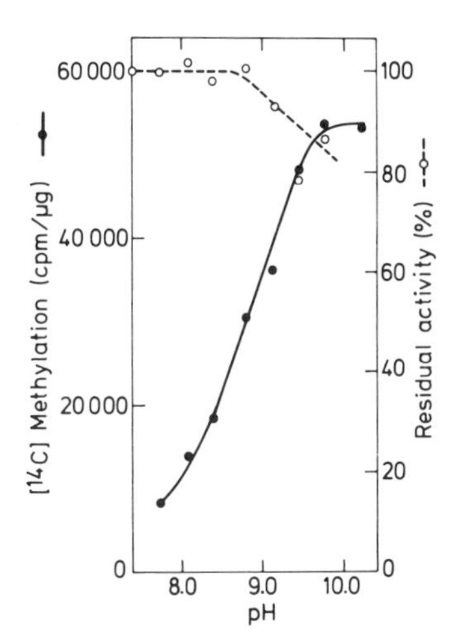


FIG. 1. Effect of pH on the extent of reductive methylation of initiation factor IF-3. The reaction was carried out in a volume of 15 $\mu$l under conditions identical to those described in the text, the only exception being the pH of the borate buffer. The indicated pH values were measured at the temperature of the reaction (1°). The extent of methylation was determined following sodium dodecyl sulfate–polyacrylamide slab gel electrophoresis of aliquots of the reaction mixtures. The stained bands of IF-3 were cut out, dissolved by a 16-hr incubation at 55° with 1 ml of Soluene 350 (Packard), and the radioactivity was determined after addition of 10 ml of toluene–PPO–POPOP. The test of IF-3 activity was performed as described in this series, Vol. 60 [18].

Another variable to be considered is the amount of formaldehyde in the reaction mixture. Although it has been shown with a model compound, butylamine, that alkylation occurs optimally with a stoichiometric amount of formaldehyde ($H_2CO$:butylamine = 2),[1] a quite different situation exists in the case of reductive methylation of proteins. As seen in Fig. 2, an approximately linear relationship exists between extent of methylation and concentration of formaldehyde until approximately 15-fold excess of formaldehyde over the total number of lysines (20)[7] is reached. Higher concentrations do not increase the extent of methylation, and excessively high concentrations may be deleterious for both the activity of the protein and the efficiency of the reaction. Since the number of lysines in a given protein is not always known and variations in the number of lysines available for chemical modification as well as the individual p*K*s of their $\epsilon$-$NH_2$ groups are expected to vary from one case to another, the extent of methylation should be checked against formaldehyde concentration for each protein, especially in light of the fact that the activity of some proteins is likely to be more susceptible than others

[7] D. Brauer and B. Wittmann-Liebold, *FEBS Lett.* **79**, 269 (1977).

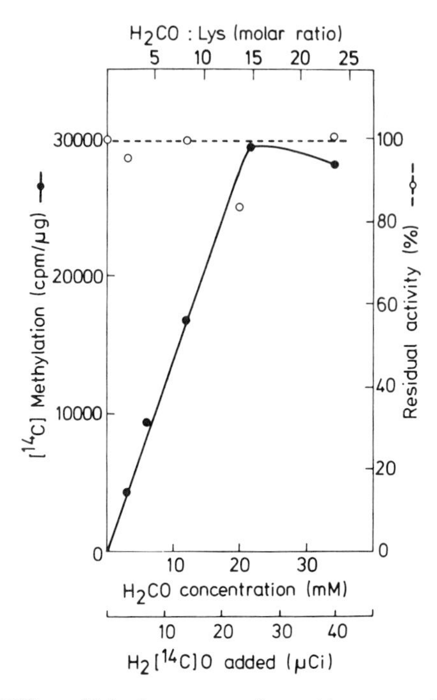


FIG. 2. Effect of [$^{14}$C]formaldehyde concentration on the extent of reductive methylation of initiation factor IF-3. All the conditions are identical to those described for Fig. 1 with the exception of the pH of the reaction (9.0) and [$^{14}$C]formaldehyde concentration. In each case the amount of $NaBH_4$ added (in a single addition, cf. text) was equal to one-fourth of the amount of $H_2CO$.

to higher formaldehyde concentrations. Provided that the amount of formaldehyde remains above saturation, the concentration of the protein to be labeled can be varied to meet individual needs without affecting the efficiency of the reaction (cf. Table I).

The amount of borohydride used as a reducing agent plays an important role in determining the efficiency of the reaction. For this reason the necessary amount of borohydride should be accurately determined from the amount of formaldehyde used in each reaction. In addition, since solutions of sodium borohydride are unstable, it is necessary to dissolve the sodium borohydride at 0° just before use. As seen in Fig. 3, the methylation displays an optimum for the amount of borohydride equivalent to the stoichiometric requirement of the formaldehyde (1 mol of $H_2CO$ = 0.25 mol of borohydride). Amounts of borohydride less than stoichiometric are obviously not enough to reduce all the Schiff bases, thus resulting in a lower yield of methylation. Increasing the amount of

TABLE I
EFFECT OF PROTEIN CONCENTRATION ON THE EXTENT OF METHYLATION

| Tube No. | Protein concentration (mg/ml) | Extent of [$^{14}$C]methylation (cpm/$\mu$g) |
|---|---|---|
| 1 | 0.64 | 19914 |
| 2 | 1.60 | 17823 |
| 3 | 3.20 | 18795 |

borohydride above the stoichiometric requirement, however, is also counterproductive and for a 3-fold excess of borohydride a 60% decrease in the yield of the reaction is observed (Fig. 3). It has been proposed that the borohydride should be added in several small aliquots.[3] We do not find any advantage in doing so, as long as the total amount of borohydride added is equivalent to the stoichiometric requirement. If this is not the case, however, the stepwise addition represents an advantage when an excess of reagent is used (since the first addition seems to be the most important), but represents a disadvantage if the total amount of borohydride added is insufficient, as shown in Table II.

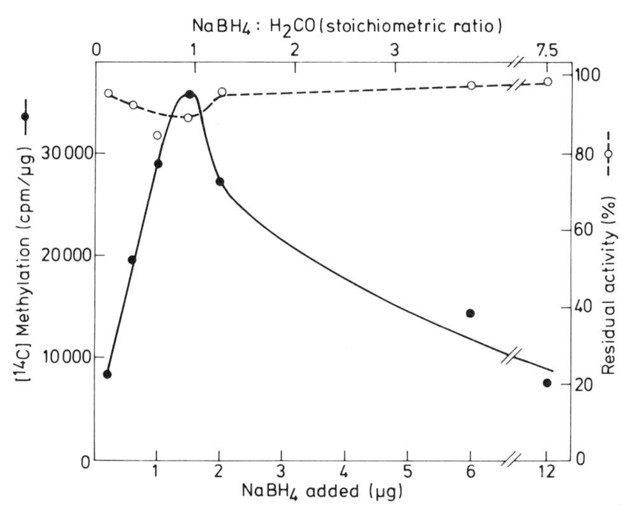


FIG. 3. Effect of $NaBH_4$ concentration on the extent of reductive methylation of initiation factor IF-3. All the conditions are identical to those described for Fig. 1 with the exception that the amount of $NaBH_4$ added was varied as indicated.

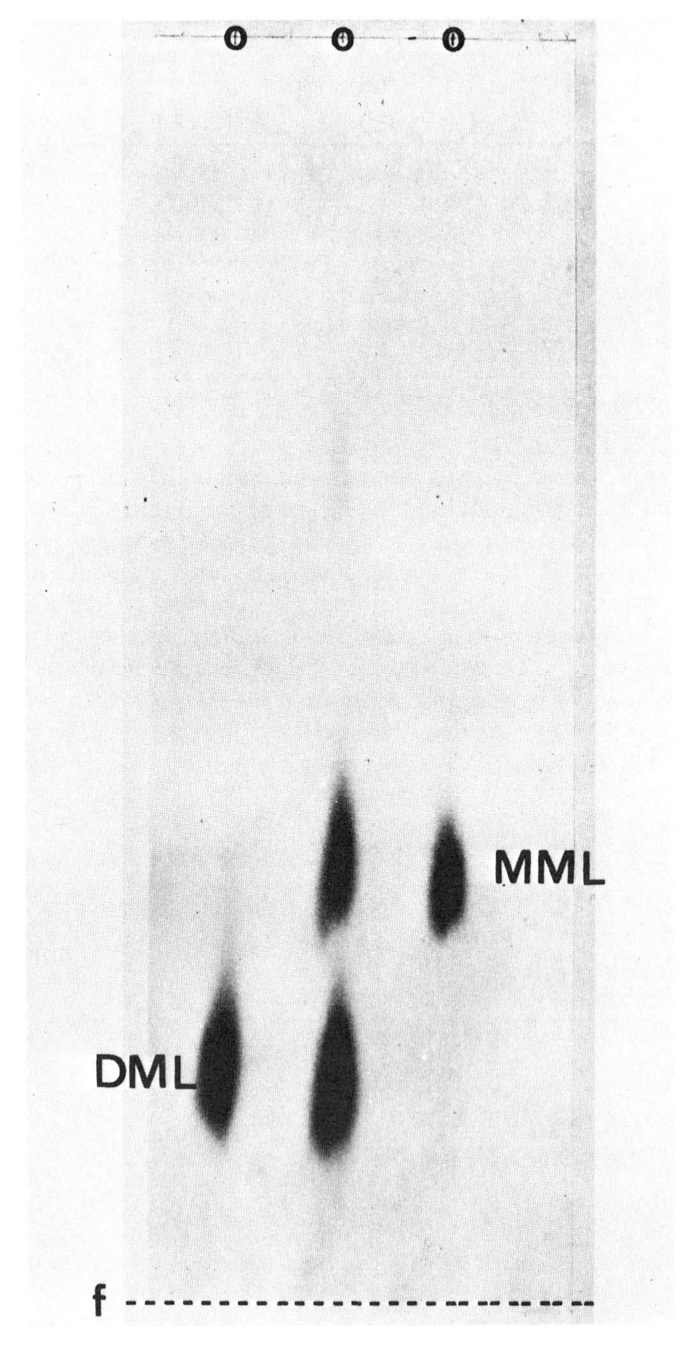


FIG. 4. Separation of monomethyl- and dimethyllysine by descending paper chromatography. The chromatographic system is similar to that described by I. Stewart [*J. Chromatogr.* **10,** 404 (1963)] and J.-H. Alix and D. Hayes [*J. Mol. Biol.* **86,** 139 (1974)]. $^{14}$C-

TABLE II
EFFECT OF STEPWISE OR SINGLE ADDITION OF $NaBH_4$ ON THE EXTENT OF METHYLATION

| Expt. No. | $NaBH_4$ added, total (μg) | $NaBH_4$:$H_2CO$ (stoichiometric ratio) | Mode of addition | Extent of [$^{14}$C]methylation (cpm/μg) |
|---|---|---|---|---|
| 1 | 0.6 | 0.37 | Single addition | 19,398 |
| | 0.6 | 0.37 | 4 Equal aliquots | 11,548 |
| 2 | 6.0 | 3.70 | Single addition | 14,455 |
| | 6.0 | 3.70 | 4 Equal aliquots | 31,787 |
| 3 | 25.0 | 15.60 | Single addition | 7,789 |
| | 25.0 | 15.60 | 4 Equal aliquots | 9,460 |

*Protein Chemical Analysis of the Labeled Product*

Reductive methylation yields primarily, if not exclusively, monomethyl- and dimethyllysines.[1,3] The relative yields of these two products can be determined after acid hydrolysis, with an amino acid analyzer[1] or, if this is not available, by descending paper chromatography (cf. Fig. 4). When applied to the labeled IF-3, this method showed 76.8% and 23.2% dimethyllysine and monomethyllysine, respectively, under labeling conditions giving the highest yield of methylation (approximately 10 pmole of $^{14}CH_3$ per picomole of IF-3).

In agreement with results reported by others,[8,9] we find that the tryptic digestion of the methylated proteins is severely affected by the modification of the lysines. The dimethyllysines are apparently more resistant to the tryptic digestion than monomethyllysines, which in turn are more resistant than unmodified lysines. This results in composite tryptic fin-

[8] L. Benoiton and J. Deneault, *Biochim. Biophys. Acta* **113,** 613 (1966).

[9] M. Goreck and Y. Shalitin, *Biochem. Biophys. Res. Commun.* **29,** 189 (1967).

labeled methylated IF-3 is subjected to acid hydrolysis in 5.7 *N* HCl containing 0.02% 2-mercaptoethanol *in vacuo* for 20 hr at 110°. After addition of monomethyl and dimethyllysine carrier, the sample is applied onto Whatman 3 MM paper (55 cm long), pretreated by dipping in 80 m*M* EDTA, pH 7, and dried. The chromatogram is developed for approximately 20 hr at 20° in a system containing phenol (110 ml), *m*-cresol (110 ml), and 64 m*M* Na borate buffer, pH 9.3 (10 ml). After chromatography the paper is allowed to dry completely in a fume hood with a steam of warm air, and finally it is sprayed with ninhydrin. The chromatogram is cut into 4–5 cm longitudinal strips, which are cut into 1.5-cm segments and placed into vials for radioactivity determination after the addition of 10 ml of Bray's solution. The radioactivity was found to be unequally distributed in two peaks coinciding with the positions of the stained spots of monomethyllysine (MML, 23.2%) and dimethyllysine (DML, 76.8%). f, front; o, origin.

gerprints that contain not only the usual peptides but also partial digestion products, so that overall a larger number of radioactive peptides than expected is obtained. This does not represent an absolute obstacle to the identification of the peptides, however, and coupled with the use of other proteolytic enzymes whose activities are not affected by the modification of lysines (i.e., *Staphylococcus* protease and thermolysin), it is possible to reach the unequivocal identification of the labeled peptides. Peptide maps of several proteins labeled *in vitro* have been analyzed in our laboratory. It seems clear that, with a few exceptions, when an isolated native protein is labeled nearly all lysine-containing peptides become radioactive to a greater or lesser extent.

*Labeling of Translational Factors and Purified Ribosomal Proteins*

*In vitro* labeling of proteins by reductive methylation can be carried out using either [$^{14}$C]- or [$^{3}$H]formaldehyde or sodium boro[$^{3}$H]hydride. While the use of radioactive formaldehyde results in the specific formation of labeled monomethyl- and dimethyllysines, it should be noted that the labeling with sodium boro[$^{3}$H]hydride may also result not only in the labeling of the methyl groups, but also in the incorporation of some exchangeable tritium.

As mentioned above, the reductive methylation procedure has been successfully applied in this and many other laboratories to label *in vitro* several translational factors without apparent loss of biological activity (cf. Table III).

A typical labeling procedure with radioactive formaldehyde is given here: protein, 0.4 mg in 0.2 ml, is dialyzed for 12 hr at 2°–4° against several changes of buffer B. The dialyzed protein (200 $\mu$l) is placed into a glass test tube in an ice bath under a fume hood. Just before the reaction is started by the addition of formaldehyde to the protein solution, 1 mg of sodium borohydride is dissolved in 1 ml of ice-cold water. A new vial of [$^{14}$C]formaldehyde (250 $\mu$Ci, 45 mCi/mmol, or an equivalent amount of [$^{3}$H]formaldehyde) is opened, and 50 $\mu$l of buffer B are added so that the radioactive formaldehyde can be quantitatively transferred into the test tube containing the protein. After mixing on a Vortex mixer for 1–2 sec, the reaction mixture is incubated on ice for 30–60 sec and the reaction is completed by the addition of 40 $\mu$l of the 1 mg/ml borohydride solution. The reaction mixture is mixed again on a Vortex mixer for 2–3 sec and immediately dialyzed against the most suitable buffer for the protein in question.

If the amount of protein to be labeled is extremely small or if it does not withstand a long period at high pH, the dialyses preceding and following the reaction can be replaced by gel filtration on Sephadex (preferably G-50).

TABLE III
TRANSLATIONAL FACTORS AND RIBOSOMAL PROTEINS[a] LABELED *in Vitro* BY REDUCTIVE METHYLATION

| Protein labeled | Source of label and approximate specific activities[b] obtained | | | |
|---|---|---|---|---|
| | [$^{14}$C]Formaldehyde (cpm/pmol) | Na boro[$^{3}$H]hydride (cpm/pmol) | Activity | Reference |
| IF-1 | 35 | 5360 | + | *d* |
| IF-2 | 820–1000 | — | + | *d,e* |
| IF-3 | 200–820 | 22727 | + | *d,f,g* |
| EF-G | — | 1100–11000 | + | *h* |
| EF-Tu | — | 1100–11000 | + | *h* |
| PPT[c] | — | 1100–11000 | + | *h* |
| S1 | 700–1400 | — | + | *d,i,j,k* |
| L7/L12 | + | — | + | *l* |

[a] Only the ribosomal proteins for which a specific functional test other than rRNA binding or ribosome reconstitution is available have been included in this table.
[b] The approximate specific activities are given for an 85% and 50% counting efficiency for $^{14}$C and $^{3}$H radioactivity, respectively.
[c] Pyrophosphoryltransferase (stringent factor).
[d] C. Gualerzi and C. L. Pon, unpublished results, 1977.
[e] G. A. J. M. van der Hofstad, J. A. Foekens, P. J. van den Elsen, and H. O. Voorma, *Eur. J. Biochem.* **66,** 181 (1976).
[f] D. A. Hawley, M. J. Miller, L. I. Slobin, and A. J. Wahba, *Biochem. Biophys. Res. Commun.* **61,** 329 (1974).
[g] R. L. Heimark, L. Kahan, K. Johnston, J. W. B. Hershey, and R. R. Traut, *J. Mol. Biol.* **105,** 219 (1976).
[h] U. Kleinert and D. Richter, *FEBS Lett.* **55,** 188 (1975).
[i] G. Jay and R. Kaempfer, *J. Mol. Biol.* **82,** 193 (1974).
[j] K. Isono and S. Isono, *Proc. Natl. Acad. Sci. U.S.A.* **73,** 767 (1976).
[k] J. E. Sobura, M. R. Chowdhury, D. A. Hawley, and A. J. Wahba, *Nucl. Acids. Res.* **4,** 17 (1977).
[l] R. Amons and W. Möller, *Eur. J. Biochem.* **44,** 97 (1974).

To label proteins with sodium boro[$^{3}$H]hydride, we have used the following procedure. Protein, 115 $\mu$g in 0.6 ml of buffer B, is treated with 5 $\mu$l of a 3.5% formaldehyde solution. After a 2-min incubation in an ice bath, 20 $\mu$l of an aqueous solution containing approximately 16 mCi of freshly dissolved sodium boro[$^{3}$H]hydride (8 Ci/mmol) are added to the reaction mixture. After 5 min at 1°, 50 $\mu$l of a 10 mg/ml solution of nonradioactive sodium borohydride are added and the sample is dialyzed against a buffer suitable for the protein. The specific activities obtained by this method for IF-1 and IF-3 were approximately $5.9 \times 10^{5}$ cpm/$\mu$g and $1 \times 10^{6}$ cpm/$\mu$g, respectively. The ribosomal binding activity of IF-1 labeled *in vitro* with sodium boro[$^{3}$H]hydride as well as by

TABLE IV
RIBOSOMAL BINDING ACTIVITY OF *in Vitro* LABELED IF-1[a]

| Additions | Factor bound to 30 S | |
|---|---|---|
| | [$^3$H]IF-1 (cpm) | [$^{14}$C]IF-1 (cpm) |
| None | 6021 | 115 |
| IF-3 | 21996 | 884 |

[a] The ribosomal binding of radioactive IF-1 was measured by sucrose gradient centrifugation under conditions identical to those described for the binding of IF-3 [C. L. Pon and C. Gualerzi, *Biochemistry* **15**, 804 (1976)]. The amount of 30 S ribosomal subunits used in each assay was 0.5 $A_{260}$ units and, when present, the amount of IF-3 was 1 μg.

[$^{14}$C]formaldehyde treatment was determined by sucrose gradient analysis. The results are shown in Table IV.

*Labeling Ribosomal Proteins in Situ or in Solution*

The ribosomal proteins of either subunit have also been labeled both *in situ* and in solution without loss of biological activity. The radioactive proteins so obtained have been found to be active in ribosome reconstitution,[10-13] 16 S or 23 S rRNA binding,[11,13,14] and other functional tests.[10,15,16] The labeling of total ribosomal proteins is a convenient tool for obtaining protein markers to be used as internal standards (e.g., in two-dimensional electrophoresis) in chemical modification experiments.[17] The reductive methylation of ribosomal proteins has also been applied to quantitate individual ribosomal proteins in ribosomal subunits made artificially protein deficient (e.g., by high-salt washing) or following partial or total reconstitution[18] (see below).

For the labeling of purified ribosomal proteins, the procedure described in the preceding section can be applied. However, to label total

[10] G. Moore and R. R. Crichton, *FEBS Lett.* **37**, 74 (1973).
[11] W. A. Held, B. Ballou, S. Mizushima, and M. Nomura, *J. Biol. Chem.* **249**, 3103 (1974).
[12] C. L. Pon, R. Brimacombe, and C. Gualerzi, *Biochemistry* **16**, 5681 (1977).
[13] E. Spicer, J. Schwarzbauer, and G. R. Craven, *Nucl. Acids Res.* **4**, 491 (1977).
[14] R. R. Crichton and H. G. Wittmann, *Proc. Natl. Acad. Sci. U.S.A.* **70**, 665 (1973).
[15] K. Isono and S. Isono, *Proc. Natl. Acad. Sci. U.S.A.* **73**, 767 (1976).
[16] J. E. Sobura, M. R. Chowdhury, D. A. Hawley, and A. J. Wahba, *Nucl. Acids Res.* **4**, 17 (1977).
[17] R. Ewald, C. Pon, and C. Gualerzi, *Biochemistry* **15**, 4786 (1976).
[18] C. L. Pon and C. Gualerzi, *Biochemistry* **15**, 804 (1976).

30 S and 50 S ribosomal proteins *in situ* or after denaturation with guanidinium HCl, we routinely apply the following procedure:

Two samples containing 100 $A_{260}$ units of either 30 S or 50 S ribosomal subunits in 1.0 ml of buffer A are placed into two glass centrifuge tubes. The ribosomal subunits are then precipitated by addition of 0.75 volume of ice-cold ethanol. After centrifugation, the pellets are thoroughly drained and resuspended in 0.5 ml of buffer C. Two 0.2-ml aliquots of 30 S and two 0.2-ml aliquots of 50 S subunits are placed in four glass centrifuge tubes, which are used for the reaction. One sample of 30 S and one of 50 S are treated with 10 μl (10 μg) of pancreatic RNase and are incubated for 2 hr at 37°. At the end of the incubation these two samples receive 0.25 ml of 10 *M* guanidinium HCl (must be warmed to dissolve) while the remaining two samples receive 0.26 ml of water. Four vials, each containing 50 μCi of [$^{14}$C]formaldehyde (10 mCi/mmol) in 0.015 ml of aqueous solution are opened, and each receives 30 μl of buffer C. The contents of each vial are quantitatively transferred to the four reaction tubes containing the samples to be labeled. After 1–2 sec of mixing on a Vortex mixer, the samples are incubated for 2 min in an ice bath. Of a freshly made (1 mg/ml) solution of $NaBH_4$ (see above), 35 μl are added. The samples are kept on ice for an additional 2 min, after which the acetic acid extraction of the ribosomal proteins[19] is started at once.

The radioactive ribosomal proteins obtained by this method are then analyzed by two-dimensional gel electrophoresis[20] on 10 × 10 cm gel slabs. The labeled ribosomal proteins obtained at the end of this procedure (1 to 2 × $10^6$ cpm) are sufficient for four to six two-dimensional gel electrophoreses if no carrier proteins are added, and if Coomassie Blue is used for staining. After destaining, the slabs are kept for at least 2 hr in deionized water, after which the stained protein spots, identified according to the nomenclature of Kaltschmidt and Wittmann,[21] are cut out and placed into low-background scintillation vials. One milliliter of Soluene 350 (Packard) is added to the vials, which are then tightly closed with screw caps and incubated for 16 hr at 55°. The radioactivity is determined after addition of 10 ml of toluene–PPO–POPOP scintillation fluid. The yield of methylation varies for individual 30 S and 50 S ribosomal proteins labeled *in situ* or in the presence of 5 *M* guanidinium HCl, but on average the counts recovered in each stained spot, range between 2000 and 20,000 cpm (each plate representing one-fifth of the total labeled protein recovered).

[19] S. J. S. Hardy, C. G. Kurland, P. Voynow, and G. Mora, *Biochemistry* **8**, 2897 (1969).
[20] E. Kaltschmidt and H. G. Wittmann, *Anal. Biochem.* **36**, 401 (1970).
[21] E. Kaltschmidt and H. G. Wittmann, *Proc. Natl. Acad. Sci. U.S.A.* **67**, 1276 (1970).

The above procedure has also been applied to quantitate the proteins contained in subparticles obtained by partial reconstitution.[18] In this case, control 30 S ribosomal subunits and particles reconstituted omitting specific ribosomal proteins were isolated by sucrose gradient centrifugation, precipitated with ethanol, and then labeled with [$^{14}$C]formaldehyde in the presence of guanidinium–HCl. The amount of individual proteins in the reconstituted subparticle was then determined after two-dimensional electrophoresis in the presence of total 30 S carrier proteins using the radioactivity incorporated into protein S4 of both the subparticle and control 30 S subunits as internal reference.[18]

### Labeling with Ethylmaleimide, *N*-[Ethyl-2-$^3$H]

A second method used in our laboratory that allows the *in vitro* labeling of initiation factor IF-3 to a satisfactory specific activity without loss of biological activity and with minimal change in its chemical structure is the reaction with radioactive *N*-ethylmaleimide (NEM).

To label IF-3 with radioactive NEM we follow the procedure given below. To 3 mg of IF-3 in 1 ml of buffer D are added 2 $\mu$l of 1 *M* dithiothreitol (DTT). After incubation for 2 hr at 37° the protein solution is exhaustively dialyzed at 2°–4° against buffer D containing no DTT under a stream of $N_2$. From a vial of ethylmaleimide, *N*-[ethyl-2-$^3$H] (150 mCi/mmol, 1 mCi/ml of pentane, New England Nuclear), 0.15 ml is removed and put in a glass test tube, which is placed in an ice bath in a desiccator attached to a vacuum pump to remove the pentane solvent. After approximately 5 min, when all pentane is evaporated, 0.2 ml of water are added to the tube and the NEM is dissolved by mixing on a Vortex for a few seconds. The 1 ml of protein solution is then added to the [$^3$H]NEM solution and the reaction mixture is incubated for 45 min at 37°. The reaction is stopped by the addition of 12 $\mu$l of 1 *M* DTT. The unbound radioactivity is removed by exhaustive dialysis against buffer D or gel filtration on a Sephadex G-50 column.

Figure 5 shows the time course of [$^3$H]NEM labeling of IF-3 under the conditions described above. As seen from the figure, the reaction is complete after approximately 30–45 min at 37° at an NEM concentration of 0.8 m*M*. Figure 5 also shows that the loss of biological activity of IF-3 after this period is negligible. Stoichiometric determination shows that after 60 min approximately 1 mol of NEM is incorporated per mole of IF-3, and peptide mapping shows that all the radioactivity is incorporated into the single SH-containing tryptic peptide T1 of IF-3. Since after this reaction all the $\epsilon$-$NH_2$ groups of IF-3 are preserved, the use of the factor

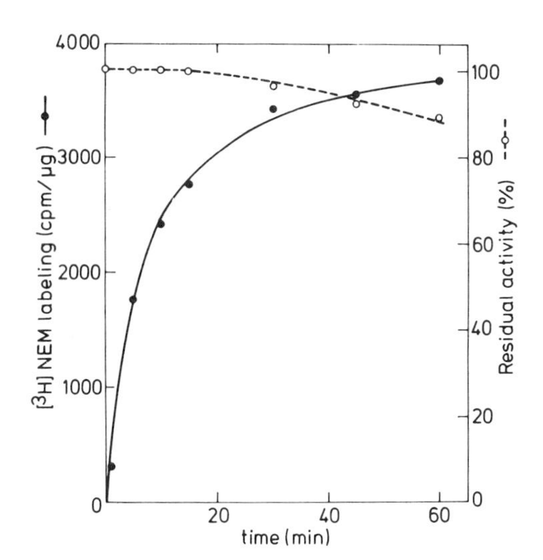


FIG. 5. Time course of IF-3 labeling with ethylmaleimide *N*-[ethyl-$^3$H]. The reaction conditions are described in the text. The extent of reaction and the IF-3 activity test are performed as described in the legend of Fig. 1.

labeled *in vitro* by NEM is ideal for those reactions for which the integrity of $\epsilon$-$NH_2$ groups of lysines is desirable (e.g., cross-linking with bisimidoesters and diazides). Cross-linking of [$^3$H]NEM-IF-3 to the 16 S rRNA via formaldehyde has been reported.[12]

In addition to IF-3, we have tried this method on ribosomal protein S1. The resulting labeled protein can be reconstituted into 30 S ribosomal subunits, which, in agreement with other results,[22] preserve at least some of their activity. Although we have not tried other ribosomal proteins, it should be possible to apply the NEM *in vitro* labeling to all those ribosomal proteins that contain nonessential SH groups.[23]

[22] A. Kolb, J. M. Hermoso, J. O. Thomas, and W. Szer, *Proc. Natl. Acad. Sci. U.S.A.* **74,** 2379 (1977).

[23] P. B. Moore, *J. Mol. Biol.* **60,** 169 (1971).

# [59] Peptidyl-Puromycin Synthesis on Polyribosomes from *Escherichia coli*

*By* SIDNEY PESTKA

Several model systems have been used to study transpeptidation on ribosomes. The reaction of puromycin with numerous derivatives of aminoacyl-tRNA as synthetic donors has provided the basis for most model systems (Fig. 1). Common synthetic donors have included fMet-tRNA, acetyl-Phe-tRNA, polylysyl-tRNA, and fragments of tRNA such as C-A-C-C-A(Ac-Phe) and C-A-A-C-C-A(fMet).[1-10] The ribosomes used generally have been washed in 1 *M* $NH_4Cl$, although ribosomes washed in low salt have also been used. Because of apparent disparate results of inhibitors on peptide bond synthesis in model systems and intact cells,[11] it was useful to study transpeptidation in a more nearly physiological system, which behaved similarly to the process in intact cells. For this reason, native polyribosomes have been prepared and used as the enzyme–substrate complex containing peptidyl-tRNA. The transfer of the peptidyl group to [$^3$H]puromycin as an acceptor was measured by determining the [$^3$H]puromycin incorporated into trichloroacetic acid-precipitable material.

## Preparation of Polyribosome Extract

*Reagents*

Medium NK: 15 g nutrient broth in 1 liter of 0.1 *M* potassium phosphate, pH 7.7; 0.25% (w/v) glucose

Sucrose-Tris: 25% (w/v) sucrose in 40 m*M* Tris·HCl, pH 8.1

Lysozyme (muraminidase): 6.4 mg of egg white lysozyme (EC

[1] B. E. H. Maden, R. R. Traut, and R. E. Monro, *J. Mol. Biol.* **35**, 333 (1968).
[2] M. E. Gottesman, *J. Biol. Chem.* **242**, 5564 (1967).
[3] I. Rychlik, *Biochim. Biophys. Acta* **114**, 425 (1966).
[4] M. S. Bretscher and K. A. Marcker, *Nature (London)* **211**, 380 (1966).
[5] A. Zamir, P. Leder, and D. Elson, *Proc. Nat. Acad. Sci. U.S.* **56**, 1794 (1966).
[6] J. Lucas-Lenard and F. Lipmann, *Proc. Nat. Acad. Sci. U.S.* **57**, 1050 (1967).
[7] H. Weissbach, B. Redfield, and N. Brot, *Arch. Biochem. Biophys.* **127**, 705 (1968).
[8] S. Pestka, *Arch. Biochem. Biophys.* **136**, 80 (1970).
[9] R. E. Monro and K. A. Marcker, *J. Mol. Biol.* **25**, 347 (1967).
[10] J. L. Lessard and S. Pestka, *J. Biol. Chem.* **247**, 690 (1972).
[11] S. Pestka, *Annu. Rev. Microbiol.* **25**, 487 (1971).



ISBN 0-12-504180-2

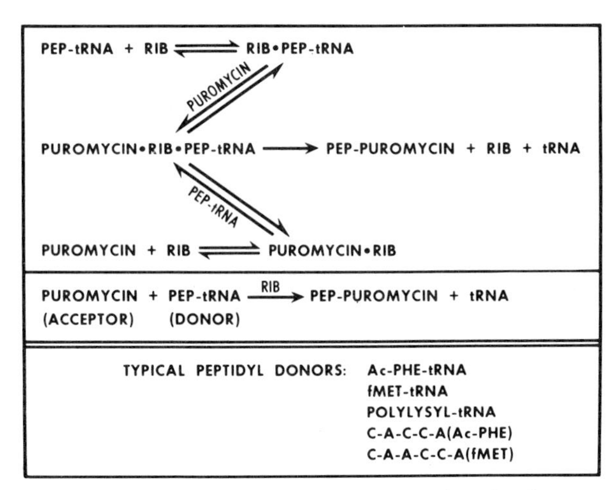


FIG. 1. Outline of the steps of the puromycin reaction. In the equations written, all starting components are considered to be free in solution. During protein synthesis, however, peptidyl-tRNA (PEP-tRNA) is a ribosomal bound intermediate and the usual acceptor is aminoacyl-tRNA. A list of several commonly used peptidyl donors is given in the figure.

3.2.1.17) in 1.0 ml of 0.25 *M* Tris·HCl, pH 8.1. This solution should be freshly made each time before use.

$Na_2EDTA$: 0.1 *M*

$MgSO_4$: 1.0 *M*

Lysing Medium: 0.5% (w/v) Brij 58; 50 m*M* $NH_4Cl$; 5 m*M* $MgSO_4$; 10 *M* Tris·HCl, pH 7.5

Deoxyribonuclease I (EC 3.1.4.5): 5 mg/ml (ribonuclease-free, Worthington)

Polysomes were prepared by modification of the method of Godson and Sinsheimer.[12] *E. coli* B was grown to about 300 Klett units (using the No. 42 blue filter), at which time cells were still in log phase in medium NK. Under these conditions 300 Klett units was approximately equivalent to $9 \times 10^8$ cells/ml. The cells were rapidly chilled to 0° to 5° within 10 seconds by swirling in a stainless steel beaker surrounded by a dry ice–acetone bath. The suspension of cells was then centrifuged for 10 minutes at 10,000 rpm in a GSA rotor of a Sorvall RC2B centrifuge. The cells were resuspended in 9.0 ml sucrose-Tris solution; 1.0 ml of a freshly made solution of lysozyme was then added. To start the action of lysozyme, 0.5 ml of 0.1 *M* $Na_2EDTA$ was added. The concentrated cell suspension was kept in an ice bath for 90 seconds with

[12] G. N. Godson and R. L. Sinsheimer, *Biochim. Biophys. Acta* **149**, 476, 489 (1967).

occasional shaking. To stop lysozyme action, 0.12 ml of 1 *M* $MgSO_4$ was added to the mixture. The cell protoplast suspension was sedimented at 10,000 rpm for 5 minutes in a SS-34 rotor of the Sorvall centrifuge. The supernatant portion was discarded and the inside of the tube carefully wiped to remove any remaining liquid. The pellet was resuspended in 9 ml of lysing medium by stirring with a glass rod: 0.1 mg of DNase in a volume of 0.02 ml was added to the suspension at this time. The suspension was kept in lysing medium 10 minutes at 0°–5°C. The viscous lysate was then centrifuged for 10 minutes in the SS-34 Sorvall rotor at 10,000 rpm. The supernatant was gently aspirated, and the pellet was discarded. The final solution contained 150–250 $A_{260}$ units/ml. The absorbance at 260 nm of each milliliter as determined by sucrose gradient centrifugation was typically distributed as follows[13]: 45% in the fraction sedimenting at less than 30 S; 28%, 15%, 7%, and 5% of the absorbance was found in polyribosomes, 70 S, 50 S, and 30 S fractions, respectively. Washed polyribosomes free of supernatant fraction and free of ribosomal particles could be prepared by sedimenting the polysome extract through a sucrose gradient as described in the legend to Fig. 7. In order to obtain purified polysomes active in forming peptidylpuromycin, the sucrose gradients used must contain 0.01 to 0.1 *M* $K^+$. At lower or higher $K^+$ concentrations, the purified polysomes are inactivated. These purified polysomes behaved essentially identically to the polysome extracts with respect to peptidyl-puromycin synthesis.[14,15] Similar procedures were used to prepare polysomes from other *E. coli* strains such as MRE 600 and Q13.

Determination of Peptidyl-[$^3$H]Puromycin Synthesis

*Reagents*

Polyribosome extracts (150–250 $A_{260}$/ml)
[$^3$H]Puromycin: 40 $\mu M$; specific activity $\geq$ 700 mCi/mmole
5× Standard Buffer: 0.50 *M* KCl; 20 m*M* $MgCl_2$; 0.25 *M* Tris·acetate, pH 7.2

After addition of 0.010, 0.005, and 0.010 ml of the polyribosome extract, [$^3$H]puromycin solution, and 5× standard buffer, respectively, each 0.050-ml reaction mixture contained the following components: 0.10 *M* KCl; 5 m*M* $MgCl_2$; 50 m*M* Tris·acetate, pH 7.2; 10 m*M* $NH_4Cl$ and 0.1% (w/v) Brij 58 contributed by the polyribosome extract; about

[13] S. Pestka and H. Hintikka, *J. Biol. Chem.* **246**, 7723 (1971).
[14] S. Pestka, *Proc. Nat. Acad. Sci. U.S.* **69**, 624 (1972).
[15] S. Pestka, *J. Biol. Chem.* **247**, 4669 (1972).

2–4 $A_{260}$ units of polyribosome extract; and 4 $\mu M$ [$^3$H]puromycin. Unless otherwise noted, the polyribosome extract was added last to start reactions, and tubes were warmed at the temperature of incubation for 1 minute prior to addition of polysomes. Reactions were incubated for 1 minute at 24° or as otherwise indicated. [$^3$H]Puromycin incorporation into nascent polypeptides was measured as follows: reactions were stopped by addition of 2 ml of cold 10% (w/v) trichloroacetic acid and permitted to sit in an ice bath for at least 5 minutes; the contents of each tube were filtered through a polyvinylchloride Millipore filter (type BDWP, 0.6 $\mu$m pore size, 25 mm diameter); each tube and filter were washed three times with 3-ml portions of 5% (w/v) trichloroacetic acid; the filter was washed eight times with 3-ml portions of absolute ethanol at room temperature.[14] The filters were then dried under an infrared lamp and placed in a scintillation fluor; radioactivity was determined as previously described.[14] Extensive washing of these filters with absolute ethanol was necessary to remove completely unreacted [$^3$H]puromycin. In the absence of puromycin, under the conditions of the assay for peptidyl-puromycin synthesis, leucine incorporation into polypeptides was negligible.

## Characteristics of Peptidyl-[$^3$H]Puromycin Formation

The kinetics of peptidyl-[$^3$H]puromycin formation at various temperatures is shown in Fig. 2. The initial rate of peptidyl-puromycin formation is nearly a linear function of temperature throughout the temperature range studied. The extent of peptidyl-puromycin synthesis at 15 minutes also increases with temperature. An Arrhenius plot provides an estimate of 5.6 kcal/mole for the activation energy of peptidyl-puromycin synthesis. The value for 0° (solid square) does not fall on the straight line as the other points representing temperatures higher than 0°.

Since the polysome extract contained 50 m$M$ $NH_4Cl$, all reaction mixtures for evaluating the effect of monovalent cations contained 10 m$M$ $NH_4^+$. Addition of $K^+$ beyond 10 m$M$ $NH_4^+$ concentration increased the rate of peptidyl-puromycin synthesis slightly; nevertheless, peptidyl-puromycin synthesis occurred in the absence of additional $K^+$ or $NH_4^+$ ions (Fig. 3). Optimum $NH_4^+$ was found to be 50 m$M$ to 0.1 $M$ $NH_4^+$. At higher concentrations of $NH_4^+$, the rate of peptidyl-puromycin synthesis was reduced. Addition of $Na^+$ or $Li^+$ was markedly inhibitory to peptidyl-puromycin synthesis.

Optimum magnesium concentration for peptidyl-puromycin synthesis was 4 m$M$ (Fig. 4). At lower or higher concentrations of $Mg^{2+}$, the rate of peptidyl-puromycin synthesis was substantially reduced. In fact, at 20 m$M$ $Mg^{2+}$ the rate of synthesis was about half the rate at 4 m$M$ $Mg^{2+}$.

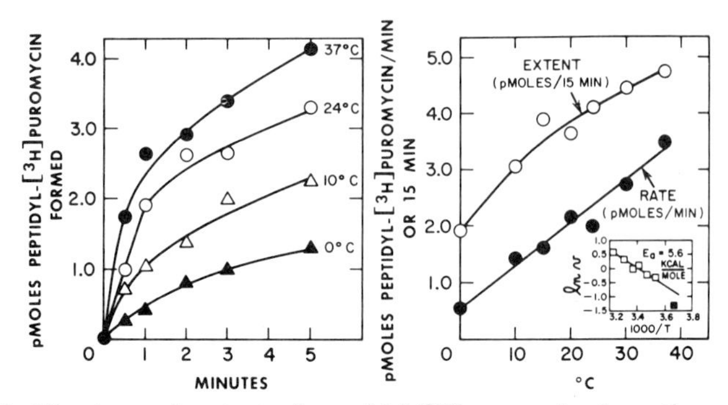


FIG. 2. Kinetics and extent of peptidyl-[$^3$H]puromycin formation at various temperatures. Each 0.050-ml reaction mixture contained the components indicated under "Determination of Peptidyl-[$^3$H]Puromycin Synthesis." Reaction mixtures were incubated at various temperatures and assayed at the times indicated on the abscissa of the left panel. For the determination of the rate of peptidyl-[$^3$H]puromycin formation as a function of temperature (right panel, filled circles, ●), the picomoles of peptidyl-puromycin synthesized per minute was determined from the initial slope of time curves (not all of which are shown in the left panel); this rate is plotted as a function of time in the right panel as picomoles of peptidyl-[$^3$H]puromycin formed per minute. In addition, the extent of peptidyl-puromycin synthesis at 15 minutes is also given at various temperatures (unfilled circles, ○). From the rate of peptidyl-puromycin synthesis at various temperatures ($v$) an Arrhenius plot is derived (inset of right panel); from the slope of the line, the activation energy ($E_a$) of peptidyl-[$^3$H]puromycin synthesis is calculated to be 5.6 kcal/mole. The data for the Arrhenius plot are given as unfilled squares (□) for all temperatures except 0° which is plotted as a solid square (■).

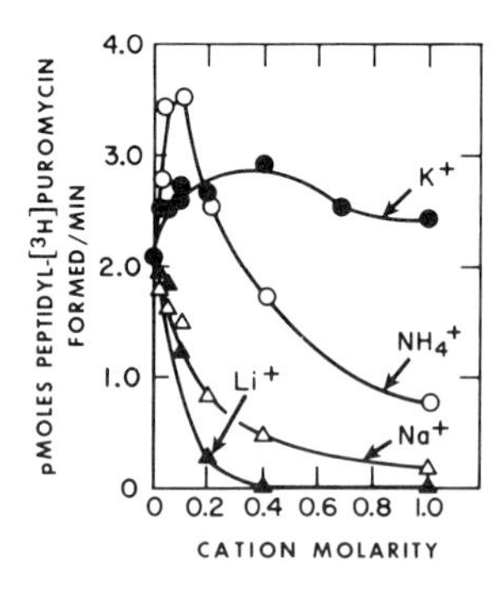


FIG. 3. Effect of monovalent cations on the rate of peptidyl-puromycin formation. Each 0.050-ml reaction mixture contained the following components: 50 m$M$ Tris·acetate, pH 7.2; 5 m$M$ $MgCl_2$; monovalent cation concentration as specified on the abscissa; 10 m$M$ $NH_4Cl$ was present in each reaction mixture as the contribution from the polysome suspension; 2.8 $A_{260}$ units of the polysome preparation; and 4 $\mu M$ [$^3$H]puromycin (977 mCi/mmole). Reactions were incubated at 37° for 1 minute and the rate of peptidyl-[$^3$H]puromycin formation was determined as described in the text.

Maximal rate of peptidyl-puromycin formation occurred at the following cation concentrations: 4 m$M$ $Mg^{2+}$, 1.3 m$M$ $Mn^{2+}$, 1.2 m$M$ $Ca^{2+}$, 10 m$M$ putrescine, and 0.4 m$M$ spermidine. The cations were effective in supporting peptidyl-puromycin synthesis in the following order: $Mg^{2+} > Mn^{2+} >$ putrescine $> Ca^{2+} >$ spermidine. All showed a stimulatory phase at low concentrations and an inhibitory phase at cation concentrations above the optimum. Although excess $Mg^{2+}$ could only partially inhibit peptidyl-puromycin synthesis, excess $Mn^{2+}$ or spermidine produced essentially total inhibition (Fig. 4).

The rate of peptidyl-puromycin formation as a function of pH is shown by the data of Fig. 5. At pH 5, the rate of the reaction is negli-

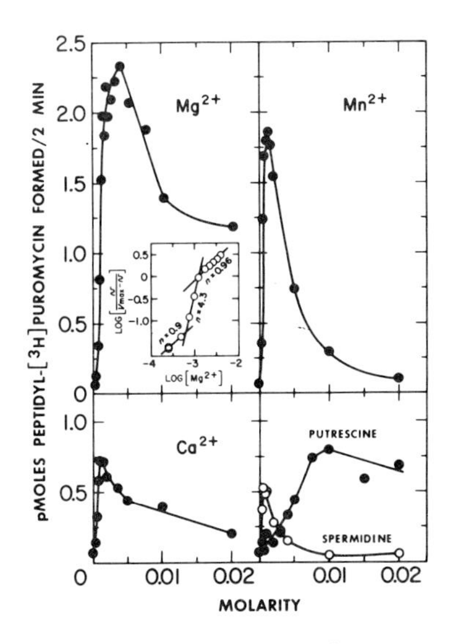


FIG. 4. Effect of divalent cations and oligoamines on rate of peptidyl-puromycin synthesis. Each reaction mixture contained the following in a total of 0.20 ml: divalent cation or oligoamine concentration as indicated on the abscissa; 0.10 $M$ KCl; 50 m$M$ Tris·acetate, pH 7.2; 2.5 $A_{260}$ units of polyribosome preparation; and 1 $\mu M$ [$^3$H]puromycin. Polysomes were added last to start the incubations and reaction mixtures were incubated at 24° for 2 minutes. Formation of peptidyl-[$^3$H]puromycin was determined as described in the text. Reaction volumes were 0.20 ml instead of the usual 0.05-ml so that the $Mg^{2+}$ in the polyribosome preparation could be sufficiently diluted to determine $Mg^{2+}$, $Mn^{2+}$, $Ca^{2+}$, putrescine, and spermidine curves. The final concentration of $Mg^{2+}$ due to $Mg^{2+}$ in the polyribosome preparation was 0.25 m$M$. The cation concentrations on the abscissa refer to the concentrations of the cations in addition to this 0.25 m$M$ $Mg^{2+}$ present in each reaction mixture; however, in the case of $Mg^{2+}$ the actual total $Mg^{2+}$ concentration is plotted. A Hill plot of the data for the $Mg^{2+}$ curve is also presented as an insert to the $Mg^{2+}$ portion of the figure. In these experiments a polyribosome extract containing 250 $A_{260}$ units/ml was used.

gible. The optimum pH for the rate of peptidyl-puromycin formation was found to be 9.1. Above pH 9.1, the velocity of the reaction decreases.

Since it is known from independent experiments that one puromycin molecule reacts with each peptidyl-tRNA (see footnote 11 for a review), the value of the interaction coefficient as 1.0 is consistent with independent determination of this parameter (Fig. 6). The value for the $K_m$ of 3 $\mu M$ is about two orders of magnitude lower than the $K_m$ determined for acetylphenylalanyl or formyl-methionyl-puromycin synthesis.[8,16]

When polyribosomes are centrifuged through a sucrose gradient containing no $K^+$ or $NH_4^+$, the resulting polyribosomes are inactive in syn-

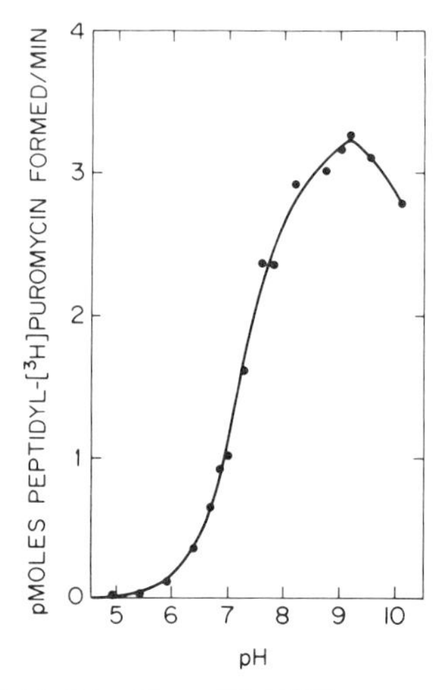


FIG. 5. Effect of pH on rate of peptidyl-[$^3$H]puromycin synthesis. Each 0.20-ml reaction mixture contained the components indicated in the text except that the buffer and volume were varied as follows: Tris concentration was 50 m$M$, and $K^+$ was present as potassium acetate (0.1 $M$); the appropriate pH was obtained by prior adjustment with acetic acid at the temperature of the reaction mixture. This Tris·acetate:potassium acetate combination was useful as a buffer throughout the pH range examined. To slow down the rate of peptidyl-puromycin synthesis, the volume of each reaction mixture was 0.20 ml with the following components present: 0.10 $M$ potassium acetate; 2.5 m$M$ $NH_4Cl$; 4.25 m$M$ $MgCl_2$; 50 m$M$ Tris·acetate, pH as given on the abscissa; 0.025% Brij 58 (w/v); 2.8 $A_{260}$ units of polyribosome extract; and 1 $\mu M$ [$^3$H]puromycin. Reaction mixtures were incubated at 15, 30, and 60 seconds at each pH. From this curve, the initial rate of peptidyl-puromycin synthesis was determined at each pH. The initial rate of peptidyl-puromycin synthesis as a function of pH is presented in the figure.

[16] S. Fahnestock, H. Neumann, V. Shashoua, and A. Rich, *Biochemistry* 9, 2477 (1970).

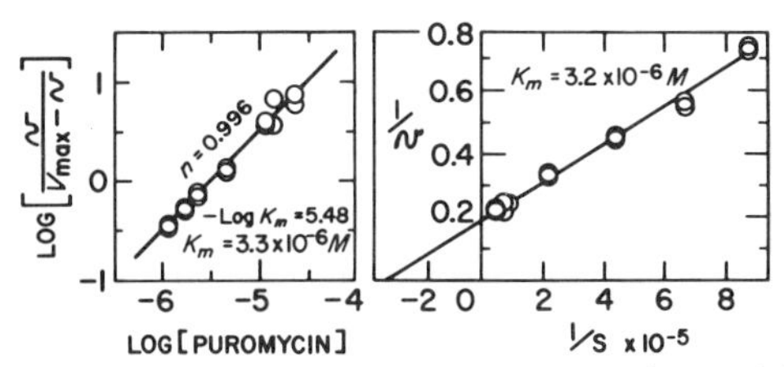


FIG. 6. Hill and double reciprocal plots for puromycin participation in peptidyl-puromycin synthesis. Each 0.050-ml reaction mixture contained the components described in the text except that the puromycin concentration was varied. The data were plotted in the left panel according to the Hill equation. In the right panel, a double reciprocal plot with respect to puromycin as a substrate is plotted. For substrates, $\log[v/(V_{max} - v)] = n \log [S] - \log K$; where $v$ is reaction rate, $V_{max}$ is maximal velocity, $n$ is the interaction coefficient, [S] is substrate concentration and $K$ is equal to $K_m$ (when $n = 1$). For the Hill plot, $K_m$ is 3.3 $\mu M$ by extrapolation to the intercept of the ordinate; and 3.0 $\mu M$ from the intercept of the abscissa (where $\log[v/(V_{max} - v)] = 0$). From the double reciprocal plot, $K_m = 3$ $\mu M$.

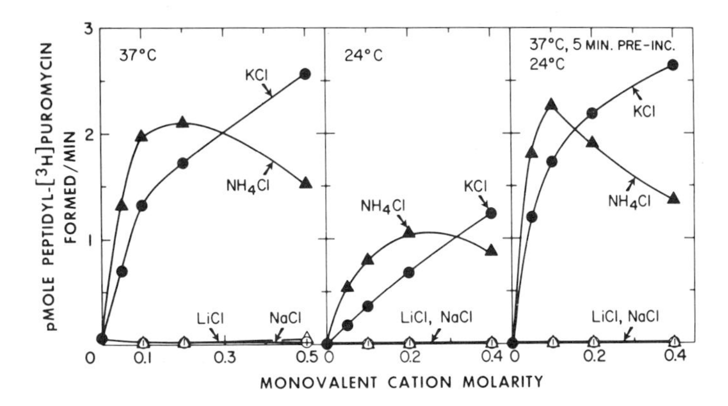


FIG. 7. Effect of monovalent cations on the rate of peptidyl-[³H]puromycin formation with washed polyribosomes. For these experiments, washed polyribosomes were used. These were prepared by layering 0.25 ml of the polyribosome preparation onto a 10% to 30% (w/v) sucrose gradient containing 5 m$M$ $MgCl_2$ and 5 m$M$ Tris·HCl, pH 7.2. The gradient was centrifuged in a Spinco SW 65 rotor at 60,000 rpm for 80 minutes at 0°. The brownish-yellow pellet of polyribosomes was resuspended in 5 m$M$ $MgCl_2$ and 5 m$M$ Tris·HCl, pH 7.2, at a concentration of 54 $A_{260}$ units/ml. Each 0.050-ml reaction mixture contained the following components: 1.0 $A_{260}$ unit of washed polyribosomes; 5 m$M$ Tris·acetate, pH 7.2; 5 m$M$ $MgCl_2$; no Brij 58 was present in reaction mixtures; 4 $\mu M$ [³H]puromycin; and monovalent cation concentration as indicated on the abscissa. Reaction mixtures were incubated at 37° for 1 minute or 24° for 1 minute, the left and center panels, respectively. In the case of the right panel, reaction mixtures containing washed polyribosomes were incubated at 37° for 5 minutes prior to addition of [³H]puromycin; after addition of [³H]puromycin, incubations were performed at 24° for 1 minute.

thesizing peptidyl-puromycin (Fig. 7). In contrast to the results with polyribosomes prepared and stored in the presence of 50 m*M* $NH_4^+$ (Fig. 3) where additional $K^+$ or $NH_4^+$ is only slightly stimulatory, these polyribosomes washed free of $NH_4^+$ are relatively inactive in forming peptidyl-puromycin unless they are activated by prior incubation with relatively high concentrations of $K^+$ or $NH_4^+$ (Fig. 7). That activation rather than a direct effect on the reaction rate is involved is suggested by comparison of the data of Figs. 3 and 7. If polyribosomes are incubated at 37° for 1 minute in the presence of $K^+$ or $NH_4^+$, substantial activity is seen (left panel, Fig. 7); in contrast, however, incubation at 24° for 1 minute produced a relatively small amount of peptidyl-puromycin synthesis (center panel, Fig. 7). If the polyribosomes are incubated at 37° for 5 minutes and then assayed for activity at 24° for 1 minute (right panel, Fig. 7), it can be seen that substantial activity was restored during the 5-minute 37° incubation. Under none of these conditions was $Li^+$ or $Na^+$ active.

| ANTIBIOTIC | SYNTHETIC MODELS FOR PEPTIDE BOND SYNTHESIS | NATIVE PEPTIDE BOND SYNTHESIS |
|---|---|---|
| CHLORAMPHENICOL | ● | ● |
| SPARSOMYCIN | ● | ● |
| AMICETIN | ● | ● |
| GOUGEROTIN | ● | ● |
| BLASTICIDIN S | ● | ● |
| LINCOMYCIN | ● | ○ |
| ERYTHROMYCIN | ◓ | ○ |
| NIDDAMYCIN | ● | ○ |
| CARBOMYCIN | ● | ○ |
| TYLOSIN | ● | ○ |
| SPIRAMYCIN III | ● | ○ |
| PA114 A | ● | ○ |
| VERNAMYCIN A | ● | ○ |
| ALTHIOMYCIN | ◓ | ● |

●, INHIBITION ○, NO EFFECT

FIG. 8. Summary of effects of antibiotics on assays. The column "model systems" refers to synthetic donors such as Ac-Phe-tRNA, fMet-tRNA, polylysyl-tRNA, C-A-C-C-A(Ac-Phe) or C-A-A-C-C-A(fMet). The column native peptide bond synthesis refers to peptidyl-[$^3$H]puromycin synthesis on native polysomes. A filled circle, ●, designates inhibition by that antibiotic; an unfilled circle, ○, designates no effect. For a few of the antibiotics as erythromycin inhibition of peptide bond synthesis in model systems depends on the assay: polylysyl-puromycin synthesis is inhibited, but fMet-puromycin or Ac-Phe-puromycin synthesis is not. For details and a review, see S. Pestka, *Annu. Rev. Microbiol.* **25**, 487 (1971).

A comparison of the effects of antibiotics on model systems for peptide bond synthesis and on peptidyl-[$^3$H]puromycin formation with native polysomes is shown in Fig. 8. Chloramphenicol, sparsomycin, amicetin, gougerotin, and blasticidin S inhibit peptide bond synthesis in both these assays. Lincomycin, the macrolides (erythromycin, niddamycin, carbomycin, tylosin, and spiramycin III), and the streptogramins A (PA114 A and vernamycin A) are potent inhibitors in appropriate model systems, but cannot inhibit peptidyl-puromycin synthesis on native polysomes.[11,15] One antibiotic, althiomycin, was found to inhibit transpeptidation on polyribosomes, but its effect on transpeptidation in model systems seemed to depend on the $K^+$ and $Mg^{2+}$ concentrations. At high $K^+$ (0.4 *M*) or $Mg^{2+}$ (0.04 *M*) concentrations the inhibitory effect of althiomycin on peptide bond formation appears to be reduced or abolished.[17] The correlation of the effects of antibiotics on native polyribosomes with their effects on intact cells or protoplasts is good.[11] Thus, studies with native polyribosomes can be used to predict the effects of antibiotics on intact cells; and these polyribosome extracts can be used to examine in detail the kinetics of transpeptidation.[14,15]

[17] S. Pestka, *in* "Antibiotics, Mode of Action" Vol. II (F. Hahn and J. Corcoran, eds.), in press.

# [60] Separation of Large Quantities of Eukaryotic Ribosomal Subunits by Zonal Ultracentrifugation

*By* Corinne C. Sherton, Ralph F. Di Camelli, and Ira G. Wool

Bacterial ribosomes dissociate into subunits if the concentration of magnesium is lowered,[1] and large quantities of pure ribosomal subunits (2 g) can be separated by centrifugation in a zonal rotor.[2] However, the ribosomes of eukaryotic cells do not dissociate completely when sus-

[1] A. Tissières, D. Schlessinger, and F. Gros, *Proc. Nat. Acad. Sci. U.S.* **46**, 1450 (1960).

[2] E. F. Eikenberry, T. A. Bickle, R. R. Traut, and C. A. Price, *Eur. J. Biochem.* **12**, 113 (1970).



ISBN 0-12-504180-2

pended in low concentrations of magnesium—complete dissociation also requires the addition of a chelating agent.[3-5] Unfortunately, subunits prepared in that way are irreversibly altered; they do not recombine to form 80 S monomers, nor are they active in protein synthesis.[3,4,6]

Eukaryotic ribosomes can be dissociated by high concentrations of potassium (0.3–1.0 *M*) and the subunits will recombine to form monomers which synthesize protein in the presence of added template RNA.[3,7-21] In some cases complete dissociation requires preincubation of the particles with puromycin to remove nascent peptide and messenger RNA.[3,10,16,18,19,22]

Smaller quantities of eukaryotic ribosomal subunits (60–500 mg) have been separated in a zonal rotor.[18,21,23-28] Attempts to separate larger amounts of mammalian subunits have hitherto not been completely suc-

[3] T. E. Martin, F. S. Rolleston, R. B. Low, and I. G. Wool, *J. Mol. Biol.* **43**, 135 (1969).
[4] H. Lamfrom and E. Glowacki, *J. Mol. Biol.* **5**, 97 (1962).
[5] Y. Tashiro and P. Siekevitz, *J. Mol. Biol.* **11**, 149 (1965).
[6] Y. Tashiro and T. Morimoto, *Biochim. Biophys. Acta* **123**, 523 (1966).
[7] T. E. Martin and I. G. Wool, *Proc. Nat. Acad. Sci. U.S.* **60**, 569 (1968).
[8] T. E. Martin and I. G. Wool, *J. Mol. Biol.* 43, 151 (1969).
[9] S. A. Bonanou and H. R. V. Arnstein, *FEBS (Fed. Eur. Biochem. Soc.) Lett.* 3, 348 (1969).
[10] G. R. Lawford, *Biochem. Biophys. Res. Commun.* **37**, 143 (1969).
[11] H. R. V. Arnstein, *Biochem. J.* **117**, 55P (1970).
[12] A. K. Falvey and T. Staehelin, *J. Mol. Biol.* **53**, 1 (1970).
[13] A. M. Reboud, J. P. Reboud, C. Wittmann, and M. Arpin, *Biochim. Biophys. Acta* **213**, 437 (1970).
[14] K. Terao and K. Ogata, *Biochem. Biophys. Res. Commun.* 38, 80 (1970).
[15] E. Bermek, H. Monkemeyer, and R. Berg, *Biochem. Biophys. Res. Commun.* **45**, 1294 (1971).
[16] E. Busiello, M. DiGirolamo, and L. Felicetti, *Biochim. Biophys. Acta* **228**, 289 (1971).
[17] M. S. Kaulenas, *Biochem. Biophys. Res. Commun.* **43**, 1081 (1971).
[18] B. Mechler and B. Mach, *Eur. J. Biochem.* **21**, 552 (1971).
[19] C. H. Faust, Jr. and H. Matthaei, *Biochemistry* **11**, 2682 (1972).
[20] A. M. Reboud, M. Arpin, and J. P. Reboud, *Eur. J. Biochem.* **26**, 347 (1972).
[21] B. A. M. van der Zeijst, A. J. Kool, and H. P. J. Bloemers, *Eur. J. Biochem.* **30**, 15 (1972).
[22] G. Blobel and D. Sabatini, *Proc. Nat. Acad. Sci. U.S.* **68**, 390 (1971).
[23] L. H. Kedes, R. J. Koegel, and E. L. Kuff, *J. Mol. Biol.* **22**, 359 (1966).
[24] E. S. Klucis and H. J. Gould, *Science* **152**, 378 (1966).
[25] S. Bonanou, R. A. Cox, B. Higginson, and K. Kanagalingam, *Biochem. J.* **110**, 87 (1968).
[26] B. M. Mullock, R. Hinton, M. Dobrota, D. Froomberg, and E. Reid, *Eur. J. Biochem.* **18**, 485 (1971).
[27] M. L. Petermann and A. Pavlovec, *Biochemistry* **10**, 2770 (1971).
[28] B. A. van der Zeijst and H. Bult, *Eur. J. Biochem.* **28**, 463 (1972).

cessful.[29] We have now adapted the use of a hyperbolic sucrose density gradient[2] in a Spinco Ti 15 zonal rotor to achieve excellent separation of up to 1.4 g of rat liver or muscle ribosomes.[30]

*Media*

Medium A: tris(hydroxymethyl)aminomethane (Tris)·HCl (20 m$M$), pH 7.8; KCl (830 m$M$); $MgCl_2$ (12.5 m$M$); 2-mercaptoethanol (MSH) (20 m$M$)
Medium B: Tris·HCl (20 m$M$), pH 7.8; KCl (500 m$M$); $MgCl_2$ (3 m$M$); MSH (20 m$M$)
Medium C: Tris·HCl (200 m$M$), pH 7.8; KCl (800 m$M$); $MgCl_2$ (125 m$M$)
Medium D: Tris·HCl (200 m$M$), pH 7.8; KCl (800 m$M$); $MgCl_2$ (30 m$M$)
Medium E: Tris·HCl (10 m$M$), pH 7.6; KCl (80 m$M$); $MgCl_2$ (12 m$M$)
Medium F: Tris·HCl (10 m$M$), pH 7.6; KCl (500 m$M$); $MgCl_2$ (5 m$M$)
Medium G: Tris·HCl (50 m$M$), pH 7.6

*Reagents*

Carbon; decolorizing, alkaline (Norit-A) (Fisher Scientific Co.)
Puromycin dihydrochloride (Nutritional Biochemical Corp.)
Sodium dodecyl sulfate (SDS), 20% in water (Fisher Scientific Co.)

*Preparation of Ribosomes.* We generally prepare muscle[3,31] and liver[3,8] ribosomes from male Sprague-Dawley rats that weigh 100–120 g. A modification[32] which we have found valuable in the preparation of skeletal muscle ribosomes is described in detail elsewhere in this volume.[33] However, any of the standard procedures for the preparation of relatively uncontaminated particles (from any cell type) will do. One should choose conditions of centrifugation likely to yield the maximum number of pure particles. Ribosome pellets can be stored at $-20°$ for several months without loss of activity.

[29] T. E. Martin, I. G. Wool, and J. J. Castles, this series, Vol. 20, p. 417.
[30] C. C. Sherton and I. G. Wool, *J. Biol. Chem.* **247**, 4460 (1972).
[31] W. S. Stirewalt, J. J. Castles, and I. G. Wool, *Biochemistry* **10**, 1594 (1971).
[32] R. Zak, J. Ettinger, and D. A. Fischmann, *in* "Research in Muscle Development and the Muscle Spindle" (R. Pizybylski, J. Vander Meullen, M. Victor, and B. Banker, eds.), p. 163 (Excerpta Med. Found. Int. Congr. Ser. No. 240). Elsevier, Amsterdam, 1971.
[33] C. C. Sherton and I. G. Wool, this volume [61].

TABLE I
PREPARATION OF SOLUTIONS FOR ZONAL SEDIMENTATION IN MEDIUM A

| Solution | Medium C (ml) | KCl (g) | 2.5 $M$ KCl (ml) | 60% sucrose[a] (ml) | 1 $M$ MSH (ml) | Volume[b] (ml) |
|---|---|---|---|---|---|---|
| Medium A (1.0347)[c] | 100 | — | 300 | — | 20 | 1000 |
| 7.4% Sucrose (1.0592) | 100 | — | 300 | 123 | 20 | 1000 |
| 38% Sucrose (1.1803) | 100 | 55 | — | 633 | 20 | 1000 |
| 45% Sucrose (1.2100) | 150 | 84 | — | 1,125 | 30 | 1500 |

[a] The sucrose (60%, w/v in water) was treated with Norit-A as described in the text to remove material absorbing in the ultraviolet.
[b] Made to volume with water.
[c] Density, g/cm³.

## Separation of Ribosomal Subunits

*Preparation of Ribosomes.* Ribosome pellets are suspended in medium A (Table I) or medium B (Table II) by gentle homogenization, and aggregates are removed by centrifugation for 20 minutes at 2000 *g*. The final volume should be 40 ml, and should contain 13,000–15,000 $A_{260}$ units (1.2–1.4 g) of ribosomes. The suspension can be kept in ice until it is loaded into the rotor.

*Preparation of the Gradient.* The sucrose used for the gradients should be freshly prepared and free of material that absorbs in the ultraviolet. Contaminants that absorb in the ultraviolet are removed from a 60% sucrose solution by heating with Norit-A (approximately 70 g per liter) and filtering twice through Whatman No. 1 filter paper; 20 m$M$ MSH is included in the gradient to preserve the activity of the subunits. Since

TABLE II
PREPARATION OF SOLUTIONS FOR ZONAL SEDIMENTATION IN MEDIUM B

| Solution | Medium D (ml) | KCl (g) | 2.5 $M$ KCl (ml) | 60% sucrose[a] (ml) | 1 $M$ MSH (ml) | Volume[b] (ml) |
|---|---|---|---|---|---|---|
| Medium B (1.0243)[c] | 100 | — | 168 | — | 20 | 1000 |
| 7.4% Sucrose (1.0496) | 100 | — | 168 | 123 | 20 | 1000 |
| 38% Sucrose (1.1680) | 100 | — | 168 | 633 | 20 | 1000 |
| 45% Sucrose (1.1938) | 150 | 47 | — | 1,125 | 30 | 1500 |

[a] The sucrose (60%, w/v in water) was treated with Norit-A as described in the text to remove material absorbing in the ultraviolet.
[b] Made to volume with water.
[c] Density, g/cm³.

the MSH is gradually oxidized, only the 45% sucrose solution (in which the ribosomes are never actually suspended during separation) can be frozen, stored, and reused.

A choice between medium A and medium B should be made on the basis of the behavior of the particular ribosome preparation that is to be used (see Remarks). Intact, active subunits can be prepared with either medium, but medium B gives better subunit separation with rat liver and skeletal muscle ribosome preparations and increases the yield 38% for the 40 S subunit and 56% for the 60 S subunit.

*Loading the Zonal Rotor.* A hyperbolic sucrose density gradient is generated in a Spinco Ti 15 zonal rotor using a Beckman Model 141 Gradient Pump with a hyperbolically shaped program cam[2] (the gears are selected so the total delivery will be 2 liters). The heavy solution line is filled with 38% sucrose. The light solution line, the three pump cylinders, and the line leading to the outer fitting of the loading head assembly (peripheral line) are equilibrated with 7.4% sucrose. The program cam is set at 0 (at which setting only 7.4% sucrose from the light solution reservoir is pumped into the rotor). It is the shape of the cam (actually its height as it rotates) which determines the proportions of the total outflow coming from the light and heavy solution lines.

The assembled rotor is placed in a specially adapted Beckman L2-65B centrifuge. With the zonal operation switch set for load and the refrigeration set at 26°, the rotor shield is put in place, and the rotor is started at 2500 rpm. When the rotor is at speed, the water-cooled loading head assembly is attached, and the gradient is generated at a speed control setting of 860 (power switch on low). Thus, the hyperbolic gradient is pumped into the rotor through the peripheral line, beginning with 7.4% and ending with 38% sucrose.

The formation of the gradient is complete when a total of 795 ml of both sucrose solutions have entered the pump, although it remains for a portion of the gradient to be transferred from the pump to the rotor. The heavy solution line is now transferred to the 45% sucrose reservoir. (The time at which the change is made can be indicated with a mark on the program cam.) While that is being done, the pump should be turned off and care taken that no air bubbles enter the line. Pumping is resumed. When the mixing of the gradient is completed (maximum height of the program cam) the program cylinder is arrested, so that only 45% sucrose enters the mixing cylinder. Pumping is continued until the rotor is full (total capacity 1665 ml), and the top of the gradient flows out the line from the center fitting of the loading head assembly (central line). The inner sector of the rotor now contains the 795 ml of hyperbolic gradient, and the outer sector the 870 ml of 45% sucrose cushion.

The peripheral line is disconnected from the pump. The continuous output and mixing cylinders and the heavy solution line are then equilibrated with medium B (the program cylinder remains arrested). A syringe containing 20 ml of 7.4% sucrose solution is attached to the end of the central line. The end of the peripheral line is placed in a beaker containing 45% sucrose solution. Bubbles that have accumulated under the outer rim of the rotor are now removed by slowly pushing them out with the 7.4% sucrose in the syringe, and then pulling 45% sucrose solution back in through the peripheral line. This operation should be repeated 2 to 3 times, until no bubbles are seen.

The suspended ribosomes which have been stored on ice are now dissociated by incubation for 15 minutes at 37° in 0.1 m*M* puromycin. The sample is introduced onto the hyperbolic gradient as a linear inverse gradient, produced with a two-chambered linear gradient maker. The mixing chamber contains 40 ml of 7.4% sucrose and the second chamber contains the 40 ml of ribosome suspension. The mixing chamber is connected to the central line, and the sample is slowly loaded onto the hyperbolic gradient by pulling 45% sucrose solution out of the rotor through the peripheral line, using a syringe.

The sample is overlayed with 680 ml of medium B pumped in through the central line (speed control 230 low for first 200 ml; 460 low for remainder), as 45% sucrose is removed from the peripheral line. The first 100 ml of the latter are combined with the 45% sucrose solution

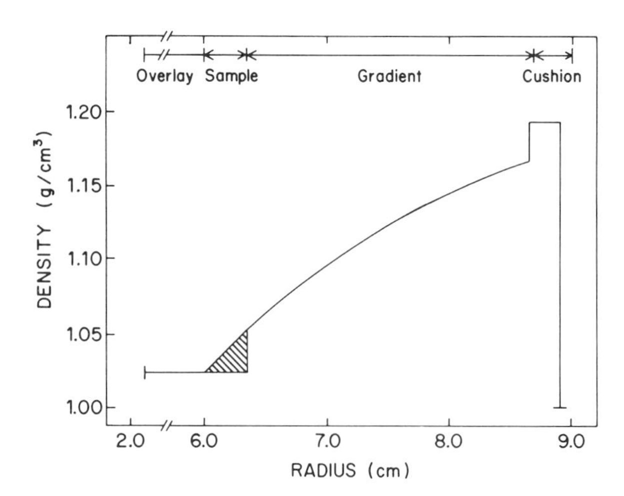


FIG. 1. Diagram of the configuration in the Spinco Ti 15 rotor prior to zonal sedimentation of ribosomal subunits. The densities are of sucrose solutions in medium B at room temperature (23°). The shaded area represents 1.35 g of ribosomes in an inverse linear sucrose gradient.

displaced during loading of the sample, and the absorption at 260 nm is determined to discover whether there has been a leak across the rotating rulon seal from the central line to the peripheral one. The loading head is then removed, and the rotor cap is attached to seal the rotor.

A continuous gradient of sucrose has been formed, starting at a radius of 6.0 cm from the rotor core, with the sample contained in the first 90 ml of linear gradient. At 6.35 cm the concentration of sucrose in the gradient is 7.36%, and at 8.66 cm it is 37.2%. Between 8.66 and 8.89 cm is a cushion of 45% sucrose solution. The relationships are shown diagrammatically in Fig. 1.

After a vacuum is formed, centrifugation is at 13,500 rpm for 17 hours at 26°. The zonal operation switch may now be placed on run, activating the centrifuge's normal safety circuits. If there has been a slight leak (2–3% of the total sample) across the central seal of the rotating rulon seal in the loading head assembly, the total running time should be increased by 10% because not all the buffer overlay was actually introduced into the rotor.

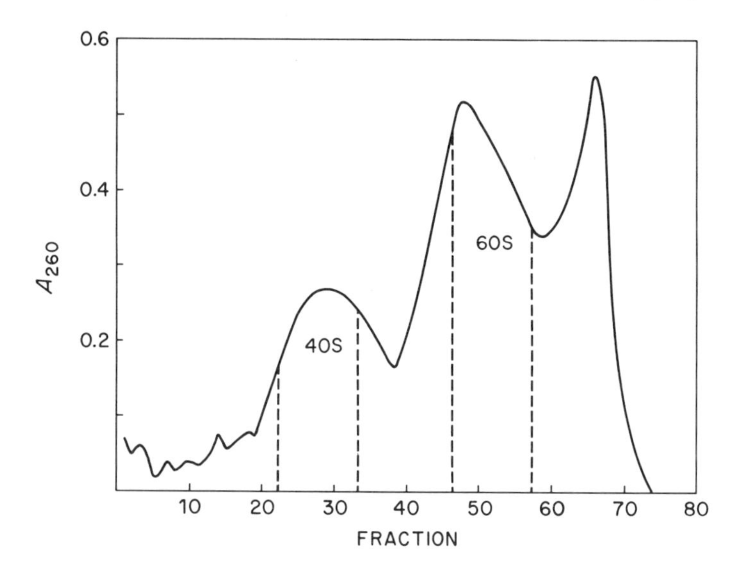


FIG. 2. Sedimentation of liver ribosomal subunits in a zonal rotor [C. C. Sherton and I. G. Wool, *J. Biol. Chem.* **247**, 4460 (1972)]. Liver ribosomal subunits (1.38 g) were separated by centrifugation for 17 hours at 13,500 rpm in a Spinco Ti 15 zonal rotor using a 7.4 to 38.0% hyperbolic sucrose density gradient [E. F. Eikenberry, T. A. Bickle, R. R. Traut, and C. A. Price, *Eur. J. Biochem.* **12**, 113 (1970)] in medium A. Fractions (10 ml) were collected, and those indicated by the interrupted lines were pooled; the other fractions were discarded; 117 mg of 40 S and 243 mg of 60 S subunits were recovered.

*Unloading the Zonal Rotor.* When the separation is completed the rotor speed is reduced to 2500 rpm; the operation switch is placed on load, the vacuum is released, and the rotor cap is removed. The pump is connected to the peripheral line and is equilibrated with untreated 60% sucrose (w/v. in water)—a total of 2.5 liters is required for unloading. The loading head is attached to the rotor, and the contents of the rotor are displaced with 60% sucrose.

A total of 1 liter of solution is removed through the central line and discarded (the first 200 ml are at a speed control setting of 230 low; the second 200 ml at 460 low; the remainder at 895 low). Now 10-ml fractions are collected (the flow rate is 40 ml per minute) until the contents of the rotor have been completely displaced by 60% sucrose (a total of some 75–80 fractions). A 1:50 dilution of each fraction is made with a 1:1 mixture of the 7.4% and 38% sucrose solutions, and the absorption at 260 nm determined.

*Precipitation of Ribosomal Subunits.* The fractions containing the subunits, either 40 S or 60 S (Figs. 2 and 3), are pooled and dialyzed against

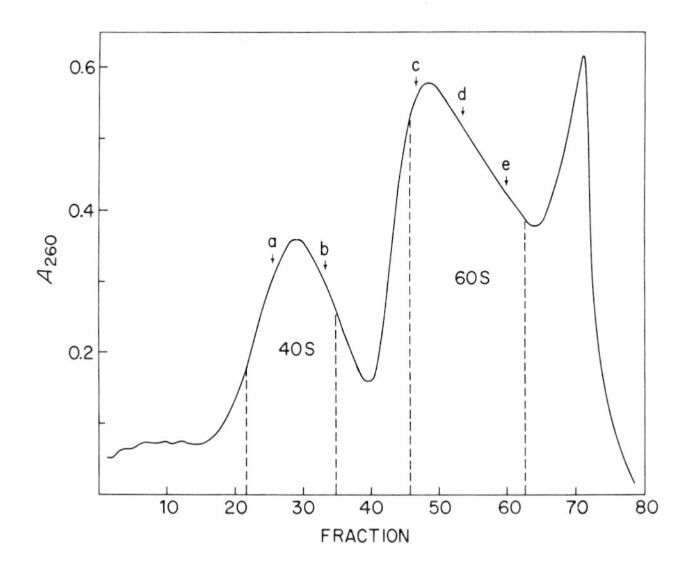


FIG. 3. Sedimentation of liver ribosomal subunits in a zonal rotor. Liver ribosomal subunits (1.35 g) were separated by centrifugation for 18.75 hours at 13,500 rpm in a Spinco Ti 15 zonal rotor using a 7.4 to 38.0% hyperbolic sucrose density gradient [E. F. Eikenberry, T. A. Bickle, R. R. Traut, and C. A. Price, *Eur. J. Biochem.* **12**, 113 (1970)] in medium B. The fractions indicated by the interrupted lines were pooled; 162 mg of 40 S and 413 mg of 60 S subunits were recovered. The analysis of fractions a–e is in Fig. 5.

8 liters of medium E for at least 36 hours with 3 to 4 changes of buffer. Dialysis is necessary because high concentrations of sucrose and potassium interfere with precipitation of ribosomes by ethanol.[34] Cold 95% ethanol (0.2 of a volume) is added to the dialyzed subunits and the suspension is kept at 0° for at least 1 hour (to ensure complete precipitation of the particles). The subunits are collected by centrifugation at 10,000 *g* for 15 minutes; they may be stored at −20°.

### Purification of 60 S Subunits

The 60 S subunits isolated in this way are contaminated with small amounts of 40 S subunits and for some purposes must be resolved further.[30] Fractions containing 60 S subunits collected from centrifugation in a zonal rotor (about 740 mg) are incubated again for 15 minutes at 37° in medium A or B containing 0.1 m*M* puromycin. They are recentrifuged in the same way in a zonal rotor using a hyperbolic sucrose

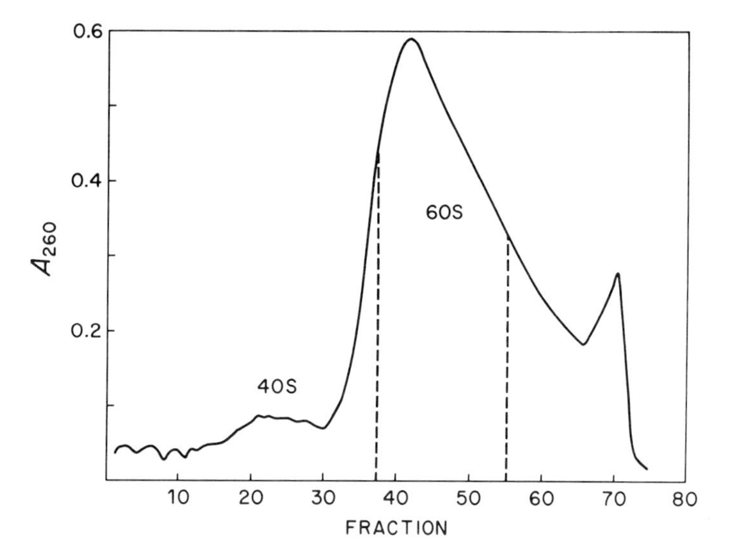


FIG. 4. Sedimentation of 60 S subunits of liver ribosomes in a zonal rotor [C. C. Sherton and I. G. Wool, *J. Biol. Chem.* **247**, 4460 (1972)]. Fractions containing 736 mg of 60 S subunits from three previous centrifugations in a zonal rotor were pooled and centrifuged again for 17 hours at 13,500 rpm in a Spinco Ti 15 zonal rotor using a 7.4 to 38.0% hyperbolic sucrose density gradient [E. F. Eikenberry, T. A. Bickle, R. R. Traut, and C. A. Price, *Eur. J. Biochem.* **12**, 113 (1970)] in medium A. The fractions indicated by the interrupted lines were pooled; 324 mg of 60 S subunits were recovered.

[34] M. S. Kaulenas, *Anal. Biochem.* **41**, 126 (1971).

gradient in either medium A or B, and fractions are collected (Fig. 4), dialyzed, and precipitated with ethanol as described above.

Analysis of the Purity of Ribosomal Subunits

The purity of the subunit fractions can be determined by zonal centrifugation (Fig. 5). Samples in medium E are layered on 5.2 ml of a 10 to 30% linear sucrose gradient in medium F. Centrifugation is at 60,000 rpm for 40 minutes at 26° in a Spinco SW 65 rotor, or at 48,000 rpm for 55 minutes at 26° in a Spinco SW 50.1 rotor. The distribution of particles is determined with an ISCO (Instrument Specialities Company) density gradient fractionator and UV analyzer.

The 40 S subunit fractions from the first centrifugation in the zonal rotor are generally free of contamination with 60 S subunits (Fig. 5,a and b). The slower sedimenting 60 S subunit fractions from the first centrifugation in the zonal rotor show some contamination with 40 S particles (Fig. 5c). The faster sedimenting 60 S subunit fractions have decreasing 40 S contamination, but contain a small amount of 90 S particles–60 S dimers (Fig. 5,d and e). The 60 S particles which have been purified by a second centrifugation in the zonal rotor are free of contamination (Fig. 6a).

Another method of determining whether subunit preparations are contaminated is analysis of their RNA.[3] Sufficient 20% SDS is added to a suspension of ribosomal particles in medium E to give a final concentration of 0.1%. The sample is incubated for 5 minutes at 37°, and then layered onto 5.2 ml of a 5 to 20% linear sucrose gradient in medium G. Centrifugation is at 4° for 2 hours at 60,000 rpm in a Spinco SW 65 rotor or for 2.75 hours at 48,000 rpm in a Spinco SW 50.1 rotor. The presence of 18 S (small subunit) and 28 S (large subunit) RNA is determined with an ISCO density gradient fractionator and UV analyzer.

The 40 S subunit fractions from the first zonal centrifugation contain only 18 S RNA (Fig. 5,f and g). The 60 S subunit fractions from the first zonal centrifugation contain predominantly 28 S RNA, but also a small amount of 18 S RNA, and usually some RNA that sediments between 18

FIG. 5. Sedimentation in sucrose gradients of subunit fractions and their RNA. The purity of the subunit fractions collected from the zonal centrifugation depicted in Fig. 3 was determined (a–e) after dialyzing the samples against medium E. The samples were analyzed on 10 to 30% linear sucrose gradients in medium F, centrifugation was at 48,000 rpm for 55 minutes at 26° in a Spinco SW 50.1 rotor. The RNA of the subunit fractions was analyzed (f–j) after suspending the particles in medium G with 0.1% SDS, and incubating for 5 minutes at 37°. The samples were centrifuged in 5 to 20% linear sucrose density gradients in medium G at 48,000 rpm for 2.75 hours at 4° in a Spinco SW 50.1 rotor.

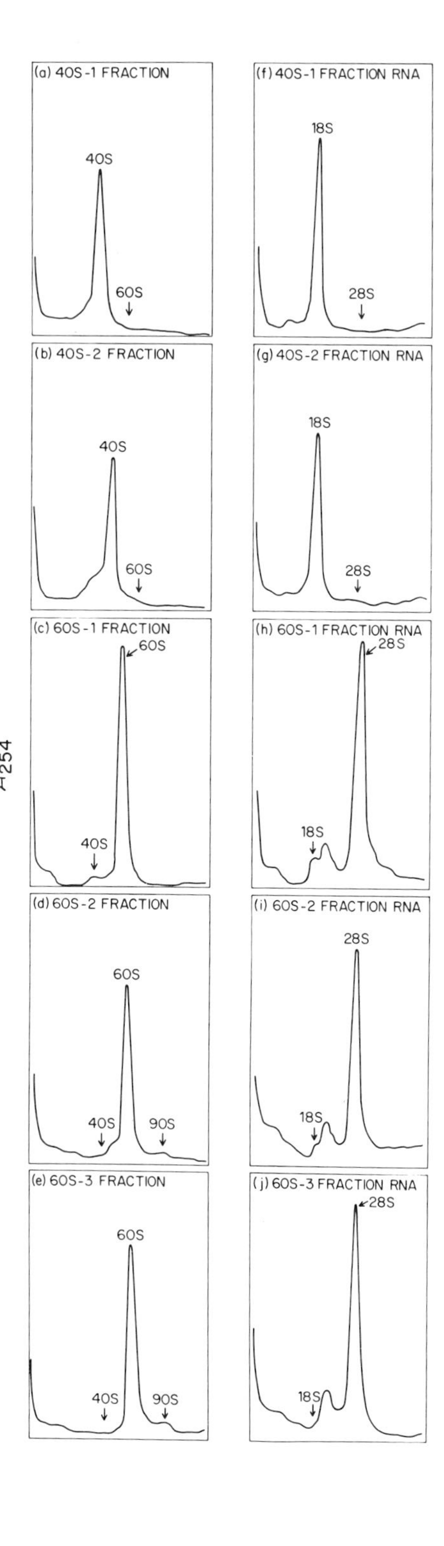

(a) 40S-1 FRACTION
40S
60S
(b) 40S-2 FRACTION
40S
60S
(c) 60S-1 FRACTION
60S
40S
(d) 60S-2 FRACTION
60S
40S
90S
(e) 60S-3 FRACTION
60S
40S
90S
(f) 40S-1 FRACTION RNA
18S
28S
(g) 40S-2 FRACTION RNA
18S
28S
(h) 60S-1 FRACTION RNA
28S
18S
(i) 60S-2 FRACTION RNA
28S
18S
(j) 60S-3 FRACTION RNA
28S
18S
$A_{254}$

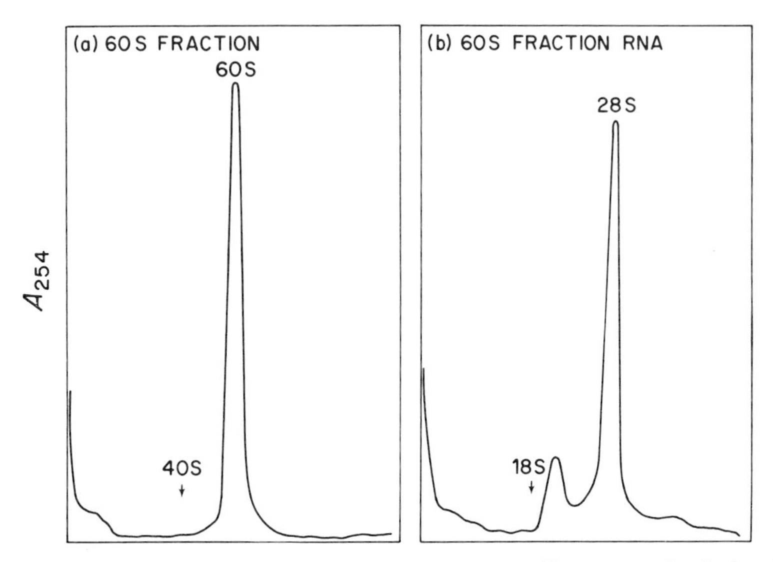


FIG. 6. Sedimentation in sucrose gradients of a purified 60 S subunit fraction and its RNA. The 60 S subunits collected from the zonal centrifugation depicted in Fig. 4 were analyzed as in Fig. 5 to determine the purity of the particles and their RNA.

and 28 S; the latter is probably from breakdown of 28 S RNA[3,20] (Fig. 5,h–j). The 60 S particles which have been purified by a second centrifugation in the zonal rotor are free of contamination with 18 S RNA (Fig. 6b).

## Remarks

Active ribosomal subunits have been obtained from a variety of eukaryotic cells, including rat liver[8,10,13,14,20,22,27] and skeletal muscle,[3,7,8] mouse liver[12] and plasmocytoma tumor,[18,19] rabbit reticulocytes[9,11,16] and skeletal muscle,[8] dog pancreas,[35] human tonsil[15] and reticulocytes,[36] HeLa cells,[37] insects,[17,34] yeast,[21] and protozoa.[8] Ribosomal subunits have been separated by unfolding in urea[27] and by ion-exchange chromatography[35]; but in all other cases, modifications of the original Martin and Wool procedure[3] of centrifugation through high concentrations of potassium were used. In selecting a method for separating large quantities of eukaryotic ribosomal subunits, it is important to consider the best means of dissociating the ribosomes, the optimal conditions for the separation of the sub-

[35] S. R. Dickman and E. Bruenger, *Biochemistry* **8**, 3295 (1969).
[36] J. E. Fuhr, C. Natta, A. Bank, and P. A. Marks, *Biochim. Biophys. Acta* **240**, 70 (1971).
[37] E. A. Zylber and S. Penman, *Biochim. Biophys. Acta* **204**, 221 (1970).

units, and the type of zonal centrifugation that will yield the largest amounts of pure subunits.

Polysomes are resistant to dissociation. Several methods have been used to reduce their numbers and hence to increase the yield of subunits. Incubation of HeLa cells at 42° greatly reduces the proportion of polysomes present[37]; incubation at 0 to 4° has the same effect on cricket nymph[17] and human reticulocyte[36] polysomes. The most efficient, and the most generally used, method of dissociating polysomes is incubation with puromycin to release the nascent peptide and probably the messenger RNA. Incubation, generally with 0.1–0.5 m*M* puromycin, may be in high concentrations of potassium[19,22,30,37] as described here, or at low monovalent cation concentrations in the presence of supernatant factors and an energy generating system.[3,10,15,16,18] Falvey and Staehelin,[12] prepared subunits from "runoff" ribosomes by first incubating mouse liver polysomes in the presence of all the components necessary for protein synthesis; they used that procedure because they found that subunits prepared with puromycin were less active. Nonetheless, our method does produce active subunits and has the advantage of not requiring the addition of extraneous proteins during the separation procedure.

The conditions of centrifugation are also important for the separation of active subunits in good yield. Centrifugation is at 26° because at 4° 40 S and 60 S rat liver and muscle subunits form 55 S and 90 S aggregates,[3,8] respectively. However, others have been able to isolate eukaryotic ribosomal subunits at lower temperatures (0 to 10°),[12,17–19,21,37] and mouse plasmocytoma subunits are less active when prepared at 20° rather than at 5°.[19]

One or more ribosomal proteins has sulfhydryl groups which must be kept reduced if the function of the particles is to be preserved.[15,38] Active subunits have been prepared without addition of either 2-mercaptoethanol or dithiothreitol,[11,12,17,18,37] even though a sulfhydryl reducing reagent has been shown to be an absolute requirement for the preparation of active subunits by others.[3,7,8,21]

The separation of subunits is generally in relatively high concentrations of potassium and low concentrations of magnesium ions (the use of other mono- and divalent cations has been investigated[37,39]). It is thought that potassium replaces magnesium ions which stabilize the interaction between the subunits. Thus, the ratio of potassium to magnesium ions must be high enough so that dissociation is complete.[10,18,19,22] Furthermore, the concentration of potassium ions must be high enough to re-

[38] B. S. Baliga and H. N. Munro, *Nature* (*London*) *New Biol.* **233**, 257 (1971).

[39] A. M. Reboud, M. Buisson, and J. P. Reboud, *Eur. J. Biochem.* **26**, 354 (1972).

move factors or supernatant proteins adventitiously bound to the surface of the subunit,[11,39,40] yet not so high as to extract ribosomal proteins.[13,18-20] On the other hand, the magnesium concentration must not be so low as to destabilize the structure of the ribosomal subunit, and thus produce irreversible unfolding and the release of the 5 S RNA from the 60 S subunit.[10,18,19] We believe that the use of 500 m*M* potassium and 3 m*M* magnesium (medium B) best fits these criteria. The ratio of potassium to magnesium concentrations in medium B is 167; high enough to produce complete dissociation of the particles and better separation of the subunit peaks than is obtained with 830 m*M* potassium and 12.5 m*M* magnesium (medium A) (cf. Figs. 2 and 3). The subunits prepared in medium B are intact (cf. Fig. 5), and do not contain proteins removed by higher concentrations of potassium.[41]

Centrifugation in a zonal rotor is the only practicable method to prepare large quantities of ribosomal subunits. Linear sucrose density gradients allow zone broadening (sectorial dilution) which may result in contamination of subunit fractions; moreover, they can only be used to separate at the most 300 mg of ribosomes.[23-27] Earlier we reported a rather unsuccessful attempt to separate 1.0 g of rat liver ribosomes using a linear gradient.[29] Equivolumetric sucrose density gradients—ones in which sample particles of like density pass through a constant volume of gradient per unit time—have been devised to eliminate the problem of sectorial dilution.[42] However, they have been successfully used only for the separation of up to 130 mg of ribosomes.[21,28] The hyperbolic sucrose density gradient developed by Eikenberry *et al.*—in which an initially stable zone, such as the ribsome sample in its inverse gradient, does not undergo sectorial dilution—has been used by them to separate up to 2 g of *E. coli* ribosomes.[2] The same gradient has also been used to separate 500 mg of mouse plasmocytoma ribosomes.[18] We have modified the concentration of ions in the hyperbolic gradient, added a sulfhydryl reducing agent, and adjusted the temperature to achieve an efficient means for the preparation of eukaryotic ribosomal subunits; the method allows for the preparation of large quantities of pure, intact subunits from rat liver or muscle ribosomes.[30]

[40] S. M. Heywood, *Cold Spring Harbor Symp. Quant. Biol.* **34**, 799 (1969).
[41] C. C. Sherton and I. G. Wool, unpublished results (1972).
[42] M. S. Pollack and C. A. Price, *Anal. Biochem.* **42**, 38 (1971).

# [61] Two-Dimensional Polyacrylamide Gel Electrophoresis of Eukaryotic Ribosomal Proteins

*By* CORINNE C. SHERTON and IRA G. WOOL

Most ribosomal proteins are basic and most lie in a narrow range of molecular sizes: they are, therefore, difficult to separate. Moreover, ribosomal proteins have no catalytic activity so their isolation, purification, and characterization cannot be followed by enzyme assay. The only feasible alternative is electrophoresis. A number of procedures have been used: two-dimensional polyacrylamide gel electrophoresis, developed by Kaltschmidt and Wittmann,[1] has expanded the resolving power of the technique and has proved valuable for analysis of prokaryotic[1-7] and eukaryotic[8-17] ribosomal proteins. For example, all 55 *Escherichia coli* ribosomal proteins can (with one exception) be displayed as individual spots on a single gel slab.[2]

We have adapted the Kaltschmidt and Wittmann procedure for our purposes and used it to analyze rat liver and skeletal muscle ribosomal proteins. We shall describe the technique in detail.

[1] E. Kaltschmidt and H. G. Wittmann, *Anal. Biochem.* **36**, 401 (1970).

[2] E. Kaltschmidt and H. G. Wittmann, *Proc. Nat. Acad. Sci. U.S.* **67**, 1276 (1970).

[3] M. Dzionara, E. Kaltschmidt, and H. G. Wittmann, *Proc. Nat. Acad. Sci. U.S.* **67**, 1909 (1970).

[4] E. Kaltschmidt, *Anal. Biochem.* **43**, 25 (1971).

[5] H. E. Homann and K. H. Nierhaus, *Eur. J. Biochem.* **20**, 249 (1971).

[6] E. Kaltschmidt and H. G. Wittmann, *Biochimie* **54**, 167 (1972).

[7] G. Stöffler and H. G. Wittmann, *J. Mol. Biol.* **62**, 407 (1971).

[8] O. H. W. Martini and H. J. Gould, *J. Mol. Biol.* **62**, 403 (1971).

[9] T. Hultin and A. Sjöqvist, *Anal. Biochem.* **46**, 342 (1972).

[10] B. L. Jones, N. Nagabhushan, A. Gulyas, and S. Zalik, *FEBS (Fed. Eur. Biochem. Soc.) Lett.* **23**, 167 (1972).

[11] H. Welfle, J. Stahl, and H. Bielka, *Biochim. Biophys. Acta* **243**, 416 (1971).

[12] H. Welfle, *Acta Biol. Med. Ger.* **27**, 547 (1971).

[13] H. Bielka, J. Stahl, and H. Welfle, *Arch. Geschwultsforsch.* **38**, 109 (1971).

[14] Huynh-Van-Tan, J. Delaunay, and G. Schapira, *FEBS (Fed. Eur. Biochem. Soc.) Lett.* **17**, 163 (1971).

[15] J. Delaunay and G. Schapira, *Biochim. Biophys. Acta* **259**, 243 (1972).

[16] C. C. Sherton and I. G. Wool, *J. Biol. Chem.* **247**, 4460 (1972).

[17] H. Welfle, J. Stahl, and H. Bielka, *FEBS (Fed. Eur. Biochem. Soc.) Lett.* **26**, 228 (1972).




ISBN 0-12-504180-2

Preparation of Ribosomes and Ribosomal Subunits

*Media*

Medium A: KCl, 100 m*M*; $MgCl_2$, 5 m*M*; EGTA, 5 m*M*; sodium pyrophosphate, 5 m*M*, pH 6.8. N.B.: The sodium pyrophosphate is added just before using.
Medium B: tris(hydroxymethyl)aminomethane (Tris)·HCl, 50 m*M*, pH 7.6; KCl, 250 m*M*; $MgCl_2$, 12.5 m*M*; EGTA, 5 m*M*; sucrose, 0.25 *M*
Medium C: Tris·HCl, 20 m*M*, pH 7.8; KCl, 80 m*M*; $MgCl_2$, 12.5 m*M*
Medium D: Tris·HCl, 20 m*M*, pH 7.8; KCl, 880 m*M*; $MgCl_2$, 12.5 m*M*; β-Mercaptoethanol (MSH), 20 m*M*, is included when the medium is used to dissociate ribosomes. The presence of a sulfhydryl-protecting agent is necessary to preserve the activity of subunits.
Medium E: Tris·HCl, 10 m*M*, pH 7.6; KCl, 80 m*M*; $MgCl_2$, 12 m*M*
Medium F: Tris·HCl, 10 m*M*, pH 7.6; KCl, 500 m*M*; $MgCl_2$, 5 m*M*
Medium G: Tris·HCl, 50 m*M*, pH 7.6
Medium H: Tris·HCl, 10 m*M*, pH 7.7; magnesium acetate, 100 m*M*

*Reagents*

Lubrol WX, 10% in 10 m*M* $MgCl_2$ (General Biochemicals)
Sodium deoxycholate (DOC), 10% in water (Nutritional Biochemicals Corp.)
Sodium dodecyl sulfate (SDS), 20% in water (Fisher)
Puromycin dihydrochloride (Nutritional Biochemicals Corp.)
Ethyleneglycol-bis(β-aminoethyl ether) *N,N'*-tetraacetic acid (EGTA), 50 m*M* (Sigma)
Sodium pyrophosphate, 50 m*M* (Fisher)
ATP, 0.1 *M* (Sigma)
GTP, 5 m*M* (Schwarz/Mann)
Solid phosphoenolpyruvate (Calbiochem Corporation)

*Preparation of Ribosomes.* We generally prepare muscle[18,19] and

[18] T. E. Martin, F. S. Rolleston, R. B. Low, and I. G. Wool, *J. Mol. Biol.* **43**, 135 (1969).
[19] W. S. Stirewalt, J. J. Castles, and I. G. Wool, *Biochemistry* **10**, 1594 (1971).

liver[18,20] ribosomes from male Sprague-Dawley rats of about 100–120 g; however, any of the standard procedures for the preparation of relatively uncontaminated particles (from any cell type) will do. One should choose conditions of centrifugation likely to yield the maximum number of pure particles.

A modification[21] that we have found valuable in the preparation of skeletal muscle ribosomes is the following: Finely minced skeletal muscle is stirred for 10 minutes in 2.5 volumes of medium A. The medium is drained through a stainless steel strainer and the muscle is stirred again for 10 minutes in 2.5 volumes of fresh medium A. The medium is removed, and the tissue is washed several times with medium B. EGTA chelates calcium ions and causes muscle to relax: the result is that the tissue is considerably softer (because of a lack of rigor) and hence easier to homogenize. The yield of ribosomes is almost doubled.

The muscle is now homogenized in 2 volumes of medium B in a Virtis 45 homogenizer—30 seconds, at a setting of 85. The homogenate is centrifuged (in a Sorvall model RC-2 centrifuge) in 250 ml polycarbonate bottles in a GSA rotor for 15 minutes at 13,000 *g*. The supernatant is filtered through glass wool and set aside; the pellet is resuspended in an amount of medium B equal to the original tissue volume and centrifuged at 13,000 *g* for 15 minutes. The supernatant is filtered through glass wool and retained. The pellet is discarded. The combined supernatant is centrifuged in a Spinco 30 rotor at 30,000 rpm for 2 hours in a Spinco Model L-2 centrifuge. The supernatant is decanted and discarded. The pellets are homogenized in medium C containing 0.25 *M* sucrose to which is added 6 ml of Lubrol WX (10% in 10 m*M* $MgCl_2$) and 12 ml of freshly prepared DOC (10% in water) for each 100 ml of medium C. Homogenization is with a motor-driven Teflon pestle and a size C glass tissue grinder. The homogenate is centrifuged in a Sorvall GSA rotor at 13,000 *g* for 15 minutes. The supernatant (7 ml) is layered on 5 ml of medium C in 0.5 *M* sucrose and centrifuged for 2.5 hours at 40,000 rpm in a Spinco 40 rotor in a Spinco L-2 centrifuge. The supernatant is carefully aspirated so that the DOC at the top does not contaminate the ribosomal pellet; the tubes are drained by inverting them, and the sides are dried. The ribosomal pellets may be stored at $-70°$ for several months without loss of activity.

The particles used to prepare total ribosomal protein for electrophoresis should be as free from contamination as possible. It is helpful to incubate ribosome preparations to remove nascent peptide and endogenous mRNA.[20]

[20] T. E. Martin and I. G. Wool, *J. Mol. Biol.* **43**, 151 (1969).

[21] R. Zak, P. Etlinger, and D. A. Fischmann, *Res. Muscle Develop. and Muscle Spindle, Excerpta Med. Found. Int. Congr. Ser.* **240**, p. 163.

(Incubation is not necessary if the ribosomes are to be used to prepare subunits—see below.) Ribosomes are suspended by gentle homogenization in medium C, and incubated in medium C (7 mg of ribosomes per milliliter) containing: puromycin (0.1 m*M*); ATP (5 m*M*); GTP (0.5 m*M*); phosphoenolpyruvate (46.5 $\mu$g/ml); MSH (10 m*M*); and liver or muscle supernatant (G-25 fraction[22]—cf. this volume [18]). It is important that muscle supernatant be prepared in medium C (with 0.25 *M* sucrose) rather than medium B.[23] The higher concentration of KCl (250 m*M*) in medium B solubilizes myosin which interferes with the preparation of G-25 fraction. The ribosome suspension is incubated for 30 minutes at 37°. Ribosomal aggregates are removed by centrifugation at 12,000 *g* for 10 minutes. The ribosomes in the supernatant are collected by sedimentation (45,000 rpm for 5.5 hours at 4° in a Spinco Ti 50 rotor) through 4 ml of 0.5 *M* sucrose in medium D. The high concentration (880 m*M*) of KCl in medium D removes extraneous proteins as well as initiation and elongation factors bound to the ribosomes. The particles are referred to as stripped ribosomes.

*Preparation of Ribosomal Subunits*. Liver ribosomes (1–1.4 g is a convenient amount) that have not been stripped are dissociated by incubation for 15 minutes at 37° in medium D containing 0.1 m*M* puromycin and 20 m*M* MSH. The ribosomal subunits are separated by centrifugation at 13,500 rpm for 17 hours at 26° in a Spinco Ti 15 zonal rotor using a hyperbolic sucrose density gradient in medium D containing 20 m*M* MSH.[24] The gradient is formed over a cushion of 45% sucrose in the same medium; the concentration of sucrose in the gradient is 7.36% at 6.35 cm, and 37.2% at 8.66 cm from the rotor core. The sample, in a linear sucrose gradient of 0 to 7.4%, is layered onto the hyperbolic gradient. We have found that the yield of subunits from muscle ribosomes is improved if the concentration of potassium in the sucrose gradient is decreased from 0.88 to 0.5 *M* and that of magnesium from 12.5 to 3 m*M*.

After centrifugation the gradient is displaced with 60% sucrose and 10 ml fractions are collected. The fractions containing the subunits, either 40 S or 60 S (Fig. 1), are pooled and dialyzed against 8 liters of medium E for at least 36 hours with 3–4 changes of buffer. Sucrose and high concentrations of potassium interfere with precipitation of ribosomes by

[22] D. P. Leader, I. G. Wool, and J. J. Castles, *Proc. Nat. Acad. Sci. U.S.* **67**, 523 (1970).

[23] J. J. Castles and I. G. Wool, *in* "Protein Biosynthesis in Non-Bacterial Systems" (J. A. Last and A. I. Laskin, eds.), p. 1. Dekker, New York, 1972.

[24] E. F. Eikenberry, T. A. Bickle, R. R. Traut, and C. A. Price, *Eur. J. Biochem.* **12**, 113 (1970).

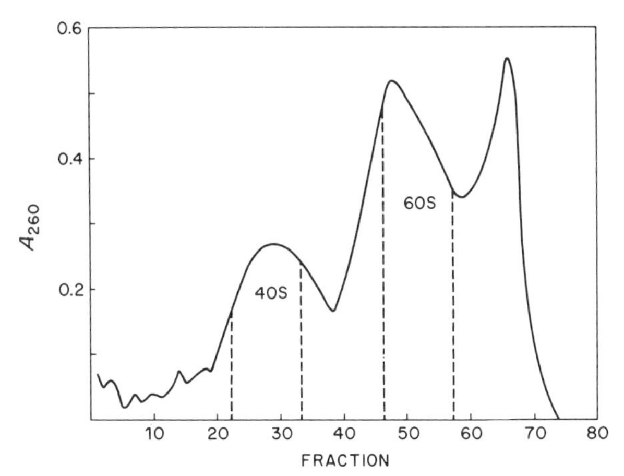


FIG. 1. Sedimentation of liver ribosomal subunits in a zonal rotor [C. C. Sherton and I. G. Wool, *J. Biol. Chem.* **247**, 4460 (1972)]. Liver ribosomal subunits (1.38 g) were separated by centrifugation for 17 hours at 13,500 rpm in a Spinco Ti 15 zonal rotor using a 7.4–38.0% hyperbolic sucrose density gradient [E. M. Eikenberry, T. A. Bickle, R. R. Traut, and C. A. Price, *Eur. J. Biochem.* **12**, 113 (1970)]. Fractions (10 ml) were collected, and those indicated by the interrupted lines were pooled; the other fractions were discarded; 117 mg of 40 S and 243 mg of 60 S subunits were recovered.

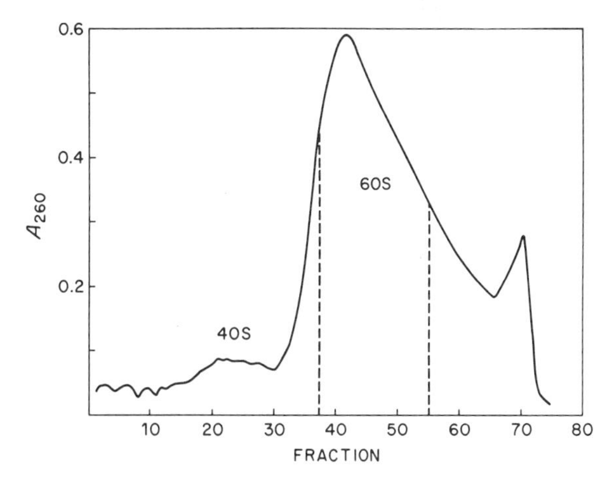


FIG. 2. Sedimentation of 60 S subunits of liver ribosomes in a zonal rotor [C. C. Sherton and I. G. Wool, *J. Biol. Chem.* **247**, 4460 (1972)]. Fractions containing 736 mg of 60 S subunits from three previous centrifugations in a zonal rotor were pooled and centrifuged again for 17 hours at 13,500 rpm in a Spinco Ti 15 zonal rotor using a 7.4–38.0% hyperbolic sucrose density gradient. [E. M. Eikenberry, T. A. Bickle, R. R. Traut, and C. A. Price, *Eur. J. Biochem.* **12**, 113 (1970)]. The fractions indicated by the interrupted lines were pooled; 324 mg of 60 S subunits were recovered.

ethanol.[25] Cold 95% ethanol (0.2 of a volume) is added. The suspension is kept at 0° for at least 1 hour (to ensure complete precipitation of the particles) and the subunits are collected by centrifugation at 10,000 *g* for 15 minutes. The ethanol-precipitated ribosomal pellets may be stored at −20°.

*Purification of 60 S Subunits.* The 60 S subunits isolated in this way are contaminated with variable amounts of 40 S subunits and must be resolved further to be suitable for a determination of the number of large subunit ribosomal proteins. The fractions containing 60 S subunits collected from centrifugation on three occasions in a zonal rotor (about 740 mg) are incubated again for 15 minutes at 37° in medium D containing 0.1 m*M* puromycin and 20 m*M* MSH. They are recentrifuged in a zonal rotor and fractions collected (Fig. 2), dialyzed, and precipitated with ethanol as described above.

Smaller amounts of 60 S subunit fractions can be purified by centrifugation in a Spinco SW 27 rotor. After suspension of the subunits in medium D, and centrifugation at 13,000 *g* for 10 minutes to remove

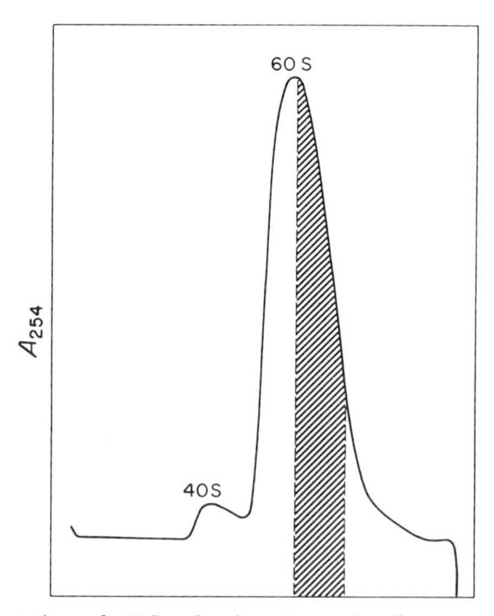


FIG. 3. Sedimentation of 60 S subunits of muscle ribosomes in a Spinco SW 27 rotor. Approximately 7.2 mg of 60 S subunits from a previous centrifugation on a zonal rotor were recentrifuged for 16 hours at 13,500 rpm using a 10–30% linear sucrose density gradient. The fractions indicated by the interrupted lines were pooled; 2.3 mg of 60 S subunits were recovered.

[25] M. S. Kaulenas, *Anal. Biochem.* **41**, 126 (1971).

aggregates, they are incubated for 5 minutes at 37° in 0.1 m*M* puromycin and 20 m*M* MSH. The suspension (0.8–1.0 ml containing 6.7–7.2 mg of ribosomes) is then layered onto a 37-ml 10 to 30% linear sucrose density gradient containing medium D and 20 m*M* MSH. Centrifugation is at 26° for 4 hours at 26,500 rpm, or 16 hours at 13,500 rpm. The distribution of the particles in the gradient is determined with an Instrument Specialties Co., Inc. (ISCO) Model D, density gradient fractionator, and the Model UA-2 UV analyzer,[18,23] and a portion of the 60 S peak is collected (cf. Fig. 3). The 60 S subunits are dialyzed and precipitated with ethanol as described above.

*Analysis of the Purity of Ribosomal Subunits.* The purity of the subunit fractions can be determined by zonal centrifugation (Fig. 4). The samples are dialyzed against medium E and layered on 5.2 ml of a 10–30% linear sucrose gradient in medium F. Centrifugation is at 60,000 rpm for 40 minutes at 26° in a Spinco SW 65 rotor, or at 48,000 rpm for 55 minutes in a Spinco SW 50.1 rotor. The distribution of particles

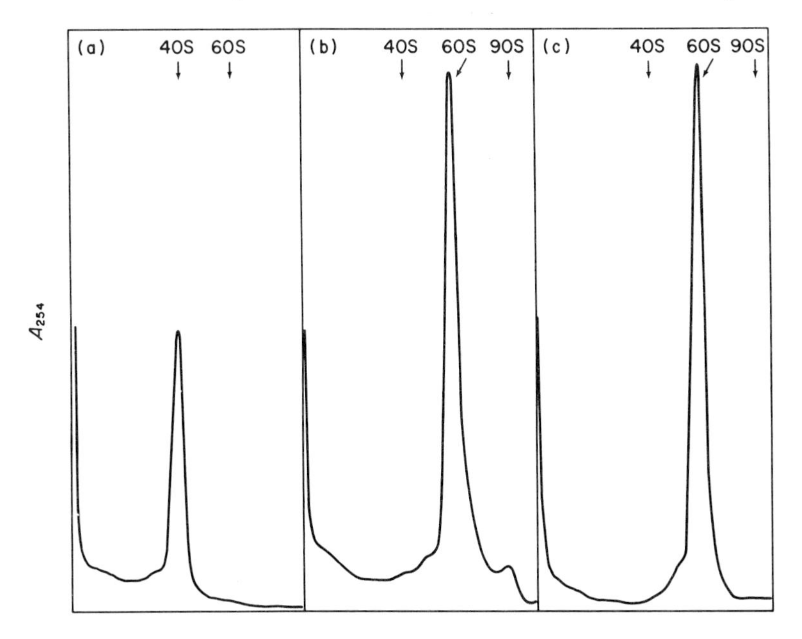


FIG. 4. Sedimentation of subunit fractions in sucrose gradients. [C. C. Sherton and I. G. Wool, *J. Biol. Chem.* **247**, 4460 (1972)]. The purity of subunit fractions was determined after dialyzing the samples against medium E. The samples were analyzed on 10–30% linear sucrose gradients in medium F; centrifugation was at 60,000 rpm for 40 minutes at 26° in a Spinco SW 65 rotor. (a) 40 S fraction (18 μg) from Fig. 1; (b) 60 S fraction (36 μg) from Fig. 1; (c) 60 S fraction (36 μg) from Fig. 2.

is determined with an ISCO density gradient fractionator and UV analyzer. The 40 S subunits from the first centrifugation in the zonal rotor are generally free of contamination with 60 S subunits (Fig. 4a); the 60 S subunits purified by either method described above are also generally free of contamination with 40 S subunits (Fig. 4c).

Another method of determining whether subunit preparations are contaminated is to analyze their constituent RNA.[18] Sufficient 20% SDS is added to a suspension of ribosomal particles in medium G so as to give a final concentration of 0.1%. The sample is incubated for 5 minutes at 37°, and then layered onto 5.2 ml of a 5 to 20% linear sucrose gradient in medium G. Centrifugation is at 4° for 2 hours at 60,000 rpm in a Spinco SW 65 rotor or for 2.75 hours at 48,000 rpm in a Spinco SW 50.1 rotor. The presence of 18 S (small subunit) and 28 S (large subunit) RNA is determined with an ISCO density gradient fractionator and UV analyzer (Fig. 5). Samples containing 60 S subunits that have not been

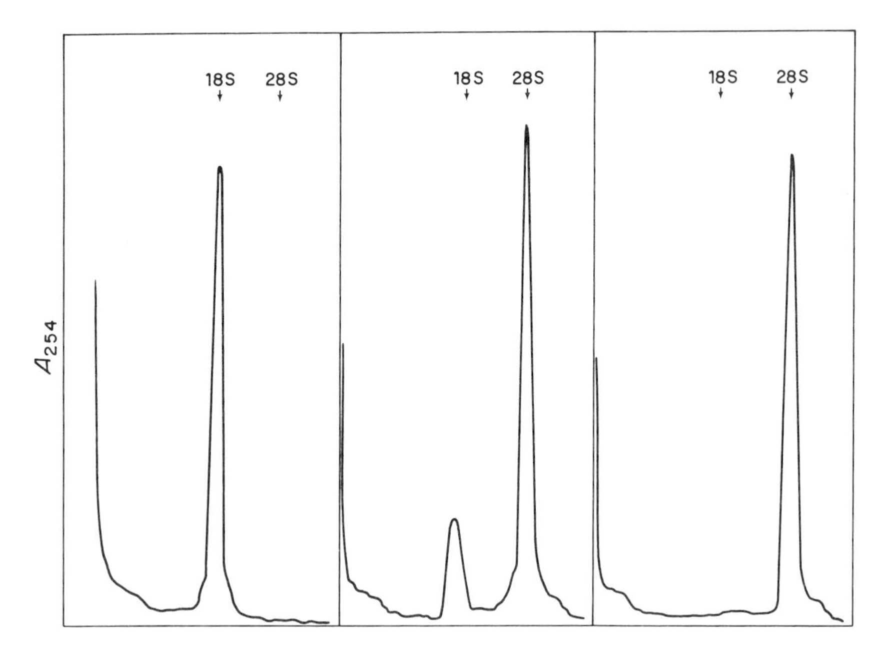


FIG. 5. Sedimentation in sucrose gradients of the RNA of subunit fractions. Subunit fractions were suspended in medium G with 0.1% SDS and incubated for 5 minutes at 37°. The samples were centrifuged in 5 to 20% linear sucrose density gradients in medium G at 60,000 rpm for 2 hours at 4° in a Spinco SW 65 rotor; (a) 40 S fraction (18 μg) from Fig. 1; (b) 60 S fraction (22 μg) from Fig. 1; (c) 60 S fraction (18 μg) from Fig. 3.

purified contain 18 S RNA (Fig. 5b); whereas purified 60 S preparations do not (Fig. 5c).

It is extremely important that ribosomal proteins to be used for two-dimensional electrophoresis be extracted from subunits that are intact and as free of contamination as possible. Subunits prepared by dissociation of ribosomes with high concentrations of KCl will recombine to form ribosomes active in protein synthesis,[18,26] whereas subparticles prepared by treatment with EDTA and other harsh methods are no longer active[18,24] and are not satisfactory for analysis of ribosomal proteins. The particles prepared by removal of magnesium from ribosomes unfold to varying degrees.[18,27-29] During zonal sedimentation, unfolded monomers may co-sediment with the large subunit, and unfolded large subunits contaminate the small subunit fraction. Contamination, caused in that way, has been responsible for erroneous reports that 60 S subunits (prepared with EDTA) contained all the proteins of the ribosome,[30] and that 40 S and 60 S subunits contained 15 proteins having identical mobilities.[11] In the latter case, cross contamination of less than 5% was reported. Obviously, even that is too much.

## Extraction of Ribosomal Proteins

### *Reagents*

Glacial acetic acid, and 1 *N*

Proteins are extracted from ribosomes and from ribosomal subunits with acetic acid by a modification of the procedure of Hardy *et al.*[31] Ribosomes are suspended (50–60 mg/ml) with gentle homogenization in medium H; 2 volumes of glacial acetic acid are added, and the mixture is stirred for 1 hour at 0°. Ribosomal RNA is removed by centrifugation at 15,000 *g* for 10 minutes; the pellet is washed with an equal volume of one-third medium H and two-thirds glacial acetic acid (v/v), and recentrifuged. The combined supernatant, containing the ribosomal proteins, is dialyzed against about 50 volumes of 1 *N* acetic acid for 48 hours with 4 changes of acid. The dialysis tubing is acetylated[32] to prevent the loss

[26] T. E. Martin and I. G. Wool, *Proc. Nat. Acad. Sci. U.S.* **60**, 569 (1968).
[27] Y. Tashiro and T. Morimoto, *Biochim. Biophys. Acta* **123**, 523 (1966).
[28] R. F. Gesteland, *J. Mol. Biol.* **18**, 356 (1966).
[29] H. Lamfrom and E. R. Glowacki, *J. Mol. Biol.* **5**, 97 (1962).
[30] M. DiGirolamo and P. Cammarano, *Biochim. Biophys. Acta* **168**, 181 (1968).
[31] S. J. S. Hardy, C. G. Kurland, P. Voynow, and G. Mora, *Biochemistry* **8**, 2897 (1969).
[32] L. C. Craig, this series, Vol. 11, p. 870.

of low molecular weight proteins. The dialyzed proteins are then lyophilized.

We have not extracted ribosomal proteins by other methods. Ford has reported[33] that acetic acid is not as efficient as lithium chloride-urea[34] or guanidine hydrochloride[35] in the extraction of proteins from *Xenopus* ribosomes. However, the method he used[36] did not include the modification (high concentration of magnesium) introduced by Hardy *et al.*,[31] which greatly increases the efficiency of the method. Kaltschmidt and Wittmann[6] extracted *E. coli* ribosomal proteins in various ways and evaluated the efficiency of the procedures by two-dimensional electrophoresis. The most complete extraction was with 2 parts glacial acetic acid and one part buffer containing 200 m*M* magnesium chloride. (The LiCl-urea and RNase methods were almost as effective.) Extraction was not complete if the concentration of magnesium in the buffer was 10 m*M*. Unfortunately, 100 m*M* magnesium (the concentration we have used in our buffer) was not tested.

If, for the moment, we take the number of ribosomal proteins as the criterion for the completeness of extraction (and there is a question whether that is an entirely valid standard), then it is obvious that the several procedures give remarkably similar results. Martini and Gould,[8] Hultin and Sjöqvist,[9] and Huynh-Van-Tan, *et al.*[14] all extracted mammalian ribosomes with LiCl-urea[34] and analyzed the proteins by two-dimensional electrophoresis: they found 62-63, 48, and 75 proteins, respectively. The number Hultin and Sjöqvist found is low, but they also found only 48 proteins when ribosomes were extracted with HCl, and from their electropherograms it is obvious that the proteins were incompletely resolved. Welfle *et al.* found 72 different proteins when they extracted ribosomes with HCl,[17] and 76 when they used 67% acetic acid and 50 m*M* magnesium chloride in the buffer.[11] With the exception of the findings of Hultin and Sjöqvist,[9] the several procedures give results quite comparable to the 68–72 different ribosomal proteins we found after extracting liver ribosomes with 67% acetic acid and 100 m*M* magnesium chloride in the buffer.[16] Although we cannot be certain that any procedure extracts all the eukaryotic ribosomal proteins, the acetic acid–magnesium chloride, urea–lithium chloride, and HCl methods seem equally efficient.

[33] P. J. Ford, *Biochem. J.* **125**, 1091 (1971).
[34] P. S. Leboy, E. C. Cox, and J. G. Flaks, *Proc. Nat. Acad. Sci. U.S.* **52**, 1367 (1964).
[35] R. A. Cox, this series, Vol. 12B, p. 120.
[36] P. B. Moore, R. R. Traut, H. Noller, P. Pearson, and H. Delius, *J. Mol. Biol.* **31**, 441 (1968).

At least that is a reasonable assumption until a systematic evaluation is carried out.

## Two-Dimensional Polyacrylamide Gel Electrophoresis

*Reagents*

Acrylamide (Eastman Organic Chemicals)
*N*,*N*′-Methylene-bisacrylamide (Eastman Organic Chemicals)
*N*,*N*,*N*′,*N*′-Tetramethylethylenediamine (TEMED) (Eastman Organic Chemicals)
Crystalline ammonium persulfate (APS), 0.6%; 1.5%; 6%; 8% (w/v) in water (Eastman Organic Chemicals)
Riboflavin (Eastman Organic Chemicals)
Naphthol Blue Black, 1% in 7.5% acetic acid (Eastman Organic Chemicals)
Solid boric acid (Sigma)
Solid disodium ethylenediamine tetraacetic acid (EDTA) (Sigma)
Solid Tris·HCl (Sigma)
Solid urea, and 8 *M*
NaOH, 2 *N*
KOH, 5 *N*
Acetic acid, 7.5%

*Equipment*

We have used the apparatus designed by Kaltschmidt and Wittmann and described by them in great detail in their first publication.[1] Comparable equipment is now available from several manufacturers.

*General Procedures.* We have adapted the technique of Kaltschmidt and Wittmann[1] for the electrophoresis of mammalian ribosomal proteins. The "standard" conditions for two-dimensional electrophoresis at room temperature are: 8% acrylamide gel, pH 8.6, for 40 hours at 90 V (initial current 2.7 mA/tube) in the first dimension; 18% acrylamide gel, pH 4.2, for 40 hours at 105 V (initial current 100 mA/gel) in the second dimension. If purified acrylamide solutions are used (see below) the voltage is reduced by 25%. The concentrations of the gels and the pH of the buffers are selected to maximize separation of the ribosomal proteins by charge in the first dimension and by size in the second, and to provide the best possible resolution of the largest number of proteins.

There is some leeway in the selection of the concentration of acrylamide, pH, voltage, etc., for the separation of mammalian ribosomal proteins.[8,9,11,14,17] Regardless of the conditions one finally selects as most suitable, it is important to perform electrophoresis with different per-

TABLE I
BUFFERS TO BE USED FOR ELECTROPHORESIS OF RIBOSOMAL PROTEINS IN THE FIRST DIMENSION

| Reagent[a] | pH 7.6 | pH 8.6 | pH 9.6 | pH 10.6 |
|---|---|---|---|---|
| Boric acid (g) | 37.1 | 28.8 | 6.3 | 1.86 |
| Tris·HCl (g) | 25.5 | 43.65 | 87.3 | 98.1 |
| EDTA (disodium salt) (g) | 7.05 | 7.2 | 11.7 | 7.05 |
| 2 $N$ NaOH (ml) | — | 55 | 52 | 32 |

[a] Sufficient water is added to give a final volume of 3 liters.

centages of acrylamide and at different pH's, in order to detect proteins which may fortuitously coelectrophorese in the "standard" conditions.[2,16] We generally vary the acrylamide percentage in the first dimension from 5% to 10% (while keeping the acrylamide: methylene–bisacrylamide ratio constant), and vary the pH of the first dimension buffer from 7.6 to 10.6 (Table I).

*First Dimension.* Soft glass tubes (inner diameter 6 mm) are cut to a length of 180 mm, the ends are fire polished, cleaned in chromic acid solution, and dried. The acrylamide and methylene-bisacrylamide used in the preparation of separation and sample gel solutions (Table II) should first be recrystallized,[37] to reduce streaking and smudging in the electropherogram. The 8 $M$ urea solution used in the preparation of the gels is passed through a column of Bio-Rad AG 501-X8(D) mixed bed resin to remove cyanate. Cyanate can lead to carbamylation of lysine residues,[38]

TABLE II
PREPARATION OF POLYACRYLAMIDE GEL SOLUTIONS TO BE USED FOR ELECTROPHORESIS OF RIBOSOMAL PROTEINS IN THE FIRST DIMENSION

| Reagent[a] | Separation gel | Sample gel |
|---|---|---|
| Acrylamide (g) | 40 | 10 |
| Methylene-bisacrylamide (g) | 1.5 | 0.5 |
| Boric acid (g) | 16 | 0.8 |
| EDTA (disodium salt) (g) | 4 | 0.21 |
| Tris·HCl (g) | 24.3 | — |
| TEMED (ml) | 1.5 | 0.15 |

[a] Sufficient water is added to give a final volume of 115 ml of separation gel or 60 ml of sample gel.

[37] U. E. Loening, *Biochem. J.* **102**, 251 (1967).
[38] G. R. Stark, W. H. Stein, and S. Moore, *J. Biol. Chem.* **235**, 3177 (1960).

and thus increase the number of protein bands or spots obtained on electrophoresis.

For each 5 gel tubes to be filled, 18.75 ml of 8 *M* urea are combined with 5.75 ml of separation gel solution (Table II). The solution is deaerated (with vacuum suction) for 15 minutes to remove oxygen and reduce bubble formation, and APS (0.5 ml of 1.5%) is added. The final ratio of 8 *M* urea:separation gel solution:1.5% APS is 75:23:2. The solution is mixed, deaerated for 20 seconds, quickly pipetted into the gel tubes to a height of 100 mm, and overlayed with water (to ensure a level surface). At least 15 minutes is allowed for polymerization.

The sample gel is made by first combining 6.0 ml of 8 *M* urea with 1.92 ml of sample gel solution (Table II); 1.98 ml of the mixture are deaerated for 15 minutes. Next, 10 $\mu$l of 0.6% APS and 10 $\mu$l of riboflavin (1 mg/ml in 6 *M* urea) are added, mixed, and deaerated for 20 seconds. The final ratio is: 8 *M* urea (75):sample gel solution (24):0.6% APS (0.5):riboflavin (0.5). The water overlaying the polymerized gel is removed and a 150-$\mu$l aliquot of the sample gel mixture is added to each tube (as spacer gel) and overlayed with water. Since the spacer gel contains only 4% acrylamide, it will, with an appropriate buffer, concentrate the proteins into thin layers before their entry into the 8% acrylamide separation gel. The spacer gel is polymerized in front of a fluorescent light for at least 20 minutes. A fluorescent light is necessary for polymerization when the gel contains riboflavin.

The ribosomal proteins are dissolved in sample gel: the concentration can be varied from 5 to 10 mg/ml and the volume from 50 to 400 $\mu$l. In general, the optimal amount for the analysis of 40 S and 60 S ribosomal proteins is 1.0 mg; and for 80 S, 1.6 mg (Fig. 6). To be able to record on film proteins which stain lightly, it is necessary to use at least twice the optimal amount of protein. On the other hand, proteins which stain intensely and which migrate closely can only be seen as separate when less than the optimal amount is used.

At the pH of the sample gel (8.2) the very basic ribosomal proteins are barely soluble. Incubating the gel solution containing the ribosomal protein sample at 37° for 10–15 minutes may help to solubilize the proteins, but with some preparations the solution remains translucent. To the sample gel solution containing ribosomal proteins are added 0.005 volume each of APS (0.6%) and riboflavin (1 mg/ml). The sample is deaerated for 20 seconds. An appropriate amount (see above) is layered onto the spacer gel and overlayed with water. Polymerization is carried out with a fluorescent light for at least 30 minutes.

Another portion of sample gel is deaerated, layered over the actual sample, and polymerized to form an upper spacer gel in the same way we described for the lower spacer gel. The upper separation gel is formed in the following way: 11.25 ml of 8 *M* urea and 3.45 ml of separation gel solution (Table II) are combined and deaerated for 15 minutes; APS (0.3 ml of 1.5%) is added and the solution is deaerated for 20 seconds and pipetted over the upper spacer gel to the top of the tubes. A small layer of water may be applied to facilitate polymerization, which is for at least 15 minutes.

Electrophoresis is carried out with 3 liters of whichever buffer is selected (Table I). We have omitted the 6 *M* urea which Kaltschmidt and Wittmann[1] used in their first dimension buffer. We find no discernible difference in the protein pattern without urea. In "standard" conditions (pH 8.6), the pH of the buffer in the chambers changes no more than 0.2 pH unit during electrophoresis. Electrophoresis is from the anode (top) to the cathode (bottom).

The first-dimension gels are removed from the tubes by first injecting glycerin to loosen the gels, which adhere to the tube wall, and then water

FIG. 6. Two-dimensional electropherograms of liver ribosomal proteins [C. C. Sherton and I. G. Wool, *J. Biol. Chem.* **247**, 4460 (1972)]. Standard conditions were used for the electrophoresis. The anode was on the left in the first dimension and at the top in the second. (a) 40 S ribosomal subunit protein (1 mg); (b) 60 S ribosomal subunit protein (1 mg); (c) 80 S ribosomal protein (1.6 mg). Some of the spots on the gels were clearly visible to the eye, but are barely discernible (S1) or cannot be seen (S12) in the photograph. Those proteins are included in the schematics (Figs. 7–9).

FIG. 6(b).

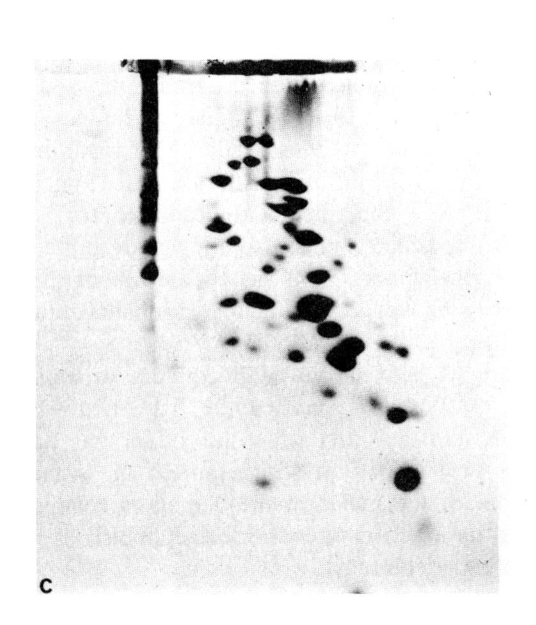


FIG. 6(c).

TABLE III
PREPARATION OF POLYACRYLAMIDE GEL AND DIALYSIS BUFFER TO BE USED FOR ELECTROPHORESIS OF RIBOSOMAL PROTEINS IN THE SECOND DIMENSION

| Reagent[a] | Polyacrylamide gel | Dialysis buffer |
|---|---|---|
| Acrylamide (g) | 360 | — |
| Methylene-bisacrylamide (g) | 10 | — |
| Urea (g) | 720 | 960 |
| TEMED (ml) | 11.6 | — |
| Glacial acetic acid (ml) | 104.6 | 1.48 |
| 5 *N* KOH (ml) | 19.2 | 4.8 |

[a] Sufficient water is added to give a final volume of 1933 ml of polyacrylamide gel solution or 2 liters of dialysis buffer.

to force them out of the tube. (They break rather easily, but may be reassembled and placed atop the second-dimension gel slab.) On removal from the tubes the gels are dialyzed for a total of 45 minutes against 3 changes of 8 *M* urea, pH 7.2 buffer (Table III). Dialysis is to reduce the pH of the gels in order to adapt them to the second-dimension buffer.

*Second Dimension.* The 18% acrylamide gel solution (Table III) is prepared, deaerated for 30 minutes, and kept at 4° for at least 1 hour prior to use. Polymerization is performed at 4° to slow the process and to dissipate the large amount of heat generated. Gel solution, 483.3 ml, is combined with 16.7 ml of 6% APS (29:1) and poured into the base of the electrophoresis apparatus to seal off the bottoms of the slabs. At least 1 hour is allowed for polymerization. The first-dimension gels are positioned in the slots; 1160 ml of gel solution are combined with 40 ml of 8% APS and poured slowly to form slabs (20 cm × 20 cm × 0.5 cm). Bubbles under the first-dimension gels are removed. As polymerization proceeds, the gels shrink, and more gel solution is added to the top. The exact amount of APS that is required for polymerization to occur in a reasonable time varies with the age of the reagent.

Again, at least 1 hour must be allowed for complete polymerization. Electrophoresis is at room temperature. The 12 liters of pH 4.2 buffer to be used for electrophoresis in the second dimension contain 168 g of glycine and 18.0 ml of acetic acid (Table III). The pH will sometimes change as much as 2 units during the 40 hours required for electrophoresis. Electrophoresis is from the anode (top) to the cathode (bottom).

*Staining and Destaining.* The gels are stained for 20 minutes in 1% Naphthol Blue Black in 7.5% acetic acid. The dye is purified by electrophoresis through 10% acrylamide gel at 100 mA for 40 hours—the dye remaining on the cathode side of the gel is used. The gels are destained for 1 hour in running distilled water, and subsequently by continuous diffusion destaining[39] in 17 liters of 7.5% acetic acid. A simple, convenient method is to use an inexpensive model 425 "Dynaflo" aquarium pump to circulate the destaining solution through a filter of activated charcoal. The charcoal should be stirred for consecutive 24-hour periods with about 3 volumes each of 4 *N* NaOH, then 2 *N* HCl, and finally distilled water; the procedure removes colloidal carbon and soluble contaminants. We have found that Fisher Activated Cocoanut charcoal (6–14 mesh) works especially well.

*Comments.* When ribosomal proteins are analyzed by two-dimensional electrophoresis shadowed streaks occur in the upper portion of the gel slabs on the cationic side, a region where one would expect to find the proteins of greatest molecular size (Fig. 6). The streaks are probably caused by protein aggregates linked by disulfide bonds. Kaltschmidt and Wittmann[2] have shown that similar streaks, which occur when two-dimensional electrophoresis of *E. coli* ribosomal proteins is carried out, can be eliminated if the proteins are oxidized with performic acid or reduced and alkylated with iodoacetamide. We have not repeated that procedure. However, we have found that purification of acrylamide and methylene-bisacrylamide by the procedure of Loening[37] diminishes the intensity of the streaks.

A much more serious difficulty is the variable amount of protein which remains at the origin during electrophoresis in the first dimension and causes the formation of bands, by migration directly down from the origin, during electrophoresis in the second dimension (Fig. 6). The formation of the bands is concentration dependent. We suspect the bands to be proteins that do not migrate in the first dimension because of the reduced solubility of ribosomal proteins at pH 8.6 (the pH in the first dimension). The proteins would be expected to be soluble at the pH (4.2) used in the second dimension, although a variable amount of material still remains at the origin. The bands appear to have mobilities in the second dimension equal to those of some of the ribosomal proteins. However, bands are not formed for all the proteins, and the bands present have different intensities. Therefore, the problem posed is whether variable amounts of each protein may be retarded at the origin, and whether, in fact, a particular protein might be lost in this way. As far as we know, we have never lost one of our proteins from the two-dimensional pattern,[16] although we have an instance in which a single protein (S9) was all but completely absent. Thus we cannot be certain that there do not exist

[39] L. J. Gathercole and L. Klein, *Anal. Biochem.* 44, 232 (1971).

ribosomal protein species which are insoluble at pH 7.6–10.6 and do not migrate from the origin.

Similar bands below the first dimension origin are apparent in all the other published electropherograms of mammalian ribosomal proteins

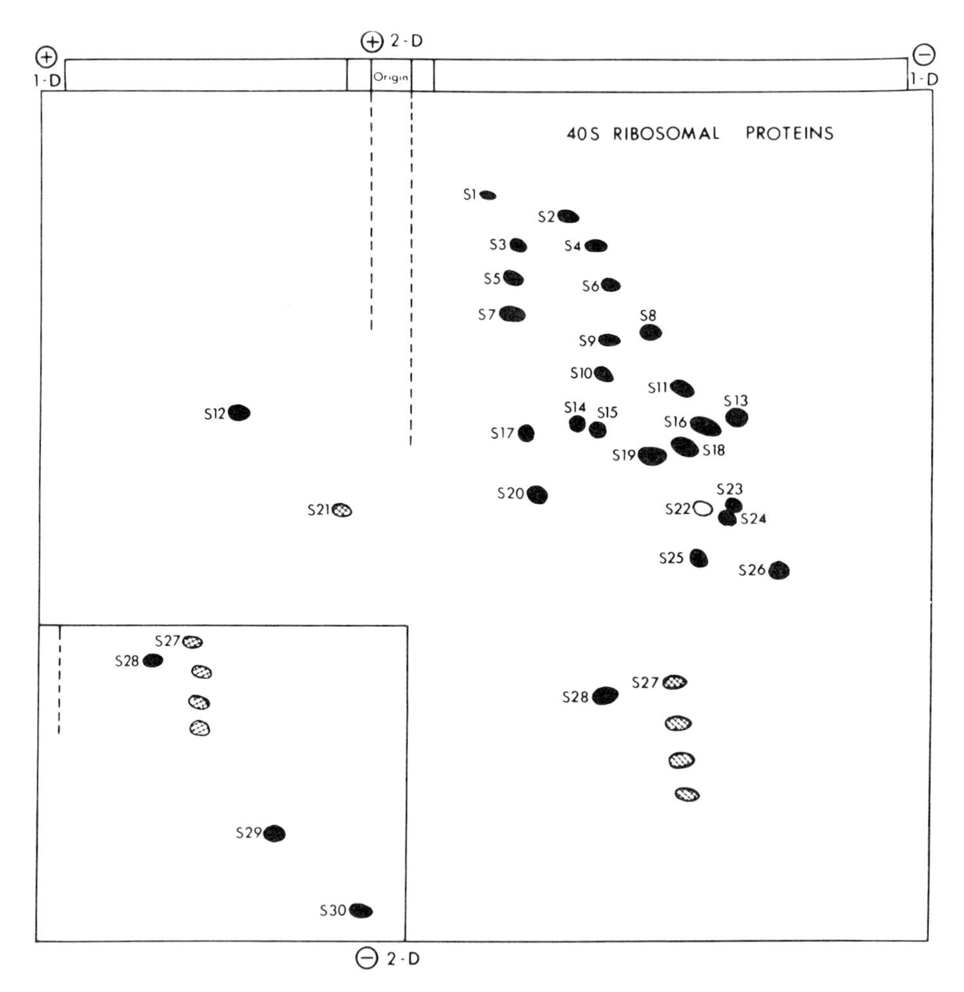


FIG. 7. Schematic of the two-dimensional electropherogram of the proteins of the 40 S subunit of liver ribosomes [C. C. Sherton and I. G. Wool, *J. Biol. Chem.* **247**, 4460 (1972)]. The solid spots were always seen; the crosshatched spots either varied in location (S21) or in intensity of staining (S27 and its three satellite spots); the open spot (S22) was seen only when the conditions of electrophoresis were changed. The diagram includes spots that are difficult to see in the photographs (Fig. 6). The intensity of the staining of the spots is not represented with fidelity.

analyzed by the Kaltschmidt and Wittmann procedure.[11-14,17] The two-dimensional electrophoresis method developed by Martini and Gould[8] (and also that used by Hultin and Sjöqvist[9]), has the great advantage of using a pH of 4.5 for electrophoresis in the first dimension. At this pH the ribosomal proteins are far more soluble, and there are no difficulties with proteins precipitating, or forming bands in the second dimension.

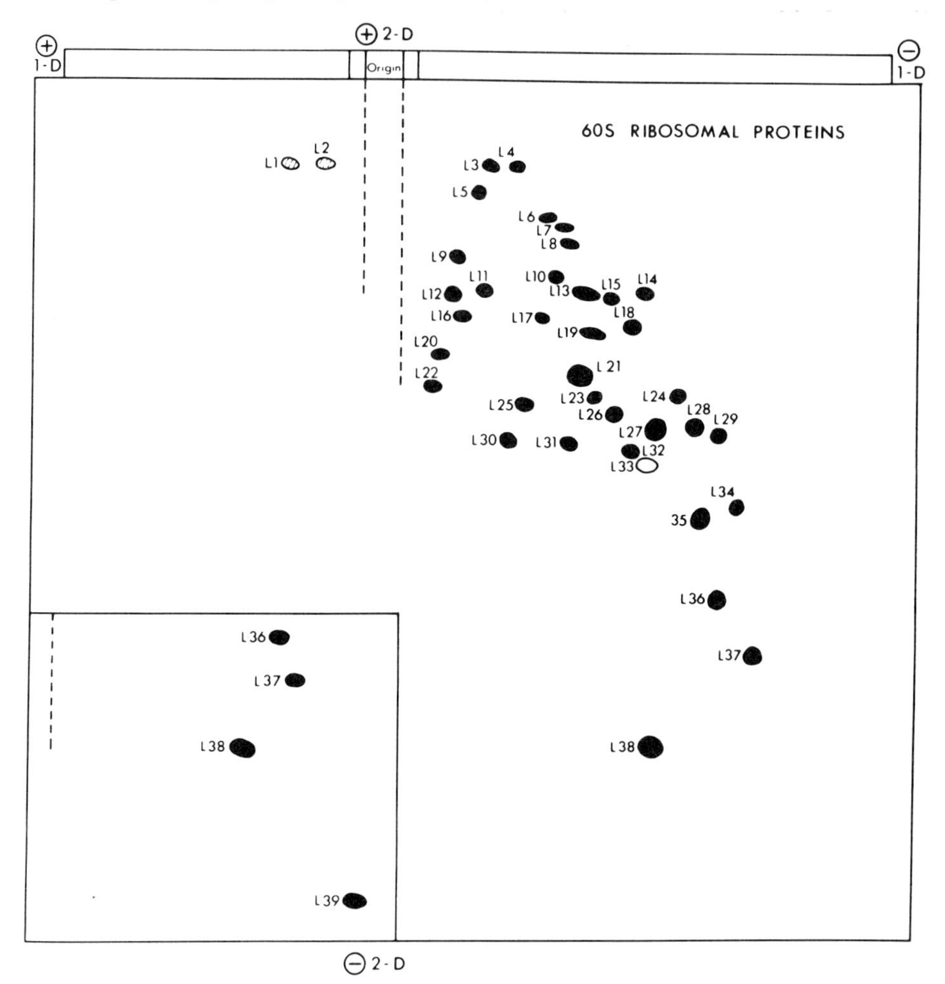


FIG. 8. Schematic of the two-dimensional electropherogram of the proteins of the 60 S subunit of liver ribosomes [C. C. Sherton and I. G. Wool, *J. Biol. Chem.* **247**, 4460 (1972)]. The solid spots were always seen; the crosshatched spots (L1 and L2) were seen only when an excess of protein was applied; the open spot (L33) was seen only when the conditions of electrophoresis were changed.

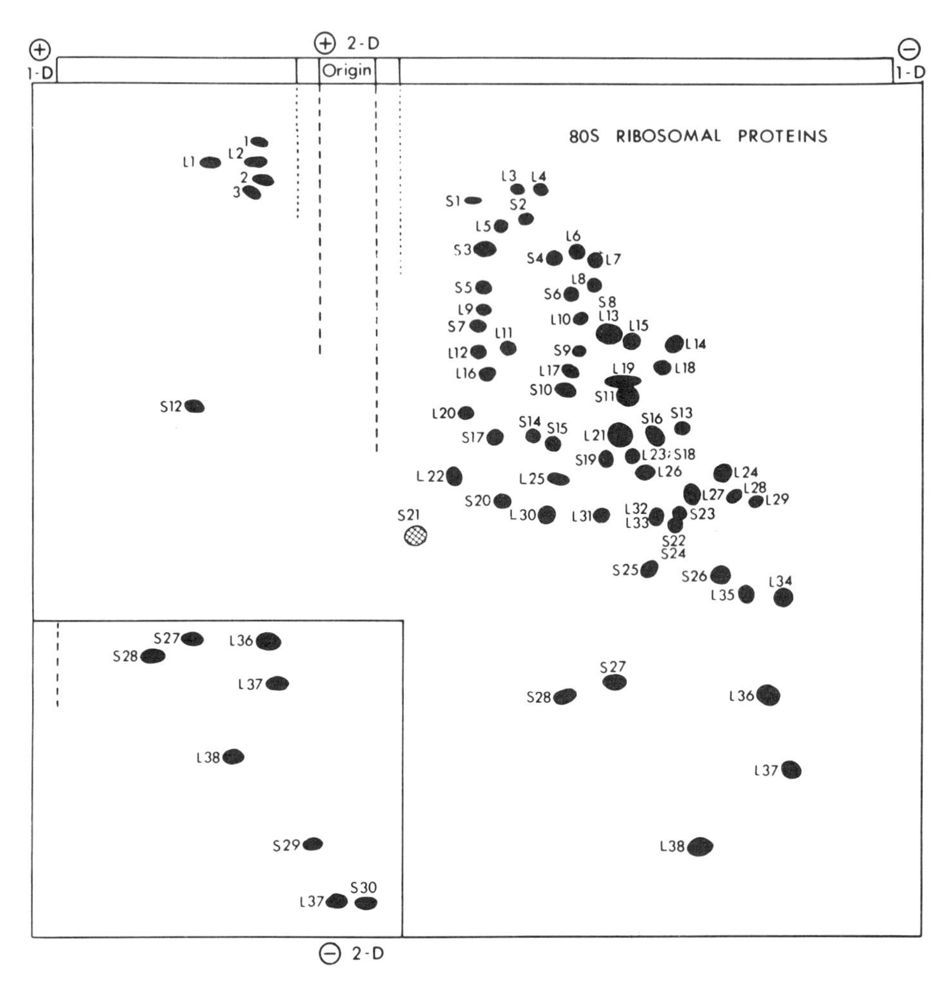


FIG. 9. Schematic of the two-dimensional electropherogram of the proteins of rat liver 80 S ribosome monomers [C. C. Sherton and I. G. Wool, *J. Biol. Chem.* **247**, 4460 (1972)]. The solid spots were always seen; the crosshatched spot (S21) varied in location; the open spots (1, 2, 3) were seen when ribosome monomer proteins were analyzed, but not when subunit proteins were electrophoresed.

Another advantage of that procedure is that electrophoresis in the second dimension is in 0.1% sodium dodecyl sulfate which enables one to determine the molecular weight of the ribosomal proteins.

## Concluding Remarks

We have used two-dimensional electrophoresis to compile a catalogue of the proteins of rat liver ribosomes and ribosomal subparticles (Figs.

7–9). It is to be noted that the schematic drawings (which depict the proteins) do not represent with fidelity the intensity of staining of the individual spots; moreover, they are composites from a large number of electropherograms in which the amount of sample and the conditions of electrophoresis were varied. We have chosen to number the protein spots of each subunit separately along horizontal lines beginning at the upper left. The 60 S subunit proteins are designated by an L (large subunit), the 40 S subunit proteins by an S (small subunit).

There are 30 proteins in the 40 S subunit. Two are variable in position (S21, S27) or amount (S27), two can be seen only when the time of electrophoresis in the first dimension is decreased (S29, S30), and one is seen only when the conditions of electrophoresis are altered (S22). There are 39 proteins in the 60 S subunit. Two are present in very small amounts (L1, L2), one can be seen only at the shorter time (L39), and one is seen only in altered conditions (L33). The proteins of the two subunits, with one possible exception (S8, L13) are unique. Preparations of 80 S ribosomes contain 3 proteins not present in either subunit; thus, we estimate that mammalian ribosomes contain 68–72 different proteins.

Two-dimensional electrophoresis has proved a useful method for the resolution of mammalian ribosomal proteins. Electropherograms can be used to display and count the proteins of ribosomal subunits and monomers; and even to determine their molecular weights (using the SDS method[8]). In addition, two-dimensional electrophoresis can be employed to compare the protein composition of ribosomes from different species or tissues and to determine the isoelectric points of the proteins. The catalog of proteins should prove a valuable reference for the isolation, purification, and characterization of individual eukaryotic ribosomal proteins.

# [62] Isotopic Labeling and Analysis of Phosphoproteins from Mammalian Ribosomes

*By* LAWRENCE BITTE and DAVID KABAT

## I. Introduction and Scope

It has been recently shown that at least five polypeptide chains from ribosomes of rabbit reticulocytes,[1,2] rat liver cells,[3] and mouse sarcoma

[1] D. Kabat, *Biochemistry* **9**, 4160 (1970).

[2] D. Kabat, *J. Biol. Chem.* **247**, 5338 (1972).

[3] J. E. Loeb and C. Blat, *FEBS (Fed. Eur. Biochem. Soc.) Lett.* **10**, 105 (1970).



ISBN 0-12-504180-2

180 tumor cells[4] are phosphoproteins rather than simple proteins. The phosphoryl groups in these proteins are present in *o*-phosphoseryl and in *o*-phosphothreonyl residues. Since these phosphoryl groups turn over intracellularly, the proteins become radioactive when the cells are incubated with [$^{32}$P]orthophosphate. Furthermore, these phosphoproteins from rabbit reticulocyte and from mouse sarcoma 180 ribosomes have the same molecular weights and may be homologous proteins related by evolution.[4] Intracellular phosphorylation of these polypeptides is markedly elevated by hormonal stimuli or by cyclic AMP.[5-7] *In vitro* phosphorylation of ribosomal proteins by cyclic AMP-dependent protein kinases (ATP:protein phosphotransferase EC 2.7.1.37) has also been reported by several groups.[3,8-10] In addition, a protein kinase which may be the physiologically active enzyme has been isolated in association with reticulocyte[11] and with fibroblast[10] ribosomes.

This is the only enzymatic modification of ribosomal proteins known to occur in mammalian cells. Although a similar metabolism is apparently absent from *Escherichia coli*,[12] cyclic AMP-dependent protein phosphorylation reactions have been increasingly implicated in the control of metabolism and growth of mammalian cells and in the regulation of physiological responsiveness. Frequently, the phosphoproteins occur in complex assemblages of macromolecules, such as in chromosomes,[13] in viruses,[14] in secretory granules,[15] in microtubules,[16] and in rhodopsin on photoreceptor membranes.[17]

The purpose of this chapter is to describe the methods which we have

[4] L. Bitte and D. Kabat, *J. Biol. Chem.* **247**, 5345 (1972).
[5] L. Bitte and D. Kabat, manuscript in preparation, 1972.
[6] M. L. Cawthon, L. Bitte, A. Krystosek, and D. Kabat, *J. Biol. Chem.* (in press) 1973.
[7] C. Blat and J. E. Loeb, *FEBS (Fed. Eur. Biochem. Soc.) Lett.* **18**, 124 (1971).
[8] C. Eil and I. G. Wool, *Biochem. Biophys. Res. Commun.* **43**, 1001 (1971).
[9] G. M. Walton, G. N. Gill, I. B. Abrass, and L. D. Garren, *Proc. Nat. Acad. Sci. U.S.* **68**, 880 (1971).
[10] C. Li and H. Amos, *Biochem. Biophys. Res. Commun.* **45**, 1398 (1971).
[11] D. Kabat, *Biochemistry* **10**, 197 (1971).
[12] J. Gordon, *Biochem. Biophys. Res. Commun.* **44**, 579 (1971).
[13] T. A. Langan, "Regulatory Mechanisms for Protein Synthesis in Mammalian Cells," p. 101. Academic Press, New York, 1968.
[14] M. Strand and J. T. August, *Nature (London) New Biol.* **233**, 137 (1971).
[15] F. Labrie, S. LeMaire, G. Poirer, G. Pelletier, and R. Boucher, *J. Biol. Chem.* **246**, 7311 (1971).
[16] D. B. P. Goodman, H. Rasmussen, and F. DiBella, *Proc. Nat. Acad. Sci. U.S.* **67**, 652 (1970).
[17] D. Bownds, J. Dawes, J. Miller, and M. Stahlman, *Nature (London) New Biol.* **237**, 639 (1972).

found most useful for the isotopic labeling and characterization of phosphoproteins from mammalian ribosomes. We describe methods for removing contaminating phosphoproteins and for establishing the reality of this ribosomal modification. These same methods should be applicable to analysis of other phosphoproteins, especially those which occur in complex subcellular organelles.

## II. Labeling of Ribosomal Phosphoproteins with [$^{32}$P]-Orthophosphate (Methods for Reticulocytes and for Sarcoma 180 Cells)

### *A. Principles*

Nearly all phosphorus in mammalian ribosomes is present in the constituent RNA molecules (approximately $6.5 \times 10^3$ P-atoms), with only a small number occurring in the proteins (approximately 7–11 P-atoms).[2] Consequently, it is desirable to $^{32}$P-label the phosphoryl groups on the proteins in conditions in which ribosomal RNA synthesis is absent. This has been accomplished by utilizing reticulocytes (which lack a nucleus and are accordingly inactive in RNA synthesis),[1,2] by utilizing tumor cells cultured in the presence of actinomycin D (an inhibitor of RNA synthesis),[4] or by labeling purified ribosomes *in vitro* with [γ-$^{32}$P]ATP in the presence of protein kinase.[3,8–11]

### *B. Solutions Used*

a. Physiological salt solution (0.13 *M* NaCl, 5 m*M* KCl, 1.5 m*M* $MgCl_2$), used for washing mammalian cells
b. Buffer A (10 m*M* KCl, 1.5 m*M* $MgCl_2$, 10 m*M* Tris·HCl, pH 7.4), a low ionic strength buffer in which ribosomes are stable
c. Buffer B (0.25 *M* KCl, 10 m*M* $MgCl_2$, 10 m*M* Tris·HCl, pH 7.4), a high ionic strength buffer in which single ribosomes may partially dissociate into subunits
d. Buffer C (0.10 *M* KCl, 40 m*M* NaCl, 5 m*M* Mg acetate, 20 m*M* Tris·HCl, pH 7.6), an intermediate ionic strength buffer in which ribosomes are stable
e. Sodium deoxycholate. A 10% solution is stored at room temperature
f. New methylene blue stain: 5.0 g Methylene blue NN (Allied Chemical, Morristown, New Jersey), 8.5 g of NaCl and 4.0 g of sodium citrate·2$H_2O$, mixed with 1 liter of $H_2O$ followed by filtration
g. Toluidine blue stain: 1% solution in physiological salt solution
h. Nutritional medium for reticulocytes (modified from Hori and

Rabinovitz[18]). The modified medium contains physiological salt solution supplemented with 1 mg/ml glucose, 0.4 mg/ml $NaHCO_3$, 0.2 m*M* ferrous ammonium sulfate, and 5% fetal calf serum. Amino acids are present in the following concentrations, expressed as millimoles per liter: L-glutamine 0.096, L-histidine 0.116, L-leucine 0.20, L-lysine 0.090, L-phenylalanine 0.080, L-serine 0.086, L-tryptophan 0.015, L-tyrosine 0.042, L-valine 0.154, L-methionine 0.040, L-arginine 0.040 and L-isoleucine 0.031. L-Cysteine, L-alanine, L-asparagine, glycine, L-proline, and L-threonine are present at 10 $\mu M$

i. Nutritional medium for sarcoma 180 cells. This medium contains Krebs bicarbonate buffer lacking inorganic phosphate (0.12 *M* NaCl, 5 m*M* KCl, 2.5 m*M* $CaCl_2$, 1 m*M* $MgSO_4$, 0.2% $NaHCO_3$, pH 7.8–8.0) supplemented with 1 mg/ml glucose, with 10% fetal calf serum, and with L-amino acids in the final concentrations recommended by Lee *et al.*[19]

*C. Procedures for Reticulocytes*

Rabbits are made anemic by subcutaneous injection for at least 7 days with 10 mg per kilogram of body weight of phenylhydrazine hydrochloride. A neutralized solution containing 25 mg/ml is prepared daily just before use. We generally make a new rabbit anemic each week and use it throughout the following week. The blood is collected into heparinized beakers which are kept chilled in ice. The cells are sedimented by centrifugation at 800 *g* for 6 minutes and are washed three times with physiological salt solution. Blood cells are routinely stained with new methylene blue stain in order to determine the proportion of reticulocytes. The washed cells are resuspended in one volume of physiological salt solution. Four drops of cell suspension and three drops of stain are mixed in a tube and are incubated at 37° for 1 hour. A smear of cells is then made on a slide, which is examined in the microscope. We routinely obtain by these methods at least 95% reticulocytes.

The washed cells are resuspended at $7 \times 10^8$ cells/ml in the nutritional medium for reticulocytes. Generally, we add 5 $\mu$g of actinomycin D per milliliter to suppress any RNA synthesis in contaminating leukocytes. The cell suspension is swirled in a water bath at 37° for 15 minutes before addition of 50 $\mu$Ci/ml [$^{32}$P]orthophosphoric acid (New England Nuclear Corp., Boston, Massachusetts). After incorporation for various time periods, the cell suspension is diluted 5-fold with ice-cooled physio-

[18] M. Hori and M. Rabinovitz, *Proc. Nat. Acad. Sci. U.S.* **59**, 1349 (1968).

[19] S. Y. Lee, V. Krsmanovic, and G. Brawerman, *Biochemistry* **10**, 895 (1971).

logical salt solution and the cells are sedimented at 800 *g* for 6 minutes. All subsequent procedures are at 2°. The packed cells are lysed by the addition of 4 volumes of buffer A. After centrifugation at 10,000 *g* for 10 minutes, the supernatant is collected into an ice-cooled beaker. A solution of 0.1 *M* acetic acid is added dropwise with swirling until the pH is reduced to pH 5.1; this causes the ribosomes and many cellular enzymes to coprecipitate, leaving the hemoglobin in solution. The precipitate is collected by centrifugation at 10,000 *g* for 10 minutes. The precipitate obtained from 1 ml of packed cells is redissolved in 1 ml of buffer A, B, or C (the choice depends on the experiment, as is described below); a few drops of 1 *M* Tris·HCl, pH 7.5, are added to the samples in order to facilitate their resolution; and the solutions are then clarified by centrifugation at 10,000 *g* for 10 minutes.

The resulting solutions contain $^{32}$P-labeled ribosomes and are used for

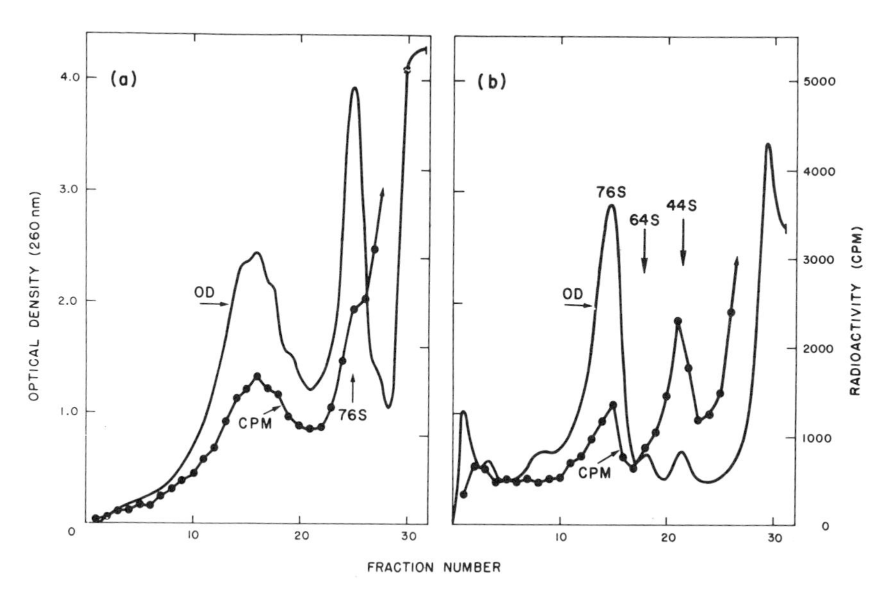


FIG. 1. Sucrose gradient sedimentation of $^{32}$P-labeled reticulocyte ribosomes. Reticulocytes were incubated for 60 minutes with [$^{32}$P]orthophosphate. The extract containing ribosomes was layered onto 29-ml linear gradients of 15–30% sucrose dissolved in buffer A in tubes for the SW 25.1 Spinco rotor. In (a), centrifugation was at 25,000 rpm for 2.5 hours. In (b) centrifugation was at 17,500 rpm for 20 hours; however, 18,000 rpm for 19 hours gives better results for preparative purposes. The gradient fractions were supplemented with 0.050 mg bovine serum albumin and were then precipitated with 10% trichloroacetic acid at 0° for 30 minutes. The precipitates were filtered onto 0.45 μm Millipore filters for radioactivity measurement from D. Kabat, *Biochemistry* 9, 4160 (1970).

further ribosome purification (see Section III). When such solutions are sedimented into sucrose gradients, results such as those in Fig. 1 are reproducibly obtained. It can be seen that radioactivity is associated with the rapidly sedimenting polyribosomes, with the single ribosomes (76 S), and with the large (66 S) and small (44 S) subribosomal particles. This radioactivity is precipitable with 10% trichloroacetic acid, suggesting that it is associated with macromolecules.

*D. Procedures for Sarcoma-180 Cells*

The sarcoma 180 cell culture is maintained by transferring 0.2 ml of ascites fluid ($1 \times 10^7$ cells) from the peritoneum of infected mice into the peritoneum of new 28–35 g female Swiss Webster mice. Cells are harvested for experimentation between 5 and 7 days postinoculation and are then washed three times by centrifugation with cold physiological salt solution. Generally, we obtain between 2 and 6 ml of ascites fluid (with 0.1–0.4 ml of packed cells per milliliter of ascites fluid) from each mouse. Cultures of ascites cells occasionally change their growth characteristics or karyotype. Consequently, in beginning work with such cultures it is desirable to store aliquots of the cells in a liquid $N_2$ freezer. We routinely use the procedure of Hauschka *et al.*[20] for sarcoma 180 cells except that dimethyl sulfoxide (10% of the final cell suspension volume) is used instead of glycerol.

For isotopic incorporation, the cells are suspended in the nutritional medium for sarcoma 180 cells in the ratio of 1 ml of packed cells to 29 ml of medium; the flasks are incubated at 37° in a rotary shaking water bath. Actinomycin D (10 μg/ml) is added to the cell suspension to inhibit RNA synthesis 10 minutes prior to addition of [$^{32}$P]orthophosphoric acid (50 μCi/ml of cell suspension). After incorporation for various time periods, the cells are chilled by dilution with four volumes of cold physiological salt solution and are collected by centrifugation at 600 *g* for 5 minutes at 2–4°. The cells are then washed twice by centrifugation in cold physiological salt solution.

Cell lysis and ribosome preparation are performed at 2°. Washed cells are suspended in three volumes of buffer C. Triton X-100 (Sigma Chemical Co., St. Louis, Missouri) is added at a concentration of 0.25% and lysis is accomplished by drawing the cell suspension into a Pasteur pipette six times. A small sample of the cell lysate is stained with toluidine blue and is observed with a phase contrast microscope in order to ensure that cell lysis is complete with no disruption of nuclei. The cell lysate is then centrifuged at 12,000 *g* for 10 minutes to produce a supernatant fraction

[20] T. S. Hauschka, J. T. Mitchell, and D. J. Niederpruem, *Cancer Res.* **19**, 643 (1959).

which we shall refer to as the cell extract. Ribosomes are precipitated from this cell extract by titration to pH 5.1 and are further treated as is described above for reticulocytes. Sedimentation of the resulting solutions containing $^{32}$P-labeled sarcoma 180 ribosomes in sucrose gradients is shown in Fig. 2. The labeling results appear to be very similar to those obtained with reticulocytes (Fig. 1).

This choice of conditions for lysis of sarcoma 180 cells with Triton X-100 deserves some comment. Lysis of rat liver or rabbit muscle cells in the presence of 1% Triton X-100 is known to solubilize certain membrane constituents[21-23] and to cause leakage of ribonucleoproteins from

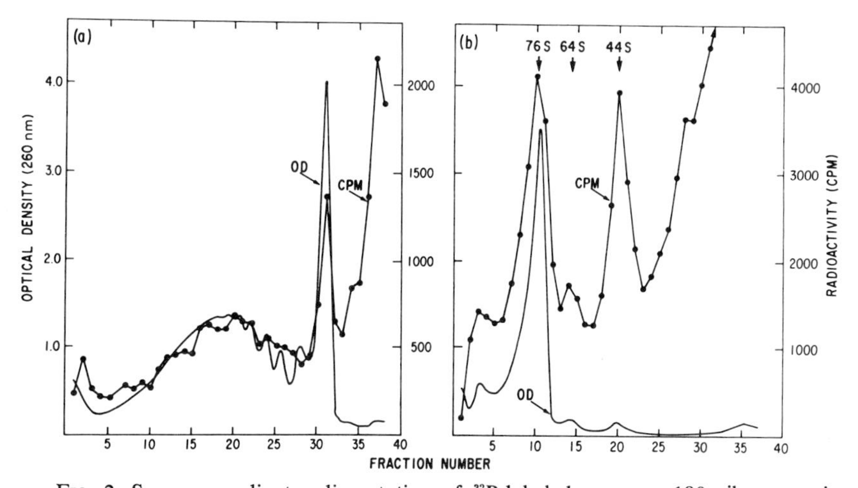


FIG. 2. Sucrose gradient sedimentation of $^{32}$P-labeled sarcoma 180 ribosomes in a Spinco SW 27 rotor. Sarcoma 180 cells were incubated for 60 minutes with [$^{32}$P]-orthophosphate in a nutrient medium containing 10 $\mu$g/ml actinomycin D. In (a), the ribosomes were concentrated by pH 5.1 precipitation and were then further purified by reprecipitation with 50 m*M* $MgCl_2$ [M. Takanami, *Biochim. Biophys. Acta* **39**, 318 (1960)]. Forty $OD_{260}$ units of ribosomes were then sedimented at 25,000 rpm for 3 hours in a linear 38 ml 15–40% sucrose gradient in buffer C. In (b), the preparative procedures were the same except that centrifugation was in buffer A at 25,000 rpm for 17 hours. In addition, the $MgCl_2$ precipitation step was eliminated because it causes a loss of subribosomal particles. Analytical procedures were as in Fig. 1, except that precipitates in 10% trichloroacetic acid were heated to 90° for 20 minutes before filtration onto membranes in order to remove nucleic acids. The experiments in frames (a) and (b) were done independently with different mice. Adapted from L. Bitte and D. Kabat, *J. Biol. Chem.* **247**, 5345 (1972).

[21] S. Olsnes, *Biochim. Biophys. Acta* **232**, 705 (1971).

[22] B. H. McFarland and G. Inesi, *Arch. Biochem. Biophys.* **145**, 456 (1971).

[23] A. L. J. Gielkens, T. J. M. Berns, and H. Bloemendal, *Eur. J. Biochem.* **22**, 478 (1971).

nuclei[24]; these can contaminate the resulting ribosome preparations. Furthermore, when this lysis is carried out in low ionic strength buffers, there occurs an artifactural adsorption of polyribosomes onto the modified membranes and released membranous proteins, resulting in a severe lowering of polysome yields. Such adsorption does not occur if the monovalent cation concentration is elevated to 150 m*M*.[21] For these reasons, we have used as low a Triton X-100 concentration as possible for complete cell lysis (0.25%) and a lysis buffer 140 m*M* in monovalent cations. Higher salt concentrations were avoided because S-180 single ribosomes are partially dissociated into subunits in 150 m*M* KCl.[19] When isolated in our conditions, polysomes are routinely obtained in high yields as compared with previous reports[19,23,25-27] and are undegraded as judged by the absence of nascent polypeptide chains from single ribosomes. We obtain between 4–7 mg of ribosomes from each milliliter of sedimented sarcoma 180 cells. Furthermore, the polysome size distribution is unaffected by dissolution of membranes in the cell extract with 0.5% sodium deoxycholate. This implies that the polysomal clusters in the extracts are not significantly aggregated onto membranes.[21]

## III. Purification of $^{32}$P-Labeled Mammalian Ribosomes

### *A. Principles*

The above methods yield solutions containing partially purified $^{32}$P-labeled reticulocyte and sarcoma 180 ribosomes. Further purification is needed in order to obtain preparations suitable for analysis of phosphorylated ribosomal components. However, there are no rigorous criteria for defining the purity of mammalian ribosomal preparations. Some true intracellular constituents may be only loosely bound and may be lost readily; conversely, many contaminants adsorb strongly to ribosomes. It has been our experience that multiple cycles of ribosome sedimentation and resolution is an ineffective means of removing such adsorbed contaminants. Furthermore, it is generally recognized by workers in this field that highly purified mammalian ribosomes are very unstable, especially when they have been repeatedly sedimented from solution and dissolved in fresh buffers.

Accordingly, we have used preparative procedures that do not require multiple cycles of ribosome sedimentation and resolution. The ribosomes

[24] S. Olsnes, *Biochim. Biophys. Acta* **213**, 149 (1970).
[25] B. L. M. Hogan and A. Korner, *Biochim. Biophys. Acta* **169**, 139 (1968).
[26] R. K. Morse, H. Hermann, and S. M. Heywood, *Biochim. Biophys. Acta* **232**, 403 (1971).
[27] I. Faiferman, L. Cornudella, and A. O. Pogo, *Nature* (*London*) **233**, 234 (1971).

are sedimented through sucrose solutions in buffers of differing ionic strengths. Although the phosphoprotein content of the preparations is highly dependent on the ionic strength used during the centrifugation, certain of the phosphoproteins cannot be extracted from the ribosomes with high ionic strength buffers, even in the presence of 0.5% sodium deoxycholate. Since these firmly bound components in reticulocyte and in sarcoma 180 ribosomes are very similar (see below), and since they occur in ribosomes in a reasonable stoichiometry,[2] we have concluded[4] that they are true ribosome constituents rather than contaminants.

### *B. Sucrose Gradient Purification*

Solutions containing ribosomes dissolved in buffers A, B, or C (see Section II, B) are fractionated by centrifugation in sucrose gradients made in the same buffers. Frequently, ribosomes in buffer B are adjusted to 0.5% sodium deoxycholate 5 minutes before layering onto the sucrose gradients. The sedimentation conditions for the Spinco SW 25.1 and SW 27 rotors are described in the legends to Figs. 1 and 2, respectively. Polyribosomes are obtained from gradients like those in Figs. 1a and 2a, whereas single ribosomes and subribosomal particles are prepared from gradients like those in Figs. 1b and 2b. After centrifugation, the gradients are pumped directly through a flow-cell in a Gilford spectrophotometer and the absorbance at 260 nm is plotted on a recorder. The appropriate regions of the sucrose gradient are collected directly into ice-cooled flasks and the ribosome fractions are then pelleted by centrifugation at 65,000 rpm for 2 hours in the Spinco 65 rotor or at 40,000 rpm for 4 hours in the Spinco 40 rotor. The $^{32}P$-labeled ribosome pellets are stored in a freezer.

### *C. Direct Sedimentation of Ribosomes*

Partially purified ribosomes dissolved in buffers A, B, or C are diluted to 7 ml and are layered carefully over 2 ml of 15% sucrose dissolved in the same buffers. Ribosomes in buffer B are often adjusted to 0.5% sodium deoxycholate before they are layered over the sucrose solutions. The ribosomes are then pelleted by centrifugation at 65,000 rpm for 45 minutes in the Spinco 65 rotor or at 40,000 rpm for 90 minutes in the Spinco 40 rotor. The supernatants are carefully decanted, and the ribosome pellets are stored in a freezer.

### *D. Uses of Ionic Strength in Preparation of $^{32}P$-Labeled Ribosomes*

Ribosomes prepared by the above two sedimentation methods (Sections B and C) are equally pure as judged by their phosphoprotein compositions. In other words, sedimentation of ribosomes through a small 2-ml cushion of sucrose solution removes the same $^{32}$P-labeled contaminants which are removed by sedimentation of the ribosomes through sucrose gradients. The more important variable which influences $^{32}$P-phosphoprotein content of isolated ribosomes is the ionic strength used for their preparation. The ionic strengths of buffers A, C, and B are 0.02, 0.17, and 0.29, respectively. Table I shows that a substantial portion of the nonnucleic acid $^{32}$P-labeled material is extracted from ribosomes by the higher ionic strength buffers.

As we will document more fully below, certain $^{32}$P-phosphoproteins remain associated with ribosomes regardless of the buffers employed for their preparation, whereas other phosphoproteins are more readily extracted. The extracted phosphoproteins may be contaminants or weakly bound ribosome constituents. Figure 3 shows an electrophoretic comparison on polyacrylamide gels of the $^{32}$P-labeled materials from reticulocyte

TABLE I
INFLUENCE OF IONIC STRENGTH OF BUFFERS ON $^{32}$P-CONTENT OF RIBOSOMAL PREPARATIONS

| Experiment | Source of ribosomes[a] | Buffer used[b] | Specific activity of ribosome preparations[c] (cpm/$OD_{260}$) | Percentage of radioactivity remaining |
|---|---|---|---|---|
| 1. | Reticulocytes | A | 2,750 | 100 |
| | | C | 1,760 | 64 |
| | | B | 1,180 | 43 |
| | | B+ | 450 | 16 |
| 2. | Reticulocytes | A | 3,470 | 100 |
| | | C | 1,820 | 54 |
| | | B+ | 880 | 26 |
| 3. | Sarcoma 180 | A | 25,400 | 100 |
| | | C | 22,800 | 90 |
| | | B+ | 12,000 | 47 |

[a] Ribosomes were all prepared by the direct sedimentation method from cells labeled for 60 minutes with [$^{32}$P]orthophosphate. The cells were labeled as described in Section II, except that the sarcoma 180 cell suspension contained 0.25 mCi/ml of [$^{32}$P]orthophosphate rather than 0.050 mCi/ml.

[b] Formulas of these buffers are given in Section II, B. Buffer "B+" is buffer B supplemented with 0.5% sodium deoxycholate.

[c] Radioactivity in ribosome preparations was measured after precipitation with 10% trichloroacetic acid. The precipitated samples in 10% trichloroacetic acid were heated at 90° for 20 minutes to hydrolyze nucleic acids and were then filtered onto 0.45 μm Millipore membranes. The membranes were rinsed thoroughly with 5% trichloroacetic acid.

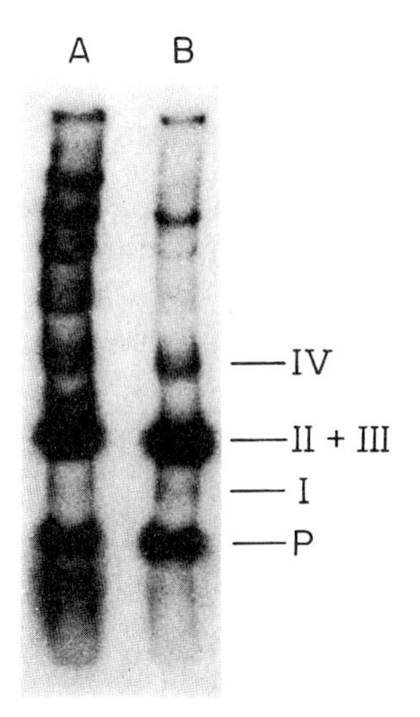


FIG. 3. Role of ionic strength in preparation of $^{32}$P-labeled ribosomes. The figure shows an electrophoretic fractionation of $^{32}$P-labeled components from reticulocyte ribosomes which had been purified either in low ionic strength buffer A or in high ionic strength buffer B. The techniques of electrophoresis in sodium dodecyl sulfate–polyacrylamide gels and of autoradiographic visualization of radioactive components are described in Section IV. Migration is toward the bottom. All resolved components have been identified as phosphoproteins except for the band labeled "P" which migrates faster than any ribosomal proteins and is extracted from ribosomes with 0.5% sodium deoxycholate. The nomenclature used to identify the phosphoproteins which remain bound to ribosomes washed with buffer B is described below (Section IV). Gel concentration is 4%. Components Ia and Ib are not separated in these 4% gels, and component III is not resolved from the more heavily labeled component II. Adapted from D. Kabat, *Biochemistry* 9, 4160 (1970).

ribosomes prepared in either buffers A or B; the radioactive components are visualized by autoradiography. Clearly, the effect of raising the ionic strength is to cause a highly selective extraction of phosphorylated components.

## IV. Electrophoresis of $^{32}$P-Labeled Ribosomal Components in Sodium Dodecyl Sulfate-Polyacrylamide Gels

### *A. Principles*

The $^{32}$P-labeled constituents of mammalian ribosomes can be conveniently analyzed by electrophoresis in polyacrylamide gels in the presence of sodium dodecyl sulfate (SDS). Many of the basic techniques have been recently described.[28] The advantages of this method are the following: (a) Ribosomes and most other subcellular assemblages of macromolecules are fully dissolved and dissociated into their components in the presence of the SDS buffer. The sample can be layered directly onto the gel without prior separation of the RNA and protein constituents. The method is therefore very convenient and simple. (b) The fractionation of proteins in SDS-gels is on the basis of their molecular weights; a plot of the mobility of proteins versus the logarithm of their molecular weights falls approximately on a straight line.[29,30] (c) The resolution of constituents is excellent; it is at least as good as other electrophoretic methods of fractionating mammalian ribosomal proteins.[31]

*B. Reagents and Solutions*

a. Acrylamide-bisacrylamide stock solutions. These are stable at 4° for several months. The acrylamide and *N,N'*-methylenebisacrylamide are obtained from Eastman Organic Chemicals (Rochester, New York) and are recrystallized before use.[32] Acrylamide is dissolved in $CHCl_3$ at 50°, and the solution is filtered. Crystallization occurs during storage overnight at −20°. The crystals are washed with heptane. Methylenebisacrylamide is recrystallized from water and then from acetone. After dissolving in $H_2O$ at 90° and filtering, the solution is stored overnight at 4° and the crystals are collected by filtration. 10 g of crystals are then dissolved in acetone at 50° and are recrystallized at −20°.

Stock 1 (for gels containing less than 5% acrylamide). This contains 15 g acrylamide and 0.75 g of *N,N'*-methylenebisacrylamide in 100 g of solution.

Stock 2. This contains 20 g of acrylamide and 0.5 g of bisacrylamide in 100 g of solution.

b. Ammonium persulfate. A 1.6% solution is kept in the refrigerator and is prepared freshly each week.

c. Stock electrophoresis buffer (10×). It is 0.36 *M* Tris, 0.30 *M* $NaH_2PO_4$, and 10 m*M* EDTA. The actual electrophoresis buffer

[28] J. V. Maizel, Jr., *in* "Methods in Virology" (K. Maramorosch and H. Koprowski, eds.), Vol. V. Academic Press, New York, 1971.

[29] A. L. Shapiro, E. Viñuela, and J. V. Maizel, Jr., *Biochem. Biophys. Res. Commun.* **28**, 815 (1967).

[30] A. K. Dunker and R. R. Reuckert, *J. Biol. Chem.* **244**, 5047 (1969).

[31] H. W. S. King, H. J. Gould, and J. J. Shearman, *J. Mol. Biol.* **61**, 143 (1971).

[32] U. E. Loening, *Biochem. J.* **102**, 251 (1967).

contains 100 ml of stock buffer, 900 ml of $H_2O$, and 6.0 g of sodium dodecyl sulfate (Matheson, Coleman and Bell).

d. TEMED: *N,N,N',N'*-Tetramethylethylenediamine (Eastman Organic Chemicals, Rochester, New York).

e. Coomassie brilliant blue stain (Schwarz/Mann, Orangeburg, New York). This contains 0.2% in destaining solution.

f. Destaining solution. This contains 100 ml of acetic acid, 200 ml of methanol, and 700 ml of $H_2O$.

g. Gel dissolving solution.[33] It must be freshly made. It contains 1% $NH_4OH$ dissolved in 30% $H_2O_2$.

h. Scintillation fluid. This is made by mixing 3 g of 2,5-diphenyloxazole (PPO), 0.3 g of 1,4-bis[2(4-methyl-5-phenyloxazolyl)]-benzene(dimethyl POPOP), 400 ml of toluene, and 800 ml of Triton X-100.

### *C. Gel Preparation*

Eight-centimeter gels are made in 6 mm (i.d.) plastic tubes. For preparation of 4% gels, 6.67 ml of stock 1 acrylamide–bisacrylamide and 1.25 ml of 1.6% ammonium persulfate are diluted to 25 ml with electrophoresis buffer lacking SDS; 20 μl of TEMED is added. The solution is then quickly mixed and placed into the plastic tubes, and the menisci are overlain with 100 μl of $H_2O$. Polymerization is complete in 30 minutes at room temperature. 8% Gels are made identically, except that we use 10 ml of the stock 2 acrylamide–bisacrylamide solution. Before use, the gels in the plastic tubes are soaked overnight at room temperature in electrophoresis buffer containing 0.6% SDS.

### *D. Sample Preparation*

The pellets containing purified ribosomes are dissolved at room temperature in a few drops of 0.6% SDS–electrophoresis buffer adjusted to 5% sucrose. Of the resulting solution, 10 μl, is added to 1 ml of $H_2O$ and the absorbance at 260 nm is measured. The absorbance should be between 0.5 and 1.0 units/ml. The solution is then adjusted to 1% 2-mercaptoethanol and is heated at 60° for 30 minutes to reduce disulfide bonds.

### *E. Electrophoresis*

The gels are prerun at 5 mA/gel for 1 hour before 1 $OD_{260}$ unit of the ribosome sample in a volume of 10–30 μl is layered onto the upper

[33] D. Goodman and H. Matzura, *Anal. Biochem.* **42**, 481 (1971).

surface of the gel. Electrophoresis is at room temperature at 5 mA/gel for approximately 2 hours.

*F. Staining of Proteins in the Gels*

Proteins in the gel can be stained with Coomassie brilliant blue. The gels are removed from the plastic tubes and are washed overnight at room temperature with 12.5% trichloroacetic acid. This is required for removing SDS from the gel, since the detergent coprecipitates with the stain. The gels are then placed into the stain for 8–16 hours and are then washed overnight with several changes of destaining solution. Destaining is continued until the gel background is clear.

*G. Autoradiographic Detection of $^{32}P$-Labeled Constituents*

Radioactive components in the gels can be visualized by autoradiography. The cylindrical gels are cut into 4 longitudinal sections with the apparatus described by Fairbanks *et al.*[34] They are then dried onto high-wet strength paper (Schleicher and Schuell, No. 497) or onto Whatman 3 MM paper. Gels with 4% polyacrylamide can be dried without special apparatus on the high-wet strength paper; the gel slices dry smoothly without shrinking or cracking in an oven at 75°. However, more concentrated gels require a special drying technique.[28,34] The dried gels on the paper backing are then pressed together with X-ray film (Kodak, single coated type SB-54) for varying time periods before the film is developed.

*H. Transverse Sectioning of Gels for Direct Measurement of Radioactivity*

We frequently transection the gels with a commercially available sectioning apparatus (Brinkman Instruments, Westbury, New York). Gels for this purpose are made as described above except that they also contain 10% glycerol; this facilitates the accurate sectioning of frozen gels.[35] The 1 mm gel sections are dissolved by incubation at room temperature in 0.8 ml of gel dissolving solution. The incubation is for 24–48 hours in tightly stoppered scintillation vials. After addition of 15 ml of scintillation fluid, radioactivity is measured in a liquid scintillation spectrometer.

*I. Uses of Electrophoresis for Analyzing $^{32}P$-Labeled Ribosomes*

We present here some representative electrophoretic analyses of the radioactive components from $^{32}P$-labeled mammalian ribosomes. On the

[34] G. Fairbanks, Jr., C. Levinthal, and R. H. Reeder, *Biochem. Biophys. Res. Commun.* **20**, 343 (1965).

[35] R. A. Weinberg, U. E. Loening, M. Willems, and S. Penman, *Proc. Nat. Acad. Sci. U.S.* **58**, 1088 (1967).

one hand, these data illustrate the reproducibility and some uses of this method. Second, the $^{32}$P-labeled phosphoproteins from rabbit reticulocyte and from mouse sarcoma 180 ribosomes coelectrophorese in the SDS-polyacrylamide gels, suggesting that they have the same molecular weights.[4] Based on this apparent similarity between different species and tissues, we anticipate that these electrophoretic patterns may be generally obtained for mammalian cells. Accordingly, this data should serve as a useful prototype for comparison with future analyses made in other laboratories.

Figure 4 shows a typical electrophoretic fractionation in an 8% polyacrylamide gel of the $^{32}$P-labeled components from reticulocyte ribosomes, the radioactive components being visualized by autoradiography. All the resolved components are phosphoproteins, as determined by the criteria described below, and they are numbered in the order of their mobilities in the gel. Components Ia and Ib migrate closely together on 4% gels and

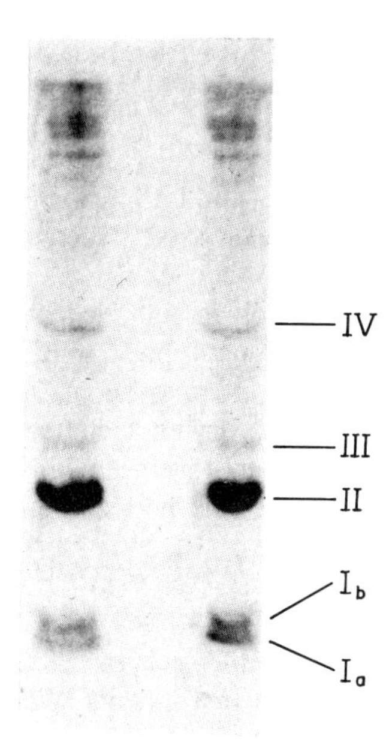


FIG. 4. Electrophoresis of $^{32}$P-labeled components from reticulocyte ribosomes in a sodium dodecyl sulfate–polyacrylamide gel. The gel concentration is 8%; visualization of radioactive components is by autoradiography; and migration is toward the bottom. The reticulocytes were labeled with [$^{32}$P]orthophosphate for 60 minutes (Section II), and the ribosomes were then purified by the direct sedimentation method in high ionic strength buffer B supplemented with 0.5% sodium deoxycholate (Section III). The two internal sections of the gel were dried next to each other on the paper backing and used for autoradiography. Both of the sections are presented to illustrate the reproducibility obtained.

are resolved slightly on 8% gels. In addition to the five numbered components (Ia, Ib, II, III, and IV) there are several more slowly migrating phosphoproteins which occur in ribosome preparations in variable amounts. The five components assigned numerals in Fig. 4 are reproducibly obtained in ribosome preparations. We have estimated their molecular weights[28-30] as 18,200, 19,500, 27,500, 33,000, and 53,000 for components Ia, Ib, II, III, and IV, respectively.

However, polyribosomes and single ribosomes contain different proportions of these radioactive phosphoproteins. Figure 5 shows a time course for the $^{32}$P-labeling of these constituents in reticulocyte poly- and single ribosomes. Single ribosomes contain relatively more of components I and III, whereas polysomes contain relatively more radioactivity in component II. Analysis of this difference has indicated[2] that mammalian ribosomes are heterogeneous in their molecular constitution.

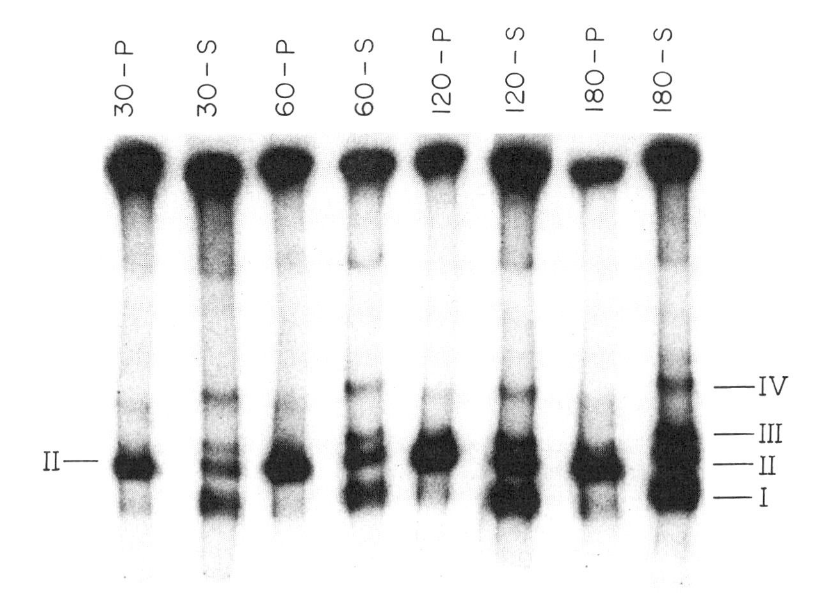


FIG. 5. Comparison of $^{32}$P-labeled components from reticulocyte polyribosomes and single ribosomes by electrophoresis in sodium dodecyl sulfate–polyacrylamide gels. Gel concentration is 4%, and the visualization of radioactive components is by autoradiography. Ribosomes were purified in buffer B containing 0.5% sodium deoxycholate (see Section III) from cells labeled for 30, 60, 120, or 180 minutes with [$^{32}$P]orthophosphate. S is single ribosomes; P is polyribosomes. These types of ribosomes have distinctive and reproducibly obtained patterns of phosphorylation. Adapted from D. Kabat, *J. Biol. Chem.* **247**, 5338 (1972).

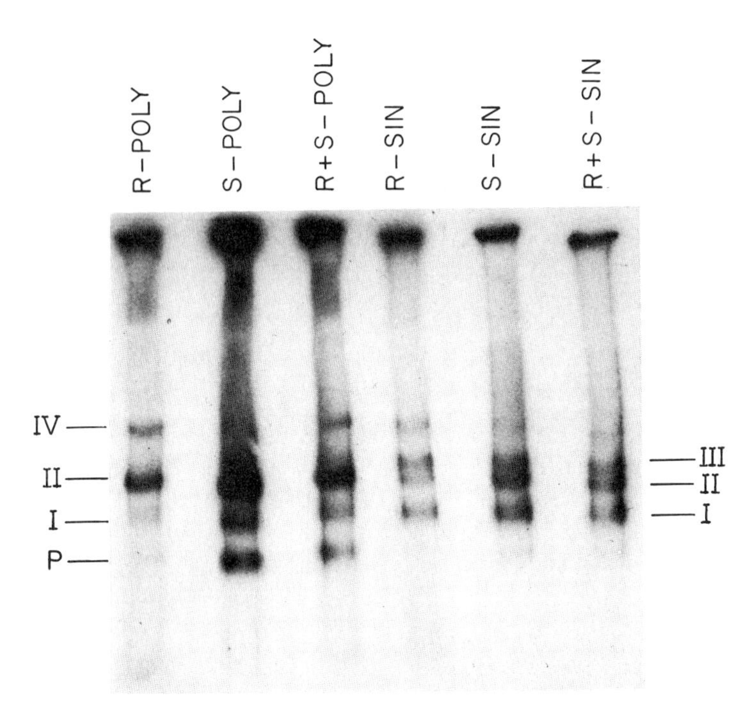


FIG. 6. Comparative electrophoresis of $^{32}$P-labeled ribosomal components from rabbit reticulocytes and from mouse sarcoma 180 cells. The cells were labeled for 60 minutes and their ribosome fractions were then prepared from sucrose gradients containing buffer C. Gel concentrations are 4%; migration is downward; and the radioactive components are visualized by autoradiography. The nomenclature of the phosphoproteins I–IV is described in the text. The "P" component occurs in polysomes from reticulocytes and from sarcoma 180 cells in variable amounts but may not be a protein (see Fig. 3). S-SIN and S-POLY are sarcoma 180 single ribosomes and polyribosomes, respectively. R-SIN and R-POLY are reticulocyte single ribosomes and polyribosomes, respectively. R + S-SIN is mixture of reticulocyte and of sarcoma 180 single ribosomes. R + S-POLY is a mixture of reticulocyte and of sarcoma 180 polyribosomes. The sarcoma 180 preparations are usually more contaminated than those from reticulocytes, as evidenced by the higher background of radioactivity on the sarcoma 180 gels. From L. Bitte and D. Kabat, *J. Biol. Chem.* **247**, 5345 (1972).

A comparison of the $^{32}$P-labeled components from rabbit reticulocyte and mouse sarcoma 180 polyribosomes and single ribosomes is shown in Fig. 6. All the labeled components except P have been identified as phosphoproteins. The reticulocyte phosphoproteins are identified by Roman numerals as was described above. Clearly, the phosphoprotein patterns from reticulocyte and from sarcoma 180 ribosomes are very similar. No electrophoretic components appear only in sarcoma 180 or only in reticulocyte ribosomes since mixtures of ribosomes from these two sources exhibit the same number of bands as do the unmixed samples. Furthermore, the distribution of these components between polysomes and single ribosomes is similar in both cells; however, significant differences between these distributions have been noted.[4]

## V. Membrane Filter Assay of Radioactivity in the Phosphoprotein Components of $^{32}$P-Labeled Ribosomes

### *A. Principles*

The quantity of radioactivity incorporated into protein components of $^{32}$P-labeled ribosomes can be analyzed following protein precipitation with 10% trichloroacetic acid (TCA). The acidified precipitates are heated to 90° for 20 minutes and are then collected onto membrane filters. Methods are described for removing three types of radioactive contaminants from the precipitated proteins.

### *B. Removal of RNA*

Two procedures are employed to eliminate contamination with radioactive RNA. First, the incorporation of [$^{32}$P]orthophosphate into ribosomes occurs in conditions in which RNA synthesis is greatly inhibited (Section II). Second, nucleic acids are removed by heating in 10% TCA at 90° for 20 minutes.[36,37] However, it should be recognized that this treatment also results in a 10–20% loss of protein from the precipitates.[37]

The amount of radioactivity lost from TCA-precipitated $^{32}$P-labeled ribosomes by heating at 90° for 20 minutes is presented in Table II. Reproducibly, this treatment removes only a small fraction of radioactivity from the precipitates.

That this treatment with hot TCA causes quantitative extraction of $^{32}$P-labeled RNA was verified using purified radioactive rRNA. In addition, we have analyzed reticulocyte ribosomes which were labeled *in vivo,* in conditions favoring the labeling of RNA. An anemic rabbit was bled 38 hours after it had been injected with 1 mCi of [$^{32}$P]orthophosphate. The polysomes had a specific activity of 31,000 cpm/mg protein, as measured after precipitation with 10% TCA at 0°. However, the specific activity was reduced to 430 cpm/mg protein after heating in 10% TCA to 90° for 20 minutes. This residual radioactivity was not further reduced by a

[36] W. C. Schneider, *J. Biol. Chem.* **161**, 293 (1945).
[37] D. Kennell, this series, Vol. 12, p. 686.

TABLE II
HOT TRICHLOROACETIC ACID LABILE RADIOACTIVITY IN PRECIPITATES OF $^{32}$P-LABELED RIBOSOMES

| Experiment | Source of ribosomes[a] | Filtering method[b] | Percentage of radioactivity lost to hot TCA treatment |
|---|---|---|---|
| 1. | Reticulocytes | A | 14 |
| 2. | Sarcoma 180 | A | 25 |
| | | B | 27 |
| | | C | 23 |

[a] Ribosomes were $^{32}$P-labeled and isolated from buffer B as described in the text (Sections II, B and C, and III, C). The resulting ribosomal pellets were dissolved in cold water and were used for 10% TCA precipitation at 0°.

[b] Filtration Method A. All ribosome samples were precipitated in cold 10% TCA in test tubes. Control samples were chilled for 60 minutes, were collected by filtration onto Millipore filters (0.45 μm), and were then rapidly rinsed 3 times with 5% TCA at room temperature. The remaining samples were allowed to stand in ice for only 10 minutes, were heated at 90° for 20 minutes and then chilled to 0° for 30 minutes, and were then collected onto Millipore filters as described above.

Filtration Method B. This method is identical to method A except that fiberglass filters (Reeve Angel, 934A-A) were substituted for Millipore filters.

Filtration Method C. The ribosomal preparations were spotted onto the center of fiberglass filters (50 μl samples) and were dried at room temperature. The filters were then placed into 50 ml of cold 10% TCA for 30 minutes. Control filters remained in the cold acid for an additional 50 minutes before they were placed in a Millipore filtration apparatus for rinsing with 5% TCA as in Methods A and B. The remaining filters were heated to 90° for 20 minutes, were chilled on ice for 30 minutes, and were then rinsed on the filtration apparatus as described above.

second treatment with hot acid, suggesting that it was not in unhydrolyzed nucleic acids. Thus, the hot TCA method is very efficient for removing RNA from $^{32}$P-labeled ribosomal proteins.

### *C. Extraction of Phospholipids*

The quantity of radioactivity in contaminating phospholipids is estimated by measuring the loss of radioactivity when the TCA-precipitate is extracted with a series of lipid solvents. The solvents and sequence of extractions used were suggested by Davidson *et al.*[38] The precipitate is extracted sequentially once with acetone, ethanol, and chloroform; twice with ethanol–ethyl ether (3:1); and then once with ethyl ether.

One method we have used[1,4] is to extract the TCA-precipitated ribosomal proteins in test tubes following the hot acid treatment (see Section

[38] J. N. Davidson, S. C. Fraser, and W. C. Hutchison, *Biochem. J.* **49**, 311 (1951).

B). The precipitate is sedimented by centrifugation, and the solvents are removed by aspiration. However, this procedure produces variable losses of protein from the precipitate and is especially difficult for the chloroform extraction in which the precipitate floats on the solvent surface.

Consequently, we have more recently used a method in which the TCA precipitation is done in a glass fiber filter (Table II, Method C) and the filter is then extracted with the series of nonaqueous solvents. We observe no loss of radioactivity when this procedure is used for samples of $^{32}P$-labeled ribosomes from reticulocytes and from sarcoma 180 cells. This is true regardless of the buffers used for ribosome preparation (see Table I). Hence, we conclude that phospholipid contamination does not contribute significantly to the radioactivity which remains in the ribosomal precipitates after the hot acid treatment.

### *D. Other Contaminants*

Greenaway[39] has shown that Mg–ATP complexes can contaminate proteins precipitated with cold 10% TCA. Dissolving the precipitates in cold 0.1 *M* NaOH and then reprecipitating with TCA was suggested as a method for removing this source of contamination. Other workers[40] have also used a 0.1 *M* NaOH resolution step as a means to remove contaminants from phosphoproteins precipitated with cold 10% TCA. However, such contamination is apparently not significant when the precipitates of $^{32}P$-labeled ribosomal phosphoproteins are heated in 10% TCA at 90° for 15–20 minutes. Resolution of such treated precipitates in cold 0.1 *M* NaOH does not cause any further reduction in measured radioactivity.

## VI. Proteolytic Digestion of Ribosomal Phosphoproteins

### *A. Principles*

Pronase (*Streptomyces griseus* protease) is highly active in solutions containing 0.6% sodium dodecyl sulfate, whereas ribonuclease is markedly inhibited in these conditions.[41] This forms the basis of a simple procedure for proteolysis of phosphoproteins in the absence of any degradation of nucleic acids. The electrophoretic mobilities of the $^{32}P$-labeled ribosomal constituents are compared before and after the proteolysis.

### *B. Proteolysis*

Sedimented pellets of $^{32}P$-labeled ribosomes are dissolved in electro-

[39] P. J. Greenaway, *Biochem. Biophys. Res. Commun.* **47**, 639 (1972).

[40] R. J. DeLange, R. Kemp, W. D. Riley, R. A. Cooper, and E. G. Krebs, *J. Biol. Chem.* **243**, 2200 (1968).

[41] D. Kabat, *Anal. Biochem.* **39**, 228 (1971).

phoresis buffer containing 0.6% sodium dodecyl sulfate as described above (Section IV, D). After the absorbance of the solution at 260 nm is adjusted to 50 units/ml, an aliquot containing 20 μl is placed into a second tube. Two microliters of a freshly prepared Pronase solution [5 mg/ml Pronase (Calbiochem, Los Angeles, California) dissolved in the 0.6% SDS-electrophoresis buffer] is added to this tube. After 30 minutes at room temperature, the Pronase-treated sample is layered onto the surface of a polyacrylamide gel for electrophoresis. Twenty microliters of the untreated control sample is layered onto another gel. Electrophoresis is then performed as described above (Section IV).

Typical electrophoretic results with Pronase digested $^{32}P$-labeled reticulocyte ribosomes have been described.[1] All the radioactive components are converted by pronase into more rapidly migrating materials which have a mobility approximately the same as the fast moving P component (see Fig. 3). Because polypeptide mobilities in SDS-gels are increased when their molecular weights are reduced,[28-30] such data suggest that the labeled components are proteins. Furthermore, it has been shown[41] that RNA is not degraded by this proteolysis procedure.

## VII. Methods for Selective Cleavage of Phosphate–Protein Bonds

### *A. Principles*

Information concerning the mode of binding of phosphate with protein can be obtained by studying the conditions which result in bond cleavage.

### *B. Alkaline Phosphatase*

*Escherichia coli* and intestinal alkaline phosphatases (orthophosphoric monoester phosphohydrolase EC 3.1.3.1) are relatively nonspecific phosphomonoesterases which also catalyze hydrolysis of acyl phosphates.[42] The phosphate which is bonded to ribosomal proteins is efficiently cleaved by the *E. coli* enzyme,[1] in conditions in which ribonucleic acid remains stable.

$^{32}P$-labeled ribosomes are dissolved in buffer B (see Section II, B) at a concentration of 8 mg/ml and the solution is separated into two portions. After warming to 37°, a small volume of *E. coli* alkaline phosphatase (ribonuclease-free type BAPF, Worthington Biochemical Corp., Freehold, New Jersey) is added to one portion of the ribosome solution to give a final enzyme concentration of 0.4 mg/ml (13 U/ml); 0.1-ml aliquots of the solutions are periodically removed into tubes containing 1.9 ml of 5% trichloroacetic acid. These tubes are warmed to 90° for 15 minutes and

[42] J. W. Sperow and L. G. Butler, *Biochim. Biophys. Acta* **146**, 175 (1971).

are then chilled to 0° for at least 30 minutes before the precipitates are filtered onto Millipore membranes for radioactivity measurement. Other samples of the ribosome solutions are cooled in ice at various times after enzyme addition and the ribosomes are subsequently pelleted by centrifugation (Section III, C). These sedimented ribosomes are then analyzed by electrophoresis in polyacrylamide gels (Section IV).

This enzymatic treatment quantitatively removes the $^{32}$P-labeled phosphate from the mammalian ribosomal proteins I–IV (Fig. 4). The half-time for removal of the ribosome-associated radioactivity is 5 minutes in these conditions. Components Ia and Ib are dephosphorylated more rapidly than is component II. Furthermore, the integrity of polyribosomal structure is not destroyed by dephosphorylation of the ribosomal proteins.

### C. *Hydroxylamine*

One common mode of bonding of phosphate with proteins is by acyl linkage to carboxyl groups. Such acyl phosphates are susceptible to rapid aminolysis at pH 5.5 with hydroxylamine. On the contrary, phosphate esters are stable in these conditions.

$^{32}$P-labeled ribosomes are dissolved in water at a concentration of 10 mg/ml. One aliquot is diluted into 9 volumes of 1 *M* succinic acid–1 *M* hydroxylamine, pH 5.5; whereas a second aliquot is diluted into 9 volumes of 1 *M* succinic acid, pH 5.5 lacking any hydroxylamine. The resulting ribosome solutions are incubated at 37°, and 0.1-ml aliquots are periodically taken into tubes containing 1.9 ml of 5% trichloroacetic acid. These tubes are heated to 90° for 15 minutes and are then chilled for at least 30 minutes before the precipitates are filtered onto Millipore membranes for radioactivity measurement. Acyl phosphate bonds in proteins are very rapidly cleaved in these conditions.[43,44] However, the bonds of phosphate with ribosomal proteins are resistant to this treatment.

### D. *The pH Stability of Phosphate–Protein Bonds*

The stability of phosphate–protein bonding is highly dependent upon the pH. Furthermore, the shape of the curve relating bond stability with pH is characteristic of the type of bonding.[44] For example, acyl phosphates are rapidly hydrolyzed at both extremes of pH, a fact that imposes a U shape on the stability curves. On the other hand, phosphate monoesterified to hydroxyl groups of seryl or threonyl residues are very stable at low pH and unstable at high pH. If the phosphoryl group is linked to a nitrogen of a histidyl or lysyl residue, the hydrolysis curve is high at low pH and low

[43] A. Martonosi, *J. Biol. Chem.* **244**, 613 (1969).

[44] R. S. Anthony and L. B. Spector, *J. Biol. Chem.* **247**, 2120 (1972).

at high pH. Phosphoryl groups attached to the oxygen of a tyrosyl residue or to a sulfur atom of a cysteinyl residue also exhibit distinctive hydrolytic behaviors.[44]

$^{32}P$-labeled ribosomes are dissolved in water at a concentration of 5 mg/ml. 0.1-ml aliquots are then diluted into tubes containing 1.9 ml of various buffers. Two tubes are prepared for each buffer. The buffers (0.10 *M* unless otherwise specified) are pH 1.2 and 2.4 (HCl and KCl), pH 3.1 ($KH_2PO_4$), pH 5.9 ($K_2HPO_4$), pH 7.2 and 8.4 (Tris·HCl), pH 9.8 ($KHCO_3$–$K_2CO_3$), pH 13.1 (0.20 *N* KOH) and pH 13.5 (1.0 *N* KOH). After incubation at 55° for 120 minutes, the samples are chilled to 0° and adjusted to 20% trichloroacetic acid. The resulting precipitates are collected by centrifugation and resuspended in cold 5% trichloroacetic acid. Of the two samples incubated in each of the buffers, one is filtered directly onto a Millipore membrane for radioactivity measurement. The other sample is heated to 90° for 20 minutes and is then chilled to 0° for 30 minutes before filtration. The latter procedure removes $^{32}P$-labeled nucleic acids which may be present in the ribosome preparations. Radioactivity on the membranes is then measured in a low background gas flow counter. The pH stability profile of sarcoma 180 ribosomal phosphoproteins is shown in Fig. 7. The shape of this profile is that expected of *o*-phosphoseryl and *o*-phosphothreonyl residues.

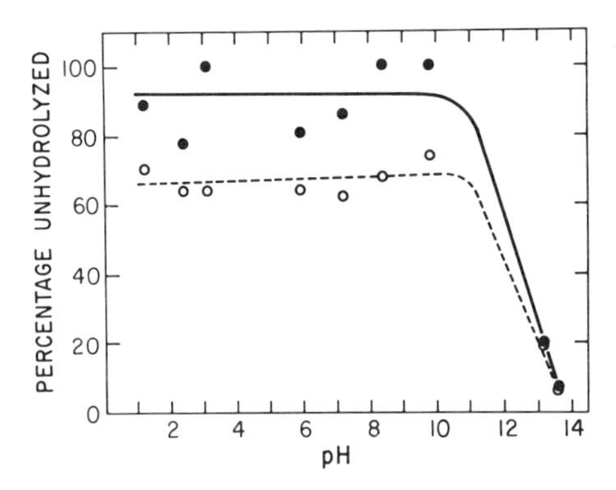


FIG. 7. Effect of pH on the stability of ribosomal protein-phosphate bonds. The $^{32}P$-labeled sarcoma 180 ribosomes were isolated by the direct sedimentation method. After incubation at 55° for 2 hours at each pH, samples were precipitated with cold 5% trichloroacetic acid and were filtered onto membranes. Other samples were heated in 5% trichloroacetic acid at 90° for 20 minutes before filtration; this removes nucleic acids and 15–20% of proteins from the precipitates. This latter portion of the data shows that the radioactivity analyzed here is not in nucleic acids. The percentage of radioactivity is measured with respect to a control sample at pH 7.2 which was kept chilled at 0°. ●—●, Cold trichloroacetic acid; ○- -○, treated with hot trichloroacetic acid.

## VIII. Detection of *o*-Phosphoserine and *o*-Phosphothreonine in Ribosomal Hydrolyzates

### *A. Principles*

Radioactive *o*-phosphoserine and *o*-phosphothreonine can be detected in acid hydrolyzates of $^{32}$P-labeled ribosomal proteins. This is possible because the hydrolysis of the phosphomonoester bonds is considerably slower in acid than the hydrolysis of peptide bonds. After hydrolysis of the protein, the amino acids are fractionated by high voltage paper electrophoresis at pH 1.85; only the phosphorylated amino acids migrate toward the cathode in these conditions.[13]

### *B. Hydrolysis*

Solutions containing $^{32}$P-labeled ribosomes are precipitated with 5% trichloroacetic acid at 0°. After heating at 90° for 15 minutes to hydrolyze nucleic acids and subsequent chilling at 0° for 30 minutes, the radioactive precipitate is collected by centrifugation. It is washed with ethanol–ethyl ether (3:1) and then with acetone. The precipitate from 1 mg of ribosomes is suspended in 2 ml of 6 *M* HCl in a sealed hydrolysis tube, which is then heated to 105° for 7 hours. The hydrolyzate is evaporated to dryness in a vacuum desiccator and the residue is dissolved in 100 μl of paper electrophoresis buffer (see below). The quantity of *o*-phosphoserine and of *o*-phosphothreonine is analyzed following electrophoretic separation of the amino acids.

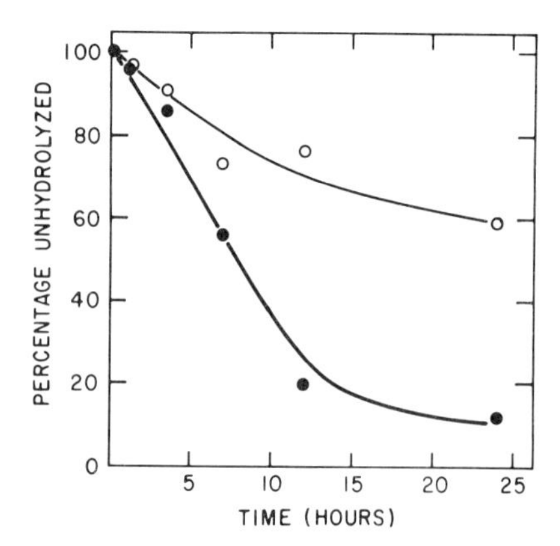


FIG. 8. Kinetics of hydrolysis of *o*-phosphoserine (●) and of *o*-phosphothreonine (○) in 6 *M* HCl at 105°. These amino acids were purified from hydrolyzates by paper electrophoresis, were staining with Cd-ninhydrin, and were then quantitatively assayed (see text).

The rates of hydrolysis of *o*-phosphoserine and of *o*-phosphothreonine in these hydrolysis conditions are shown in Fig. 8. Phosphothreonine is relatively resistant to hydrolysis. We selected 7 hours as the optimal time for phosphoprotein hydrolysis in order to allow complete polypeptide hydrolysis with only minimal loss of these phosphorylated amino acids.

*C. Paper Electrophoresis*

Electrophoresis is in a water-cooled flat plate apparatus on Whatman 3 MM paper.[13] The paper strip (10 inches wide by 20 inches long) is wet with paper electrophoresis buffer (2.5% formic acid, 7.8% acetic acid) and is blotted so that it is only slightly damp. Radioactive sample (10 $\mu$l) is applied at the origin; 10-$\mu$l control samples containing 2 $\mu$g of *o*-phosphoserine and of *o*-phosphothreonine are applied to adjacent positions on the origin. A control sample containing [$^{32}$P]orthophosphate is also analyzed. A voltage of 3000 V is then applied across the electrodes for 120 minutes. The paper is dried in an oven at 75°, then sprayed thoroughly with a Cd-ninhydrin stain[45] (made by mixing 200 ml of acetone, 20 ml of $H_2O$, 4.0 ml of acetic acid, 200 mg of Cd acetate and 2.0 g of ninhydrin; this stain is stable in dark bottles at 4°); it is then heated at 75° for 3 hours. Positions of the marker phosphoserine and phosphothreonine are circled with a pencil and the paper is placed together with X-ray film for autoradiographic localization of radioactive compounds. Such regions of the paper can be cut out and analyzed quantitatively for radioactivity in a liquid scintillation spectrometer.

The color which develops on the paper after Cd-ninhydrin staining can be quantitatively eluted with methanol. Its absorbance at 500 nm is proportional to the quantity of the amino acid.[45]

Figure 9 shows a typical electrophoretic analysis of an acid hydrolyzate of $^{32}$P-labeled reticulocyte ribosomes. The radioactive spots are visualized by autoradiography; and the positions of electrophoresis of *o*-phosphoserine, *o*-phosphothreonine, and orthophosphate are also indicated. Of the total radioactivity in the hydrolyzates, we routinely observe that approximately 13% coelectrophoreses with *o*-phosphoserine, 2% with *o*-phosphothreonine, and the remainder with orthophosphate. These data support other evidence that these amino acids are present in mammalian ribosomes; however, the data do not exclude the possibility that other phosphorylated amino acids might occur in ribosomes. Furthermore, the actual percentages of phosphoserine and phosphothreonine cannot be determined by this method because the rates of phosphomonoester hydrolyses are likely to

[45] W. J. Dreyer and E. Bynum, this series, Vol. 11, p. 32.

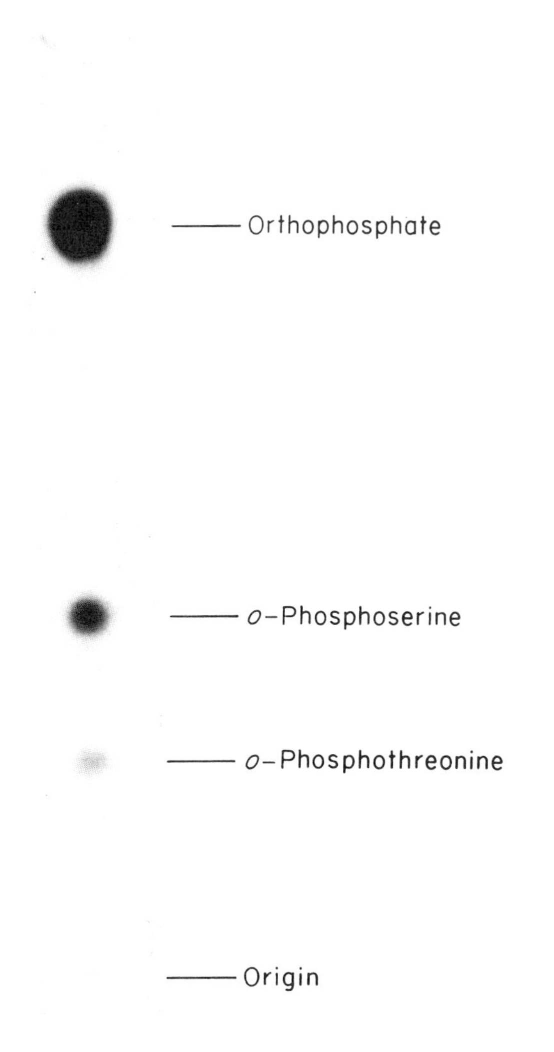


FIG. 9. Separation of $^{32}$P-labeled *o*-phosphoserine, *o*-phosphothreonine, and orthophosphate by paper electrophoresis. The radioactive hydrolyzate (12,000 dpm) is from $^{32}$P-labeled reticulocyte polyribosomes, and the compounds are visualized by autoradiography. The positions of electrophoresis of standard marker compounds are indicated.

be different (from Fig. 8) when the amino acids are present in peptide chains.

## IX. Quantitative Analysis of Phosphate in Ribosomal Proteins

### *A. Principles*

Nucleic acids are removed from ribosomes by heating in 5% trichloroacetic acid (TCA) at 90° for 30 minutes and from phospholipids by extraction. The residual protein fraction contains phosphoseryl and phosphothreonyl residues. Such phosphate is released by incubation with 1 *M* NaOH at 37° for 16 hours, and it is analyzed by a chemical method. Such analysis indicates that the quantity of phosphate bonded to ribosomal proteins is appreciable (approximately 14 phosphates per ribosome).

### *B. Extraction of Nonprotein Phosphate*

Solutions containing 10 mg of ribosomes are adjusted to 10% TCA, and the precipitates are collected by centrifugation. They are then washed twice by centrifugation with 5 ml of cold 5% TCA. The precipitates are resuspended in 5 ml of 5% TCA, then the suspensions are heated to 90° for 15 minutes; after chilling to 2° for 30 minutes, the precipitates are resedimented. The latter step is repeated, and the resulting precipitates are washed once again with 5 ml of cold 5% TCA. The precipitates are then extracted with ethanol–ethyl ether (3:1) at 60° for 15 minutes and finally with cold acetone.

### *C. Alkaline Hydrolysis of Protein–Phosphate Bonds*

The resulting protein precipitates are transferred into plastic tubes and are dissolved in 0.50 ml of 1 *M* NaOH (reagent grade, low carbonate type from Merck and Company, Inc., Rahway, New Jersey). Use of plastic tubes is necessary because NaOH causes a leaching of phosphate from glass. A control tube is incubated without any protein. The solutions are incubated at 37° for 16 hours, and the protein is reprecipitated by addition of 0.33 ml of 50% trichloroacetic acid. After chilling to 0° for 1 hour, the precipitates are removed by sedimentation and are saved for protein determination. The supernatants are transferred into separate tubes and are analyzed for orthophosphate concentration.

### *D. Phosphate Determination*

Phosphate is analyzed by a modification of Sumner's method.[46] To the phosphate-containing solutions resulting from step C (see above) are added 0.10 ml of 7.5 *N* $H_2SO_4$, 0.1 ml of 6.6% ammonium molybdate $[(NH_4)_6Mo_7O_{24} \cdot 4H_2O]$, and 0.08 ml of freshly prepared ferrous sulfate solution (made by mixing 5 g of $FeSO_4 \cdot 7H_2O$, 50 ml $H_2O$, and 1 ml of 7.5 *N* $H_2SO_4$). The quantity of phosphate is determined by comparing the

[46] J. B. Sumner, *Science* **100**, 413 (1944).

TABLE III
QUANTITATIVE ANALYSIS OF PHOSPHATE IN RIBOSOMAL PROTEINS

| Cells used | Proteins analyzed[a] | Nanomoles $P_i$ per milligram protein |
|---|---|---|
| Reticulocytes | A-ribosomes | 9.6 |
| | C-ribosomes | 9.1 |
| | B+-ribosomes | 7.1 |
| Sarcoma 180 Cells | C-ribosomes | 17 |
| | B-ribosomes | 12.5 |
| | Cell-extract proteins[b] | 14 |
| Sea urchin eggs[c] | Cell-extract proteins | 2.8[d] |
| Sea urchin Blastulae[c] | Cell-extract proteins | 1.8[d] |

[a] Ribosomes were prepared by the direct sedimentation method, A-ribosomes from buffer A, B-ribosomes from buffer B, C-ribosomes from buffer C, B+-ribosomes from buffer B supplemented with 0.5% sodium deoxycholate. Formulas of these buffers are given in Section II, B.

[b] This fraction is the supernatant remaining after the cell lysate is centrifuged at 12,000 *g* for 10 minutes (see Section II, D).

[c] Generously provided by Lawrence Kedes, Stanford University Medical School, Palo Alto, California.

[d] These relatively low values for sea urchin proteins are maximal estimates of their $P_i$ content and may reflect the presence of some contaminating material which contributes to the $P_i$ assay.

absorbance at 700 nm of these latter solutions with similar solutions prepared with known amounts of orthophosphate. If the ribosomal proteins were $^{32}$P-labeled, the specific activity of orthophosphate can be determined by measuring the radioactivity in an aliquot of the solutions after their absorbance has been recorded.

### *E. Protein Determination*

The protein precipitates from step C are dissolved in 20 ml of 1 *M* NaOH, and the quantity of protein is determined by the procedure of Lowry *et al.*,[47] using bovine serum albumin as the standard.

### *F. Results with Mammalian Ribosomes*

Some representative data for mammalian ribosomes are presented in Table III. The quantities of $P_i$ in proteins from ribosomes prepared with high ionic strength buffer B are slightly lower than the value of 16.1 nmoles of $P_i$ per milligram of protein measured[3] for ribosomes from rat liver cells; however, the purity of these latter ribosomes was not demonstrated. Our

[47] O. H. Lowry, N. J. Rosebrough, A. L. Farr, and R. J. Randall, *J. Biol. Chem.* **193**, 265 (1951).

data suggest that there are an average of approximately 14 phosphoryl groups in the proteins of reticulocyte ribosomes prepared from buffer B containing 0.5% sodium deoxycholate. [This calculation assumes that these ribosomes are 50% protein and have a molecular weight of $4.0 \times 10^6$. Thus, each nanomole of ribosomes contains 2.0 mg of protein and, accordingly, 14.2 nmoles of $P_i$.] Furthermore, the sarcoma 180 ribosome preparations contain even more $P_i$; however, they are also more contaminated with nonribosomal proteins. These analyses of phosphate content are consistent with the previous conclusion, based on another technique, that there are 7–11 protein-associated phosphoryl groups which are turning over in each reticulocyte ribosome.[2]

Acknowledgments

Report of work supported by U.S. Public Health Service Grants CA-11347 and HL-CA-14960-04. We thank M. Laurence Cawthon and Janet Ploss for their assistance and advice.

# [63] Phosphorylation of Ribosomal Proteins: Preparation of Rat Liver Ribosomal Protein Kinases. Assay of the Phosphorylation of Ribosomal Subunit Proteins. Assay of the Function of Phosphorylated Ribosomes

*By* CHARLES EIL and IRA G. WOOL

The proteins of eukaryotic ribosomes are phosphorylated *in vivo* and *in vitro*.[1–8] In the latter case the reaction is catalyzed by protein kinases present in the cytosol[1,4,5,8] or associated with ribosomes.[3,5,7] Two such enzymes, both stimulated by cyclic adenosine 3′,5′-monophosphate (cyclic AMP), can be isolated from rat liver cytosol; the protein kinases transfer

[1] J. E. Loeb and C. Blat, *FEBS (Fed. Eur. Biochem. Soc.) Lett.* **10**, 105 (1970).
[2] D. Kabat, *Biochemistry* **9**, 4160 (1970).
[3] D. Kabat, *Biochemistry* **10**, 197 (1971).
[4] G. M. Walton, G. N. Gill, I. B. Abrass, and L. D. Garren, *Proc. Nat. Acad. Sci. U.S.* **68**, 880 (1971).
[5] C. Eil and I. G. Wool, *Biochem. Biophys. Res. Commun.* **43**, 1001 (1971).
[6] C. Blat and J. E. Loeb, *FEBS (Fed. Eur. Biochem. Soc.) Lett.* **18**, 124 (1971).
[7] C. C. Li and H. Amos, *Biochem. Biophys. Res. Commun.* **45**, 1398 (1971).
[8] H. Yamamura, Y. Inoue, R. Shimomura, and Y. Nishizuka, *Biochem. Biophys. Res. Commun.* **46**, 589 (1972).



ISBN 0-12-504180-2

the terminal phosphate from ATP[5] to serine and threonine residues of specific proteins of rat liver ribosomal subunits.[9]

We shall describe techniques we have devised (or adapted) for the preparation of rat liver ribosomal protein kinases, for the assay of the phosphorylation of the proteins of ribosomal subunits, for analysis of the product of the reaction, for assay of ribosomal phosphoprotein phosphatase, and for the comparison of the function of phosphorylated and nonphosphorylated ribosomes.

## Preparation of Rat Liver Cytosol Ribosomal Protein Kinases I and II

### *Reagents*

$\beta$-Mercaptoethanol (MSH), 1 *M*
Glacial acetic acid
Glycerol
KOH (0.5 *M*)
$KH_2PO_4$, 1 *M*
$K_2HPO_4$, 1 *M*
Solid $(NH_4)_2SO_4$
Solid NaCl
Aged calcium phosphate gel (10.5% solids) (Sigma)
DEAE-cellulose (DE-52 resin) (Whatman)
Hydroxylapatite (Bio-Rad)

### *Buffers*

TM: tris(hydroxymethyl)aminomethane (Tris)·HCl, 10 m*M*, pH 7.7; MSH, 10 m*M*
TMG: Tris·HCl, 10 m*M*, pH 7.7; MSH, 10 m*M*; glycerol, 10%
PMG: potassium phosphate buffer, pH 7.5, containing: MSH, 10 m*M*; glycerol, 10%; and potassium phosphate at concentrations of 0.05 and 0.25 *M*
Potassium phosphate, 250 m*M*, pH 8.1; MSH, 10 m*M*

*Separation of Protein Kinases I and II.* Rat liver ribosomal protein kinases are prepared by a modification of the method Gill and Garren[10] used to purify a histone kinase from bovine adrenal supernatant. All procedures are carried out at 4°. The 100,000 *g* post-microsomal supernatant from rat liver (prepared as described by Martin and Wool[11]) is made

[9] E. Eil and I. G. Wool, *J. Biol. Chem.* **248**, 5123 (1973).
[10] G. N. Gill and L. D. Garreñ, *Biochem. Biophys. Res. Commun.* **39**, 335 (1970).
[11] T. E. Martin and I. G. Wool, *J. Mol. Biol.* **43**, 151 (1969).

10 m*M* in MSH with the 1 *M* stock solution. The pH is adjusted to 5.0 with glacial acetic acid and the precipitate is removed by centrifugation at 10,000 *g* for 20 minutes. The pH of the supernatant is made 7.0 with 0.5 *M* KOH and brought to 45% saturation by the slow addition of solid ammonium sulfate. The precipitate is collected by centrifugation at 27,000 *g* for 20 minutes, dissolved in TM, and dialyzed against the same buffer for 16 hours.

Aged calcium phosphate gel, adjusted to 3% in TM (w/v), is added slowly to the protein solution (about 12 mg/ml) and stirred for 35 minutes. The ratio of gel to protein should be 1:1 (w/w). The gel is collected by centrifugation at 3000 *g* for 5 minutes and washed twice with 75 ml of TM. The protein kinase activity is eluted three times with 50 ml each of 0.25 *M* potassium phosphate buffer (pH 8.1). The combined eluates are dialyzed against TMG for 16 hours.

Generally about 550 mg of dialyzed protein is applied to a DEAE-cellulose column (22 × 3 cm) that has been equilibrated with TMG. The contents of the column are washed with two column volumes of TMG at a flow rate of 30 ml/hour and the protein kinase eluted (at the same rate) with 1 liter of a linear gradient of 0.04 to 0.40 *M* NaCl in TMG. Fractions (10 ml) are collected and the absorbance at 280 nm is determined. The protein kinase activity of the fractions are assayed (see below). Generally,

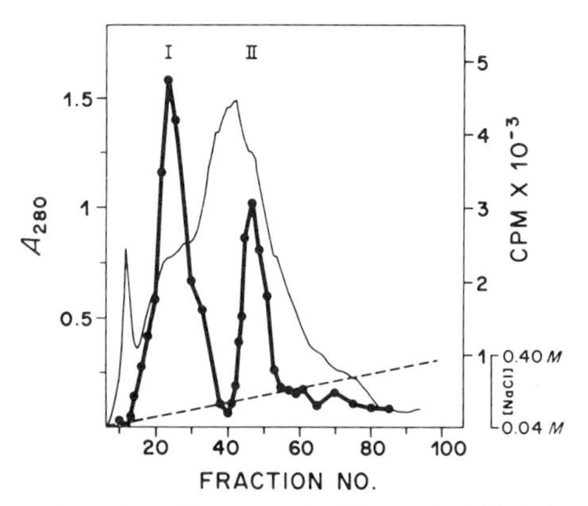


FIG. 1. The separation of rat liver protein kinases by DEAE-cellulose chromatography [C. Eil and I. G. Wool, *Biochem. Biophys. Res. Commun.* 43, 1001 (1971)]. Selected fractions were assayed for ribosomal protein kinase activity in the presence of cyclic AMP using 60 S ribosomal subunits as substrate. The radioactivity incorporated into 60 S ribosomes in the absence of added protein fractions was 375 cpm. Each 1000 cpm incorporated into ribosomal protein is equivalent to 9.1 pmoles of phosphate. The continuous thin line is the $A_{280}$ of the eluate; the thicker line is the protein kinase activity of selected fractions (cpm of $^{32}P$ incorporated into protein of 60 S ribosomal subunits); the interrupted line is the concentration of NaCl.

protein kinase I eluates at 0.1 *M* and protein kinase II at 0.2 *M* NaCl (Fig. 1).[5]

*Purification of Protein Kinase I.* Protein kinase I can be further purified by a modification of the procedure of Yamamura *et al.*[12] The peak fractions of protein kinase I from the DEAE-cellulose eluate are pooled and brought to 65% saturation by slow addition of solid ammonium sulfate. The precipitate is collected by centrifugation, dissolved in TMG, and dialyzed overnight against 50 m*M* PMG.

Approximately 25 mg of dialyzed protein is applied to a hydroxyapatite column (2.3 $\times$ 2.5 cm) that has been equilibrated with 50 m*M* PMG. The column is washed with 40 ml of 50 m*M* PMG at a flow rate of 12 ml/hour and the protein kinase activity eluted (at the same rate) with 200 ml of a linear gradient of 50 m*M* to 0.25 *M* potassium phosphate buffer (pH 7.5). Fractions (4 ml) are collected and the absorbance at 280 nm is determined. The protein kinase activity of the fractions are assayed (see below). The fractions from the peak containing the cyclic AMP-dependent enzyme activity (generally eluted at 0.1 *M* potassium phosphate) are pooled and the protein precipitated by the slow addition of solid ammonium sulfate to 65% saturation. After centrifugation, the precipitate is dissolved in TMG and dialyzed overnight against TMG to remove ammonium sulfate and potassium phosphate. We designate the sample purified protein kinase I.

## Preparation of Ribosomes and Ribosomal Subunits

*Reagents*

Puromycin dihydrochloride, 1 m*M* (Nutritional Biochemicals Corporation)
Ethanol

*Media*

Medium A: Tris·HCl, 50 m*M*, pH 7.6; KCl, 80 m*M*; $MgCl_2$, 12.5 m*M*; MSH, 20 m*M*
Medium B: Tris·HCl, 50 m*M*, pH 7.6; KCl, 880 m*M*; $MgCl_2$, 12.5 m*M*; MSH, 10 m*M*

We generally prepare liver ribosomes from male Sprague-Dawley rats of about 100–120 g,[11] however, any of the standard procedures for preparation of relatively uncontaminated particles (from any eukaryotic cell type) will do. The ribosomes are suspended in medium B and incubated for 15 minutes at 37° with 0.1 m*M* puromycin—to remove nascent peptide

[12] H. Yamamura, M. Takeda, A. Kumon, and Y. Nishizuka, *Biochem. Biophys. Res. Commun.* **40**, 675 (1970).

and dissociate the ribosomes into subparticles.[13,14] After incubation the ribosomal suspension is layered on a 10–30% linear sucrose gradient and centrifuged for 4 hours at 25° in a Spinco SW 27 rotor at 27,000 rpm. The gradients are displaced with 50% sucrose using an Instrument Specialities Co., Inc. (ISCO) Model D density gradient fractionator, and the effluent is analyzed at 254 nm with an ISCO model UA-2 UV analyzer. Subunit fractions are collected[13] and dialyzed overnight against medium A. (Sucrose and high concentrations of potassium interfere with precipitation of ribosomes by ethanol.[15]) Cold 95% ethanol (0.2 of a volume) is added to the subunit fractions and the suspension is kept at 0° for at least 1 hour (to ensure complete precipitation of the particles). The subunits are collected by centrifugation at 10,000 *g* for 10 minutes, carefully drained free of residual ethanol, and suspended in medium suitable for assay of ribosomal protein kinase activity (see below). The concentration of ribosomes and ribosomal subunits is calculated from their absorption at 260 nm[16]; 1 $A_{260}$ unit is taken to be the equivalent of 45 $\mu$g of ribosomal RNA (rRNA).

## Assay of the Phosphorylation of the Proteins of 40 S and 60 S Ribosomal Subunits by Protein Kinase

### *Reagents*

- Cyclic AMP, $10^{-5}$ *M* (Sigma)
- Caffeine, $10^{-3}$ *M* (Nutritional Biochemicals Corporation)
- [$\gamma$-$^{32}$P]ATP, 0.1 m*M*, 2–10 $\times$ $10^{-4}$ cpm/nmole; prepared by the method of Glynn and Chappell[17]
- Trichloroacetic acid (TCA), 5 and 10%
- Formic acid, 88%
- BBOT scintillation fluid: 5 g of 2,5-bis[2-(5-*tert*-butylbenzoxazolyl)]thiophene (Packard) in 500 ml of toluene and 500 ml of Methyl Cellosolve[16]

### *Media*

- Medium C: Tris·HCl, 40 m*M*, pH 7.7; $MgCl_2$, 5 m*M*; MSH, 10 m*M*; glycerol, 5%

*Assay.* The reaction requires only a substrate protein that will accept phosphate, ATP, magnesium, and protein kinase. Variations in the condi-

[13] T. E. Martin, I. G. Wool, and J. J. Castles, this series, Vol. 20, p. 417 (1971).

[14] W. S. Stirewalt, J. J. Castles, and I. G. Wool, *Biochemistry* **10**, 1594 (1971).

[15] M. S. Kaulenas, *Anal. Biochem.* **41**, 126 (1971).

[16] I. G. Wool and P. Cavicchi, *Proc. Nat. Acad. Sci. U.S.* **56**, 991 (1966).

[17] I. M. Glynn and J. B. Chappell, *Biochem. J.* **90**, 147 (1964).

tions are frequently necessary, but we describe a typical assay. Ribosomal subunits—generally 28 μg (rRNA) of 40 S or 70 μg of 60 S subunits—are incubated in 0.1 ml of medium C containing a preparation of protein kinase (the amount varies with the purity and activity of the enzyme: 10 μg of purified protein kinase I would be adequate) and [$\gamma$-$^{32}$P]ATP (0.1 m$M$; 2 to 10 $\times$ $10^4$ cpm/nmole). When cyclic AMP is added, the concentration is $10^{-5}$ $M$. If the enzyme preparation has not been purified by chromatography on DEAE-cellulose, and if cyclic AMP is included, it is best to add caffeine (1 m$M$) to inhibit contaminating phosphodiesterase.[18] The reaction is started by adding [$\gamma$-$^{32}$P]ATP and the tubes are stirred (on a Vortex mixer) and incubated for 15 minutes at 37°. The reaction is stopped by adding 1 ml of 10% TCA and mixing.

*Determination of the Incorporation of Radioactivity from [$\gamma$-$^{32}$P]ATP into Ribosomal Proteins.* A convenient procedure is to collect the radioactive phosphorylated ribosomal proteins on glass fiber filters (grade 934 AH, 2.4 cm, Reeve Angel). The samples (in TCA) are heated in a boiling-water bath for 15 minutes, to hydrolyze nucleic acids, and then cooled at 0°. The protein is collected on glass fiber disks held in a Millipore multiple sample manifold. The filters are washed with 30 ml of 5% TCA and transferred to scintillation vials containing 0.5 ml of 88% formic acid to dissolve the protein. BBOT scintillation fluid (10 ml) is added, the samples are stirred on a Vortex mixer and the radioactivity is determined (with an efficiency of 95%) in Packard Tri-Carb liquid scintillation spectrometer.

## Characterization of Ribosomal Phosphoproteins by Electrophoresis in Polyacrylamide Gels and by Radioautography

### *Reagents*

$MgCl_2$, 0.5 $M$
KCl, 2.5 $M$
Magnesium acetate, 1 $M$
Acetic acid, glacial and 7%
Formic acid, 88%
Urea, 7 $M$ containing 0.1 $M$ MSH
Solid Tris

### *Media*

Medium D: Tris·HCl, 40 m$M$, pH 7.7; MSH, 10 m$M$; glycerol, 5%
Medium E: Tris·HCl, 10 m$M$, pH 7.7; magnesium acetate, 100 m$M$

[18] T. W. Rall and E. W. Sutherland, *J. Biol. Chem.* **232**, 1077 (1958).

*Autoradiography of Phosphorylated Ribosomal Proteins.* The identity of the individual ribosomal proteins phosphorylated by protein kinase can be determined if the proteins are extracted from the ribosomal subunits,[19] separated by electrophoresis in polyacrylamide gels,[20,21] and autoradiographs of the gels made.[5]

Ribosomal subunits, approximately 400 $\mu$g (rRNA) of 40 S or 1.0 mg of 60 S, are phosphorylated in 12-ml conical Corex tubes in 1.6 ml of medium D containing 0.1 m*M* [$\gamma$-$^{32}$P]ATP and protein kinase, with or without cyclic AMP. Neither magnesium nor potassium are essential for phosphorylation of ribosomal proteins,[9] but it is at times important to add the cations (see below) and that can be done with appropriate amounts of concentrated solutions. Incubation is for 30 minutes at 37°. The reaction is stopped by the addition of 0.18 ml of ice-cold 1 *M* magnesium acetate (to precipitate subunits) and the samples kept at 0° for at least 15 minutes. The ribosomes are collected by centrifugation at 10,000 *g* for 10 minutes, resuspended in approximately 1 ml of medium A, and reprecipitated by the addition of 1/9 volume of 1 *M* magnesium acetate. After centrifugation the ribosomes are suspended with a glass rod in 0.1–0.2 ml of medium E; 2 volumes of glacial acetic acid are added, and the mixture is stirred with a glass rod for 1 hour at 0° to extract the ribosomal proteins.[19] The ribosomal RNA is removed by centrifugation at 15,000 *g* for 10 minutes; the pellet is washed with approximately the same volume of 1/3 medium C and 2/3 glacial acetic acid and recentrifuged. The combined supernatant, containing the ribosomal proteins, is dialyzed against 1000 volumes of 7% acetic acid for 3–4 hours. Dialysis is in acetylated tubing[22] to prevent loss of low molecular weight proteins. The concentration[23] and the radioactivity (see above) of the protein in the extract is determined. The dialyzate is lyophilized (to reduce the volume and to remove acetic acid) and the protein dissolved in 0.1–0.2 ml of 7 *M* urea containing 0.1 *M* MSH. The pH of the samples is adjusted to 8 with a small amount of solid Tris and incubated for 3 hours at 37° to assure that the proteins are reduced.[21] The ribosomal proteins (80 $\mu$g of 40 S or 120 $\mu$g of 60 S) are separated by electrophoresis in discontinuous polyacrylamide gels at pH 4.5 in 6 *M* urea.[20,21] The gels are stained, destained, and sliced into several 1 mm

[19] S. J. S. Hardy, C. G. Kurland, P. Voynow, and G. Mora, *Biochemistry* **8**, 2897 (1969).

[20] P. S. Leboy, E. C. Cox, and J. G. Flaks, *Proc. Nat. Acad. Sci. U.S.* **52**, 1367 (1964).

[21] R. B. Low and I. G. Wool, *Science* **155**, 330 (1967).

[22] L. C. Craig, this series, Vol. XI, p. 870 (1967).

[23] O. H. Lowry, J. T. Rosebrough, A. L. Farr, and R. J. Randall, *J. Biol. Chem.* **193**, 265 (1951).

longitudinal sections. The slices are dried under strong vacuum on filter paper on a Buchner funnel; drying is speeded by covering the slices on the funnel with Saran Wrap and heating with an infrared lamp. X-ray film (Kodak, NS-54T) is placed over the dried gels on filter paper and exposed 1–8 days.

### Determination of Phosphoserine and Phosphothreonine in Hydrolyzed Phosphorylated Ribosomal Proteins

*Reagents*

HCl, 6 *N*
Perchloric acid, 70%
Solid ammonium molybdate
Phosphoserine, 2 mg/ml $H_2O$
Phosphothreonine, 2 mg/ml $H_2O$
$KH_2PO_4$, 2 mg/ml $H_2O$

It is important to establish that the phosphate incorporated into ribosomal protein is covalently bound to amino acid residues, for $^{32}P$-labeled RNA can contaminate ribosome proteins. Since the RNA might be far more radioactive than the protein a small amount of contamination could be very misleading.

Phosphorylation of ribosomal proteins by protein kinase leads to the formation of phosphoserine and phosphothreonine.[1–4,9] To detect those amino acids ribosomal subunits are phosphorylated with [$\gamma$-$^{32}P$]ATP and the ribosomal proteins are extracted. The proteins are then hydrolyzed in 6 *N* HCl and the hydrolyzate is subjected to paper electrophoresis at pH 1.9. At that pH, phosphoserine and phosphothreonine migrate to the anode whereas the other amino acids migrate to the cathode. The radioactivity is located by radioautography of the electrophoretogram.

Ribosomal subunits, 40 S and 60 S, are phosphorylated, and the proteins are extracted as described above for the preparation of radioactive ribosomal proteins for gel electrophoresis. Lyophilized 40 S proteins (90 $\mu$g, about 25,000 cpm) and 170 $\mu$g of lyophilized 60 S proteins (about 75,000 cpm) are dissolved separately in 0.5–1.0 ml of 6 *N* HCl. The tubes are evacuated, sealed, and heated at 110° for approximately 7 hours. Hydrolysis of the proteins for longer periods leads to extensive breakdown of the ester bonds in phosphoserine and phosphothreonine.[3] The samples are dried in a rotary evaporator. The hydrolyzates are dissolved in 100 $\mu$l of 2.5% formic acid–7.8% acetic acid, pH 1.9. The samples to be analyzed, authentic phosphoserine (50 $\mu$l), phosphothreonine (50 $\mu$l), and phosphate (50 $\mu$l), are spotted onto Whatman 3 MM paper (46 × 57 cm)

10 cm from the end to be attached to the cathode. The sample is applied in several small portions, with drying after each application, to prevent spreading. Drying is speeded with a hair dryer. Both ends of the paper are moistened with the formic acid–acetic acid mixture (pH 1.9) and then applied to the electrophoresis apparatus (Savant; Flat Plate). Electrophoresis is for 2 hours at 2750 V. The paper is dried in an oven at 60° and cut to separate the radioactive samples from the authentic markers. The radioactive samples on the paper are covered with X-ray film (Kodak NS-2T) and exposed for 2 weeks. The portion of the paper with the authentic samples of phosphoserine, phosphothreonine, and inorganic phosphate is sprayed with a mixture of 3% perchloric acid, 0.1 *N* HCl, and 1% w/v ammonium molybdate to locate inorganic and organic phosphate.[24] The paper is dried in an oven and developed with strong ultraviolet light.[24]

*Comment*

*Effect of Cations.* Magnesium is generally required for kinase reactions having ATP as a substrate; nonetheless, maximum phosphorylation of ribosomal proteins occurs when subunits are incubated without addition of magnesium. Magnesium is bound to the ribosome and no doubt is the source of sufficient cation to support catalytic activity of the enzyme. If ribosomes are treated with EDTA to remove the bound magnesium and then assayed without added magnesium, no phosphorylation occurs. With ordinary preparations of ribosomal subunits increasing concentrations of magnesium (up to 20 m*M*) inhibit protein phosphorylation.

Potassium also inhibits ribosomal protein phosphorylation, though to a lesser extent than magnesium. Increments in the concentration of potassium, to about 50 m*M*, decrease phosphorylation of ribosomal subunits. The effects may be a consequence of the influence of the cation on the structure of the ribosomal subparticles. However, potassium also inhibits the phosphorylation of histone by protein kinase.[9] The structure of histones is not known to be affected by potassium; hence, the effect of the cation may be in part on the enzyme.

## Assay of Ribosomal Phosphoprotein Phosphatase

A prerequisite for the study of the function of phosphorylated ribosomes, is assurance that the preparation of factors used to measure ribosome activity does not contain ribosomal phosphoprotein phosphatases. If phosphorylated ribosomes are incubated with unresolved cytosol (from ascites cells or from rat liver) in circumstances used to assay protein syn-

[24] R. M. C. Dawson, *in* "Lipid Chromatography Techniques" (G. V. Marinetti, ed.), Vol. I, p. 179. Dekker, New York, 1967.

thesis, as much as 70% of the phosphate is released from the particles.[25] It is obvious that an inhibitor of ribosomal phosphoprotein phosphatase would be a great convenience (provided, of course, that it did not interfere with ribosome function). Unfortunately, we have not discovered such a reagent, although we have tried a number of inhibitors of other phosphatases. What is required then is that the factor preparations be sufficiently resolved so as to be relatively free of ribosomal phosphoprotein phosphatase. We describe an assay that allows one to assess the extent of contamination of factor preparations with ribosomal phosphoprotein phosphatase.

*Reagents*

ATP, 0.1 *M* (Pabst)
GTP, 5 m*M* (Pabst)
Creatine phosphate (Boehringer Mannheim Corporation)
Creatine phosphokinase (Boehringer Mannheim Corporation)
Aminoacyl-tRNA–*E. coli* B tRNA aminoacylated[26,27] with 20 different amino acids
Polyuridylic acid (Miles Laboratories Incorporated)

*Medium*

Medium F: Tris·HCl, 20 m*M*, pH 7.5; KCl, 125 m*M*, magnesium acetate, 5 m*M*; MSH, 10 m*M*

*Preparation of Radioactive Phosphorylated Ribosomes.* The substrate for assay of ribosomal phosphoprotein phosphatase is $^{32}$P-labeled ribosomes. They are prepared by incubating ten times the usual amount of ribosomal subunits—280 $\mu$g (rRNA) of 40 S or 720 $\mu$g (rRNA) of 60 S subparticles—in 1 ml of medium C in Corex centrifuge tubes with saturating amounts of protein kinase, $10^{-5}$ *M* cyclic AMP, and 0.1 m*M* [$\gamma$-$^{32}$P] ATP. Incubation is for 25 minutes at 37°. The concentration of $MgCl_2$ is raised to 12.5 m*M* (by addition of sufficient 0.5 $MgCl_2$) and potassium to 80 m*M* (with sufficient 2.5 *M* KCl). The ribosomes are precipitated by the addition of 0.2 volume of cold 95% ethanol (see above) and collected by centrifugation at 10,000 *g* for 10 minutes. The phosphorylated ribosomes are suspended in 0.5 ml of medium C and insoluble material is removed by centrifugation at 3000 *g* for 5 minutes.

*Ribosomal Phosphoprotein Phosphatase Assay.* The ribosomal phosphoprotein phosphatase activity of various factor preparations can be deter-

[25] C. Eil and I. G. Wool, *J. Biol. Chem.* **248**, 5130 (1973).
[26] G. von Ehrenstein and F. Lipmann, *Proc. Nat. Acad. Sci. U.S.* **47**, 941 (1961).
[27] I. G. Wool and P. Cavicchi, *Biochemistry* **6**, 1231 (1966).

mined by incubating them with $^{32}P$-labeled subparticles—2.5 μg (rRNA) of 40 S and 6.25 μg (rRNA) of 60 S subunits—in 0.1 ml of medium C. However, our purpose has generally been to test the function of phosphorylated ribosomes, so we have assayed phosphatase in the same circumstances in which we measure protein synthesis. The $^{32}P$-labeled subunits (in the amount specified above) are incubated in 0.1 ml of medium F containing: ATP, 1 m*M*; GTP, 0.1 m*M*; polyuridylic acid, 10 μg; creatine phosphate, 5 m*M*; creatine phosphokinase, 20 μg; and aminoacyl-tRNA, 50 μg. The reaction is started by adding the radioactive ribosomes. Incubation is at 37° for 10–45 minutes depending on the amount of ribosome phosphoprotein phosphatase. The reaction is stopped by adding 1 ml of 10% TCA, and the mixture is heated at 90–95° for 15 minutes. The precipitate is collected on glass fiber disks, and the radioactivity is measured as described for the protein kinase assay (see above). Ribosome phosphoprotein phosphatase activity of protein fractions is calculated from the loss of radioactivity from $^{32}P$-labeled ribosomes after allowance is made for loss of radioactivity from ribosomes incubated without the protein fractions. The loss of radioactivity in the latter case does not exceed 15% of the amount originally present.

## Assay of the Function of Phosphorylated Ribosomes

### *Medium*

Medium G: Tris·HCl, 40 m*M*, pH 7.7; KCl, 10 m*M*; $MgCl_2$, 5 m*M*; MSH, 10 m*M*; glycerol, 5%

There are two requisites for the study of the function of phosphorylated ribosomes: the particles should be fully phosphorylated at the start of the experiment and the loss of phosphate during the assay should be kept to a minimum. The first requirement is met by choosing optimal conditions (concentrations of ions, of enzymes, etc.) for the phosphorylation of ribosomal proteins by protein kinase. For ribosomes to retain activity, phosphorylation of the subunits must be carried out in medium containing magnesium and potassium. Since these ions also inhibit protein kinase, they should be kept to the minimal concentration consonant with maintenance of ribosome function. If subunits are phosphorylated in medium containing 5 m*M* $MgCl_2$ and 10 m*M* KCl, protein kinase activity is not appreciably depressed, and the capacity of the subunits to synthesize protein (during which assay the potassium is usually 75–125 m*M*) is hardly decreased. No synthesis of protein occurs if potassium is omitted from the preincubation medium.[9,28] Although protein kinase I does not require a reducing agent for its activity, MSH is included in

[28] P. H. Näslund and T. Hultin, *Biochim. Biophys. Acta* **204**, 237 (1970).

the protein kinase reaction to preserve the function of the ribosomes.[29,30]

Ribosomal subunits, 40 S (20–30 μg of rRNA) and 60 S (50–75 μg of rRNA), are preincubated separately in 0.1 ml of medium G containing: ATP, 1 m*M*—the ATP ordinarily need not be radioactive; cyclic AMP, 10 μ*M*; and protein kinase. Incubation is for 25 minutes at 37°. The reaction is stopped by cooling the samples on ice, and aliquots of the subunits are used to assay their function.

We have used this procedure to compare the ability of phosphorylated and nonphosphorylated ribosomal subunits to synthesize polyphenylalanine, to translate encephalomyocarditis virus RNA, to bind aminoacyl-tRNA, and to reassociate to form 80 S monomers.[25] While we have not examined all the possibilities, we have not yet discovered an appreciable difference in the activity of ribosomal subunits as a result of phosphorylation.[25]

*Remarks*

A reliable assay of the activity of ribosomal protein kinase requires that the enzyme be resolved from inhibitors present in crude rat liver cytosol. The eluate from calcium phosphate gel contains detectable ribosomal protein kinase activity, but catalyzes considerable phosphorylation in the absence of ribosomes. The protein in the enzyme preparation that is phosphorylated is removed by DEAE-cellulose column chromatography. That procedure also separates two protein kinases: neither shows a substrate preference; both phosphorylate 40 S subunits, 60 S subunits, and calf histone.[5] Moreover, protein kinase I also phosphorylates casein and protamine sulfate as well as 80 S ribosomes.[9] However, the 80 S ribosomes accept less phosphate than subunits and are contaminated with protein kinase since they can be phosphorylated in the absence of added enzyme. Apparently, the enzyme is removed when subunits are prepared by treating ribosomes with 0.88 *M* KCl and centrifuging them through a sucrose gradient.

At least three proteins of the 40 S subunit of liver ribosomes and nine of the 60 S subunit are phosphorylated by ribosomal protein kinases.[5] The extent of phosphorylation varies with the kinase preparation used, the presence of cyclic AMP, and the ionic conditions, although none of these variations leads to the phosphorylation of different ribosomal proteins.[9] The reaction involves the transfer of the terminal phosphate of ATP and not the exchange of phosphates already present in ribosomal proteins.[9] GTP will not serve as a phosphate donor.[9]

[29] T. E. Martin, F. S. Rolleston, R. B. Low, and I. G. Wool, *J. Mol. Biol.* 43, 135 (1969).

[30] H. C. McCallister and R. S. Schweet, *J. Mol. Biol.* 34, 519 (1968).

# [64] Protein–RNA Interactions in the Bacterial Ribosome

*By* ROBERT A. ZIMMERMANN

The assembly, function, and stability of the ribosome are all assured by a complex network of specific associations among its protein and RNA constituents. Over one-third of the 50 to 55 ribosomal proteins from *Escherichia coli* have been shown to bind directly and independently to homologous 16 S, 23 S, and 5 S RNAs.[1] These interactions are believed to play a major role in the early phases of 30 S and 50 S subunit assembly by initiating the formation of a series of protein nuclei within localized, and relatively antonomous, segments of the RNA. At the same time, certain proteins appear to provoke appreciable alterations in RNA configuration, which may stimulate contacts between more distant portions of the nucleic acid chains. As additional components are integrated into the nascent subunits and the particles become increasingly compact, the number of protein–RNA associations undoubtedly multiplies. Such interactions must therefore make a substantial contribution to molding the functional sites required in protein synthesis as well as to maintaining the mature subunits in their active conformation throughout many cell generations.

The importance of direct protein–RNA interaction in ribosome reconstitution was clearly delineated by the derivation of an assembly map for the *E. coli* 30 S subunit which showed that the binding of a small group of proteins to the 16 S RNA was a prerequisite for subsequent assembly reactions.[2] Since that time, the specificity, stoichiometry, and stability of primary protein–RNA complexes formed from components of both subunits have been the subject of numerous investigations, and the binding sites for many of the proteins have been located within the ribosomal RNAs.[1,3–5] In addition, equilibrium constants and thermodynamic pa-

[1] R. A. Zimmermann, *in* "Ribosomes" (M. Nomura, A. Tissières, and P. Lengyel, eds.), p. 225. Cold Spring Harbor Laboratory, Cold Spring Harbor, New York, 1974.

[2] S. Mizushima and M. Nomura, *Nature (London)* **226,** 1214 (1970).

[3] R. A. Zimmermann, G. A. Mackie, A. Muto, R. A. Garrett, E. Ungewickell, C. Ehresmann, P. Stiegler, J.-P. Ebel, and P. Fellner, *Nucl. Acids Res.* **2,** 279 (1975).

[4] C. Branlant, A. Krol, J. Sriwidada, and J.-P. Ebel, *J. Mol. Biol.* **116,** 443 (1977).

[5] V. A. Erdmann, *Prog. Nucl. Acid. Res. Mol. Biol.* **18,** 45 (1976).





ISBN 0-12-504180-2

rameters have been evaluated in a few instances, and the analysis of cooperative changes in RNA secondary and tertiary structure that result from protein binding is underway. Despite these advances, our understanding of ribosomal protein–RNA interaction is still in many ways quite elementary and a host of critical problems remains to be resolved.

The following sections are intended to provide an overview of techniques currently used in the study of interactions among protein and RNA constituents of bacterial ribosomes. Since the methods employed to prepare the ribosomal components frequently have an important bearing on the feasibility of the experiments and on the quality of the results, they will be considered in some detail. Procedures for the formation, identification and enzymic fragmentation of ribonucleoprotein complexes will then be described. Finally, the application of several specialized techniques—both conventional and novel—to ribosomal protein–RNA association will be surveyed, with particular attention to the kinds of information they provide.

## Isolation of Ribosomal Components

*Escherichia coli,* strain MRE600, is the usual source of ribosomal components for studies on protein–RNA interactions (Table I). The preparative techniques described in this section are equally applicable to other strains of *E. coli,* however, and, with only slight modifications, to a wide variety of other bacteria as well.[6]

### *Growth and Harvesting of Bacteria*

Unlabeled bacteria are grown to a density of 2 to 5 $\times$ $10^9$ cells/ml at 37° in a medium composed of 10 g of yeast extract (Difco), 34 g of $KH_2PO_4$, 8.4 g of KOH, and 10 g of glucose per liter.[7] Cells are harvested with a continuous-flow centrifuge, frozen in Dry Ice without washing, and stored at −70°. Uniform labeling of cellular constituents is accomplished by incubating exponentially growing bacteria in the presence of radioactive precursors for at least 5 generations at 37° in a medium containing 7 g of $NaH_2PO_4 \cdot 2H_2O$, 3 g of $K_2HPO_4$, 0.5 g of NaCl, 1.0 g of $NH_4Cl$, 1.32 g of $(NH_4)_2SO_4$, 0.2 g of $MgCl_2 \cdot 6\ H_2O$, 0.018 g of $CaCl_2 \cdot 4\ H_2O$, and 4 g of glucose per liter.[7] Labeling of proteins is performed by the addition of 1 mCi of $^3H$-labeled amino acid mixture and 20 mg of vitamin-free Casamino acids (Difco), or of 20 mCi $H_2[^{35}S]O_4$ and 1 mg of each of the commonly occurring amino acids except methionine and cysteine, per 100-ml culture. In the latter case, the medium is also de-

[6] D. L. Thurlow and R. A. Zimmermann, *Proc. Natl. Acad. Sci. U.S.A.,* **75**, 2859 (1978).
[7] A. Muto, C. Ehresmann, P. Fellner, and R. A. Zimmermann, *J. Mol. Biol.* **86,** 411 (1974).

mixture. After electrophoresis, the gels are fixed in 5.6% (v/v) acetic acid, stained with 0.04% (w/v) methylene blue in 0.2 *M* Na acetate, pH 4.7, and scanned at 600 nm in a spectrophotometer equipped with a linear transport accessory.

*Alternative Procedures.* Dissociation of 70 S ribosomes with 100 m*M* LiCl–0.5% SDS in buffer F has proved to be preferable to the phenol method in terms of both yield and purity for the isolation of small amounts of $^{32}P$-labeled ribosomal RNAs.[12]

Ribosomal RNA has also been obtained by precipitation from a solution of 30 S or 50 S subunits in 15 m*M* Tricine, pH 8.0, 800 m*M* Mg acetate, 4 *M* urea by the addition of 3 volumes of glacial acetic acid. The 16 S RNA prepared in this way has been reported to bind several ribosomal proteins that cannot associate directly with phenol-extracted 16 S RNA.[13]

An alternative to the purification of ribosomal RNAs by zonal centrifugation consists in the use of chromatography on lysine-Sepharose (Pharmacia).[14] The RNA is bound to this substance in 20 m*M* Tris·HCl, pH 7.5, 10 m*M* $MgCl_2$ and eluted according to size with a linear salt gradient in the same buffer. Up to 15 mg of total cellular RNA has been resolved into discrete tRNA, 5 S RNA, 16 S RNA, and 23 S RNA fractions on a 1.6 cm i.d. × 7.5 cm column. Successful separation of 5 S RNA from tRNA on Sephadex G-100 (Pharmacia) has also been described.[15]

### *Separation of Ribosomal Proteins*

*Buffers.* Compositions of the buffers are as follows:

- Buffer G: 50 m*M* $NaH_2PO_4$, 10 m*M* methylamine, 6 *M* urea, 4 m*M* 2-mercaptoethanol, pH 6.5
- Buffer H: 50 m*M* Na acetate, 6 *M* urea, 4 m*M* 2-mercaptoethanol, pH 5.6
- Buffer I: 10 m*M* $NaH_2PO_4$, 40 m*M* $Na_2HPO_4$, 10 m*M* methylamine, 6 *M* urea, 4 m*M* 2-mercaptoethanol, pH 8.4

*Preparation of Ion-Exchange Columns.* Phosphocellulose (Schwarz/Mann, regular high capacity)[16] or carboxymethyl cellulose (Whatman,

[12] P. Fellner, *Eur. J. Biochem.* **11,** 12 (1969).

[13] H.-K. Hochkeppel, E. Spicer, and G. R. Craven, *J. Mol. Biol.* **101,** 155 (1976).

[14] D. S. Jones, H. K. Lundgren, and F. T. Jay, *Nucl. Acids Res.* **3,** 1569 (1976).

[15] R. Monier and J. Feunteun, this series, Vol. 20, p. 494.

[16] This grade of phosphocellulose is no longer available from Schwarz/Mann; suitable replacements include Whatman P-11 and Brown Selectacel phosphate, standard grade.

CM 52) is precycled by washing with 0.5 *N* NaOH, distilled $H_2O$, and 0.5 *N* HCl. The slurry is collected on a paper filter (Whatman, 3 MM) by suction, washed extensively with distilled $H_2O$, and resuspended in either buffer G (phosphocellulose) or buffer H (carboxymethyl cellulose). Fine particles comprising 20–40% of the wet volume are decanted after allowing the celluloses to settle from 5–10 volumes of the appropriate buffer for 30–60 min in a graduated cylinder. Removal of fines is crucial to the maintenance of adequate flow rates during column chromatography, especially in the case of carboxymethyl cellulose. The washed, settled ion exchanger is stirred gently in 2–3 volumes of buffer and poured into a glass tube fitted with a disk of fine-mesh nylon cloth that retains the slurry. Packing is allowed to proceed by gravity alone until a height of 10–20 cm is attained in order to prevent overcompression and clogging of the bottom portion of the column. Thereafter, the effluent clamp is opened slightly, permitting a slow flow of buffer to pass through the packed cellulose. When the desired height is reached, the column is thoroughly equilibrated with the running buffer using a peristaltic pump.[17]

*Extraction and Chromatography.* From one to several hundred milligrams of ribosomal proteins can be conveniently prepared for chromatography by the following procedure.[18] As an example, a relatively large-scale preparation will be described. One gram of 30 S or 50 S subunits is suspended in 50 ml of buffer A. The $Mg^{2+}$ concentration is adjusted to 100 m*M* and two volumes of glacial acetic acid are added. The mixture is stirred for 60 min on ice and then centrifuged for 10 min at 20,000 *g*. The supernatant, which contains the bulk of the protein, is decanted and saved, while residual protein is recovered from the insoluble pellet by extraction with 75 ml of 67% acetic acid (v/v) in buffer A with 100 m*M* $MgCl_2$. The combined supernatants, containing about 300 mg of ribosomal protein, are dialyzed against two 2-liter portions of buffer I followed by three 2-liter portions of buffer G over a period of 36–48 hr until they reach equilibrium with the starting buffer. Proportionate reductions in volumes are made when smaller amounts of ribosomes are processed by this method.

Initial chromatography is performed on phosphocellulose at pH 6.5.[18] Column sizes and elution rates are adjusted to the amount of protein to be fractionated. Satisfactory separation of 10–15 mg of total 30 S or 50 S subunit proteins can be achieved with 0.5 cm i.d. × 40 cm columns, flow rates of 5 ml per hour, and salt gradients of 300 ml. For larger preparations, up to 300 mg of protein may be loaded on a 2.5 cm i.d. × 60 cm

[17] R. A. Zimmermann and G. Stöffler, *Biochemistry* **15,** 2007 (1976).

[18] S. J. S. Hardy, C. G. Kurland, P. Voynow, and G. Mora, *Biochemistry* **8,** 2897 (1969).

column and eluted with a 6000-ml salt gradient at 50 ml per hour. As a rule of thumb, the ratios of protein to column cross section and to elution volume should be maintained within a factor of 2 or 3 of these indicated in the examples above to avoid the poor separation that results from overloading and the poor yields that result from underloading. Unresolved protein mixtures can generally be fractionated by a second cycle of chromatography on carboxymethyl cellulose at pH 5.6,[19] scaling column cross section, gradient volume, and flow rate to roughly one-third of those used for the phosphocellulose column.

In a typical run, the dialyzed protein solution is applied to a phosphocellulose column, washed with 2 volumes of buffer G, and eluted with a linear gradient of 0 to 0.5 *M* NaCl in the same buffer. A constant flow rate is assured during loading, washing, and elution by the use of a peristaltic pump. From 300 to 400 fractions are collected and analyzed individually for absorbance at 230 nm and/or radioactivity in order to locate the proteins. Peak fractions are pooled and concentrated in an ultrafiltration cell (Amicon) fitted with a UM-2 membrane. The identity and purity of the proteins in each pool is assessed by polyacrylamide gel electrophoresis. Mixtures are consolidated, dialyzed against 100 volumes of buffer H, and applied to a carboxymethyl cellulose column. After washing with two volumes of buffer H, proteins are eluted with a linear gradient of 0.05 to 0.35 *M* Na acetate in 6 *M* urea, 4 m*M* 2-mercaptoethanol, pH 5.6. Fractions are collected and processed as for the phosphocellulose column. Protein concentration is determined by the method of Lowry *et al.*[20] using egg-white lysozyme and bovine serum albumin as standards.

Chromatograms illustrating the isolation of unlabeled 30 S subunit proteins and $^3$H-labeled 50 S subunit proteins are presented in Figs. 1 and 2. RNA-binding proteins are underlined. The profiles are highly reproducible for a given column although small variations in elution position may occur with different batches of phosphocellulose or carboxymethyl cellulose. The purified components, which are generally recovered in 75–85% yield, are stored at −20° in 6 *M* urea. Such preparations have been found to retain their RNA-binding capacities for at least 6 months, and often for several years, when maintained at concentrations of 1 mg/ml or above. A decline in the binding activity of S8, L2, and, on occasion, certain other proteins has been found to occur after chromatography on carboxymethyl cellulose. In such cases, pure protein frac-

[19] E. Otaka, T. Itoh, and S. Osawa, *J. Mol. Biol.* **33,** 93 (1968).

[20] O. H. Lowry, N. J. Rosebrough, A. L. Farr, and R. J. Randall, *J. Biol. Chem.* **193,** 265 (1951).

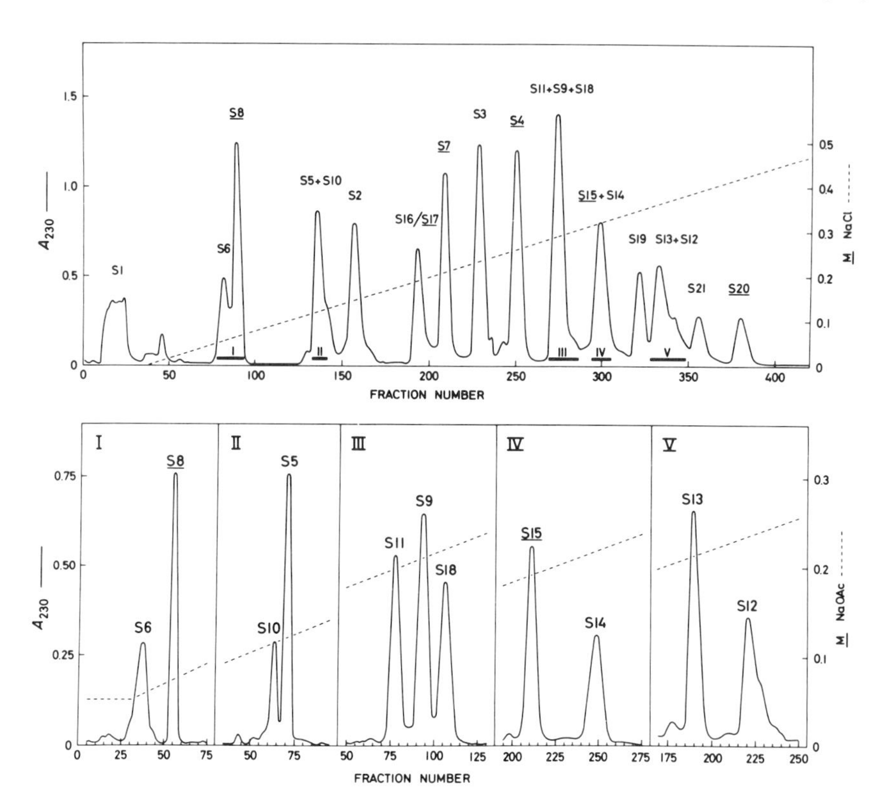


FIG. 1. Chromatographic fractionation of unlabeled 30 S subunit proteins from *Escherichia coli*. *Upper panel*: 200 mg of 30 S subunit proteins were applied to a 2.0 cm i.d. × 55 cm column of phosphocellulose at pH 6.5 and eluted with a linear 4000-ml salt gradient at a flow rate of 30 ml/hr. *Lower panel*: Protein mixtures from the first column were resolved on a 1.5 cm i.d. × 50 cm column of carboxymethyl cellulose at pH 5.6. Elution was accomplished with a 2000-ml gradient of Na acetate at 14 ml/hr. RNA-binding proteins are underlined. ——, $A_{230}$; - - -, salt concentration. Adapted from A. Muto, C. Ehresmann, P. Fellner, and R. A. Zimmermann, *J. Mol. Biol.* **86,** 411 (1974).

tions can usually be selected from the phosphocellulose column on the basis of polyacrylamide gel analysis of each tube across the relevant peak.

Analytical electrophoresis of ribosomal proteins is performed on both one- and two-dimensional polyacrylamide gels. One-dimensional gels, consisting of 7.5% acrylamide and 0.2% *N,N'*-methylene bisacrylamide in 8 *M* urea at pH 4.5,[21] are adequate for preliminary characterization. Proteins are visualized by staining with 0.05% Coomassie Brilliant Blue in 50% (v/v) methanol–7.5% (v/v) acetic acid, and their purity is estimated

[21] P. S. Leboy, E. C. Cox, and J. G. Flaks, *Proc. Natl. Acad. Sci. U.S.A.* **52,** 1367 (1964).

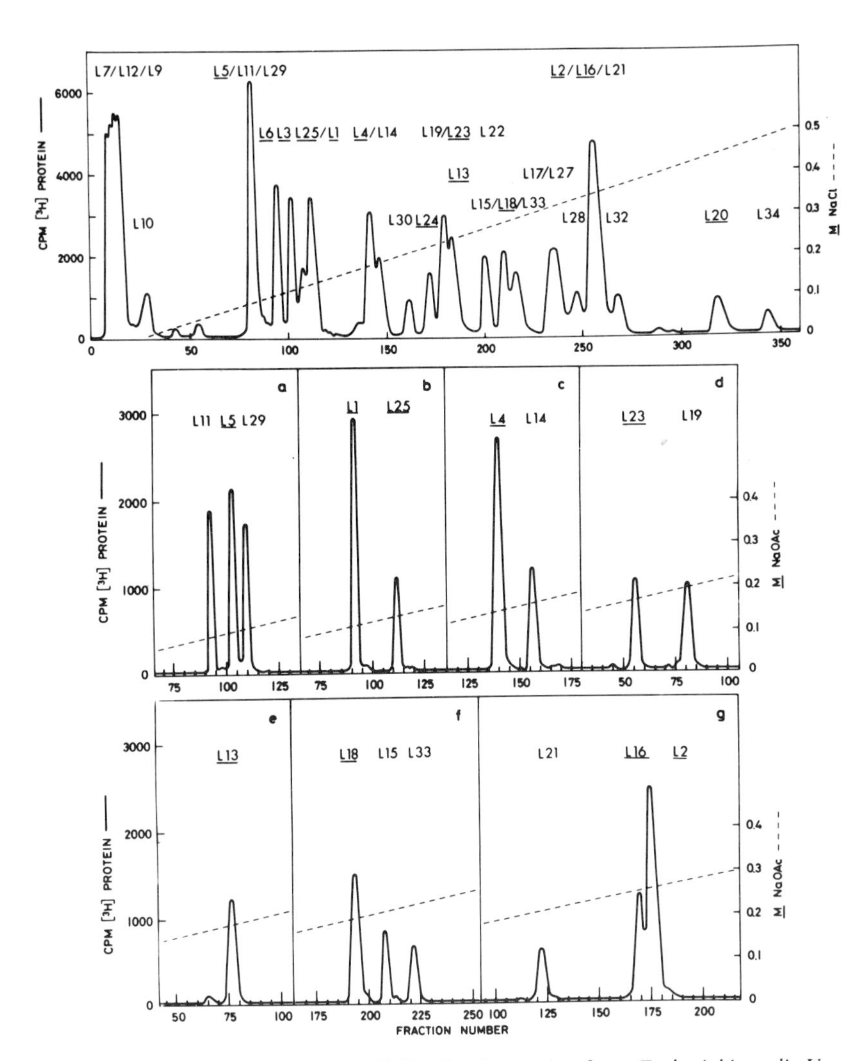


FIG. 2. Separation of $^{3}$H-labeled 50 S subunit proteins from *Escherichia coli*. *Upper panel*: 10 mg of total 50 S proteins were chromatographed on a 0.5 cm i.d. × 40 cm phosphocellulose column. Proteins were eluted with a 300-ml NaCl gradient at a flow rate of 4.5 ml/hr. *Middle and lower panels*: Proteins unresolved on phosphocellulose were fractionated by chromatography on a 0.3 cm i.d. × 40 column cm of carboxymethyl cellulose using 120-ml Na acetate gradients and an elution rate of 1.8 ml/hr. RNA-binding proteins are underlined.——, $^{3}$H-labeled protein; - - -, salt concentration. Adapted from R. A. Zimmermann and G. Stöffler, *Biochemistry* **15**, 2007 (1976).

by scanning the stained gels at 560 nm in a spectrophotometer. More definitive identifications can be obtained with the two-dimensional system of Kaltschmidt and Wittmann.[22]

*Alternative Procedures.* A number of techniques for the fractionation of *E. coli* 30 S and 50 S subunit proteins by ion-exchange chromatography, gel filtration, and electrophoresis have been worked out in several different laboratories. Most of the procedures for isolating 50 S subunit components entail some degree of prefractionation by high-salt or urea treatment prior to chromatographic or electrophoretic separation. Advantages and disadvantages of the various approaches to ribosomal protein purification have been discussed.[23]

Two methods have been developed for the special task of isolating the 5 S RNA-binding proteins, L5, L18, and L25. One employs the selective sequestration of these proteins by affinity chromatography on columns containing 5 S RNA covalently attached to an agarose matrix.[24] The other takes advantage of the fact that a complex of 5 S RNA with L5, L18, and L25 can be liberated from 50 S subunits treated with solid-phase pancreatic RNase A by sucrose gradient centrifugation in the presence of EDTA.[25] Proteins can then be dissociated from the RNA and further purified by ion-exchange or gel-filtration chromatography. In both cases, the recovery of L5 is found to be considerably lower than that of L18 and L25.

Ribosomal proteins may also be isolated by a new procedure that involves extraction and chromatography in concentrated LiCl solutions and thereby circumvents the use of acetic acid and urea.[26] Some of these proteins have been found to associate directly with 16 S and 23 S RNAs under conditions in which their counterparts prepared by more conventional methods are unable to do so.

## Formation and Analysis of Protein–RNA Complexes

### *Formation of Complexes*

*Buffers.* Compositions of the buffers are as follows:

Buffer J: 50 m*M* Tris·HCl, pH 7.6, 20 m*M* $MgCl_2$, 300–350 m*M* KCl, 6 m*M* 2-mercaptoethanol

[22] E. Kaltschmidt and H. G. Wittmann, *Anal. Biochem.* **36,** 401 (1970).
[23] H. G. Wittmann, *in* "Ribosomes" (M. Nomura, A. Tissières, and P. Lengyel, eds.), p. 93. Cold Spring Harbor Laboratory, Cold Spring Harbor, New York, 1974.
[24] H. B. Burrell and J. Horowitz, *FEBS Lett.* **49,** 306 (1975).
[25] U. Chen-Schmeisser and R. A. Garrett, *FEBS Lett.* **74,** 287 (1977).
[26] J. Littlechild and A. L. Malcolm, *Biochemistry* **17,** 3363 (1975).

Buffer K: 50 m$M$ $NaH_2PO_4$·methylamine, pH 7.6, 1 $M$ KCl, 1 $M$ urea, 6 m$M$ 2-mercaptoethanol

*Standard Assay.* From 1 to 2 molar equivalent(s) of one or more ribosomal protein(s) is mixed with an appropriate quantity of 16 S, 23 S or 5 S ribosomal RNA in 100 $\mu$l of buffer J containing 25 $\mu$g of bovine serum albumin at 0°. Fresh polypropylene tubes or detergent-washed polystyrene tubes are suitable for this purpose, but glass should be avoided whenever possible since many ribosomal proteins are strongly adsorbed to vitreous surfaces. Reaction mixtures are routinely heated to 40° for 30 min, chilled in an ice bath, and analyzed by one of the several techniques described below.

*Solution Conditions.* Solution conditions for the formation of specific protein–RNA complexes were originally based upon those used for reconstitution of 30 S ribosomal subunits.[27] The dependence of interaction upon pH and ionic environment has been systematically investigated for the association of S4 and S8 with 16 S RNA, of L24 with 23 S RNA, and of L5, L18, and L25 with 5 S RNA.[10, 28, 29] In all cases, optimal binding occurs between pH 7.5 and 8.5, although substantial association is observed from pH 6 to 9 for most of the complexes. The L18– and L25–5 S RNA interactions, however, appear to be unstable below pH 7.[10] The dependence of association upon $K^+$ concentration is relatively weak. The various optima occur between 100 to 400 m$M$ KCl and nonspecific binding may become a problem at concentrations less than 100 m$M$ KCl. By contrast, the influence of $Mg^{2+}$ ions is quite striking. Whereas the protein-binding capacity of the 16 S and 23 S RNAs is negligible below 1 m$M$ $MgCl_2$, it rises sharply to its maximum level at about 10 m$M$ $MgCl_2$.[28,29] Protein–5 S RNA interactions do not exhibit such a strong $Mg^{2+}$ dependence, and appreciable binding of L5 and L18 occurs at 1 m$M$ $MgCl_2$ or less.[10] The L25–5 S RNA complex remains completely stable at 0.2 m$M$ $MgCl_2$.

It is worth commenting on several other solution constituents—bovine serum albumin, 2-mercaptoethanol, and urea—that are present in the incubations either by design or by accident. The addition of bovine serum albumin has been found to reduce the nonspecific adsorption of ribosomal proteins to tubes, pipettes, and filters during the binding assays and is therefore beneficial in quantitative analyses. The usefulness of mercaptoethanol is open to question, however. The reducing agent can usually be omitted without adverse results even though a number of the RNA-

[27] P. Traub and M. Nomura, *Proc. Natl. Acad. Sci. U.S.A.* **59,** 777 (1968).

[28] C. Schulte and R. A. Garrett, *Mol. Gen. Genet.* **119,** 345 (1972).

[29] C. Schulte, C. A. Morrison, and R. A. Garrett, *Biochemistry* **13,** 1032 (1974).

binding proteins contain unique cysteine residues.[30] Finally, small amounts of urea are introduced into the reaction mixtures along with the ribosomal proteins. Although the binding of certain proteins is unaffected by the presence of 1 *M* urea, others are much more sensitive to this substance. In order to prevent the urea concentration from exceeding 0.2 *M*, where it does not appear to hinder any of the individual protein-RNA associations, a portion of the protein stock solutions is dialyzed into buffer K and stored at −20°. Most of the proteins have been found to remain soluble under these conditions. Alternatively, the reaction mixture itself may be dialyzed against buffer J prior to the heating step.

*Temperature.* In many cases, maximum levels of protein-RNA interaction can be obtained only after the reaction mixture has been incubated at 40°. The complexes themselves are actually more stable at low temperatures,[29,31] however, and components heated separately have been found to interact efficiently at 0°.[28,32] The temperature-dependent step appears to be necessary for the elimination of structural heterogeneities that arise in both protein and RNA molecules during purification.[28,29,32]

*Component Concentrations.* Detection of protein-RNA association depends upon the stability of the complexes, the solubilities of the components, and the sensitivity of the analytical procedures. Association constants for several such interactions have been found to lie between $10^6$ and $10^9$ $M^{-1}$.[31,33,34] Complex formation therefore requires component concentrations of $10^{-9}$ to $10^{-6}$ *M* and, since many of the detection methods themselves displace the equilibrium through dilution, pressure, or other perturbation, a more practical range is $10^{-8}$ to $10^{-5}$ *M*. In general, there is no problem in maintaining ribosomal components at $10^{-5}$ *M*, although this value does represent the limit of solubility of certain proteins, such as L5, in buffer J.[10] A wide variety of analytical techniques may be used at the higher concentrations since reaction mixtures will contain tens or even hundreds of micrograms of material that can be readily quantitated by standard absorbance and colorimetric assays. This is particularly advantageous when analysis entails gel electrophoresis. Below $10^{-6}$ *M*, however, it is more convenient to use radioactive RNAs and proteins. A rough guide to the concentration ranges in which various

[30] G. Stöffler and H. G. Wittmann, *in* "Molecular Mechanisms of Protein Biosynthesis" (H. Weissbach and S. Pestka, eds.), p. 117. Academic Press, New York, 1977.

[31] P. Spierer, A.A. Bogdanov, and R.A. Zimmermann, *Biochemistry* **17**, in press.

[32] A. Muto and R.A. Zimmermann, *J. Mol. Biol.* **121**, 1 (1978).

[33] E. Spicer, J. Schwarzbauer, and G.R. Craven, *Nucl. Acids Res.* **4**, 491 (1977).

[34] J. Feunteun, R. Monier, R. Garrett, M. Le Bret, and J. B. Le Pecq, *J. Mol. Biol.* **93**, 535 (1975).

kinds of unlabeled and labeled components are applicable is given in Table II. The choice of radioisotope has been made so as to ensure the presence of at least 1000 cpm per component in each assay, assuming the specific activities of labeling indicated in the first section. It is important to note that the suggestions provided in Table II apply to 100-$\mu$l incubation mixtures and may be extrapolated to smaller or larger volumes as appropriate.

*Analysis of Complexes*

A broad spectrum of methods has been used to separate protein-RNA complexes from unbound protein or RNA either for quantitative analysis or for purification of specific fragment complexes. The ensuing discussion is intended to describe applications of the different techniques. Although the most important experimental parameters will be indicated in each case, further details will be found in the original literature.

*Sucrose Gradient Centrifugation.* Reaction mixtures of 100 $\mu$l are layered onto linear 3% (w/v) to 15% (w/v) sucrose gradients in buffer J.[7] Four-milliliter gradients, used routinely for samples containing intact 5 S, 16 S, or 23 S RNAs, are centrifuged in the Spinco SW 60 rotor at 50,000 rpm for 10, 3, or 2 hr, respectively, at 2°. Twelve-milliliter gradients, employed mainly in the fractionation of RNA mixtures and RNase digests, are centrifuged in the Spinco SW 41 rotor at 40,000 rpm for 6–24 hr at 2°, depending on the requirements of the experiment. Gradient effluents are collected in 10 (SW 60) or 20 (SW 41) fractions, each of which is analyzed for absorbance and/or radioactivity. Labeled samples are precipitated in 5% (w/v) trichloroacetic acid and collected on glass-fiber filters, which are then dried and counted in a liquid scintillation spectrometer using a 5-ml cocktail containing 4 g of Omnifluor (New England Nuclear) per liter of toluene.[7]

*Gel Filtration.* Gel filtration media composed of polyacrylamide, agarose, or dextran with molecular exclusion limits of 100,000 to 500,000 may be used to separate ribonucleoprotein complexes containing 16 S and 23 S RNAs from unbound protein. The gel is equilibrated with buffer J and packed in a 0.5 cm i.d. × 40 cm glass column over a disk of fine-mesh nylon cloth. The reaction mixture is loaded onto the top of the column and eluted with buffer J at a flow rate of 2–3 ml/hr with the aid of a peristaltic pump. Fractions of 100–200 $\mu$l are collected and analyzed. BioGel P-150 (Bio-Rad Laboratories) has been used by Schaup *et al.*[35] for the study of interactions between radioactively labeled 30 S subunit

[35] H. W. Schaup, M. Green, and C. G. Kurland, *Mol. Gen. Genet.* **109,** 193 (1970).

TABLE II

CHOICE OF RIBOSOMAL COMPONENTS FOR USE IN DIFFERENT CONCENTRATION RANGES[a]

| Concentration ($M$) | Protein | | 16 S and 23 S RNAs | | 5 S RNA | |
|---|---|---|---|---|---|---|
| | $\mu$g/assay | Form | $\mu$g/assay | Form | $\mu$g/assay | Form |
| $10^{-6}$ to $10^{-5}$ | 1–30 | Unlabeled | 50–1000 | Unlabeled | 4–40 | Unlabeled |
| $10^{-7}$ to $10^{-6}$ | 0.1–3 | $^{3}$H-labeled | 5–100 | Unlabeled | 0.4–4 | $^{14}$C-Labeled |
| $10^{-8}$ to $10^{-7}$ | 0.01–0.3 | $^{35}$S-Labeled | 0.5–10 | $^{14}$C-Labeled | 0.04–0.4 | $^{14}$C-Labeled |
| $<10^{-8}$ | <0.01 | — | <0.5 | $^{14}$C-Labeled or $^{32}$P-Labeled | 0.04 | $^{32}$P-Labeled |

[a] The table indicates the minimum concentrations at which various unlabeled and radioactively labeled ribosomal components can be conveniently detected and quantitated. A 100-$\mu$l reaction mixture is assumed, as are specific activities of 5000 cpm/$\mu$g for $^{3}$H-labeled protein, 30,000 cpm/$\mu$g for $^{35}$S-labeled protein, 3000 cpm/$\mu$g for [$^{14}$C]RNA and $3 \times 10^{6}$ cpm/$\mu$g for [$^{32}$P]RNA. Quantitation of very small amounts of protein is difficult by any technique. Somewhat higher protein specific activities can be attained by subjecting purified proteins to reductive methylation [G. Moore and R. R. Crichton, *FEBS Lett.* **37,** 74 (1973)]. Since this procedure leads to the modification of lysine residues, however, the functional activity of the proteins may be altered as a result.

proteins and 16 S RNA. Garrett and his colleagues have used filtration on BioGel A-0.5 *M* (Bio-Rad Laboratories) to isolate complexes containing unlabeled ribosomal proteins and RNA.[36] In this case, RNA was estimated by absorbance at 260 nm. Protein was analyzed by electrophoresing the complexes into polyacrylamide gels, which were then stained with Coomassie Brilliant Blue and quantitated by densitometry.

*Polyacrylamide Gel Electrophoresis.* Electrophoresis on polyacrylamide gels has been used both for the analysis of complexes of protein with intact ribosomal RNA[35,36] and for the isolation of protein-RNA fragment complexes from RNase digests.[37-39] Gels are made from a 19:1 (w/w) mixture of acrylamide:*N,N'*-methylene bisacrylamide in 10 m*M* Tris·acetate, pH 7.2, 1 to 10 m*M* Mg acetate and are cast in 0.6 cm i.d. × 12 cm glass tubes or as a slab of the desired size.[40] For complexes of 16 S or 23 S RNAs, gels containing 2–3% (w/v) acrylamide mixture and 0.5% agarose are used. When protein–5 S RNA or protein–RNA fragment complexes are to be separated, the gels are made up in 3.5–10% (w/v) acrylamide mixture without agarose. After polymerization, the gels are preelectrophoresed for 1 hr, the samples are loaded in 10–20% (w/v) sucrose and 0.002% bromphenol blue, and electrophoresis is continued for several hours until the tracking dye is 1–2 cm from the bottom of the gel. Electrophoretic separations are carried out at 4° and at 8 to 10 mA per square centimeter of gel cross section, using the same buffer in the reservoirs as in the gel. Depending on the purpose of the experiment, the gels can be processed in a number of different ways. Unlabeled RNA can be detected by spectrophotometric scanning at 260 nm before staining or at 600 nm after fixing in 5.6% (v/v) acetic acid and staining with a 0.04% (w/v) solution of methylene blue in 0.2 *M* Na acetate, pH 4.7 Protein can be selectively colored with Coomassie Brilliant Blue and quantitated by scanning in a similar fashion. When radioactive components are present, the gel is frozen and sliced at regular intervals. Each slice is then placed in a vial, 5 ml of toluene containing 3% (v/v) Protosol (New England Nuclear) and 4 g of Omnifluor per liter are added, and the samples are assayed for radioactivity in a scintillation counter. If the object of the run is to recover specific $^{32}$P-labeled RNA fragments, the gel is subjected to autoradiography and the exposed film is used as a guide to excise the bands from which RNA is eluted for further analysis.[40]

[36] R. A. Garrett, K. H. Rak, L. Daya, and G. Stöffler, *Mol. Gen. Genet.* **114,** 112 (1974).
[37] C. Branlant, A. Krol, J. Sriwidada, and P. Fellner, *FEBS Lett.* **35,** 265 (1973).
[38] G. A. Mackie and R. A. Zimmermann, *J. Biol. Chem.* **250,** 4100 (1975).
[39] E. Ungewickell, R. Garrett, C. Ehresmann, P. Stiegler, and P. Fellner, *Eur. J. Biochem.* **51,** 165 (1975).
[40] R. De Wachter and W. Fiers, this series, Vol. 21, p. 167.

*Membrane Filter Assay.* The retention of ribonucleoprotein complexes on membrane filters provides a convenient means for the quantitative analysis of protein–RNA interaction under equilibrium conditions. The technique is well suited to the study of protein–5 S RNA complexes since L5, L18, and L25 all bind to the filter whereas 5 S RNA is not retained unless specifically associated with one of the proteins.[31,41] Reaction mixtures containing radioactively labeled proteins and 5 S RNA in 100 $\mu$l of buffer J are filtered through 13-mm cellulose acetate/cellulose nitrate membranes (Millipore, type HA) previously equilibrated in buffer J under vacuum. The filtration apparatus may be jacketed so that its temperature can be varied with the aid of a thermostatted circulating water system. A flow rate of about 500 $\mu$l/min is established by gentle suction. The filters are washed with 100 $\mu$l of buffer J, dried, and analyzed for radioactivity. Less than 5% of the 5 S RNA applied to the filter remains bound in the absence of protein whereas roughly 65% of the protein is retained.[31] The specific retention of 5 S RNA in the presence of protein generally reaches a maximum when about 50% of the protein is complexed.

*Affinity Chromatography.* Two variants of affinity chromatography have been used in the investigation of ribosomal protein–RNA interaction. In the first, 30 S subunit protein S20 was covalently bound to agarose and shown to selectively bind 16 S RNA in the presence of 23 S RNA.[42] Second, a 5 S RNA–agarose column has been used with success in the isolation of proteins that bind to this nucleic acid molecule.[24]

*Applications.* Unless large quantities of purified material are available, it is generally advantageous to employ radioactive components in studies of complex formation. Besides opening a wide concentration range to investigation, the use of protein and RNA labeled with different isotopes permits rapid and accurate estimation of both reactants at any stage of the analysis by scintillation counting. Components of high specific activity can also be diluted with their unlabeled counterparts in order to obtain a convenient counting rate of any given concentration. Finally, the availability of $^{32}$P-labeled RNA facilitates nucleotide sequence determination.

Recoveries of protein and RNA following sucrose gradient centrifugation, gel filtration, or polyacrylamide gel electrophoresis are all reasonably good, ranging from 70 to 100% for both kinds of molecules. All these techniques are thus applicable to the quantitative analysis of com-

[41] R. S. T. Yu and H. G. Wittmann, *Biochim. Biophys. Acta* **324**, 375 (1973).

[42] L. Gyenge, V. A. Spiridonova, and A. A. Bogdanov, *FEBS Lett.* **20**, 209 (1972).

plex formation but they are not in general suitable for equilibrium studies.

Although a somewhat smaller proportion of the starting material is recovered on cellulose membranes, the filter assay affords an opportunity for analyzing protein–RNA association in mixtures that are close to or at equilibrium. Owing to this characteristic, it has been possible to measure association constants and thermodynamic parameters for the three protein–5 S RNA interactions.[31] The filter assay has also been adapted to the investigation of protein–16 S RNA complexes.[33] Since proteins bound to the large ribosomal RNAs are not retained by the membrane, the extent of interaction must be inferred from the amounts of protein and RNA in the filtrate. This approach should be used with caution, however, because a significant fraction of the protein is known to pass through the filter in the absence of RNA at the concentrations normally used for such experiments. Steps must therefore be taken to ensure that all the protein in the filtrate is actually bound to RNA.

Affinity chromatography offers greater promise as a preparative technique than as an analytical technique. The potential utility of 5 S RNA–agarose columns in the fractionation of L5, L18, and L25 has been mentioned in an earlier section. Immobilized ribosomal proteins, on the other hand, could likely be exploited in the isolation of protein-binding fragments from enzymically digested ribosomal RNAs.

*Binding Stoichiometry and Criteria of Specificity.* Two criteria are generally used to establish that RNA molecules or RNA fragments contain specific protein binding sites. First, a given protein is expected to interact with only one species of RNA in the presence of others.[36] There will of course be exceptions to this rule, since certain proteins may serve to unite two RNA molecules and hence possess binding sites on each. In the same vein, other proteins may associate with two or more distinct fragments derived from a single RNA molecule. A second and more rigorous test of specificity demands that binding sites in the RNA or protein become saturated in the presence of excess protein or RNA, respectively.[35,36] In practice, it is usually easier to measure saturation of the RNA because complexes can be separated from unbound protein more readily than from free RNA. If the RNA contains a single specific binding site for the protein, the molar protein:RNA ratio should not surpass 1:1. Saturation curves are constructed by mixing increasing amounts of protein with a fixed amount of RNA at concentrations at least an order of magnitude greater than the presumed equilibrium constant so that interaction is strongly favored. The complexes are then isolated, the quantities of protein and RNA in each one are determined, and the molar protein:RNA ratio is computed. Calculations are facilitated by the use of different radioisotopes in protein and RNA, as a single assay provides all necessary information assuming that the specific activities and molecular weights of the components are known. The molecular weights of the RNAs and RNA-binding proteins of the *E. coli* ribosome are given in Table I. Examples of saturation curves for the association of L5, L18, and L25 with 5 S RNA are presented in Fig. 3. The relatively large excess of L5 required to establish the plateau could indicate that the protein is not fully active whereas the rather low molar protein:RNA ratio at saturation might be taken to mean that a substantial fraction of the RNA does not possess intact binding sites for the protein. When L5 binding is cooperatively stimulated by L18, however, a binding ratio of 1:1 is reached at an L5:5 S RNA input ratio of 1:1.[10] These results suggest that saturation data should be interpreted with circumspection.

*Fragmentation of RNA and Protein*

*Digestion of Ribosomal RNA*. Specific fragments of the 5 S, 16 S, and 23 S RNAs can be produced by limited enzymic hydrolysis of free RNA or ribonucleoprotein complexes. RNase $T_1$ and pancreatic RNase A have proved to be particularly useful for this purpose because they are active under solution conditions that promote optimal protein–RNA interaction. In fact, it is usually possible to recover large, discrete frag-

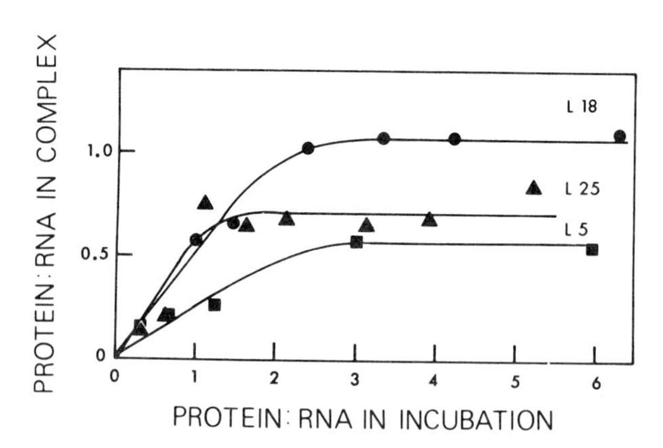


FIG. 3. Saturation curves for the binding of L5, L18, and L25 to the 5 S RNA. Increasing amounts of $^{3}H$-labeled proteins were incubated with fixed amounts of $^{14}C$-labeled 5 S RNA in buffer J. The complexes were separated by sucrose gradient centrifugation, and the molar ratios of protein and RNA in the 5 S peak were calculated from the specific radioactivities and molecular weights of the components. ■—■, L5-5 S RNA; ●—●, L18-5 S RNA: ▲—▲, L25-5 S RNA. Adapted from P. Spierer and R. A. Zimmermann, *Biochemistry* **17**, 2474 (1978).

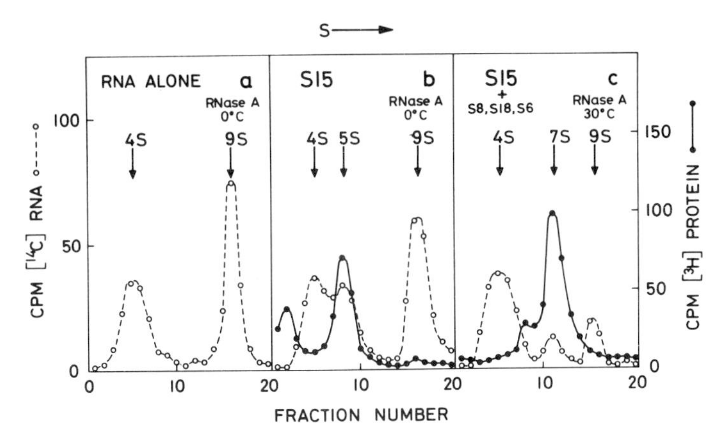


FIG. 4. Limited RNase digestion of 16 S RNA and of protein-16 S RNA complexes. $^{14}C$-Labeled 16 S RNA was mixed with (a) no addition, (b) $^{3}H$-labeled S15 and (c) $^{3}H$-labeled S15 in the presence of unlabeled S6, S8, and S18 in buffer J. After incubation at 40°, the mixtures were chilled and pancreatic RNase A was added at an enzyme:RNA (w/w) ratio of 1:5. Tubes (a) and (b) remained at 0° whereas tube (c) was heated to 30° for 5 min and then quickly chilled. The reaction mixtures were resolved by sedimentation on sucrose gradients. ○—○, $^{14}C$-Labeled RNA; ●—●, $^{3}H$-labeled protein. R. A. Zimmermann, unpublished results.

ments from the RNA only when digestion is carried out at high $Mg^{2+}$ concentrations.[7] The divalent cations apparently contribute to the intrinsic stability of the RNA by promoting compact folding of the nucleic acid chain. In addition, fragments can often be isolated in the presence of protein that are absent from digests of free RNA. In such cases, it is assumed that the protein protects the RNA segment from hydrolysis because of its proximity to labile bonds in the nucleic acid chain.

Ribonuclease digestion is usually carried out immediately after formation of the ribonucleoprotein complex. Depending on the extent of hydrolysis desired, RNase $T_1$ (Sankyo) or RNase A (Worthington) is added at an enzyme:RNA ratio of 1:5 to 1:1000 (w/w) and the mixture is incubated for 15–30 min at 0°.[7,38,39] The sample is then fractionated by sucrose gradient centrifugation in buffer J or by polyacrylamide gel electrophoresis in 10 m*M* Tris·acetate, pH 7.8, containing 1 to 10 m*M* Mg acetate. When several proteins are simultaneously bound to the RNA, homogeneous digestion products can sometimes be obtained only by heating the reaction mixture briefly after addition of RNase.

A number of the points discussed above are illustrated in Fig. 4. Hydrolysis of free 16 S RNA with pancreatic RNase A at 0° in 20 m*M* $MgCl_2$ produces fragments that sediment at 4 S and 9 S (Fig. 4a). Digestion of the protein S15–16 S RNA complex under identical conditions results in the formation of a third, protected fragment of about 5 S with which the protein remains associated (Fig. 4b). When a complex of 16 S RNA with S6, S8, S15, and S18 is treated with RNase A at 0°, the proteins are recovered with fragments ranging in size from 7 S to 13 S. After a 5-min incubation at 30°, however, the yield of higher molecular weight material is greatly reduced and the bound proteins are retained exclusively by the 7 S RNA component (Fig. 4c). Sequence analysis of $^{32}$P-labeled RNA showed that the 9 S fragment encompasses 550 residues from the 5′ end of the 16 S RNA whereas the 4 S material comprises a mixture of segments from the central and 3′ regions of the parent molecule.[7] The 5 S RNA, by contrast, is homogeneous, consisting of 150 nucleotides from the middle of the 16 S RNA; the 7 S fragment contains roughly 300 nucleotides, including all those present in the 5 S fragment as well as a sequence of 150 residues adjacent to its 3′ end.[3]

Although the sucrose gradient technique has been quite effective for the purification of RNA fragments containing one to several hundred residues, polyacrylamide gels offer better resolution of fragments encompassing 20–100 nucleotides.[38,39] Optical scans of gels used in the isolation of S4-protected sequences of the 16 S RNA are reproduced in Fig. 5. The gel method has also been widely exploited in conjunction with autoradiography for the fractionation of $^{32}$P-labeled fragments.

One of the first steps in the analysis of isolated RNA fragments, whether derived from free RNA or from ribonucleoprotein complexes, is to evaluate their protein-binding capacity.[43] Sucrose gradient fractions or polyacrylamide gel eluates are extracted with phenol and the RNA recovered by precipitation with ethanol. Carrier tRNA is frequently added to a final concentration at 50 $\mu$g/ml at this stage to obtain quantitative precipitation of the ribosomal RNA. The fragments are tested for their ability to interact with one or more ribosomal proteins by standard procedures. Binding saturation assays are especially useful in determining the proportion of the RNA that retains functional binding sites for the protein(s) in question.[38,44]

Sequences present in the RNA fragment are determined by the methods of fingerprinting and secondary oligonucleotide analysis originally developed by Sanger and co-workers[45] and recently extended by Uchida *et al.*[46] Ribosomal RNA fragments isolated by the procedures described

[43] R. A. Zimmermann, A. Muto, P. Fellner, C. Ehresmann, and C. Branlant, *Proc. Natl. Acad. Sci. U.S.A.* **69**, 1282 (1972).

[44] R. A. Zimmermann, A. Muto, and G. A. Mackie, *J. Mol. Biol.* **86**, 433 (1974).

[45] F. Sanger and G. G. Brownlee, this series, Vol. 12A, p. 361.

[46] T. Uchida, L. Bonen, H. W. Schaup, B. J. Lewis, L. Zablen, and C. Woese, *J. Mol. Evol.* **3**, 63 (1974).

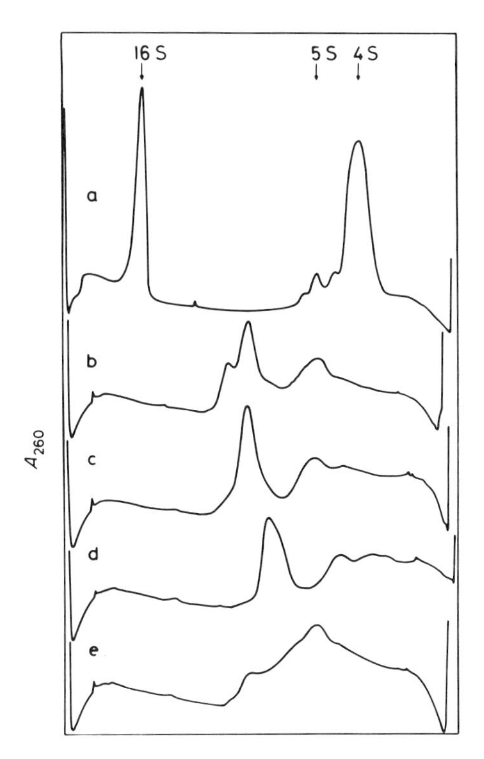


FIG. 5. Hydrolysis of S4–16 S RNA complexes with pancreatic RNase. Complexes between S4 and unlabeled 16 S RNA were formed in buffer J and treated with various amounts of pancreatic RNase. The products were separated by electrophoresis in 3.5% polyacrylamide gels, which were scanned in a spectrophotometer at 260 nm after the run. Migration is from left to right. (a) 16 S RNA and tRNA standards, no RNase added; (b) through (d) S4–S16 RNA complexes digested with increasing amounts of RNase; (e) RNase hydrolyzate of 16 S RNA in the absence of S4. Adapted from G. A. Mackie and R. A. Zimmermann, *J. Biol. Chem.* **250,** 4100 (1975).

above almost invariably contain a number of internal scissions, and on occasion substantial internal deletions as well, that are masked by secondary and tertiary interactions within the RNA molecule. Important information about the structure of the protein-binding sites can often be inferred from the pattern of such "hidden breaks" or excisions. The interactions that mask the discontinuities in the nucleic acid molecule are usually stable to normal manipulations such as centrifugation, electrophoresis, phenol extraction, and the like. However, intact subfragments can be resolved by electrophoresis in polyacrylamide gels containing denaturing agents such as urea.[47] To this end, the appropriate amount of

[47] C. Ehresmann, P. Stiegler, P. Fellner, and J.-P. Ebel, *Biochimie* **54,** 901 (1972).

a 19:1 (w/w) mixture of acrylamide:*N,N'*-methylene bisacrylamide is dissolved in 40 m*M* Tris·HCl, pH 7.2, 2 m*M* $Na_2EDTA$, 6 *M* urea and polymerized either in cylindrical tubes or in a slab-gel apparatus.[38,47] Discontinuous gels in which the lower two-thirds are composed of 15% (w/v) acrylamide and the upper one-third of 10% (w/v) acrylamide, are particularly convenient since fragments varying in length from 10 to 300 nucleotides can be fractionated on a single 40-cm slab. Samples in gel buffer with 10 to 20% (w/v) sucrose and 0.002% (w/v) bromphenol blue are loaded after a 1-hr preelectrophoresis, and the run is continued until the marker dye is a few centimeters from the bottom of the gel. After visualization of the bands by autoradiography, the RNA is eluted or subjected to further purification on a second urea gel containing a higher concentration of acrylamide.[47] The primary structure of this material may be analyzed in the usual way, and the binding capacity of the subfragments may be tested as well if so desired.

Figure 6 shows how a combination of gel techniques was used to isolate and analyze segments of the 23 S RNA that are protected from RNase attacked by protein L1.[48] Digestion of the L1–23 S RNA complex with RNase $T_1$ generated a ribonucleoprotein fragment, designed RNP 2, that migrated to a position in the gel that contained no products derived from hydrolysis of 23 S RNA alone (Fig. 6a). The RNA component of RNP 2 was subsequently resolved into a set of continuous, partially overlapping subfragments by electrophoresis under denaturing conditions (Fig. 6b). Fingerprinting of the subfragments revealed that they span a region of approximately 150 nucleotides at the 3' end of the 23 S RNA. The RNA extracted from RNP 2 was also shown to reassociate with L1 by a polyacrylamide gel assay (Fig. 6c).

Application of the methods described in this section has generated an appreciable amount of information on the distribution and characteristics of protein binding sites in both 16 S and 23 S RNAs.[3,4] This was possible largely because small, protein-specific fragments are not difficult to isolate on an analytical scale. Further chemical and physical studies on protein–RNA interaction would be greatly facilitated by the availability of means for the large-scale preparation of such sequences. Reasonable alternatives must therefore be found to replace sucrose gradient centrifugation, where the resolution is inadequate, and polyacrylamide gel electrophoresis, where the yield is very low. Fractionation of fragmented RNA by DEAE-cellulose[15] or reversed-phase[49] chromatography could well meet the combined demands of high resolution and high capacity in future investigations.

[48] P. Sloof, R. Garrett, A. Krol, and C. Branlant, *Eur. J. Biochem.* **70,** 447 (1976).

[49] R. L. Pearson, J. F. Weiss, and A. D. Kelmers, *Biochim. Biophys. Acta* **228,** 770 (1971).

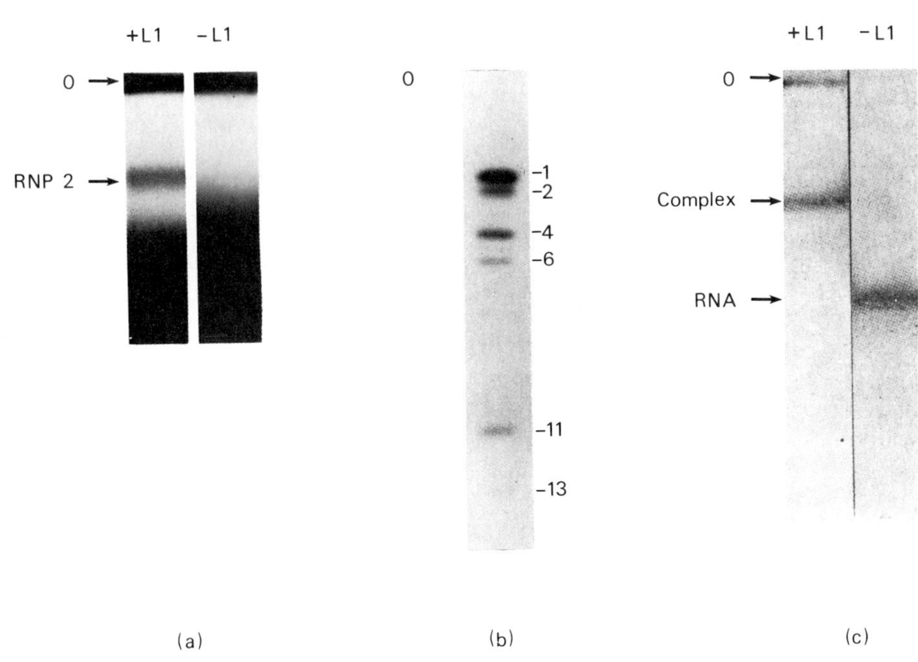


FIG. 6. Isolation and properties of a ribonucleoprotein fragment derived from the L1–23 S RNA complex. A complex of 50 S subunit protein L1 with $^{32}P$-labeled 23 S RNA was formed at 42°, chilled on ice, and treated with RNase $T_1$ at an enzyme:RNA (w/w) ratio of 1:8. (a) Fractionation of digests of the complex (left) and of free 23 S RNA (right) was carried out on an 8% polyacrylamide slab gel under nondenaturing conditions. (b) RNA subfragments present in the gel slice containing RNP 2 were dissociated from protein and separated from one another by electrophoresis into a second gel composed of 0.1% sodium dodecyl sulfate, 8 *M* urea, and 12–15% acrylamide. (c) The RNA moiety of RNP 2 was extracted from the initial gel with phenol, dialyzed, and incubated with protein L1 at 40°. The protein-fragment complex was then subjected to electrophoresis in an 8% polyacrylamide (left) in parallel with a sample of uncomplexed RNA fragment (right). Adapted from P. Sloof, R. Garrett, A. Krol, and C. Branlant, *Eur. J. Biochem.* **70,** 447 (1976).

*Enzymic Hydrolysis of Ribosomal Proteins.* The isolation of protein fragments capable of specifically binding to ribosomal RNA would clearly be of benefit in the study of protein–RNA association. With a few exceptions, however, little progress has been made in this area. Partial digestion of the S4–16 S RNA complex with trypsin has led to the isolation of a fragment comprising roughly 80% of the protein that was capable of reattaching to the 16 S RNA.[50,51] RNA-binding fragments have also been derived from S8 both by treatment of the isolated protein with

[50] C. Schulte, E. Schiltz, and R. A. Garrett, *Nucl. Acids Res.* **2,** 931 (1975).

[51] L.-M. Changchien and G. R. Craven, *J. Mol. Biol.* **108,** 381 (1976).

cyanogen bromide and by digestion of the S8–16 S RNA complex with proteinase K.[52] A systematic approach to the production, purification, and analysis of protein fragments still awaits development.

## Application of Specialized Techniques

Although the methodology discussed above has led to the acquisition of a substantial body of data on protein–RNA interaction, a number of more specialized techniques have also provided important insights into such processes. A number of these are identified in the following paragraphs and a brief description of their application to the ribosomal system will be presented. Particular emphasis will be placed upon the experimental conditions, the amounts of material required, and the nature of the information they can furnish.

### *Chemical Procedures*

*Establishment of Protein–RNA Cross Links.* Several RNA-binding proteins have been photochemically linked to their respective RNAs by irradiation of reconstituted complexes,[53–55] or of intact subunits,[56,57] with ultraviolet light. Covalent S4– and S7–16 S RNA complexes were isolated under conditions that dissociate noncovalently attached material, and the cross-linked segments of both protein and RNA were analyzed by fingerprinting.[53–55,57] Positive identification of cross-linked peptides was not possible, however, since conclusions were based on the absence of particular spots from standard tryptic peptide maps.[53,54] While such studies can provide valuable information on the proximity of specific protein and RNA segments in the complexes, it is still not clear whether these regions contain the actual recognition sites.

Cross-linking by affinity labeling or with bifunctional reagents has not yet been applied successfully to individual ribosomal protein–RNA complexes, despite the great potential of these techniques for defining structural relationships among interacting macromolecules.

*Chemical Modification of Protein and RNA.* Little use has been made of chemical modification as a probe of protein–RNA interaction. One drawback to such methods is that even in free components, identical

[52] J. Bruce, E. J. Firpo, and H. W. Schaup, *Nucl. Acids Res.* **4,** 3327 (1977).

[53] B. Ehresmann, J. Reinbolt, and J.-P. Ebel, *FEBS Lett.* **58,** 106 (1975).

[54] B. Ehresmann, J. Reinbolt, C. Backendorf, D. Tritsch, and J.-P. Ebel, *FEBS Lett.* **67,** 316 (1976).

[55] B. Ehresmann, C. Backendorf, C. Ehresmann, and J.-P. Ebel, *FEBS Lett.* **78,** 261 (1977).

[56] K. Möller and R. Brimacombe, *Mol. Gen. Genet.* **141,** 343 (1975).

[57] J. Rinke, A. Yuki, and R. Brimacombe, *Eur. J. Biochem.* **64,** 77 (1976).

residues in dissimilar environments differ in their accessibility to various reagents. Therefore, it is frequently difficult to evaluate the specificity of changes in reactivity that occur when the components are incorporated into ribonucleoprotein complexes. Nonetheless, the binding of S4 to 16 S RNA has been demonstrated to completely protect from reductive methylation two lysine residues that are available in the free protein.[58] One of them falls within a peptide believed to be covalently linked to the 16 S RNA by ultraviolet irradiation.[53] The influence of prior modification on the binding capacities of S4, S8, S15, and S20 has also been assessed.[59] Interaction was little affected by blockage of cysteine, tryptophan, and lysine residues, but severely impaired after oxidation of methionine residues.

Kethoxal treatment of 5 S RNA, whether free in solution or a constituent of either the 50 S subunit or the 70 S ribosome, results in the modification of two specific guanine residues.[60–62] Since the modified 5 S RNA can be reintegrated into 50 S subunits with full restoration of activity, its ability to interact with ribosomal proteins is apparently not curtailed.[60,62] By contrast, treatment of free 5 S RNA with the cytosine-specific reagent methoxyamine or with nitrous acid leads to its rapid inactivation.[60] Kethoxal has also been used to investigate accessible guanine residues in both 16 S and 23 S RNAs in intact ribosomal particles.[63,64] However, none of the techniques for modifying specific classes of nucleotide bases has been systematically exploited in the study of protein–RNA interaction, even though an analysis of the pattern of residues protected by bound protein would appear to be useful in defining the topography of the binding sites.

*Physical Procedures*

*Thermodynamic and Kinetic Parameters.* Apparent association constants ($K'$) for complexes of ribosomal RNA with individual ribosomal proteins can be measured by any technique that permits discrimination of bound and unbound components at equilibrium. In addition, the components should be readily quantifiable at concentrations comparable to the reciprocal of $K'$ since the greatest sensitivity is obtained in this range. The membrane filter assay provides a simple method for making such

[58] R. Amons, W. Möller, E. Schiltz, and J. Reinbolt, *FEBS Lett.* **41,** 135 (1974).
[59] L. Daya-Grosjean, J. Reinbolt, O. Pongs, and R. A. Garrett, *FEBS Lett.* **44,** 253 (1974).
[60] G. Bellemare, B. R. Jordan, J. Rocca-Serra, and R. Monier, *Biochimie* **54,** 1453 (1972).
[61] H. F. Noller and W. Herr, *J. Mol. Biol.* **90,** 181 (1974).
[62] N. Delihas, J. J. Dunn, and V. Erdmann, *FEBS Lett.* **58,** 76 (1975).
[63] H. F. Noller, *Biochemistry* **13,** 4694 (1974).
[64] W. Herr and H. F. Noller, *Biochemistry* **17,** 307 (1978).

measurements as long as the process of filtration does not perturb the equilibrium. Precautions required to ensure the validity of the filter assay have been discussed elsewhere.[65,66] The apparent standard free-energy change for complex formation may be calculated from the equilibrium constant at any given temperature ($\Delta G^{\circ\prime} = -RT \ln K'$) and, if the variation of $\ln K'$ with temperature is linear, the apparent standard enthalpy change may be derived by application of the van't Hoff equation ($[\partial \ln K'/\partial(1/T)]_p = -\Delta H^{\circ\prime}/R$). Finally, the corresponding entropy change can be computed from $\Delta G^{\circ\prime}$ and $\Delta H^{\circ\prime}$ ($\Delta G^{\circ\prime} = \Delta H^{\circ\prime} - T\Delta S^{\circ\prime}$). Parameters for the interaction of L5, L18 and L25 with 5 S RNA have recently been measured in this fashion.[31] In principle, the same kinds of techniques can be employed to determine the association and dissociation rate constants as long as these processes can be adequately resolved as a function of time.

*Fluorescence Techniques.* The association of certain proteins with the 5 S and 16 S RNAs leads to the displacement of the fluorescent dye ethidium bromide from the nucleic acid molecules.[34,67] It is possible to take advantage of this phenomenon for the detection of protein–RNA complex formation at concentrations as low as $10^{-7}$ $M$. In particular, the ability of L18 to stimulate the release of the dye from 5 S RNA has been used to estimate an affinity constant for the L18–5 S RNA interaction.[34] Since some proteins do not cause appreciable displacement of ethidium bromide, however, the generality of the approach is limited. The application of singlet–singlet energy transfer[68] to the measurement of distances between fluorescent labels at specific locations in interacting protein and RNA molecules has not to date been reported.

*Scattering Techniques.* Information on the size and shape of macromolecules may be obtained from a number of different light-scattering techniques. Low-angle X-ray scattering is particularly appropriate for particles whose dimensions are on the order of 50 to 200 Å. Analysis of isolated ribosomal components, or of their complexes, by this method provides a means for estimating the corresponding radii of gyration, volumes, and molecular weights.[69–74] These parameters, as well as other

[65] M. Yarus and P. Berg, *Anal. Biochem.* **35,** 450 (1970).
[66] A. D. Riggs, H. Suzuki, and S. Bourgeois, *J. Mol. Biol.* **48,** 67 (1970).
[67] A. Bollen, A. Herzog, A. Favre, J. Thibault, and F. Gros, *FEBS Lett.* **11,** 49 (1970).
[68] K.-H. Huang, R. H. Fairclough, and C. R. Cantor, *J. Mol. Biol.* **97,** 443 (1975).
[69] P. G. Connors and W. W. Beeman, *J. Mol. Biol.* **71,** 31 (1972).
[70] R. Österberg, B. Sjöberg, and R. A. Garrett, *FEBS Lett.* **65,** 73 (1976).
[71] R. Österberg, B. Sjöberg, and R. A. Garrett, *Eur. J. Biochem.* **68,** 481 (1976).
[72] R. Österberg, B. Sjöberg, R. A. Garrett, and J. Littlechild, *FEBS Lett.* **73,** 25 (1977).
[73] R. Österberg and R. A. Garrett, *Eur. J. Biochem.* **79,** 67 (1977).
[74] R. Österberg, B. Sjöberg, R. A. Garrett, and E. Ungewickell, *FEBS Lett.* **80,** 169 (1977).

known constraints, are used to construct molecular models which are generally based on ellipsoids, ellipsoidal segments, or even more complex structures. Theoretical scattering curves are then calculated from the models and compared with the angular dependence of scattering determined by experiment in order to achieve the best fit. Data on L18, 5 S RNA, and the L18–5 S RNA complex, for instance, suggest that both molecules are elongated and that they interact with one another in a highy asymmetric fashion.[70,71,73] Low-angle X-ray scattering can also be used to analyze equilibria in multicomponent systems. One disadvantage of the method is its requirement for protein and RNA concentrations in the range of $10^{-5}$ to $10^{-4}$ *M*, where insolubility and aggregation can pose substantial problems.

Another optical technique that has been applied to the characterization of protein–RNA interaction is laser light scattering, which permits the accurate determination of translational diffusion constants.[75,76] This approach provides a particularly convenient and sensitive probe of changes in macromolecular conformation because diffusion constants can be measured on very small amounts of material in roughly 1 minute with a precision of ±1%. Experiments are routinely performed with samples of 20 μl at concentrations between $10^{-6}$ and $10^{-5}$ *M* for ribosomal proteins and 5 S RNA, and as little as $10^{-7}$ *M* for the large ribosomal RNAs, so that aggregation is usually avoided. In addition, the technique is appropriate for the investigation of concentration- and temperature-dependent dissociation phenomena since measurements are made under equilibrium conditions. Rapid kinetic assays are not feasible, however. Determination of diffusion constants for the 16 S RNA and a number of protein–16 S RNA complexes by laser light scattering has revealed that S4, S7, S8, and S15, either alone or in combination, stabilize the nucleic acid chain against unfolding at low $Mg^{2+}$ concentration and actually promote the formation of a more compact conformation at the higher $Mg^{2+}$ concentrations necessary for 30 S subunit reconstitution.[76]

*Circular Dichroism.* The circular dichroism (CD) of RNA molecules is related to the extent of base-pairing and base-stacking within the nucleic acid chain. Alterations in either will provoke changes in the position and intensity of the main positive CD band, which lies between 260 and 275 nm, as well as in other portions of the spectrum. The sensitivity of the method is roughly equivalent to that of absorbance at 260 nm, so that measurements may be carried out at 5 S RNA concentrations of about $10^{-6}$ *M* and 16 S or 23 S concentrations of $10^{-7}$ *M* or

[75] N. C. Ford, Jr., *Chemica Scripta* **2,** 193 (1972).

[76] A. A. Bogdanov, R. A. Zimmermann, C.-C. Wang, and N. C. Ford, Jr., *Science*, in press.

less. For the detection of changes in RNA secondary structure that accompany the association of protein, attention is focused on the spectral region above 240 nm, where the circular dichroism of protein is negligible. Below 240 nm, the CD spectrum contains contributions from both protein and RNA, which cannot be easily distinguished. In a practical sense, the application of CD methods to ribosomal protein–RNA interaction is restricted to 5 S RNA or to RNA fragments of comparable size, since perturbations in secondary structure will generally occur only within limited portions of the RNA and the corresponding CD changes must be detected over the signal from the molecule as a whole. In the case of protein L24, association with fragments from the 23 S RNA has been found to produce a slight decrease in the CD maximum at 265 nm.[77] By contrast, the binding of L18 to 5 S RNA has been found to stimulate a 15–20% increase in the main positive band at 268 nm[31,78] A similar change is observed when L5, L18, and L25 are added together.[31,79] Uncertainties in the interpretation of CD results are all too apparent in these reports, however, since similar spectral changes in the 5 S RNA have been variously taken to indicate an increase in secondary structure,[78] an increase in the stacking of single-stranded regions,[79] and an alteration in the regularity of existing helical segments.[31]

*Thermal Transitions* Direct measurement of the energy absorbed by marcomolecules undergoing thermal order–disorder transitions may be accomplished by differential scanning calorimetry.[80] In practice, the heat capacity of the sample is recorded as a function of temperature as it is continuously heated from 20° to near 100°. The melting of various structural domains within the solute is delineated by changes in the heat capacity at the corresponding transition temperatures. The enthalpies of melting are obtained by integrating the areas under the peaks and can be used to derive a number of other thermodynamic parameters. The greatest utility of the technique lies in the analysis of small RNAs and their complexes where the total number of independent transitions is not too large.[81,82] In addition, relatively high concentrations of material are required since the quantities of heat absorbed are minute. For tRNA or 5 S RNA, a 1-ml sample containing roughly 1–3 mg of solute ($5 \times 10^{-5}$ to $10^{-4}$ *M*) is necessary for each run. By the proper choice of solution

[77] T. R. Tritton and D. M. Crothers, *Biochemistry* **15,** 4377 (1976).

[78] D. G. Bear, T. Schleich, H. F. Noller, and R. A. Garrett, *Nucl. Acids Res.* **4,** 2511 (1977).

[79] J. W. Fox and K.-P. Wong, *J. Biol. Chem.* **253,** 18 (1978).

[80] P. L. Privalov, *FEBS Lett.* **40,** S140 (1974).

[81] H.-J. Hinz, V. V. Filimonov, and P. L. Privalov, *Eur. J. Biochem.* **72,** 79 (1977).

[82] T. L. Marsh, J. F. Brandts, and R. A. Zimmermann, unpublished results.

conditions, it should be possible to distinguish a series of discrete peaks that correspond to the melting of different elements of secondary and tertiary structure. If two or more transitions occur at about the same temperature, however, it may be hard to accurately decompose the melting curve into its component parts. A more serious problem lies in the identification of the calorimetric peaks with particular domains in the molecule. In the case of 5 S RNA, whose secondary structure is uncertain and whose tertiary structure is totally unknown, assignment of transitions must await the analysis of specific fragments, despite the fact that reasonably well-resolved calorimetric melting curves have been obtained.[82] Nonetheless, when these difficulties are overcome, scanning microcalorimetry will undoubtedly prove to be of considerable value in characterizing the intramolecular structure of 5 S RNA and small fragments of the larger ribosomal RNAs, as well as the manner in which various structural domains are affected by the presence of bound proteins.

Further dissection of thermal transitions within individual structural domains of small RNA molecules or RNA fragments may be achieved by temperature-jump relaxation spectroscopy, which permits kinetic analysis of the melting process following a limited, but very rapid increase in temperature.[83] T-jump measurements are therefore of utility in defining the effects of bound ligands on specific elements of RNA structure and can also provide insights into the cooperativity of the corresponding transitions. In the binding of coat protein to a short segment of phage R17 RNA, for example, it was possible to demonstrate that complex formation produces a dramatic reduction in the melting rate of a single hairpin loop without altering its melting temperature.[84] This technique has also been used in the ribosomal system for studies on the interaction of L24 with fragments derived from 23 S RNA.[77]

*Electron Microscopy.* Direct visualization of protein–RNA complexes by electron microscopy has been achieved in a number of instances.[85–87] Complexes of S4 and S8 with 16 S RNA, and of L23 and L24 with 23 S RNA have been examined after spreading in 80% dimethyl sulfoxide.[85,86] In this method, a considerable fraction of the RNA appears to condense around the protein, making it difficult to precisely locate the site of attachment or to discern the structural features of the binding region.

[83] D. M. Crothers, *in* "Procedures in Nucleic Acid Research" (G. L. Cantoni and D. R. Davies, eds.), Vol. 2, p. 369. Harper & Row, New York, 1971.
[84] J. Gralla, J. A. Steitz, and D. M. Crothers, *Nature (London)* **248,** 204 (1974).
[85] N. Nanninga, R. A. Garrett, G. Stöffler, and G. Klotz, *Mol. Gen. Genet.* **119,** 175 (1972).
[86] P. Sloof, R. A. Garrett, and N. Nanninga, *Mol. Gen. Genet.* **147,** 129 (1976).
[87] M. D. Cole, M. Beer, T. Koller, W. A. Strycharz, and M. Nomura, *Proc. Natl. Acad. Sci. U.S.A.* **75,** 270 (1978).

More recently, S4- and S8-16 S RNA complexes were prepared for electron microscopy by fixation with formaldehyde and spreading in the presence of benzyldimethylalkylammonium chloride.[87] In this case, the proteins did not cause the RNA to clump but, rather, stabilized characteristic loops that could be positioned relative to the two termini of the otherwise extended nucleic acid molecule. The 5′ and 3′ ends were not distinguished in this work, however, although specific labeling of the termini is now technically feasible. Electron microscopic studies lend support to the binding site locations established by biochemical techniques and offer considerable promise as means to investigate tertiary interactions within the large ribosomal RNA molecules.

## Concluding Remarks

During the past several years, the study of ribosomal protein-RNA interactions has been directed toward two main objectives. First, considerable effort has been devoted to locating protein binding sites along the 16 S, 23 S, and 5 S ribosomal RNAs. Second, an attempt has been made to discover the elements of sequence or structure within the nucleic acids that comprise the signals for protein recognition. The former task has been successfully accomplished for most of the proteins that bind early in 30 S and 50 S subunit assembly, but the existence of a general recognition code remains in doubt owing to the great structural diversity of the attachment sites themselves. Since some proteins interact with structurally complex regions spanning several hundred nucleotides and others, with relatively short helical segments, it seems more likely that specific protein-RNA associations represent the concerted action of a variety of different binding mechanisms.

In order to better understand the process of protein-RNA complex formation, investigations must now be extended to a broader range of phenomena. The ability of certain proteins to cooperatively stimulate the binding of others has been demonstrated, for instance, but the structural basis for cooperativity, and in particular its relationship to protein-induced alterations in RNA structure, await elucidation. Moreover, there is not a single instance in which the base or amino acid sequences involved in protein-RNA recognition have been positively identified. Similarly, the relative contributions of hydrogen-bonding, stacking, and electrostatic interaction to the specificity and stability of the complexes are unknown. Information on the size and shape of the interacting molecules is also scant, and physicochemical parameters for the associations are just beginning to emerge. The resolution of these and other questions will undoubtedly require the deployment of fresh analytical approaches.

The methods enumerated in the preceding section do not in any way comprise an exhaustive listing. Rather, they are representative of procedures that have already proved to be of some utility in the ribosomal system. While a number of others, such as singlet–singlet fluorescence energy transfer, would appear to be immediately applicable, the use of nuclear magnetic resonance and X-ray crystallography, which offer great potential for revealing the structure of macromolecular complexes, is presently beset by a variety of unresolved technical problems. For the time being, it is likely that progress in the understanding of protein–RNA interaction will be achieved mainly by the judicious exploitation of analytical methods already in use.

### Acknowledgments

The author gratefully acknowledges receipt of a U.S. Public Health Service Research Career Development Award (GM-00129) as well as support by grants from the National Science Foundation (PCM74-00392) and from the National Institutes of Health (GM-22807).

# [65] The Use of Membrane Filtration to Determine Apparent Association Constants for Ribosomal Protein–RNA Complex Formation

*By* JEAN SCHWARZBAUER and GARY R. CRAVEN

Progress toward understanding the rules governing protein-nucleic acid interactions in the bacterial ribosome has been impeded by the lack of convenient techniques for measuring the equilibrium constants of protein-RNA complex formation. The procedures previously used for the isolation of r-protein-rRNA complexes (velocity sedimentation,[1] gel filtration,[2] and polyacrylamide–agarose gel electrophoresis[3]) completely disrupt the equilibrium between unbound protein, uncomplexed RNA, and the protein-RNA complex. With this limitation it has been possible to study only strong protein-RNA associations. Weak, yet specific, binding interactions may remain unobserved by these methods. The technique

[1] H.-K. Hochkeppel, E. Spicer, and G. R. Craven, *J. Mol. Biol.* **101,** 155 (1976).
[2] H. W. Schaup, M. Green, and C. G. Kurland, *Mol. Gen. Genet.* **109,** 193 (1970).
[3] R. A. Garrett, K. H. Rak, L. Daya, and G. Stöffler, *Mol. Gen. Genet.* **114,** 112 (1971).



ISBN 0-12-504180-2

described here, that of nitrocellulose membrane filtration,[4] allows accurate measurement of the selective association of ribosomal proteins with RNA without disturbing the equilibrium.

We have observed that the ribosomal proteins can be adsorbed, at appropriately low concentrations, to nitrocellulose membranes under conditions where specific protein–RNA complexes fail to be bound by the membrane. This observation offers the opportunity to rapidly separate unbound, or "free," protein from the protein involved in a protein–RNA complex. This method permits the determination of the amount of protein bound to RNA and the amount of free protein in reaction mixtures containing different relative concentrations of protein and RNA. The resultant data can be used to calculate the number of binding sites for the protein on the RNA and the apparent association constant for the binding reaction.

## Apparatus

The design for the original apparatus used in our studies of protein–RNA complex formation is an adaptation of one developed by Paulus to measure the binding of small molecules to proteins.[5] Figure 1 is a photograph of our apparatus, which contains twelve channels, each 5 mm in diameter. The nitrocellulose membranes are triangular sections cut from filters saturated with the buffer used in the binding reaction. The use of membranes that are not prewetted with buffer leads to a significant loss of material. The membranes are placed between the upper block and lower block across the channels. The lower/block side channels are sealed with tape. The samples, usually 300-400 μl in volume, are placed in the channels of the upper block above the membrane. The sample channels are sealed with screw plugs and nitrogen pressure is applied equally to all 12 channels through the central adaptor. Using nitrocellulose filters (type HA, pore size 0.45 μm, Millipore Corp.) a pressure of approximately 2 psi pushes the sample through the membrane in less than 10 sec. The apparatus is inverted while maintaining nitrogen pressure. Aliquots are removed from the filtrate and assayed for protein and RNA concentrations.

## Preparation of Materials

Most investigations of ribosomal protein–RNA interactions have employed proteins purified by phosphocellulose or carboxymethylcellulose

[4] E. Spicer, J. Schwarzbauer, and G. R. Craven, *Nucl. Acids Res.* **4,** 491 (1977).
[5] H. Paulus, *Anal. Biochem.* **32,** 91 (1969).

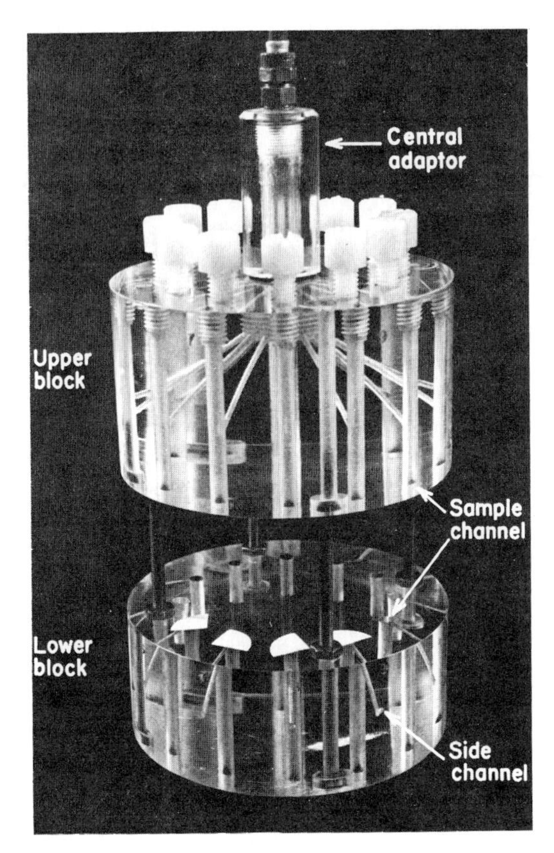


FIG. 1. The apparatus used for isolation of protein–RNA complexes shown with nitrocellulose filter sections in place over the lower block sample channels.

chromatography followed by Sephadex gel filtration when needed.[6,7] The solutions used in the fractionation procedures usually contain 6 *M* urea as a denaturant and a reducing agent such as dithiothreitol (DTT) to prevent unwanted protein–protein aggregation. After fractionation the proteins are concentrated, dialyzed against fresh Tris–urea–DTT buffer, quick frozen in aliquots, and stored at −70°.

In general, it is wise to prepare and store the ribosomal proteins with as much care as possible. It is not uncommon to find different batches of a given protein to have widely different capacities to specifically bind

[6] S. J. S. Hardy, C. G. Kurland, P. Voynow, and G. Mora, *Biochemistry* **8,** 2897 (1969).
[7] R. A. Traut, H. Delius, C. Ahmad-Zadeh, T. A. Bickle, D. Pearson, and A. Tissières, *Cold Spring Harbor Symp. Quant. Biol.* **34,** 25 (1969).

RNA.[8] However, one of the salient advantages of the nitrocellulose membrane filtration technique is that the competency of any protein preparation for RNA binding can be readily determined. This is done using the procedure described below and should be done on every batch of protein prior to detailed binding experiments.

The individual proteins can be made radioactive by reductive methylation of lysine residues with [$^3$H]$NaBH_4$ and formaldehyde.[9] Care should be taken to treat the protein as lightly as possible (approximately one hit per molecule) to avoid any potential inactivation of binding activity. The radiopurity of the $^3$H-labeled proteins can be judged by polyacrylamide gel electrophoresis.[10] The specific activity of the protein must be accurately determined. We have found that the Lowry method[11] gives a satisfactory measure of ribosomal protein concentration using lysozyme as the standard. The final radioactive preparation of protein can be tested for competency in binding relative to nonradioactive protein by doing a binding experiment with a mixture of reductively methylated and unmodified protein. If the radioactive protein is unaltered in its binding ability, there should be an equal competition between the labeled and the unlabeled protein for binding to the RNA. In our experience, the reductive methylation does not decrease the binding capacity of the ribosomal proteins.

The RNA to be used in binding experiments can be prepared either by the traditional phenol-dodecyl sulfate method[12] or the newer technique using acetic acid and urea.[1] The latter procedure yields 16 S RNA capable of binding at least six new proteins that are not bound by the phenol-SDS preparation. The expression of this increased binding capacity by acetic acid-urea RNA is rapidly lost upon freezing and thawing; therefore the RNA should be used, if possible, on the same day it is prepared. The RNA is conveniently stored in Tricine [*N*-tris(hydroxymethyl)methylglycine] or phosphate buffers, pH 7.0–7.8, without $Mg^{2+}$.

## Preparation of Protein–RNA Complexes

Approximately one $A_{260}$ unit of 16 S RNA ($0.75 \times 10^{-10}$ mol) is diluted into reconstitution buffer (RB; 30 m*M* Tricine, pH 7.4, 0.4 *M* KCl, 20 m*M* Mg acetate, 1 m*M* DTT) and preincubated for 10 min at 42° to

[8] J. Littlechild, J. Dijk, and R. A. Garrett, *FEBS Lett.* **74**, 292 (1977).

[9] G. Moore and R. R. Crichton, *FEBS Lett.* **37**, 74 (1973).

[10] P. Voynow and C. G. Kurland, *Biochemistry* **10**, 517 (1971).

[11] O. H. Lowry, N. J. Rosebrough, A. L. Farr, and R. J. Randall, *J. Biol. Chem.* **198**, 265 (1951).

[12] P. Traub, S. Mizushima, C. V. Lowry, and M. Nomura, this series, Vol. 20, p. 399.

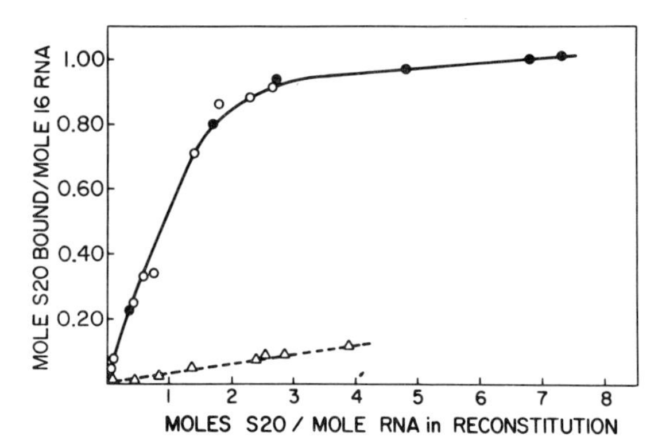


FIG. 2. Saturation binding curve for S20–16 S RNA interaction. Increasing concentrations of $^3$H-labeled S20 were reconstituted with 16 S RNA (——) or yeast RNA ( - - - ). The solid curve is a composite of two experiments (○ and ●) utilizing the same preparations of protein but different preparations of RNA [E. Spicer, J. Schwarzbauer, and G. R. Craven, *Nucl. Acids Res.* **4**, 491 (1977)].

remove any aggregation. The $^3$H-labeled, purified proteins are dialyzed against 2 × RB. The protein solution is then added to the RNA along with sufficient water to bring the final concentration of buffer and salt to that of RB. Less than $4 \times 10^{-10}$ mol of protein are added to prevent saturation of the filter with unbound protein (see Spicer *et al.*[4]). Twelve separate binding reactions are carried out, each approximately 400 μl in volume and each in a separate reaction vessel. The reaction mixtures are incubated for 1 hr at 42°.

The relatively low concentration of protein used in this method does have the disadvantage that there is an increased percentage of the protein that adheres to solid surfaces. We have found that the loss of proteins is decreased if glass reaction vessels and pipettes are avoided.

After completion of the binding reaction, the mixture is cooled in ice; an aliquot is taken, diluted appropriately, and measured for absorbance at 260 nm to precisely determine the RNA concentration. A second aliquot is taken to measure radioactivity to determine the total protein concentration in each sample. Finally, a 300-μl aliquot is taken, placed in the apparatus, and filtered as described above. Aliquots are removed from the filtrate and analyzed for RNA recovery and the concentration of bound protein.

## Typical Results

Varying the ratio of moles of protein to moles of RNA in twelve individual reaction mixtures gives data that can be plotted as a typical saturation binding curve, as in Fig. 2. The data in Fig. 2 are for the

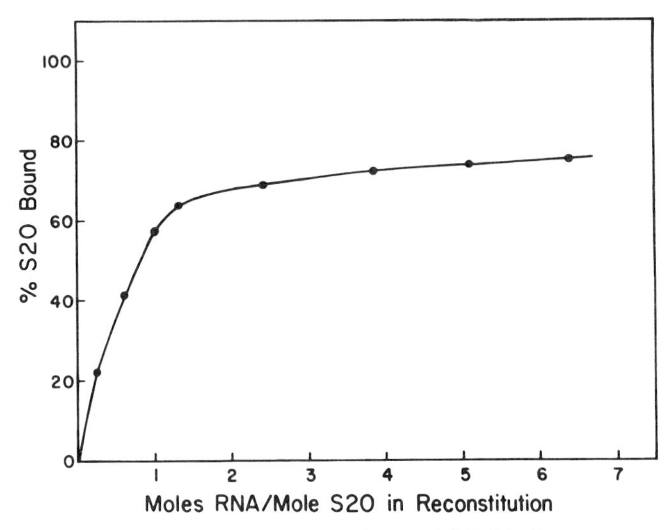


FIG. 3. Competency of protein S20 for binding to 16 S RNA. A constant amount of $^3$H-labeled S20 was incubated with increasing concentrations of 16 S RNA. At saturation, 75% of the total S20 was bound to the RNA.

binding of protein S20 to phenol-extracted 16 S RNA. Included in this figure are the data obtained from the important control experiment of binding S20 to a nonspecific RNA, in this case yeast RNA. As can be seen, the amount of protein "carried through" the filter by the nonspecific RNA is relatively minor; however, this control must be done for every new situation examined to prove that the effects observed are attributable to specific macromolecular interactions.

The actual RNA binding efficiency of a given protein preparation can be readily determined by conducting a binding experiment similar to that in Fig. 2 except that the protein concentration is kept constant and the RNA concentration is increased until all competent protein molecules are bound. Such an experiment for protein S20 is shown in Fig. 3. In this preparation only 75% of the protein molecules are active in binding 16 S RNA. This means that a fraction of the protein population apparently cannot take part in the binding reaction and that fraction should therefore be excluded in the calculations of association constants. However, it is not necessary to correct for the competency of the RNA preparation as the methods used for data analysis (see below) take this into account.

## Data Analysis and Interpretation

The binding of a ligand (in this case, a protein, P) to a macromolecule such as RNA (R) to give the complex R–P can be characterized by Eq.

(1)[13]

$$\bar{v} = nk(A)/[1 + K(A)] \quad (1)$$

where $\bar{v}$ is the number of moles of protein bound per mole of RNA, $K$ is the apparent association constant and equals (R–P)/[(R) (P)], $n$ is the number of identical and independent sites on the RNA, and $(A)$ is the molarity of free protein. The expression in Eq. (1) can be rewritten as

$$1/\bar{v} = 1/[nK(A)] + 1/n \quad (2)$$

A double reciprocal plot of $1/\bar{v}$ versus $1/(A)$ gives a line with a slope of $1/nK$ and a $y$ intercept of $1/n$.

Equation (1) can also be written in an alternative form as shown by Scatchard.[14]

$$\bar{v}/(A) = Kn - \bar{v}K \quad (3)$$

Using the Scatchard expression, $K$ and $n$ can be calculated from a plot of $\bar{v}/(A)$ versus $\bar{v}$. Thus the apparent association constant for the binding reaction and the number of independent sites on the RNA can be determined from a knowledge of $\bar{v}$ and (A) employing either of these equations. The data obtained from the nitrocellulose membrane filtration technique yield the concentration of RNA, concentration of total protein, and concentration of bound protein (concentration of protein–RNA complex). The ratio (bound protein):(RNA) is $\bar{v}$ and $(A)$ equals (total protein) minus (bound protein). Figure 4a is the double-reciprocal plot of the data on protein S20 originally displayed in Fig. 2. Figure 4b shows the same data plotted by the Scatchard method. Both methods yield comparable $K$ and $n$ values by linear regression analysis. However, the Scatchard method is capable of revealing characteristics of the reaction system which are difficult to analyze by the double-reciprocal plot. Any curvature in the line is more readily observed in a Scatchard plot. Curvature indicates either ligand–ligand interaction or noncooperative binding of the ligand at more than one site on the macromolecule, depending upon the direction of the curvature.[15]

## Concluding Comments

The filtration of protein–RNA complexes through nitrocellulose is a rapid and convenient procedure for the analysis of the binding reaction

[13] K. E. van Holde, "Physical Biochemistry." Prentice-Hall, Englewood Cliffs, New Jersey, 1971.

[14] G. Scatchard, *Ann. N.Y. Acad. Sci.* **51,** 660 (1949).

[15] J. D. McGhee and P. H. von Hoppel, *J. Mol. Biol.* **86,** 469 (1974).

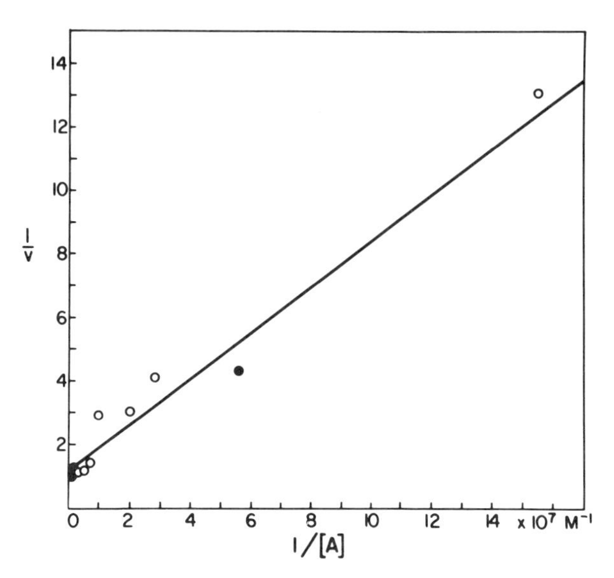


FIG. 4a. Double-reciprocal plot analysis of the binding of S20 to 16 S RNA (from Fig. 2); $n$ = 0.93 sites, $K$ = 1.4 × $10^7$ $M^{-1}$. From E. Spicer, J. Schwarzbauer, and G. R. Craven, *Nucl. Acids Res.* **4,** 491 (1977).]

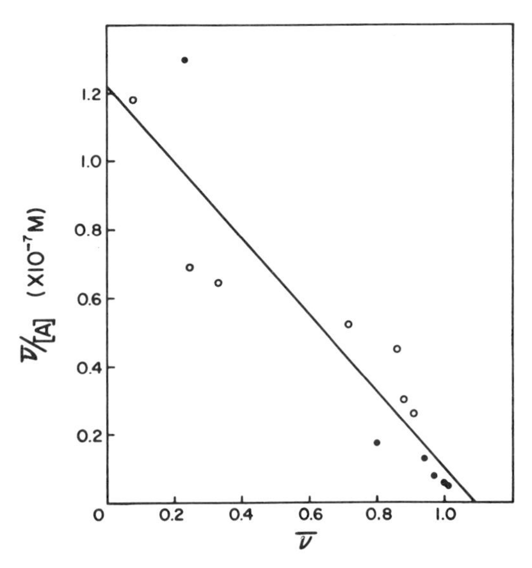


FIG. 4b. Scatchard plot for the association of S20 with 16 S RNA (from Fig. 2); $n$ = 1.09 sites, $K$ = 1.22 × $10^7$ $M^{-1}$.

at equilibrium. The method has numerous potential applications. For example, we have found that protein–RNA complexes involving up to twelve different proteins and 16 S RNA will readily pass through the membrane in RB solution. Thus, quantitative studies on protein binding to preformed protein–RNA complexes could be performed by this method. Other applications might include the examination of different solvent conditions and their effect on the strength of the association. Besides being applied to the study of ribosome structure, this fractionation technique could be adapted for use with other protein–nucleic acid systems.

# [66] Preparation and Analysis of Mitochondrial Ribosomes

*By* ALAN M. LAMBOWITZ

Mitochondria contain a distinct species of ribosome that functions in the synthesis of specific polypeptides of the inner mitochondrial membrane.[1–3] The mitochondrial (mit) ribosomes of microorganisms and higher plants have sedimentation coefficients (*s*) of 70–80 S, whereas those of animal cells have sedimentation coefficients of 55–60 S. Distinctive features of mit ribosomes include their sensitivity to inhibitors of bacterial protein synthesis, dissociation of monomers at relatively high $Mg^{2+}$ concentrations (10 m*M* $Mg^{2+}$ at 200 m*M* KCl) and, in fungi and animal cells, the lack of 5 S and 5.8 S RNA components.[1–3] Mit rRNAs are characterized by low GC content (25–40% in microorganisms compared to 50–60% for cytosolic rRNAs), a low degree of secondary structure, and a paucity of methylated nucleotides (less than 0.1% compared to 1–2% for cytosolic rRNAs).[1,4–7]

Mit ribosomes have been isolated from many organisms (listed in references cited in footnotes 1 and 2) by procedures that include the

[1] P. Borst and L. A. Grivell, *FEBS Lett.* **13,** 73 (1971).

[2] N.-H. Chua and D. J. L. Luck, *in* "Ribosomes" (M. Nomura, A. Tissières, and P. Lengyel, eds.), 519. Cold Spring Harbor Laboratory, Cold Spring Harbor, New York, 1974.

[3] G. Schatz and T. L. Mason, *Annu. Rev. Biochem.* **43,** 51 (1974).

[4] C. Vesco and S. Penman, *Proc. Natl. Acad. Sci. U.S.A.* **62,** 218 (1969).

[5] D. T. Dubin, *J. Mol. Biol.* **84,** 257 (1974).

[6] A. M. Lambowitz and D. J. L. Luck, *J. Mol. Biol.* **96,** 207 (1975).

[7] J. Klootwijk, I. Klein, and L. A. Grivell, *J. Mol. Biol.* **97,** 337 (1975).





ISBN 0-12-504180-2

following steps: (a) isolation of mitochondria, (b) lysis of mitochondria using either deoxycholate or a nonionic detergent, (c) preparation of a mit ribosomal pellet, and (d) separation of monomers or subunits on sucrose gradients. The major difficulties are contamination by cytosolic 80 S ribosomes and degradation of mit rRNAs by nucleases that are released or activated during mitochondrial lysis. The procedures described below were developed during studies of mit ribosome assembly in *Neurospora* to permit the isolation of highly purified mit ribosomal subunits containing intact rRNAs.[8] This review emphasizes two features of the procedures which may be adapted to other organisms: purification of the mitochondria by flotation gradient centrifugation[9] and substitution of $Ca^{2+}$ for $Mg^{2+}$ during mitochondrial lysis to inhibit nuclease activity.[8] The reader is referred to Grivell *et al.*[10] for the preparation of active yeast mit ribosomes and to Greco *et al.*[11] and Ibrahim and Beattie[12] for the preparation of active mit ribosomes from mammalian liver.

## Purification of Mitochondria

*Neurospora* mitochondria virtually free of cytosolic ribosome contamination can be prepared by flotation gradient centrifugation.[9] All steps are carried out at 0–3°. Cells are disrupted in 15% sucrose, 10 m*M* Tricine·KOH, 0.2 m*M* EDTA, pH 7.5. Nuclei and cell debris are pelleted at 1000 *g* (Sorvall GSA rotor, 2500 rpm, 10 min), and mitochondria are pelleted at 15,000 *g* (Sorvall GSA rotor, 10,000 rpm, 30 min). The initial mitochondrial pellet is resuspended in 5–10 ml of 20% sucrose, 10 m*M* Tricine·KOH, 0.1 m*M* EDTA, pH 7.5, and the mitochondria are pelleted again at 15,000 *g* (30 min). Care is taken to remove excess 20% sucrose from the final pellet, which is then resuspended in 4–5 ml of 60% sucrose, 10 m*M* Tricine·KOH, 0.1 m*M* EDTA, pH 7.5, using a loose-fitting Teflon pestle. Flotation gradient centrifugation is carried out in an SW 41 rotor with gradients containing 44 to 55% sucrose, 10 m*M* Tricine·KOH, pH 7.5. Each gradient accommodates *ca* 0.5 ml of packed mitochondria and larger rotors can be used to scale-up the procedure. When linear gradients are used, the gradients are prepared first and the mitochondrial suspension is underlaid with a long-stem Pasteur pipette. Satisfactory results can also be obtained using step gradients, and in this case the mitochondrial suspension is simply transferred to an SW 41 tube and then overlaid

[8] A. M. Lambowitz and D. J. L. Luck, *J. Biol. Chem.* **251,** 3081 (1976).
[9] P. M. Lizardi and D. J. L. Luck, *Nature (London), New Biol.* **229,** 140 (1971).
[10] L. A. Grivell, L. Reijnders, and P. Borst, *Biochim. Biophys. Acta* **247,** 91 (1971).
[11] M. Greco, P. Cantatore, G. Pepe, and C. Saccone, *Eur. J. Biochem.* **37,** 171 (1973).
[12] N. G. Ibrahim and D. S. Beattie, *FEBS Lett.* **36,** 102 (1973).

with 4 ml of the 55% sucrose solution followed by 2–3 ml of the 44% sucrose solution. After centrifugation at 40,000 rpm (90 min), the mitochondria form a tight band at the interface of the 44% and 55% sucrose layers. The mitochondria are removed with a Pasteur pipette, diluted with the appropriate buffer, and pelleted prior to use (see below).

The concentration of EDTA during the initial differential centrifugation steps is critically important for the flotation gradient centrifugation. The optimal concentration must be determined for each cell type and, in our experience, even for different strains of the same organism. Excessive EDTA damages the mitochondria, causing them to band diffusely within the 55% sucrose layer, whereas too little EDTA leads to contamination by cytosolic 80 S ribosomes, which bind to the outer mitochondrial membrane. It should be emphasized that the exclusion of EDTA or the addition of $Mg^{2+}$ during mitochondrial purification invariably leads to contamination by cytosolic ribosomes. It is not surprising, therefore, that studies in which mitochondria are prepared in $Mg^{2+}$-containing media have produced controversial results.[13,14]

## Inhibition of Nuclease Activity in Mitochondrial Lysates

The problem of mitochondrial nuclease activity is eliminated by substituting $Ca^{2+}$ for $Mg^{2+}$ in the mitochondrial lysis medium. The suppression of nuclease activity by $Ca^{2+}$ may result from inactivation of mitochondrial nuclease(s)[15] and/or direct protection of the RNA by $Ca^{2+}$-binding.[16] Since it is known that $Ca^{2+}$ can replace $Mg^{2+}$ in stabilizing ribosome structure,[17] the substitution can form the basis of ribosome isolations.

Figure 1 shows an experiment to determine the $Ca^{2+}$ concentration required to inhibit nuclease activity in *Neurospora* mitochondrial lysates. Mitochondria were lysed by Nonidet in buffers containing 5–50 m*M* $CaCl_2$ in combination with low or high salt (10 and 500 m*M* KCl, respectively). The lysates were incubated for 1 hr at 3°, after which the RNAs were extracted and analyzed by gel electrophoresis. As shown in Fig. 1, mit rRNAs are recovered intact from both the low- and high-salt buffers in the presence of 50 m*M* $Ca^{2+}$. As the $Ca^{2+}$ concentration is decreased

[13] R. Datema, E. Agsteribbe, and A. M. Kroon, *Biochim. Biophys. Acta* **335,** 386 (1974).

[14] R. Michel, G. Hallermayer, M. A. Harmey, F. Miller, and W. Neupert, *Biochim. Biophys. Acta* **478,** 316 (1977).

[15] S. Linn and I. R. Lehman, *J. Biol. Chem.* **241,** 2694 (1966).

[16] K. Cremer and D. Schlessinger, *J. Biol. Chem.* **249,** 4730 (1974).

[17] F.-C. Chao and H. K. Schachman, *Arch. Biochem. Biophys.* **61,** 220 (1956); see also A. S. Spirin, *FEBS Lett.* **40** (Suppl.), 38 (1974).

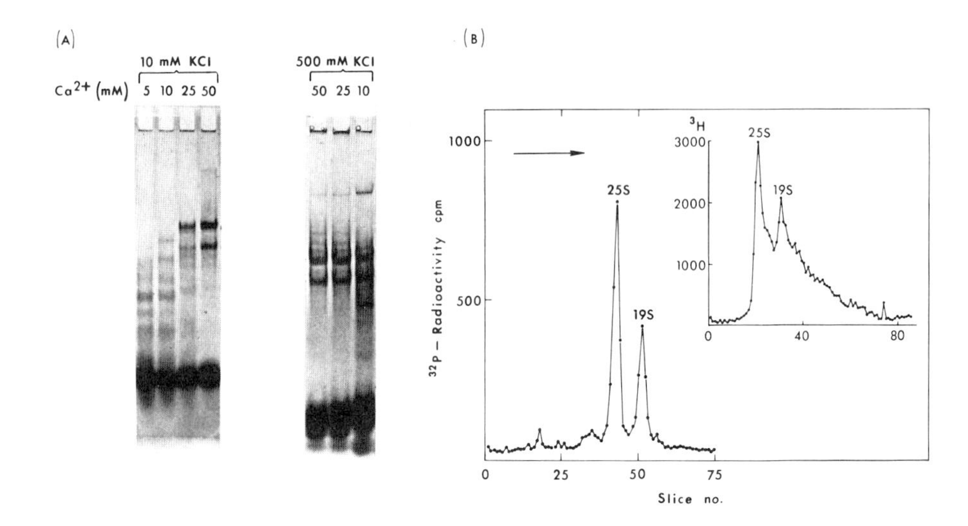


FIG. 1. (A) Gel electrophoresis of RNAs from mitochondrial lysates incubated in $Ca^{2+}$-containing buffers. Mitochondria from wild-type strain Em 5256A were purified by flotation gradient centrifugation and then divided into several parts. Each of the final mitochondrial pellets was resuspended at a density of less than 4 mg of protein per milliliter in 2.0 ml of buffers containing 10 or 500 m*M* KCl, 5-50 m*M* $CaCl_2$, 25 m*M* Tris·HCl, pH 7.5, and 5 m*M* dithiothreitol as indicated in the figure. The mitochondria were then lysed by addition of 0.1 ml of 20% Nonidet P-40, the lysates were incubated for 1 hr at 3°, and the RNAs were extracted using the SDS-diethylpyrocarbonate method [P. M. Lizardi and D. J. L. Luck, *Nature (London), New Biol.* **229,** 140 (1971)]. Yeast tRNA was added as carrier during the extraction. The extracted RNA was precipitated twice with ethanol, and aliquots were taken for electrophoretic analysis. The direction of electrophoresis is from top to bottom. The heavily stained material near the bottom of the gels is 4 S RNA. (B) Gel electrophoresis of RNAs extracted from mitochondrial ribosomal pellets prepared in $Ca^{2+}$-containing buffers ($HKCTD_{500/50}$ and 1.85 *M* sucrose in $HKCTD_{500/25}$) as described in the text. RNAs were extracted from the pellet, and an aliquot was taken for electrophoretic analysis. The inset shows the same experiment but with a lysis medium containing 400 m*M* KCl, 30 m*M* $MgCl_2$, 20 m*M* Tricine·KOH, pH 7.9, and 40 m*M* dithiothreitol. Other experimental details are given in [A. M. Lambowitz and D. J. L. Luck, *J. Biol. Chem.* **251,** 3081 (1976)]. The arrow indicates the direction of electrophoresis.

from 50 m*M*, there is progressively more degradation of the RNAs, but the degradation is less pronounced in the high-salt medium, where binding of nucleases to RNA is electrostatically inhibited. The advantage of $Ca^{2+}$-containing medium is illustrated in Fig. 1B by the integrity of mit rRNAs which can be extracted from ribosomal pellets prepared by sedimentation of the lysate through a cushion of 1.85 *M* sucrose. The inset shows the same experiment carried out in $Mg^{2+}$-containing buffers.

Preparation of Mitochondrial Ribosomes

Based on the above results, the following procedure was developed for the preparation of mit ribosomes. All steps are carried out at 0 to 3° unless otherwise specified. Mitochondria are removed from flotation gradients with a Pasteur pipette, diluted with 3–4 volumes of $HKCTD_{500/50}$ (500 m*M* KCl, 50 m*M* $CaCl_2$, 25 m*M* Tris·HCl, pH 7.5, 5 m*M* dithiothreitol), and centrifuged in a Beckman type 40 rotor (25,000 rpm, 10 min). After the supernatant is carefully removed with a Pasteur pipette, the pellet containing a maximum of 0.5 ml of packed mitochondria is resuspended in 3.8 ml $HKCTD_{500/50}$ and lysed by addition of 0.2 ml of 20% Nonidet P-40 (Particle Data Laboratories, Elmhurst, Illinois). To separate mit ribosomes from membrane contaminants, the lysate is layered over a 1.85 *M* sucrose cushion containing $HKCTD_{500/25}$ (500 m*M* KCl, 25 m*M* $CaCl_2$, 25 m*M* Tris·HCl, pH 7.5, 5 m*M* dithiothreitol) and centrifuged in a Beckman type 65 rotor (55,000 rpm, 17 hr, 3°). The lysate and the top of the sucrose layer are then carefully withdrawn with a Pasteur pipette, and the sides of the tube are washed three times with distilled water to remove contaminating nucleases. The remainder of the 1.85 *M* sucrose layer is then removed, and the translucent ribosomal pellet is rinsed quickly with approximately 1 ml of ice cold, distilled water. The final mit ribosomal pellet should be virtually free of membrane contamination provided that (a) the mitochondria are resuspended without visible clumps prior to lysis, and (b) the pH of the lysis buffer is carefully adjusted to 7.5 (3°). (If membrane fragments are present, a short clarifying spin (15,000 *g*, 15 min) can be included at this point.) The procedure yields 20–40 μg of mit ribosomes per milligram of mitochondrial protein.

Since "clean" mit ribosomal pellets contain very little nuclease activity, it is possible to substitute $Mg^{2+}$ for $Ca^{2+}$ in subsequent steps. Ribosomal subunits are routinely separated by centrifugation through sucrose gradients containing 500 m*M* KCl. The mit ribosomal pellet is suspended in one volume of cold distilled water followed quickly by one volume of 2× buffer (1.0 *M* KCl, 50 m*M* $MgCl_2$, 50 m*M* Tris·HCl, pH 7.5, 10 m*M* dithiothreitol). Monomers are dissociated by addition of 1 m*M* puromycin·KOH, pH 7.5 and incubation at 35° for 15 min.[18] Then 0.3–0.4 ml of the suspension containing up to 5 $OD_{260}$ units are layered over linear gradients of 5 to 20% sucrose containing $HKMTD_{500/25}$ (500 m*M* KCl, 25 m*M* $MgCl_2$, 25 m*M* Tris·HCl, pH 7.5, 5 m*M* dithiothreitol). The gradients are centrifuged in an SW 41 rotor at 40,000 rpm (3 hr, 3°). Similar

[18] G. Blobel and D. Sabatini, *Proc. Natl. Acad. Sci. U.S.A.* **68**, 390 (1971).

procedures, but with different gradient buffers, are used to prepare mit ribosomal monomers[19] and mit ribosomal precursor particles.[20]

## Isolation of Mitochondrial Polysomes

Isolation of mit polysomes is a serious problem since in most cases it has been difficult to distinguish "true" mit polysomes from aggregated monosomes. Putative mit polysomes have been isolated from *Euglena*,[21,22] yeast,[23] and HeLa cells.[24] However, only the *Euglena* preparations have all the expected polysome characteristics: i.e., sedimentation as a series of peaks on sucrose gradients with the higher order polysomes dissociated at low $Mg^{2+}$ concentration or by treatment with RNase. By contrast, both the yeast and HeLa cell mit polysomes sediment more amorphously on sucrose gradients and the HeLa cell mit polysomes possess atypical properties (e.g., resistance to dissociation by EDTA or RNase). Kuriyama and Luck[25] isolated puromycin-dissociable, membrane-bound ribosomes from *Neurospora* mitochondria, but did not determine whether these included higher-order polysomes. We find that at least 80% of the ribosomes isolated from *Neurospora* mitochondria are in the form of monomers or subunits, independent of ionic conditions, the presence of chloramphenicol to inhibit "runoff" or precautions to inhibit nuclease activity.

## Protein Synthetic Activities

There appears to be no general method that gives active mit ribosomes from different organisms. Mit ribosomes isolated from yeast and mammalian liver carry out poly(U)-directed polyphenylalanine synthesis at rates near 2000 pmol per milligram of RNA per 30 min, comparable to rates for *E. coli* ribosomes.[10–12] However, mit ribosomes isolated from other organisms by similar procedures have rates of only 4–300 pmol per milligram of RNA per 30 min.[1] Manipulations that enhance activity in one system diminish activity in other systems.[cf. 10,11]

*Neurospora* mit ribosomes as prepared by Küntzel carried out poly(U)-directed polyphenylalanine synthesis at relatively low rates, 4–

[19] R. J. LaPolla and A. M. Lambowitz, *J. Mol. Biol.* **116**, 189 (1977).
[20] A. M. Lambowitz, N.-H. Chua, and D. J. L. Luck, *J. Mol. Biol.* **107**, 223 (1976).
[21] N. G. Avadhani and D. E. Buetow, *Biochem. Biophys. Res. Commun.* **46**, 773 (1972).
[22] N. G. Avadhani and D. E. Buetow, *Biochem. J.* **128**, 353 (1972).
[23] H. R. Mahler and K. Dawidowicz, *Proc. Natl. Acad. Sci. U.S.A.* **70**, 111 (1973).
[24] D. Ojala and G. Attardi, *J. Mol. Biol.* **65**, 273 (1972).
[25] Y. Kuriyama and D. J. L. Luck, *J. Cell Biol.* **59**, 776 (1973).

100 pmol per milligram of RNA per 30 min.[26] We obtain somewhat higher rates, at least 100–400 pmol per milligram of RNA per 30 min, for *Neurospora* mit ribosomes prepared in high salt, $Ca^{2+}$-containing buffers. The rates are decreased if $Ca^{2+}$ is replaced by $Mg^{2+}$ during the isolation and the lowest rates, less than 100 pmol per milligram of RNA per 30 min, are obtained for mit ribosomes isolated by the method of Küntzel. Polyphenylalanine synthesis by *Neurospora* mit ribosomes has a $K^+$ optimum of 60 m$M$, a $Mg^{2+}$ optimum of 10 m$M$ and a temperature optimum between 25° and 30°.

In some respects, the initiation and elongation steps of mit protein synthesis are analogous to those in bacterial systems. *Neurospora* mit ribosomes, for example, can recognize, bind, and translocate fMet-tRNA in response to AUG. This capacity is lost when the ribosomes are washed in 1 $M$ $NH_4Cl$, a procedure that removes initiation factors, and restored by initiation factors from *E. coli*.[27] Similarly, elongation factors T and G appear to be interchangeable with those from *E. coli*, but not with those from cytosolic ribosomes.[28] In terms of practical significance, protein synthetic activities of mit ribosomes are often assayed using *E. coli* supernatant fractions, taking advantage of their relatively low nuclease activity compared to mitochondrial supernatants.

### Isolation of RNA from Whole Mitochondria

Mit RNA free of cytosolic RNA contamination can be obtained directly from flotation gradient mitochondria by extraction using the sodium dodecyl sulfate (SDS)-diethylpyrocarbonate method.[9,29] Individual RNA species can then be separated by sucrose gradient centrifugation or by electrophoresis through composite agarose–acrylamide gels.[30] The latter procedure is used for analytical work, for example in the identification of r-precursor RNA species.[31] For preparative purposes, the purity of individual RNA species must be carefully checked. In our experience, highly purified mit rRNAs are most easily prepared from purified mit ribosomal subunits (see below) whereas mit rRNAs obtained by sucrose gradient centrifugation of whole mit RNAs may be contaminated, probably by mRNA species. The same point was made previously by Borst

[26] H. Küntzel, *FEBS Lett.* **4,** 140 (1969).

[27] F. Sala and H. Küntzel, *Eur. J. Biochem.* **15,** 280 (1970).

[28] M. Grandi and H. Küntzel, *FEBS Lett.* **10,** 25 (1970).

[29] F. Solymosy, I. Fedorcsak, A. Gulyas, G. L. Farkas, and L. Ehrenberg, *Eur. J. Biochem.* **5,** 520 (1968).

[30] A. C. Peacock and C. W. Dingman, *Biochemistry* **7,** 668 (1968).

[31] Y. Kuriyama and D. J. L. Luck, *J. Mol. Biol.* **73,** 425 (1973).

and Grivell.[1] It is worth noting in addition that even gross contamination may be camouflaged on the gradients.

Mit rRNAs are known to display anomalous electrophoretic and sedimentation behavior that complicates molecular weight estimates.[1] The electrophoretic mobility of mit rRNA is unusually low compared to *E. coli* rRNA standards and is also strongly influenced by temperature and ionic strength.[1] This behavior is thought to reflect an open conformation and relatively little secondary structure.

## Isolation of rRNAs from Ribosomal Subunits

Ribosomal subunits isolated after dissociation of monomers with puromycin are the best source of highly purified rRNAs which may be required for hybridization studies or fingerprint analysis. Ribosomal subunits from pooled gradient fractions are centrifuged overnight (50 Ti rotor, 50,000 rpm, 3°) or precipitated by addition of 2.3 volumes of ethanol and incubation at −20° overnight. If the concentration of ribosomal subunits is less than 1 $OD_{260}$ unit/ml, the subunits are ethanol precipitated in the presence of carrier yeast tRNA (added to bring the final RNA concentration to 1 $OD_{260}$ unit/ml). RNA can be extracted from subunits using either the SDS-/diethylpyrocarbonate method[9,29] or the phenol-Pronase-SDS method.[8] Treatment with Pronase may be required to remove residual protein in some types of experiments. Figure 2 shows gel profiles of rRNAs extracted from ribosomal subunits of two *Neurospora* strains.

## Analysis of Mit Ribosomal Proteins

A number of conventional electrophoretic systems have been adapted for the analysis of mit ribosomal proteins. Figure 3 shows the separation of *Neurospora* mit ribosomal proteins (molecular weight ratio, 10,000-60,000) using a highly resolving one-dimensional system consisting of SDS gels with a 7 to 15% gradient of polyacrylamide.[20,32] Preparation of the gels has been described by Chua and Bennoun.[32] Sample preparation is carried out as follows: mit ribosomal subunits are recovered from pooled gradient fractions by overnight centrifugation or by ethanol precipitation as described above. If the concentration of ribosomal subunits is less than 1 $OD_{260}$ unit/ml, ethanol precipitation is carried out in the presence of carrier yeast tRNA. Pellets containing 0.2 to 1.0 $OD_{260}$ units of ribosomal subunits are dissolved in 50 m*M* $NaCO_3$, 50 m*M* dithiothreitol, 2% (w/v) SDS, 12% (w/v) sucrose, and 0.04% (w/v) bromophenol

[32] N.-H. Chua and P. Bennoun, *Proc. Natl. Acad. Sci. U.S.A.* **72,** 2175 (1975).

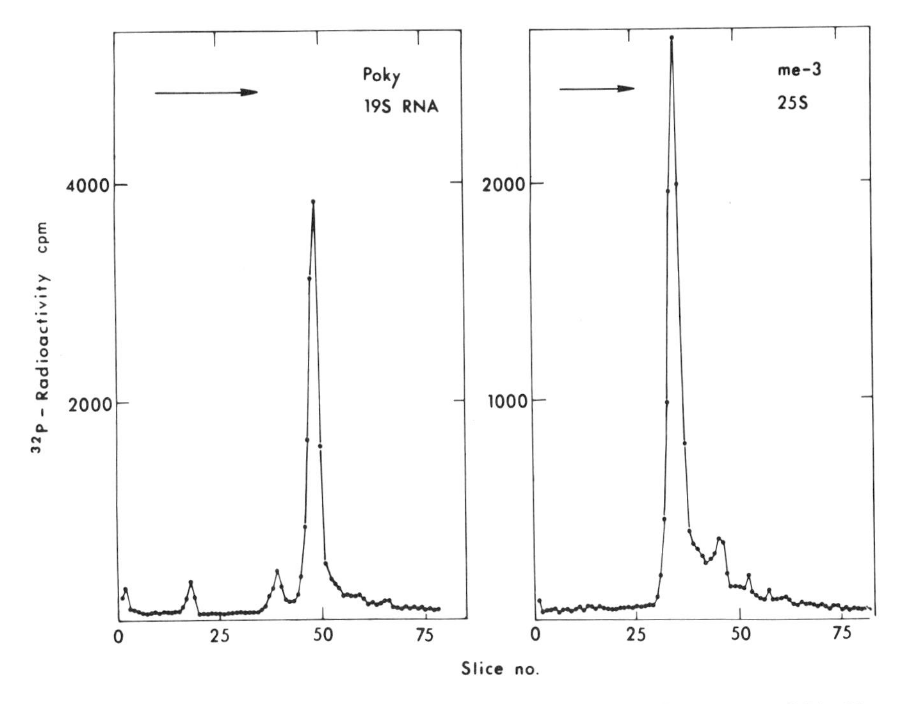


FIG. 2. Gel electrophoresis of purified 19 S and 25 S RNAs from the me-3 [A. M. Lambowitz and D. J. L. Luck, *J. Mol. Biol.* **96,** 207 (1975)] and *poky* strains of *Neurospora*. $^{32}$P-Labeled RNAs were isolated from purified ribosomal subunits using the phenol–Pronase–SDS method [A. M. Lambowitz and D. J. L. Luck, *J. Biol. Chem.* **251,** 3081 (1976)] and separated by gel electrophoresis as described in the text.

blue and applied directly to the gels for analysis. Since prior separation of the RNA and protein moieties is not required and since carrier tRNA does not interfere, this method is suited for the analysis of small amounts of material, a situation often encountered in subunit-deficient mutants.

Figure 4 shows two-dimensional gel analysis of *Neurospora* mit ribosomal proteins using a modification of the system of Mets and Bogorad.[33] In this case, sample preparation requires that subunit pellets be prepared by overnight centrifugation. Proteins are extracted by a modification of the acetic acid method.[34] Pellets containing 2–5 $OD_{260}$ units of subunits are rinsed quickly and resuspended in 100 $\mu$l of cold distilled water. The proteins are then extracted by addition of 0.5 ml of a mixture containing 80% (v/v) acetic acid, 40 m$M$ Mg acetate and 4 m$M$ Tris·HCl, pH 7.5. The suspension is drawn up and down in a Pasteur pipette about

[33] L. J. Mets and L. Bogorad, *Anal. Biochem.* **57,** 200 (1974).

[34] S. J. S. Hardy, C. G. Kurland, P. Voynow, and G. Mora, *Biochemistry* **8,** 2897 (1969).

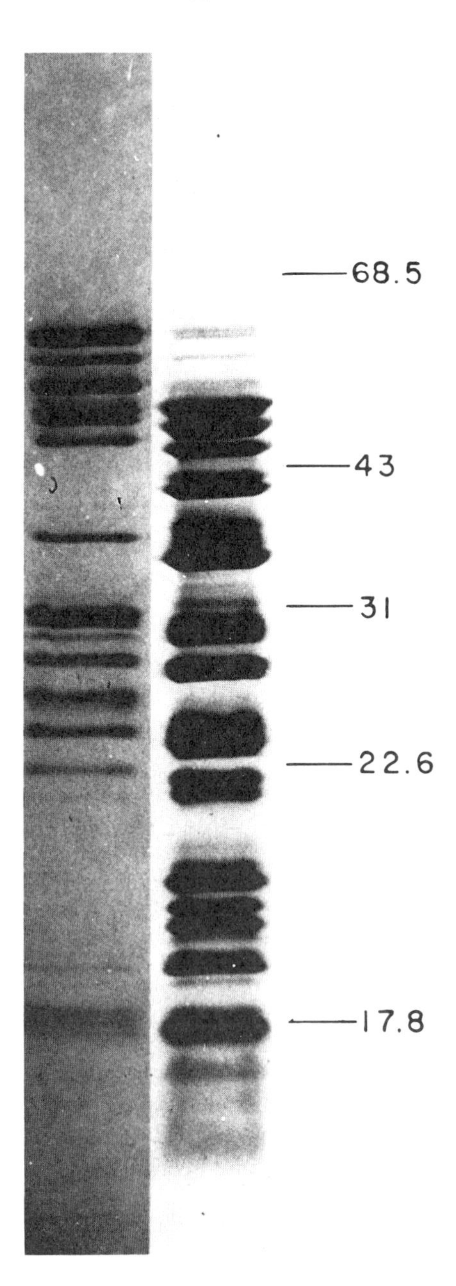
Small subunit
Large subunit
68.5
43
31
22.6
17.8

20 times, and the RNA residue is pelleted in a Sorvall SS 34 rotor (10,000 rpm, 15 min, 3°). The protein-containing supernatant is recovered using a Pasteur pipette with the tip drawn to a fine capillary to exclude RNA fragments. The RNA residue is reextracted, the supernatants are pooled, and proteins are precipitated by addition of 8 volumes of cold acetone. After overnight incubation at −20°, the precipitated proteins are pelleted at 3000 rpm (15 min, 3°) using a clinical swinging-bucket centrifuge. The protein pellet is washed twice with cold acetone to remove residual acetic acid. It is then dried under a stream of filtered air and finally dissolved in 40 μl of Mets and Bogorad sample buffer containing 8 *M* urea, 10 m*M* Bistris, 7 mM mercaptoethanol, 10 m*M* dithiothreitol, adjusted to pH 4.0 with acetic acid. Because of the scarcity of material, gel loads are quantitated by the amount of subunits present at the beginning of the extraction.

Electrophoresis in the first dimension is carried out on thin, slab gels (19 cm × 15 cm × 0.8 mm) with 1.5-cm slots. The gels contain 8 *M* urea, 4% acrylamide, 0.1% bisacrylamide, 57 m*M* Bistris adjusted to pH 5.0 with acetic acid. Polymerization is by addition of TEMED (3 μl/ml) and ammonium persulfate (0.3 mg/ml). The upper buffer is 10 m*M* Bistris adjusted to pH 4 with acetic acid and the lower buffer is 0.18 *M* potassium acetate, adjusted to pH 5 with acetic acid. Electrophoresis is toward the cathode at a constant current of 35 mA for about 5.5 hr until the pyronine Y tracking dye has migrated 1.3 times through the gel. Slots for the first dimension are cut out, rinsed with transfer buffer (55 m*M* Tris·$SO_4$, pH 6.1), placed over the second-dimension gels, rinsed again with transfer buffer, and then electrophoresed without additional equilibration. Second dimensions are the same SDS-polyacrylamide gradient gels that are used for one-dimensional analysis (see Fig. 3). The gel dimensions are 34 cm × 25 cm × 1 mm, so that two first-dimension slots can fit over a single second-dimension gel. Electrophoresis is toward the anode at constant current of 40 mA until the bromophenol blue tracking dye reaches the bottom of the gel.

Two-dimensional gel electrophoresis can be used to define the protein composition of ribosomal subunits and to look for altered proteins in mutant strains. In the case of the Mets and Bogorad system, it is usually assumed that charge differences in mutant proteins will be detected by altered mobility in the first dimension and size differences by altered mobility in the second dimension. The sensitivity of the first dimension was tested directly by carbamylating mit ribosomal proteins to produce

---

FIG. 3. One-dimensional gel electrophoretic analysis of small and large subunit proteins of wild-type strain Em 5256*A*. The arrows indicate the positions of molecular weight standards. The direction of electrophoresis is from top to bottom.

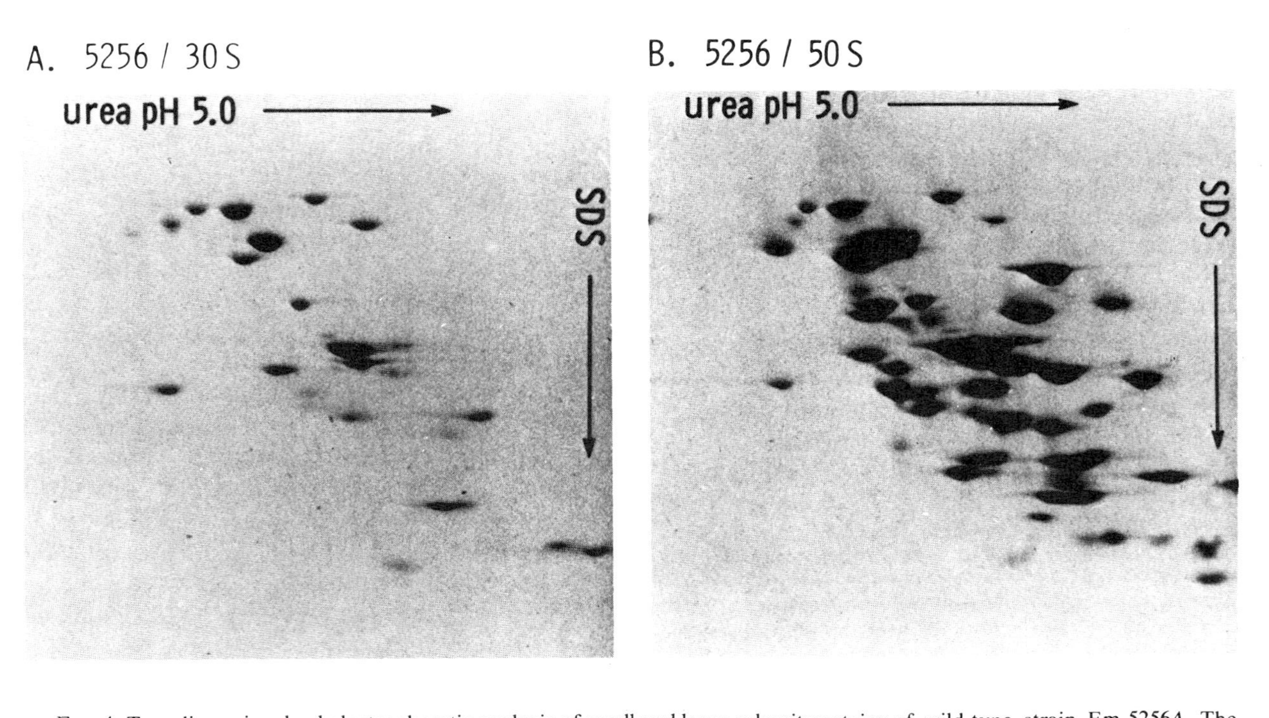


FIG. 4. Two-dimensional gel electrophoretic analysis of small and large subunit proteins of wild-type strain Em 5256*A*. The large subunit pattern shows a background of small subunit proteins presumably due to small subunit dimers that cosediment with large subunits on the gradient. Electrophoresis was carried out using the modified Mets and Bogorad system described in the text.

a series of modified proteins differing in charge (see method of Steinberg *et al.*[35]). In fact, the arrays of modified proteins were found to decrease in size with increasing molecular weight, a result suggesting that mobility in the first dimension is dependent on both charge and molecular weight and that the contribution of molecular weight increases for larger proteins. The minimum conclusion is that it would be difficult to detect single charge differences in high-molecular-weight proteins using this system. Since isoelectric focusing systems that give satisfactory resolution of very basic proteins have not yet been described, total analysis of ribosomal protein mutations may require a combination of several different gel systems and/or supplementary protein fingerprinting techniques.[36]

## Acknowledgments

The author thanks Richard A. Collins, Robert J. LaPolla, and Carmen A. Mannella for critically reading the manuscript. The data on polyphenylalanine synthesis by *Neurospora* mitochondrial ribosomes are from a manuscript in preparation by Robert J. LaPolla, Julian Scheinbuks, and Alan M. Lambowitz. The author is supported by N.I.H. Grant GM 23961 and a Basil O'Connor Starter Research Grant from the National Foundation—March of Dimes.

[35] R. A. Steinberg, P. H. O'Farrell, U. Friedrich, and P. Coffino, *Cell* **10,** 381 (1977).

[36] D. W. Cleveland, S. G. Fischer, M. W. Kirschner, and U. K. Laemmli, *J. Biol. Chem.* **252,** 1102 (1977).

# Index

## A

## F

## G



## H



## I

L

## M



## N



## O

## Q



## R

## U



## V



## W



## X



## Y



## Z